25 *Years*
Previous Solved Papers

ESE 2025

UPSC ENGINEERING SERVICES EXAMINATION

Preliminary Examination

CIVIL ENGINEERING

Objective Solved Papers

Volume-II

- Topicwise presentation
- Thoroughly revised & updated

Also useful for

State Engineering Services Examinations, Public Sector Examinations & Other Competitive Examinations

MADE EASY PUBLICATIONS

www.madeeasypublications.org

MADE EASY Publications Pvt. Ltd.

Corporate Office: 44-A/4, Kalu Sarai (Near Hauz Khas Metro Station), New Delhi-110016

Contact: 9021300500

E-mail: infomep@madeeasy.in

Visit us at: www.madeeasypublications.org

ESE-2025 : Preliminary Examination
Civil Engineering : Volume-II
Topicwise Objective Solved Questions : (2000-2024)

1st Edition : 2006
2nd Edition : 2007
3rd Edition : 2008
4th Edition : 2009
5th Edition : 2010
6th Edition : 2011
7th Edition : 2012
8th Edition : 2013
9th Edition : 2014
10th Edition : 2015
11th Edition : 2016
12th Edition : 2017
13th Edition : 2018
14th Edition : 2019
15th Edition : 2020
16th Edition : 2021
17th Edition : 2022
18th Edition : 2023

19th Edition : 2024

Director's Message

B. Singh (Ex. IES)

Engineering is one of the most chosen graduating field. Taking engineering is usually a matter of interest but this eventually develops into "purpose of being an engineer" when you choose engineering services as a career option.

Train goes in tunnel we don't panic but sit still and trust the engineer, even we don't doubt on signalling system, we don't think twice crossing over a bridge reducing our travel time; every engineer has a purpose in his department which when coupled with his unique talent provides service to mankind.

I believe *"the educator must realize in the potential power of his pupil and he must employ all his art, in seeking to bring his pupil to experience this power"*. To support dreams of every engineer and to make efficient use of capabilities of aspirant, MADE EASY team has put sincere efforts in compiling all the previous years' ESE-Pre questions with accurate and detailed explanation. The objective of this book is to facilitate every aspirant in ESE preparation and so, questions are segregated chapterwise and topicwise to enable the student to do topicwise preparation and strengthen the concept as and when they are read.

I would like to acknowledge efforts of entire MADE EASY team who worked hard to solve previous years' papers with accuracy and I hope this book will stand up to the expectations of aspirants and my desire to serve student fraternity by providing best study material and quality guidance will get accomplished.

B. Singh (Ex. IES)
CMD, MADE EASY Group

Volume-II

CIVIL ENGINEERING

Objective Solved Questions

of UPSC Engineering Services Examination

Contents

UNIT

I

Fluid Mechanics & Hydraulic Machines

Syllabus

Fluid Mechanics, Open Channel Flow, Pipe Flow: Fluid properties; Dimensional Analysis and Modeling; Fluid dynamics including flow kinematics and measurements; Flow net; Viscosity, Boundary layer and control, Drag, Lift, Principles in open channel flow, Flow controls. Hydraulic jump; Surges; Pipe networks.

Hydraulic Machines and Hydro power: Various pumps, Air vessels, Hydraulic turbines–types, classifications & performance parameters; Power house–classification and layout, storage, pondage, control of supply.

Contents

■■■■

1 Fluid Properties

1.1 Match **List-I** with **List-II** and select the correct answer using the codes given below the lists:

List-I
A. Concentrated sugar solution
B. Sewage sludge
C. Blood
D. Air

List-II
1. Dilatant fluid
2. Bingham plastic fluid
3. Pseudo-plastic fluid
4. Newtonian fluid

Codes:

	A	B	C	D
(a)	1	2	3	4
(b)	1	2	4	3
(c)	2	1	3	4
(d)	2	1	4	3

[ESE : 2001]

1.2 Match **List-I** (Definitions) with **List-II** (Properties) and select the correct answer using the codes given below the lists:

List-I
A. Newtonian fluid
B. Ideal fluid
C. Thixotropic fluid
D. Rheological fluid

List-II
1. Frictionless and incompressible
2. Viscosity is invariant with shear stress
3. Viscosity decreases at higher shear stress
4. Viscosity increases at higher shear stress

Codes:

	A	B	C	D
(a)	2	4	1	3
(b)	3	1	4	2
(c)	2	1	3	4
(d)	3	4	1	2

[ESE : 2002]

1.3 Which one of the following statements is correct?
(a) Dynamic viscosity of water is nearly 50 times that of air.
(b) Kinematic viscosity of water is 30 times that of air.
(c) Water in soil is able to rise a considerable distance above the groundwater table due to viscosity.
(d) Vapour pressure of a liquid is inversely proportional to the temperature.

[ESE : 2003]

1.4 Which of the following fluids can be classified as non-Newtonian?
1. Kerosene oil
2. Diesel oil
3. Human blood
4. Toothpaste
5. Water

Select the correct answer using the codes given below:
(a) 1 and 2
(b) 3 and 4
(c) 2 and 5
(d) 1 and 5

[ESE : 2003]

1.5 **Assertion (A):** At the standard temperature, the kinematic viscosity of air is greater than that of water at the same temperature.
Reason (R): The dynamic viscosity of air at standard temperature is lower than that of water at the same temperature.
(a) both A and R are true and R is the correct explanation of A
(b) both A and R are true but R is not a correct explanation of A
(c) A is true but R is false
(d) A is false but R is true

[ESE : 2004]

1.6 The velocity distribution for flow over a plate is given by $u = 0.5y - y^2$ where 'u' is the velocity in m/s at a distance 'y' meter above the plate. If the dynamic viscosity of the fluid is 0.9 N-s/m^2, then what is the shear stress at 0.20 m from the boundary?

(a) $0.9\ N/m^2$ (b) $1.8\ N/m^2$
(c) $2.25\ N/m^2$ (d) $0.09\ N/m^2$

[ESE : 2005]

1.7 A flat plate of $0.15\ m^2$ is pulled at 20 cm/s relative to another plate, fixed at a distance of 0.02 cm from it with a fluid having $\mu = 0.0014$ $N\text{-}s/m^2$ separating them. What is the power required to maintain the motion?

(a) 0.014 W (b) 0.021 W
(c) 0.035 W (d) 0.042 W

[ESE : 2006]

1.8 Which one of the following expresses the height of rise or fall of a liquid in a capillary tube?

(a) $\dfrac{4wd}{\sigma\cos\alpha}$ (b) $\dfrac{\sigma\cos\alpha}{4w\alpha}$

(c) $\dfrac{4\sigma\cos\alpha}{wd}$ (d) $\dfrac{wd}{4\sigma\cos\alpha}$

where,
w = Specific weight of the liquid
α = Angle of contact of the liquid surface
σ = Surface tension

[ESE : 2007]

1.9 Match **List-I** (Curve identification in figure) with **List-II** (Nature of fluid) and select the correct answer using the codes given below the lists:

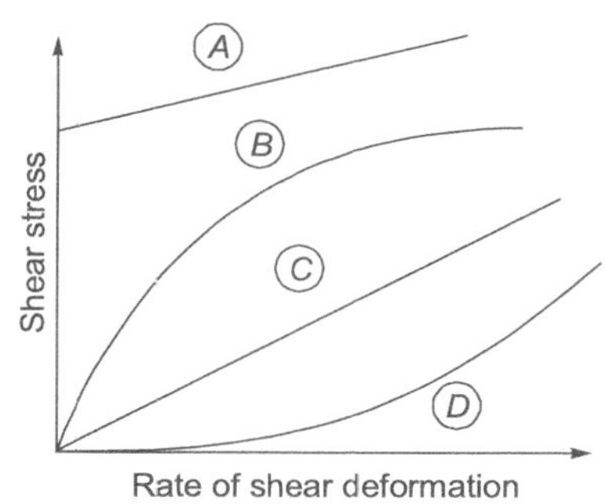

	List-I		**List-II**
A.	Curve *A*	1.	Newtonian
B.	Curve *B*	2.	Dilatant
C.	Curve *C*	3.	Ideal bingham plastic
D.	Curve *D*	4.	Pseudo-plastic

Codes:

	A	B	C	D
(a)	3	4	1	2
(b)	2	4	1	3
(c)	3	1	4	2
(d)	2	1	4	3

[ESE : 2010]

1.10 Assertion (A): The movement of two blocks of wood weltec with hot glue requires greater and greater effort as the glue is drying up.

Reason (R): Viscosity of liquids varies inversely with temperature.

(a) both A and R are true and R is the correct explanation of A
(b) both A and R are true but R is not a correct explanation of A
(c) A is true but R is false
(d) A is false but R is true

[ESE : 2010]

1.11 Match **List-I** with **List-II** and select the correct answer using the code given below the lists:

List-I
A. Specific Gravity
B. Coefficient of viscosity
C. Kinematic viscosity
D. Stress

List-II
1. $M^0L^2T^{-1}$
2. $M^0L^0T^0$
3. $ML^{-1}T^{-1}$
4. $ML^{-1}T^{-2}$

Code:

	A	B	C	D
(a)	2	3	1	4
(b)	4	3	1	2
(c)	2	1	3	4
(d)	4	1	3	2

[ESE : 2011]

1.12 Poise has the unit of

(a) Dyne-cm/s^2 (b) Dyne-cm/s
(c) Dyne-s/cm (d) Dyne-s/cm^2

[ESE : 2011]

1.13 Which one of the following statements is correct?

(a) For water at 100° Celsius at sea level, the vapour pressure is equal to atmospheric pressure.
(b) Surface energy (or tension) is caused by the force of adhesion between liquid molecules.
(c) Viscosity of a fluid is the property exhibited by it both in static and in dynamic conditions.
(d) Air is 50, 000 times more compressible than water.

[ESE : 2011]

1.14 Which one of the following statements is correct?

(a) Dynamic viscosity is the property of a fluid which is not in motion

(b) Surface energy is fluid property giving rise to the phenomenon of capillarity in water

(c) Cavitation results from the action of very high pressure

(d) Real fluids have lower viscosity than ideal fluids

[ESE : 2011]

1.15 **Statement (I):** As temperature increases, viscosity of air decreases.

Statement (II): As temperature increases, activity of the air molecules increases.

(a) Both Statement (I) and Statement (II) are individually true and Statement (II) is the correct explanation of Statement (I)

(b) Both Statement (I) and Statement (II) are individually true but Statement (II) is NOT the correct explanation of Statement (I)

(c) Statement (I) is true but Statement (II) is false

(d) Statement (I) is false but Statement (II) is true

[ESE : 2013]

1.16 The surface tension in a soap bubble of 50 mm diameter with its inside pressure being 2.5 N/m^2 above the atmospheric pressure is

(a) 0.0125 N/m (b) 0.0156 N/m

(c) 0.2 N/m (d) 0.0312 N/m

[ESE : 2015]

1.17 The surface tension of water at 20°C is 75×10^{-3} N/m. The difference in water surfaces within and outside an open-ended capillary tube of 1 mm internal bore, inserted at the water surface, would nearly be

(a) 7 mm (b) 11 mm

(c) 15 mm (d) 19 mm

[ESE : 2016]

1.18 **Statement (I):** The shear strain graph for a Newtonian fluid is linear.

Statement (II): The coefficient of viscocity μ of the fluid is not constant.

(a) Both Statement (I) and Statement (II) are individually true and Statement (II) is the correct explanation of Statement (I)

(b) Both Statement (I) and Statement (II) are individually true but Statement (II) is ***not*** the correct explanation of Statement (I)

(c) Statement (I) is true but Statement (II) is false

(d) Statement (I) is false but Statement (II) is true

[ESE : 2016]

1.19 The surface tension in a soap bubble of 20 mm diameter, when the inside pressure is 2.0 N/m^2 above atmospheric pressure, is

(a) 0.025 N/m (b) 0.0125 N/m

(c) 5×10^{-3} N/m (d) 4.25×10^{-3} N/m

[ESE : 2018]

1.20 A jet of water has a diameter of 0.3 cm. The absolute surface tension of water is 0.072 N/m and atmospheric pressure is 101.2 kN/m^2. The absolute pressure within the jet of water will be

(a) 101.104 kN/m^2 (b) 101.152 kN/m^2

(c) 101.248 kN/m^2 (d) 101.296 kN/m^2

[ESE : 2018]

1.21 A glass tube of 2.5 mm internal diameter is immersed in oil of mass density 940 kg/m^3 to a depth of 9 mm. If a pressure of 148 N/m^2 is needed to form a bubble which is just released, what is the surface tension of the oil?

(a) 0.041 N/m (b) 0.043 N/m

(c) 0.046 N/m (d) 0.050 N/m

[ESE : 2018]

1.22 A plate 0.025 mm distant from a fixed plate moves at 60 cm/s and requires a force of 0.2 kgf/m^2 to maintain this speed. The dynamic viscosity of the fluid between the plates will be nearly

(a) 9.2×10^{-10} kgfs/cm^2

(b) 8.3×10^{-10} kgfs/cm^2

(c) 7.4×10^{-10} kgfs/cm^2

(d) 6.5×10^{-10} kgfs/cm^2

[ESE : 2019]

1.23 What is the minimum size of glass tube that can be used to measure water level if the capillary rise in the tube is to be restricted to 2 mm? (Take surface tension of water in contact with air as 0.073575 N/m)

(a) 1.5 cm (b) 1.0 cm

(c) 2.5 cm (d) 2.0 cm

[ESE : 2021]

1.24 The pressure outside the droplet of water of diameter 0.04 mm is 10.32 N/cm^2 (atmospheric pressure). What is the pressure within the droplet if surface tension is 0.0725 N/m of water?
(a) 11.045 N/cm^2 (b) 10.32 N/cm^2
(c) 9.45 N/cm^2 (d) 8.595 N/cm^2
[ESE : 2022]

1.25 What is the viscosity of a liquid whose kinematic viscosity is 6 stokes and specific gravity is 1.90?
(a) 1.14 poise (b) 11.40 poise
(c) 0.114 Ns/m^2 (d) 11.40 Ns/m^2
[ESE : 2022]

1.26 How does the bulk modulus of elasticity of a fluid change with increasing pressure?
(a) It remains constant
(b) It decreases with increase in pressure
(c) It increases with increase in pressure
(d) It becomes zero
[ESE : 2024]

■■■■

Answers Fluid Properties

1.1 (a)	1.2 (c)	1.3 (a)	1.4 (b)	1.5 (b)	1.6 (d)	1.7 (d)	1.8 (c)	1.9 (a)
1.10 (a)	1.11 (a)	1.12 (d)	1.13 (a)	1.14 (b)	1.15 (d)	1.16 (b)	1.17 (c)	1.18 (c)
1.19 (c)	1.20 (c)	1.21 (a)	1.22 (b)	1.23 (a)	1.24 (a)	1.25 (b)	1.26 (c)	

Explanations Fluid Properties

1.1 (a)

Newtonian Fluids: Air, water, mercury, glycerine, kerosene and other engineering fluids under normal circumstances.

Pseudoplastic: Fine particle suspension, gelatine, blood, milk, paper pulp, polymeric solutions such as rubbers, paints.

Dilatant fluids: Ultra fine irregular particle suspension, sugar in water, aqueous suspension of rice starch, quicksand, butter.

Ideal plastics or Bingham fluids: Sewage sludge, drilling muds.

Visco-elastic fluids: Liquid solid combination in pipe flow, bitumen, tar, asphalt, polymerized fluids with drag reduction features.

Thixotropic: Printer's ink, crude oil, lipstick, certain paints and enamels.

Rheopectic fluids: Very rare liquid-solid suspensions, gypsum suspension in water and bentonite solutions.

1.2 (c)

For thixotropic fluid $\tau = \tau_0 + \mu\left(\frac{du}{dy}\right)^n \quad n < 1$

As shear stress τ increases viscosity decreases with higher rates of shear strain.

1.3 (a)

- The dynamic viscosity of water is approximately 50 to 55 times of air at 20°C.
- The kinematic viscosity of air is nearly 15 times of water at 20°C.
- Water in soil rises due to capillary (not due to viscosity)
- Vapour pressure of a liquid is directly proportional to the temperature.

1.4 (b)

Example of Newtonian fluids: Air, Water, Kerosene oil, Diesel oil

Example of non-Newtonian fluids: Toothpaste, Human blood, Paint, Shampoo.

1.5 (b)

Kinematic viscosity of air is greater because the density of air is very small compared to that of water.

1.6 (d)

Shear stress,

$$\tau = \mu\frac{du}{dy} = \mu(0.5-2y)$$

at $y = 0.2$ m

$$\tau = 0.9 \times (0.5 - 2 \times 0.2) = 0.09 \text{ N/m}^2$$

1.7 (d)

Force, $F = \tau A$

$$F = \mu\frac{du}{dy}\times A$$

$$= 0.0014\times\frac{0.2}{0.02\times10^{-2}}\times0.15$$

$$= 0.21 \text{ N}$$

Power required

$$= F \times v$$

$$= 0.21 \times 0.2$$

$$= 0.042 \text{ W}$$

1.8 (c)

Let us suppose that the level of liquid has risen (or fallen) by 'h' above (or below) the general liquid surface when a tube of diameter 'd' is inserted in the liquid. For equilibrium of vertical forces acting on the mass of liquid lying above (or below) the general liquid level, the weight of liquid column 'h' (or the total internal pressure in the case of capillary depression) must be balanced by the force, at surface of the liquid, due to surface tension σ. Thus, equating these two forces we have

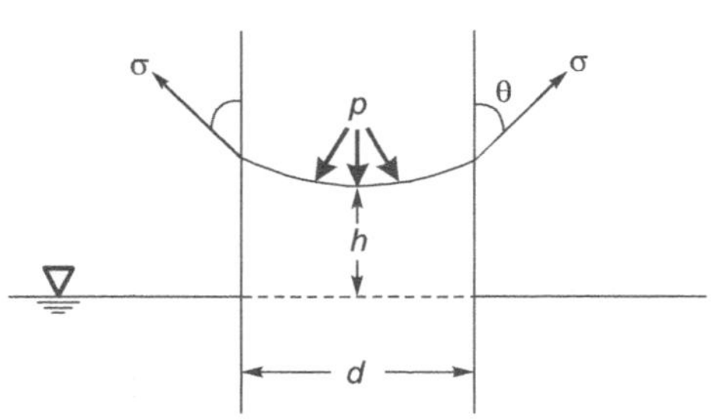

$$\frac{w\pi d^2}{4}\times h = \pi d\,\sigma\cos\theta$$

$$\Rightarrow \quad h = \frac{4\sigma\cos\theta}{wd}$$

1.9 (a)

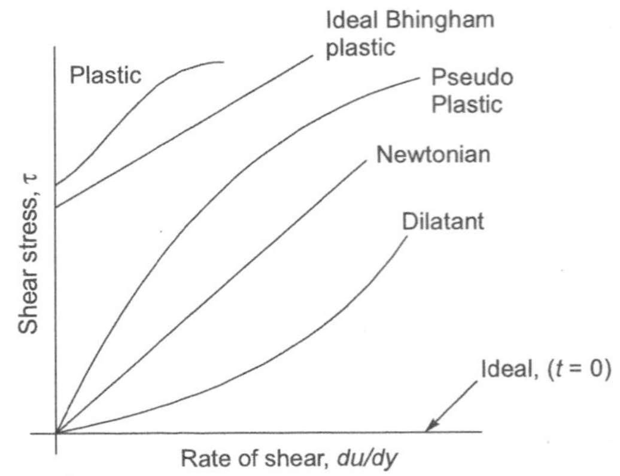

1.10 (a)

Viscosity of liquids varies inversely with temperature so as the glue is dries up and cools down, its viscosity increases. Thus movement of two blocks of wood welted with hot glue would require greater effort.

1.11 (a)

Specific gravity is the ratio of specific weight of a fluid to the specific weight of a standard fluid. Being a ratio of two quantities with same units, specific gravity has no unit.

$\therefore M^oL^oT^o$

Coefficient of viscocity = μ

unit of μ = Ns/m^2

$= (MLT^{-2})T/L^2$

$= ML^{-1}T^{-1}$

Kinematic viscocity :

$$\nu = \frac{\mu}{\rho}$$

unit of ν = $\text{m}^2\text{/s}$

$= L^2T^{-1}$

$= M^oL^2T^{-1}$

$$\text{Stress} = \frac{\text{Force}}{\text{Area}} = \frac{MLT^{-2}}{L^2}$$

$$= ML^{-1}T^{-2}$$

1.12 (d)

Poise is unit of dynamic viscocity.

$$\text{Poise} = \frac{\text{Dyne}-\text{s}}{\text{cm}^2}$$

$$1 \text{ Poise} = 0.1 \text{ Ns/m}^2$$

1.13 (a)

- At boiling point the vapour pressure of a fluid becomes equal to the atmospheric pressure. Water boils at 100°C therefore at 100°C at sea level the vapour pressure, is equal to the atmospheric pressure.
- Surface tension is caused by the force of cohesion between liquid molecules.
- Air is about 20000 times more compressible than water.

$$k_{air} = 1.03\times 10^5 \text{ N/m}^2$$
$$k_{water} = 2.06 \times 10^9 \text{ N/m}^2$$

1.14 (b)

Dynamic viscosity is the property of fluid in motion in which one layer of fluid exerts viscous force on the other layer. Cavitation occurs due to negative pressure.

Ideal fluids have no viscosity and surface tension and they are incompressible.

1.15 (d)

As the temperature increases, molecules move faster and there is an increase in molecular interaction. Molecules start colliding with each other, retarding the motion of gases, resulting in increase in viscosity of gases.

1.16 (b)

$$p\times\frac{\pi d^2}{4} = \sigma\times 2(2\pi r)$$
$$= \sigma\times 2\pi d$$
$$\sigma = \frac{pd}{8} = \frac{2.5\times 0.05}{8} = 0.0156 \text{ N/m}$$

1.17 (c)

In this problem, the bore is not defined properly. Here bore is used for radius.

$$h = \frac{2\sigma\cos\theta}{\rho\cdot g\cdot R}$$
$$= \frac{2(75\times 10^{-3})}{10^3\times 9.81\times(1\times 10^{-3})}$$
$$= 15.29 \text{ mm}$$

1.19 (c)

For soap bubble

$$\Delta P = \frac{4\sigma}{R}$$
$$2 = \frac{4\sigma}{0.01}$$
$$\sigma = 5 \times 10^{-3} \text{ N/m}$$

1.20 (c)

For jet of water

$$\Delta P = \frac{\sigma}{R}$$
$$\Delta P = \frac{0.072}{0.15\times 10^{-2}}$$
$$\Delta P = 0.048 \text{ kPa}$$

Absolute pressure within the jet

= 101.2 + 0.048 = 101.248 kPa

1.21 (a)

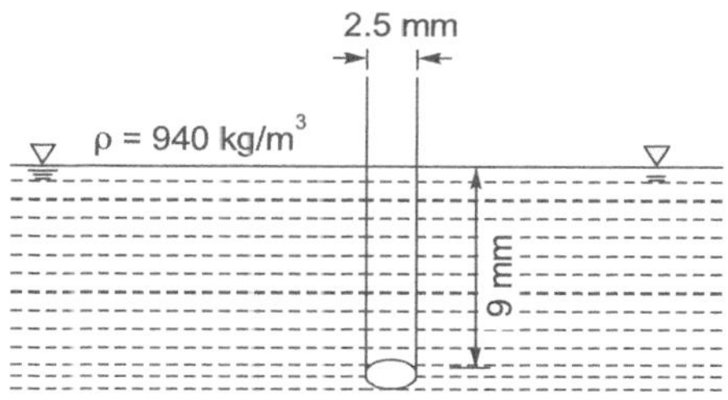

$$\Delta P = \frac{2\sigma}{R}$$
$$\Delta P = 148 - (940 \times 9.81 \times 0.009)$$
$$= 148 - 82.99 = 65.01$$

Now, $65.01 = \dfrac{2\sigma}{1.25\times 10^{-3}}$

$$\sigma = 0.04063 \text{ N/m}$$

1.22 (b)

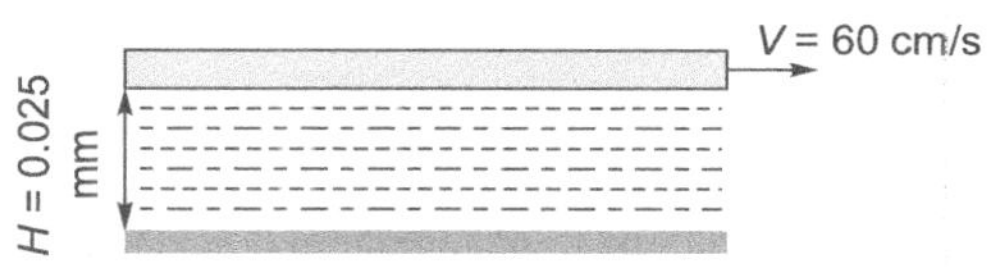

$$\frac{F}{A} = 0.2 \text{ kgf/m}^2 = \frac{0.2}{10^4} \text{ kgf/cm}^2$$

We know that

$$\tau = \mu.\frac{V}{H}$$
$$\frac{F}{A} = \mu.\frac{V}{H}$$
$$\frac{0.2}{10^4} = \mu\frac{(60)}{0.025\times 10^{-1}}$$
$$\mu = 8.3 \times 10^{-10} \text{ kgfs/cm}^2$$

1.23 (a)

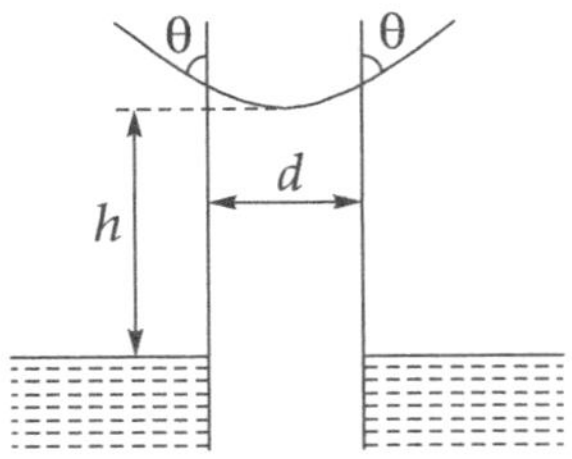

Capillary rise

$$h = \frac{4\sigma\cos\theta}{\rho g.d}$$

Given: Capillary rise

h = 2 mm

Surface tension

σ = 0.073575 N/m

For glass tube – water contact,

Angle of contact,

$\theta = 0°$

$$\therefore \quad 2\times10^{-3} = \frac{4\times0.073575\times1}{9.81\times1000\times d}$$

$$d = \frac{4\times0.073575}{9.81\times2}$$

$$= \frac{3}{200} \text{ m} = 1.5 \text{ cm}$$

∴ Size of the glass tube

d = 1.5 cm

1.24 (a)

Excess pressure,

$$\Delta P = \frac{2\sigma}{R} = \frac{2(.0725)}{0.02\times10^{-3}}$$

$$= 7250 \text{ N/m}^2 = 0.7250 \text{ N/cm}^2$$

$$= \frac{2(.0725)}{0.02\times10^{-3}}$$

$$= 7250 \text{ N/m}^2 = 0.7250 \text{ N/cm}^2$$

Now, $P_i - P_o$ = 0.725 N/cm²

$\Rightarrow P_i$ = 10.32 + 0.725 = 11.045 N/cm²

1.25 (b)

Given, $\nu = 6$ stokes $= 6\times10^{-4}$ m²/s

$S_l = 1.9$

$\therefore \quad \rho_l = 1.90\times10^3$ kg/m³

Now, $\nu = \dfrac{\mu}{\rho}$

$\mu = 6\times10^{-4}\times1.9\times10^3$

μ = 1.14 Ns/m² = 11.4 poise.

1.26 (c)

Bulk modulus, $k = \dfrac{\Delta p}{\dfrac{\Delta V}{V}}$

$\therefore \quad k \propto \Delta P$

$\Rightarrow$ Bulk modulus increases with increase in pressure.

■■■■

2 Manometry

2.1 Which one of the following expresses the difference in the pressure at the floors of the tank shown above in the figure?

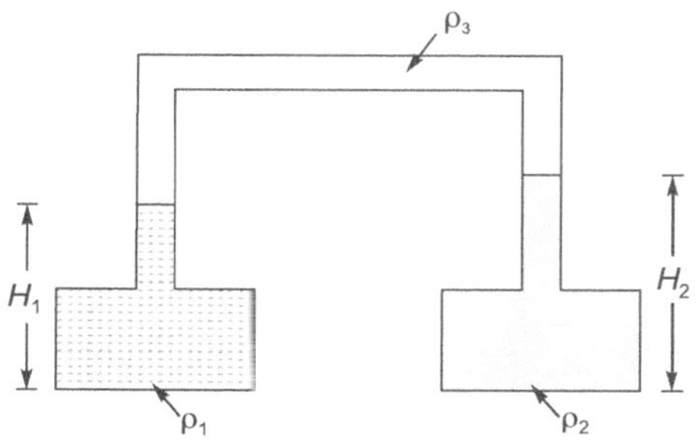

(a) $(\rho_2 - \rho_1)gH_2$
(b) $(\rho_2 - \rho_1)gH_1$
(c) $\rho_1 gH_1 + \rho_3 gH_2 - \rho_2 gH_2$
(d) $\rho_1 gH_1 + \rho_3 g(H_2 - H_1) - \rho_2 gH_2]$

where ρ_1, ρ_2 and ρ_3 are the densities of the different fluids.

[ESE : 2006]

2.2 Multi U-tube manometers with different fluids are used to measure
(a) low pressures
(b) medium pressures
(c) high pressures
(d) very low pressures

[ESE : 2006]

2.3 A pressure gauge reads 57.4 kPa and 80 kPa, respectively at heights of 8 m and 5 m fitted on the side of a tank filled with liquid. What is the approximate density of the liquid in kg/m^3?

(a) 393 (b) 768
(c) 1179 (d) 7530

[ESE : 2008]

2.4 In the below figure the pressure gauge will record a gauge pressure equivalent to

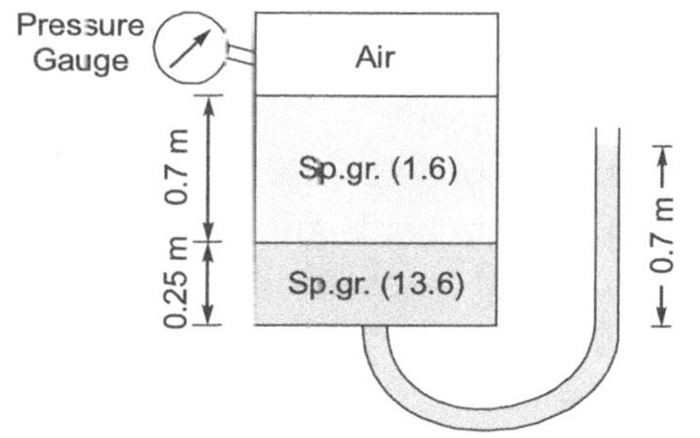

(a) 6.12 m of water
(b) 1.21 m of mercury
(c) 0.5 bar
(d) 34 000 Pa

[ESE : 2011]

2.5 The absolute pressure at a point 2.5 m below the clear water surface is measured as 125.703 kN/m^2. If the atmospheric pressure is taken as 101.325 kN/m^2, the gauge pressure in kN/m^2 at this point would be

(a) 113.514 (b) 24.378
(c) 45.401 (d) 56.757

[ESE : 2013]

2.6 The standard atmospheric pressure is 101.32 kPa. The local atmospheric pressure is 91.52 kPa. If a pressure at a flow path is recorded at 22.48 kPa (gauge), it is equivalent to

(a) 69.04 kPa (abs) (b) 88.4 kPa (abs)
(c) 114.0 kPa (abs) (d) 123.0 kPa (abs)

[ESE : 2014]

2.7 What is the depth of a point below water surface in sea, where pressure intensity is 1.006 MN/m^2? (Specific gravity of sea water is 1.025)

(a) 60 m (b) 80 m
(c) 100 m (d) 120 m

[ESE : 2021]

2.8 Two pressure points in a water pipe are connected to a manometer which has the form of an inverted U-tube. The space above the water in the two limbs of the manometer is filled with toluene (specific gravity is 0.875). If the difference of level of water columns in the two limbs reads 12.0 cm, what is the corresponding difference of pressure? (Take g = 9.81 m/s^2)

(a) 110.49 N/m^2 (b) 128.12 N/m^2
(c) 131.34 N/m^2 (d) 147.15 N/m^2

[ESE : 2021]

2.9 If pressure head of water is 100 m and specific gravity of kerosene is 0.81, what is the pressure head of kerosene?

(a) 123.5 m of kerosene
(b) 241.3 m of kerosene
(c) 75.1 m of kerosene
(d) 52.4 m of kerosene [ESE : 2021]

2.10 If V_s is the velocity of the vector, r is the radius of cuvature and $\frac{\partial V_n}{\partial t}$ is the local normal acceleration; what is the expression for total normal acceleration of fluid particles?

(a) $\frac{dV_n}{dt} = V_s^2 + r\frac{\partial V_n}{\partial t}$ (b) $\frac{dV_n}{dt} = \frac{V_s^2}{r} + \frac{\partial V_n}{\partial t}$

(c) $\frac{dV_n}{dt} = r + V_s^2\frac{\partial V_n}{\partial t}$ (d) $\frac{dV_n}{dt} = \frac{r}{V_s^2} + \frac{\partial V_n}{\partial t}$

[ESE : 2024]

■■■■

Answers Manometry

2.1 (d) 2.2 (c) 2.3 (b) 2.4 (c) 2.5 (b) 2.6 (c) 2.7 (c) 2.8 (d) 2.9 (a) 2.10 (b)

Explanations Manometry

2.1 (d)

$$p_1 - \rho_1 gH_1 - \rho_3 g(H_2 - H_1) + \rho_2 gH_2 = p_2$$

$$\therefore\ p_1 - p_2 = \rho_1 gH_1 + \rho_3 g(H_2 - H_1) - \rho_2 gH_2$$

2.2 (c)

To measure a very high pressure, a single U-tube manometer used may become too tall. So a number of U-tubes of reasonable sizes are used with same or different manometric fluids.

2.3 (b)

The pressure difference may be given as

$$p_1 - p_2 = \rho g\,(h_2 - h_1)$$

$\Rightarrow$ $80 \times 10^3 - 57.4 \times 10^3 = \rho \times 9.81 \times (8 - 5)$

$\Rightarrow$ $22600 = \rho \times 29.43$

$\Rightarrow$ $\rho = 768 \text{ kg/m}^3$

2.4 (c)

Let gauge pressure be = P

$P + (1.6 \times \rho_w) \times g \times 0.7 + (13.6 \times \rho_w) \times g \times 0.25 - (13.6 \times \rho_w) \times g \times 0.7 = 0$

$P = 5 \times g \times \rho_w$

$$= 50000\frac{\text{N}}{\text{m}^2} = \frac{50000}{10^5}\text{bar} = 0.5 \text{ bar}$$

2.5 (b)

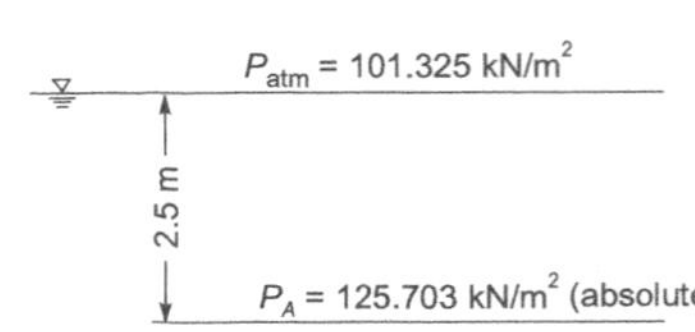

Gauge pressure at 2.5 m depth level

$\therefore$ $= 125.703 - 101.325$

$= 24.378 \text{ kN/m}^2$

2.6 (c)

Absolute pressure

= Local atmospheric pressure + Gauge pressure

= 91.52 + 22.48

= 114 kPa

2.7 (c)

Given: $P_B = 1.006 \text{ MN/m}^2 = 1.006 \times 10^6 \text{ N/m}^2$

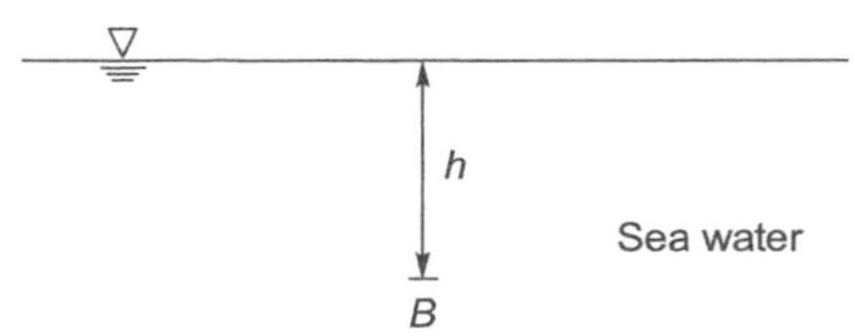

Special gravity,

$$s = 1.025$$

Pressure at any depth 'h' below water surface:

$$P_B = \gamma_{sea} h = \rho_{sea}\, gh = s\rho_w gh$$

$$1.006 \times 10^6 = 1.025 \times 10^3 \times 9.81 \times h$$

$$h = 100.05 \text{ m} \simeq 100 \text{ m}$$

2.8 (d)

Inverted U-tube manometer

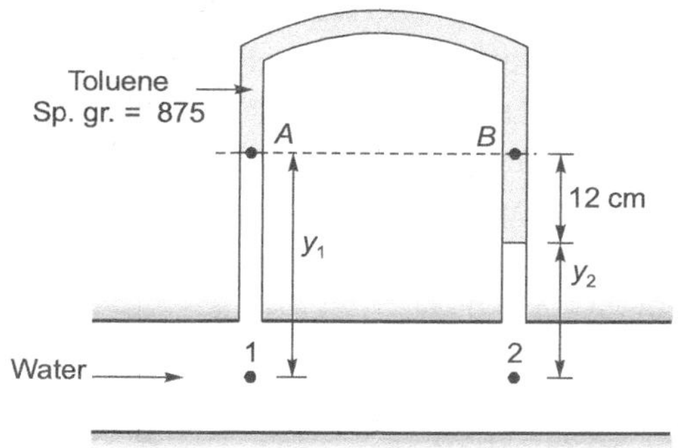

From figure, we write

$$P_A = P_B$$

$$P_1 - \rho g y_1 = P_2 - \rho g y_2 - \rho_{toluene}\, g\,(0.12)$$

$$P_1 - P_2 = \rho g y\,(y_1 - y_2) - \rho_{toulene} \times g \times 0.12$$

$$= 1000 \times 9.81 \times 0.12 - 0.875 \times 1000 \times 9.81 \times 0.12$$

$$= 1000 \times 9.81 \times 0.12\,(1 - 0.875)$$

$$\Rightarrow \quad P_1 - P_2 = 147.15 \text{ N/m}^2$$

2.9 (a)

For equal pressure at any point in two liquid column

$$(P)_{water} = (P)_{kerosene}$$

$$\rho_w g h_{water} = S\rho_w g h_{kerosene}$$

$$100 = 0.81 \times h_{kerosene}$$

$$\therefore \quad h_{kerosene} = \frac{100}{0.81} = 123.456 \text{ m} \simeq 123.5 \text{ m}$$

2.10 (b)

Total normal acceleration,

$$\frac{dV_n}{dt} = \underbrace{\frac{V_s^2}{r}}_{\text{convective}} + \underbrace{\frac{\partial V_n}{\partial t}}_{\text{local}}$$

■■■■

3 Hydrostatic Forces

3.1 When the water surface coincides with the top edge of a rectangular vertical gate 40 m (wide) × 3 m (deep), then the depth of centre of pressure is

(a) 1 m (b) 1.5 m
(c) 2 m (d) 2.5 m

[ESE : 2000]

3.2 An isosceles triangular plate of base 3 m and altitude 3 m is immersed vertically in an oil of specific gravity 0.8. The base of the plate coincides with the free surface of oil. The centre of pressure will lie at a distance of (from free surface)

(a) 2.5 m (b) 2 m
(c) 1.5 m (d) 1 m

[ESE : 2002]

3.3 As the depth of immersion of a vertical plane surface increases, the location of centre of pressure

(a) falls closer to the centre of gravity of the area
(b) moves away from the centre of gravity of the area
(c) ultimately coincides with the centre of gravity of the area
(d) falls much below the centre of gravity of the area

[ESE : 2003]

3.4 A rectangular tank 10 m × 5 m in plan and 3 m deep is divided by a partition wall parallel to the shorter wall of the tank. One of the compartments contains water to a depth of 3 m, and the other a lighter liquid of specific gravity 0.75 to a depth of 2 m. The resultant pressure thrust on the partition wall is

(a) 1000 kg (b) 1500 kg
(c) 2000 kg (d) 2500 kg

[ESE : 2012]

3.5 A square gate, 1.5 m × 1.5 m, one of the vertical sides of a fully filled water tank, has one side on the free water surface. It is hinged on the lower horizontal side and is held in position by a force applied on the vertical central line at a depth of 0.75 m below the free surface. The right magnitude of this force is

(a) 500 × 9.81 N
(b) 600 × 9.81 N
(c) 750 × 9.81 N
(d) 1000 × 9.81 N

[ESE : 2015]

3.6 A tank is 1.8 m deep and square length of 4.5 m at the top and square length of 3 m at the bottom. The four sides are plane and each has the same trapezoidal shape. The tank is completely full of oil weighing 936 kg/m^3. What is the resultant pressure on each side?

(a) 5750 kgf (b) 5500 kgf
(c) 5250 kgf (d) 5140 kgf

[ESE : 2016]

3.7 An oil of specific gravity 0.9 contained in a vessel. At a point the height of oil is 40 m and for the density of water = 1000 kg/m^3, the corresponding height of water at the point will be

(a) 28 m (b) 32 m
(c) 36 m (d) 40 m

[ESE : 2020]

3.8 A hydraulic press has a ram of 300 mm diameter and a plunger of 45 mm diameter. When the force applied at the plunger is 50 N, the weight lifted by the hydraulic press will be nearly

(a) 2133 N (b) 2223 N
(c) 2316 N (d) 2406 N

[ESE : 2020]

3.9 A 2 m wide tank contains water upto a height of 0.50 m above its base. An immiscible liquid of specific gravity 0.80 is filled on the top of the water upto 1 m height. What is the total pressure force on one side of the tank? (Take density of water as 1000 kg/m^3 and g = 9.81 m/s^2)

(a) 7.85 kN (b) 24.52 kN
(c) 10.3 kN (d) 18.15 kN

[ESE : 2022]

3.10 The pressure distribution in a liquid subjected to a constant horizontal acceleration is

(a) same as hydrostatic pressure distribution
(b) less than the hydrostatic pressure distribution
(c) more than the hydrostatic pressure distribution
(d) equal to hydrostatic pressure distribution minus weight of liquid

[ESE : 2023]

Answers Hydrostatic Forces

3.1 (c) 3.2 (c) 3.3 (a) 3.4 (*) 3.5 (*) 3.6 (a) 3.7 (c) 3.8 (b) 3.9 (d)
3.10 (a)

Explanations Hydrostatic Forces

3.1 (c)

Depth of centre of pressure,

$$\bar{h}_{cp} = \frac{2d}{3} = \frac{2\times 3}{3} = 2\text{m}$$

3.2 (c)

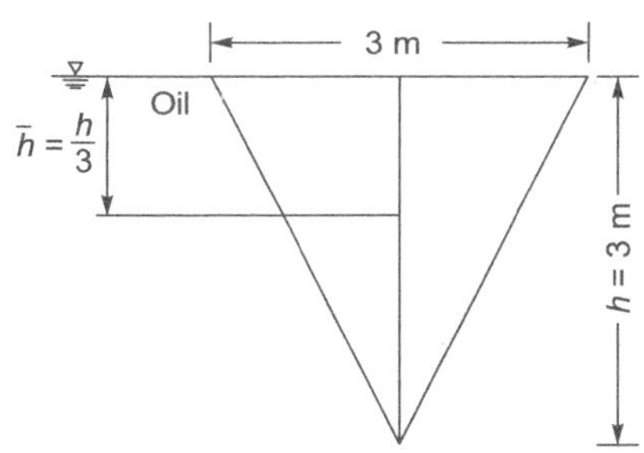

Centre of Pressure depth

$$\bar{h}_{cp} = \bar{h} + \frac{I_{Gxx}\sin^2\theta}{\bar{h}A}$$

$$= \frac{3}{3} + \left(\frac{3\times 3^3}{36\times\frac{1}{2}\times 3\times 3\times\frac{3}{3}}\right)$$

$$= 1 + 0.5 = 1.5 \text{ m}$$

3.4 (*)

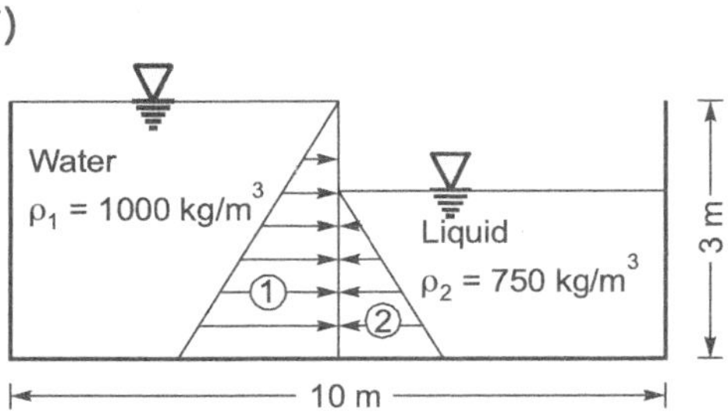

$$P_1 = \rho_1\times\frac{3}{2}\times(3\times 5)$$

$$= 1000\times\frac{3}{2}\times 3\times 5 = 22500 \text{ kg}$$

$$P_2 = \rho_2\times\frac{2}{2}\times(2\times 5)$$

$$= 750\times 1\times 2\times 5 = 7500 \text{ kg}$$

Resultant pressure thrust,

$$P_R = P_1 - P_2 = 15000 \text{ kg}$$

3.5 (*)

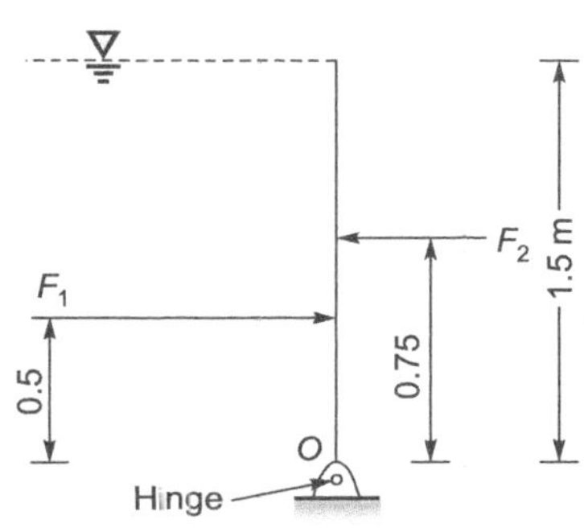

Taking moment about hinge O

$$F_1\times 0.5 = F_2\times 0.75$$

$$(\rho g A\bar{h})0.5 = F_2\times 0.75$$

$$1000\times 9.81\times(1.5\times 1.5)\times 0.75\times 0.5 = F_2\times 0.75$$

$$F_2 = 1125\times 9.81 \text{ N}$$

3.6 (a)

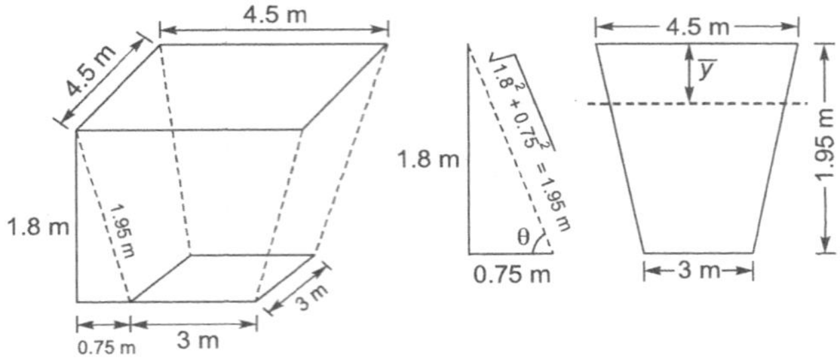

$$\sin\theta = \frac{1.8}{1.95}$$

$$\bar{y} = \left[\frac{2(3)+4.5}{3+4.5}\right]\times\frac{1.95}{3} = 0.91$$

Now, $\sin\theta = \dfrac{\bar{h}}{0.91}$

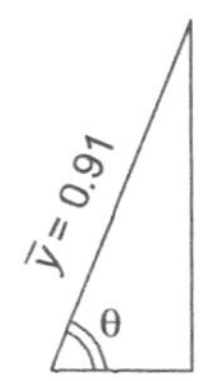

$$\frac{1.8}{1.9} = \frac{\bar{h}}{0.91}$$

$$\bar{h} = 0.84$$

$$F = \rho\cdot g\cdot \bar{h}\cdot A$$

$$= (936)(9.81)(0.84)\times\left[\frac{1}{2}\times(3+4.5)\times 1.95\right]\text{N}$$

$$= 5749.38 \text{ kgf}$$

3.7 (c)

Pressure due to oil column height
= Pressure due to water column height

$(900)g\,40 = (1000)gh$

h = 36 m of water column

3.8 (b)

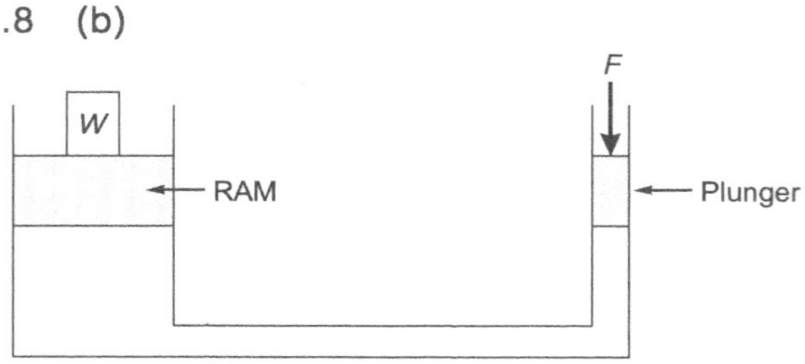

According to Pascal law,

$$\frac{W}{A} = \frac{F}{a}$$

A = Area of Ram, a = Area of plunger

$$\frac{W}{\frac{\pi}{4}\times(0.3)^2} = \frac{50}{\frac{\pi}{4}\times(0.045)^2}$$

$$W = 2222.22 \text{ N} \simeq 2223 \text{ N}$$

3.9 (d)

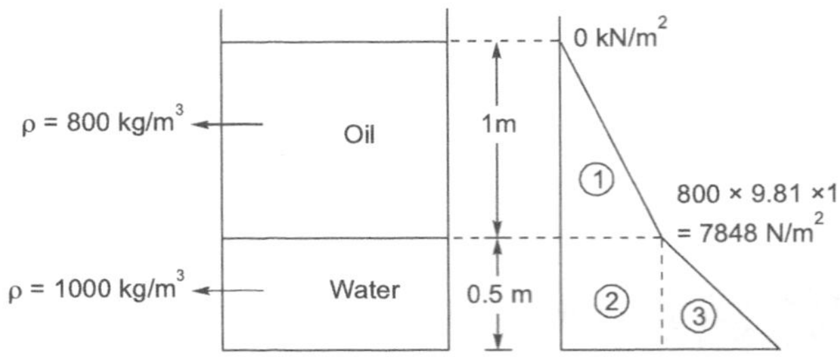

Pressure diagram:

$800\times 9.81\times 1 + 1000\times 9.81\times 0.5$
$= (7848 + 4905)$ N/m² $= 12753$ N/m²

On one side of tank

Total pressure force tank,

$$F = \left[\frac{1}{2}\times 1\times 7848 + 0.5\times 7848 + \frac{1}{2}\times 0.5\times(12753-7848)\right]\times 2$$

$\Rightarrow F$ = 18148.5 N = 18.15 kN

3.10 (a)

Concept of acceleration vessel containing liquid:

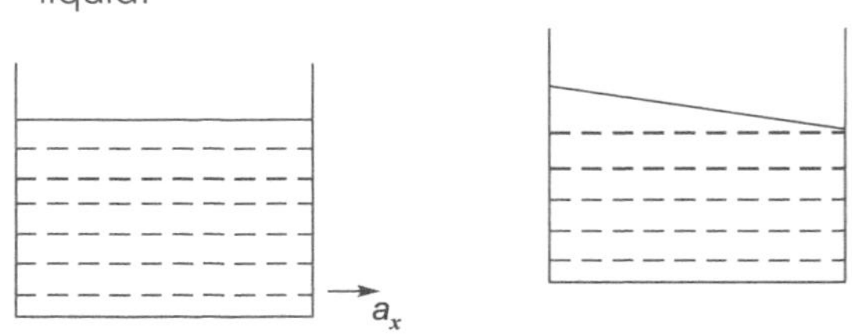

$$dP = \rho a_x\,dx - \rho(a_z + g)dz$$

$$dP = -\rho g dz$$

$$\frac{dP}{dz} = -\rho g$$

From the above equation, it is clear that pressure head at any point in a liquid subjected to a constant horizontal acceleration is equal to the height of the liquid column above that point. Therefore, the pressure distribution in a liquid subjected to a constant horizontal acceleration is same as hydrostatic pressure distribution. Planes of constant pressure are therefore, parallel to the inclined surface.

■■■■

4 Buoyancy & Floatation

4.1 A body is floating as shown in the given figure. The centre of buoyancy, centre of gravity and metacentre are labelled respectively as *B*, *G* and *M*. The body is

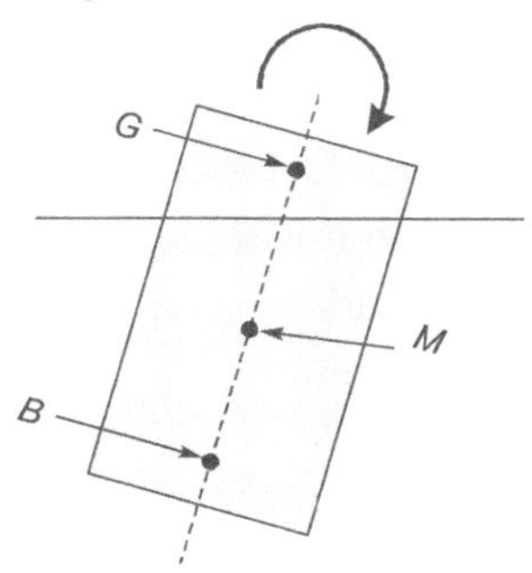

(a) vertically stable
(b) vertically unstable
(c) rotationally stable
(d) rotationally unstable [ESE : 2000]

4.2 If a water tank, partially filled with water is being carried on a truck, moving with a constant horizontal acceleration, the level of liquid will
(a) rise and fall alternately on the front side of the tank
(b) fall on the rear side of the tank
(c) remain the same on both sides of the tank
(d) rise on the rear side and fall on the front side of the tank

[ESE : 2000]

4.3 A metal cube of size 15 cm × 15 cm × 15 cm and specific gravity 8.6 is submerged in a two-layered liquid, the bottom layer being mercury and the top layer being water. The percentage of the volume of the cube remaining above the interface will be, approximately
(a) 68 (b) 63
(c) 40 (d) 25 [ESE : 2001]

4.4 Consider the following statements related to buoyancy in fluid statics:
1. Principle of buoyancy is applicable both to floating bodies and to submerged bodies.
2. Archimedes formulated the first theory of buoyancy.
3. In analyzing buoyancy of a floating body it is assumed that the resultant vertical force passes through centre of pressure.
4. In a free-body diagram of a floating body summation of all horizontal forces is taken as zero.

Which of these statements are correct?
(a) 1, 3 and 4 (b) 1, 2 and 4
(c) 1, 2 and 3 (d) 2, 3 and 4

[ESE : 2003]

4.5 A solid cylinder of length *L*, diameter *D* and specific gravity 0.6 floats in neutral equilibrium in water with its axis vertical. What is the ratio of *L* to *D*?

(a) $\frac{\sqrt{3}}{2}$ (b) $\frac{2\sqrt{3}}{5}$

(c) $\frac{4}{5\sqrt{3}}$ (d) $\frac{5}{4\sqrt{3}}$ [ESE : 2004]

4.6 Consider the figure below relating to buoyancy in water.

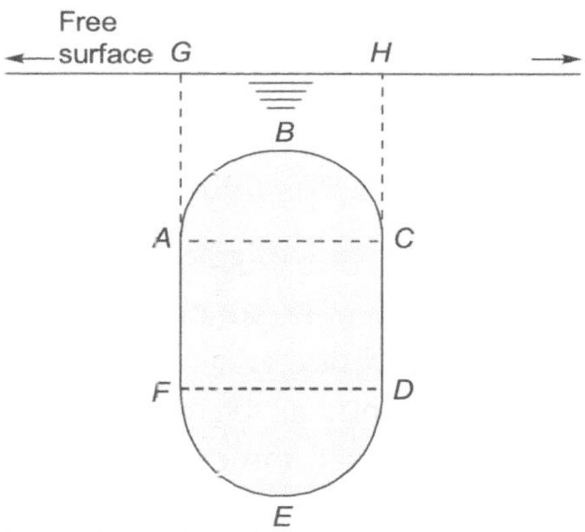

What will be the downward force upon the top of the body *ABCDEF*?
(a) The weight of the liquid column *ABCHG*
(b) The weight of the liquid column *DEFGH*
(c) The weight of the liquid column *ABCHG* – the weight of the liquid column *DEFGH*

(d) The weight of the liquid column *ABCHG* + the weight of the liquid column *DEFGH*

[ESE : 2009]

4.7 **Statement (I):** Fluid pressure is a scalar quantity.
Statement (II): Fluid thrust always acts downwards.

(a) Both Statement (I) and Statement (II) are individually true and Statement (II) is the correct explanation of Statement (I)
(b) Both Statement (I) and Statement (II) are individually true but Statement (II) is NOT the correct explanation of Statement (I)
(c) Statement (I) is true but Statement (II) is false
(d) Statement (I) is false but Statement (II) is true

[ESE : 2013]

4.8 A rectangular pontoon has a width of 6 m, a length of 12 m, and a draught of 1.5 m in fresh water (density = 1000 kg/m^3). Its draught in sea water having density of 1025 kg/m^3 is

(a) 1.04 m (b) 1.24 m
(c) 1.46 m (d) 1.50 m [ESE : 2014]

4.9 A ship has a metacentric height of 0.90 m and its period of rolling is 20 seconds. The relevant radius of gyration is nearly

(a) 5.5 m (b) 7.5 m
(c) 9.5 m (d) 11.5 m

[ESE : 2015]

4.10 A solid cylinder of length *H*, diameter *D* and of relative density *S* floats in neutral equilibrium in water with its axis vertical. What is the ratio of *H* to *D* if *S* = 0.6?

(a) 0.86 (b) 0.72
(c) 0.52 (d) 0.46 [ESE : 2016]

4.11 A solid cylinder of circular section of diameter *d* is of material with specific gravity S_s. This floats in a liquid of specific gravity S_l. What is the maximum length of the cylinder if equilibrium is to be stable with the cylinder axis vertical?

(a) $\dfrac{dS_s}{2\sqrt{S_s(S_l - S_s)}}$ (b) $\dfrac{dS_l}{\sqrt{8S_s(S_l - S_s)}}$

(c) $\dfrac{dS_l}{\sqrt{2S_s(S_l - S_s)}}$ (d) $\dfrac{d}{\sqrt{8(S_l - S_s)}}$

[ESE : 2016]

4.12 An open rectangular tank of dimensions 4m × 3m × 2m contains water to a height of 1.6m. It is then accelerated along the longer side. What is the maximum acceleration possible without spilling the water? If this acceleration is then increased by 10%, what amount of water will be spilt off?

(a) 1.472 m/s^2 and 0.48m^3
(b) 1.962m/s^2 and 0.48m^3
(c) 1.472 m/s^2 and 0.52m^3
(d) 1.962m/s^2 and 0.52m^3 ESE : 2016]

4.13 An ocean liner, 240 m long and 24 m wide, displaces 654 MN of sea-water (ρ = 1025 kgf/m^3). The second moment of inertia of the water plane about its fore-aft axis is 2/3 of that of the circumscribing rectangle. The position of the centre of buoyance is 2.30 m below the centre of gravity. How high is metacentre above the centre of buoyancy (to the nearest cm)?

(a) 49 cm (b) 53 cm
(c) 58 cm (d) 65 cm [ESE : 2017]

4.14 A ship weighs 127 MN. On filling the ship's boats on one side with water weighing 600 kN with the mean distance of the boats from the centre line of the ship being 10 m, the angle of displacement of the plumb line is 2°16′. The metacentric height will be nearly
(Take sin 2°16′ = 0.04, cos 2°16′ = 0.9992 and tan 2°16′ = 0.04)

(a) 1.73 m (b) 1.42 m
(c) 1.18 m (d) 0.87 m

[ESE : 2019]

4.15 Consider the following statements regarding stability of floating bodies:

1. If metacentre (M) lies above centroid (G), then the body is said to be in unstable equilibrium.
2. If metacentre (M) lies below centroid (G), then the body is said to be in unstable equilibrium.
3. If metacentre (M) coincides with centroid (G), then the body is said to be in stable equilibrium.

Which of the above statements is/are correct?

(a) 1 and 3 (b) 2 and 3
(c) 2 only (d) 1 only

[ESE : 2024]

■■■■

Answers Buoyancy & Floatation

4.1 (d) 4.2 (d) 4.3 (c) 4.4 (b) 4.5 (d) 4.6 (a) 4.7 (c) 4.8 (c) 4.9 (c)
4.10 (b) 4.11 (b) 4.12 (b) 4.13 (b) 4.14 (c) 4.15 (c)

Explanations Buoyancy & Floatation

4.1 (d)

Since metacentre is below centre of gravity the body is in rotationally unstable equilibrium in rotation.

4.2 (d)

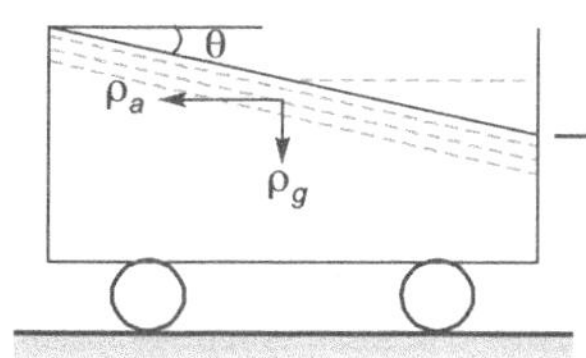

Horizontal acceleration

$$\theta = \tan^{-1}\left(\frac{a}{g}\right)$$

Water rises by θ angle on rear side and falls on the front end.

4.3 (c)

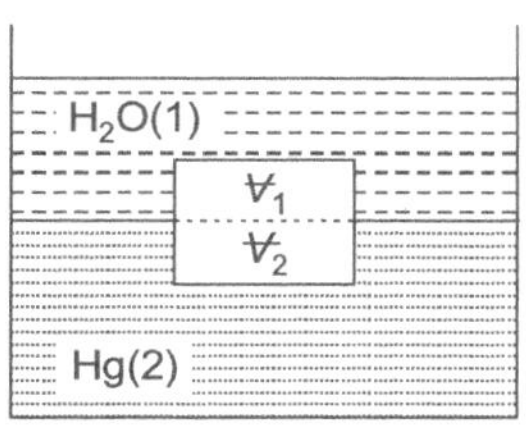

Specific gravity of block = 8.6

Force analysis

$$M_b g = F_{B_1} + F_{B_2}$$

$$\rho_b \forall_b g = \rho_w \forall_1 g + \rho_{Hg} \forall_2 g$$

Divide by ρ_w

$$\frac{\rho_b}{\rho_w}\forall_b = \forall_1 + \frac{\rho_{Hg}}{\rho_w}\forall_2 \qquad (\forall_2 = \forall_b - \forall_1)$$

$$8.6\forall_b = \forall_1 + 13.6(\forall_b - \forall_1)$$

$$\frac{\forall_1}{\forall_b}\times 100 = \frac{5}{12.6}\times 100 = 39.7$$

4.4 (b)

In analysing buoyancy of a floating body it is assumed that the resultant vertical force passes through centre of buoyancy so statement 3 is incorrect.

4.5 (d)

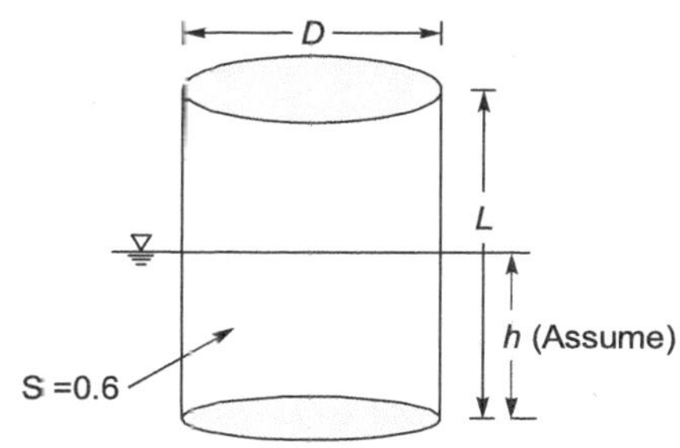

$$GM = 0$$

$$\frac{I}{V_{Displaced\ liquid}} - BG = 0$$

$$\frac{\frac{\pi D^4}{64}}{\frac{\pi D^2}{4}\times h} = \frac{L}{2} - \frac{h}{2}$$

$$\frac{D^2}{8h} = L - h \qquad \text{...(i)}$$

Force balancing

Weight of body = F_B

$$(600)\frac{\pi D^2}{4}\times L\times g = 10^3\times\frac{\pi D^2}{4}\times h\times g$$

$$h = 0.6\,L \qquad \text{...(ii)}$$

From (i) and (ii)

$$\frac{D^2}{8(0.6L)} = L - 0.6\,L$$

$$\frac{D^2}{L^2} = 8(0.6)(0.4)$$

$$\frac{D}{L} = \frac{4\sqrt{3}}{5}$$

or

$$\frac{L}{D} = \frac{5}{4\sqrt{3}}$$

4.6 (a)

The downward force upon the top of the body *ABCDEF* will be equal to the weight of the liquid contained in the portion *ABC* extending vertically above the curved surface upto the free surface of the liquid.

4.7 (c)

Fluid pressure is a scalar quantity. Its magnitude is defined but its direction is not. Fluid thrust can act in any direction.

4.8 (c)

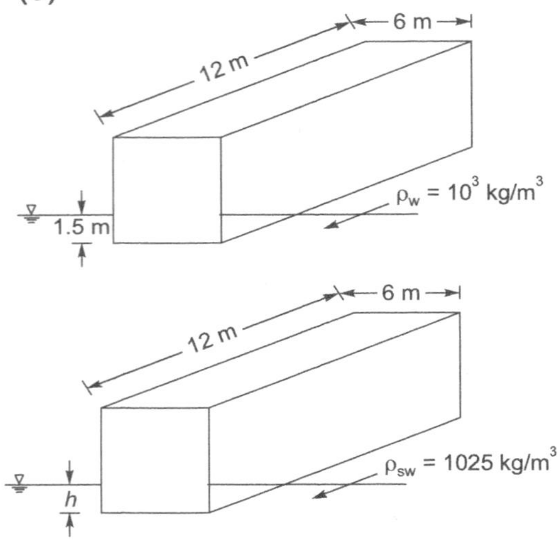

wt. of body = Buoyancy force in fresh water (F_{B_1})

= Buoyancy force in sea water (F_{B_2})

So, $F_{B_1} = F_{B_2}$

$$\rho_w g \forall_1 = \rho_{sw} g \forall_2$$

$$(10)^3 (12 \times 6 \times 1.5) = (1025)(12 \times 6 \times h)$$

$$h = 1.46 \text{ m}$$

4.9 (c)

Time period,

$$T = 2\pi\sqrt{\frac{k^2}{gGM}} \text{ seconds}$$

k = radius of gyration

GM = metacentric height = 0.90 m

$$20 = 2\pi\sqrt{\frac{k^2}{9.81 \times 0.9}}$$

$$k = 9.5 \text{ m}$$

4.10 (b)

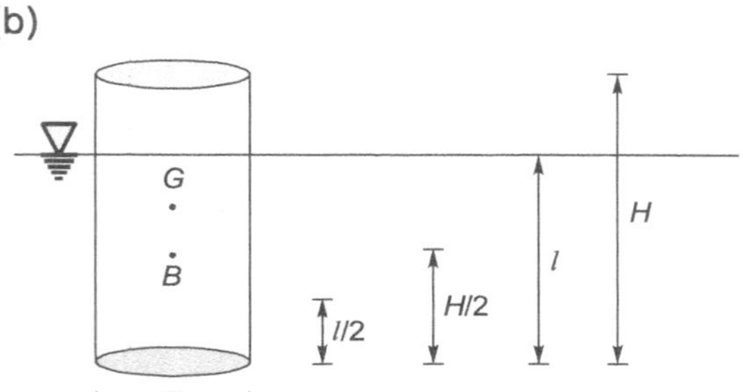

$$F_B = M_b \cdot g$$

$$(10^3)\left(\frac{\pi D^2}{4}\right) \cdot l \cdot g = (600) \cdot g \cdot H\left(\frac{\pi D^2}{4}\right)$$

$$l = 0.6\,H \quad \text{...(i)}$$

For Neutral equilibrium,

$$GM = 0$$

$$\frac{I}{\forall} - BG = 0$$

$$\frac{\frac{\pi D^4}{64}}{\frac{\pi D^2 \cdot l}{4}} - \frac{1}{2}[H - l] = 0$$

$$\frac{D^2}{16} - \frac{1}{2}[H - l] = 0$$

$$\frac{D^2}{8} = H - l$$

$$\frac{D^2}{8l} = H[1 - 0.6]$$

$$\frac{H^2}{D^2} = \frac{1}{8 \times 0.6 \times 0.4}$$

$$\frac{H}{D} = 0.721$$

4.11 (b)

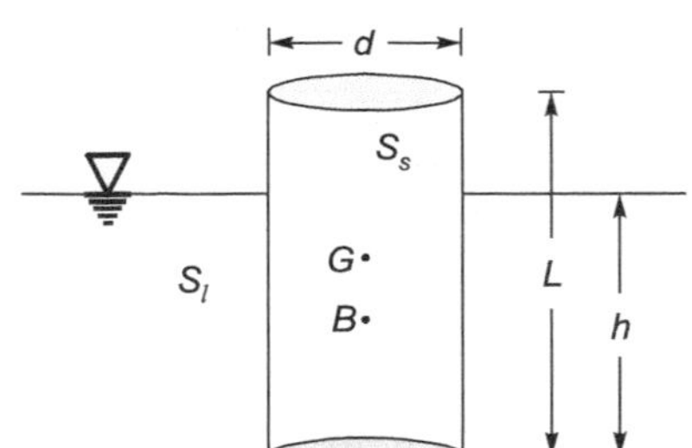

$$M_b g = F_B$$

$$\rho_s\left[\frac{\pi d^2}{4}\right]Lg = \rho_l\left[\frac{\pi d^2}{4}\right]hg$$

$$\rho_s L = \rho_l h$$

$$h = \frac{\rho_s}{\rho_l} \times L$$

or $$h = \frac{S_s}{S_l} \times L \quad \text{...(i)}$$

Take $$GM = 0$$

$$\frac{I}{\forall} - BG = 0$$

$$\frac{\frac{\pi d^4}{64}}{\frac{\pi d^2}{4}\times h}-\frac{1}{2}(L-h)=0$$

$$\frac{d^2}{16h}-\frac{1}{2}[L-h]=0$$

$$\frac{d^2}{8h}=L-h$$

$$d=\sqrt{8h(L-h)}$$

$$=\sqrt{8\frac{S_l}{S_l}L\left(L-\frac{S_sL}{S_l}\right)}=L\sqrt{8\frac{S_s}{S_l}\left[1-\frac{S_s}{S_l}\right]}$$

$$=\frac{L}{S_l}\sqrt{8S_s\,[S_l-S_s]}$$

$$L=\frac{d\times S_l}{\sqrt{8S_s\,(S_l-S_s)}}$$

4.12 (b)

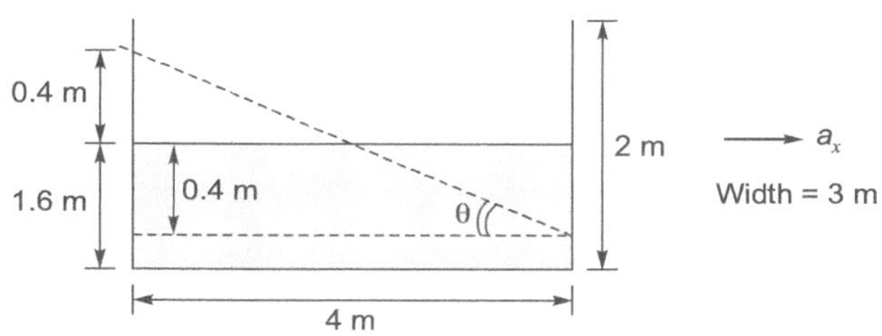

$$\tan\theta=\frac{a_x}{g}$$

$$\frac{0.8}{4}=\frac{a_x}{9.81}$$

$a_x = 1.962$ m/s^2

Now acceleration is increased by 10%

$a_x = 1.962 + 0.1962 = 2.158$ m/sec^2

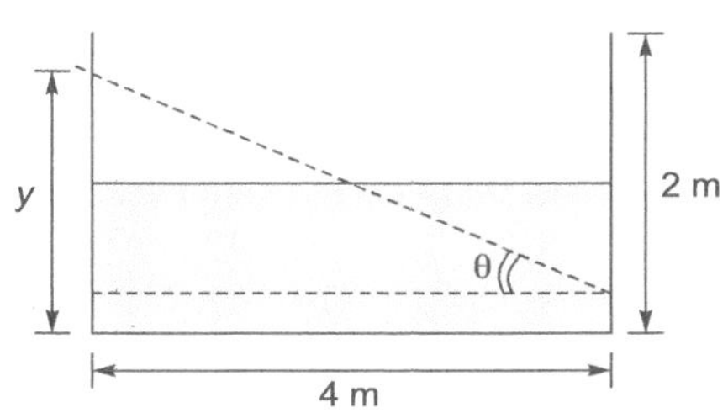

$$\tan\theta=\frac{y}{4}$$

$$\frac{2.158}{9.81}=\frac{y}{4}$$

$y = 0.8799$ m

Final volume

$$=(4\times3\times2)-\left[\frac{1}{2}\times0.88\times4\times3\right]=18.72\text{ m}^3$$

Initial volume = 4 × 3 × 1.6 = 19.2 m^3

Spillout volume = 19.2 − 18.72 = 0.48 m^3

4.13 (b)

$\rho_{sea\ water}\times\forall_{displaced}\times g$ = wt. of displaced water

$$1025\times\forall_{displaced}\times9.81=654\times10^6$$

$$\forall_{displaced}=65040.65\text{ m}^3$$

$$\frac{I}{\forall_{displaced}}-BG=GM$$

$$\frac{\frac{2}{3}\times240\times\frac{24^3}{12}}{65040.65}-2.3=GM$$

$$2.833-2.3=GM$$

So, GM = 0.533 m = 53.3 cm ≃ 53 cm

4.14 (c)

According to experimental method of determination of metacentric height

$$GM=\frac{wx}{W\tan\theta}$$

where, w = 600 kW, W = 127 MN = 127 × 10^3 kN, θ = 2°16″, x = 10 m

$$GM=\frac{600\times10}{127\times10^3\tan2°16''}$$

$$=\frac{6000}{127\times10^3\times0.04}$$

$$=1.18\text{ m}$$

4.15 (c)

- If matecentre (M) lies above centroid (G), then the body is said to be in stable equilibrium.
- If metacentre (M) coincides with centroid (G), then the body is said to be in neutral equilibrium.

Therefore only statement 2 is correct.

5 Fluid Kinematics

5.1 A two-dimensional flow is described by velocity components $u = 2x$ and $v = -2y$. The discharge between points (1, 1) and (2, 2) is equal to

(a) 9 units (b) 8 units
(c) 7 units (d) 6 units **[ESE : 2000]**

5.2 The stream function for a two-dimensional flow is given by $\psi = 2xy$. The resultant velocity at a point $P(2, 3)$ is

(a) 8.45 units/s (b) 7.21 units/s
(c) 6.44 units/s (d) 5.18 units/s

[ESE : 2003]

5.3 Consider the following parameters related to fluid flow:

1. Vorticity
2. Velocity potential
3. Stream function

Which of these parameters exist both in rotational and irrotational flows?

(a) 1 and 2 (b) 2 and 3
(c) 1 and 3 (d) 1, 2 and 3

[ESE : 2003]

5.4 Consider the following statements related to concept of continuity equation and the concept of control volume in deriving the equation:

1. Continuity equation relates velocity component and density of the fluid at a point in a fluid flow.
2. Continuity equation assumes that no void occurs in the fluid and fluid mass is neither created nor destroyed.
3. The shape of control volume for deriving the equation of continuity is assumed to be a parallelepiped.
4. For incompressible fluids the equation of continuity does not contain the viscosity terms.

Which of these statements are correct?

(a) 1, 2, 3 and 4 (b) 1 and 2
(c) 2, 3 and 4 (d) 1 and 4

[ESE : 2003]

5.5 Match **List-I** (Format of representation) with **List-II** (Context/Relevant to) and select the correct answer using the codes given below the lists:

List-I

A. $\frac{\partial u}{\partial x} + \frac{\partial v}{\partial y}$

B. $\frac{\partial v}{\partial x} - \frac{\partial u}{\partial y}$

C. $u\frac{\partial u}{\partial x} + v\frac{\partial u}{\partial y}$

D. $\frac{q\theta}{2\pi}$

List-II

1. Relevant to a velocity potential
2. Rate of rotation about a relevant axis
3. Pressure gradient in a relevant direction
4. Continuity of flow

Codes:

	A	B	C	D
(a)	3	2	4	1
(b)	4	1	3	2
(c)	3	1	4	2
(d)	4	2	3	1

[ESE : 2007]

5.6 If the stream function $\psi = 3x^2 - y^3$, what is the magnitude of velocity at point (2, 2)?

(a) 9 (b) 13
(c) 15 (d) 17 **[ESE : 2007]**

5.7 Consider the following statements:

1. Fluids of low viscosity are all irrotational.
2. Rotation of the fluid is always associated with shear stress.

Which of these statements is/are correct?

(a) 1 only (b) 2 only
(c) Both 1 and 2 (d) Neither 1 nor 2

[ESE : 2008]

5.8 Consider the following statements in respect of two-dimensional incompressible flow with velocity components u and v in x and y directions respectively:

1. The continuity equation is $\frac{\partial u}{\partial x} = \frac{\partial v}{\partial y}$
2. The acceleration in x-direction is $a_x = \frac{\partial u}{\partial t} + u\frac{\partial u}{\partial x} + v\frac{\partial u}{\partial y}$
3. The condition of irrotationality is $\frac{\partial u}{\partial y} = \frac{\partial v}{\partial x}$
4. The equation of a streamline is $udy = -vdx$

Which of these statements are correct?

(a) 2 and 3 only (b) 1 and 2 only
(c) 1, 2 and 3 (d) 3 and 4

[ESE : 2009]

5.9 Match **List-I** (Name of curve) with **List-II** (Curve in figure) and select the correct answer using the codes given below the lists:

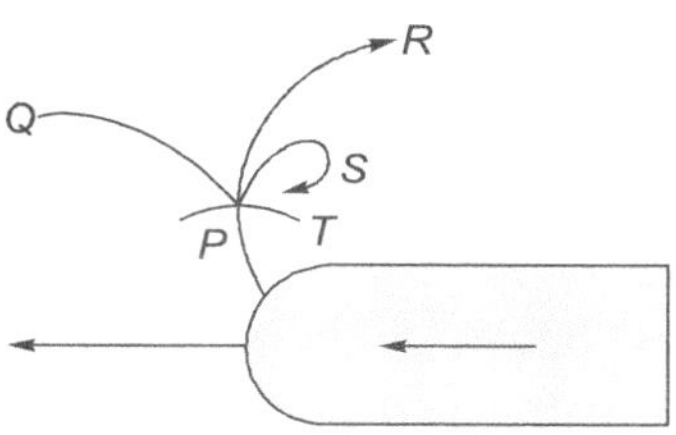

List-I	List-II
A. Equipotential line	1. *PQ*
B. Pathline	2. *PR*
C. Streakline	3. *PS*
D. Streamline	4. *PT*

Codes:

	A	B	C	D
(a)	2	3	1	4
(b)	4	1	3	2
(c)	2	1	3	4
(d)	4	3	1	2

[ESE : 2010]

5.10 Consider the following statements:

1. A streamline is an imaginary line within the flow for which the normal at any point relates to the acceleration at that point.
2. Convective acceleration is the change in velocity with respect to distance only.
3. Temporal acceleration expresses variation of velocity with respect to time only.
4. Both convective acceleration and temporal acceleration can coexist.

Which of these statements are correct related to fluid kinematics?

(a) 1, 2, 3 and 4 (b) 1, 2 and 3 only
(c) 2 and 3 only (d) 2, 3 and 4 only

[ESE : 2010]

5.11 Consider the following statements:

1. There is no flow across a streamline.
2. Streamline spacing varies directly with velocity at the section.
3. Streamlines do not cross.
4. In steady flow, streamline pattern does not change with time.

Which of these statements in respect of stream flow pattern are correct?

(a) 1, 2, 3 and 4 (b) 1 and 2 only
(c) 1, 3 and 4 only (d) 2, 3 and 4 only

[ESE : 2011]

5.12 X-component of velocity in a two dimensional incompressible flow is given by $u = 2y^2 + 6xy$. If the Y-component of velocity v is zero at $y = 0$, the expression for Y-component of velocity is given by

(a) $v = 3y^2 + f(y)$ (b) $v = 3y^2 + f(x)$
(c) $v = -3y^2 + f(y)$ (d) $v = -3y^2 + f(x)$

[ESE : 2013]

5.13 In a two-dimensional flow, with its stream function $\psi = 2xy$, the velocity at a point (3, 4) is

(a) 12.0 units (b) 10.0 units
(c) 8.0 units (d) 6.0 units

[ESE : 2016]

5.14 A fluid flow is described by a velocity fluid

$\bar{U} = 4x^2 i - 5x^2 yj + 1k$.

What is the absolute velocity (in magnitude) at the point (2, 2, 1)?

(a) $\sqrt{1802}$ (b) $\sqrt{1828}$
(c) $\sqrt{1840}$ (d) $\sqrt{1857}$

[ESE : 2016]

5.15 A line source of strength 15π m/s is situated within a uniform stream flowing at –12 m/s (i.e. right to left). At a distance of 0.6 m downstream from the source is an equal sink. How far will the stagnation points be from the nearest source/sink?

(a) 0.38 m (b) 0.46 m
(c) 0.52 m (d) 0.58 m

[ESE : 2016]

5.16 A fluid flow field is given by

$$U = 2xyi + yzj - \left(2yz + \frac{z^2}{2}\right)k.$$

1. The flow is viscous.
2. The flow is steady.
3. The flow is incompressible.
4. The magnitude of the total velocity vector at a point (1, 4, 3) is nearest to 27 units.

Which of the above statements are correct?

(a) 1 and 3 (b) 1 and 4
(c) 2 and 3 (d) 2 and 4

[ESE : 2017]

5.17 A steady, two dimensional, incompressible flow field is represented by

$u = x + 3y + 3$ and $v = 2x - y - 8$

In this flow field, the stagnation point is

(a) (3, 2) (b) (–3, 2)
(c) (–3, –2) (d) (3, –2)

[ESE : 2018]

5.18 The stream function of a doublet with horizontal axis and of strength μ is

(a) $\frac{\mu}{2\pi}r$ (b) $\frac{\mu}{2\pi r}\cos\theta$
(c) $\frac{\mu}{2\pi}r\sin\theta$ (d) $\frac{\mu}{2\pi}\frac{\sin\theta}{r}$

[ESE : 2018]

5.19 Which one of the following is the use of flow net analysis in fluid mechanics?

(a) To determine the streamlines and equipotential lines.
(b) To determine, downward lift pressure above hydraulic structure.
(c) To determine the viscosity for given boundaries of flow.
(d) To design the hydraulic structure.

[ESE : 2020]

5.20 A stream function is given by $\psi = 3x^2 - y^3$. What is the magnitude of velocity components at the point (2,1)?

(a) 8.52 (b) 9.17
(c) 10.81 (d) 12.37

[ESE : 2021]

5.21 The stream function is given by the expression $\psi = 2x^2 - y^2$. What is the resultant velocity at a point denoted by $x = 2$ and $y = 3$?

(a) 10 (b) 12
(c) 15 (d) 18

[ESE : 2022]

5.22 **Statement I** : The negative derivative of velocity potential function with respect to any direction gives the fluid velocity in that direction.

Statement II : The partial derivative of stream function in any direction gives the velocity component in that direction.

(a) Both Statement (I) and Statement (II) are individually true and Statement (II) is the correct explanation of Statement (I)
(b) Both Statement (I) and Statement (II) are individually true but Statement (II) is ***not*** the correct explanation of Statement (I)
(c) Statement (I) is true but Statement (II) is false
(d) Statement (I) is false but Statement (II) is true

[ESE : 2023]

Answers Fluid Kinematics

5.1 (d)	5.2 (b)	5.3 (c)	5.4 (a)	5.5 (d)	5.6 (d)	5.7 (b)	5.8 (a)	5.9 (d)
5.10 (d)	5.11 (c)	5.12 (d)	5.13 (b)	5.14 (d)	5.15 (a)	5.16 (c)	5.17 (d)	5.18 (d)
5.19 (d)	5.20 (d)	5.21 (a)	5.22 (c)					

Explanations Fluid Kinematics

5.1 (d)

$$v = \frac{\partial \psi}{\partial x} = -2y \quad \text{...(i)}$$

$$u = -\frac{\partial \psi}{\partial y} = -2x \quad \text{...(ii)}$$

Integrating (i), we get

$$\psi = -2xy + f(y) \quad \text{..(iii)}$$

Differentiating (iii) w.r.t. y, we get

$$\frac{\partial \psi}{\partial y} = -2x + f'(y) \quad \text{...(iv)}$$

Equating (ii) and (iv), we get

$$f'(y) = 0 \quad \text{...(v)}$$

Inegrating (v), we get

$$f(y) = C$$

where C is a numerical constant which can be treated as zero.

From (iii), we get

$$\psi = -2xy$$

At (1, 1); $\psi_1 = -2$ units

At (2, 2); $\psi_2 = -8$ units

$$Q_{\text{per unit width}} = |\psi_2 - \psi_1| = 8 - 2 = 6 \text{ units}$$

5.2 (b)

$$u = -\frac{\partial \psi}{\partial y} = -2x = -4\text{units}$$

$$v = \frac{\partial \psi}{\partial x} = 2y = 6\text{units}$$

$$V = \sqrt{u^2 + v^2} = \sqrt{16 + 36} = 7.21 \text{ units}$$

5.3 (c)

Velocity potential function can not be defined for rotational flow.

5.4 (a)

- Continuity equation is based on the conservation of mass. The continuity equation for 3D flow in Cartesian system is $\frac{\partial u}{\partial x} + \frac{\partial v}{\partial y} + \frac{\partial w}{\partial z} + \frac{\partial \rho}{\partial t} = 0$.
- Continuity equation assumes that no void occurs in the fluid and fluid mass is neither created nor destroyed.
- In deriving the continuity equation a control volume of parallelepiped is taken

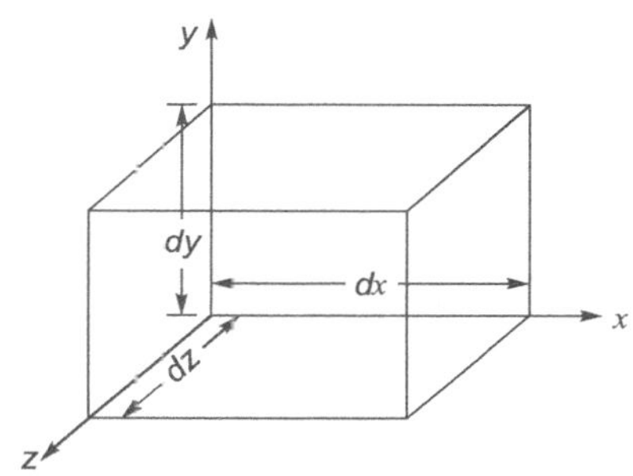

- In incompressible flow, the viscosity is not involved in continuity equation.

5.5 (d)

- Continuity equation is $\frac{\partial u}{\partial x} + \frac{\partial v}{\partial y} = 0$.
- Rate of rotation about a relevant axis

$$w_z = \frac{1}{2}\left[\frac{\partial v}{\partial x} - \frac{\partial u}{\partial y}\right]$$

- $\frac{q\theta}{2\pi}$ is the stream function for source. As there is a relation between stream function and potential function so $\frac{q\theta}{2\pi}$ is related to potential function also.

5.6 (d)

Stream function, $\psi = 3x^2 - y^3$

$$\therefore \quad \frac{\partial \psi}{\partial x} = 6x \text{ and } \frac{\partial \psi}{\partial y} = -3y^2$$

But $u = -\frac{\partial \psi}{\partial y}$

$= -(-3y^2) = 3 \times 2^2 = 12$

$$v = \frac{\partial \psi}{\partial x} = 6x = 6 \times 2 = 12$$

$\therefore$ Resultant velocity $= \sqrt{u^2 + v^2}$

$= \sqrt{(12)^2 + (12)^2}$

$= \sqrt{288} = 16.97 \approx 17$

5.7 (b)

- Statement 1 is incorrect since the fluids having low viscosity may be rotational. Example flow of air (inside the boundary layer) over a flat plate.
- Rotation of fluid is always associated with shear stress.

5.8 (a)

For two dimensional incompressible flow,

- the continuity equation is $\frac{\partial u}{\partial x} + \frac{\partial v}{\partial y} = 0$, so statement 1 is wrong.
- the equation of a stream line is $vdx - udy = 0$, so statement 4 is wrong.

5.9 (d)

- Equipotential line and stream line are orthogonal to each other. In the figure PT shown as equipotential line and perpendicular to PT, PR is shown as stream line.
- Path line is represented as PS.
- Streak line is represented as PQ. (It is the locus of different particles released from the same location at an instant).

5.10 (d)

- Statement 1 is wrong.
- Statement 2 and 3 are correct.
- Statement 4 is correct. For a nonuniform and unsteady flow both convective and temporal acceleration exist.

5.11 (c)

- Statement 1 is correct. Since the flow is along the streamline not across the streamline.
- Statement 2 is wrong. Streamline spacing is $|\psi_1 - \psi_2|$ which represent discharge per unit width area.

$$Q = AV \qquad (Q \text{ is constant})$$

$$A \propto \frac{1}{V}$$

So streamline spacing is inversely proportional to velocity at any section.

- Statement 3 is correct, streamlines never cut each other in a flow.
- Statement 4 is correct.

5.12 (d)

$$u = 2y^2 + 6xy$$

From continuity equation

$$\frac{\partial u}{\partial x} + \frac{\partial v}{\partial y} = 0$$

$$\frac{\partial}{\partial x}\left(2y^2 + 6xy\right) + \frac{\partial v}{\partial y} = 0$$

$$\Rightarrow \qquad -6y = \frac{\partial v}{\partial y}$$

$$6y + \frac{\partial v}{\partial y} = 0$$

Int. it $\quad v = -3y^2 + F(x)$

5.13 (b)

$\psi = 2xy$

$u = -\frac{\partial \psi}{\partial y} = -2x = -(2)(3)$

$= -6$ units

$v = \frac{\partial \psi}{\partial x} = 2y = 2(4) = 8$ units

Velocity of (3, 4),

$V = \sqrt{(-6)^2 + (8)^2} = 10$ units

5.14 (d)

$$\vec{V} = 4x^2\hat{i} - 5x^2y\hat{j} + 1\hat{k}$$

At point (2, 2, 1)

$u = 4(2)^2 = 16$

$$v = -5x^2y = -5(2)^2(2) = -40$$
$$w = 1$$
$$V = \sqrt{16^2 + (-40)^2 + 1^2}$$
$$= \sqrt{1857} \text{ m/sec}$$

5.15 (a)

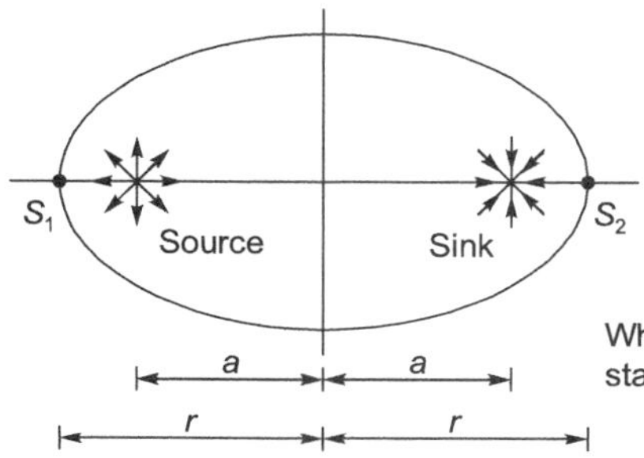

Where S_1 and S_2 are stagnation point.

Given, $a = \dfrac{0.6}{2} = 0.3$ m

$u = 12$ m/sec

Now, $r = a\sqrt{\dfrac{Q}{\pi a U_\infty} + 1}$

$$0.3 = \sqrt{\frac{15\pi}{\pi(0.3)(12)} + 1}$$
$$= 0.6819 \text{ m}$$

So, $r - a = 0.6819 - 0.3 = 0.381$ m

5.17 (d)

$u = x + 3y + 3$
$v = 2x - y - 8$

At stagnation point the velocity of net flow becomes to zero.

$$|\vec{V}| = 0$$

means $u = 0$ and $v = 0$

$x + 3y + 3 = 0$
$2x - y - 8 = 0$

By eq. (i) and (ii)

$x = 3$
$y = -2$

Point = 3, −2

5.20 (d)

Given stream function,

$$\psi = 3x^2 - y^3$$

Velocity component:

In x-direction, $u = -\dfrac{\partial\psi}{\partial y} = -\dfrac{\partial}{\partial y}(3x^2 - y^3)$

$$= -(0 - 3y^2)$$
$$u = 3y^2$$

In y-direction, $v = \dfrac{\partial\psi}{\partial x} = \dfrac{\partial}{\partial x}(3x^2 - y^3)$

$$= (3 \times 2x - 0)$$
$$v = 6x$$

At (2, 1)

$u = 3(1)^2 = 3$ units
$v = 6(2) = 12$ units

$\therefore$ $\vec{V} = u\hat{i} + v\hat{j} = 3\hat{i} + 12\hat{j}$

$$|\vec{v}| = \sqrt{u^2 + v^2}$$
$$= \sqrt{3^2 + 12^2} = \sqrt{153} \text{ m/s}$$
$$|\vec{v}| = 12.37 \text{ m/s}$$

5.21 (a)

Given: $\psi = 2x^2 - y^2$

$\therefore$ $u = -\dfrac{\partial\psi}{\partial y} = -\dfrac{\partial}{\partial y}(2x^2 - y^2) = 2y$

$\therefore$ $v = \dfrac{\partial\psi}{\partial x} = \dfrac{\partial}{\partial x}(2x^2 - y^2) = 4x$

At $(x, y) = (2, 3)$,

$\Rightarrow$ $u = 2 \times 3 = 6$ units

$\Rightarrow$ $v = 4 \times 2 = 8$ units

$\therefore$ $V_R = \sqrt{u^2 + v^2} = \sqrt{6^2 + 8^2}$

$\Rightarrow$ $V_R = \sqrt{100} = 10$ units

■■■■

6 Fluid Dynamics, Flow Measurements

6.1 A 30 cm diameter, 90° elbow has one limb vertical. Average velocity of flow of water through the elbow is 5 m/s and the pressure intensity is 4 kPa. The vertical component of force to keep the elbow in position will be

(a) 0.28 kN (b) 1.49 kN
(c) 1.77 kN (d) 2.05 kN

[ESE : 2001]

6.2 One orifice is located at a distance 'x' from the free surface while another orifice is located at the same distance 'x' from the bottom of the tank as shown in the figure.

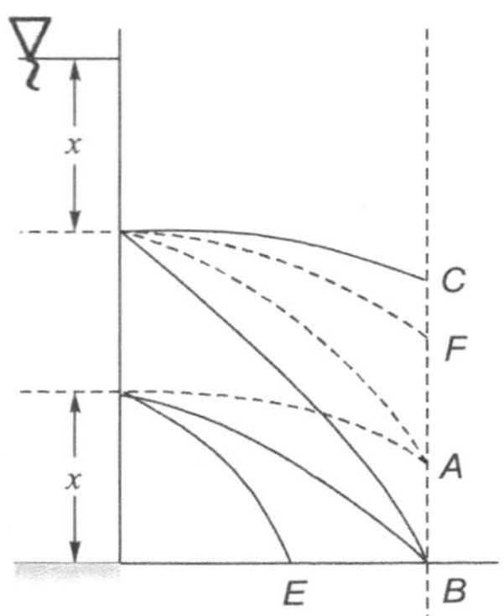

The water jets through the orifices

(a) intersect at point A
(b) intersect at point B
(c) strike the plane at point C and B respectively
(d) would be striking at E and F only

[ESE : 2002]

6.3 Match **List-I** (Equation/Principle) with **List-II** (Property) and select the correct answer using the codes given below the lists:

List-I
A. Energy equation
B. Continuity equation
C. Moment of momentum principle
D. Impulse-momentum principle

List-II
1. Force on a moving vane
2. Lawn sprinkler
3. Pressure at a point in a pipeline
4. Flow at two sections of a tapering pipe

Codes:

	A	B	C	D
(a)	4	3	2	1
(b)	3	4	1	2
(c)	4	3	1	2
(d)	3	4	2	1

[ESE : 2003]

6.4 Consider the following statements related to a horizontal venturimeter:

1. The velocity of flow in the main pipe is greater and the pressure is lesser than that at the throat section.
2. The velocity of flow in the main pipe is lower and the pressure is larger than that at the throat section.
3. The pressure difference between the main pipe and throat section is positive.
4. The pressure difference between the main pipe and throat section is negative.

Which of these statements are not correct?

(a) 1 and 4 (b) 1 and 2
(c) 2 and 3 (d) 3 and 4

[ESE : 2003]

6.5 When a body moves through still water at a constant velocity of 4.5 m/s, the velocity of water at 0.8 m ahead of the nose of the body is 3.0 m/s. What will be the difference in pressure between the nose and the point 0.8 m ahead of it?

(a) 875 N/m^2 (b) 1000 N/m^2
(c) 1125 N/m^2 (d) 1250 N/m^2

[ESE : 2003]

6.6 Match **List-I** with **List-II** and select the correct answer using the codes given below the lists:

List-I
A. The type of vortex flow in which no external torque rotates the fluid mass

B. Flow of liquid inside the impeller of a centrifugal pump
C. Free surface of forced vortex flow
D. In forced vortex, the rise of liquid level at ends

List-II
1. The fall of liquid at the axis of rotation
2. Free vortex flow
3. Parabolic
4. Forced vortex flow

Codes:

	A	B	C	D
(a)	4	2	3	1
(b)	2	4	3	1
(c)	4	2	1	3
(d)	2	4	1	3

[ESE : 2003]

6.7 Match **List-I** (Causes) with **List-II** (Effects) pertaining to rotation and vertical/horizontal motion of liquid masses at constant acceleration and select the correct answer using the codes given below the lists:

List-I
A. The form of the free surface of liquid in a rotating open vessel
B. The form of surface of the liquid in a tanker moving in the direction of its length
C. The pressure at any point in a container's base moving with upward acceleration
D. The pressure at any point in a container's base moving with downward acceleration

List-II
1. An inclined plane
2. A paraboloid of revolution
3. Greater than the hydrostatic value
4. Less than the hydrostatic value

Codes:

	A	B	C	D
(a)	4	1	3	2
(b)	4	3	2	1
(c)	2	3	1	4
(d)	2	1	3	4

[ESE : 2004]

6.8 Match **List-I** (Features) with **List-II** (Type of manometers) and select the correct answer using the codes given below the lists:

List-I
A. Open ended manometer for positive pressure
B. Negative pressure manometer
C. For measuring pressure in liquids or gases
D. For measuring pressure in liquids only

List-II

1. 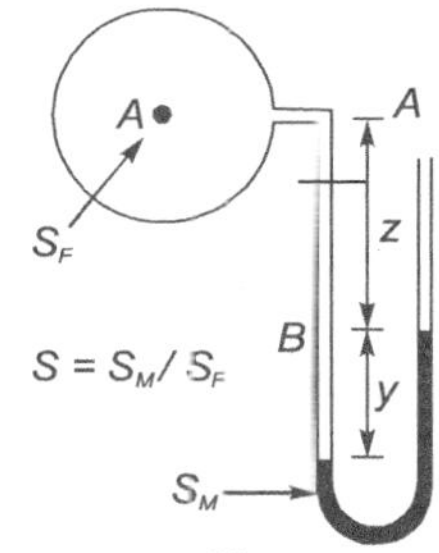

2. 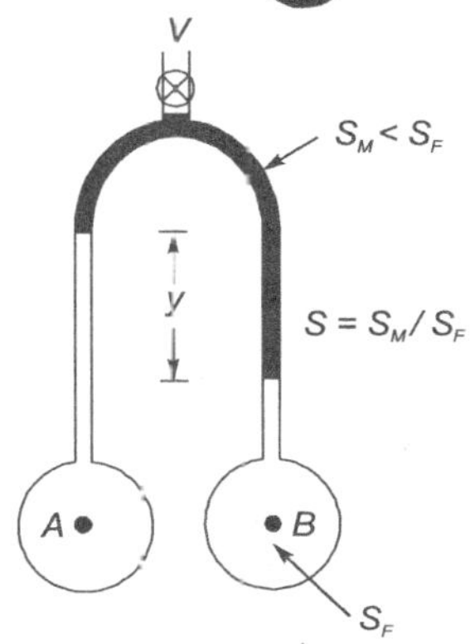

3. 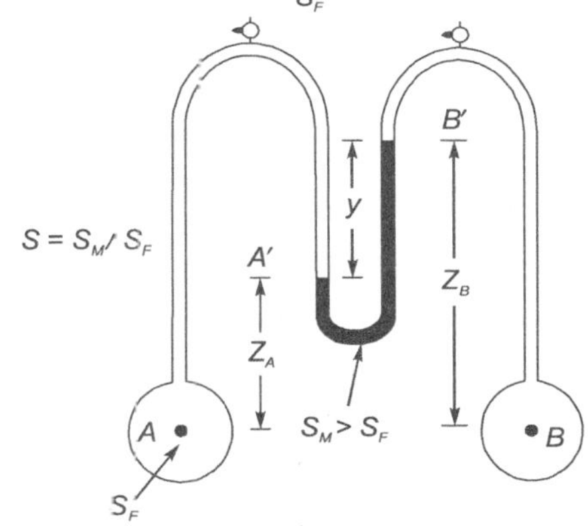

4. 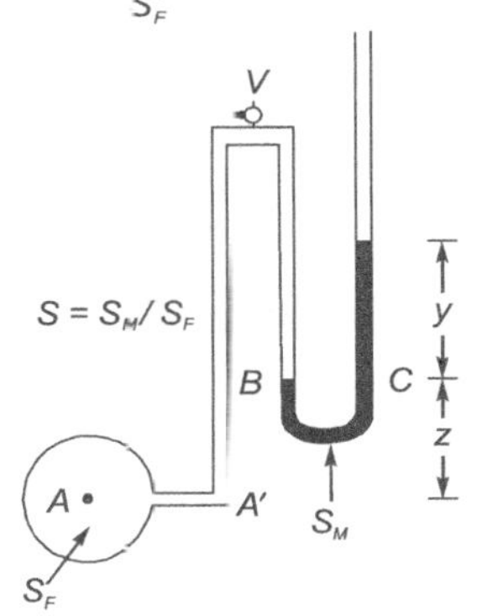

Codes:

	A	B	C	D
(a)	3	4	1	2
(b)	1	2	3	4
(c)	3	2	1	4
(d)	1	4	3	2

[ESE : 2004]

6.9 Match **List-I** with **List-II** and select the correct answer using the codes given below the lists:

List-I

A. Equation of motion along a streamline
B. Euler's equation
C. Pressure exerted by a free jet
D. Rotating lawn sprinkler

List-II

1. Principle of moment of momentum
2. Bernoulli's equation
3. Equation for conservation of momentum
4. Momentum equation

Codes:

	A	B	C	D
(a)	4	1	2	3
(b)	2	3	4	1
(c)	4	3	2	1
(d)	2	1	4	3

[ESE : 2005]

6.10 A Pitot-static tube, with a coefficient of 0.98 is used to measure the velocity of water in a pipe. The stagnation pressure recorded is 3 m and the static pressure is 0.5 m. What is the velocity of flow?

(a) 7.2 m/s (b) 6.8 m/s
(c) 5.9 m/s (d) 5.2 m/s [ESE : 2006]

6.11 Assertion (A): It takes longer to drain a reservoir with a long vertical pipe taken down from its bottom discharging into atmosphere than with an orifice at the bottom.

Reason (R): The relative height of the water surface elevation in the reservoir compensates for the friction loss in the pipe besides the entry and exit losses in the pipe.

(a) both A and R are true and R is the correct explanation of A
(b) both A and R are true but R is not a correct explanation of A
(c) A is true but R is false
(d) A is false but R is true [ESE : 2006]

6.12 In an air flow the velocity is measured by a Pitot tube (coefficient = 1.0). The mass density of air can be taken as 1.2 kg/m^3. If the head difference in a vertical U-tube holding water is 12 mm, then what is the velocity of air in m/s?

(a) 10 (b) 14
(c) 17 (d) 20 [ESE : 2007]

6.13 Consider the following:

1. Force on pipe bends and transitions
2. Jet propulsion
3. Flow velocities in open channel
4. Vortex flow

Which of the above admit employing the moment of momentum equation?

(a) 1 and 2 only (b) 1, 2 and 3
(c) 1 and 3 only (d) 2, 3 and 4

[ESE : 2009]

6.14 Consider the following statements:

The moment of momentum equation in fluid dynamics can be used

1. to find the torque exerted on sprinklers by water
2. to determine the force in a flow passage, when stream changes direction/magnitude.

Which of these statements is/are correct?

(a) 1 only (b) 2 only
(c) Both 1 and 2 (d) Neither 1 nor 2

[ESE : 2010]

6.15 The movement of air mass in the case of Tornado can be described as

(a) Forced vortex throughout
(b) Free vortex throughout
(c) Forced vortex at the core and free vortex outside
(d) Free vortex at the core and forced vortex outside

[ESE : 2011]

6.16 Consider the following statements:

Euler's equation of motion

1. can be derived from Navier-Strokes equations.
2. refers to energy balance.
3. develops into Bernoulli's equation under appropriate conditions.
4. is applicable to rotational as well as irrotational flows.

Which of these statements are correct?

(a) 1, 2, 3 and 4 (b) 1 and 2 only
(c) 1 and 3 only (d) 3 and 4 only

[ESE : 2013]

6.17 At a distance of 20 cm from the axis of a whirlpool in an ideal liquid, the velocity is 10 m/s. At a radius of 50 cm, what will be the depression of the free surface of the liquid below that at a very large distance?
(Take $1/g = 0.102\ s^2/m$)
(a) 0.408 m (b) 0.612 m
(c) 0.816 m (d) 1.224 m
[ESE : 2013]

6.18 The Bernoulli's equation is applicable to:
(a) Both steady and unsteady flows
(b) Real fluids
(c) All fluids and flows along a stream tube
(d) Steady flow of ideal fluids along a stream tube [ESE : 2013]

6.19 **Statement (I):** The integration of differential form of Euler's equation of motion yields to Bernoulli's equation.
Statement (II): Euler's equation is based on the assumption of conservation of mass.
(a) Both Statement (I) and Statement (II) are individually true and Statement (II) is the correct explanation of Statement (I)
(b) Both Statement (I) and Statement (II) are individually true but Statement (II) is NOT the correct explanation of Statement (I)
(c) Statement (I) is true but Statement (II) is false
(d) Statement (I) is false but Statement (II) is true
[ESE : 2013]

6.20 Flow measurements with a Prandtl-Pitot tube showed that the tip readings varied only across the flow while the side-opening readings varied only in the direction of flow. The type of flow is
(a) Uniform irrotational
(b) Uniform rotational
(c) Non-uniform irrotational
(d) Non-uniform rotational [ESE : 2013]

6.21 An open cylindrical tank of 75 cm diameter and 1.5 m high contains water upto 1.2 m depth. If the cylinder is rotated about its vertical axis, what is the maximum angular velocity (in radians per second) that can be attained without spilling any water?
(a) 7.55 (b) 8.08
(c) 9.15 (d) 10.02 [ESE : 2013]

6.22 A horizontal venturimeter with inlet diameter of 30 cm and throat diameter of 15 cm is used to measure the flow of water. The reading on a differential manometer connected to the inlet and the throat is 20 cm of mercury. If $C_d = 0.98$, the rate of flow is nearly
(a) 12.5 *l*/s (b) 25 *l*/s
(c) 125 *l*/s (d) 250 *l*/s
[ESE : 2014]

6.23 A mercury water manometer has a gauge difference of 0.8 m. The difference in pressure measured in metres of water is
(a) 0.8 (b) 1.06
(c) 10.05 (d) 8.02 [ESE : 2015]

6.24 An orifice is located in the side of a tank with its centre 10 cm above the base of the tank. The constant water level is 1.0 m above the centre of orifice. The coefficient of velocity is 0.98. On the issuing jet, the horizontal distance from the vera-contracta to where the jet is 10 cm below vera-contracta is
(a) 1.62 m (b) 1.00 m
(c) 0.62 m (d) 0.32 m [ESE : 2016]

6.25 The velocity of water at the outer edge of a 60 cm diameter whirlpool, where the water level is horizontal is 2.5 m/s. The velocity of water at a level where the diameter of the whirlpool is 15 cm, is
(a) 1 m/s (b) 5 m/s
(c) 8 m/s (d) 10 m/s
[ESE : 2016]

6.26 A cylindrical vessel with closed bottom and open top is 0.9 m in diameter. What is the rotational speed about its vertical axis (with closed bottom below and open top above) when the contained incompressible fluid will rise 0.5 m at the inner circumference of the vessel and a space of 0.4 m diameter at the bottom will have no fluid thereon? Take $g = 10\ m/s^2$.
(a) 650 rpm (b) 600 rpm
(c) 580 rpm (d) 470 rpm
[ESE : 2016]

Directions: The following items consists of two statements; one labelled as **'Statement (I)'** and the other as **'Statement (II)'**. You are to examine these two statements carefully and select the answers to these items using the codes given below:

Codes:

(a) Both Statement (I) and Statement (II) are individually true and Statement (II) is the correct explanation of Statement (I)
(b) Both Statement (I) and Statement (II) are individually true but Statement (II) is NOT the correct explanation of Statement (I)
(c) Statement (I) is true but Statement (II) is false
(d) Statement (I) is false but Statement (II) is true

6.27 **Statement (I):** When flow through a pipeline is measured through fixing a venturimeter, the computed flow will not be sensitive to the alignment of the centre line of the set-up—horizontal or sloping, up or down, along the flow direction.

Statement (II) : The difference in the readings on the manometer limbs is by itself always adjusted for the ratio of the densities of the two liquids—the manometer liquid and the liquid whose flow rate is being measured—in the development of the formula for computing the discharge.

[ESE : 2017]

6.28 **Statement (I)) :** Bernoulli's equation is applicable to any point in the flow field provided the flow is steady and irrotational.

Statement (II) : The integration of Euler's equation of motion to derive Bernoulli's equation involves the assumptions that velocity potential exists and that the flow conditions do not change with time at any point.

[ESE : 2018]

6.29 A vertical cylindrical tank, 2 m diameter has, at the bottom, a 5 cm diameter, sharp-edged orifice, for which $C_d = 0.6$. Water enters the tank at a constant rate of 9 l/sec. At what depth above the orifice will the level in the tank become steady?

(a) 2.95 m (b) 2.75 m
(c) 2.60 m (d) 2.50 m

[ESE : 2018]

6.30 What is the rotational speed in rpm of a 0.8 m diameter cylindrical container, held with axis vertical, if the fluid contained in it rises to 0.6 m height at the sides and leaves a circular space 0.3 m diameter on the bottom uncovered?

(a) 90.2 rpm (b) 88.4 rpm
(c) 86.0 rpm (d) 83.7 rpm

[ESE : 2018]

6.31 For frictionless adiabatic flow of compressive fluid, the Bernoulli's equation with usual notations is

(a) $\frac{k}{k-1}\frac{p_1}{w_1}+\frac{v_1^2}{2g}+z_1=\frac{k}{k-1}\frac{p_2}{w_2}+\frac{v_2^2}{2g}+z_2+h_L$

(b) $\frac{k}{k-1}\frac{p_1}{w_1}+\frac{v_1^2}{2g}+z_1=\frac{k}{k-1}\frac{p_2}{w_2}+\frac{v_2^2}{2g}+z_2$

(c) $\frac{p_1}{w_1}+\frac{v_1^2}{2g}+z_1+H_m=\frac{p_2}{w_2}+\frac{v_2^2}{2g}+z_2$

(d) $\frac{k}{k-1}\frac{p_1}{w_1}+\frac{v_1^2}{2g}+z_1+H_m=\frac{p_2}{w_2}+\frac{v_2^2}{2g}+z_2+h_L$

[ESE : 2019]

6.32 Which one of the following is a device used for measuring the velocity of flowing water in pipes or open channels?

(a) Pitot tube (b) Piezometer
(c) Venturimeter (d) Venturi tube

[ESE : 2021]

6.33 Which one of the following is the correct assumption for the derivation of Bernoulli's equation?

(a) The flow is compressible
(b) Viscosity is zero
(c) The flow is unsteady
(d) The flow is rotational

[ESE : 2023]

6.34 Consider the following statements regarding ultrasonic flowmeters :

1. There are no moving parts.
2. They cannot measure flow quantities in reverse flow.
3. There is no direct contact with the fluid, there is no danger of corrosion or clogging.

Which of the above statements are correct?

(a) 1 and 2 only (b) 1 and 3 only
(c) 2 and 3 only (d) 1, 2 and 3

[ESE : 2024]

Answers Fluid Dynamics, Flow Measurements

6.1 (d)	6.2 (b)	6.3 (d)	6.4 (a)	6.5 (*)	6.6 (b)	6.7 (d)	6.8 (d)	6.9 (b)
6.10 (b)	6.11 (d)	6.12 (b)	6.13 (d)	6.14 (a)	6.15 (c)	6.16 (c)	6.17 (c)	6.18 (d)
6.19 (c)	6.20 (d)	6.21 (c)	6.22 (c)	6.23 (c)	6.24 (c)	6.25 (d)	6.26 (*)	6.27 (a)
6.28 (a)	6.29 (a)	6.30 (b)	6.31 (b)	6.32 (a)	6.33 (b)	6.34 (b)		

Explanations Fluid Dynamics, Flow Measurements

6.1 (d)

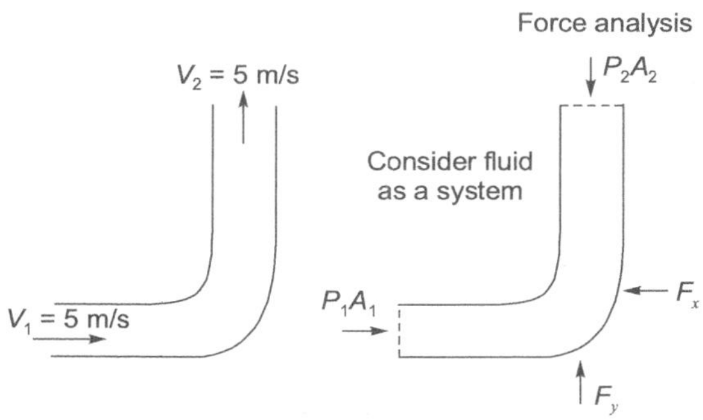

$$P_1 = P_2 = 4000 \text{ Pa}$$
$$D_1 = D_2 = 30 \text{ cm}$$
F_x, F_y = Force exerted on the fluid

Momentum eq. in y-direction.

$$F_y - P_2A_2 = \dot{m}V_2 - \dot{m}(0)$$
$$F_y = \dot{m}V_2 + P_2A_2$$
$$= (\rho A_2 V_2)V_2 + P_2A_2$$
$$= \left[(1000)(5)^2\right] + 4000\left[\frac{\pi}{4}(0.3)^2\right] = 2.05 \text{ kN}$$

6.2 (b)

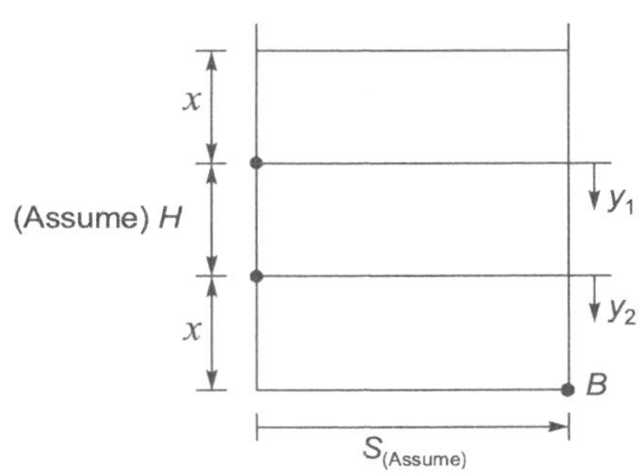

Assume C_v is same for both the orifice and the horizontal distance travelled by both the jets is same (S).

$$C_v = \frac{S}{2\sqrt{x.y_1}} = \frac{S}{2\sqrt{(x+H)y_2}}$$

$$\frac{1}{\sqrt{x.y_1}} = \frac{1}{\sqrt{(x+H)y_2}}$$

$$\frac{y_1}{y_2} = \frac{x+H}{x}$$

It is possible if the jet intersect at 'B'.

6.4 (a)

The cross-section area of main pipe is more than that of throat section. Therefore, from continuity equation velocity of flow in main pipe will be lower than that at the throat section. The flow from main pipe to throat takes place due to pressure gradient and therefore pressure in main pipe is more.

So statement 1 is incorrect.

6.5 (*)

$$\frac{P_1 - P_2}{\rho g} = \frac{V_2^2 - V_1^2}{2g}$$

$$P_1 - P_2 = \frac{1000}{2}\left[4.5^2 - 3^2\right]$$

$$= 500\left[4.5^2 - 3^2\right] = 5625 \text{ N/m}^2$$

No any option is given in the problem.

6.6 (b)

In forced vortex flow:

1. In free vortex flow no external torque rotates the fluid mass.
2. Flow of liquid inside impeller of a centrifugal pump is forced vortex flow.
3. The free surface of liquid is in parabolic shape.
4. Rise of liquid levels at the ends is equal to the fall of liquid at axis of rotation.

6.7 (d)

Form of surface of liquid in

1. Rotating open vessel- Paraboloid of revolution
2. Tanker moving along length- inclined plane

Container moving in the vertical direction

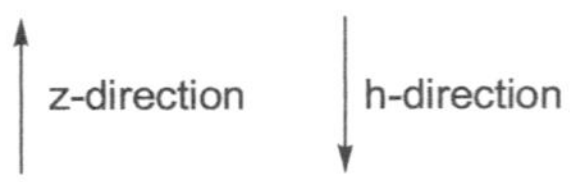

$$\frac{dP}{dh} = \rho(a_z + g)$$

For upward motion a_z is +ve so pressure is greater than the hydrostatic value. For downward motion a_z is – ve so pressure is lesser than the hydrostatic value.

6.9 (b)

The moment of momentum equation can be applied to analyze rotating systems, such as turbines, sprinklers, and fans. Bernoulli's equation is used to applicable to steady incompressible flow along a streamline. Euler's equation is based on conservation of momentum. Pressure exerted by free jet can be analyzed by momentum equation.

6.10 (b)

$$V = C_V\sqrt{2gh}$$

For horziontal

$$h = \frac{P_{stag}}{\rho g} - \frac{P_{static}}{\rho g}$$

$$= 3 - 0.5 = 2.5 \text{ m}$$

$$= 0.98 \times \sqrt{2 \times 9.81 \times 2.5} = 6.86 \text{ m/s}$$

6.11 (d)

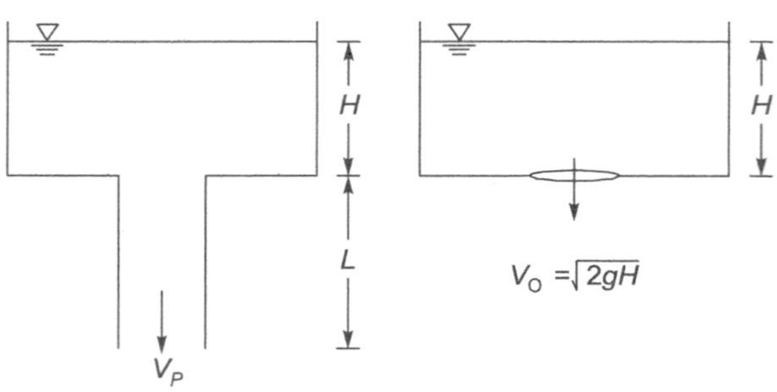

Reservoir with a long vertical pipe of length L

$V_P = \sqrt{2g(H+L) - h_f}$

Since V_P is greater than V_O. So reservoir with long vertical pipe of length L will take less time to empty then the reservoir having an orifice at the bottom. Hence assertion is wrong.

6.12 (b)

$$V = C_V\sqrt{2gh}$$

$$h = x\left[\frac{\rho_w}{\rho_a} - 1\right]$$

$$= 0.012\left[\frac{10^3}{1.2} - 1\right]$$

$$= 9.988 \text{ m}$$

$$V = 1 \times \sqrt{2 \times 9.81 \times 9.988}$$

$$= 13.99 \text{ m/s}$$

$$\approx 14 \text{ m/s}$$

6.13 (d)

- Force on pipe bends employ the impulse momentum equation.
- Vortex flow employ the moment of momentum equation.

6.15 (c)

A combination of free and forced vortex is known as compound vortex or Rankine vortex. In a tornado inner core region experiences forced vortex while outer core region experiences free vortex.

6.17 (c)

Whirlpool is a example of free vortex flow.

So, $v \propto \frac{1}{r}$ i.e. vr = constant

Now, $v_1 = 10$ m/s, $r_1 = 20$ cm

When $r_2 = 50$ cm

$$v_2 = \frac{v_1 r_1}{r_2} = \frac{10 \times 20}{50} = 4 \text{ m/s}$$

According to fundamental equation of vortex flow,

$$dP = \frac{\rho v^2 dr}{r} - \rho g dz \left\{v = \frac{c}{r}\right\}$$

$$0 = \frac{\rho C^2}{r^2 r} dr - \rho g dz$$

$$dz = \frac{\rho C^2}{r^3 g} dr$$

Int. it

$$z = \frac{c^2}{g} \times -\frac{1}{2} \times \left[\frac{1}{r^2}\right]_{0.5}^{\infty}$$

$$[c = vr = 0.2 \times 10]$$

$$z = 0.816 \text{ m}$$

6.18 (d)

Bernoulli's equation is applicable to steady, incompressible flow along a streamline when the effect of viscous effect is neglected.

6.19 (c)

Statement (2) is wrong.

- Euler's equation based on the assumption of conservation of momentum.

6.20 (d)

As tip reading varied only across the flow while side-opening reading varied only in the direction of flow, velocity of flow is not same across the flow. Hence the flow is non-uniform rotational.

6.21 (c)

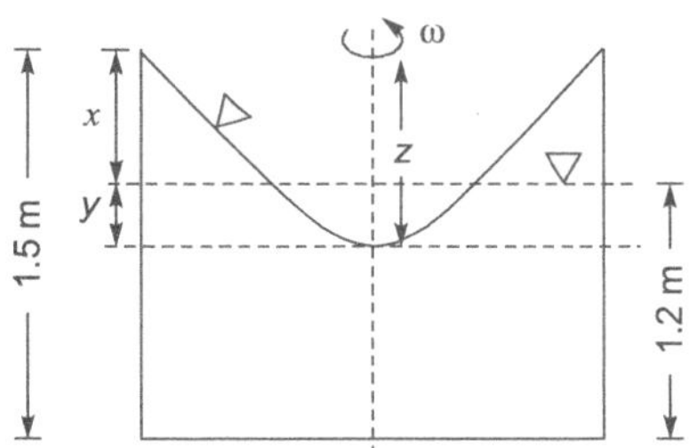

$$R = \frac{75}{2}\text{cm} = 0.375 \text{ m}$$

For conservation of volume

$$x = y$$

$$x = y = 1.5 - 1.2 = 0.3$$

$$z = 0.3 + 0.3 = 0.6 \text{ m}$$

$$z = \frac{\omega^2 R^2}{2g}$$

$$0.6 = \frac{\omega^2 (0.375)^2}{2g}$$

$$\omega = 9.15 \text{ rad/s}$$

6.22 (c)

In Venturimeter,

Rate of flow,

$$Q = \frac{C_d A_1 A_2}{\sqrt{A_1^2 - A_2^2}}\sqrt{2gh}$$

Here, $h = 20\left(\frac{\rho_{Hg}}{\rho_w} - 1\right)$

$$= 20\left(\frac{13.6 \times 10^3}{10^3} - 1\right)$$

$$= 20\,(13.6 - 1) = 252 \text{ cm}$$

$$\therefore \quad Q = \frac{0.98 \times \frac{\pi}{4} \times 30^2 \times \frac{\pi}{4} \times 15^2}{\sqrt{\left[\frac{\pi}{4} \times 30^2\right]^2 - \left[\frac{\pi}{4} \times 15^2\right]^2}} \times \sqrt{2 \times 981 \times 252}$$

$$= \frac{0.98 \times 30^2 \times \frac{\pi}{4} \times 15^2}{\sqrt{30^4 - 15^4}} \times \sqrt{2 \times 981 \times 252}$$

$$= 125.76 \; l/\text{sec}$$

6.23 (c)

$$h = x\left(\frac{\rho_{Hg}}{\rho_w} - 1\right) = 0.8\left(\frac{13.6}{1} - 1\right) = 10.08 \text{ m}$$

6.24 (c)

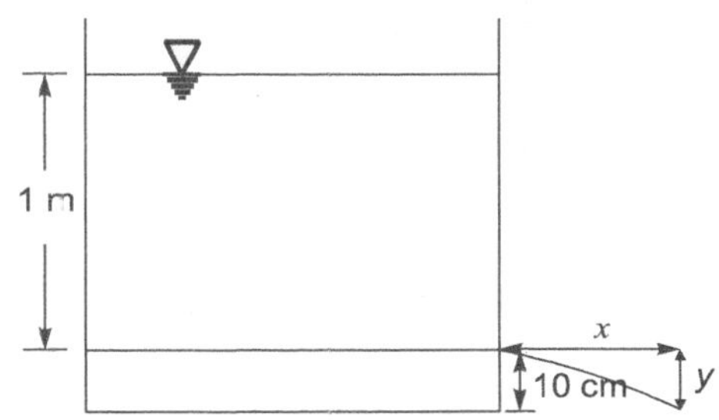

Given, $y = 10$ cm $= 0.1$ m

We know that

$$C_v = \frac{x}{2\sqrt{y \cdot H}}$$

$$0.98 = \frac{x}{2\sqrt{0.1 \times 1}} = 0.62 \text{ m}$$

6.25 (d)

Whirlpool is an example free vortex.

So, $\quad V \cdot r = \text{constant}$

$$2.5 \times 30 = V \times 7.5$$

$$V = 10 \text{ m/sec}$$

6.26 (*)

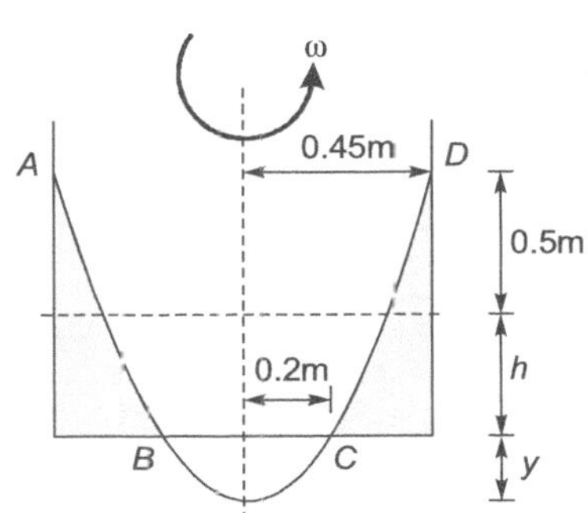

We know that

$$y + h + 0.5 = \frac{\omega^2 (0.45)^2}{2g} \quad ...(i)$$

$$y = \frac{\omega^2 (0.20)^2}{2g} \quad ...(ii)$$

By eq. (i) and (ii)

$$h + 0.5 = \frac{\omega^2}{2g}\left[0.45^2 - 0.2^2\right] \quad ...(iii)$$

Since there is no fluid.

Spillout so initial and final volume remains same.

So $V_{Initial} = V_{Final}$

$$\pi(0.45)^2 h = \pi(0.45)^2 (0.5 + h) - V_{ABCD}$$

$$V_{ABCD} = \pi(0.45)^2 (0.5)$$

$$\frac{\pi(0.45)^2 (0.5 + h + y) - \pi(0.2)^2 y}{2}$$

$$= \pi(0.45)^2 (0.5)$$

$$\pi(0.45)^2 (0.5) + \pi(0.45)^2 h + \pi(0.45)^2 y - \pi(0.2)^2 y$$
$$= 2\pi(0.45)^2 (0.5)$$

$$(0.45)^2 h + [(0.45)^2 - (0.2)^2]y = (0.45)^2 (0.5)$$

$$h + \left[\frac{0.45^2 - 0.2^2}{0.45^2}\right]y = 0.5$$

$$h + 0.8y = 0.5$$

$$h = 0.5 - 0.8y \qquad y = \frac{\omega^2 (0.2)^2}{2g}$$

$$h = 0.5 - 0.8\left[\frac{\omega^2 (0.2)^2}{2g}\right]$$

Put the value in eq. (3)

$$0.5 - \frac{0.8\omega^2 (0.2)^2}{2g} + 0.5$$

$$\omega = 10.14 \text{ rad/s}$$

$$\frac{2\pi N}{60} = 100.14$$

$$N = 96.834 \text{ rad/s}$$

6.29 (a)

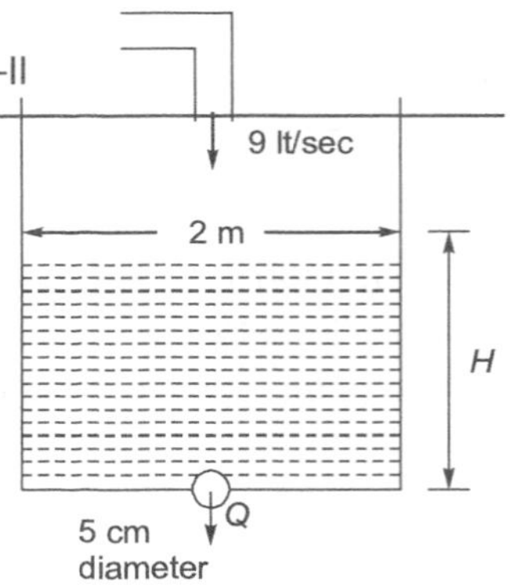

For steady level in the tank discharge released from the tank is equal to the water enter in the tank.

$$Q = 9 \times 10^{-3}$$

$$C_d\left(\frac{\pi}{4}d^2\right)\sqrt{2gH} = 9 \times 10^{-3}$$

$$C_d a\sqrt{2gh} = 9 \times 10^{-3}$$

$$0.6 \times \frac{\pi}{4}(0.05)^2 \times \sqrt{2 \times 9.81 \times H} = 9 \times 10^{-3}$$

$$H = 2.97 \text{ m}$$

6.30 (b)

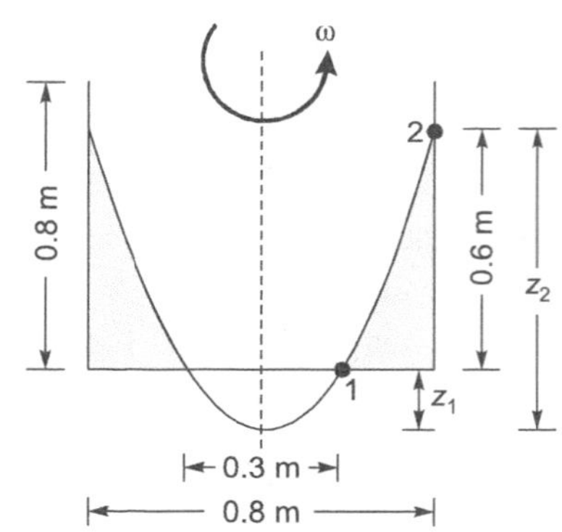

$$z_2 - z_1 = \frac{\omega^2}{2g}\left[R_2^2 - R_1^2\right]$$

$$0.6 = \frac{\omega^2}{2g}\left[0.4^2 - 0.15^2\right]$$

$$\omega = 9.253 \text{ rad/sec}$$

$$\frac{2\pi N}{60} = 9.253$$

$$N = 88.36 \text{ rpm}$$

6.31 (b)

As we know that,

$$\int \frac{dP}{\rho g} + \frac{V^2}{2g} + z = \text{constant} \quad \text{...(i)}$$

For a compressible flow, undergoing an adiabatic process

$$\frac{P}{\rho^k} = c \text{ (constant)}$$

$$dP = K \,.\, C \,.\, \rho^{k-1} \,.d\rho$$

By equation (i)

$$\int \frac{K.C.\rho^{k-1}.d\rho}{\rho g} + \frac{V^2}{2g} + z = \text{constant}$$

$$\frac{KC}{g}\int \rho^{k-2}\, d\rho + \frac{V^2}{2g} + z = \text{constant}$$

$$\frac{K.C.\rho^{k-1}}{g(k-1)} + \frac{V^2}{2g} + z = \text{constant}$$

$$\frac{K.C\rho^{k-1}}{g(k-1)}.\frac{\rho}{\rho} + \frac{V^2}{2g} + z = \text{constant}$$

$$\frac{K}{K-1}.\frac{C.\rho^k}{\rho.g} + \frac{V^2}{2g} + z = \text{constant} (\because P = C.\rho^k)$$

$$\frac{K}{K-1}.\frac{\rho}{\rho.g} + \frac{V^2}{2g} + z = \text{constant}$$

6.32 (a)

Pitot tube working

The liquid flows up the tube and when equilibrium is attained, the liquid reaches a height above the free surface of the water stream.

Since the static pressure, under this situation, is equal to the hydrostatic pressure due to its depth below the free surface, the difference in level between the liquid in the glass tube and the free surface becomes the measure of dynamic pressure. Therefore, we can write, neglecting friction.

$$p_0 - p = \frac{pV^2}{2} = h\rho g$$

where p_0, p and V are the stagnation pressure, static pressure and velocity respectively at point A.

$$V = \sqrt{2gh}$$

Such a tube is known as a Pitot tube and provides one of the most accurate mean of measuring the fluid velocity.

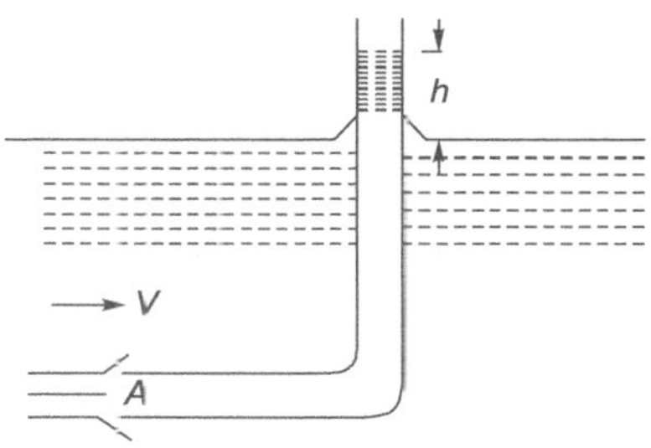

- Venturimeter is used to measure flow rate.
- Piezometer is used to measure pressure head.

■■■■

7 Viscous Flow of Incompressible Fluid

7.1 Consider the following velocity profiles in a pipeline:

1. 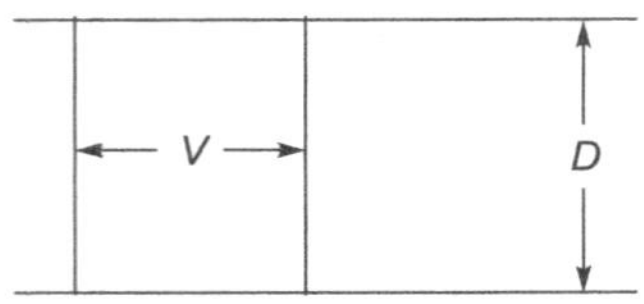

2. 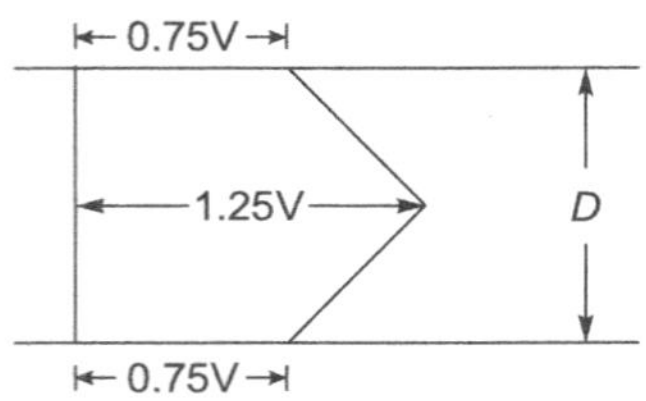

3. 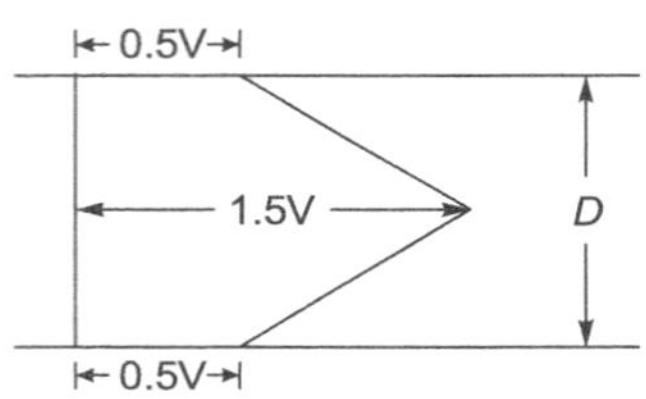

4. 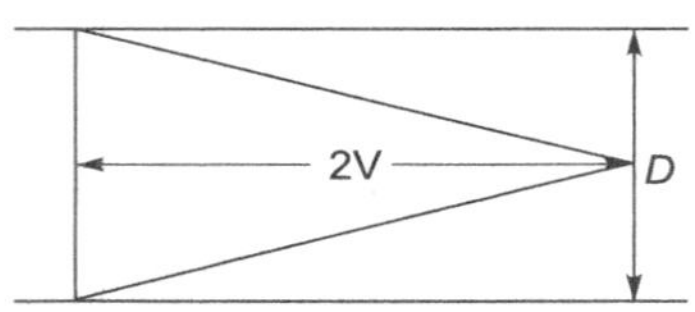

Among these profiles, the momentum correction factor would be
(a) least in 4
(b) highest in 1
(c) more in 3 than that for 2
(d) the same in 1, 2, 3 and 4

[ESE : 2000]

7.2 If the velocity profile in laminar flow is parabolic, then the shear stress profile must be
(a) a hyperbola (b) a parabola
(c) a straight line (d) an ellipse

[ESE : 2000]

7.3 In a laminar flow between two static parallel plates, the velocity at mid-point is found to be 2.0 m/s. If the space between the plates is 10 cm, then the discharge per unit width (in $m^3/s/m$) will be
(a) 0.01 (b) 0.02
(c) 0.10 (d) 0.20

[ESE : 2001]

7.4 An oil of viscosity 9 poise is flowing through a horizontal pipe of 60 mm diameter. If the flow is laminar, the pressure drop in 100 m length of the pipe is 1800 kN/m^2. The average velocity of flow of the oil is
(a) 2.25 m/s (b) 0.55 m/s
(c) 5.5 m/s (d) 22.5 m/s

[ESE : 2002]

7.5 Oil (μ = 0.44 Pa-s, r = 888 kg/m^3) is filled in the space between two parallel plates which are 18 mm apart. The upper plate is moving at a velocity of 4 m/s. What is the shear stress on the plate?
(a) 97.8 Pa (b) 48.9 Pa
(c) 79.8 Pa (d) 87.9 Pa

[ESE : 2004]

7.6 The pressure gradient in the direction of flow is equal to the
(a) shear gradient parallel to the direction of flow
(b) shear gradient normal to the direction of flow
(c) velocity gradient parallel to the direction of flow
(d) velocity gradient normal to the direction of flow

[ESE : 2005]

7.7 A circular pipe of radius R carries a laminar flow of a fluid. The average velocity is indicated as the local velocity at what radial distance, measured from the centre?
(a) 0.50R (b) 0.71R
(c) 0.67R (d) 0.29R

[ESE : 2007]

7.8 In a 4 cm diameter pipeline carrying laminar flow of a liquid with $\mu = 1.6$ centipoise, the velocity at the axis is 2 m/s. What is the shear stress midway between the wall and the axis?

(a) 0.01 N/m^2 (b) 0.0125 N/m^2
(c) 0.0175 N/m^2 (d) 0.02 N/m^2

[ESE : 2007]

7.9 Match **List-I** (Principle) with **List-II** (Effect) and select the correct answer using the codes given below the lists:

List-I
- **A.** Hele Shaw Flow
- **B.** Stoke's law
- **C.** Hagen-Poiseuille flow
- **D.** Pascal's law

List-II
1. Surface equal pressure
2. Settling of fine particles
3. Laminar flow between parallel plates
4. Laminar flow in tubes

Codes:

	A	B	C	D
(a)	2	3	4	1
(b)	3	2	4	1
(c)	2	3	1	4
(d)	3	2	1	4

[ESE : 2008]

7.10 Consider the following statements:
1. Shear stress is maximum at the centre line.
2. Maximum velocity is 3/2 times the average velocity.
3. Discharge varies inversely with the coefficient of viscosity.
4. Slope of hydraulic gradient line increases linearly with the velocity of flow.

Which of the above statements are correct in connection with a steady laminar flow through a circular pipe?

(a) 1, 3 and 4 (b) 3 and 4 only
(c) 1 and 3 only (d) 2 and 4

[ESE : 2009]

7.11 Consider the following with respect to the application of the Navier-Stokes equations:
1. Laminar flow in circular pipes
2. Laminar flow between concentric rotating cylinders
3. Laminar unidirectional flow between stationary parallel plates
4. Laminar unidirectional flow between parallel plates having relative motion

Which of these statements is/are correct?

(a) 1 only (b) 2 and 3 only
(c) 3 and 4 only (d) 1, 2, 3 and 4

[ESE : 2009]

7.12 Match **List-I** (Velocity distribution) with **List-II** (Types of flow) and select the correct answer using the codes given below the lists:

List-I **List-II**

A. 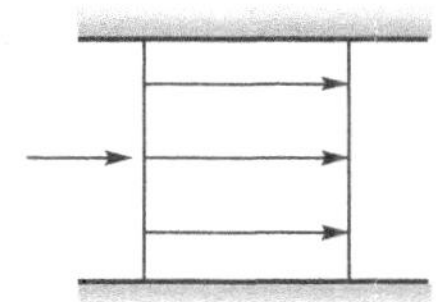 **1.** Couette

B. 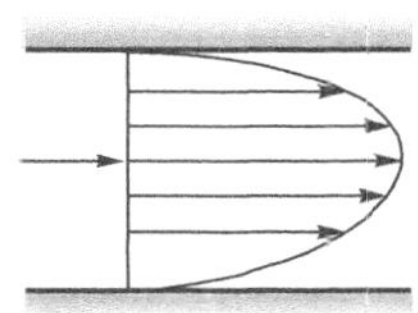 **2.** Ideal fluid

C. 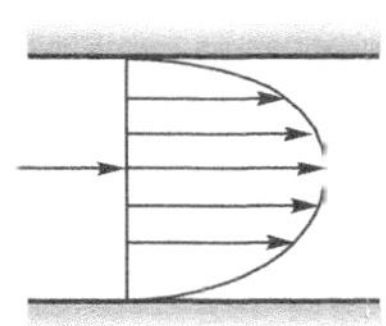 **3.** Poiseuille

D. 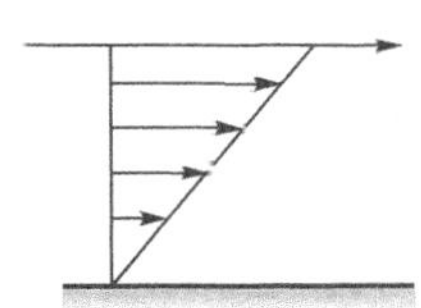 **4.** Turbulent

Codes:

	A	B	C	D
(a)	1	3	4	2
(b)	2	3	4	1
(c)	1	4	3	2
(d)	2	4	3	1

[ESE : 2010]

7.13 Assertion (A): Reynolds number of a fluid flow is indicative of the relative dominance of the effects of momentum transfer between adjacent layers of the flow over the viscous stresses.

Reason (R): For flow at high Reynolds numbers, the velocity profile is logarithmic.

(a) both A and R are true and R is the correct explanation of A
(b) both A and R are true but R is not a correct explanation of A
(c) A is true but R is false
(d) A is false but R is true

[ESE : 2010]

7.14 The pressure drop in a 30 cm diameter horizontal pipe is 60 kPa in a distance of 15 m. The wall shear stress in kPa is

(a) 0.1 (b) 0.2
(c) 0.3 (d) 0.4

[ESE : 2012]

7.15 Consider the following statements in respect of steady laminar flow through a circular pipe:

1. Shear stress is zero on the central axis of the pipe.
2. Discharge varies directly with the viscosity of the fluid.
3. Velocity is maximum at the centre of the pipe.
4. Hydraulic gradient varies as the square of the mean velocity of flow.

Which of these statements are correct?

(a) 1, 2, 3 and 4 (b) 1 and 3 only
(c) 2 and 4 only (d) 3 and 4 only

[ESE : 2012]

7.16 A wall shear stress of 40 Pa exists in a laminar flow in a 12 cm diameter pipe. At a distance of 2 cm from the wall, what will be the shear stress?

(a) 22.2 Pa (b) 24.0 Pa
(c) 26.7 Pa (d) 28.0 Pa

[ESE : 2013]

7.17 Consider the following statements regarding laminar flow through a circular pipe:

1. The friction factor is constant.
2. The frictional factor depends upon the pipe roughness.
3. The friction factor varies inversely with the Reynolds number of flow.
4. The velocity distribution is parabolic.

Which of these statements are correct?

(a) 1 and 3 (b) 3 and 4
(c) 2 and 3 (d) 2 and 4

[ESE : 2013]

7.18 In a laminar flow through a circular pipe of diameter 200 mm, the maximum velocity is found to be 1 m/s. The velocity at a radial distance of 50 mm from the axis of the pipe will be

(a) 0.5 m/s (b) 0.25 m/s
(c) 0.75 m/s (d) 1.25 m/s

[ESE : 2014]

7.19 In a laminar flow between two fixed plates held parallel to each other at a distance *d*, the shear stress is:

1. Maximum at plane *d*/2 away from each plate and zero at the plate boundaries.
2. Zero throughout the passage.
3. Maximum at the plate boundaries and zero at a plane *d*/2 away from each plate.

Which of the above statements is/are correct?

(a) 1 only (b) 3 only
(c) 2 only (d) 1, 2 and 3

[ESE : 2016]

7.20 Water flows through a pipe of 200 mm in diameter 60 m long with a velocity of 2.5 m/s. What is the head loss due to friction using Darcy's formulae? (Assume $f = 0.005$)

(a) 1.52 m (b) 1.79 m
(c) 1.85 m (d) 1.91 m

[ESE : 2023]

■■■■

Answers Viscous Flow of Incompressible Fluid

7.1 (c)	7.2 (c)	7.3 (c)	7.4 (a)	7.5 (a)	7.6 (b)	7.7 (b)	7.8 (*)	7.9 (b)
7.10 (b)	7.11 (d)	7.12 (b)	7.13 (b)	7.14 (c)	7.15 (b)	7.16 (c)	7.17 (b)	7.18 (c)
7.19 (b)	7.20 (d)							

Explanations Viscous Flow of Incompressible Fluid

7.1 (c)

More uniform flow gives less moment correction factor.

The average velocity in all the cases is V. The momentum correction factor is defined as

$$\beta = \frac{1}{A}\int\left(\frac{v}{V}\right)^2 dA = \frac{1}{D}\int_0^D\left(\frac{v}{V}\right)^2 dy$$

Figure 1: $\beta_1 = 1.0$

Figure 2: $\beta_2 = \frac{2}{D}\int_0^{D/2}\left(\frac{1.25v - 0.5v\,y/D}{V}\right)^2 dy$

$$= 2\int_0^{1/2}\left(1.25 - 0.5\frac{y}{D}\right)^2 d(d/D)$$

$$= 1.27$$

Figure 3: $\beta_3 = 2\int_0^{1/2}\left(1.5 - \frac{y}{D}\right)^2 d(y/D)$

$$= 1.583$$

Figure 4: $\beta_4 = 2\int_0^{1/2}\left(2 - 2\frac{y}{D}\right)^2 d(y/D)$

$$= 2.33$$

Thus $\beta_1 < \beta_2 < \beta_3 < \beta_4$

7.2 (c)

If the velocity profile for Laminar flow is parabolic, then the shear stress profile must be linear.

$$\tau = \mu\frac{du}{dy}$$

7.3 (c)

$$u_{max} = 1.5\bar{u}$$

$$2 = 1.5\bar{u}$$

$$\bar{u} = \frac{4}{3}\text{ m/s}$$

The discharge per unit width = $\bar{u}A$

$$= \frac{4}{3}\times(0.1\times 1) = 0.13\text{ m}^3\text{/s/m}$$

7.4 (a)

$$1\text{ poise} = 0.1\text{ N-s/m}^2$$

$$\therefore \quad \mu = 0.9\text{ N-s/m}^2$$

$$L = 100\text{ m}$$

Apply Hagen-Poiseuille equation

$$\frac{P_1 - P_2}{\rho g} = \frac{32\mu\bar{U}L}{\rho g D^2}$$

$$\Delta P = 1800\text{ kN/m}^2$$

$$1800\times 10^3 = \frac{32(0.1)\bar{U}(100)}{(0.06)^2}$$

$$\bar{U} = 2.25\text{ m/s}$$

7.5 (a)

Shear stress, $\tau = \mu\frac{du}{dy} = 0.44\times\frac{4}{0.018}$

$$= 97.8\text{ Pa}$$

7.6 (b)

The pressure gradient in the direction of flow is equal to

$$\frac{dp}{dx} = \frac{d\tau}{dy}$$

the shear gradient in the transverse direction of flow.

7.7 (b)

For laminar flow in a circular pipe, the local velocity may be given as

$$u = U_{max}\left[1 - \frac{r^2}{R^2}\right] \quad \text{...(i)}$$

$$\bar{u} = U_{max}\left[1 - \frac{r^2}{R^2}\right]$$

(For Laminar pipe flow)

$$U_{max} = 2\bar{U}$$

$$\frac{1}{2} = 1 - \frac{r^2}{R^2}$$

$$\frac{r^2}{R^2} = \frac{1}{2}$$

$$r = 0.71\,R$$

7.8 (*)

$$D = 4 \text{ cm}$$

$$\mu = 1.6 \text{ centipoise}$$

$$u_{max} = 2 \text{ m/s}$$

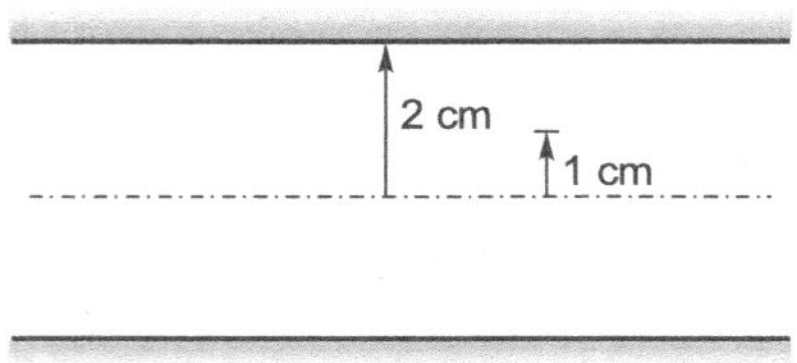

$$\tau = -\frac{r}{2}\left(\frac{\partial P}{\partial x}\right) \quad \ldots\text{(i)}$$

We know that

$$u_{max} = -\frac{1}{4\mu}\left(\frac{\partial P}{\partial x}\right)R^2$$

$$2 = -\frac{1}{4 \times 1.6 \times 10^{-3}}\left(\frac{\partial P}{\partial x}\right)(0.02)^2$$

$$\frac{\partial P}{\partial x} = -32 \text{ Pa/m}$$

By eq. (i)

$$\tau = -\frac{r}{2}(-32)$$

At $r = 1$ cm

$$\tau_{r = 1\text{ cm}} = -\frac{0.01}{2}(-32)$$

$$= 0.16 \text{ N/m}^2$$

7.9 (b)

Hele Shaw Flow: It is is defined as low Reynond's number flow between parallel plates seperated by an infiltensimally small gap. It is important to micro flows.

7.10 (b)

- Statement 1 wrong.
  For Laminar flow through a circular pipe shear stress is zero at the centre.

$$\tau = -\frac{r}{2}\left(\frac{\partial p}{\partial x}\right)$$

- Statement 2 is wrong.
  For Laminar flow through a circular pipes

$$U_{max} = 2\bar{U}$$

- Statement 3 is correct.
  According to Hagen-Poiseuille equation

$$h_f = \frac{32\mu\bar{U}L}{\rho g D^2} = \frac{128}{\pi}\frac{\mu QL}{\rho g D^4}$$

So, $$Q \propto \frac{1}{\mu}$$

- Statement 4 is correct.
  According to Hagen-Poiseuille equation

$$h_f = \frac{32\mu\bar{U}L}{\rho g D^2}$$

$$\frac{h_f}{L} = \frac{32\mu\bar{U}}{\rho g D^2}$$

$$\frac{h_f}{L} \propto \bar{U}$$

7.11 (d)

In Navier's-Stoke equation viscous force term is considered and in all the above mentioned flow viscous force can't be neglected.

7.12 (b)

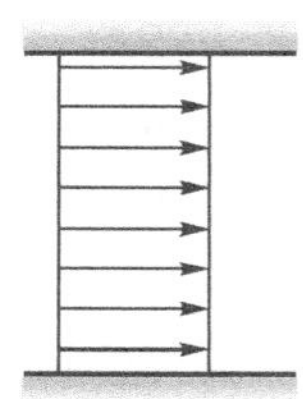

Ideal flow: The velocity profile is uniform.

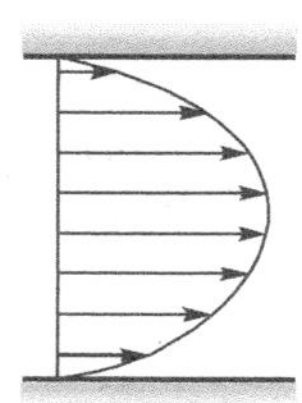

Poiseuille flow: The velocity profile is parabolic.

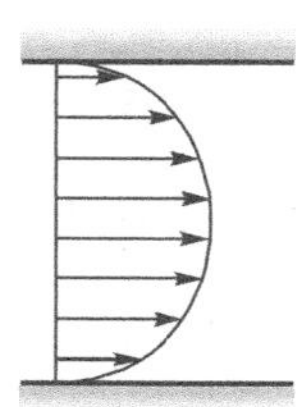

Turbulent flow: The velocity profile is logarithmic. (It is more flatter than laminar flow)

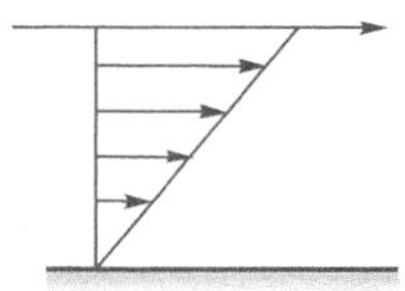

Couette flow: It is velocity distribution of simple or plain couette flow.

7.13 (b)

- In laminar flow (low Reynold's number) momentum transfer between adjacent layer of flow is negligible whereas In turbulent flow (high Reynold's number) momentum transfer between adjacent layer of flow is significant.
- At high Reynold's number the flow becomes turbulent and for turbulent flows the velocity distribution is logarithmic.

7.14 (c)

The wall shear stress, τ, is given by

$$\tau = -\left(\frac{\partial P}{\partial x}\right)\frac{R}{2}$$

$$\tau = -\left(\frac{-60}{15}\right)\times\frac{15}{100\times 2}$$

$$\tau = 0.3 \text{ kPa}$$

7.15 (b)

- Statement 1 is correct.

$$\tau = -\frac{r}{2}\left(\frac{\partial P}{\partial x}\right)$$

At $r = 0$, $\tau = 0$

- Statement 2 is wrong.
  According to Hagen Poiseuille equation

$$h_f = \frac{128}{\pi}\frac{\mu QL}{\rho g D^4}$$

$$Q \propto \frac{1}{\mu}$$

- Statement 3 is correct.
  Velocity is maximum at the centre.

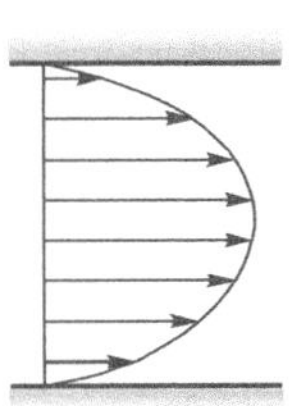

- Statement 4 is wrong.

$$h_f = \frac{32\mu\bar{U}L}{\rho g D^2}$$

$$\frac{h_f}{L} = \frac{32\mu}{\rho g D^2}\bar{U}$$

$$\frac{h_f}{L} \propto \bar{U}$$

7.16 (c)

We know that

$$\tau = -\frac{r}{2}\left(\frac{\partial P}{\partial x}\right) \quad \text{...(i)}$$

At wall $r = R$, $\tau = \tau_0$

$$\tau_0 = -\frac{R}{2}\left(\frac{\partial P}{\partial x}\right) \quad \text{...(ii)}$$

Divide eq. (i) and (ii)

$$\frac{\tau}{\tau_0} = \frac{r}{R}$$

$$\tau = \frac{r}{R}\tau_0 \quad \text{...(iii)}$$

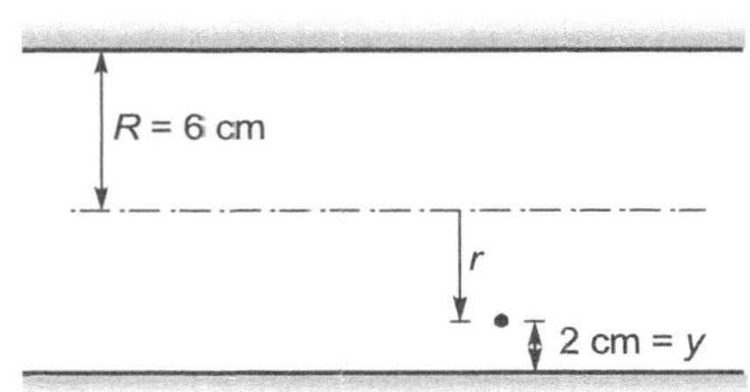

$$r = R - y$$
$$= 6 - 2$$
$$= 4 \text{ cm}$$

By equation (iii)

$$\tau = 4\frac{4}{6}\times 40 = 26.7 \text{ Pa}$$

7.17 (b)

For Laminar flow through pipe

- Statement 1 is wrong.

$$f = \frac{64}{\text{Re}}$$

  f is dependent on Re.
- Statement 2 is wrong.
  f only dependent on Re.
- Statement 3 is correct.
- Statement 4 is correct.

$$u = -\frac{1}{4\mu}\left(\frac{\partial P}{\partial x}\right)\left(R^2 - r^2\right)$$

Velocity distributionis parabolic.

7.18 (c)

$D = 200$ mm

$u_{max} = 1$ m/s

$u_{r=50\text{ mm}} = ??$

We know that

$$u = -\frac{1}{4\mu}\left(\frac{\partial P}{\partial x}\right)\left(R^2 - r^2\right)$$

At $r = 0$, $u = u_{max}$

$$u_{max} = -\frac{1}{4\mu}\left(\frac{\partial P}{\partial x}\right)R^2$$

$$1 = -\frac{1}{4\mu}\left(\frac{\partial P}{\partial x}\right)(0.1)^2$$

$$-\frac{1}{4\mu}\left(\frac{\partial P}{\partial x}\right) = \frac{1}{(0.1)^2} \quad \text{...(i)}$$

At $r = 50$ mm

$$u = -\frac{1}{4\mu}\left(\frac{\partial P}{\partial x}\right)\left(R^2 - r^2\right)$$

$$= \frac{1}{(0.1)^2}\left[0.1^2 - 0.05^2\right]$$

$= 0.75$ m/s

7.19 (b)

For Laminar flow between two parallel fixed plates.

- Statement 1 is wrong.
  Shear stress is maximum at wall and zero at $\frac{d}{2}$ (centre line).

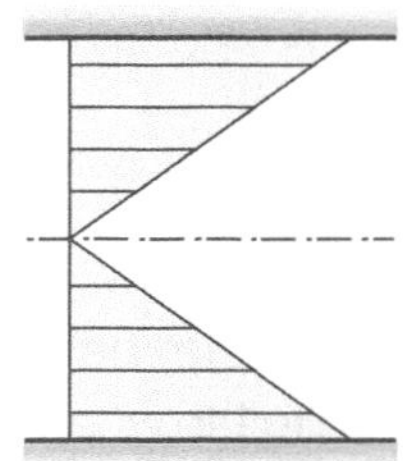

- Statement 2 is wrong.
- Statement 3 is correct.

7.20 (d)

Head loss due to friction, $h_f = \dfrac{flv^2}{2gd}$

$$h_f = \frac{0.005 \times 60 \times (2.5)^2}{2 \times 9.81 \times 0.2} = 0.477 \text{ m}$$

No answer is matching.

Considering, 0.005 as Fannings friction coefficient, then darcy's friction factor 'f' = 4 × 0.005 = 0.02

$$\therefore\ h_f = \frac{0.02 \times 60 \times (2.5)^2}{2 \times 9.81 \times 0.2} = 1.91 \text{ m}$$

■■■■

8 Flow Through Pipes

8.1 Consider the following conditions for the pipe network shown in the given figure (Notations have the usual meaning with suffixes 1, 2 and 3 referring to respective pipes):

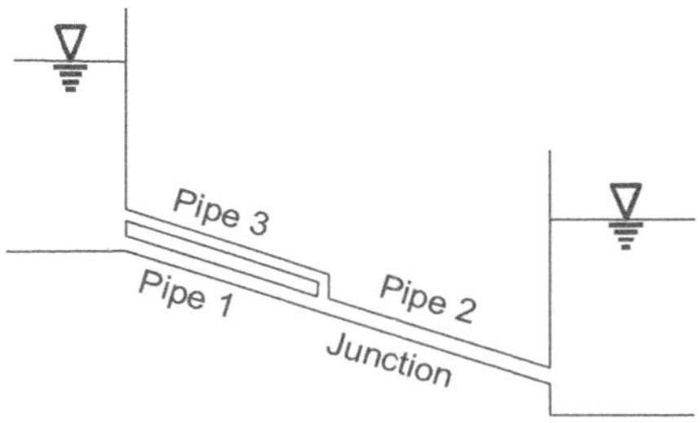

1. $Q_1 = Q_3$ 2. $Q_2 = Q_1 + Q_3$
3. $h_{f1} = h_{f3}$ 4. $h_{f1} = h_{f2} = h_{f3}$

Which of these conditions must be satisfied by this pipe network?

(a) 1 and 3 (b) 2 and 3
(c) 1 and 4 (d) 2 and 4

[ESE : 2000]

8.2 Maximum pressure rise due to water hammer in a pipeline (a = area of the pipe; V_0 = Velocity, g = acceleration due to gravity; t = time period; L = length of the pipeline) is

(a) $\dfrac{aV_0}{2g}$

(b) $\dfrac{aV_0^2}{2g}$

(c) $\dfrac{LV_0}{gt}$

(d) Independent of the dimensions of the pipe

[ESE : 2000]

8.3 A and B are at the same elevation of 2.5 m above datum in the valve and pipe line shown in the given figure. Velocity head of 0.5 m, head loss in valve of 0.2 m and pressure head of 2.5 m are the parameters at A. Piezometric head at B is equal to

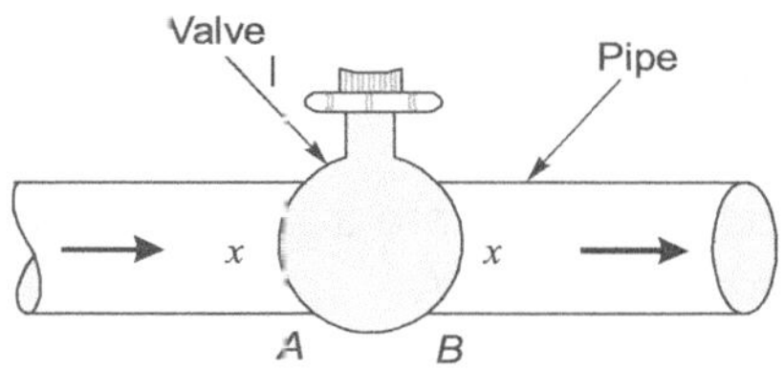

(a) 5.5 m (b) 5.3 m
(c) 5.0 m (d) 4.8 m

[ESE : 2000]

8.4 Two tanks are connected in parallel by two pipes A and B of identical friction factors and lengths. If the size of pipe A is double than that of pipe B, then their discharges will be in the ratio of

(a) 2 (b) 4
(c) 5.66 (d) 32

[ESE : 2001]

8.5 Water of ν = 1 centistoke flows through a 1 cm diameter pipe. Critical flow will correspond to a discharge of approximately

(a) 0.008 l/s (b) 0.013 l/s
(c) 0.016 l/s (d) 0.025 l/s

[ESE : 2001]

8.6 The procedure to be followed in solving for discharge in a simple pipe problem, when h_f, L, D, V and ε are given is to

[**Note :** Re is Reynolds number,
f = friction coefficient,
ε/D = relative roughness
V = velocity]

(a) assume f, compute : V, Re, ε/D and calculate f; and repeat if necessary
(b) assume Re, compute f, check ε/D
(c) assume V, compute Re and calculate f. V again
(d) assume ε, compute V, Re and calculate f

[ESE : 2002]

8.7 The Chezy's coefficient C is related to Darcy-Weisbach friction factor 'f' as

(a) $C = \sqrt{(g/8f)}$ (b) $C = \sqrt{(8g/f^{1/4})}$

(c) $C = \sqrt{(8g/f)}$ (d) $C = \sqrt{(f/8g)}$

[ESE : 2002]

8.8 **Assertion (A):** Abrupt rise in pressure in a pipe line results from sudden closure of a valve.

Reason (R): Pressure wave propagates through elastic medium.

(a) both A and R are true and R is the correct explanation of A
(b) both A and R are true but R is not a correct explanation of A
(c) A is true but R is false
(d) A is false but R is true

[ESE : 2002]

8.9 Match **List-I** with **List-II** and select the correct answer using the codes given below the lists:

List-I	List-II
A. Flow development	1. Surge tank
B. Pipe network	2. Entrance length
C. Water hammer	3. Darcy-Weisbach equation
D. Friction loss	4. Hardy-Cross method

Codes:

	A	B	C	D
(a)	2	4	1	3
(b)	4	2	3	1
(c)	2	4	3	1
(d)	4	2	1	3

[ESE : 2003]

8.10 Which one of the following conditions must **NOT** be satisfied in the Hardy-Cross analysis of pipe network?

(a) Continuity principle demands that flow into a network junction is equal to the flow out of it.
(b) Momentum equation must be satisfied so that the force in each loop is balanced.
(c) Darcy-Weisbach head loss equation is to be used in computing head loss in elementary circuits. The equation is of the form $h_L = kQ^n$.
(d) $\Sigma h_L = \Sigma kQ^n$ must be equated to zero. If not, a flow correction factor ΔQ is to be made for arriving at a solution.

[ESE : 2003]

8.11 Which one of the following phenomena in a pipe flow is termed as water hammer?

(a) The sudden rise of pressure in a long pipe due to sudden closure of valve.
(b) The rise of a pressure in a pipe flow due to gradual closure of valve.
(c) The rise of negative pressure.
(d) The zero pressure in a pipe flow.

[ESE : 2003]

8.12 Which one of the following statements is not correct?

(a) Cavitation forms where the local pressure drops up to the vapour pressure conditions due to increase in local velocity.
(b) Cavitation takes place in turbine blades, runners of pumps, spillways where local velocity increases considerably.
(c) Cavitation may be prevented by maintaining the ambient pressure relatively higher at probable points of occurrence by raising their levels.
(d) Boundaries likely to be affected by cavitation can be protected against pitting by covering it with tough materials like stainless steel or rubber.

[ESE : 2003]

8.13 Two long pipes in parallel are used to carry water between two reservoirs. The diameter of one pipe is twice that of the other. Both the pipes have the same value of friction factor. Neglect minor losses. What is the ratio of flow rates through the two pipes?

(a) 2.8 (b) 5.6
(c) 8 (d) 11.3

[ESE : 2004]

8.14 Which one of the following statements is **NOT** correct?

(a) Water hammer is a phenomenon which occurs due to oscillation of water column away from the valve.
(b) Rigid water column theory specifies that water is incompressible and pipe material does not expand.
(c) Water hammer occurs under unsteady conditions of flow.

(d) Rigid water column theory is used less in practice as it does not reflect field situation.
[ESE : 2006]

8.15 Which one of the following statements is **NOT** correct?

(a) Solution of pipe network by Hardy Cross method is a trial and error solution.
(b) At a junction of pipes, total inflow is equal to the total outflow.
(c) Loss of head due to flow in a clockwise direction should be equal to loss of head in flow in a counter clockwise direction.
(d) Hardy Cross method can be extended to open channel flow also.

[ESE : 2006]

8.16 Water flows into junction J from reservoirs *A* and *B* through connecting pipes, the head loss through these being, respectively $10.Q_A^2$ and $4.Q_B^2$. The water level elevations at the reservoirs at *A* and *B* are 25.9 m and 18 m, respectively. The inflow at *J* is discharged out at *C* into the atmosphere. The head loss through pipe *JC* is $1.Q_C^2$. The gauge pressure at *J* is 9 m. What is the residual gauge pressure of the outflow at *C*?

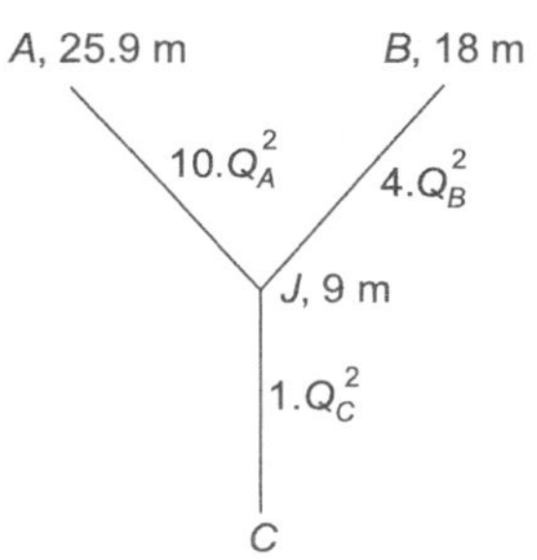

(a) 2.56 m
(b) 1.86 m
(c) 1.16 m
(d) 0.46 m [ESE : 2006]

8.17 While using Darcy-Weisbach equation for estimating head loss in a pipe flow, the friction factor was misjudged by + 20% for this case, the error in estimating discharge is

(a) + 10%
(b) + 40%
(c) – 40%
(d) – 10%

[ESE : 2008]

8.18 What is the energy loss through a length of pipe from which all the discharge is taken out uniformly along its length, with respect to the case when all the discharge is delivered to the end?

(a) 2/3
(b) 1/3
(c) 3/2
(d) 1/2

[ESE : 2008]

8.19 The power transmitted through a pipeline is maximum when the head lost due to friction in the pipe is equal to

(a) the total supply head
(b) half of the total supply head
(c) one-third of the total supply head
(d) one-fourth of the total supply head

[ESE : 2009]

8.20 In a siphon system employed for carrying water from a reservoir *A* at a higher elevation to another reservoir *B* at lower elevation, both being separated by a higher hill, what will be the pressure at the 'Summit' (*S*)?

(a) Equal to the pressure at the water surface of reservoir *A*
(b) Higher than the pressure at the water surface of reservoir *A*
(c) Equal to the pressure at the water surface of reservoir *B*
(d) Less than the pressure at both *A* and *B* above

[ESE : 2009]

8.21 Consider the following statements:
Cavitation generally results from a combination of several influences

1. by reduction of pressure intensity below a limiting value
2. by increase in either elevation or the velocity of flow
3. by reduction of pressure load in the system
4. by decrease in the velocity of flow

Which of these statements are correct?

(a) 1, 2 and 3
(b) 1 and 2 only
(c) 2 and 3 only
(d) 3 and 4

[ESE : 2009]

8.22 A sluice inlet in a dam with streamlined entrance is shown in figure below.

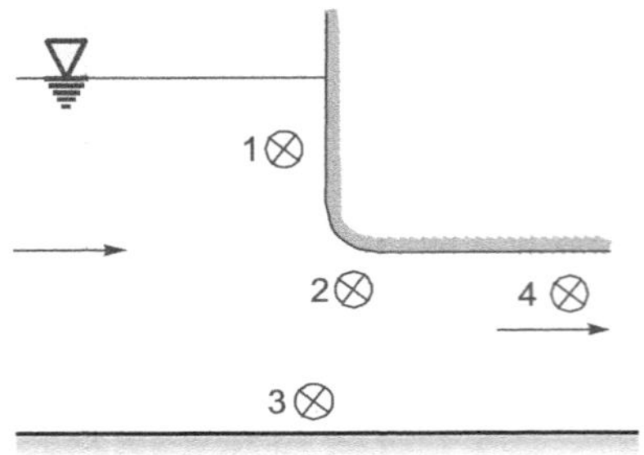

If cavitation is expected to occur, it will first appear at the point

(a) 1 (b) 2
(c) 3 (d) 4 [ESE : 2010]

8.23 Consider the following statements which are related to the phenomenon of cavitation in fluid flow:

1. Cavitation occurs when local velocity is decreased so that local pressure increases to a high degree.
2. Cavitation occurs if elevation is high thereby decreasing ambient pressure.
3. Cavitation occurs if local velocity is increased so that the local pressure decreases.
4. Cavitation is dependent on vapour pressure of the fluid.

Which of these statements are correct?

(a) 1, 2 and 3 (b) 1, 2 and 4
(c) 1, 3 and 4 (d) 2, 3 and 4

[ESE : 2010]

8.24 In a pipe network of municipal water supply, a parallel pipe is sometimes installed over a portion of the pipe mainly for

(a) reducing water hammer pressure
(b) decreasing the pumping power need
(c) increasing the head available at the node
(d) increasing the discharge [ESE : 2010]

8.25 Two identical pumps, each capable of delivering 0.2 cumec, against a head of 30 m, are connected in parallel. The resulting discharge is

(a) 0.4 cumec against a head of 30 m
(b) 0.4 cumec against a head of 60 m
(c) 0.2 cumec against a head of 30 m
(d) 0.2 cumec against a head of 60 m

[ESE : 2011]

8.26 Assertion (A): Loss of head at a sudden contraction in a pipe is smaller than the loss at a sudden expansion in the pipe.

Reason (R): Increase in turbulence level is higher at a sudden expansion than at a sudden contraction.

(a) both A and R are true and R is the correct explanation of A
(b) both A and R are true but R is not a correct explanation of A
(c) A is true but R is false
(d) A is false but R is true [ESE : 2011]

8.27 In a pressure penstock 4500 m long, water is flowing at a velocity of 4 m/s. If the velocity of the pressure wave travelling in the pipe, due to sudden complete closure of a valve at the downstream end, is given as 1500 m/s, what would be the period of oscillation in seconds under frictionless conditions?

(a) 6 (b) 8
(c) 9 (d) 11

[ESE : 2011]

8.28 Three pipes *A*, *B* and *C* have the following basic geometries:

Pipe	*A*	*B*	*C*
Diameter	*D*	*D*/2	2*D*
Length	*L*	*L*	4*L*

If these pipes are connected in series, by assuming the value of friction factor *f* to be same for all the three pipes and the equivalent pipe, this set of pipes in series is equivalent to a pipe of length L_e and diameter *D* and friction factor *f* with the equivalent length L_e being equal to

(a) $5\frac{1}{8}L$ (b) $4\frac{1}{8}L$
(c) $26\frac{1}{8}L$ (d) $33\frac{1}{8}L$

[ESE : 2012]

8.29 The velocity of pressure wave in water of infinite extent is 1414 m/s. The velocity of propagation of water hammer pressure in a pipe carrying water and having diameter = 40 cm, pipe thickness = 4 mm, with *E* (Modulus of elasticity) of the pipe material = 2.1×10^{11} Pa, and *K* (Bulk modulus of water) = 2.1×10^{9} Pa, is

(a) 1410 m/s (b) 2000 m/s
(c) 1000 m/s (d) 700 m/s
[ESE : 2012]

8.30 If the velocity of flow as well as the diameter of the flowing pipe are respectively doubled through pipe system in use since long, the head loss will thereafter be
(a) Halved (b) Doubled
(c) Increased 4 times (d) No change
[ESE : 2012]

8.31 Let C_1 be the velocity of pressure wave travelling along rigid pipe carrying water with its bulk modulus 2.16×10^9 N/m^2. Let C_2 be the velocity of pressure wave travelling along a rigid pipe carrying oil of relative density 0.600 with its bulk modulus as 1.296×10^9 N/m^2 through a similar pipe. What will be the ratio $\frac{C_1}{C_2}$?
(a) 0.01 (b) 0.1
(c) 1.0 (d) 10.0
[ESE : 2012]

8.32 Which of the following pairs are correctly matched?
1. Piezometric head : Sum of datum head and pressure head
2. Dynamic head : Sum of datum head and velocity head
3. Stagnation head : Sum of Piezometric head and velocity head
4. Total head : Sum of Piezometric head and dynamic head

(a) 1, 2 and 3 only (b) 1, 3 and 4 only
(c) 2, 3 and 4 only (d) 1, 2, 3 and 4
[ESE : 2012]

8.33 It is proposed to increase the discharge by 20% in circular pipe carrying oil in laminar regime. If all other factors remain unchanged, power consumption to maintain the modified flow relative to the original flow would increase by
(a) 10% (b) 20%
(c) 44% (d) 52% [ESE : 2013]

8.34 In a pipe network:
(a) The algebraic sum of discharges around each elementary circuit must be zero
(b) The head at each node must be the same
(c) The algebraic sum of the drop in piezometric head around each elementary circuit is zero
(d) The piezometric head loss in each line of each circuit is the same
[ESE : 2013]

8.35 In a pipeline 920 m long, the velocity of propagation of pressure wave is 960 m/s. Rapid closure of a downstream valve will entail, when the maximum time for the closure is:
(a) 8 s (b) 6 s
(c) 4 s (d) 2 s [ESE : 2013]

8.36 Statement (I): Gravity is a determinant in estimating pipe flow magnitudes.
Statement (II): The term 'g' appears explicitly in Darcy-Weisback formula for pipe flow resistance.
(a) Both Statement (I) and Statement (II) are individually true and Statement (II) is the correct explanation of Statement (I)
(b) Both Statement (I) and Statement (II) are individually true but Statement (II) is ***not*** the correct explanation of Statement (I)
(c) Statement (I) is true but Statement (II) is false
(d) Statement (I) is false but Statement (II) is true
[ESE : 2013]

8.37 A liquid of density ρ and bulk modulus K flows with a mean velocity V in a long rigid pipe of diameter D. A sudden closure of a valve at the downstream end of the pipe will produce a maximum water hammer head of
(a) $\frac{V\sqrt{\rho}}{g\sqrt{K}}$ (b) $\frac{V\sqrt{g}}{\rho\sqrt{K}}$
(c) $\frac{V\sqrt{K}}{g\sqrt{\rho}}$ (d) $\frac{V\rho}{\sqrt{gK}}$ [ESE : 2014]

8.38 Two reservoirs are connected by two pipes P and Q. The pipes have the same diameter and length and are placed in parallel. If the friction factor of P is 9 times that of Q, then the discharge in P to that in Q is
(a) 0.5 (b) 0.45
(c) 0.33 (d) 0.27 [ESE : 2015]

8.39 Assertion (A): Total energy line and the hydraulic gradient line for a pipe flow cannot intersect each other.

Reason (R): The vertical differences between these two lines must equal the velocity head.

(a) Both A and R are true and R is the correct explanation of A
(b) Both A and R are true but R is not a correct explanation of A
(c) A is true but R is false
(d) A is false but R is true [ESE : 2015]

8.40 What is the maximum power available at the downstream end of a pipeline 3 km long, 20 cm in diameter, if water enters at the upstream end at a pressure of 720 m of water, with taking pipe friction coefficient as 0.03 and g as 10 m/s^2?

(a) 770 mhp (b) 740 mhp
(c) 700 mhp (d) 660 mhp

[ESE : 2016]

8.41 Two tanks A and B, of constant cross-sectional area of 10 m^2 and 2.5 m^2, respectively, are connected by a 5 cm pipe, 100 m long, with $f = 0.03$. If the initial difference of water levels is 3 m, how long will it take for 2.5 m^3 of water to flow from A to B? Considering entry and exit losses, it can be grossly assumed that the flow velocity, in m/s, through the pipe is $1.75\sqrt{h}$, where h is in m, taking g = 10 m/sec^2, also, may take area of pipe as 2×10^{-3} m^2.

(a) 535 sec (b) 516 sec
(c) 485 sec (d) 467 sec [ESE : 2016]

8.42 A pipe of 324 mm diameter, having friction coefficient as 0.04, connects two reservoirs with 15 m difference in their water levels through a 1500 m long pipe. What will be the discharge through the pipe?

(a) 104 lps (b) 134 lps
(c) 165 lps (d) 196 lps

[ESE : 2016]

8.43 Statement (I) : Pipes carrying water are anchored at bends and other points of unbalanced thrusts.

Statement (II): Pipes are anchored by firmly embedding in massive blocks of concrete or masonry to counter side thrusts due to hydrodynamic forces exerted on the joints.

(a) Both Statement (I) and Statement (II) are individually true and Statement (II) is the correct explanation of Statement (I)
(b) Both Statement (I) and Statement (II) are individually true but Statement (II) is ***not*** the correct explanation of Statement (I)
(c) Statement (I) is true but Statement (II) is false
(d) Statement (I) is false but Statement (II) is true

[ESE : 2016]

8.44 Cavitation is likely to occur if

1. Pressure becomes very high.
2. Temperature becomes low.
3. Pressure at the specific point falls below vapour pressure.
4. Energy is released with the onset of a high intensity wave due to noise and vibration of the machine.

Which of the above statements are correct?

(a) 1 and 3 (b) 2 and 3
(c) 3 and 4 (d) 2 and 4

[ESE : 2016]

8.45 Two reservoir maintain a constant difference of water levels of 11.25 m and are connected by a 10 cm diameter pipeline of 294.3 m length. The total of all head losses, by friction valve losses, bend losses, inlet and exit losses, and velocity head can be taken as 98.1 v^2/2g (in m) where v is the flow velocity through the pipe (in m/sec). Assuming that the valve at the downstream end is suddenly opened so that there is no pressure wave, what will be the time taken for the velocity of flow in the pipe to attain 95% of the steady terminal velocity? Take

$$\frac{1}{9.81} = 0.102.$$

(a) $2.25 \log_e 19$ (b) $2 \log_e 19$
(c) $2.25 \log_e 39$ (d) $2 \log_e 39$

[ESE : 2017]

8.46

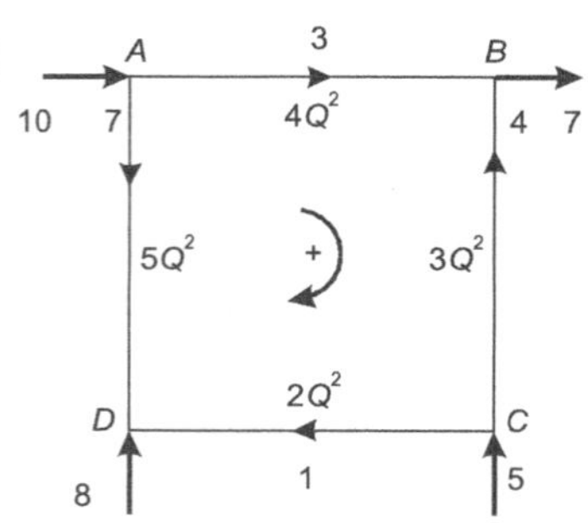

A pipe network is shown with all needful input data to compute the first iteration improved magnitudes of the initially assumed flows in the branches. What will be the such improved flow magnitudes in branches AB and CD? Consider to first decimal accuracy.

(a) A to B : 5.1; C to D : 3.1
(b) A to B : 5.7; C to D : 2.8
(c) A to B : 4.9; C to D : 3.4
(d) A to B : 5.5; C to D : 3.8 [ESE : 2018]

8.47 In a siphon, the summit is 5 m above the water level in the tank from which the flow is being discharged. If the head loss from the inlet to the summit is 2.5 m and the velocity head at the summit is 0.5 m, (taking γ = 10 appropriate units) the pressure head at the summit is

(a) –80 kPa (b) –3 m of water (abs)
(c) 5 m of water (abs) (d) 18 m of water (abs)

[ESE : 2018]

8.48 A network of pipes conveying water to a city has the following specifications. The diameter of a main pipe is 30 cm and it branches into two pies of diameters 20 cm and 15 cm respectively. If the average velocity in the main pipe is 2.5 m/s and the average velocity in the 20 cm pipe is measured as 2 m/s, what is the velocity in the 15 cm pipe?

(a) 8.84 m/s (b) 7.44 m/s
(c) 5.84 m/s (d) 6.44 m/s

[ESE : 2021]

8.49 A pipeline of uniformly varying cross-section carries an oil of specific gravity 0.87. The diameter of pipe is 200 mm at end A and 500 mm at end B. The end B is located at 4 m higher than A. What is the loss of head in the pipeline if the pressure reading at A is 9.81 N/cm^2 and at B is 5.886 N/cm^2? (Take discharge is 200 litres/s and g = 9.81 m/s^2)

(a) 2.609 m (b) 26.09 cm
(c) 2.109 m (d) 21.09 cm

[ESE : 2022]

8.50 The water is flowing with a velocity of 1.5 m/s in a pipe of length 2500 m and diameter 500 mm. A valve is provided at the end of the pipe. What is the rise in pressure if the valve is closed in 25 seconds? (Take velocity of pressure wave is 1460 m/s)

(a) 15 N/cm^2 (b) 1500 N/cm^2
(c) 150 N/m^2 (d) 15 kN/m^2

[ESE : 2022]

8.51 If the pressure heads at the different sections of the pipe are plotted to scale as vertical ordinates above the axis of the pipe and all these points are joined by a slopping straight line, which is known as

(a) Hydraulic gradient line
(b) Total energy line
(c) Total head line
(d) Energy grade line [ESE : 2023]

8.52 The difference in the water surface level of two reservoirs which are connected by a siphon is 8 m. The length of siphon is 600 m and its diameter 0.3 m. If the siphon is running full, then what is the discharge? (Take f = 0.02)(in m^3/s)

(a) 3.242 (b) 2.117
(c) 1.842 (d) 0.137 [ESE : 2023]

8.53 **Statement I** : In Hardy cross method, a trial distribution of discharges is made arbitrary.
Statement II : The continuity equation must satisfy at each junction. [ESE : 2023]

8.54 Consider the following statements regarding flow measurement through pipes :

1. The reduction in constriction diameter causes velocity to increase.
2. High velocities in constriction cause low pressures in the system.
3. The reduction in constriction diameter enables lesser accuracy in its measurement.

Which of the above statements is/are correct :

(a) 1 and 2 (b) 2 and 3
(c) 2 only (d) 1 and 3

[ESE : 2024]

■■■■

Answers Flow Through Pipes

8.1 (b)	8.2 (c)	8.3 (d)	8.4 (c)	8.5 (c)	8.6 (a)	8.7 (c)	8.8 (a)	8.9 (a)
8.10 (b)	8.11 (a)	8.12 (c)	8.13 (b)	8.14 (a)	8.15 (d)	8.16 (c)	8.17 (d)	8.18 (b)
8.19 (c)	8.20 (d)	8.21 (b)	8.22 (b)	8.23 (d)	8.24 (d)	8.25 (a)	8.26 (a)	8.27 (a)
8.28 (d)	8.29 (c)	8.30 (b)	8.31 (c)	8.32 (b)	8.33 (c)	8.34 (c)	8.35 (c)	8.36 (d)
8.37 (c)	8.38 (c)	8.39 (a)	8.40 (d)	8.41 (d)	8.42 (a)	8.43 (a)	8.44 (c)	8.45 (d)
8.46 (a)	8.47 (a)	8.48 (d)	8.49 (a)	8.50 (a)	8.51 (a)	8.52 (d)	8.53 (b)	8.54 (a)

Explanations Flow Through Pipes

8.1 (b)

As per the continuity equation at the junction,
$Q_1 + Q_3 = Q_2$

Since connect between pipe 1 and pipe 3 is parallel, $h_{f1} = h_{f3}$

Since diameter is not given no relation can be established between discharges.

8.2 (c)

In gradual closure of pipe valve, the pressure head rise is

$$\frac{P}{\rho g} = \frac{V_0}{g}\frac{L}{T}$$

8.3 (d)

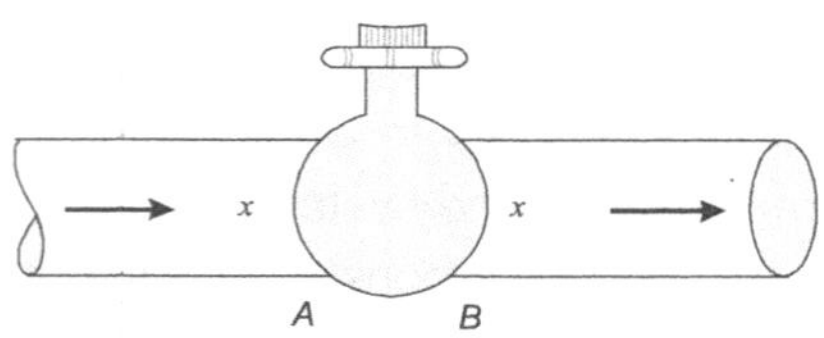

Apply energy equation between (A) and (B)

$$\frac{P_A}{\rho g} + \frac{V_A^2}{2g} + z_A = \frac{P_B}{\rho g} + \frac{V_B^2}{2g} + z_B + h_f$$

$$2.5 + 0.5 + 2.5 = \frac{P_B}{\rho g} + 0.5 + z_B + 0.2$$

$$\frac{P_B}{\rho g} + z_B = 4.8 \text{ m} \qquad (V_A = V_B)$$

8.4 (c)

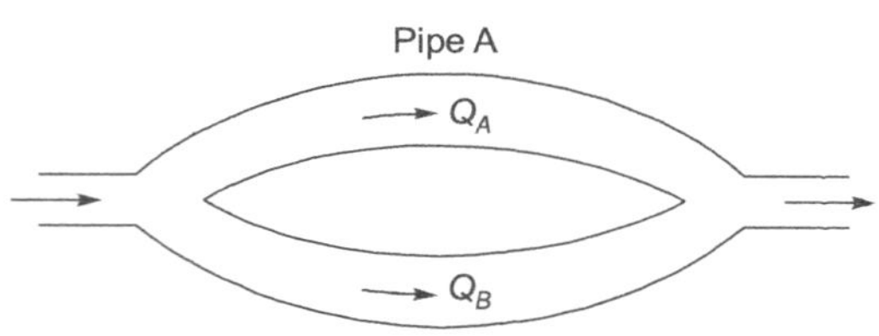

Given: $D_A = 2D_B$

L & f is same for both the pipes.

For parallel pipes

$$h_{f_A} = h_{f_B}$$

$$\frac{8Q_A^2}{\pi^2 g}\frac{fL}{D_A^5} = \frac{8Q_B^2}{\pi^2 g}\frac{fL}{D_B^5}$$

$$\frac{Q_A^2}{(2D_B)^5} = \frac{Q_B^2}{D_B^5}$$

$$\frac{Q_A}{Q_B} = (2)^{5/2}$$

$$= 5.66$$

8.5 (c)

$$\text{Re}_c = 2000 \text{ for critical flow}$$

Viscosity, $\nu = 0.01$ Stokes

$$\frac{VD}{\nu} = 2000$$

Velocity of flow,

$$V = \frac{2000 \times 0.01}{1} = 20 \text{ cm/s}$$

Flow at critical Reynolds number

$$Q = V \times \frac{\pi D^2}{4} = 20 \times \frac{\pi}{4} \times 1$$

$$= 15.7 \text{ cm}^3\text{/s} = 0.0157 \text{ } l\text{/s}$$

8.6 (a)

(i) Assume f (It is function of Re and ε/D)

Head Loss, $h_f = f\dfrac{L}{D}\dfrac{V^2}{2g}$

(ii) From head loss formula compute V.

(iii) Find ε/D from given data.

Reynolds number, $Re = \frac{VD}{\nu}$

(iv) Now again find 'f'. (say from Colebrook equation)

(v) If the computed 'f' is same as assumed 'f', then stop, otherwise repeat the procedure with new value of 'f'.

8.7 (c)

According to Chezy's formula

$$V = C\sqrt{mi}$$

m = Hydraulic mean depth

$$m = \frac{A}{P} = \frac{\pi D^2/4}{\pi D}$$

$$m = \frac{D}{4}$$

$$i = \frac{h_f}{L}$$

C = Chezy's constant

$$V = C\sqrt{\frac{D}{4}\frac{h_f}{L}}$$

$$h_f = \frac{4LV^2}{C^2D} \quad \text{...(i)}$$

According to Darcy Weisbach equation

$$h_f = \frac{fLV^2}{2gD} \quad \text{...(ii)}$$

By eq. (i) and (ii)

$$\frac{4LV^2}{C^2D} = \frac{fLV^2}{2gD}$$

$$C = \sqrt{\frac{8g}{f}}$$

8.8 (a)

Abrupt rise in pressure in a pipe line results from sudden closure of a value because pressure wave propagates through elastic medium. (flowing fluid).

8.9 (a)

- Surge tanks are used to absorb the increase in pressure due to water hammer phenomenon.
- Entrance length in a pipe is the length where boundary layer increases and the flow is established or developed fully. Entrance length (L_e)

$\frac{L_e}{D} = 0.07Re(0.06Re)$ For laminar flow given by **Rouse**

$\frac{L_e}{D} = 50$ For turbulent flow

where D is the diameter of the pipe.

Entrance length for turbulent flow is much smaller than laminar flow.

8.10 (b)

- It is necessary to satisfy continuity equation.
- The loss of head due to flow in clockwise direction must be equal to the loss of head due to flow in anticlockwise direction.
- Darcy-Weisbach equation must be satisfied

$$h_f = kQ^n$$

8.11 (a)

A water hammer commonly occurs when a valve closes suddenly at an end of a pipeline system, and there is sudden rise in the pressure. As a result a high pressure wave propagates in the pipe.

8.12 (c)

- Increase in local velocity decreases the pressure and therefore cavitation may occur.
- Raising the levels of the point will cause higher gradient and flow velocities will increase. This may cause cavitation.

8.13 (b)

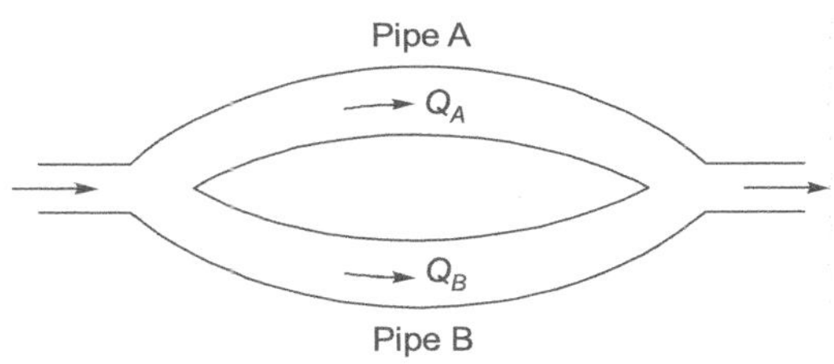

Given: $D_A = 2D_B$

L & f is same for both the pipes.

For parallel pipes

$$h_{f_A} = h_{f_B}$$

$$\frac{8Q_A^2}{\pi^2 g}\frac{fL}{D_A^5} = \frac{8Q_B^2}{\pi^2 g}\frac{fL}{D_B^5}$$

$$\frac{Q_A^2}{(2D_B)^5} = \frac{Q_B^2}{D_B^5}$$

$$\frac{Q_A}{Q_B} = (2)^{5/2} = 5.66$$

8.14 (a)

Water hammer phenomenon occurs near to the valve.

8.15 (d)

The Hardy Cross method is an iterative method for determining the flow in pipe network systems where the inputs and outputs are known, but the flow inside the network is unknown. This method is not applicable to open channel flow.

8.16 (c)

Using continuity equation,

$$Q_C = Q_A + Q_B$$

Head loss in pipe,

$$AJ = 10Q_A^2 = 25.9 - 9 = 16.9\text{m}$$

$$\therefore \quad Q_A = 1.3 \text{ m}^3/\text{s}$$

Head loss in pipe,

$$BJ = 4Q_B^2 = 18 - 9 = 9\text{m}$$

$$Q_B = 1.5 \text{ m}^3/\text{s}$$

$$\therefore \quad Q_C = 1.3 + 1.5 = 2.8 \text{ m}^3/\text{s}$$

Head loss in pipe,

$$JC = 1Q_C^2 = 7.84\text{m}$$

Residual head at,

$$C = 9 - 7.84 = 1.16 \text{ m}$$

8.17 (d)

The head loss through the pipe is given by

$$h_L = \frac{fLQ^2}{12.1D^5}$$

If all the parameters remain same other than friction factor and discharge then

$$fQ^2 = \text{constant}$$

Let this constant be C, then

$$fQ^2 = C$$

$$\Rightarrow \quad Q^2 = \frac{C}{f} \qquad \text{...(i)}$$

Differentiating Q w.r.t f, we get

$$2QdQ = C \times \left(-\frac{1}{f^2}\right) df \qquad \text{...(ii)}$$

Dividing (ii) by (i), we get

$$\frac{2QdQ}{Q^2} = \frac{C \times \left(-\frac{1}{f^2}\right) df}{\frac{C}{f}}$$

$$\Rightarrow \quad \frac{2dQ}{Q} = -\frac{df}{f}$$

$$\Rightarrow \quad \frac{dQ}{Q} = -\frac{1}{2} \times \frac{df}{f}$$

$$\Rightarrow \quad dQ = -\frac{Q}{2} \times \frac{df}{f}$$

$$\Rightarrow \quad dQ = -\frac{100}{2} \times \frac{20}{100}$$

$$\Rightarrow \quad dQ = -10\%$$

8.18 (b)

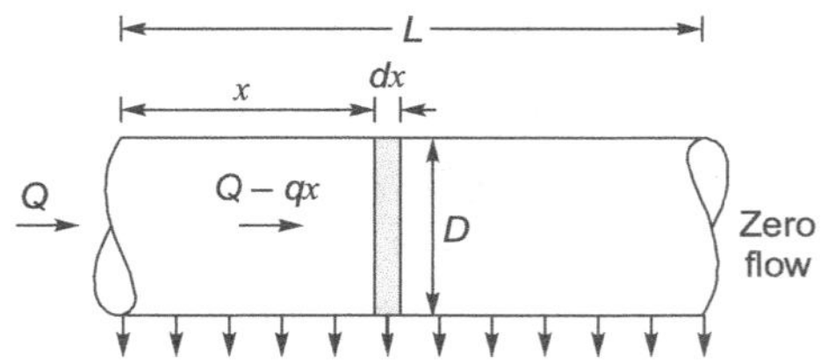

Assume q = uniform discharge per unit length ($Q = qL$)

$$dh_f = \frac{8(Q - qx)^2}{\pi^2 g} \frac{f dx}{D^5}$$

Int. it

$$h_f = \frac{8}{\pi^2 g} \frac{f}{D^5} \int_0^L (Q - qx)^2 dx$$

$$= \frac{8}{\pi^2 g} \frac{f}{D^5} \int_0^L \left[Q^2 + q^2x^2 + 2qQx\right] dx$$

$$= \frac{8}{\pi^2 g} \frac{f}{D^5} \left[Q^2 x + \frac{q^2x^3}{3} - \frac{2qQx^2}{2}\right]_0^L$$

$$= \frac{8}{\pi^2 g} \frac{f}{D^5} \left[Q^2 L + \frac{q^2L^3}{3} - qQL^2\right]$$

(Since $Q = qL$)

$$= \frac{8}{\pi^2 g} \frac{f}{D^5} \left[Q^2 L + \frac{Q^2 L}{3} - Q^2 L\right]$$

$$= \frac{8Q^2}{\pi^2 g} \frac{fL}{D^5} = \frac{1}{3}$$

8.19 (c)

If H is the total supply head and h_f is the loss of head due to friction, then head available at the outlet of pipe (neglecting the minor losses) is $H - h_f$. If Q is the discharge through the pipe, V is the velocity of flow and L and D are the length and the diameter of the pipe respecting, then from

Dercy-Weisbach equation we have

$$h_f = \frac{fLV^2}{2gD} \text{ and } Q = \frac{\pi D^2}{4} \times V$$

The power available at the outlet of pipeline is given by

$$P = \gamma Q(H - h_f)$$

$$\Rightarrow \quad P = \gamma \times \frac{\pi D^2}{4} \times V \times \left(H - \frac{fLV^2}{2gD}\right)$$

For a given pipe and head at the entrance, the condition for the maximum power transmitted through the pipe line may be obtained by differentiating P w.r.t. V and equating it to zero

$$\therefore \quad \frac{dP}{dV} = \gamma \times \frac{\pi D^2}{4} \times \left(H - \frac{3fLV^2}{2gD}\right) = 0$$

$$\Rightarrow H - \frac{3fLV^2}{2gD} = 0$$

$$\Rightarrow H - 3h_f = 0$$

$$\Rightarrow h_f = \frac{H}{3}$$

8.20 (d)

Pressure at summit i.e. at the highest point of siphon is always negative. Therefore there is a minimum limit for pressure at the summit. It can never be allowed to go below vapour pressure. Pressure at siphon will be less than at A and B.

8.21 (b)

When the pressure in any part of the flow passage reaches the vapor pressure (limiting value) of the flowing liquid, its starts vaporizing and small bubbles of vapor form in large numbers. The bubbles when reach to the high pressure zones, suddenly collapse as the vapor condenses to liquid again. Due to sudden collapsing of the bubbles or cavities the surrounding liquid rushes in to fill them. The liquid moving from all directions collides at the centre of the cavity, thus giving rise to very high local pressures. Any solid surface in the vicinity is also subjected to these intense pressures. The alternate formation and collapse of vapor bubbles may cause severe damage to the surface which ultimately fails by fatigue and the surface becomes badly scoured and pitted. This phenomenon is known as cavitation which is found to occur in turbines and pumps as well as in various hydraulic structures such as penstocks, gates valves, spillways etc. Since the total energy remains constant in the system thus if the pressure is reduced to limiting value, the velocity or elevation might have been increased and vice-versa.

8.22 (b)

Velocity will be maximum at point 2 and then there will be momentum change and possibility of pressure drop. Bubbles will generate at point 2 and collapse a little while flowing towards point 4.

8.23 (d)

Cavitation occurs when increase in local velocity causes local pressure to reduce below vapour pressure.

8.24 (d)

When Pumps are in parallel individual volume flow rates are added and head remains constant.

8.25 (a)

Pumps in Parallel: Total discharge is cumulative of discharge of individual pump while head remains same.

Pumps in Series: Discharge remains same while the total head is cumulative head of individual pump.

8.26 (a)

In the case of sudden expansion the fluid emerging from the smaller pipe being unable to take a sharp turn at the corners follows a path of streamlines. Consequently, the pockets of turbulent eddies are formed in the corners and result in dissipation of energy in the form of head. The lost hydraulic energy is thus converted into the thermal energy (raising slightly the fluid temperature) which is in turn lost to the surrounding medium.

Loss of energy at sudden enlargement is given by

$$h_L = \frac{V_1^2}{2g}\left(1 - \frac{A_1}{A_2}\right)^2 = \frac{V_2^2}{2g}\left(\frac{A_2}{A_1}\right)^2$$

Loss of energy at sudden contraction is given by

$$h_L = \frac{V_2^2}{2g}\left(\frac{1}{C_c}-1\right)^2$$

C_c = Coefficient of contraction.

The turbulence level and formation of eddies is more prominent in the case of sudden expansion than at sudden contraction.

8.27 (a)

Time period of oscillation refers to the total time that the wave takes to start from the closed valve and reach back at the same location after traversing the entire length of penstock twice.

$$T = \frac{2\times 4500}{1500} = 6\text{s}$$

8.28 (d)

The head loss in a circular pipe of diameter D, and length L, carrying a discharge Q is given by

$$h_f = \frac{f\times L\times Q^2}{12.1\times D^5}$$

When pipes are connected in series total head loss is given by

$$h_f = h_{f_A} + h_{f_B} + h_{f_C}$$

$$\Rightarrow \frac{f\times L_e\times Q^2}{12.1\times D^5} = \frac{f\times L\times Q^2}{12.1\times D^5} + \frac{f\times L\times Q^2}{12.1\times\left(\frac{D}{2}\right)^5} + \frac{f\times 4L\times Q^2}{12.1\times(2D)^5}$$

$$\Rightarrow \frac{L_e}{D^5} = \frac{L}{D^5}+\frac{32L}{D^5}+\frac{L}{8D^5}$$

$$\therefore L_e = L+32L+\frac{L}{8}$$

$$\Rightarrow L_e = 33\frac{1}{8}L$$

8.29 (c)

$K = 2.1\times 10^9$ Pa; $E = 2.1\times 10^{11}$ Pa

$\rho = 1000$ kg/m^3; $D = 400$ mm

$t = 4$ mm

Velocity of propagation of water hammer pressure

$$= \sqrt{\frac{K/\rho}{1+\frac{KD}{Et}}} = \sqrt{\frac{2.1\times\frac{10^9}{1000}}{1+1}} = 1024.7 \text{ m/s}$$

8.30 (b)

According to Darcy Weisbach

$$h_f = \frac{fLV^2}{d\times 2g}$$

Now, $d' = 2d$, $V' = 2V$

$$\therefore \quad h_f' = \frac{fL(2V)^2}{2d\times 2g} = 2\times\frac{fLV^2}{d\times 2g}$$

$$\therefore \quad h_f' = 2h_f$$

8.31 (c)

Velocity of pressure wave is given by

$$C = \sqrt{\frac{K}{\rho}}$$

$$C_1 = \sqrt{\frac{2.16\times 10^9}{1000}}$$

Now, $$C_2 = \sqrt{\frac{1.296\times 10^9}{0.6\times 1000}}$$

$$\therefore \quad \frac{C_1}{C_2} = 1$$

8.32 (b)

As per Bernoulli's equation total head,

$$H = \frac{P}{\rho g}+Z+\frac{V^2}{2g}$$

Where,

$\frac{P}{\rho g}$ is pressure head or static head

Z is datum or elevation head

$\frac{V^2}{2g}$ is velocity head or dynamic head or kinetic head

$\frac{P}{\rho g}+Z$ is also called piezometric head

$\frac{P}{\rho g}+\frac{V^2}{2g}$ is called stagnation head

8.33 (c)

$$P = \dot{m}gh_L = \rho Qgh_L$$

Now $$h_L = \frac{fLQ^2}{12\times D^5}$$

$$\therefore \quad P \propto Q^2$$

$$\Rightarrow \quad P_2 = \left(\frac{Q_2}{Q_1}\right)^2 \times P_1$$

$$P_2 = (1.44\,P_1)$$

$\therefore$ Power will increase by 44%.

8.34 (c)

To maintain continuity at junction,

Total Q_{in} = Total Q_{out}

In a network, algebraic sum of the drop in piezometric head is zero.

8.35 (c)

$$t = \frac{2L}{C} = \frac{2(1920)}{960} = 4 \text{ sec}$$

8.36 (d)

In pipe flow, flow occurs under pressure and not through gravity.

8.37 (c)

Pressure head rise due to water hammer for rigid pipe

$$\frac{P}{\rho g} = \frac{VC}{g} \qquad \text{where } C = \sqrt{\frac{k}{\rho}}$$

8.38 (c)

Pipes are placed parallel

$$\therefore \quad h_{f_P} = h_{f_Q}$$

$$\left(\frac{fLQ^2}{12.1D^5}\right)_P = \left(\frac{fLQ^2}{12.1D^5}\right)_Q$$

$$9f\,Q_P^2 = f.Q_Q^2$$

$$Q_P = \frac{Q_Q}{3} = 0.33\,Q_Q$$

8.39 (a)

Total energy line and the hydraulic gradient line for a pipe flow cannot intersect each other. Hydraulic gradient line is sloping due to friction loss and other losses.

8.40 (d)

As per the question no options is matching with the solution.

$$h_f = \frac{H}{3} = \frac{720}{3} = 240 \text{ m}$$

$$H_{exit} = \frac{2H}{3} = 480 \text{ m}$$

$$h_f = \frac{8Q^2 \times f \times L}{\pi^2 g D^5}$$

$$240 = \frac{8Q^2 \times 0.12 \times 3000}{\pi^2 g \times 0.2^5}$$

$$\Rightarrow \quad Q = 0.0508 \text{ m}^3/\text{s}$$

$$P_{exit} = \rho Q g H_{exit}$$

$$= \frac{1000 \times 0.508 \times 9.81 \times 480}{735.5}$$

$$= 325.31 \text{ mhp}$$

Alternatively

In this problem, the value of friction factor is not defined properly. Friction coefficient is given as point 0.03 if we take $f = 0.03$, where f is friction factor then, we can choose (d) as nearest answer.

$$h_f = \frac{8Q^2 \times f \times L}{\pi^2 g D^5}$$

$$240 = \frac{8Q^2 \times 0.03 \times 3000}{\pi^2 g \times 0.2^5}$$

$$\Rightarrow \quad Q = 0.102 \text{ m}^3/\text{s}$$

$$P_{exit} = \rho Q g H_{exit}$$

$$= \frac{1000 \times 0.102 \times 9.81 \times 480}{735.5}$$

$$= 653.02 \text{ mhp}$$

8.41 (d)

Let at any instant the difference in the water levels in the two reservoirs be h and in time dt the water level in the tank A fall by an amount dH, Then, the water level will rise in tank B by an amount $\left(dH\frac{A_1}{A_2}\right)$. Now if dh is the change in difference in head causing the flow

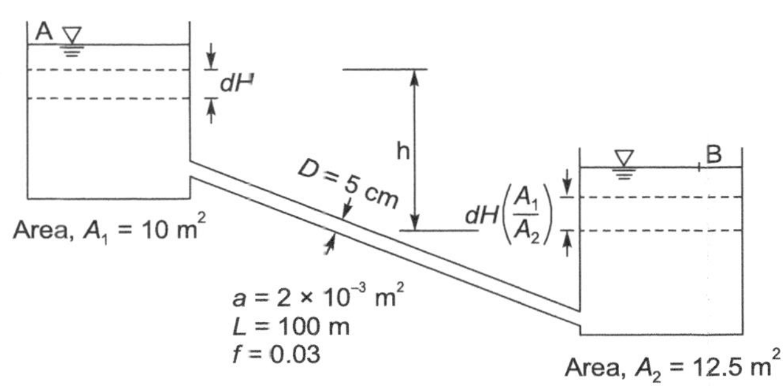

$$dh = dH + dH\frac{A_1}{A_2}$$

$$dH = \frac{dh}{1+\frac{A_1}{A_2}}$$

Now, $-A_1\, dH = a.v.dt \qquad V = 1.75\sqrt{h}$

$$-A_1 \frac{dh}{1+\frac{A_1}{A_2}} = a \times 1.75\sqrt{h}\,dt$$

$$\frac{-A_1A_2}{(A_1+A_2)a(1.75)}\int_{h_1}^{h_2}\frac{dh}{\sqrt{h}} = \int_0^T dt$$

$$T = \frac{2A_1A_2}{a(A_1+A_2)(1.75)}\left[\sqrt{h_1}-\sqrt{h_2}\right]$$

$h_1 \Rightarrow$ The difference of water levels in the reservoirs before flow commences

$h_1 = 3$ m

$h_2 \Rightarrow$ The difference of water levels in the reservoirs after flow commences

Level fall in tank A $= \frac{2.5}{10} = 0.25$ m

Level rise in tank B $= \frac{2.5}{2.5} = 1$ m

So, $h_2 = h_1 - (1 + 0.25)$
$= 3 - (1.25) = 1.75$ m

Now,

$$T = \frac{2(10)(2.5)}{(2\times10^{-3})(10+2.5)(1.75)}\left[\sqrt{3}-\sqrt{1.75}\right]$$

$$= 1142.86\left(\sqrt{3}-\sqrt{1.75}\right) = 467.66 \text{ sec}$$

8.42 (a)

In this problem, the value of friction factor is not defined properly. Actually the value of friction factor is given but written as friction coefficient.

$$15 = \frac{8Q^2}{\pi^2 g}\times\frac{(0.04)(1500)}{(0.324)^5}$$

$Q = 103.9$ *lps*

8.44 (c)

Cavitation occurs when at a specific point fluid pressure becomes less than vapor pressure or temperature become too high. Energy is released with the onset of pressure wave of high intensity wave due to noise and vibration of machine.

8.45 (d)

$$H = \frac{V_0^2}{2g} + k\cdot\frac{V_0^2}{2g}$$

where $V_0 \Rightarrow$ steady flow velocity

$$11.25 = \frac{V_0^2}{2g}[1+k]$$

$$11.25 = \frac{V_0^2}{2g}[1+98.1]$$

$$11.25 = V_0^2 \times 5.051 = 1.49 \text{ m/sec}$$

Time take for establishment of flow

$$t = \frac{L}{(1+k)V_0}\ln\frac{V_0+V}{V_0-V}$$

$$= \frac{294.3}{(99.1)(1.49)}\ln\left(\frac{V_0+V}{V_0-V}\right) = 2\ln\left(\frac{V_0+V}{V_0-V}\right) \qquad \text{...(1)}$$

when $\frac{Q}{Q_0} = 0.95$

Then $\frac{V}{V_0} = 0.95$

Put the value in equation

$$t = 2\ln\left(\frac{1+0.95}{1-0.95}\right) = 2\log_e 39$$

8.46 (a)

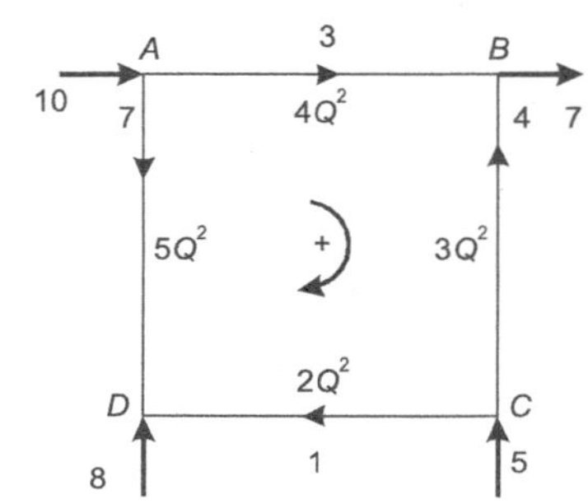

S.No.	Pipe	r	Q_0	rQ_0^2	$\lvert 2rQ_0\rvert$
1	AB	4	3	+36	24
2	BC	3	4	–48	24
3	CD	2	1	+2	4
4	DA	5	7	–245	70
				$\Sigma rQ_0^2 = -255$	$\Sigma\lvert 2rQ_0\rvert = 122$

$$\Delta Q = \frac{-\Sigma rQ_0^2}{\Sigma|2rQ_0|} = \frac{-(-255)}{122} = 2.09$$

Pipeline $AB = 3 + 2.09 = 5.09$
$\simeq 5.1$

Pipeline $CD = 3 + 2.09 = 3.09 \simeq 3.1$

8.47 (a)

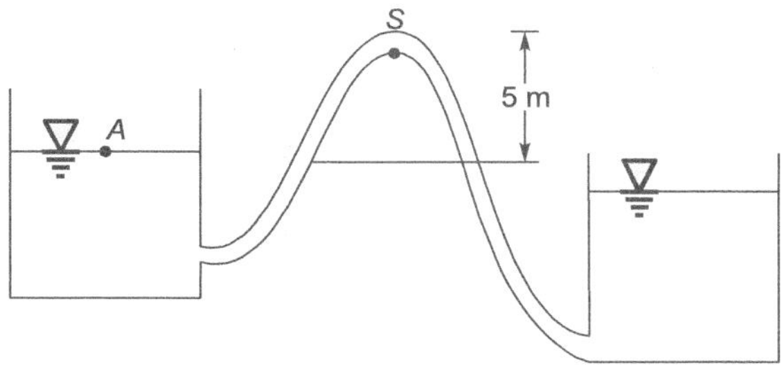

Apply energy equation between (A) & (S)

$$\frac{P_A}{\rho g}+\frac{V_A^2}{2g}+z_A = \frac{p_s}{\rho g}+\frac{v_s^2}{2g}+z_s+h_{fs}$$

$$0+0+0 = \frac{p_s}{\rho g}+0.5+5+2.5$$

$$\frac{p_s}{\rho g} = -8 \text{ m of } H_2O$$

$$p_s = -(10^3)\times(10)\times 8$$

$$= -80 \text{ kPa}$$

8.48 (d)

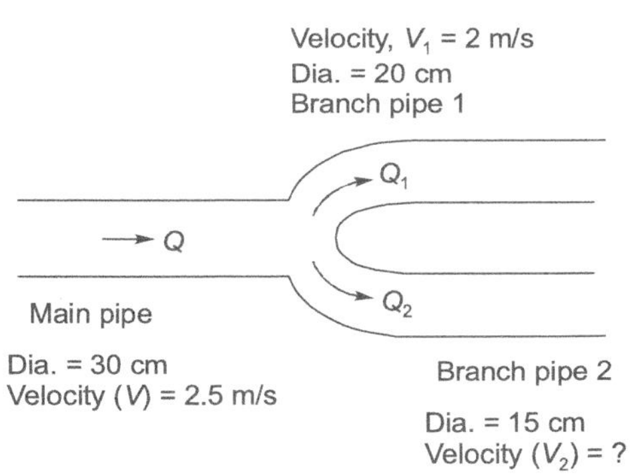

Let V_2 be the velocity in the branch pipe 2, Applying continuity equation

$$Q = Q_1 + Q_2$$

$$AV = A_1V_1 + A_2V_2$$

$$\frac{\pi}{4}d^2V = \frac{\pi}{4}d_1^2V_1+\frac{\pi}{4}d_2^2V_2$$

$$\Rightarrow \quad V_2 = \left(\frac{d}{d_2}\right)^2 V-\left(\frac{d_1}{d_2}\right)V_1$$

$$= \left(\frac{30}{15}\right)^2\times 2.5-\left(\frac{20}{15}\right)^2\times 2$$

$$= 10-\frac{32}{9} = \frac{58}{9} = 6.44 \text{ m/s}$$

8.49 (a)

Given, $P_A = 9.81$ N/cm^2,

$P_B = 5.886$ N/cm^2,

$d_A = 200$ mm $= 0.2$ m, $d_B = 500$ mm $= 0.5$ m

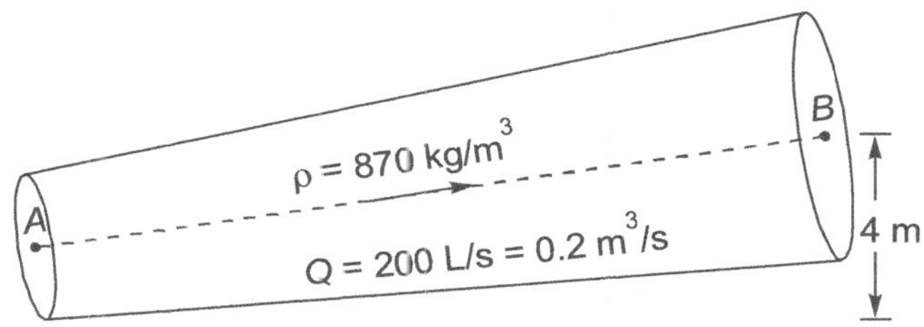

For end A, $$\frac{P_A}{\rho g} = \frac{9.81\times10^4}{870\times9.81} = 11.494 \text{ m}$$

$$\frac{V_A^2}{2g} = \frac{1}{2\times9.81}\left\{\frac{0.2}{\frac{\pi}{4}(0.2)^2}\right\}^2 = 2.066 \text{ m}$$

Given, $Z_B - Z_A = 4$ m

For end B, $$\frac{P_B}{\rho g} = \frac{5.886\times10^4}{870\times9.81} = 6.896 \text{ m}$$

$$\frac{V_B^2}{2g} = \frac{1}{2\times9.81}\left\{\frac{0.2}{\frac{\pi}{4}\times(0.5)^2}\right\}^2 = 0.053 \text{ m}$$

Head loss for given pipeline

$$= \left(\frac{P_A}{\rho g}+\frac{V_A^2}{2g}+Z_A\right)-\left(\frac{P_B}{\rho g}+\frac{V_B^2}{2g}+Z_B\right)$$

$= [(11.494 + 2.066 + Z_A) - (6.896 + 0.053 + Z_B)]$

$= 2.611 + Z_A - Z_B = (6.611 - 4)$ m

$= 2.611$ m

8.50 (a)

Actual time for valve closure,

$$T_a = 25 \text{ seconds}$$

Time taken by pressure wave to reach valve

$$= \frac{2L}{C} = \frac{2\times2500}{1460} = 3.4246 \text{ seconds}$$

Since, $\frac{2L}{C} < T_a \Rightarrow$ Gradual closure of valve

$$\therefore \quad \frac{P}{\rho g} = \frac{L}{g}\times\frac{V}{T_a}$$

$$\Rightarrow \quad P = \rho\frac{LV}{T_a} = (10^3)\times\frac{2500(1.5)}{25}$$

$$\Rightarrow \quad P = 15\times10^4 \text{ N/m}^2$$

$$\Rightarrow \quad P = 15 \text{ N/cm}^2$$

8.52 (d)

$L = 600$ m
$D = 0.3$ m
$f = 0.02$

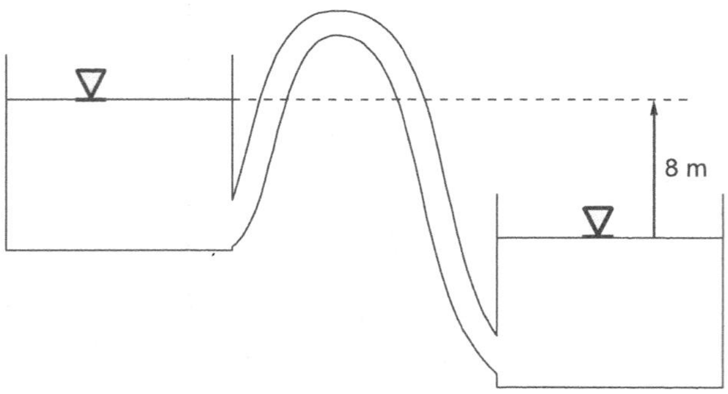

$$h_f = \frac{8Q^2}{\pi^2 g} \times \frac{f \cdot L}{D^5}$$

$$\Rightarrow \quad 8 = \frac{8Q^2}{\pi^2 g} \times \frac{(0.02)(600)}{(0.3)^5}$$

$$\Rightarrow \quad Q = 0.14 \text{ m}^3/\text{s}$$

8.54 (a)

For a constant discharge, reduction in area increases velocity and thereby, causing low pressure in the system.

■■■■

9 Drag and Lift Force

9.1 **Assertion (A):** Total drag is reduced if the boundary layer on the surface of a cylinder separates further downstream of the leading point.
Reason (R): As the separation point moves further downstream, form drag is reduced and the skin drag is only marginally increased.
(a) both A and R are true and R is the correct explanation of A
(b) both A and R are true but R is not a correct explanation of A
(c) A is true but R is false
(d) A is false but R is true [ESE : 2001]

9.2 Which one of the following statements relating to vortex flow is **INCORRECT**?
(a) In the formation of a free vortex, stream-lines are axisymmetric and irrotational.
(b) In a forced vortex, work transfer between the fluid and the surroundings takes place and the flow is rotational.
(c) In a free vortex, radial motion towards the core takes place due to variation of depth of water in the whirlpool or due to the difference of pressure resulting from higher velocity near the core.
(d) In a free vortex, there is no variation of energy from streamline to streamline and irrotationality is not deviated from near the core.
[ESE : 2001]

9.3 Magnus effect may be used advantageously in games such as cricket, tennis, table tennis and golf. In order to obtain a lift, i.e. a rising curve for the trajectory of the ball, from left to right, the nature of the spin to be given is
(a) anticlockwise in the vertical plane
(b) anticlockwise in the horizontal plane
(c) clockwise in the vertical plane
(d) clockwise in the horizontal plane
[ESE : 2001]

9.4 Match **List-I** (Flow pattern past a cylinder) with **List-II** (Pressure distribution) and select the correct answer using the codes given below the lists:

List-I

A. 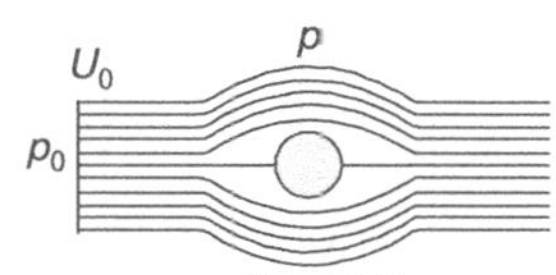

Re < 0.5

B. 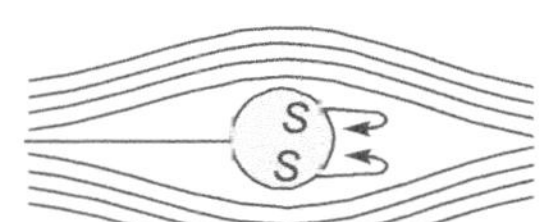

2 < Re < 30

C. 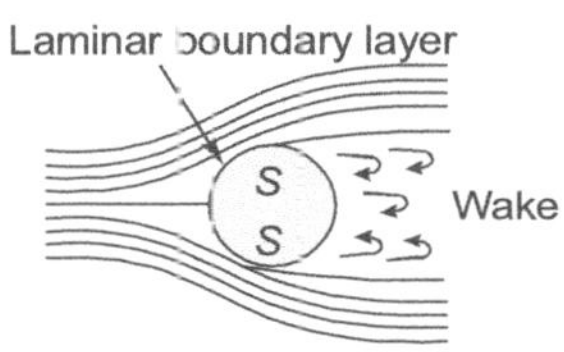

$10^3 \le Re \le 2 \times 10^5$

D. 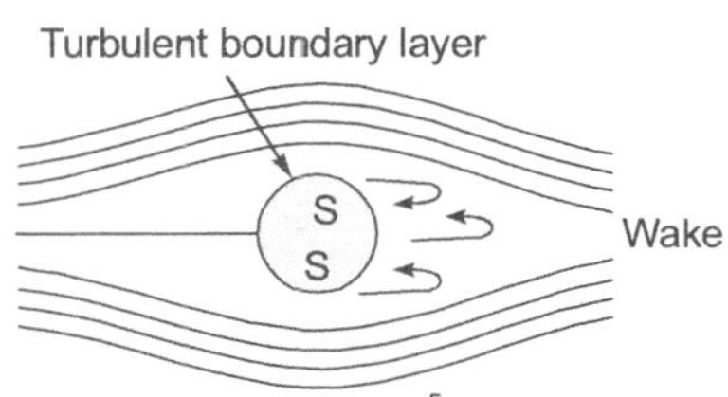

$Re > 2 \times 10^5$

List-II

1. 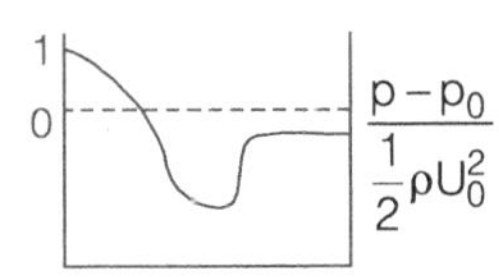

2. 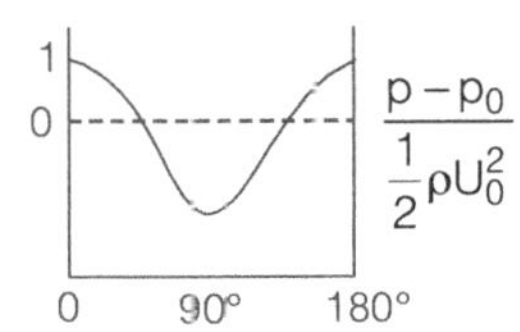

3. 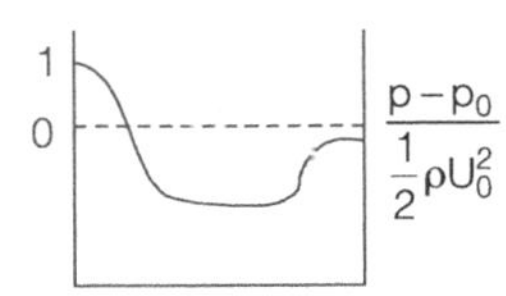

4. 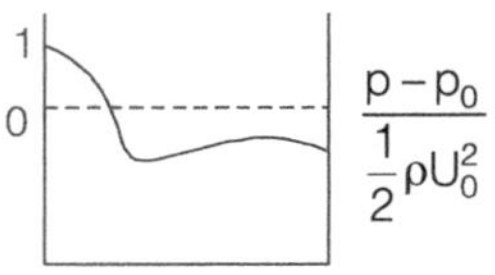

5. 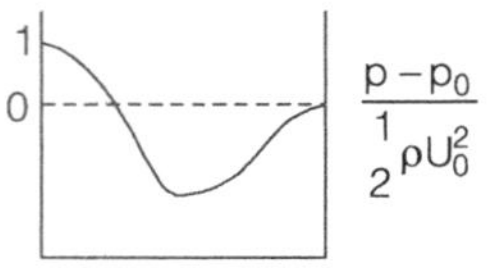

Codes:

	A	B	C	D
(a)	2	5	1	3
(b)	4	3	5	2
(c)	2	1	5	4
(d)	4	3	1	2

[ESE : 2002]

9.5 Which one of the following gives the lift force F_L produced on a rotating circular cylinder per unit length in a uniform flow?

(a) $F_L = \frac{u\Gamma}{\rho}$ (b) $F_L = \rho u \Gamma$

(c) $F_L = \frac{\rho\Gamma}{u}$ (d) $F_L = \frac{\rho u}{\Gamma}$

where,
u = free stream velocity
Γ = circulation
ρ = density of the fluid [ESE : 2006]

9.6 Drag force on a cylinder for turbulent flow compared to laminar flow is

(a) same (b) more
(c) less (d) very high

[ESE : 2006]

9.7 A streamlined body is one for which the

(a) skin friction is zero
(b) thickness of the body is less than 1/100 of its length
(c) corners are rounded off
(d) separation occurs, if at all, at the farthest downstream part of the body [ESE : 2007]

9.8 Body *M* has twice the weight, twice the projected area and twice the drag coefficient of body *N*. How many times is the terminal velocity of the body *M* in air, compared to that of the body *N*?

(a) 8 (b) 2
(c) $\sqrt{2}$ (d) $1/\sqrt{2}$ [ESE : 2007]

Directions: The following items consists of two statements; one labelled as **'Assertion (A)'** and the other as **'Reason (R)'**. You are to examine these two statements carefully and select the answers to these items using the codes given below:

Codes:

(a) both A and R are true and R is the correct explanation of A
(b) both A and R are true but R is not a correct explanation of A
(c) A is true but R is false
(d) A is false but R is true

9.9 **Assertion (A):** If a cylindrical body is placed in a fluid stream and is rotated, then a lift force is produced on the body.

Reason (R): Rotation of the cylinder disturbs the symmetrical pattern which in turn alters the pressure distribution on the body.

[ESE : 2009]

9.10 **Assertion (A):** The drag force on a sphere is more in laminar flow than in turbulent flow.

Reason (R): C_D of a sphere is more in laminar flow than in turbulent flow.

[ESE : 2009]

9.11 The terminal velocity of a small sphere settling in a viscous fluid varies as the

(a) inverse square of the diameter
(b) inverse of the diameter
(c) first power of its diameter
(d) inverse of the fluid viscosity

[ESE : 2010]

9.12 The terminal velocity of a sphere settling in a viscous fluid varies as

(a) the Reynolds number
(b) the square of its diameter
(c) directly proportional to the viscosity of fluid
(d) its diameter

[ESE : 2011]

9.13 Singing of telephone wires in the wind occurs due to

(a) Vibrations caused by birds as they sit on, or get off, the wires
(b) Tensioning at the ends
(c) Magnus effect
(d) Generation of Karman Vortex street

[ESE : 2011]

9.14 A high tension cable 5 cm in diameter is strung-out between two towers. At a wind velocity of 22.22 m/s (corresponding Reynolds number being 7.4×10^4) the frequency of vortex shedding is
(a) 100 Hz (b) 9.33 Hz
(c) 93.3 Hz (d) 10.0 Hz
[ESE : 2012]

9.15 A very small spherical particle of sand settles in a column of viscous liquid. If the Reynolds number corresponding to its terminal velocity is 0.25, the drag coefficient of the sand particle is:
(a) 256 (b) 128
(c) 96 (d) 64 **[ESE : 2013]**

9.16 When small sphere falls freely in Stokes range through a water column, which of the following statements are correct?
1. Viscosity, surface tension and gravitational forces act on the sphere.
2. There is no turbulence possible in the wake of the falling sphere.
3. Drag coefficient, $C_d = R/24$, where R is Reynolds number.
4. Proximity of the boundary moderates the drag and lift coefficients.

(a) 1 and 2 (b) 2 and 3
(c) 3 and 4 (d) 2 and 4
[ESE : 2013]

9.17 Which of the following statements are correct?
1. Friction and pressure drag depend upon the shape and position of immersed body.
2. If a thin plate is kept parallel to the direction of flow, then the pressure drag is zero.
3. If a thin plate is kept perpendicular to the moving fluid, then friction drag is zero.

(a) 1 and 2 only (b) 1 and 3 only
(c) 2 and 3 only (d) 1, 2 and 3
[ESE : 2013]

Directions: Each of the next items consists of two statements, one labeled as the 'Statements (I)' and the other as 'Statement (II)'. You are to examine these two statements carefully and select the answers to these items using the codes given below :

Codes:
(a) Both Statement (I) and Statement (II) are individually true and Statement (II) is the correct explanation of Statement (I)
(b) Both Statement (I) and Statement (II) are individually true but Statement (II) is NOT the correct explanation of Statement (I)
(c) Statement (I) is true but Statement (II) is false
(d) Statement (I) is false but Statement (II) is true

9.18 **Statement (I):** A tennis ball with a rough surface can be made to curve more markedly than the one with a smooth surface.
Statement (II): Circulation is increased in the case of a rough ball.
[ESE : 2013]

9.19 **Statement (I):** A rectangular bridge pier experiences lateral force during flood flow.
Statement (II): The bridge pier sheds vortices alternately from its either side into the wake flow.
[ESE : 2013]

9.20 For circular cylinders, with Reynolds number greater than 1000, how would the Strouhal number behave
(a) Varies as $Re^{1/3}$ (b) Varies as $Re^{1/4}$
(c) Almost ≈ 0.16 (d) Constant at 0.21
[ESE : 2014]

9.21 An airfoil of surface area 0.1 m^2 is tested for lift L in a wind tunnel. (Conditions can be considered as incompressible flow). At an angle of attack of 5°, with standard air of density 1.22 kg/m^3, at a speed of 30 m/sec, the lift is measured to be 3.2 kgf. What is the lift coefficient? For a prototype wing of area 10 m^2, what is the approximate lift at an air speed of 160 kmph at the same angle of attack of 5°?
(a) 0.572 and 700 kgf
(b) 0.603 and 700 kgf
(c) 0.572 and 570 kgf
(d) 0.603 and 570 kgf **[ESE : 2016]**

9.22 A man, 65 kg, descends to the ground with the helps of a parachute, 18 kg. The parachute is hemispherical in shape, 2 m diameter. Density of air can be taken as 0.00125 g/cm^3 and its kinematic viscosity as 0.15 stoke. What is the terminal velocity of the parachute? [Take C_D = 1.5 and g = 1000 cm/sec^2]
(a) 16.6 m/sec (b) 15.8 m/sec
(c) 15.0 m/sec (d) 14.1 m/sec
[ESE : 2017]

9.23 A transmitter antenna is of a vertical pipe, 20 cm diameter and 25 m height, on top of a tall structure. It is subjected to wind speed of 20 m/sec. Density of air is 1.22 kg/m^3; its viscosity is 1.8×10^{-5} Ns/m^2. Drag coefficient of a (tall) circular cylinder is tabulated as

R_e	10^2	10^3	1.3×10^3	10^4	1.5×10^4	1.06×10^5	1.2×10^5	4.5×10^5
C_D	1.6	1.05	0.95	1.0	1.08	1.0	0.89	0.26

What is the drag experienced?
(a) 737 N (b) 700 N
(c) 670 N (d) 63 N

[ESE : 2018]

9.24 The phenomenon of generation of lift by rotating an object placed in a free stream is known as
(a) Coanda effect (b) Magnus effect
(c) Scale effect (d) Buoyancy effect

[ESE : 2019]

9.25 An airfoil is a streamlined body as shown in the figure below. Because of the streamlining of the body, the separation occurs only at the extreme rear of the body, resulting in

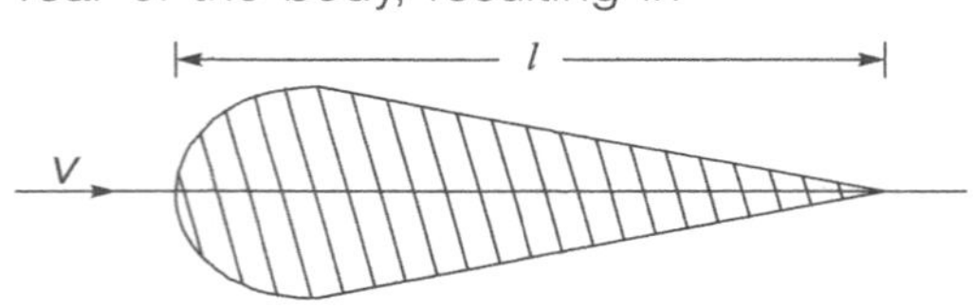

(a) A very high pressure drag
(b) A small wake and consequently small pressure drag
(c) A moderate pressure drag
(d) No pressure drag

[ESE : 2019]

9.26 When the drag force becomes equal to the weight of the body, the acceleration ceases and the net external force acting in the body becomes.
(a) Zero and the body will move at constant speed.
(b) Light and the body will move forward
(c) Zero and the body will move fast.
(d) High and the body will move at constant speed.

[ESE : 2020]

9.27 Which one of the following is an example of bodies where both drag and lift forces are produced?
(a) Hydrofiles
(b) A tall chimney exposed to wind
(c) Flow of water past a bridge pier
(d) Motion of aeroplanes, submarines, torpedoes

[ESE : 2020]

9.28 A semi-tubular cylinder of 75 mm radius with concave side upstream (drag coefficient = 2.3) is submerged in flowing water of velocity 0.6 m/s. If the cylinder is 7.2 m long and density of water is 1000 kg/m^3, what is the drag?
(a) 150 N (b) 173 N
(c) 955 N (d) 223 N

[ESE : 2021]

9.29 A flat plate 1.5 m × 1.5 m moves at 45 km/hour in stationary air of specific weight 11.3 N/m^3. If the coefficient of lift is 0.75, then what is the lift force?
(a) 120.4 N (b) 151.9 N
(c) 180.4 N (d) 225.9 N

[ESE : 2023]

9.30 A submarine which may be supposed to approximate a cylinder 3 m in diameter and 15 m long travels submerged at 1.54 m/s in sea water at 4°C. What is the drag exerted on it? (Take C_d = 0.7 and ρ = 1025 kg/m^3)
(a) 30129 N (b) 34517 N
(c) 37112 N (d) 38287 N

[ESE : 2023]

Answers Drag and Lift Force

9.1 (a)	9.2 (d)	9.3 (a)	9.4 (a)	9.5 (b)	9.6 (c)	9.7 (d)	9.8 (d)	9.9 (a)
9.10 (a)	9.11 (d)	9.12 (b)	9.13 (d)	9.14 (c)	9.15 (c)	9.16 (d)	9.17 (d)	9.18 (a)
9.19 (a)	9.20 (d)	9.21 (a)	9.22 (a)	9.23 (a)	9.24 (b)	9.25 (b)	9.26 (a)	9.27 (d)
9.28 (d)	9.29 (b)	9.30 (d)						

Explanations Drag and Lift Force

9.1 (a)

As the fluid velocity increases the separation point moves further downstream. Skin friction is increased marginally but there is reduction in form drag resulting in decrease in total drag and form drag is generally more transkin drag.

9.2 (d)

In real fluid, in the central region of a free vortex, the viscous effects becomes quite predominant, due to which a central cave of fluid with rotational motion is developed, where $V \propto r$ similar to forced vortex motion.

9.3 (a)

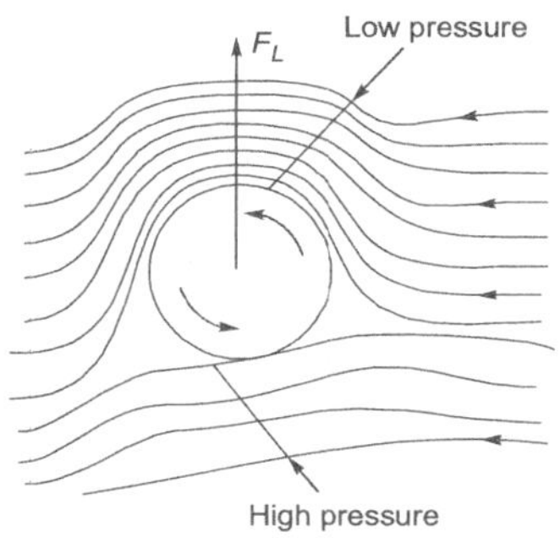

To obtain a lift nature of spin should be anticlockwise in vertical plane.

9.4 (a)

With Re < 0.5, the pressure distribution is symmetrical on both upstream and downstream of cylinder, when the Reynolds number is very small. So A–2

With increase in Re, the flow separates at downstream. The flow pattern with respect to an axis perpendicular to the direction of flow becomes unsym-metrical. At Re ranging from 2 to 30, very weak vortices are formed on the downstream of the cylinder. So B–5

For $10^3 < Re < 2 \times 10^5$, the boundary layer is laminar and the points of separation are located on the upstream half of the cylinder. The pressure remains negative with a constant value on the downstream.

For Re > 2×10^5, the turbulent boundary layer forms and the points of separation shift further downstream. The pressure distribution is negative on considerable portion of upstream and downstream end. The maximum negative pressure occurs on the upstream side.

9.5 (b)

$F_L = \rho u \Gamma$

This equation is commonly known as Kutta-Joukowski equation for lift force.

9.6 (c)

Drag coefficient as a function of Reynolds number for smooth circular cylinders and smooth spheres.

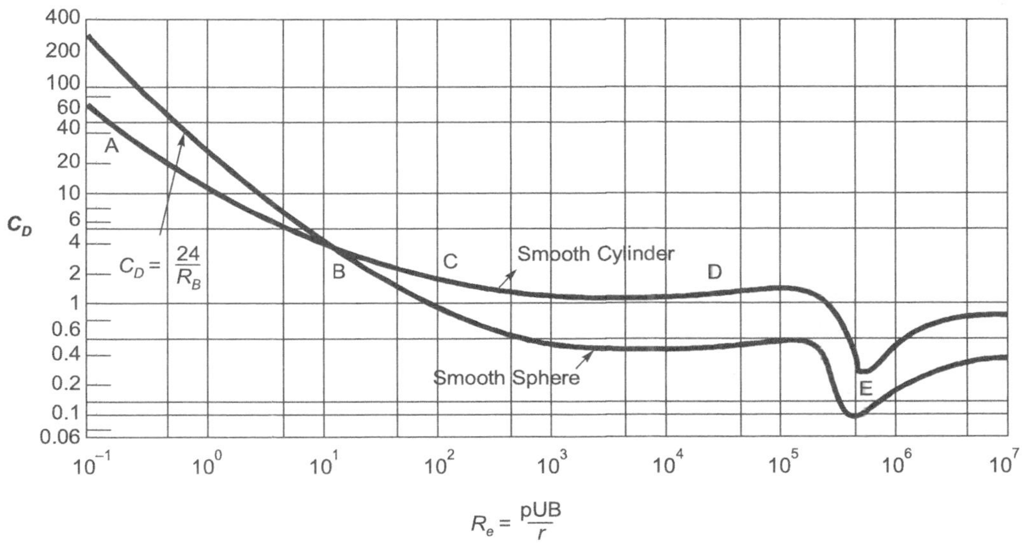

Turbulent flow occurs at higher Re. At higher Re, C_D decreases so F_D also decreases.

9.7 (d)

A streamlined body is defined as that body whose surface coincides with the streamlines, when the body is placed in a flow. For a well streamlined body the separation occurs only at the downstream end. As such the wake in this case is very small and consequently the pressure drag is also small. Thus, the total drag on the stream lined body will be due to friction only.

9.8 (d)

Using subscript 1 and 2 for body N and M respectively. Now, to attain the terminal velocity, the drag force should be equal to the weight of the body.

$$F_D = C_D A \times \frac{\rho V^2}{2}$$

But $$W = F_D$$

$$\Rightarrow \quad \frac{W_1}{W_2} = \frac{C_{D_1} A_1 \times \frac{\rho V_1^2}{2}}{C_{D_2} A_2 \times \frac{\rho V_2^2}{2}}$$

$$\Rightarrow \quad \frac{W}{2W} = \frac{C_{D_1} A_1 \times V_1^2}{2C_{D_1} \times 2A_1 \times V_2^2}$$

$$\Rightarrow \quad \frac{V_2}{V_1} = \frac{1}{\sqrt{2}}$$

9.9 (a)

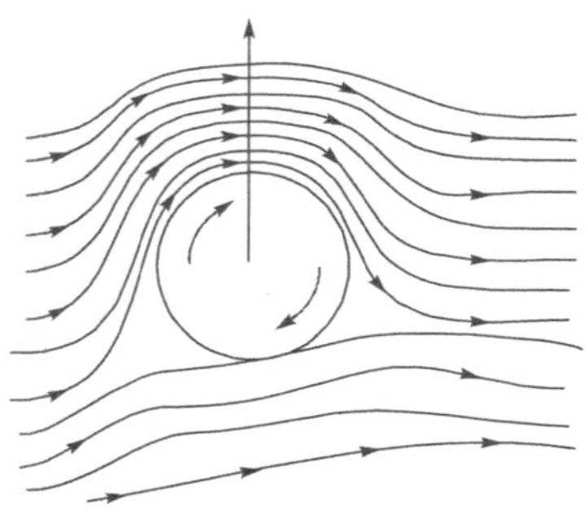

Rotation of cylinder causes crowded streamlines in top portion indicating increase in velocity while in bottom portion lines are further apart. Pressure on upper portion is lower than that at bottom so upward force will act on cylinder.

9.10 (a)

For a sphere $C_D \propto \frac{1}{\text{Re}}$

In Laminar flow Re is low.

So C_D is more.

9.11 (d)

Terminal velocity of small sphere setting in viscous fluid.

$$V_t = \frac{gd^2(\rho_s - \rho_u)}{18\mu}; \; V_t \propto \frac{1}{\mu}$$

9.12 (b)

Terminal velocity of a sphere settling in a viscous fluid is expressed by Stoke's equation.

$$V = \frac{1}{18}\frac{gd^2(\rho_s - \rho)}{\mu}$$

9.13 (d)

With Re > 90, eddies are shed alternately from the two sides of the cylinder. As a result, two rows of vortices rotating in opposite direction are formed in the wake. This is known as Von-Karmann vortex street. Alternate shedding of vortices from the two sides of a cylinder gives rise to the phenomenon of singing of telephones or power lines in the mind.

9.14 (c)

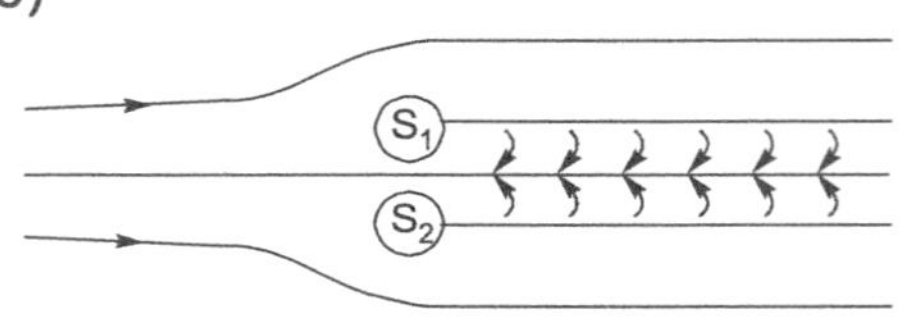

The frequency of vortex shedding is given by the shrouhal No. (S)

$$S = \frac{nD}{u_\infty}$$

n = frequency of vortex shedding
D = Dia. of cylinder
u_∞ = free stream velocity

$$S = 0.198\left[1 - \frac{19.7}{\text{Re}}\right]$$

For 250 < Re < 2 × 10^5

$$\frac{n \times 0.05}{22.22} = 0.198\left[1 - \frac{19.7}{17.4 \times 10^4}\right]$$

$$n = 87.967\text{Hz}$$

So, the nearest option is (c).

9.15 (c)

$$C_D = \frac{24}{\text{Re}} = \frac{24}{0.25} = 96$$

9.16 (d)

Surface tension does not act on falling sphere and Drag coefficient $C_d = 24/R_e$.

9.17 (d)

- Friction drag and pressure drag depend on shape and position of immersed body.
- When thin plate is held parallel to direction of flow pressure drag is zero.
- When thin plate is held perpendicular to direction of flow total drag is equal to pressure drag and friction drag is zero.

9.18 (a)

Tennis balls are designed to have rough surfaces that grab the air a little bit. So if the ball spins as it moves it will drag the air around it a little, which leads to air rotating around the ball, which makes the air on one side move faster than the air on the other, which leads to a lower pressure on one side compared to the other, which creates a net force and thus pushes the ball around.

9.19 (a)

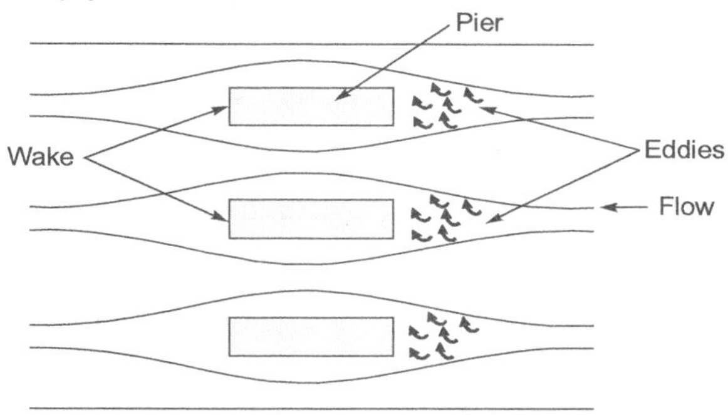

9.20 (d)

An interesting feature of flow past two dimensional bodies is the formation and alternate release of vortices behind the cylinder for Re > 30. This vortex shedding leads to lateral vibration of two-dimensional bodies. The frequency of vortex shedding n is expressed as Strouhal Number

$$S = \frac{nD}{V_o}$$

For cylinder, the value of strouhal number S can be estimated by the formula

$$S = 0.198\left[1 - \frac{19.7}{Re}\right]$$

Where, $250 < Re < 2 \times 10^5$

9.21 (a)

$$F_L = C_L \frac{\rho u_0^2 A}{2}$$

$$3.2 \times 9.81 = \frac{C_L \times 1.22 \times 30^2 \times 0.1}{2}$$

$$C_L = 0.572$$

$$F_L = 0.572 \times \frac{1}{2} \times 1.222$$

$$10 \times \left(\frac{160 \times 1000}{3600}\right)^2 = 6892.25 \text{ N} = 702.57 \text{ kfg}$$

9.22 (a)

$$Mg = F_D$$

$$10 \times (65 + 18) = C_D \times \frac{1}{2} \times \rho \times A \times V_T^2$$

$$10 \times 83 = 1.5 \times \frac{1}{2} \times 1.25 \times \frac{\pi}{4}(2)^2 \times V_T^2$$

$$V_T = 16.79 \text{ m/sec}$$

9.23 (a)

$$Re = \frac{\rho V_\infty D}{\mu} = \frac{1.22 \times 20 \times 0.2}{1.8 \times 10^{-5}} = 2.71 \times 10^5$$

$$F_D = C_D \times \frac{1}{2} \times \rho \times A \times U_\infty^2$$

$C_D = 0.602$ [On the basis of linear interpolation]

$$= 0.602 \times \frac{1}{2} \times 1.22 \times (0.2 \times 25) \times (20)^2$$

$$= 734.44 \text{ N}$$

9.25 (b)

In an airfoil separation of boundary layer occurs only at the extreme rear of the body. So, the wake formation is minimum and pressure difference between front and back end will be very less. Therefore, the form drag is comparatively very small in airfoil.

9.28 (d)

Given: For a semi-tubular cylinder

Radius, $R = 75$ mm

Drag coefficient,

$$C_D = 2.3$$

Velocity, $V = 0.6$ m/s

Length of cylinder,

$$L = 7.2 \text{ m}$$

$$\rho_w = 1000 \text{ kg/m}^3$$

Concave frontal area,

$$A = 0.075 \times 7.2 = 0.54 \text{ m}^2$$

∴ Drag force,

$$F_D = \frac{1}{2} C_D \rho A V^2$$

$$= \frac{1}{2} \times 2.3 \times 1000 \times 0.54 \times (0.6)^2$$

$$= 223.56 \text{ N}$$

Option (d) is close.

9.29 (b)

Given: $\rho g = 11.3 \text{ N/m}^3$

$$\rho = \frac{11.3}{9.81} \text{kg/m}^3$$

$$\therefore \text{ Lift force, } F_L = C_L \times \frac{1}{2} \times \rho \times A \times U_\infty^2$$

$$= 0.75 \times \frac{1}{2} \times \frac{11.3}{9.81} \times (1.5 \times 1.5) \times \left(45 \times \frac{5}{18}\right)^2$$

$$F_L = 151.86 \text{ N} \simeq 151.9 \text{ N}$$

9.30 (d)

$$\text{Drag force, } F_D = C_D \times \frac{1}{2} \times \rho A U_\infty^2$$

$$= 0.7 \times \frac{1}{2} \times 1025 \times \frac{\pi}{4} \times 3^2 \times 1.54^2 = 6014.03 \text{ N}$$

But, since none of the options matching.

$$\text{Drag force, } F_D = C_D \times \frac{1}{2} \times \rho \times A \times U_\infty^2$$

$$= 0.7 \times \frac{1}{2} \times 1025 \times (3 \times 15) \times 1.54^2$$

$$F_D = 38286.516 \text{ N}$$

Thus, most suitable answer is option (d).

■■■■

10 Boundary Layer Theory

10.1 A fluid with kinematic viscosity ν flows in laminar stage along a flat plate with free-stream velocity V. At a distance x from the leading edge, the Reynolds number of the flow is given by $R = \frac{Vx}{\nu}$. The thickness of the boundary layer at x will be proportional to

(a) $xR^{1/2}$ (b) $xR^{-1/2}$
(c) $R^{1/2}$ (d) $R^{-1/2}$

[ESE : 2001]

10.2 Match **List-I** with **List-II** and select the correct answer using the codes given below the lists:

List-I

A. $\left(\frac{\partial u}{\partial y}\right)_{y=0}$ is zero

B. $\left(\frac{\partial u}{\partial y}\right)_{y=0}$ is +ve

C. Displacement thickness

D. Momentum thickness

List-II

1. The flow is attached flow
2. The flow is on the verge of separation
3. $\int_0^\delta \frac{u}{U}\left(1-\frac{u}{U}\right)dy$
4. $\int_0^\delta \left(1-\frac{u}{U}\right)dy$

Codes:

	A	B	C	D
(a)	1	2	3	4
(b)	2	1	3	4
(c)	1	2	4	3
(d)	2	1	4	3

[ESE : 2002]

10.3 A thin smooth plate 1 m wide and 2 m long is towed through water at a velocity of 2 m/s. Assuming that boundary remains laminar, then drag on both sides of the plate is (kinematic viscosity = 10^{-6} m^2/s)

(a) 5.3 N (b) 26.6 N
(c) 53 N (d) 72.5 N

[ESE : 2003]

10.4 For a turbulent boundary layer (under zero pressure gradient), the velocity profile is described by the one-fifth power law. What is the ratio of displacement thickness to boundary layer thickness?

(a) 1/7 (b) 1/6
(c) 1/5 (d) 1/4

[ESE : 2004]

10.5 In a laminar boundary layer, the velocity distribution can be assumed to be given, in usual notations, as

$$\frac{u}{V} = \frac{y}{\delta}$$

Which one of the following is the correct expression for the displacement thickness (δ^*) for this boundary layer?

(a) $\delta^* = \delta$ (b) $\delta^* = \delta/2$
(c) $\delta^* = \delta/4$ (d) $\delta^* = \delta/6$

[ESE : 2005]

10.6 Which one of the following assumptions in deriving the boundary layer equation of flow past a flat plate at zero incidence is not correct?

(a) Uniform flow = 0
(b) Outside boundary layer velocity is v_0 throughout
(c) The boundary layer thickness δ is very small compared to distance x
(d) Pressure remains constant throughout the flow both within and outside the boundary layer

[ESE : 2006]

10.7 **Assertion (A):** Boundary layer theory is applicable only in the vicinity of the leading edge of the plate.

Reason (R): Boundary layer theory is based on the assumption that its thickness is small when compared to other linear dimensions.

(a) both A and R are true and R is the correct explanation of A
(b) both A and R are true but R is not a correct explanation of A
(c) A is true but R is false
(d) A is false but R is true

[ESE : 2006]

10.8 A flat plate with a sharp leading edge is placed along a free stream of fluid flow. Local Reynolds number at 3 cm from the leading edge is 10^5. What is the thickness of the boundary layer?
(a) 0.47 mm (b) 0.35 mm
(c) 0.23 mm (d) 0.12 mm

[ESE : 2007]

10.9 What is the momentum thickness for the boundary layer with velocity distribution $\frac{u}{U} = \frac{y}{\delta}$?
(a) $\delta/6$ (b) $\delta/2$
(c) $3\delta/2$ (d) 2δ

[ESE : 2008]

10.10 **Assertion (A):** At the point of boundary layer separation, the shear stress is zero.
Reason (R): The point of separation demarcates between zones of forward and reverse flow close to the wall.
(a) both A and R are true and R is the correct explanation of A
(b) both A and R are true but R is not a correct explanation of A
(c) A is true but R is false
(d) A is false but R is true

[ESE : 2011]

10.11 The thickness of the laminar boundary layer over a flat plate at two different sections *P* and *Q* are 0.8 cm and 2.4 cm respectively. If the section *Q* is 3.6 m downstream of *P*, the distance of section *P* from the leading edge of the plate is
(a) 0.32 m (b) 0.22 m
(c) 0.40 m (d) 0.53 m

[ESE : 2012]

10.12 **Statement-I:** Boundary layer theory is applicable only in the vicinity of the leading edge of a flat plate.
Statement-II: Boundary layer theory is based on the assumption that its thickness is small when compared to other linear dimensions in the flow.
(a) Both Statement (I) and Statement (II) are individually true and Statement (II) is the correct explanation of Statement (I)
(b) Both Statement (I) and Statement (II) are individually true but Statement (II) is NOT the correct explanation of Statement (I)
(c) Statement (I) is true but Statement (II) is false
(d) Statement (I) is false but Statement (II) is true

[ESE : 2012]

10.13 The average drag coefficient for a laminar boundary layer over a flat plate was obtained as 0.018. If all other factors remain unchanged, and length of the plate is increased by 4 times its original value, this average drag coefficient would change to
(a) 0.0036 (b) 0.0056
(c) 0.008 (d) 0.009

[ESE : 2013]

10.14 The velocity distribution in a laminar flow adjacent to a solid wall is given by $u = 3.0 \sin(5\pi y)$. The viscosity of the fluid is 5 poise. What is the shear stress at a section (*i*) $y = 0.05$ m; (*ii*) $y = 0.12$ m?
(a) 16.7 N/m^2 and 6.1 N/m^2
(b) 33.4 N/m^2 and Zero
(c) 16.7 N/m^2 and 12.3 N/m^2
(d) 16.7 N/m^2 and Zero

[ESE : 2014]

10.15 The velocity distribution in the boundary layer is given by $\frac{u}{U} = \frac{y}{\delta}$, where *u* is the velocity at a distance of *y* from the boundary and $u = U$ at $y = \delta$, δ being boundary layer thickness. Then the value of momentum thickness will be
(a) $\frac{\delta}{2}$ (b) $\frac{\delta}{4}$
(c) $\frac{\delta}{6}$ (d) $\frac{\delta}{8}$ [ESE : 2016]

10.16 What are the values of coefficients *a* and *c* if velocity distribution in a laminar boundary layer on a flat plat is:

$$f(n) = \frac{u}{U_0} = a + bn + cn^2 + dn^3 \text{ where, } n = \frac{y}{\delta}$$

(a) $\frac{1}{2}$ and 1 (b) 0 and 1

(c) $-\frac{1}{2}$ and 0 (d) 0 and 0

[ESE : 2016]

10.17 A smooth flat plate with a sharp leading edge is placed along a free stream of water flowing at 2.5 m/sec. At what distance from the leading edge will the boundary layer transition from laminar to turbulent flow? Take density of water as 1000 kg/m^3 and its viscosity as 1 centipoise. Also, what will be the boundary layer thickness at that distance?

(a) 12.8 cm and 0.113 cm
(b) 14.2 cm and 0.113 cm
(c) 12.8 cm and 0.125 cm
(d) 14.2 cm and 0.125 cm

[ESE : 2018]

10.18 If δ_1 and δ_2 are the laminar boundary layer thicknesses at a point M distant x from the leading edge when the Reynolds number of the flow are 100 and 484, respectively, then the ratio $\frac{\delta_1}{\delta_2}$ will be

(a) 2.2 (b) 4.84
(c) 23.43 (d) 45.45

[ESE : 2018]

10.19 **Statement (I):** The boundary layer thickness decreases as distance from the leading edge increases.

Statement (II): Greater is the kinematic viscosity of the fluid greater is the boundary layer thickness.

(a) Both Statement (I) and Statement (II) are individually true and Statement (II) is the correct explanation of Statement (I)
(b) Both Statement (I) and Statement (II) are individually true but Statement (II) is **NOT** the correct explanation of Statement (I)
(c) Statement (I) is true but Statement (II) is false
(d) Statement (I) is false but Statement (II) is true

[ESE : 2022]

10.20 Air at standard conditions flows over a flat plate. The free stream speed is 3 m/s. What is the thickness of boundary layer at a distance of 1 m from the leading edge of the flat plate? (Take the kinematic viscosity of air is 1.5×10^{-5} m^2/s and density is 1.23 kg/m^2)

(a) 1.80 mm (b) 1.80 cm
(c) 10.3 cm (d) 10.3 mm

[ESE : 2022]

■■■■

Answers Boundary Layer Theory

10.1 (b) 10.2 (d) 10.3 (a) 10.4 (b) 10.5 (b) 10.6 (a) 10.7 (d) 10.8 (a) 10.9 (a)
10.10 (b) 10.11 (c) 10.12 (d) 10.13 (d) 10.14 (d) 10.15 (c) 10.16 (d) 10.17 (a) 10.18 (a)
10.19 (d) 10.20 (a)

Explanations Boundary Layer Theory

10.1 (b)

For Laminar boudary layer over flat plate.

$$\delta = \frac{5x}{\sqrt{Re}} \quad \text{(Blasius solution)}$$

10.2 (d)

- $\left(\frac{du}{dy}\right)_{y=0} = 0$, the flow is on the verge of seperation.
- $\left(\frac{du}{dy}\right)_{y=0} = +ve$, the flow is attached.
- Displacement thickness = $\int_0^\delta \left(1-\frac{u}{u_\infty}\right)dy$.
- Momentum thickness = $\int_0^\delta \frac{u}{u_\infty}\left(1-\frac{u}{u_\infty}\right)dy$.

10.3 (a)

Friction coefficient, $\overline{C}_f = \dfrac{1.328}{\sqrt{Re_L}}$

$$Re_L = \frac{VL}{\nu} = \frac{2\times 2}{10^{-6}} = 4\times 10^6$$

$$\therefore \overline{C}_f = \frac{1.328}{\sqrt{4\times 10^6}} = 6.64\times 10^{-4}$$

The drag force on both sides of the plate,

$$F_D = 2\overline{C}_f\left(\frac{1}{2}\rho V^2\right)(BL)$$

$$= 2\times 6.64\times 10^{-4}\times\frac{1}{2}\times 1000\times (2)^2\times 1\times 2$$

$$= 5.31 \text{ N}$$

10.4 (b)

$$\frac{u}{u_\infty} = \left(\frac{y}{\delta}\right)^{1/5}$$

$$\delta^* = \int_0^\delta\left(1-\frac{u}{u_\infty}\right)dy$$

$$= \int_0^\delta\left[1-\left(\frac{y}{\delta}\right)^{1/5}\right]dy$$

$$= \delta - \frac{5\delta}{6} = \frac{\delta}{6}$$

$$\frac{\delta^*}{\delta} = \frac{1}{6}$$

10.5 (b)

Displacement thickness

$$\delta^* = \int_0^\delta\left(1-\frac{u}{V}\right)dy = \int_0^\delta\left(1-\frac{y}{\delta}\right)dy$$

$$= \left[y-\frac{y^2}{2\delta}\right]_0^\delta = \delta-\frac{\delta}{2} = \frac{\delta}{2}$$

10.6 (a)

Assumptions in deriving boundary layer equation

- Velocity outside boundary layer is constant throughout.
- Boundary layer thickness is very small compared to other linear dimensions.
- Pressure is constant both within and outside the boundary layer.

10.7 (d)

Statement (A) is wrong.

Since boundary layer theory is applicable not only in the vicinity of leading edge but also along the length of plate.

10.8 (a)

Since the Reynolds number is less than 5×10^5, the boundary layer is laminar. The thickness of the laminar boundary layer may be given by the Blasius' equation i.e.

$$\delta = \frac{5x}{\sqrt{Re_x}} = \frac{5\times 3\times 10}{\sqrt{10^5}}$$

$$\Rightarrow \quad \delta = 0.47 \text{ mm}$$

10.9 (a)

The momentum thickness is given by

$$\theta = \int_0^\delta\frac{u}{U}\left(1-\frac{u}{U}\right)dy = \int_0^\delta\frac{y}{\delta}\left(1-\frac{y}{\delta}\right)dy$$

$$= \int_0^\delta\left(\frac{y}{\delta}-\frac{y^2}{\delta^2}\right)dy = \frac{1}{\delta^2}\int_0^\delta(\delta y - y^2)dy$$

$$= \frac{1}{\delta^2}\left[\frac{\delta y^2}{2}-\frac{y^3}{3}\right]_0^\delta = \frac{1}{\delta^2}\left[\frac{\delta^3}{2}-\frac{\delta^3}{3}\right]$$

$$= \delta\left[\frac{3-2}{6}\right] = \frac{\delta}{6}$$

10.10 (b)

At a point of boundary layer separation, shear stress is zero because $\dfrac{du}{dy}$ at the surface bounday is zero.

10.11 (c)

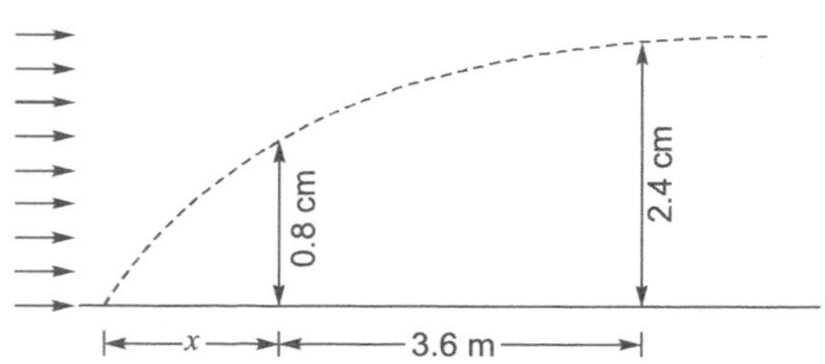

For Laminar boundary layer Over a flat plate

$$\frac{\delta}{x} = \frac{5}{\sqrt{Re_x}}$$

$$\delta \propto \sqrt{x}$$

So, $\dfrac{0.8}{2.4} = \sqrt{\dfrac{x}{x+3.6}}$

$$x = 0.45 \text{ m}$$

Since no any option is given.

So nearest answer is 0.4 m.

10.13 (d)

For Laminar boundary layer over a flat plate.

$$C_D = \frac{1.328}{\sqrt{Re_L}} = \frac{1.328}{\sqrt{\dfrac{\rho v L}{\mu}}}$$

$$\therefore \quad C_D \propto \frac{1}{\sqrt{L}}$$

$$\therefore \quad \frac{C_{D_1}}{C_{D_2}} = \frac{\sqrt{L_2}}{\sqrt{L_1}}$$

$$\Rightarrow \frac{0.018}{C_{D_2}} = \frac{\sqrt{4L_1}}{\sqrt{L_1}}$$

$$\Rightarrow C_{D_2} = 0.009$$

10.14 (d)

Shear stress, $\tau = \mu . \dfrac{\partial u}{\partial y}$

$$= 0.5\frac{\partial}{\partial y}[3\sin(5\pi y)]$$

[Here $\mu = 0.5$ Pa-s]

$$= 0.5 \times 3 \times 5\pi \cos(5\pi y)$$

$$= 23.56 \cos(5\pi y) \text{ N/m}^2$$

So, At $y = 0.05$ m

$$\tau = 23.56\cos(5\pi \times 0.05)$$

$$= 16.7 \text{ N/m}^2$$

and At $y = 0.12$ m

$$\tau = 23.56\cos(5\pi \times 0.12)$$

$$= -7.28 \text{ N/m}^2$$

Since Negative value is not acceptable.

∴ It is taken as zero.

10.15 (c)

$$\frac{u}{U} = \frac{y}{\delta}$$

$$\theta = \int_0^\delta \frac{u}{U}\left[1-\frac{u}{U}\right]dy = \int_0^\delta \frac{y}{\delta}\left[1-\frac{y}{\delta}\right]dy$$

$$= \frac{\delta}{2} - \frac{\delta}{3} = \frac{\delta}{6}$$

10.16 (c)

Apply boundary condition,

$$a = 0, \quad b = \frac{3}{2}$$

$$c = 0, \quad d = -\frac{1}{2}$$

10.17 (a)

$$u_\infty = 2.5 \text{ m/sec}$$

$$\rho_w = 10^3 \text{ kg/m}^3$$

$$\mu_w = 10^{-3} \text{ Pa-s}$$

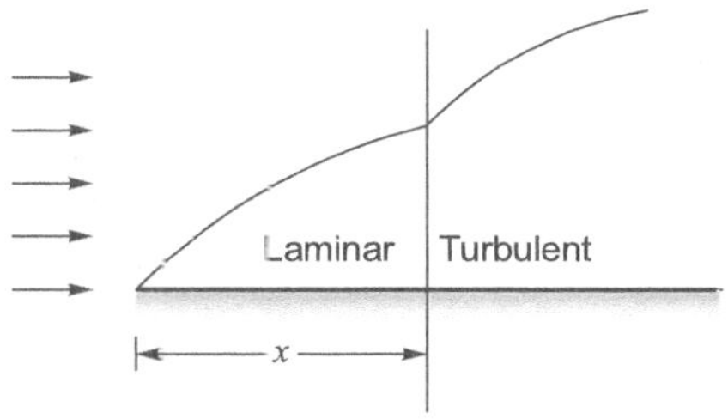

For laminar, critical Reynolds No(Re_{cr}) = 5×10^5

$$\therefore \quad Re_x = \frac{\rho \cdot U_\infty \cdot x}{\mu}$$

$$5 \times 10^5 = \frac{(10^3)\times(2.5)\times x}{10^{-3}}$$

$$x = 0.2 \text{ m}$$

$$\frac{\delta}{x} = \frac{5}{\sqrt{Re_x}}$$

$$\frac{\delta}{0.2} = \frac{5}{\sqrt{5\times 10^5}}$$

$$\delta = 1.414 \text{ mm}$$

Note: *By using Re_{cr} = 5 lakh no option is matching. If we take Re_{cr} = 3.2 lakh. Although it is not mentioned.*

Assuming transition occurs for Re between 1.3×10^5 to 4×10^6. If we take

$$Re_{cr} = 3.2 \times 10^5$$

Then $Re_x = \dfrac{\rho U_\infty x}{\mu}$

$$3.2 \times 10^5 = \frac{(10)^3(2.5)x}{10^{-3}}$$

$$x = 12.8 \text{ cm}$$

$$\frac{\delta}{x} = \frac{5}{\sqrt{Re_x}}$$

$$\frac{\delta}{0.128} = \frac{5}{\sqrt{3.2\times 10^5}}$$

$$\delta = 0.113 \text{ cm}$$

10.18 (a)

For laminar boundary layer

Blasius results

$$\frac{\delta}{x} = \frac{5}{\sqrt{Re_x}}$$

$$\delta \propto \frac{1}{\sqrt{Re_x}}$$

Now, $$\frac{\delta_1}{\delta_2} = \frac{\sqrt{Re_{x_2}}}{\sqrt{Re_{x_1}}} = \sqrt{\frac{484}{100}} = 2.2$$

10.19 (d)

The boundary layer thickness increases with the increase in distance from the leading edge.

10.20 (a)

Reynold's number,

$$Re_x = \frac{\rho U_\infty x}{\mu} = \frac{U_\infty x}{\nu}$$

$$\Rightarrow \quad Re_x = \frac{3 \times 1}{2 \times 10^{-5}} = 2 \times 10^5$$

Since, $Re_x < 5 \times 10^5$

$\Rightarrow$ Boundary layer over flat plate with laminar flow, so applying Blasius equation,

$$\therefore \quad \frac{\delta}{x} = \frac{5}{\sqrt{Re_x}}$$

$$\Rightarrow \quad \frac{\delta}{1000} = \frac{5}{\sqrt{2 \times 10^5}}$$

$$\Rightarrow \quad \delta = 11.18 \text{ m} = 1.118 \text{ cm}$$

■■■■

11 Turbulent Flow

11.1 Consider the following statements regarding friction factor variation in a pipe flow:

1. In hydrodynamically rough boundaries, friction factor 'f' is related to relative roughness ε/D, (ε is roughness and D is the diameter of pipe).
2. In hydrodynamically smooth boundaries, friction factor 'f' depends both on Reynolds number 'Re' and relative roughness.
3. If the friction factor, $f = \dfrac{0.316}{\text{Re}^{1/4}}$, then the value of the friction factor 'f' relates to that of the region of turbulent flow in smooth boundary.
4. Friction factor is directly related to the nature of boundary layer in a pipe flow.

Which of these statements are correct?

(a) 1, 2 and 4 (b) 1, 2 and 3
(c) 2, 3 and 4 (d) 1, 3 and 4

[ESE : 2000]

11.2 Consider the following statements:
Due to aging of pipes in a pipe network,

1. the roughness increases linearly with time.
2. the pipes get rusted and bent.
3. the pipes become smoother with time.

Which of these statements is/are correct?

(a) 1 and 3 (b) 2 and 3
(c) 3 only (d) 1 only

[ESE : 2008]

11.3 Match **List-I** with **List-II** and select the correct answer using the code given below the lists:

List-I

A. Uniform flow
B. Laminar flow
C. Turbulent flow
D. Irrotational flow

List-II

1. Flow through a water supply pipe
2. Flow through a straight tube of uniform diameter and uniform roughness
3. Flow above the drain hole of a wash basin
4. Flow of blood in veins and arteries

Code:

	A	B	C	D
(a)	3	1	4	2
(b)	2	1	4	3
(c)	3	4	1	2
(d)	2	4	1	3

[ESE : 2011]

11.4 The maximum velocity U_m, the mean velocity U and shear velocity u_* in the case of turbulent flow through circular pipes are related as $\dfrac{(U_m - U)}{u_*} =$

(a) 2.5 for rough boundary flow only
(b) 5.75 for smooth boundary flow only
(c) 3.75 for both smooth and rough boundary flows
(d) 5.75 for both smooth and rough boundary flows

[ESE : 2012]

11.5 The loss of head in a pipe carrying turbulent flow varies:

(a) Inversely as the square of the velocity of flow
(b) Inversely as the square of the diameter of pipe
(c) Directly as the square of the velocity of flow
(d) Directly as the velocity of flow

[ESE : 2013]

11.6 In turbulent flows through rough pipes, the ratio of the maximum velocity to the mean velocity is

(a) 2
(b) 4/3
(c) 1.1
(d) Dependent on the friction factor

[ESE : 2015]

11.7 Consider the following statements:

1. In flow through hydro-dynamically smooth pipes, the friction factor f is always a constant.
2. In flow through hydro-dynamically smooth pipes, the friction factor f is always a function of the flow Reynolds number.
3. In a fully developed rough turbulent pipe flow, the friction factor f is a function of relative roughness only.
4. In a fully developed rough turbulent pipe flow, the friction factor f is a function of the flow Reynolds number and relative roughness.

Which of the above statements are correct?

(a) 1 and 3 (b) 2 and 3
(c) 2 and 4 (d) 1 and 4

[ESE : 2016]

11.8 In the design of pipeline the usual practice is to assume that due to ageing of pipelines:

1. The effective roughness increases linearly with time.
2. The friction factor increases linearly with time.
3. The flow through the pipe becomes linearly lesser with time.

Which of the above statements is/are correct?

(a) 1, 2 and 3 (b) 3 only
(c) 2 only (d) 1 only

[ESE : 2016]

11.9 When the average flow is steady, which one of the following causes significant fluctuations in velocity, temperature and pressure?

(a) Streamlines in turbulent flow
(b) Molecular diffusion in turbulent flow
(c) Eddy motion in turbulent flow
(d) Orderly flow in turbulent flow

[ESE : 2024]

■■■■

Answers Turbulent Flow

11.1 (d) 11.2 (d) 11.3 (d) 11.4 (c) 11.5 (c) 11.6 (d) 11.7 (b) 11.8 (d) 11.9 (c)

Explanations Turbulent Flow

11.1 (d)

The friction factor in laminar flow, $f = \dfrac{64}{\text{Re}}$

For turbulent flow the pipe boundary can be classified as:

(i) Smooth boundary where f is a function of Reynolds number (Re) only
(ii) Transition boundary where f is a function of Re and relative roughness (ε/D)
(iii) Rough boundary where f is a function of ε/D only.

11.2 (d)

Due to aging of pipes, the roughness increases linearly with time as :

$$k_s = k_{s_0} + \alpha t$$

k_{s_0} is the initial roughness.

α is the rate of increase in roughness and t is the time period.

11.3 (d)

In a straight tube of uniform diameter and uniform roughness the flow properties does not vary across the length of the pipe, hence uniform flow.

Flow of blood in veins and arteries occurs as a viscous flow, hence laminar

Flow above the drain hole of a wash basin is a free vortex motion. The molecules of water does not rotate about their centre of mass hence irrotational flow.

A water supply pipe carries water at high speed, leading to rapid mixing which causes highly turbulent condition.

11.4 (c)

Common mean velocity distribution equation

$$\frac{u-\bar{u}}{v^*} = 5.75\log_{10}\frac{y}{R}+3.75$$

At $y = R$, $u = u_{max}$

$$\frac{u_{max}-\bar{u}}{v^*} = 3.75$$

11.5 (c)

$$h_f = \frac{fLV^2}{2gD} \Rightarrow h_f \propto V^2$$

11.6 (d)

Common mean velocity distribution on equation.

$$\frac{u-\bar{u}}{v^*} = 5.75\log_{10}\frac{y}{R}+3.75$$

At $y = R$, $u = u_{max}$

$$\frac{u_{max}-\bar{u}}{\bar{u}\sqrt{\frac{f}{8}}} = 3.75$$

$$\frac{u_{max}}{\bar{u}}-1 = \frac{3.75}{\sqrt{8}}\times\sqrt{f}$$

$$\frac{u_{max}}{\bar{u}} = 1+1.33\sqrt{f}$$

By Nikurase experimental results.

$$\frac{u_{max}}{\bar{u}} = 1+1.43\sqrt{f}$$

11.8 (d)

$$k_s = k_{s_0}+\alpha t$$

k_s increases linearly with time.

11.9 (c)

Eddy motion in turbulent flow.

12 Dimensional Analysis & Model Analysis

12.1 Euler number or pressure coefficient may be represented as (notations have their usual meanings)

(a) $\frac{\Delta p}{\gamma H}$ (b) $\frac{\Delta p}{(\rho V^2/2)}$

(c) $\frac{\Delta p}{\mu V/L}$ (d) $\frac{\Delta p}{\rho L g}$

[ESE : 2000]

12.2 A spillway of an irrigation project is to be studied by means of a model constructed to a scale of 1 : 9. The prototype discharge is 1000 m^3/s. Neglecting the viscous and surface tension effects, the required flow rate for the model is

(a) 12.35 m^3/s (b) 111.11 m^3/s

(c) 4.11 m^3/s (d) 1.37 m^3/s

[ESE : 2002]

12.3 The distorted models, used in the hydraulic studies, are those which

(a) have exaggeration of the vertical scale and horizontal scale

(b) are not geometrically similar to the prototypes

(c) have same vertical and horizontal scale as that of the prototypes

(d) have same Froude number as that for the prototypes

[ESE : 2003]

12.4 A 1: 30 scale model of a submarine is to be tested in a wind tunnel for its drag when it is operating at 15 km/h in ocean. The kinematic viscosity of air is 1.51×10^{-5} m^2/s and for water 1.02×10^{-6} m^2/s. What is the velocity of air in wind tunnel that should be maintained for kinematic similarity?

(a) 1500 m/s (b) 2000 m/s

(c) 1850.5 m/s (d) 2500.5 m/s

[ESE : 2004]

12.5 Which one of the following statements is not correct?

(a) Models are always smaller than the prototypes.

(b) Dynamic similarity between a model and a prototype can be verified by equating Reynolds number in a viscous flow.

(c) Mach number achieves significance when the velocity of fluid approaches or exceeds the sonic velocity.

(d) Distorted models are always exaggerated on a vertical scale.

[ESE : 2005]

12.6 Match **List-I** with **List-II** and select the correct answer using the codes given below the lists:

List-I

A. Reynolds number

B. Froude number

C. Mach number

D. Weber number

List-II

1. Formation of liquid droplet

2. High speed flow of a gas

3. Flow in closed conduits

4. Sloping interface exists between fluids of different densities

Codes:

	A	B	C	D
(a)	3	4	2	1
(b)	2	1	3	4
(c)	3	1	2	4
(d)	2	4	3	1

[ESE : 2006]

12.7 Match **List-I** with **List-II** and select the correct answer using the codes given below the lists:

List-I

A. Froude number

B. Euler number

C. Reynolds number
D. Mach number

List-II
1. Related to inertia force and elastic force
2. Related to inertia force and viscous force
3. Related to inertia force and pressure force
4. Related to inertia force and gravity force

Codes:

	A	B	C	D
(a)	4	1	3	2
(b)	2	3	4	1
(c)	4	3	2	1
(d)	2	1	4	3

[ESE : 2006]

12.8 In the model of a highway bridge constructed to a scale of 1 : 25, the force of water on the pier was measured as 5 N. What is the force (approximate) on the prototype pier?
(a) 15.6 kN (b) 25.3 kN
(c) 78 kN (d) 90 kN
[ESE : 2007]

12.9 A river model is constructed to a horizontal scale of 1 : 1000 and a vertical scale of 1 : 100. A model discharge of 0.1 m^3/s would correspond to a discharge in the prototype, of what magnitude?
(a) 10^2 m^3/s (b) 10^3 m^3/s
(c) 10^4 m^3/s (d) 10^5 m^3/s
[ESE : 2007]

12.10 In model similarity, if gravitational and inertial forces are the only important forces, then what is the discharge ratio?
(a) $L_r^{3/2}$ (b) $L_r^{1/2}$
(c) $L_r^{5/2}$ (d) $L_r^{1/3}$
where L_r = ratio of length dimension.
[ESE : 2008]

12.11 An airplane is cruising at a speed of 800 km/h at an altitude where the air temperature is 0° C. The flight Mach number at this speed is nearly
(a) 1.3 (b) 0.67
(c) 0.25 (d) 2.4
[ESE : 2010]

12.12 Distorted models are needed for:
1. Rivers
2. Dams across wide rivers
3. Harbours
(a) 1 and 2 only (b) 2 and 3 only
(c) 1 and 3 only (d) 1, 2 and 3
[ESE : 2011]

12.13 In a 1 in 50 model of a spillway, discharge is 3 m^3/sec. Find the corresponding discharge in the prototype in m^3/sec.
(a) 1061 (b) 7500
(c) 53033 (d) 375000
[ESE : 2013]

12.14 A model of a boat is built to a scale of 1/100. It experiences a resistance of 0.10 N when simulating a speed of 5 m/s of the prototype. Water is the fluid in both the cases. Neglecting frictonal forces, corresponding resistance in the prototype is:
(a) 1000 kN (b) 100 kN
(c) 10 kN (d) 1 kN
[ESE : 2013]

12.15 The performance of a hydraulic structure during a flood has been investigated in a 1/25 model based on Froude law of similarity. A flood wave passing through the model in 2 hrs. corresponds to a prototype period of:
(a) 50 hr (b) 12.5 hr
(c) 10 hr (d) 0.4 hr
[ESE : 2013]

12.16 A 1 : 25 model of a naval ship is towed at a velocity of 1.2 m/s to determine each of the two components of the total drag that may be experienced by the prototype. The wave-making drag of the model is duly estimated to be 0.37 kgf. What would be the wave making drag on the prototype ship?
(a) 1250 kgf (b) 1875 kgf
(c) 3750 kgf (d) 4625 kgf
[ESE : 2013]

12.17 A river whose discharge is 10^5 cumecs is to be studied by a 1 : 100 scale model in a laboratory. The discharge required in the model is:
(a) 0.5 m^3/s (b) 10 m^3/s
(c) 1 m^3/s (d) 5 m^3/s
[ESE : 2013]

12.18 In order to estimate the energy loss in a pipeline of 1 m diameter through which kerosene of specific gravity 0.80 and dynamic viscosity 0.02 poise is to be transported at the rate of 2 m^3/s, model tests were conducted on a 0.1 m diameter pipe using water at 20°C. If the absolute viscosity of water at 20°C is 1.00×10^{-2} poise, then the discharge required for the model pipe would be

(a) 60 *l*/s (b) 80 *l*/s
(c) 120 *l*/s (d) 160 *l*/s

[ESE : 2014]

12.19 A sphere is moving in water with a velocity of 1.6 m/s. Another sphere of twice the diameter is placed in a wind tunnel and tested with air which is 750 times less dense and 60 times less viscous (dynamically) than water. The velocity of air that will model dynamically similar conditions is

(a) 5 m/s (b) 20 m/s
(c) 10 m/s (d) 40 m/s

[ESE : 2015]

12.20 The flow in a river is 1500 cumecs. A distorted model is built with horizontal scale of 1/150 and vertical scale of 1/25. The flow rate in the model should be

(a) 0.04 m^3s^{-1} (b) 0.06 m^3s^{-1}
(c) 0.08 m^3s^{-1} (d) 0.10 m^3s^{-1}

[ESE : 2015]

12.21 **Statement (I):** Reynolds number must be the same for model and the prototype when both are tested as immersed in a subsonic flow.

Statement (II): A model should be geometrically similar to the prototype.

(a) Both Statement (I) and Statement (II) are individually true and Statement (II) is the correct explanation of Statement (I)
(b) Both Statement (I) and Statement (II) are individually true but Statement (II) is ***not*** the correct explanation of Statement (I)
(c) Statement (I) is true but Statement (II) is false
(d) Statement (I) is false but Statement (II) is true

[ESE : 2016]

12.22 Which of the following statements are correct?

1. Depression of mercury in a capillary tube is dependent on density and surface tension.
2. Modelling of flow-induced drag on a ship is done invoking both of Froude number and Reynolds number.
3. Flow of a fluid in a narrow pipe is relatable to both Reynolds number and Cauchy number.
4. Formation and collapse of a soap bubble is analyzed through employing surface tension and external pressure.
5. Flow over the downstream slope of an ogee spillway can be affected by surface tension.

Select the correct answer using the codes given below:

(a) 1, 2 and 4 (b) 1, 3 and 5
(c) 2, 3 and 4 (d) 3, 4 and 5

[ESE : 2017]

12.23 Which of the following factors are non-dimensional?

1. *C* in Chezy's equation
2. 11.6 as a measure of the sub-layer
3. $\frac{H}{N^2D^2}$ employed in comparing performance of pumps
4. $\frac{Q^2}{D^5}$ employed in computations in pipe networks
5. $\frac{U}{\sqrt{gL}}$ used in estimating wave-making drag

Select the correct answer using the codes given below:

(a) 2 and 5 (b) 2, 4 and 5
(c) 1 and 5 (d) 1, 3 and 4

[ESE : 2017]

12.24 A jet propelled aircraft is flying at a speed of 1100 km/hour at t = 20°C, k = 1.4 and R = 287 J/kgK. The Mach number at a point on the jet will be nearly

(a) 0.3 (b) 0.5
(c) 0.7 (d) 0.9

[ESE : 2020]

■■■■

Answers Dimensional Analysis & Model Analysis

12.1 (b)	12.2 (c)	12.3 (b)	12.4 (c)	12.5 (a)	12.6 (a)	12.7 (c)	12.8 (c)
12.9 (d)	12.10 (c)	12.11 (b)	12.12 (d)	12.13 (c)	12.14 (b)	12.15 (c)	12.16 (*)
12.17 (c)	12.18 (b)	12.19 (c)	12.20 (c)	12.21 (b)	12.22 (a)	12.23 (a)	12.24 (d)

Explanations Dimensional Analysis & Model Analysis

12.1 (b)

The Euler number (E_u) is a dimensionless number used in fluid flow calculations. It expresses the relationship between a local pressure drop e.g. over a restriction and the kinetic energy per volume, and is used to characterize losses in the flow, where a perfect frictionless flow corresponds to an Euler number of 0.

$$E_u = \frac{\Delta P}{(\rho V^2)/2}$$

12.2 (c)

Discharge ratio,

$$Q_r = L_r^{2.5}$$

$$\frac{Q_m}{Q_p} = L_r^{2.5}$$

$$\frac{Q_m}{1000} = \left(\frac{1}{9}\right)^{2.5}$$

$$Q_m = 4.11 \text{ m}^3/\text{s}$$

12.3 (b)

Distorted model can have vertical scale exaggeration or slope exaggeration. Therefore they are not geometrically similar. It may have either geometrical distortion, or material distortion or distortion of hydraulic quantities or a combination of these.

12.4 (c)

The Reynolds model law shall be applied in the case of fully submerged objects

$$\frac{V_m L_m}{\nu_m} = \frac{V_p L_p}{\nu_p}$$

$$\therefore \quad V_m = V_p\left(\frac{\nu_m}{\nu_p}\right)\left(\frac{L_p}{L_m}\right)$$

$$= 15 \times \frac{1.5 \times 10^{-5}}{1.02 \times 10^{-6}} \times 30$$

$$= 6661.8 \text{ km/h} = 1850.5 \text{ m/s}$$

The Reynolds model law ensures dynamic, kinematic and geometric similarities.

12.5 (a)

Models may be larger than prototypes also. e.g., model of carburettor. However in general models are smaller than the prototypes.

12.6 (a)

Weber model law is applicable when surface tension force are predominant along with inertia forces.

12.7 (c)

The Froude number is a ratio of inertial and gravitational forces. The *Euler Number* can be interpreted as a measure of the ratio of the *pressure* forces to the *inertial* forces. Reynolds is the ratio of *Inertia* force to viscous force. Mach number is ratio of *inertia force* to *elastic* force.

12.8 (c)

In this case Froude model law is applicable

Force ratio, $F_r = \rho_r L_r^3 g_r$

$$\rho_r = 1$$

$$g_r = 1$$

$$F_r = L_r^3$$

$$\frac{F_m}{F_p} = \left(\frac{1}{25}\right)^3$$

$$\frac{5}{F_p} = \left(\frac{1}{25}\right)^3$$

$$F_p = 78.125 \text{ kN}$$

12.9 (d)

Horizontal scale, $L_r = \frac{L_m}{L_p} = \frac{1}{1000}$

Vertical scale, $h_r = \frac{y_m}{y_p} = \frac{1}{100}$

By Froude model law, we have

$$\frac{V_m}{\sqrt{gy_m}} = \frac{V_p}{\sqrt{gy_p}}$$

$$\Rightarrow \quad \frac{V_m}{V_p} = \sqrt{\frac{y_m}{y_p}}$$

$$\Rightarrow \quad \frac{V_m}{V_p} = \sqrt{h_r}$$

$$\Rightarrow \quad V_r = \sqrt{h_r}$$

∴ Discharge ratio,

$$Q_r = \frac{Q_m}{Q_p} = A_r \times V_r$$

$$\Rightarrow \quad \frac{Q_m}{Q_p} = L_r h_r \times \sqrt{h_r}$$

$$\Rightarrow \quad \frac{Q_m}{Q_p} = L_r\, h_r^{3/2}$$

$$\Rightarrow \quad \frac{0.1}{Q_p} = \frac{1}{1000} \times \left(\frac{1}{100}\right)^{3/2}$$

$$\Rightarrow \quad Q_p = 0.1 \times 1000 \times 1000$$

$$\Rightarrow \quad Q_p = 10^5\ \text{m}^3/\text{s}$$

12.10 (c)

If gravitational and inertial forces are the only important forces, then the Froude number must be the same in the model and prototype. Thus

$$\frac{V_m}{\sqrt{gL_m}} = \frac{V_p}{\sqrt{gL_p}}$$

But g is same for both model and prototype

$$\therefore \quad \frac{V_m}{V_p} = \sqrt{\frac{L_m}{L_p}}$$

$$\Rightarrow \quad V_r = \sqrt{L_r}$$

Now, Discharge = Area × Velocity

$$\Rightarrow \quad Q_r = A_r \times V_r$$

$$\Rightarrow \quad Q_r = L_r^2 \times (L_r)^{1/2} = L_r^{5/2}$$

12.11 (b)

$$\text{Speed of aeroplane} = 800 \times \frac{5}{18} = 222.22\ \text{m/sec}$$

Velocity of sound in air = 332 m/sec

$$\text{Mach No.} = \frac{222.22}{332} = 0.67$$

12.12 (d)

Some of the typical examples for which distorted models are required to be prepared are rivers, dams across very wide rivers, harbours, estuaries etc., for which the horizontal dimensions are large in proportion to the vertical ones.

12.13 (c)

In the case of a flow in spillway, Froude number will dominate

$$\frac{V_m}{V_p} = \sqrt{L_r}$$

$$\frac{Q_m}{Q_p} = L_r^{5/2} = \left(\frac{1}{50}\right)^{5/2}$$

$$\Rightarrow \quad Q_p = 50^{5/2}(3) = 53033\ \text{m}^3/\text{s}$$

12.14 (b)

$$\frac{F_m}{F_p} = \left(\frac{1}{100}\right)^3$$

$$F_p = 100\ \text{kN}$$

12.15 (c)

According to Froude Lane

$$(F_r)_m = (F_r)_p$$

$$\left(\frac{V}{\sqrt{Lg}}\right)_m = \left(\frac{V}{\sqrt{Lg}}\right)_p$$

$$\frac{V_m}{V_p} = \sqrt{\frac{L_m}{L_p}} \qquad (g_m = g_p)$$

$$V_r = \sqrt{L_r}$$

$$\frac{L_r}{T_r} = \sqrt{L_r}$$

$$T_r = \sqrt{L_r}$$

$$\frac{T_m}{T_p} = \sqrt{\frac{1}{25}}$$

$$\frac{2}{T_p} = \sqrt{\frac{1}{25}}$$

$$T_p = 10\ \text{hrs}$$

12.16 (*)

$$\frac{F_m}{F_p} = \left(\frac{1}{25}\right)^3$$

$$\Rightarrow \quad F_p = 5781.25\ \text{kgf}$$

12.17 (c)

$$\frac{Q_m}{Q_p} = (L_r)^{5/2}$$

$$\Rightarrow \quad Q_m = \frac{10^5}{100^{5/2}} = 1\text{ m}^3/\text{s}$$

12.18 (b)

For dynamic similarity, Reynolds model law will be used.

$$\frac{\rho_P V_P L_P}{\mu_P} = \frac{\rho_m V_m L_m}{\mu_m}$$

$$\Rightarrow \frac{0.8 \times \rho_w \times Q_P \times 1}{A_P \times 2 \times 10^{-2}} = \frac{\rho_w \times Q_m \times 0.1}{A_m \times 1 \times 10^{-2}}$$

$$\Rightarrow Q_m = \frac{0.8 \times 1 \times \frac{\pi}{4}(0.1)^2 \times 2}{\frac{\pi}{4} \times 1^2 \times 2 \times 0.1}$$

$$\Rightarrow \quad Q = 0.08\text{ m}^3/\text{sec.} = 80\ l/\text{sec.}$$

12.19 (c)

According to Reynold's Model law

$$(\text{Re})_m = (\text{Re})_p$$

$$\frac{\rho_r V_r L_r}{\mu_r} = \text{Constant}$$

$$\frac{\rho_w \times 1.6 \times D}{\mu_w} = \frac{\frac{\rho_w}{750} \times V \times 2D}{\mu_w/60}$$

$$V = \frac{750 \times 1.6}{60 \times 2} = 10\text{ m/s}$$

12.20 (c)

Flow rate, $Q_r = (L_r)_H (L_r)_V^{3/2}$

$$\frac{Q_m}{Q_p} = \left(\frac{1}{150}\right)\left(\frac{1}{25}\right)^{3/2}$$

$$\frac{Q_m}{1500} = \frac{1}{150} \times \left(\frac{1}{25}\right)^{3/2}$$

$$Q_m = \frac{1500}{150 \times (25)^{3/2}} = 0.08\text{ m}^3/\text{s}$$

12.22 (a)

- Capillary tube depression $h = \dfrac{2\sigma\cos\theta}{\rho g R}$, dependent on surface tension, density and diameter of tube.
- In modeling of flow induced drag is done invoking both Froude no. and Reynold's no.
- Flow of a fluid in a narrow pipe mainly relatable to Reynold's no. (Cauchy number is square of Mach no. and mainly used for compressible liquids).
- Flow over the downstream slope of an Ogee spillway is affected by Froude no.

12.23 (a)

- Dimension of Chezy's constant is $L^{1/2}T^{-1}$
- Laminar sublayer thickness (δ')

$$\delta' = \frac{11.6\nu}{V^*}$$

ν = kinematic viscosity

V^* = shear velocity

11.6 = constant (dimensionless)

- $\dfrac{H}{N^2D^2} = \dfrac{L}{\left(\frac{1}{T}\right)^2 \times L^2} = T^2 L^{-1}$
- $\dfrac{Q^2}{D^5} = \dfrac{\left(\frac{L^3}{T}\right)^2}{(L)^5} = LT^{-2}$
- $\dfrac{V}{\sqrt{gL}} = \dfrac{\left(\frac{L}{T}\right)}{\sqrt{\left(\frac{L}{T^2} \times L\right)}} = 1$ (Non-dimensional)

12.24 (d)

For adiabatic process

$$K = \gamma P$$

$$C = \sqrt{\frac{K}{\rho}}, \quad P = \rho RT$$

$$= \sqrt{\gamma RT}$$

$$= \sqrt{1.4(287)(273+20)}$$

$$= 343.114\text{ m/s}$$

$$M_a = \frac{V}{C} = \frac{1100 \times \frac{5}{18}}{343.114} = 0.89$$

13 Notches and Weirs

13.1 If the error in the measurement of head in a '*V*' notch is 1%, then the error in the measurement of discharge will be

(a) 1% (b) 1.5%
(c) 2% (d) 2.5%

[ESE : 2000]

13.2 Consider the following measuring structures:

1. Submerged broad-crested weir
2. Free broad-crested weir
3. Free sharp-crested weir
4. Free ogee spillway

The correct sequence of the increasing order of the discharge coefficient of these structures is

(a) 1, 2, 4, 3 (b) 2, 1, 3, 4
(c) 2, 1, 4, 3 (d) 1, 2, 3, 4

[ESE : 2001]

13.3 A Cipolletti weir discharges water with the head of water over the crest being 250 mm. If the head due to velocity of approach is 0.01 m, what will be the excess percentage of flow, when corrected for velocity of approach as compared to when not so corrected?

(a) 3.2% (b) 4.2%
(c) 5.3% (d) 6.3%

[ESE : 2003]

13.4 The approximate discharge over a 4 m long rectangular weir (with suppressed end contractions) with head over the crest as 0.36 m is

(a) 4.2 litres/s (b) 2.4 m^3/s
(c) 6.1 litres/s (d) 1.6 m^3/s

[ESE : 2003]

13.5 The head on a sharp-crested rectangular weir of height 1.6 m and crest length 1.2 m was incorrectly observed to be 0.13 m when it was actually 0.15 m. What was the percentage error in the computed value of flow rate?

(a) 0.5% (b) 1.0%
(c) 2.0% (d) 1.5%

[ESE : 2009]

13.6 Match **List-I** with **List-II** and select the correct answer using the codes given below the lists:

List-I

A. Rehbock formula
B. Francis formula
C. A special trapezoidal weir
D. Linear proportional weir

List-II

1. Sutro weir
2. Rectangular suppressed weir
3. Rectangular side-contracted weir
4. Cippolletti weir

Codes:

	A	B	C	D
(a)	1	3	4	2
(b)	2	4	3	1
(c)	1	4	3	2
(d)	2	3	4	1

[ESE : 2009, 2011]

13.7 The absolute percentage error in the computed discharge over a rectangular weir corresponding to an absolute error of 1.5% in the measurement of head over the sill of the weir would be

(a) 1.5 (b) 2.25
(c) 2.5 (d) 3.75

[ESE : 2011]

13.8 Consider the following devices

1. Orifice
2. Borda's mouthpiece running free
3. Bell-mouthed orifice
4. External mouthpiece

What is the correct sequence of these devices by decreasing magnitude of coefficient of discharge?

(a) 2, 3, 1 and 4 (b) 4, 3, 1 and 2
(c) 4, 1, 3 and 2 (d) 2, 1, 3 and 4

[ESE : 2011]

13.9 A depth-discharge relationship of the canal section is maintained at a notch fall because the sill of the notches is
(a) Level with downstream canal bed
(b) Below the upstream canal bed
(c) Level with upstream canal bed
(d) Above the upstream canal bed
[ESE : 2011]

13.10 For accurate flow measurement in open channels, an arrangement for aeration under the nappe is necessary when the weir used is
(a) Contracted rectangular
(b) Suppressed rectangular
(c) Submerged contracted rectangular
(d) Triangular [ESE : 2013]

13.11 Among weirs used in open flow metrology, critical depth of the flow is the relevant criterion in:
(a) Broad-crested weir
(b) Cippoletti weir
(c) Parabolic weir
(d) Sharp-crested weir [ESE : 2013]

13.12 The head over a V-notch at the end of a channel is 75 cm. If an error of 0.15 cm is possible in the measurement of the head, then the percentage error in computing the discharge is
(a) 0.25 (b) 0.5
(c) 0.75 (d) 1.0 [ESE : 2015]

13.13 While conducting the flow measurement using a triangular notch, an error of 2% in head over the notch is observed. The percentage error in the computed discharge would be
(a) +7% (b) –3%
(c) +5% (d) – 4% [ESE : 2016]

13.14 In a 90° triangular notch, the error in the estimated discharge for a given head due to an error of 1% in cutting the vertex angle is
(a) zero (b) 1%
(c) $\frac{\pi}{2}$% (d) π% [ESE : 2018]

13.15 A cipolletti weir has length of 2.0 m and head over the weir is 1 m. What is the discharge over the weir? (Take C_d = 0.62)
(a) 10.35 m³/s (b) 3.66 m³/s
(c) 45.21 m³/s (d) 75.68 m³/s
[ESE : 2023]

13.16 Which one of the following is an advantage using a triangular notch over a rectangular notch?
(a) Ventilation of a triangular notch is necessary
(b) The same triangular notch cannot measure a wide range of flows accurately
(c) For heavy discharges, a triangular notch gives more accurate results than a rectangular notch.
(d) In a given triangular notch, only one reading is required to be taken for the measurement of discharge
[ESE : 2023]

13.17 A convergent mouthpiece is discharging water under a constant head of 5 metres. If the diameter of the mouthpiece is 75 mm, then what is the discharge?
(a) 38.2 litres/sec (b) 40.4 litres/sec
(c) 43.8 litres/sec (d) 46.1 litres/sec
[ESE : 2023]

■■■■

Answers Notches and Weirs

13.1 (d) 13.2 (d) 13.3 (c) 13.4 (d) 13.5 (*) 13.6 (d) 13.7 (b) 13.8 (b)
13.9 (d) 13.10 (b) 13.11 (c) 13.12 (b) 13.13 (c) 13.14 (c) 13.15 (b) 13.16 (d)
13.17 (c)

Explanations Notches and Weirs

13.1 (d)

$$\frac{dQ}{Q} = 2.5\frac{dH}{H} = 2.5\%$$

13.2 (d)

For ogee spillway $C_{d1} = 1.19\ C_{ds}$
C_{ds} is the discharge coefficient for sharp – crested weir. So 4 will come after 3.

The classification of weir is:

$H_1/B_w \leq 0.1$	Long crested weir
$01 < H_1/B_w < 0.35$	Broad-crested weir
$0.35 \leq H_1/B_w < 1.5$	Narrow-crested weir
$H_1/B_w \geq 1.5$	Sharp-crested weir

Further submerged weirs have lesser coefficient of discharge compared to free weirs. So 2 will come after 1.

13.3 (c)

Discharge when approach velocity is neglected

$$Q = \frac{2}{3} C_d L \sqrt{2g}\, H^{3/2}$$

If velocity of approach, V_a is to be taken into consideration

$$Q_a = \frac{2}{3} C_d L \sqrt{2g}\left[(H + h_a)^{3/2} - h_a^{3/2}\right]$$

Excess percentage of flow

$$= \left(\frac{Q_a - Q}{Q}\right) \times 100$$

$$= \left(\frac{Q_a}{Q} - 1\right) \times 100$$

$$= \left[\frac{(0.25 + 0.01)^{3/2} - 0.01^{3/2}}{(0.25)^{3/2}} - 1\right] \times 100$$

$= 5.3\%$

13.4 (d)

$$Q = \frac{2}{3} C_d \sqrt{2g} L_e H^{3/2}$$

$= 1.84\, L_e H^{3/2}$ (Long rectangular weir)

$L_e = (L - 0.1\, nH)$

$= 4 - 0.1 \times 2 \times 0.36$

$= 3.928$ m

$\therefore\ Q = 1.84 \times 3.928 \times (.36)^{3/2} = 1.56\ \text{m}^3/\text{s}$

13.5 (*)

For rectangular notch or weir

$$\frac{dQ}{Q} = \frac{3}{2}\frac{dH}{H}$$

$$\Rightarrow \frac{dQ}{Q} = 1.5 \times \left(\frac{0.15 - 0.13}{0.15}\right)$$

$$\Rightarrow \frac{dQ}{Q} = 0.2$$

$\therefore$ Percentage error in flow rate $= 0.2 \times 100$

$= 20\%$

13.6 (d)

(i) Rehbock's formula is an empirical formula for calculating discharge in rectangular suppressed weir. It is given by

$$Q = \frac{2}{3}\left(0.605 + 0.08\frac{H}{Z}\right) + \frac{0.001}{H}\sqrt{2g} L H^{\frac{3}{2}}$$

Where L is the length of weir, H is head in meters and Z is the crest height in meters.

(ii) Francis' formula is one of the commonly used empirical formula for computing the discharge over sharp or narrow crested weirs with or without end contraction. Francis proposed the following formulae for the discharge over rectangular weir.

(a) When the velocity of approach is not considered

$Q = 1.84(L - 0.1\, nH)H^{3/2}$

(b) With the velocity of approach taken into account

$Q = 1.84(L - 0.1\, nH_1)(H_1^{3/2} - h_a^{3/2})$

Where $H_1 = H + h_a = H + \frac{V_a^2}{2g}$, H is height of water surface above the crest of weir, and n is the number of end contractions.

13.7 (b)

For a rectangular weir

$$Q = \left(\frac{2}{3} C_d L \sqrt{2g}\right) H^{3/2} = KH^{3/2}$$

$$dQ = \frac{2}{3} K H^{1/2} \cdot dH$$

$$\Rightarrow \frac{dQ}{Q} = \frac{\frac{3}{2} K H^{1/2}}{K H^{3/2}} \cdot dH = \frac{3}{2}\frac{dH}{H}$$

$\therefore$ For an error of 1.5% in measurement of head, corresponding error in discharge $= 1.5 \times 1.5 = 2.25\%$

13.8 (b)

The coefficient of discharge of a orifice varies from 0.61 to 0.65 depending on the size and the shape of the orifice, and the head of liquid under which the flow takes place. For general purposes for small sharp-edged orifices discharging water of liquids of similar viscosity, an average value of 0.62 or 0.63 may be adopted for C_d.

A Borda's mouthpiece is a short cylindrical tube attached to a circular orifice in the side of a reservoir or tank such that it projects inwardly.
Coefficient of discharge, $C_d = C_c \times C_v$.
The value of C_v for a Borda's mouthpiece running free is equal to 1 and $C_d = C_c = 0.5$
C_d for Borda's mouth piece running full $= 0.707$.
In a bell-mouthed orifice loss of energy as the water discharges out is less.
$\therefore$ C_d for bell-mouthed orifice is more than sharp-edged orifice.

13.9 (d)

Depth-discharge relationship of canal section is maintained at notch fall because crest (sill level) of notches is above the upstream canal bed and there is no heading up or drawdown of water level in canal near the fall.

13.10 (b)

In a contracted weir (one who's notch does not span the full channel), the nappe is automatically aerated and springs free of the weir crest. In a suppressed weir (where the sides of the channel act as the ends of the weir notch), the nappe may, or may not, be aerated. So, here the arrangement for aeration under the nappe is necessary.

13.12 (b)

Percentage error in discharge with head, in V-notch is given by

$$\frac{dQ}{Q} = \frac{5}{2}\times\frac{dH}{H} = \frac{5}{2}\times\frac{0.15}{75}$$

$$\therefore \quad \frac{dQ}{Q}\times 100\% = 0.5\%$$

13.13 (c)

$$\frac{dQ}{Q} = \frac{5}{2}\frac{dH}{H} = \frac{5}{2}\times 2 = +5\%$$

13.14 (c)

$$Q = \frac{8}{15}C_d H^{5/2}\sqrt{2g}\tan\frac{\theta}{2}$$

$$\frac{dQ}{d\theta} = \frac{8}{15}C_d H^{5/2}\sqrt{2g}\sec^2\frac{\theta}{2}$$

$$= \frac{Q}{\tan\frac{\theta}{2}}\times\sec^2\frac{\theta}{2}$$

$$\frac{dQ}{d\theta} = \frac{d\theta}{\tan 45}\times\sec^2 45$$

$$= \frac{1\%}{1}\times\frac{\pi}{2}\times 2 = \frac{\pi}{2}\%$$

13.15 (b)

The discharge over a cipolletti weir is calculated by using the suppressed weir formula,

$$Q_1 = \frac{2}{3}C_d\sqrt{2g}\,LH^{3/2}$$

$$Q = \frac{2}{3}\times 0.62\times\sqrt{2\times 9.81}\times 2\times(1)^{3/2}$$

$$Q = 3.66 \text{ m}^3/\text{s}$$

13.16 (d)

A triangular notch is preferred to a rectangular notch due to following reasons:

1. The expression for discharge for a right angled V-notch is very simple.
2. For measuring low discharge, a triangular notch gives more accurate results than a rectangular notch.
3. In case of triangular notch, only one reading, i.e., H is required for the computation of discharge.
4. Ventilation of triangular notch is not necessary.

13.17 (c)

Discharge through mouthpiece

$$= \frac{\pi}{4}\times(0.075)^2\times\sqrt{2\times 9.81\times 5}$$

$= 0.04375 \text{ m}^3/\text{sec} = 43.75$ lit/sec

$\simeq 43.8$ lit/sec

■■■■

14 Impact of Jets and Turbines

14.1 Given that atmospheric pressure head = 9 m, vapour pressure head (max.) = 1 m, failure head = 40 m and cavitation coefficient $\sigma = 0.15$, the height at which the turbine can be set above the tail race level is

(a) 6 m (b) 4 m
(c) 3 m (d) 2 m

[ESE : 2000]

14.2 Consider the following statements:

1. Specific speed uniquely determines the type, shape and design of the turbine.
2. Both unit speed 'n_{11}' and unit discharge 'Q_{11}' are necessary to uniquely determine the type and design of the turbine.

Which of these statements is/are correct in respect of a hydraulic turbine?

(a) 2 alone (b) 1 alone
(c) Both 1 and 2 (d) Neither 1 nor 2

[ESE : 2000]

14.3 The ideal efficiency of a single-jet Pelton wheel is given by

$$\frac{2}{V^2}[Vv + v(V - v)\cos\phi - v^2]$$

(symbols have their usual meanings)

If friction on the buckets reduces the relative velocity by a factor k_1 and if windage losses reduce the effectiveness of the affected parameters by a factor k_2, then the resulting efficiency multiplied by V^2 is given by

(a) $2k_1k_2[Vv + v(V - v)\cos\phi - v^2]$
(b) $2[k_1Vv + k_2 v(V - v)\cos\phi - v^2]$
(c) $2[k_2Vv + k_1k_2 v(V - v)\cos\phi - v^2]$
(d) $2[k_1Vv + k_1k_2 v(V - v)\cos\phi - k_2v^2]$

[ESE : 2001]

14.4 For a hydro-electric project with reaction turbine, the draft tube at the exit from the turbine is

(a) always immersed in water
(b) always above the water
(c) may either be above or below the water
(d) above or below the water depending on the unit speed of the turbine

[ESE : 2002]

14.5 Consider the following statements in case of impulse turbine:

1. The pressure and velocity both change as the water flows through the turbine.
2. Pressure throughout remains atmospheric and the velocity alone changes.
3. All the available energy is converted into kinetic energy.
4. Water is admitted over the entire circumference of the runner.
5. Water is admitted only through the nozzle.

Which of these statements are correct?

(a) 2, 3 and 5 (b) 2 and 3
(c) 1, 4 and 5 (d) 1 and 4

[ESE : 2002]

14.6 In case of semi-circular vanes, the theoretical maximum efficiency of the wheel can be

(a) 50% (b) 67%
(c) 75% (d) 100%

[ESE : 2002]

14.7 Without residual whirl in the flow at the entr-ance to a draft tube, the best cone angle is

(a) $<6°$ (b) $9°$ to $11°$
(c) $18°$ (d) $24°$

[ESE : 2002]

Directions: The following items consists of two statements; one labelled as **'Assertion (A)'** and the other as **'Reason (R)'**. You are to examine these two statements carefully and select the answers to these items using the codes given below:

Codes:
(a) both A and R are true and R is the correct explanation of A
(b) both A and R are true but R is not a correct explanation of A
(c) A is true but R is false
(d) A is false but R is true

14.8 **Assertion (A):** Selection of turbine by specific speed criterion is more scientific.
Reason (R): Specific speed is independent of dimensions, size of both actual and specific turbine.

[ESE : 2002]

14.9 **Assertion (A):** The relative velocity of water of a Pelton turbine equals the difference between the absolute velocities of water and vane.
Reason (R): For a Pelton turbine, the inlet blade angle is 180°.

[ESE : 2003]

14.10 Pondage in a hydropower station is defined as
(a) impounding of considerable amount of excess water during seasons of surplus flow
(b) a regulating body of water in the form of relatively small amount of run-off to regulate flow variation in daily or weekly power requirements
(c) excess run-off to last for years
(d) excess run-off for a few hours only

[ESE : 2003]

14.11 The specific speed of a turbine under a head of 150 m to develop 2000 HP while running at 300 rpm is
(a) 10-35 (b) 35-60
(c) 60-300 (d) 300-1000

[ESE : 2003]

14.12 A pelton wheel operates at 630 rpm taking 3 m^3/s of water under a head of 256 m with a speed ratio of 0.48 (Given $\sqrt{19.62} = 4.43$) What is the diameter of the impeller?
(a) 0.90 m (b) 1.03 m
(c) 1.42 m (d) 1.80 m **[ESE : 2004]**

14.13 An impulse turbine of 3 m diameter is rated at 10000 kW at 300 rpm under a head of 500 m. The turbine is operated under the head of 400 m. What is the power developed?
(a) 15000 kW (b) 14000 kW
(c) 13000 kW (d) 12000 kW

[ESE : 2004]

14.14 An impulse turbine of 3 m diameter is rated at 1000 kW at 300 rpm under a head of 500 m. The turbine is operated under the head of 400 m. What is the speed at which it would run?
(a) 324 rpm (b) 336 rpm
(c) 348 rpm (d) 364 rpm

[ESE : 2004]

14.15 A forebay in a hydel system is provided at the junction of
(a) the power channel and the tail race channel
(b) the tail race channel and the penstock
(c) the penstock and the turbine
(d) the power channel and the penstock

[ESE : 2005]

14.16 Which one of the following statements is not correct?
(a) Draft tubes decrease efficiency of a turbine.
(b) Wheels of a reaction turbine must always run full.
(c) When the load on a reaction turbine is decreased, the system reduces quantity of flow of water in an immersed flow reaction turbine and reduces the power produced.
(d) Pelton wheels are impulse turbine.

[ESE : 2006]

14.17 Match **List-I** with **List-II** and select the correct answer using the codes given below the lists:

	List-I		List-II
A.	Pelton	**1.**	Draft tube is used
B.	Kaplan	**2.**	Draft tube is not used
C.	Bank	**3.**	Operates under high head and low discharge
D.	Francis	**4.**	Operates under low head and high discharge

Codes:

	A	B	C	D
(a)	2	1	3	4
(b)	3	4	2	1
(c)	2	4	3	1
(d)	3	1	2	4

[ESE : 2006]

14.18 In the selection of turbine by specific speed or head, which one of the following statements is not correct?
(a) For specific speed 10-35, Kaplan turbines
(b) For specific speed 60-300, Francis turbines
(c) For head 50-150 m, Francis turbines
(d) For head above 300 m, Pelton wheel

[ESE : 2006]

14.19 To generate 8100 kW under a head of 81 m while working at a speed of 540 rpm, what type of turbine is suitable?
(a) Pelton (b) Kaplan
(c) Bulb (d) Francis

[ESE : 2007]

14.20 Water is supplied a height of 2.8 m at the rate of 35 *l*ps to a hydraulic ram which delivers 2 *l*ps to a height of 28 m above the ram. What is the Rankine's efficiency?
(a) 58% (b) 55%
(c) 52% (d) 44%

[ESE : 2008]

14.21 Match **List-I** (Type of power house) with **List-II** (Feature) and select the correct answer using the codes given below the lists:

List-I
A. Tidal Plant
B. Storage Plant
C. Pumped Storage Plant
D. Run-of-river plant

List-II
1. Utilizes reversible turbines and improves utilization factor
2. Utilizes current flow in stream, and has pondage to improve load factor
3. Head on turbine varies with time cyclically
4. Utilizes annually average flow with storage

Codes:

	A	B	C	D
(a)	3	4	1	2
(b)	3	2	1	4
(c)	1	2	3	4
(d)	1	4	3	2

[ESE : 2008]

14.22 The shape number in the case of pumps refers to which one of the following?
(a) Ratio of sizes of suction pipe and delivery pipe
(b) Ratio of diameter and thickness of impeller
(c) Non dimensional form of specific speed
(d) Ratio of blade angle at the inlet to the blade angle at the exit

[ESE : 2008]

14.23 Match **List-I** (Component) with **List-II** (Function) and select the correct answer using the codes given below the lists:

List-I
A. Spiral casing
B. Stay ring
C. Guide vane
D. Runner

List-II
1. To allow flow of water through it to produce a torque for the rotation of the runner
2. To direct the water on the runner at an appropriate angle
3. To distribute the flow over the periphery of the runner
4. To act as column helping to support the electrical generator above the turbine

Codes:

	A	B	C	D
(a)	3	2	4	1
(b)	1	2	4	3
(c)	3	4	2	1
(d)	1	4	2	3

[ESE : 2009]

14.24 In all reaction turbines, maximum efficiency is obtained if the
(a) guide vane angle is 90°
(b) blade angle is 90° at the inlet
(c) blade angle is 90° at the outlet
(d) angle of the absolute velocity vector at the outlet is 90°

[ESE : 2010]

14.25 For attaining maximum efficiency, a Francis turbine runner is so designed as to result in radial discharge at exit. This is done by
(a) Providing runner vane angle at inlet as 90°
(b) Providing guide vane angle at inlet as 90°
(c) Providing runner vane angle at exit as 90°
(d) Designing for absolute velocity at outlet to be inclined at 90° to direction of vane there

[ESE : 2011]

14.26 Assume that water vaporizes at an absolute pressure of 1.5 m, barometric pressure head is 9.5 m, cavitation coefficient is 0.1. A turbine operates under a head of 40 m. The safe height of the runner above the tail water level, in meters, is

(a) 6 (b) 4
(c) 3 (d) 2

[ESE : 2011]

14.27 Match **List-I** with **List-II** and select the correct answer using the code given below the lists:

List-I

A. Pelton turbine
B. Francis turbine
C. Kaplan turbine
D. Banki turbine

List-II

1. Mixed flow reaction turbine
2. Operating under low head and large discharge
3. Operating under high head and large discharge
4. No draft tube

Code:

	A	B	C	D
(a)	4	2	1	3
(b)	3	1	2	4
(c)	4	1	2	3
(d)	3	2	1	4

[ESE : 2011]

14.28 **Assertion (A):** In the case of water power plants, it is advisable to provide the surge tank as close to the turbine unit as possible.

Reason (R): Purpose of surge tank is to provide the intended protection for the portion of the penstock which lies on the upstream of it.

(a) both A and R are true and R is the correct explanation of A
(b) both A and R are true but R is not a correct explanation of A
(c) A is true but R is false
(d) A is false but R is true

[ESE : 2011]

14.29 A fire hose has a nozzle attached to it, and the nozzle discharges a jet of water into the atmosphere at a velocity of 20 m/s. This causes the joint of the nozzle with the hose to be in

(a) Tension (b) A state of zero stress
(c) Compression (d) Bending stress

[ESE : 2011]

14.30 Consider the following assumptions made in the analysis of a jet impinging normally on a moving plate to introduce the principle of moment of momentum:

1. Friction between jet and plate is neglected.
2. Flow is steady.
3. Impinging momentum of jet is unchanged.
4. Plate moves at a constant velocity.

Which of these statements are relevant?

(a) 1, 2 and 4 only (b) 1, 2 and 3 only
(c) 2, 3 and 4 only (d) 1, 2, 3 and 4

[ESE : 2012]

14.31 Consider the following statements:

1. The specific speed for turbines is directly proportional to $H^{5/4}$.
2. The specific speed for turbines is inversely proportional to $H^{5/4}$.
3. The specific speed for pumps is directly proportional to $H^{3/4}$.
4. The specific speed for pumps is inversely proportional to $H^{3/4}$.

Which of these statements are correct?

(a) 1 and 3 (b) 2 and 4
(c) 1 and 4 (d) 2 and 3 [ESE : 2012]

14.32 A turbine discharging 10 m^3/s is to be designed so that a torque of 1600 kg-m is to be exerted on the impeller turning at 200 rpm under the condition that the exiting liquid exerts no moment in spite of its momentum. The tangential component of the velocity at the outer periphery of the impeller of radius 1.0 m is

(a) 0.98 m/s (b) 1.57 m/s
(c) 2.10/s (d) 2.26 m/s

[ESE : 2012]

14.33 Two Pelton turbines *A* and *B* have the same specific speed and are working under the same head. Turbine *A* produces 400 kW at 1000 rpm. If turbine *B* produces 100 kW, then its rpm is

(a) 4000 (b) 2000
(c) 1500 (d) 3000 [ESE : 2012]

14.34 Consider the following types of turbines:

1. Francis
2. Pelton with a single jet
3. Kaplan

The correct sequence of these turbines in increasing order of their specific speeds is

(a) 1, 3 and 2 (b) 2, 1 and 3
(c) 1, 2 and 3 (d) 2, 3 and 1

[ESE : 2012]

14.35 A turbine develops 7000 kW power under a head of 25 m at 145 rpm. Calculate the specific speed of turbine and state the type of turbine.

(a) 485, Pelton (b) 217, Francis
(c) 217, Pelton (d) 485, Francis

[ESE : 2013]

14.36 In all reaction turbines, for maximum efficiency, the velocity of:

(a) Swirl at outlet must be zero
(b) Flow at entrance must be zero
(c) Swirl at entrance must be zero
(d) Flow at outlet must be zero [ESE : 2013]

14.37 A turbine develops 2516 kW at 240 rpm. The torque in the shaft is approximately:

(a) 400 kNm (b) 3336 kNm
(c) 1000 kNm (d) 100 kNm

[ESE : 2013]

14.38 A Kaplan turbine has a runner of diameter 4.0 m. The diameter of the hub is 1.6 m. If the velocity of flow and the swirl velocity at the inlet side of the blade at the hub are 6.0 m/s and 10.0 m/s, respectively, the flow and swirl velocities at the inlet side of the tip are, respectively:

(a) 15.0 m/s and 10.0 m/s
(b) 6.0 m/s and 25.0 m/s
(c) 6.0 m/s and 4.0 m/s
(d) 2.5 m/s and 10.0 m/s [ESE : 2013]

14.39 A jet of oil of relative density 0.8 and of area 0.02 m^2 issues at 10 m/s. If it strikes normally on a stationary plate, the force exerted on the plate is

(a) 1906 N (b) 15606 N
(c) 1600 N (d) 19581 N

[ESE : 2013]

14.40 A two-dimensional water jet strikes a fixed two-dimensional plate at 30° to the normal to the plate. This causes the jet to split into two streams in which the ratio of larger discharge to smaller discharge is

(a) 14.0 (b) 3.0
(c) 2.0 (d) 1.0 [ESE : 2014]

14.41 In a Francis turbine, the runner blades are radial at the inlet and the discharge leaves the runner radially at the exit. For this turbine

(a) The relative velocity is radial at the outlet
(b) The absolute velocity is radial at the outlet
(c) The guide vane angle is 90°
(d) The velocity of flow is constant

[ESE : 2014]

14.42 **Statement (I):** The effective head of a turbo-machine is equal to: Gross head minus the head loss in penstock and the velocity head at the turbine inlet.

Statement (II): A turbo-machine becomes susceptible to cavitation if pressure falls below the vapour pressure of the liquid.

(a) Both Statement (I) and Statement (II) are individually true and Statement (II) is the correct explanation of Statement (I).
(b) Both Statement (I) and Statement (II) are individually true but Statement (II) is not the correct explanation of Statement (I).
(c) Statement (I) is true but Statement (II) is false.
(d) Statement (I) is false but Statement (II) is true.

[ESE : 2014]

14.43 A hydraulic turbine has an output of 6000 kW when it works under a head of 25 m runs at 100 rpm. Then the type of turbine used is

(a) Pelton wheel (b) Francis
(c) Kaplan (d) Propeller

[ESE : 2015]

14.44 The velocity heads of water at the inlet and outlet sections of a draft tube are 3.0 m and 0.20 m, respectively. The frictional and other losses in the draft tube are 0.4 m. What is the efficiency of the draft tube?

(a) 15% (b) 67%
(c) 80% (d) 86% [ESE : 2015]

14.45 Assertion (A): Possibility of cavitation is an important consideration in the selection of a turbine for a given head and a range of corresponding specific speed.

Reason (R): High-speed turbines are used for high heads.

(a) Both A and R are true and R is the correct explanation of A
(b) Both A and R are true but R is not a correct explanation of A
(c) A is true but R is false
(d) A is false but R is true **[ESE : 2015]**

14.46 At the point of operation with maximum efficiency, a turbine indicated unit power of 12 units and unit speed of 98 units and operates with 3300 kgf/s of flow. What are the speed in rpm and the specific speed of the machine respectively when its design head is 8.5 m?

(a) 285 rpm and 339
(b) 270 rpm and 360
(c) 285 rpm and 360
(d) 270 rpm and 339 **[ESE : 2016]**

14.47 A hydraulic turbine develops 8000 kW when running at 300 rpm under a head of 45 m. The speed of the same turbine under a head of 60 m is

(a) 224.4 rpm (b) 346.4 rpm
(c) 424.8 rpm (d) 485.8 rpm **[ESE : 2016]**

14.48 The important parameters describing the performance of a hydraulic machine are: P the power input, H the head produced across the machine and the efficiency, η. For a given geometrical design of the machine, the performance is characterized by the variables: H – the head increase across the machine, ρ – the fluid density, ω – the angular velocity of the rotor, D – the diameter of the rotor, μ – the fluid viscosity and Q – the flow rate; and both P and η are expressed through these variables. How many non-dimensional parameters are involved herein? Gravitational acceleration g has also to be considered necessarily

(a) 7 (b) 6
(c) 5 (d) 4 **[ESE : 2016]**

14.49 A Pelton wheel works under a head of 400 m. Frictional loss through the pipe flow is limited to 10%. The coefficient of velocity for the jet is 0.98. What is the velocity of the jet? Take, g = 10 m/s^2.

(a) 83 m/s (b) 71 m/s
(c) 65 m/s (d) 56 m/s **[ESE : 2016]**

14.50 Consider the following statements :

1. In an Impulse turbine, the pressure of the flowing water remains unchanged and is equal to atmospheric pressure.
2. In Impulse turbines, the water impinges on the buckets with 'pressure energy'.
3. In a Reaction turbine, the pressure of the flowing water remains unchanged and is equal to atmospheric pressure.

Which of the above statements is/are correct?

(a) 1 only (b) 2 only
(c) 3 only (d) 1,2 and 3 **[ESE : 2016]**

14.51 A Pelton wheel with single jet rotates at 600 rpm. The velocity of the jet from the nozzle is 100 m/s. If the ratio of the bucket velocity to jet velocity is 0.44 and the speed ratio is 0.43, what is the coefficient of velocity of the nozzle?

(a) 0.817 (b) 0.882
(c) 0.913 (d) 0.977 **[ESE : 2016]**

14.52 Statement (I): The speed of a hydraulic turbine has to be maintained constant irrespective of the load on the machine for keeping the electrical power generation frequency constant.

Statement (II): Governing of hydraulic turbines can be done by controlling the discharge through the turbines by adjusting the spear valve in Pelton turbines and the wicket in Francis or Kaplan turbines.

(a) Both Statement (I) and Statement (II) are individually true and Statement (II) is the correct explanation of Statement (I)
(b) Both Statement (I) and Statement (II) are individually true but Statement (II) is ***not*** the correct explanation of Statement (I)
(c) Statement (I) is true but Statement (II) is false
(d) Statement (I) is false but Statement (II) is true **[ESE : 2016]**

14.53 Consider the following statements regarding a turbine:

1. Specific speed plays an important role in the selection of the type of turbine.
2. An increase in specific speed of the turbine is accompanied by higher maximum efficiency.
3. The runner of too high specific speed with high available head increases the cost of the turbine on account of the high mechanical strength required.

Which of the above statements are correct?

(a) 1, 2 and 3 (b) 1 and 2
(c) 1 and 3 (d) 2 and 3

[ESE : 2017]

14.54 In a hydraulic machine, the moment of momentum of water is reduced by 15915 N.m, when the machine is rotating to 600 rpm. The power developed is

(a) 1000 kW (b) 1500 kW
(c) 2000 kW (d) 2500 kW

[ESE : 2017]

14.55 Consider the following statements in connection with hydraulic turbines:

1. The Kaplan turbine is a radial flow turbine in which the guide vane angles as well as the runner vane angles are adjustable.
2. Francis and Kaplan turbines are provided with draft tubes which carry water from exit of runner to tailwater to increases the gross head across the turbine.
3. The parts of turbines susceptible to cavitation are the guide vanes and runner vanes - drastically reducing the turbine efficiency.
4. The specific speed of a 4-jet Pelton turbine will be 28 if the specific speed of a single -jet Pelton turbine is 14.

(a) 2 and 3 (b) 1 and 3
(c) 1 and 4 (d) 2 and 4

[ESE : 2017]

14.56 Consider the following statements:

1. Surge tanks are not substitutes for forebays.
2. Pumped storage power plants are a boon to power generation.
3. Water hammer in penstocks is not dangerous.
4. Kaplan turbines are used in low head power plants.

Which of the above statements are correct?

(a) 2 and 4 (b) 1 and 4
(c) 2 and 3 (d) 1 and 3

[ESE : 2017]

14.57 **Statement (I):** In the computations of velocity triangles, or the torque of well-encased roto-dynamic flow machines, whether of outward, or inward, flow type, the tangential component of velocity at the smaller radius flow section can be almost neglected.

Statement (II): This follows from the very concept of 'outward', or 'inward', flow—to a large extent—in visualizing the working features of the machine.

(a) Both Statement (I) and Statement (II) are individually true and Statement (II) is the correct explanation of Statement (I)
(b) Both Statement (I) and Statement (II) are individually true but Statement (II) is ***not*** the correct explanation of Statement (I)
(c) Statement (I) is true but Statement (II) is false
(d) Statement (I) is false but Statement (II) is true

[ESE : 2017]

14.58 If the energy present in a jet of water of 20 cm diameter and having a velocity of 25 m/s could be extracted by a device with 90% efficiency, the power extracted would be nearly

$$\left(\text{taking } \frac{1}{2g} = 0.051\frac{\text{sec}^2}{\text{m}}\right)$$

(a) 180 kW (b) 225 kW
(c) 260 kW (d) 300 kW

[ESE : 2018]

14.59 The work done by a kN of water jet moving with a velocity of 60 m/sec when it impinges on a series of vanes moving in the same direction with a velocity of 9 m/sec is

(a) 60.2 kNm (b) 55.6 kNm
(c) 46.8 kNm (d) 45.0 kNm

[ESE : 2018]

14.60 The velocity heads of water at the inlet and outlet sections of a draft tube are 3.5 m and

0.3 m, respectively. The frictional and other losses in the draft tube can be taken as 0.5 m. What is the efficiency of the draft tube?

(a) 84.4% (b) 80.0%
(c) 77.1% (d) 74.4%

[ESE : 2018]

14.61 Consider the following regarding turbines:

1. The main function of a governor is to maintain a constant speed even as the load on the turbine fluctuates.
2. In the case of Pelton turbines, the governor closes or opens the wicket gates.
3. In the case of Francis turbines, the governor opens or closes the needle valve.
4. In the case of a Kaplan turbine, the governor swings the runner blades appropriately in addition to further closing or further opening of the wicket gates.

Which of the above statements are CORRECT?

(a) 1 and 3 only (b) 2 and 4 only
(c) 2 and 3 only (d) 1 and 4 only

[ESE : 2018]

14.62 A hydraulic turbine develops 5000 kW under a head of 30 m when running at 100 rpm. This turbine belongs to the category of

(a) Pelton wheel (b) Francis turbine
(c) Kaplan turbine (d) Propeller turbine

[ESE : 2018]

14.63 Consider the following data from a test on Pelton wheel?

Head at the base of the nozzle = 32 m
Discharge of the nozzle = 0.18 m^3/s
Area of the jet = 7500 mm^2
Power available at the shaft = 44 kW
Mechanical efficiency = 94%

The power lost in the nozzle will be nearly

(a) 3.9 kW (b) 4.7 kW
(c) 3.5 kW (d) 2.3 kW

[ESE : 2019]

14.64 Which of the following are component parts for an oil pressure governor in modern turbines?

1. Servomotor, known as relay cylinder
2. Oil sump
3. Oil pump which is driven by belt connected to turbine main shaft
4. Draft tube

(a) 1, 2 and 3 only (b) 1, 2 and 4 only
(c) 1, 3 and 4 only (d) 2, 3 and 4 only

[ESE : 2019]

14.65 Hydraulic efficiency of Francis turbine is

(a) Directly proportional to velocity of whirl at inlet and inversely proportional to net head on turbine.
(b) Directly proportional to velocity of whirl at inlet and net head on turbine.
(c) Inversely proportional to velocity at inlet and net head on turbine.
(d) Inversely proportional to velocity of whirl at inlet and directly proportional to net head on turbine.

[ESE : 2020]

14.66 A turbine develops 7225 kW power under a head of 25 m at 135 rpm. The specific speed of the turbine will be nearly

(a) 245 rpm (b) 225 rpm
(c) 205 rpm (d) 185 rpm

[ESE : 2020]

14.67 A 7.5 cm diameter jet of water strikes a curved plate at its centre with a velocity of 20 m/s. The curved plate is moving with a velocity of 8 m/s in the direction of the jet. The jet is deflected through an angle of 165°. By assuming the plate as smooth, what is the angle made by the relative velocity at the outlet of the plate?

(a) 45° (b) 30°
(c) 15° (d) 0°

[ESE : 2021]

14.68 A reservoir has a head of 40 m and a channel leading from the reservoir permits a flow rate of 34 m^3/s. If the rotational speed of the rotor is 150 rpm, what is the power of the turbine? (Take g = 9.81 m/s^2)

(a) 14.34 MW (b) 13.34 MW
(c) 12.34 MW (d) 11.34 MW

[ESE : 2021]

14.69 Full load is supplied by the turbine shaft when the diameter of jet issuing from the nozzle is 150 mm. If the load suddenly drops to 36% of the full load, what diameter of the jet should be attained by regulating the spear rod?

(a) 15 mm (b) 45 mm
(c) 90 mm (d) 180 mm

[ESE : 2021]

14.70 Which one of the following statements is correct with respect to Kaplan Turbine?

(a) The peripheral velocity at inlet is more than peripheral velocity at outlet.
(b) Velocity of flow at inlet is more than velocity of flow at outlet.
(c) The peripheral velocity at inlet and outlet are equal.
(d) Velocity of flow at outlet is more than velocity of flow at inlet.

[ESE : 2022]

14.71 The speed of the generator can be maintained constant only if the speed of the turbine runner is constant equal to the one given by equation $N = \frac{60f}{p}$ and it is known as

(a) Synchronous speed
(b) Asynchronous sped
(c) Derived speed
(d) Measured variable speed

[ESE : 2022]

14.72 Match the following lists:

List I (Specific speed (rpm))
P. 8.5 to 30
Q. 50 to 340
R. 255 to 860

List II (Type of turbine)
1. Francis turbine
2. Kaplan turbine
3. Pelton wheel turbine with single jet

Select the correct answer using the code given below :

	P	Q	R
(a)	2	3	1
(b)	3	1	2
(c)	2	1	3
(d)	1	2	3

[ESE : 2024]

14.73 Consider the following statements regarding Kaplan turbine:

1. The runner blades of Kaplan turbine runner are warped.
2. The blade angle is being greater at the hub than at the outer tip.
3. The peripheral velocity of the blades is being directly proportional to radius.

Which of the above statements is/are correct?

(a) 1 and 3 only (b) 2 and 3 only
(c) 3 only (d) 1,2 and 3

[ESE : 2024]

14.74 Consider the following statements regarding draft tube of Francis turbine :

1. It permits a suction head to be established at the runner exit.
2. It makes possible to install the turbine above the tail race level without loss of head.
3. It converts a large proportion of velocity energy rejected from the runner into useful pressure energy.

Which of the above statements is/are correct?

(a) 1 and 2 only (b) 2 and 3 only
(c) 3 only (d) 1, 2 and 3

[ESE : 2024]

■■■■

Answers Impact of Jets and Turbines

14.1 (d)	14.2 (b)	14.3 (d)	14.4 (a)	14.5 (a)	14.6 (d)	14.7 (a)	14.8 (a)
14.9 (c)	14.10 (b)	14.11 (a)	14.12 (b)	14.13 (*)	14.14 (*)	14.15 (d)	14.16 (a)
14.17 (b)	14.18 (a)	14.19 (d)	14.20 (b)	14.21 (a)	14.22 (c)	14.23 (c)	14.24 (d)
14.25 (d)	14.16 (b)	14.27 (c)	14.28 (a)	14.29 (c)	14.30 (a)	14.31 (b)	14.32 (b)
14.33 (b)	14.34 (b)	14.35 (b)	14.36 (a)	14.37 (d)	14.38 (c)	14.39 (c)	14.40 (b)
14.41 (b)	14.42 (d)	14.43 (b)	14.44 (c)	14.45 (c)	14.46 (a)	14.47 (b)	14.48 (b)
14.49 (a)	14.50 (a)	14.51 (d)	14.52 (b)	14.53 (c)	14.54 (a)	14.55 (d)	14.56 (a)
14.57 (*)	14.58 (b)	14.59 (c)	14.60 (c)	14.61 (d)	14.62 (b)	14.63 (b)	14.64 (a)
14.65 (a)	14.66 (c)	14.67 (c)	14.68 (b)	14.69 (c)	14.70 (c)	14.71 (a)	14.72 (b)
14.73 (a)	14.74 (d)						

Explanations Impact of Jets and Turbines

14.1 (d)

Thoma's cavitation parameter,

$$\sigma = \frac{H_a - H_v - H_s}{H}$$

$\therefore \quad H_s = (9-1) - 0.15 \times 40 = 2 \text{ m}$

Thus turbine can be set 2 m above tail race level.

14.2 (b)

Unit speed and unit discharge are are used to predict the performance of turbine and specific speed uniquely determines the design and shape of turbine.

14.3 (d)

Jet velocity = V

Bucket velocity = v

Relative velocity = $V - v$

Ideal efficiency is given by

$$\eta_{ideal} = \frac{2v(V-v)(1+\cos\phi)}{V^2}$$

$$= \frac{2}{V^2}\left[Vv - v(V-v)\cos\phi - v^2\right]$$

Loss of energy due to friction on bucket reduces relative velocity between jet and bucket by a factor k_1.

$$\therefore \quad \eta_1 = \frac{2}{V^2}\left[Vv + v(V-v)k_1\cos\phi - v^2\right]$$

Windage losses are expressed as $k_2\dfrac{v^2}{2g}$. So,

$$\eta_2 = \frac{2}{V^2}[k_2Vv + k_1k_2v(V-v)\cos\phi - k_2v^2]$$

14.4 (a)

Draft tube must be air-tight and under all-conditions of its operation its lower end must be submerged below the level of water in the tail race. It has two purposes:

(i) Permits a negative or suction head to be established at the runner exit.

(ii) Converts a large portion of velocity energy rejected from the runner into useful pressure energy.

14.5 (a)

The machine for which the change in static head in the runner is zero is known as impulse turbine. In these machines, the energy transfer in the runner takes place only by the change in the dynamic head of the fluid. All the available energy is converted into kinetic energy or velocity head by passing water through a contracting nozzle provided at the end of the penstock. Throughout its action water is at atmospheric pressure. The jet of water strikes at bucket attached on the circumference of the wheel.

In the reaction turbine, at the entrance to the runner, only a part of the available energy of water is converted into kinetic energy and a substantial part remains in the form of pressure energy. Thus pressure and velocity both change as the water flows through the turbine. For this gradual change of pressure to be possible the runner should be completely enclosed in an air tight casing and the passage is entirely full of water throughout the operation of the turbine.

14.6 (d)

Maximum Efficiency, $\eta_{max} = \left(\frac{1+\cos\theta}{2}\right)$

For $\theta = 0°$, the curved vanes will become semicircular and $\eta_{max} = 1$ or 100%.

14.7 (a)

For straight divergent type draft tube the central cone angle should not be more than 6° because if the angle is more than 6°, the water flowing through the draft tube will not remain in contact with the inner surface, with the result that eddies are formed and the efficiency of the draft tube is reduced.

14.8 (a)

Specific speed is independent of size and dimensions of both actual and specific turbine so comparing and selecting turbines on basis of specific speed criterion is more scientific.

14.9 (c)

For Pelton turbine, $\alpha = 0°$ and $\theta = 0°$.

θ is inlet blade angle.

14.10 (b)

Pondage usually refers to the comparably small water storage body. Such regulating body has considerably less storage than the reservoirs of large dams and conventional hydroelectric stations which can store water for long periods such as a dry season or year. Ample pondage allows a power plant meet hourly load fluctuations for a period of a week or more.

14.11 (a)

$$N_s = \frac{N\sqrt{P}}{H^{5/4}}$$

$$N_s = \frac{300\sqrt{2000}}{(150)^{5/4}}$$

$$= 25.55 \text{ (in MKS)}$$

14.12 (b)

$$\text{Speed ratio} = \frac{u}{\sqrt{2gH}}$$

$$\therefore \quad u = 0.48\times\sqrt{2\times 9.81\times 256}$$

$$= 0.48 \times 4.43 \times 16$$

$$= 34.02 \text{ m/s}$$

$$\text{But} \quad u = \left(\frac{\pi DN}{60}\right)$$

$$\therefore \quad D = \frac{60\times 34.02}{\pi\times 630} = 1.03\text{ m}$$

14.13 (*)

The power $P \propto H^{3/2}$

$$\therefore \quad P_2 = P_1\left(\frac{H_2}{H_1}\right)^{3/2}$$

$$= 10{,}000\times\left(\frac{400}{500}\right)^{3/2}$$

$$= 7155.4 \text{ kW}$$

The power developed will be reduced when the head is reduced.

However the answer is given by

$$P_2 = P_1\left(\frac{H_1}{H_2}\right)^{3/2} = 13975.4\text{ kW}$$

So one can choose 14000 kW as the answer although it is incorrect.

14.14 (*)

Speed, $N \propto \sqrt{H}$

$$\therefore \quad N_2 = N_1\sqrt{\left(\frac{H_2}{H_1}\right)}$$

$$= 300\times\sqrt{\left(\frac{400}{500}\right)}$$

$$= 268.3 \text{ rpm}$$

However the answer is given by

$$N_2 = N_1\sqrt{(H_1/H_2)} = 335.4 \simeq 336 \text{ rpm}$$

So one can choose 336 rpm as the answer although it is incorrect.

14.15 (d)

The sequence of units of hydel system are power channel – forebay – intake – short penstocks – power house (turbine) – tail race.

Thus forebay is between power channel and penstock.

14.16 (a)

Draft tube convert kinetic energy at the exit of runner into useful pressure thereby increasing the efficiency.

14.17 (b)

In impulse turbine (Pelton wheel and Banki) draft tube is not used.

Pelton wheel is suitable for high head and low discharge. Kaplan turbines operates under low head and high discharge.

14.18 (a)

Turbine type	Specific Speed	Head
Pelton	10-60	> 250 m
Francis	60-300	370 m
Kaplan	300-1000	40-50 m

14.19 (d)

Specific speed,

$$N_s = \frac{N\sqrt{P}}{H^{5/4}} = \frac{540\sqrt{8100}}{81^{5/4}} = 200$$

Francis turbine is suitable.

14.20 (b)

The Rankine's efficiency may be given as

$$\eta_r = \frac{qH}{Qh}$$

where q is rate of flow of water actually lifted

H is level of water in delivery tank w.r.t. supply tank

Q is rate of flow of water flowing past the waste valve

h is level of water in supply tank

$$\therefore \quad \eta_r = \frac{2\times(28-2.8)}{(35-2)\times 2.8} = 54.54\% \approx 55\%$$

14.21 (a)

Tidal plants: These work on the principle that there is a rise in sea water during high tide period and fall during low tide period. Thus the water rises and falls twice a day. Hence head on turbine varies with time cyclically.

Pumped Storage Plant: This generates power during peak hours, but during the off-peak hours, water is pumped back from the tail water pool to head water pool. So these plants use reversible turbines which function both as turbine as well as pump.

Storage Plant: Utilizes annually average flow with storage.

Runoff River Plant: Utilizes current flow in stream and has pondage to improve load factor.

14.22 (c)

The specific speed of a pump is defined as the speed in revolution per minute of a geometrically similar pump of such a size that under corresponding conditions it would deliver 1 litre of liquid per second against a head of 1m. It is represented by N_S and expressed as

$$N_S = \frac{N\sqrt{Q}}{H^{3/4}}$$

Where N is rotational speed in rpm for maximum efficiency

Q is discharge in m^3/s

H is net head developed in m.

It may be noted that the specific speed N_S has the dimensions of $\left[L^{3/4}\,T^{-3/2}\right]$.

The non-dimensional form of specific speed is given by

$$N_S = \frac{N\sqrt{Q}}{(gH)^{3/4}}$$

This parameter n_S is known as dimensionless specific speed or shape number of the pump.

14.23 (c)

Spiral casing: The spiral casing around the runner of the turbine is known as the volute casing or scroll case. Throughout its length, it has numerous openings at regular intervals to allow the working fluid to impinge on the blades of the runner. These openings convert the pressure energy of the fluid into momentum energy just before the fluid impinges on the blades. This maintains a constant flow rate despite the fact that numerous openings have been provided for the fluid to enter the blades, as the cross-sectional area of this casing decreases uniformly along the circumference.

Stay ring or Speed ring: It directs water from spiral casing to guine wane or wicket gates and resists the load upon it by internal pressure of water

Guide or stay vanes: The primary function of the guide or stay vanes is to convert the pressure energy of the fluid into the momentum energy. It also serves to direct the flow at design angles to the runner blades.

Runner: Runner blades are the heart of any turbine. These are the centers where the fluid strikes and the tangential force of the impact causes the shaft of the turbine to rotate, producing torque.

14.24(d)

Reaction turbine efficiency is maximum when whirling velocity at the outlet is zero and absolute velocity vector at the outlet is 90°.

14.25(d)

Francis turbine efficiency is maximum when whirling velocity at the outlet is zero and discharge is radial at exit. This is done by designing runner such that the absolute velocity vector at the outlet is inclined at 90° to direction of vane.

14.26 (b)

NPSH (Net positive suction head)

= Caviation coefficient × Manometric head

= 0.1 × 40 = 4 m

Now, NPSH = $\frac{p_{atm} - p_v}{\rho g} - h_s - h_{f_s}$

Safe height of runner

$= h_s + h_{f_s} = \frac{p_{atm} - p_v}{\rho g} - \text{NPSH}$

= 9.5 – 1.5 – 4 = 4 m

14.27 (c)

A draft tube is a pipe or passage of gradually increasing cross-sectional area which connects the runner exit to the tail race. In a pelton turbine a casing made of cast iron or fabricated steel plates is usually provided to lead water to the tail race level.

14.28 (a)

Surge Tanks: A surge tank is a cylindrical open-topped storage tank which is connected to the penstock at a suitable point. Surge tanks are provided in the case of hydroelectric developments having long penstocks. These are provided to relieve the penstocks of excess pressure caused by water hammer and to provide additional supply of water when the turbines are in need of more water on account of increased load. A surge tank however, provides protection against water hammer pressure for only that portion of the penstock which lies on the upstream of it. As such surge tank should be provided as close to the power house as is possible according to the site conditions.

14.30 (a)

After the jet impinges normally on a moving plate the direction of velocity changes. Hence the direction of momentum changes.

14.31 (b)

The specific speed for turbines is given by

$$N_S = \frac{N\sqrt{P}}{H^{5/4}}$$

The specific speed for pumps is given by

$$N_S = \frac{N\sqrt{Q}}{H^{3/4}}$$

14.32 (b)

$$T = \dot{m}V_{w_1}r_1 + \dot{m}V_{w_2}r_2$$

As given at exit water does not have moment mean $V_{w_2} = 0$.

Therefore, $T = \dot{m}V_{w_1}r_1$

$$1600 \times 9.81 = (1000 \times 10)\ V_{w_1} \times 1$$

$$V_{w_1} = 1.57 \text{ m/s}$$

14.33 (b)

$$N_{S_A} = N_{S_B}$$

$$\frac{N_1\sqrt{P_1}}{H^{5/4}} = \frac{N_2\sqrt{P_2}}{H^{5/4}}$$

$\Rightarrow$ $1000 \times \sqrt{400} = N_2 \times \sqrt{100}$

$\therefore$ $N_2 = 2000$ rpm

14.34 (b)

The type of turbine for different specific sped is given in table as

Specific speed	Type of turbine
8.5 to 30	Pelton wheel with single jet
30 to 51	Pelton wheel with two or more jet
51 to 255	Francis turbine
255 to 860	Kaplan or Propeller turbine

14.35 (b)

$$N_S = \frac{N\sqrt{P}}{H^{5/4}} = \frac{145\sqrt{7000}}{(25)^{5/4}} = 217$$

Low head ⇒ Francis turbine

14.36 (a)

$$\text{Efficiency of reaction turbine } \eta\% = \frac{V_{w_1u_1} - V_{w_2u_2}}{gH}$$

V_{w_1} = velocity of whirl at inlet

V_{w_2} = velocity of whirl at outlet

To maximize efficiency $V_{w_2u_2} = 0$

Since $u_2 \neq 0$, V_{w_2} has to be zero.

i.e. to maximize efficiency swirl at outlet must be zero.

14.37 (d)

$$P = \frac{2\pi NT}{60}$$

$$\Rightarrow \quad 2516 \times 10^3 = 2\pi(240)\frac{T}{60}$$

$$\Rightarrow \quad T = 100 \text{ kNm}$$

14.38 (c)

Velocity of flow is constant at every point.

$$V_f = \frac{Q}{\frac{\pi}{4}\left(D_0^2 - D_b^2\right)}$$

D_0 = runner/outer diameter

D_b = hub diameter

∴ at tip, velocity of flow = 6 m/s

Also, $V_{w1}u_1 = V_{w2}u_2$

At hub, $V_{w1} = 10$ m/s, $u_1 = \frac{\pi \times 1.6 \times N}{60}$

At tip, $V_{w2} = ?$, $u_2 = \frac{\pi \times 4 \times N}{60}$

$$\therefore \quad V_{w2} = \frac{10 \times 1.6}{4} = 4 \text{ m/s}$$

14.39 (c)

$\rho = 800$ kg/m³

$F = \rho AV(V_{w1})$

$A = 0.02$ m² $= 800(0.02)10(10)$

$V = 10$ m/s $= 1600$ N

14.40 (b)

On splitting of jet into two streams, the larger discharge would be

$$Q_1 = \frac{Q}{2}(1+\cos\theta)$$

and, smaller discharge would be

$$Q_2 = \frac{Q}{2}(1-\cos\theta)$$

$$\text{So, } \frac{\text{Larger discharge}}{\text{Smaller discharge}} = \frac{\frac{Q}{2}(1+\cos\theta)}{\frac{Q}{2}(1-\cos\theta)}$$

$$= \frac{1+\cos\theta}{1-\cos\theta}$$

$$= \frac{1+\cos(90^\circ+30^\circ)}{1-\cos(90^\circ-30^\circ)} = \frac{1+\frac{1}{2}}{1-\frac{1}{2}} = \frac{3}{1} = 3$$

14.42 (d)

A turbo machine is a device in which energy transfer occurs between a flowing fluid and rotating element due to dynamic action. This results in change of pressure and momentum of the fluid. Effective head: It is the head available at the inlet of the turbine. It is obtained by considering all head losses in penstock. If h is the total loss, then the effective head above the turbine is $H = Hg - hf$.

14.43 (b)

Specific speed,

$$N_s = \frac{N\sqrt{P}}{H^{5/4}} = \frac{100\sqrt{6000}}{25^{5/4}} = 138.56$$

Hence, it is Francis turbine because its specific speed varies from 60-300.

14.44 (c)

Efficiency of draft tube,

$$\eta_d = \frac{\frac{V_i^2}{2g} - \frac{V_e^2}{2g} - h_f}{\frac{V_i^2}{2g}}$$

$$= \frac{3-0.2-0.4}{3} = 0.8 \text{ i.e. } 80\%$$

14.45 (c)

High speed turbine like Kaplan turbine are used for low or medium head, while low speed turbine, like Pelton turbine are used for high head.

14.46 (a)

$$\frac{P}{H^{3/2}} = 12$$
$$P = 12 \times (8.5)^{3/2}$$
$$\frac{N}{\sqrt{H}} = 98$$
$$N = 98\sqrt{8.5} = 285$$
$$N_s = \frac{N\sqrt{P}}{H^{5/4}} = \frac{285\sqrt{12\times(8.5)^{3/2}}}{(8.5)^{5/4}}$$
$$= 339$$

14.47 (b)

$$\frac{N}{\sqrt{H}} = \text{constant}$$
$$\frac{300}{\sqrt{45}} = \frac{N}{\sqrt{60}}$$
$$N = 346.4 \text{ rpm}$$

14.49 (a)

Head at the nozzle = 0.9 × 400 = 360 m

Velocity of jet,

$$V = \sqrt{2gH} = 0.98\sqrt{2\times 9.81\times 360}$$
$$\simeq 83 \text{ m/s}$$

14.51 (d)

$$N = 600 \text{ rpm}$$
$$V_1 = 100 \text{ m/s}$$
$$\frac{u_1}{v_1} = 0.44$$
$$\Rightarrow \quad u = 44 \text{ m/s}$$
$$\frac{u}{\sqrt{2gH}} = 0.43$$
$$\Rightarrow \quad H = 533.6658 \text{ m}$$
$$V_1 = 100$$
$$c_v = \sqrt{2gH}$$
$$100 = c_v\left(\frac{44}{0.43}\right) = 0.977$$

14.54 (a)

Power $= T\omega$

$$= 15915\times\left(\frac{2\pi\times 600}{60}\right) = 1000 \text{ kW}$$

14.58 (b)

Power extracted

$$= 0.9\times\frac{1}{2}\dot{m}v^2 = 0.9\times\frac{1}{2}(\rho.a.v.).v^2$$
$$= 0.9\times\frac{1}{2}(10^3)\times\left[\frac{\pi}{4}(0.2)^2\right](25)^2$$
$$= 220.89\times 10^3 \text{ W} = 220.89 \text{ kW}$$

14.59 (c)

Work done, $W = \dot{m}(V-u)u$

$$\left(\therefore \dot{m} = \frac{\text{Weight of water}}{g}\right)$$
$$= \frac{1000}{9.81}(60-9)\times 9 \simeq 46.8 \text{ kN-m}$$

14.60 (c)

$$\eta_{DT} = \frac{\frac{V_2^2}{2g} - \frac{V_3^2}{2g} - h_f}{V_2^2/2g}$$
$$= \left[\frac{3.5-0.3-0.5}{3.5}\right]\times 100$$
$$\eta_{DT} = 77.14\%$$

14.62 (b)

$$N_S = \frac{N\sqrt{P}}{H^{5/4}} = \frac{100\sqrt{5000}}{30^{5/4}}$$
$$N_S = 100.71(\text{SI})$$

Francis turbine

14.63 (b)

Power at the base of nozzle $= \dot{m}.g.H$

$$= \rho.Q.\ g.\ H$$
$$= (10)^3\ (0.18)\ (9.81)\ (32) = 56.51 \text{ kW}$$
$$V_1 = \frac{Q}{a} = \frac{0.18}{7500\times 10^{-6}}$$

Power at the exit of nozzle $= \frac{1}{2}\dot{m}V_1^2$

$$= \frac{1}{2}\rho Q V_1^2 = \frac{1}{2}(10^3)(0.18)(24)^2$$
$$= 51.84 \text{ kW} = 24 \text{ m/s}$$

Power lost in the nozzle

$$= 56.51 - 51.84 = 4.67 \simeq 4.7 \text{ kW}$$

14.64 (a)

Draft tube is not a part of governor.

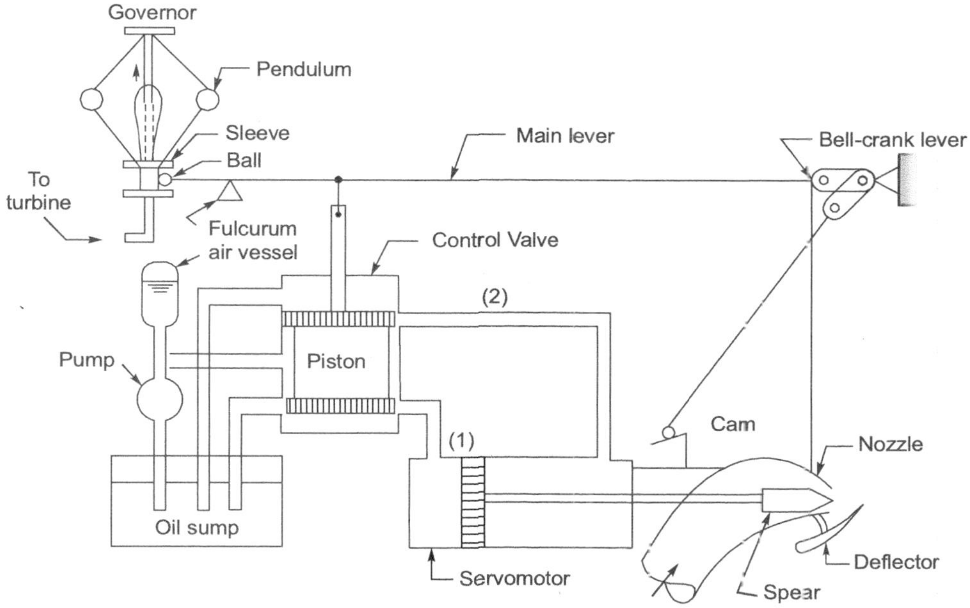

The components parts for an oil pressure governor in modern turbine are as following:

1. Servomotor also know as relay cylinder.
2. Relay valve also known as control valve or distribution valve.
3. Actuator or pendulum which is belt driven from the turbine main shaft.
4. Oil sump.
5. Oil pump which is driven by belt connected to turbine main shaft.
6. A system of oil supply pipes connecting the oil sump with the relay valve and the relay valve with the servomotor.

14.66 (c)

$$N_s = \frac{N\sqrt{P}}{H^{5/4}}$$

$$= \frac{135\sqrt{7225}}{(25)^{5/4}} = 205.27 \text{ (SI units)}$$

14.67 (c)

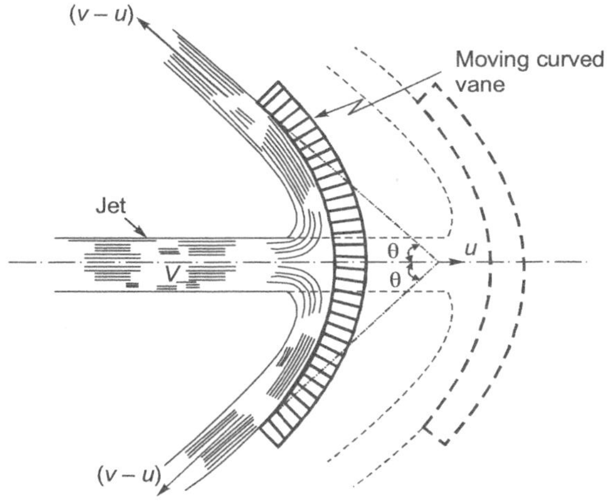

Fig. above shows the jet striking the symmetrical moving curved vane at the centre.

Let the curved vane of section be allowed to move with a velocity u in the direction of jet.

Given: Velocity of jet,

$$V = 20 \text{ m/s}$$
$$u = 8 \text{ m/s}$$

Relative velocity of jet with which it strikes the vane will be the velocity of jet relative to the plate i.e., $V - u$ which makes an angle, $\theta = 180° - 165° = 15°$ on each side as shown in figure. As plate is smooth, the relative velocity should be tangential to plate at outlet.

14.68 (b)

Given:

$$H = 40 \text{ m}$$
$$Q = 34 \text{ m}^3/\text{s}$$
$$N = 150 \text{ rpm}$$
$$g = 9.81 \text{ m/s}^2$$

Power of turbine

$$= \eta_0 (\rho g Q H) \quad (\eta_0 = 100\%)$$
$$= 1 \times (10^3)(34)(9.81)(40)$$
$$= 13.34 \times 10^6 \text{ W}$$
$$\simeq 13.34 \text{ MW}$$

14.69 (c)

Case-1: At full load,

Dia. of jet, $d_1 = 150$ mm

Discharge, $Q_1 = Q$

Case-2: When load is reduced by 36%.

i.e., $Q_2 = 0.36Q$

Dia. of jet $d_2 = ?$

Upon change in load, discharge is varied by varying the area of jet but velocity remain constant.

i.e., $Q = AV$

$Q \propto A$

$$\frac{Q_1}{Q_2} = \frac{\frac{\pi}{4} d_1^2}{\frac{\pi}{4} d_2^2}$$

$$\Rightarrow \quad \frac{Q}{0.36Q} = \frac{(150)^2}{d_2^2}$$

$$\therefore \quad d_2^2 = (150)^2 \times 0.36$$

$$d_2 = 150 \times 0.6 = 90 \text{ mm}$$

14.73 (a)

- Peripheral velocity, $u = \frac{\pi D N}{60}$

$\therefore \quad u \propto D$

- Blade angle is lesser at hub than at the outer tip.

$\therefore$ Statements 1 and 3 are correct.

14.74 (d)

All three statements are correct.

■■■■

15 Pumps

15.1 A pump running at 1414 rpm delivers 256 *l*ps of water against a head of 16 m. The pump is of the
(a) normal speed radial type
(b) double suction type
(c) mixed flow type
(d) axial flow type
[ESE : 2000]

15.2 Consider the following statements:
The head-discharge characteristics curve of two centrifugal pumps in
1. parallel can be obtained by doubling the ordinates.
2. series can be obtained by doubling the abscissa.
Which of these statements is/are correct?
(a) 2 alone (b) 1 alone
(c) Neither 1 nor 2 (d) Both 1 and 2
[ESE : 2000]

15.3 If the radius of the centrifugal pump impeller is reduced from 10 cm, the head developed by the pump will change from 10 m to
(a) 9 m of water (b) 8.1 m of water
(c) 9.487 m of water (d) 11.111 m of water
[ESE : 2000]

15.4 Reciprocating pumps are suitable for
(a) low discharge and high head
(b) high discharge and low head
(c) low discharge and low head
(d) high discharge and high head
[ESE : 2000]

15.5 Two geometrically similar pumps are running at the same speed of 100 rpm and lifting water against the heads of 25 m and 16 m respectively. First pump is having an impeller diameter of 300 mm. The impeller diameter of second pump shall be
(a) 192 mm (b) 240 mm
(c) 300 mm (d) 469 mm
[ESE : 2002]

15.6 Which one of the following statements is correct?
(a) Reciprocating pumps are less efficient than centrifugal pumps.
(b) Delivery from a reciprocating pump is pulsating.
(c) Reciprocating pumps are suitable for large discharges and smaller heads.
(d) For a negative slip to occur, a reciprocating pump must have a coefficient of discharge less than unity. [ESE : 2002]

15.7 **Assertion (A):** The shape of the impeller of rotodynamic pumps is such as to force the water in a radial/mixed/ or axial flow patterns.
Reason (R): A rotodynamic pump has a wheel or a rotating element which rotates the water in a casing and, thus, imparts energy to the water.
(a) both A and R are true and R is the correct explanation of A
(b) both A and R are true but R is not a correct explanation of A
(c) A is true but R is false
(d) A is false but R is true [ESE : 2002]

15.8 Consider the following statements relating to centrifugal pumps:
1. In centrifugal pumps discharge is proportional to speed of the impeller.
2. In centrifugal pumps discharge is proportional to the cube of the impeller diameter.
3. Specific speed is a criterion for selection of centrifugal pumps.
4. Cavitation in centrifugal pumps can be eliminated by avoiding sharp bends, lowering velocity in suction pipe and by lowering temperature.

Which of these statements are correct?
(a) 1, 2 and 3 (b) 1, 2 and 4
(c) 2, 3 and 4 (d) 1, 3 and 4
[ESE : 2003]

15.9 A centrifugal pump discharges 260 litres of water per second when running at 600 rpm. The impeller diameter at the outlet is 80 cm. It develops a head of 15.3 m. What is the approximate minimum starting speed?
(a) 425 rpm (b) 450 rpm
(c) 475 rpm (d) 500 rpm
[ESE : 2004]

15.10 A combination of centrifugal pumps of specific speed 20 and overall efficiency 80%, running at 800 rpm is to be used to pump 40 *l*ps of water to a height of 75 m. What should be the arrangement?
(a) 3 pumps in series
(b) 3 pumps in parallel
(c) 4 pumps in series
(d) 4 pumps in parallel [ESE : 2004]

15.11 Match **List-I** (Machines) with **List-II** (Associated with) and select the correct answer using the codes given below the lists:

List-I	List-II
A. Centrifugal pump	1. Per cent slip
B. Reciprocating pump	2. Bucket
C. Francis turbine	3. Guide blade
D. Pelton wheel	4. Volute chamber

Codes:

	A	B	C	D
(a)	4	3	1	2
(b)	4	1	3	2
(c)	2	3	1	4
(d)	2	1	3	4

[ESE : 2004]

15.12 The function of an air vessel in a reciprocating pump is to obtain
(a) reduction of suction head
(b) rise in delivery head
(c) continuous supply of water at uniform rate
(d) increase in supply of water
[ESE : 2005]

15.13 A high efficiency pump is required for low discharge, high head and low maintenance cost. Delivery of water need not be continuous. The pump need not run at high speed. Which one of the following is the correct choice?
(a) Centrifugal pump (b) Reciprocating pump
(c) Air lift pump (d) Hydraulic ram
[ESE : 2005]

15.14 Which one of the following inferences is not drawn by studying performance curves of centrifugal pumps?
(a) Discharge increases with speed
(b) Power decreases with speed
(c) Head increases with speed
(d) Manometric head decreases with discharge
[ESE : 2006]

15.15 The centrifugal pump should be installed above the water level in the sump such that
(a) its height is not more than 10.3 m at room temperature of liquid
(b) its height is not allowed to exceed 6.7 m
(c) the negative pressure does not reach as low as the vapour pressure
(d) the negative pressure is not allowed to develop in the impeller
[ESE : 2006]

15.16 Which one of the following statements is correct?
(a) Pumps operating in series boost the discharge whereas pumps operating in parallel boost the head.
(b) Pumps operating in parallel boost the discharge whereas pumps operating in series boost the head.
(c) In both the above cases there would be a boost in discharge only.
(d) In both the above cases there would be a boost in head only.
[ESE : 2006]

15.17 At a rated capacity of 44 cumec, a centrifugal pump develops 36 m of head when operating at 725 rpm. What is its specific speed?
(a) 327 (b) 255
(c) 350 (d) 45
[ESE : 2007]

15.18 A single-acting reciprocating pump has a plunger diameter 25 cm and stroke 35 cm. The speed of the pump is 60 rpm and it delivers 16.5 litres/second of water. What is the value of the theoretical discharge?

(a) 16.8 *l*ps (b) 17.2 *l*ps
(c) 18.0 *l*ps (d) 18.4 *l*ps

[ESE : 2007]

15.19 A fast centrifugal pump impeller has which of the following?

(a) Forward facing blades
(b) Radial blades
(c) Backward facing blades
(d) Propeller type blades [ESE : 2008]

15.20 Consider the following statements:

An air vessel is fitted on the suction side of a reciprocating pump to

1. achieve higher speed without separation
2. reduce work in overcoming frictional resistance
3. avoid excessive vibration
4. have uniform discharge

Which of these statements are correct?

(a) 1, 2 and 4 (b) 1 and 2 only
(c) 3 and 4 only (d) 2, 3 and 4

[ESE : 2008]

15.21 While selecting a centrifugal pump for your requirement of head and discharge on the basis of its performance characteristics, which one of the following criteria is to be adopted?

(a) Head, discharge and efficiency
(b) Head and discharge only
(c) Discharge only
(d) Head only [ESE : 2009]

15.22 Consider the following statements:

The function of the impeller in a centrifugal pump is to

1. convert the pressure energy into hydraulic energy
2. convert the hydraulic energy into mechanical energy
3. convert the mechanical energy into hydraulic energy
4. transform most of the kinetic energies to pressure energy

Which of these statements is/are correct ?

(a) 1 only (b) 3 only
(c) 1, 2 and 3 (d) 2, 3 and 4

[ESE : 2009]

15.23 A centrifugal pump gives maximum efficiency when its impeller blades are

(a) bent forward (b) bent backward
(c) straight (d) wave-shaped

[ESE : 2009]

15.24 By which one of the following, a small quantity of water may be lifted to a great height?

(a) Hydraulic ram
(b) Hydraulic crane
(c) Hydraulic lift
(d) Hydraulic coupling [ESE : 2009]

15.25 Consider the following devices as likely to concern with water hammer phenomenon in their design/operation:

1. Hydraulic ram
2. Hydraulic accumulator
3. Penstock
4. Surge tank
5. Draft tube

Which of the above devices is/are so concerned?

(a) 1, 2, 3, 4 and 5 (b) 1, 3 and 4 only
(c) 2 and 3 only (d) 3 only

[ESE : 2010]

15.26 Consider the following statements:

In reciprocating pumps, the air vessels are used

1. to get continuous supply of liquid at a uniform rate.
2. to save the power required to drive the pump.
3. to run at much higher speed without any danger of separation.

Which of these statements are correct?

(a) 1 and 2 only (b) 1 and 3 only
(c) 1, 2 and 3 (d) 2 and 3 only

[ESE : 2010]

15.27 A centrifugal pump will start delivering the liquid only when

(a) manometer head is greater than total head
(b) head developed tends to exceed the manometric head
(c) head developed in a centrifugal pump is due to pressure head only
(d) head developed is negligible

[ESE : 2010]

15.28 Consider the following statements:

1. Pumps in series operation can be the head to increase.

2. Pumps in series operation increase the flow rate.
3. Pumps in parallel operation increase the flow rate.
4. Pumps in parallel operation can be the head to increase.

Which of these statements are correct?

(a) 1 and 3 (b) 1 and 2
(c) 2 and 3 (d) 3 and 4 [ESE : 2010]

15.29 For a homologous model of a pump built to a scale ratio of 1 : 2, fluid and speed being the same in model and prototype, the ratio of model power to prototype power is

(a) 1/2.82 (b) 1/4
(c) 1/8 (d) 1/32 [ESE : 2010]

15.30 Conversion of dynamic velocity head into static pressure head in a centrifugal pump is the result of

(a) increasing area of flow between adjacent vanes from inlet to outlet
(b) difference in pressure between suction and delivery ends
(c) radial thrust in pumps
(d) stuffing box [ESE : 2011]

15.31 Match **List-I** with **List-II** and select the correct answer using the code given below the lists:

List-I

A. Shallow open well
B. Deep open well
C. Shallow tube
D. Deep tube well

List-II

1. Submersible pump
2. Wind power (mill)
3. Persian wheel
4. Centrifugal pump

Code:

	A	B	C	D
(a)	3	2	4	1
(b)	1	2	4	3
(c)	3	4	2	1
(d)	1	4	2	3

[ESE : 2011]

Directions: The following items consists of two statements; one labelled as **'Assertion (A)'** and the other as **'Reason (R)'**. You are to examine these two statements carefully and select the answers to these items using the codes given below:

Codes:

(a) both A and R are true and R is the correct explanation of A
(b) both A and R are true but R is not a correct explanation of A
(c) A is true but R is false
(d) A is false but R is true

15.32 Assertion (A): The efficiency of a reciprocating pump is 10-20 percent higher than that of a centrifugal pump for comparable discharge-head conditions.

Reason (R): The discharge from a reciprocating pump is dependent upon speed.

[ESE : 2011]

15.33 Assertion (A): In centrifugal pumps, flow takes place from low pressure zone to high pressure zone.

Reason (R): Possibility of separation occurring in pumps can be more and characterizing efficiency of pumps is less than that of turbines.

[ESE : 2011]

15.34 When a pump primes and works but not up to its capacity and pressure, the attributable reasons are as follows:

1. Speed may be too low.
2. Suction lift is too high.
3. Total static head is much higher than as designed.
4. Foot value has been removed.

Which of these reasons can be valid?

(a) 1, 2, 3 and 4 (b) 1, 2 and 4 only
(c) 2, 3 and 4 only (d) 1, 2 and 3 only

[ESE : 2012]

15.35 Determine the number of impellers required for a multistage pump to lift 3840 liters per minute against a total head of 80 m at a speed of 700 r.p.m., given that the limiting N_s for each impeller is 700 units.

(a) 6 (b) 5
(c) 4 (d) 3

[ESE : 2012]

15.36 A 20 cm centrifugal pump delivers 13.2 lps at a head of 45 m when running at a speed of 1350 r.p.m. A similarly designed pump of 15 cm

size runs at the same speed. What are the most likely values of discharge and delivery head serviced by this second pump?
(a) 5.57 lps and 25.3 m
(b) 7.20 lps and 25.3 m
(c) 8.27 lps and 27.8 m
(d) 8.27 lps and 29.4 m [ESE : 2012]

15.37 Consider the following statements:
Air vessels are fitted on the suction and delivery sides of a reciprocating pump to
1. Achieve higher speed without separation
2. Reduce work in overcoming frictional resistance
3. Avoid excessive vibration permanently.
4. Have nearly uniform discharge.

Which of these statements are correct?
(a) 1, 2 and 4 only (b) 1, 2 and 3 only
(c) 2, 3 and 4 only (d) 1, 2, 3 and 4
[ESE : 2012]

15.38 Given below are two lists. Which of these are properly matched?

Types of pump		Head-Discharge performance
1. Propeller pump	:	Large discharges and low heads with negligible percentage variation in head
2. Single stage centrifugal pump with backward curved blades	:	Medium heads, with decreasing head as discharge increases
3. Turbine pump	:	Medium to high heads with low, but constant, discharges.

(a) 1 and 2 (b) 2 and 3
(c) 2 only (d) 3 only [ESE : 2012]

15.39 Consider the following statements:
The total head against which a pump has to work must include, besides any other items,
1. the suction lift.
2. the delivery head
3. the head lost due to friction at entrance in the rising main
4. the head lost due to friction at exit in the rising main.

Which of these statements are correct?
(a) 1, 2 and 3 only (b) 2 and 3 only
(c) 1, 2, 3 and 4 (d) 3 and 4 only
[ESE : 2012]

15.40 **Assertion (A):** For a hydraulic ram, D'Aubuisson's efficiency is always more than Rankine's efficiency.
Reason (R): By definition, efficiency is always less than unity in any system of mechanics; and addition of a small value to both numerator and denominator in the ratio of such a case always improves the value.
(a) both A and R are true and R is the correct explanation of A
(b) both A and R are true but R is not a correct explanation of A
(c) A is true but R is false
(d) A is false but R is true
[ESE : 2012]

15.41 Consider the following statements:
To maximize the Net Positive Suction Head Available for a reciprocating pump, one must
1. Reduce the suction pipe friction loss.
2. Increase the ambient pressure at the supply reservoir.
3. Increase the pump speed.
4. Increase the static suction head.

Which of these statements are correct?
(a) 1 and 2 (b) 2 and 3
(c) 3 and 4 (d) 1 and 4
[ESE : 2013]

15.42 A hydraulic ram utilizes a supply under 4.0 m head, and delivers to an effective head of 20 m above the unit. The ratio of water wasted to the water delivered is 15. The efficiency of the pump, with neglecting frictional losses in the pipes, is:
(a) 17.5% (b) 20.0%
(c) 28.6% (d) 33.3%
[ESE : 2013]

15.43 A centrifugal pump delivers a manometric head of 12 m when pumping a liquid of relative density 0.8. If all other factors remain the same but the liquid has a relative density of 1.2, the new manometric head would be:
(a) 8 m (b) 10 m
(c) 12 m (d) 18 m [ESE : 2013]

15.44 A centrifugal pump lifts 100 l/s of a liquid against a net head of 150 kPa. The brake power required is 18.0 kW when the liquid is water with a specific weight of 9.8 kN/m^3. What would be the brake power if the liquid is a solvent having a relative density of 0.8 instead of water, with all other factors remaining the same as before?

(a) 14.4 kW (b) 18.0 kW
(c) 22.5 kW (d) 17.2 kW

[ESE : 2014]

15.45 A centrifugal pump has its impeller of 50 cm diameter at inlet, and it rotates at 1200 rpm. The tangential velocity of the impeller at the inlet is

(a) 16π m/s (b) 10π m/s
(c) 12π m/s (d) 100π m/s

[ESE : 2014]

15.46 The number of impellers required for a multistage pump to lift 4500 litres/minute against a total head 190 m at a speed of 750 rpm with specific speed not to exceed 700 is

(a) 6 (b) 8
(c) 10 (d) 12

[ESE : 2015]

Directions: The following items consists of two statements; one labelled as **'Assertion (A)'** and the other as **'Reason (R)'**. You are to examine these two statements carefully and select the answers to these items using the codes given below:

Codes:

(a) Both A and R are true and R is the correct explanation of A
(b) Both A and R are true but R is not a correct explanation of A
(c) A is true but R is false
(d) A is false but R is true

15.47 Assertion (A): In a reciprocating pump, the piston is considered to be moving with simple harmonic motion on assumption that the connecting rod is very large compared to be crank length.

Reason (R): There is acceleration at the beginning and retardation at the end of each stroke.

[ESE : 2015]

15.48 Assertion (A): By providing Air Vessels on the suction and delivery sides of a reciprocating pump, it is possible to increase the delivery head of the pump.

Reason (R): The Air Vessel terminates the acceleration head and contributes to the outgoing discharge becoming reasonably steady and uniform.

[ESE : 2015]

15.49 Assertion (A): The specific speed (N_s) of a centrifugal pump is defined as the speed (in rpm) at which it works most efficiently.

Reason (R): The specific speed is a characteristic of pumps that can be used as a basis for comparing the performance of centrifugal pumps.

[ESE : 2015]

15.50 Two identical centrifugal pumps are connected in parallel to a common delivery pipe of a system. The pump curve of each of the pumps is represented by $H = 20 - 60Q^2$ where H is manometric head of the pump and Q is the discharge of the pump. The head loss equation when two such fully-similar pumps jointly deliver the same discharge Q will be

(a) $H = 40 - 15Q^2$ (b) $H = 20 - 60Q^2$
(c) $H = 40 - 60Q^2$ (d) $H = 20 - 15Q^2$

[ESE : 2016]

15.51 In a single-acting reciprocating pump, the acceleration head at the beginning of the suction stroke is 3.5 m. If the pump is 1.5 m below the water level in the supply reservoir, the pressure head at the cylinder at that instant, considering the atmospheric pressure as 10.0 m is

(a) 2 m(abs) (b) 4 m(abs)
(c) 8 m(abs) (d) 16 m(abs)

[ESE : 2016]

15.52 Statement (II): Multistage centrifugal pumps are used to produce very high delivery heads.

Statement (II): Roto-dynamic pumps must have to be centrifugal rather than centripetal, from the very basic principles of hydrodynamics. Also, the stages are in series.

(a) Both Statement (I) and Statement (II) are individually true and Statement (II) is the correct explanation of Statement (I)
(b) Both Statement (I) and Statement (II) are individually true but Statement (II) is ***not*** the correct explanation of Statement (I)
(c) Statement (I) is true but Statement (II) is false
(d) Statement (I) is false but Statement (II) is true

[ESE : 2016]

15.53 A centrifugal pump has a impeller of 30 cm diameter and runs at 1000 rpm giving best efficiency. It delivered 1.2 m^3/minute against a head of 25 m. What is its non-dimensional specific speed (based on flow expressed in *l.p.s*)?

Take $\sqrt{9.81} = 3.132$; and $\sqrt{\sqrt{9.81}} = 1.177$.

(a) 55 (b) 63
(c) 72 (d) 80 **[ESE : 2017]**

15.54 The approximate value of BHP of a pump to supply 1 m^3/s water to a town at RL 400 m from a source, whose water level is at RL 320 m, would be

(a) 940 (b) 1000
(c) 1070 (d) 1120

[ESE : 2017]

15.55 A 20 m centrifugal pump runs at 1400 rpm delivering 0.09 m^3/sec against a head of 45 m with an efficiency of 87%. What is its non-dimensional specific speed using rps as the relevant data component?

(a) 0.482 (b) 0.474
(c) 0.466 (d) 0.458

[ESE : 2018]

15.56 Two identical centrifugal pumps are connected in parallel to a common delivery pipe of a system. The discharge performance curve of each of the pumps is represented by $H = 30 - 80Q^2$. The discharge head equation of the parallel duplex pump set is

(a) $H = 30 - 80Q^2$ (b) $H = 15 - 20Q^2$
(c) $H = 30 - 20Q^2$ (d) $H = 15 - 80Q^2$

[ESE : 2018]

15.57 Consider the following data relating to the performance of a centrifugal pump: speed = 1200 rpm, flow rate = 30 *l*/s, head = 20 m, and power = 5 kW. If the speed of the pump is increased to 1500 rpm, assuming the efficiency is unaltered, the new flow rate and head, respectively, will be

(a) 46.9 *l*/s and 25.0 m
(b) 37.5 *l*/s and 25.0 m
(c) 46.9 *l*/s and 31.3 m
(d) 37.5 *l*/s and 31.3 m

[ESE : 2018]

15.58 Water is to be pumped out of a deep well under a total head of 95 m. A number of identical pumps of design speed 1000 rpm and specific speed 900 rpm with a rated capacity of 150 *l*/s are available. The number of pumps required will be

(a) 1 (b) 3
(c) 5 (d) 7

[ESE : 2019]

15.59 A double-acting reciprocating pump having piston area 0.1 m^2 has a stroke 0.30 m long. The pump is discharging 2.4 m^2 of water per minute at 45 rpm through a height of 10 m. The slip of the pump and power required to drive the pump will be nearly

(a) 0.005 m^3/s and 4.8 kW
(b) 0.003 m^3/s and 4.8 kW
(c) 0.005 m^3/s and 4.4 kW
(d) 0.003 m^3/s and 4.4 kW

[ESE : 2019]

15.60 When speed changes in case of centrifugal pump, which of the following points are correct?

1. The shape of the velocity triangle will remain same.
2. Various angles will remain same.
3. Magnitude of velocities will change proportionately.

(a) 1 and 2 only (b) 1 and 3 only
(c) 2 and 3 only (d) 1, 2 and 3

[ESE : 2020]

15.61 Which one of the following statement is correct regarding centrifugal pumps?

(a) The discharge is fluctuating and pulsating

(b) It is used for large discharge through smaller heads
(c) The efficiency is low
(d) It runs at low

[ESE : 2020]

15.62 A centrifugal pump delivers water against a net head of 14.5 m and a design speed of 1000 rpm. The vanes are curved back to an angle of 30° with the periphery. The impeller diameter is 300 mm and the outlet width is 50 mm. What is the tangential velocity of impeller at outlet?

(a) 15.7 m/s (b) 13.2 m/s
(c) 9.7 m/s (d) 11.2 m/s

[ESE : 2021]

15.63 A double acting reciprocating pump having piston area 0.1 m^2 has a stroke length 0.30 m. The pump is discharging 2.4 m^3 of water per minute at 45 rpm through a height of 10 m. What is the slip of the pump?

(a) 0.005 m^3/s (b) 0.015 m^3/s
(c) 0.025 m^3/s (d) 0.035 m^3/s

[ESE : 2021]

15.64 Consider the following statements related to negative slip of the reciprocating pump:

1. The actual discharge of a reciprocating pump is more than the theoretical discharge.
2. The co-efficient of discharge will be more than unity.
3. When the suction pipe is short and delivery pipe is long and pump is running at slow speed, then negative slip of the pump occurs.

Which of the above statements are correct?

(a) 1 and 2 only (b) 1 and 3 only
(c) 2 and 3 only (d) 1, 2 and 3

[ESE : 2022]

15.65 A single acting reciprocating pump has a plunger of diameter 250 mm and stoke of 350 mm. If the speed of the pump is 60 rpm and it delivers 16.5 lit/sec of water against a suction head of 5 m and a delivery head of 20 m, what is the co-efficient of discharge?

(a) 0.72 (b) 0.79
(c) 0.86 (d) 0.96

[ESE : 2022]

15.66 The air vessel is not required for multicylinder pump because it has

(a) much smaller fluctuations of velocity in both the delivery and suction pipes
(b) large fluctuations of velocity in delivery pipes
(c) large fluctuations of the velocity in suction pipes
(d) moderate variation of speed of fluid reciprocating pump

[ESE : 2023]

15.67 Consider the following statements regarding positive-displacement pumps :

1. Positive-displacement pump is better able to handle shear sensitive liquids.
2. A well-sealed positive-displacement pump can create a significant vacuum pressure at its inlet, even when dry.
3. The rotor(s) of a positive displacement pump run at higher speeds than the rotor (impeller) of a dynamic pump at similar loads.

Which of the above statements are correct?

(a) 1 and 2 only (b) 1 and 3 only
(c) 2 and 3 only (d) 1, 2 and 3

[ESE : 2024]

15.68 Which one of the following heads is defined as "the head against which a centrifugal pump has to work"?

(a) Suction head (b) Delivery head
(c) Static head (d) Manometric head

[ESE : 2024]

■■■■

Answers Pumps

15.1 (c)	15.2 (c)	15.3 (b)	15.4 (a)	15.5 (b)	15.6 (b)	15.7 (b)	15.8 (b)
15.9 (c)	15.10 (c)	15.11 (b)	15.12 (c)	15.13 (b)	15.14 (b)	15.15 (c)	15.16 (b)
15.17 (a)	15.18 (b)	15.19 (c)	15.20 (a)	15.21 (a)	15.22 (b)	15.23 (b)	15.24 (a)
15.25 (b)	15.26 (c)	15.27 (b)	15.28 (a)	15.29 (d)	15.30 (a)	15.31 (a)	15.32 (b)
15.33 (b)	15.34 (a)	15.35 (b)	15.36 (a)	15.37 (a)	15.38 (a)	15.39 (c)	15.40 (a)
15.41 (a)	15.42 (d)	15.43 (a)	15.44 (a)	15.45 (b)	15.46 (c)	15.47 (b)	15.48 (a)
15.49 (d)	15.50 (d)	15.51 (c)	15.52 (a)	15.53 (c)	15.54 (c)	15.55 (d)	15.56 (c)
15.57 (d)	15.58 (b)	15.59 (c)	15.60 (d)	15.61 (b)	15.62 (a)	15.63 (a)	15.64 (a)
15.65 (d)	15.66 (a)	15.67 (a)	15.68 (d)				

Explanations Pumps

15.1 (c)

The specific speed, $N_s = \frac{N\sqrt{Q}}{H^{3/4}}$

$$N_s = \frac{1414 \times \sqrt{256}}{(16)^{0.75}} = 2828$$

Type of pump	Specific speed
Slow speed radial flow	300-900
Medium speed radial flow	900-1500
High speed radial flow	1500-2400
Mixed flow or screw type	2400-5000
Axial flow or propeller type	3400-15,000

15.2 (c)

For parallel connection discharge will get doubled, so abscissa should be doubled.

For series connection head will get doubled, so ordinate should be doubled.

15.3 (b)

$\frac{ND}{\sqrt{H}}$ will remain same even when diameter is reduced $H_2 = \left(\frac{D_2}{D_1}\right)^2 H_1 = (0.9)^2 \times 10 = 8.1$ m of water

15.4 (a)

Reciprocating pump gives constant discharge under variable heads. It is used for low discharge under variable high heads.

15.5 (b)

Equating head coefficients, we get

$$\frac{N_1D_1}{\sqrt{H_1}} = \frac{N_2D_2}{\sqrt{H_2}}$$

$$\therefore \quad D_2 = \left(\frac{1000}{1000}\right) \times \sqrt{\frac{16}{25}} \times 300 = 240 \text{ mm}$$

15.6 (b)

Reciprocating pump: Suitable for less discharge and high head. More efficient than centrifugal pump. Delivery from pump is pulsating.

$$\text{Slip} = (Q_{th} - Q_a)$$

$$Q_{th} < Q_a \quad \text{or} \quad C_d > 1$$

If coefficient of discharge is more than one, negative slip would occur.

15.7 (b)

The pumps based on the principle of fluid dynamics (angular momentum change) are called rotodynamic pump. The rotar consist of number of blades and there is relative motion between the rotar vanes and fluid. These are classified as:

(i) Centrifugal pumps with radial flow

(ii) Half axial or screw or mixed flow pumps with axial and racial flow

(iii) Axial flow or propeller pumps with axial flow.

15.9 (c)

Minimum starting speed,

$$\left[\left(\frac{\pi N_{min}}{60}\right)^2 \left(D_2^2 - D^2\right)\right] = 2gH_m$$

Given $D_1 = 80 \text{ cm} = 0.8 \text{ m}$
$H_m = 15.3 \text{ m}$
$Q = 260 \; l/s = 0.26 \text{ m}^3/s$
$N = 600 \text{ rpm}$

Inner diameter of the impeller is approximately half of outer diameter

$$\therefore \quad \frac{\pi N_{min}}{60} = \sqrt{\frac{2 \times 9.81 \times 15.3}{(0.8^2 - 0.4^2)}}$$

$$\Rightarrow \quad N_{min} = 477.6 \simeq 475 \text{ rpm}$$

15.10 (c)

Specific speed, $N_s = \dfrac{N\sqrt{Q}}{H^{3/4}}$

For parallel arrangement, $H = 75$ m

$$Q = \left(\frac{20 \times (75)^{3/4}}{800}\right)^2$$

$= 0.405 \text{ m}^3/s = 405 \; lps.$

Since the discharge of single pump is more than 40 lps so parallel arrangement is not possible.

For series arrangement, $Q = 0.04 \text{ m}^3/s$

$$H^{3/4} = \frac{800 \times \sqrt{0.04}}{20}$$

$\Rightarrow \quad H = 16$ m

Therefore number of pumps required to develop 75 m head,

$n = \dfrac{75}{16} = 4.67 \simeq 5$ pumps in series.

The closest answer is 4 pumps in series.

15.11 (b)

Guide blades are provided in Francis tube to guide flow of water.

In Pelton wheel water jet strikes the blade.

Slip in reciprocating pump is defined as the difference between theoretical discharge and actual discharge.

Volute chamber is spiral shaped casing around the impeller in volute pump which is a centrifugal pump.

15.12 (c)

Air vessel ensure continuous supply of water at a uniform rate in a reciprocating pump. It results in saving of work against friction.

Percentage of work saved during stroke in

(i) single acting pump is 84.8%

(ii) double acting pump is 39.2%

The air vessels change the indicator diagram of the reciprocating pump.

15.13 (b)

Air lift pump, centrifugal pump and hydraulic ram have low efficiency. Centrifugal pump is suitable for low head. Reciprocating pump is correct choice.

15.14 (b)

$Q \propto N; \qquad P \propto N^3; H \propto N^2$

Power increases with cube of speed

15.15 (c)

Centifugal pump should be installed above water level in sump such that negative pressure does not reach as low as vapour pressure and thereby not cause cavitation.

15.16 (b)

Pumps in parallel: individual volume flow rates are added, head remains constant.

Pumps in series: total head is sum of individual pimp heads.

Therefore pumps in parallel boost discharge whereas pumps in series boost head.

15.17 (a)

For a centrifugal pump, the specific speed is given by

$$N_S = \frac{N\sqrt{Q}}{H^{3/4}} = \frac{725 \times \sqrt{44}}{(36)^{3/4}} = 327.22$$

15.18 (b)

Theoretical discharge for single acting reciprocating pump is given by

$$Q_t = \frac{ALN}{60} = \frac{\pi \times (25 \times 10^{-2})^2 \times 35 \times 10^{-2} \times 60 \times 10^3}{4 \times 60}$$

$= 17.2 \; lps$

15.19 (c)

Impeller of a centrifugal pump is provided with a series of backward curved blades or vanes. It is mounted on a shaft which is coupled to an external source of energy (usually an electric motor) which imparts the required energy to the impeller thereby making it to rotate.

15.20 (a)

An air vessel is a close chamber having an opening at its base through which the liquid may flow into the vessel or it may flow out from the vessel. It is obvious that the difficulty with the reciprocating pups is that the supply of liquid is quite intermittent. Air vessel is fitted to obtain a uniform rate of flow of liquid in both suction and delivery pipe of a reciprocating pump. Air vessel also makes an allowance to achieve higher speed, so that the flow rate of liquid entering cylinder may be more than the mean discharge. Also, when a pump is equipped with air vessels, the fluctuations in the velocity of flow in suction and delivery pipes are eliminated, which results in reducing the head loss due to friction in these pipes and hence certain amount of energy is saved. However, air vessels do not contribute in avoiding vibrations in the pump.

15.21 (a)

Head and discharge are based on utility and greater efficiency is good economically.

15.22 (b)

Mechanical energy is converted into hydraulic energy by impeller.

Electrical energy (Input) → MOTOR → Mechanical Energy (shaft power) → IMPELLER → Hydraulic Energy (Output)

15.23 (b)

In backward bent impeller blades the outlet tip curves in direction opposite to motion and offer less resistance helping in achieving high speed leading to higher efficiency.

15.24 (a)

The hydraulic ram is a type of pump in which the energy of large quantity of water falling through small height is utilized to lift a small quantity of this water to greater height.

15.25 (b)

Design of Hydraulic ram, Penstock and surge tank is done taking into consideration the water hammer phenomenon.

15.26 (c)

Air vessels are fitted to get continuous flow of liquid at uniform rate and to achieve higher speed. It reduces fluctuations in velocity and reduces head loss thereby saving power.

15.27 (b)

Centrifugal pump will start delivering liquid when the head developed exceeds manometric head Manometric head is the sum of the actual lift (Static head) + the frictional losses in the pipes + the discharge velocity head.

15.28 (a)

Pumps in parallel: Individual volume flow rates are added, head remains constant.

Pumps in series: Total head is sum of individual pimp heads.

Therefore pumps in parallel increase discharge whereas pumps in series increase head.

15.29 (d)

Power, $P \propto L_T^5 N_s^3$

$$(N_s)_m = (N_s)_p$$

$$\frac{P_m}{P_p} = \left(\frac{L_m}{L_p}\right)^5 \times 1 = \left(\frac{1}{2}\right)^5 = \frac{1}{32}$$

15.31 (a)

The Persian wheel is a mechanical water lifting device operated usually by draught animals like bullocks, buffaloes or camels. It is used to lift water from water sources typically open wells (mainly shallow open wells).

For deep tube wells, submersible pump is suitable and for Shallow tube well, centrifugal jet pump is suitable.

Wind power (mill) can be used to lift water from deep open well.

15.32 (b)

A reciprocating pump is a positive displacement pump in which the liquid is sucked and then it is actually pushed or displaced due to the thrust exerted on it by a moving member, which results in lifting the liquid to the required height. These pumps usually have one or more chambers which are alternately filled with the liquid to be pumped and then emptied again. As such the discharge of liquid pumped by these pumps almost wholly depends on the speed of the pump.

15.33 (b)

The efficiency of a pump is generally less than that of a turbine. The flow passages in a centrifugal pump are diverging (flow is decelerating) wherease those of a turbine are coverging (flow is accelerating). The losses in a pump are more because of possiblity of flow separation and formation of eddies in a diverging flow.

15.34 (a)

When a pump primes and works but not up to its capacity and pressure the reasons are

(i) Air may be leaking into pump through suction line.
(ii) Speed may be too low.
(iii) Discharge head may be higher than anticipated.
(iv) Suction lift may be too high.
(v) Foot valve or end of suction pipe may not have sufficient submergence.
(vi) Impeller may be partly clogged or too small in diameter.
(vii) Rotation may be in the wrong direction.

15.35 (b)

If H_m is the manometric head developed by each pump, then as

$$N_s = \frac{N\sqrt{Q}}{H_m^{3/4}}$$

Given, N_s = 700 rpm, N = 700 rpm
Q = 3840 litres/minute = 64 lps

$$\therefore \quad 700 = \frac{700 \times \sqrt{64}}{H_m^{3/4}}$$

$$\therefore \quad (H_m)^{3/4} = 8$$

$$\therefore \quad H_m = (8)^{4/3} = 16$$

$\therefore$ Number of impellers required $= \dfrac{80}{16} = 5$

15.36 (a)

H_1 = 45 m; Q_1 = 13.2 lps
D_1 = 20 cm; N_1 = 1350 rpm
D_2 = 15 cm; $N_2 = N_1$ = 1350 rpm
Q_2 = ? H_2 = ?

Head coefficient

$$C_H = \frac{H}{N^2D^2}$$

$$\Rightarrow \quad \frac{H_1}{N_1^2D_1^2} = \frac{H_2}{N_2^2D_2^2}$$

$$\Rightarrow \quad \frac{45}{(20)^2} = \frac{H_2}{(15)^2} \quad (\text{As } N_2 = N_1)$$

$$\Rightarrow \quad H_2 = \frac{45 \times 225}{400}$$

$$= \frac{45 \times 9}{16} = \frac{405}{16} = 25.3 \text{ m}$$

Discharge coefficient

$$C_Q = \frac{Q}{ND^3}$$

$$\Rightarrow \quad \frac{Q_1}{N_1D_1^3} = \frac{Q_2}{N_2D_2^3}$$

$$\Rightarrow \quad \frac{13.2}{(20)^3} = \frac{Q_2}{(15)^3}$$

$$\Rightarrow \quad Q_2 = \left(\frac{15}{20}\right)^3 \times 13.2$$

$$= \frac{27}{64} \times 13.2 = 5.568 \ l\text{ps} = 5.57 \ l\text{ps}$$

15.37 (a)

An air vessel is fitted to the suction pipe and to the delivery pipe at a point close to the cylinder of a single-acting reciprocating pump.

(i) to obtain a continuous supply of liquid at a uniform rate
(ii) to save a considerable amount of work in overcoming the frictional resistance in the suction and delivery pipes, and
(iii) to run the pump at a high speed without separation.

15.38 (a)

Turbine pumps run at low speeds as they do not have backward curved blades and thus are not capable of high heads.

15.39 (c)

The total head against which a pump has to work are

(i) Static head H_s, which is the vertical distance between the liquid surfaces in the sump and the tank to which the liquid is delivered by the pump.
(ii) Velocity head $(V^2/2g)$ at the point of discharge.
(iii) Head loss, H_L, which includes the head loss due to friction in suction and delivery pipes as well as head loss in valves, bends and all

such appurtenances both on the suction and the delivery sides.

15.40 (a)

$$\text{D' Aubuisson's efficiency} = \frac{w \times H}{W \times h}$$

$$\text{Rankine's efficiency} = \frac{w \times (H - h)}{(W - w) \times h}$$

For hydraulic ram, D' Aubuisson's efficiency is more than Rankine's efficiency.

$$\frac{a + \Delta a}{b + \Delta b} > \frac{a}{b}$$

$$\frac{a + \Delta a}{b + \Delta b} - \frac{a}{b} > 0$$

adding small value to numerator and denominator improves the value of ratio.

15.41 (a)

Net positive suction head is the minimum pressure required at the suction port of the pump to keep the pump from cavitating.

$$\text{NPSH} = \frac{P_t}{\rho g} - \frac{P_v}{\rho g} - h_f$$

Increase in pressure at supply reservoir increases NPSH

Reduction in suction friction reduces h_f thereby increasing NPSH.

15.42 (d)

$$\frac{\text{water delivered}}{\text{water wasted}} = \frac{1}{15}$$

$$\text{Eff. of pump} = \frac{\text{Output}}{\text{Input}} = \frac{20}{4 \times 15} \times 100$$

$$= 33.33\%$$

15.43 (a)

In both the cases, the power output is same

$$P_1 = P_2$$

$$\rho_1 Q g h_1 = \rho_2 Q g h_2$$

$$800\,(12) = (1200)\,h_2$$

$$P = \rho Q g h$$

$$\therefore\ 800\,Qg(12) = 1200\,Qg(h)$$

$$\Rightarrow\ h = 8\text{m}$$

15.44 (a)

Considering all other factors constant,

Brake power $\propto$ specific weight

$\therefore$ Brake power required is 18 kW when liquid is water with a specific weight of 9.8 kN/m^3. Then brake power in liquid with solvent having specific density 0.8.

$$= 18\text{ kW} \times 0.8 = 14.4\text{ kW}$$

15.45 (b)

Tangential velocity

$$= \frac{\pi DN}{60} = \frac{\pi \times 50 \times 10^{-2} \times 1200}{60} = 10\,\pi\text{ m/sec}$$

15.46 (c)

$$N_s = \frac{N\sqrt{Q}}{H_m^{3/4}}$$

For n no. of pumps,

$$700 = \frac{750\sqrt{75}}{H_m^{3/4}}$$

For each pump

$$H_m = 19.49$$

$$\text{Total head} = 190$$

$$\text{No. of pump} = \frac{190}{19.49} = 9.74 \simeq 10\text{ pumps}$$

15.48 (a)

Air vessel on the suction and delivery sides of a reciprocating pump are used to:

1. Achieve higher speed without separation
2. Reduce work in overcoming frictional resistance
3. Have nearly uniform discharge
4. Increase the delivery head of the pump

15.49 (d)

The specific speed (N_S) of a pump is the speed of a geometrically similar turbine that would develop 1 kW power when working under unit head.

15.50 (d)

$H = 20 - 60Q^2$

When two pumps are joined in parallel discharge becomes half. So head loss equation now is

$$H = 20 - 60\left(\frac{Q}{2}\right)^2$$

$$H = 20 - 15Q^2$$

15.51 (c)

Suction head created by pump= –3.5 m

Available head at the entrance of cylinder = 10 + 1.5 = 11.5 m

Pressure head at cylinder = 11.5 – 3.5 = 8 m

15.53 (c)

Non-dimensional specific speed

$$N_S = \frac{N\sqrt{Q}}{(gH)^{3/4}} = \frac{100\sqrt{\frac{1.2}{60}\times 1000}}{(9.81\times 25)^{3/4}} = 72.166$$

15.54 (c)

γ_w = 10 kN/m³ = 10000 N/m³

$$\text{Power (in BHP)} = \frac{10000\times 1\times(400-320)}{746}$$

$= 1072.386 \simeq 1070$

15.55 (d)

$$N_S = \frac{N\sqrt{Q}}{(gH_m)^{3/4}} = \frac{\frac{1400}{60}\sqrt{0.09}}{(9.81\times 45)^{3/4}}$$

N_S = 0.072 revolution

$= 0.072 \times 2 \times \pi = 0.458$ radian

15.56 (c)

For total quantity of discharge (Q) the discharge from each pump will be ($Q/2$).

$$H = 30 - 80\left(\frac{Q}{2}\right)^2$$

$$H = 30 - 20\,Q^2$$

15.57 (d)

$$\left.\frac{H_m}{D^2N^2}\right|_1 = \left.\frac{H_m}{D^2N^2}\right|_2$$

$$\left.\frac{20}{1200^2}\right|_1 = \left.\frac{H_m}{1500^2}\right|_2$$

$H_{m2} = 31.3$ m

$$\left.\frac{Q}{D^3N}\right|_1 = \left.\frac{Q}{D^3N}\right|_2$$

$$\frac{30}{1200} = \left.\frac{Q}{1500}\right|_2$$

$Q_2 = 37.5$ lps

15.58 (b)

H_{total} = 95 m, N = 1000 rpm

N_S = 900, Q = 150 L/s

$$N_S = \frac{N.\sqrt{Q}}{H_m^{3/4}}$$

$$900 = \frac{1000\sqrt{150}}{H_m^{3/4}}$$

$H_m = 32.49$ m

$$\text{No. of pumps} = \frac{H_{total}}{H_m} = \frac{95}{32.49}$$

$= 2.92 \simeq 3$

15.59 (c)

Piston area (A) = 0.1 m²

Stroke (L) = 0.3 m

$$Q_{actual} = \frac{2.4}{60}\,\text{m}^3/\text{s}$$

N = 45 rpm

$H = h_s + h_d = 10$ m

$$Q_{theo} = \frac{2A.L.N}{60}\ \text{m}^3/\text{s}$$

$$= \frac{2(0.1)(0.3)(45)}{60}\ \text{m}^3/\text{s}$$

$$= \frac{2.7}{60}\ \text{m}^3/\text{s}$$

$\text{Slip} = Q_{Theo} - Q_{Act}$

$$= \frac{2.7}{60} - \frac{2.4}{60} = 0.005\ \text{m}^3/\text{s}$$

$\text{Power required} = \dot{m}.g.H = \rho.Q_{theo}.g.H$

$$= (10^3)\left(\frac{2.7}{60}\right)(9.81)(10) = 4.41\ \text{kW}$$

15.62 (a)

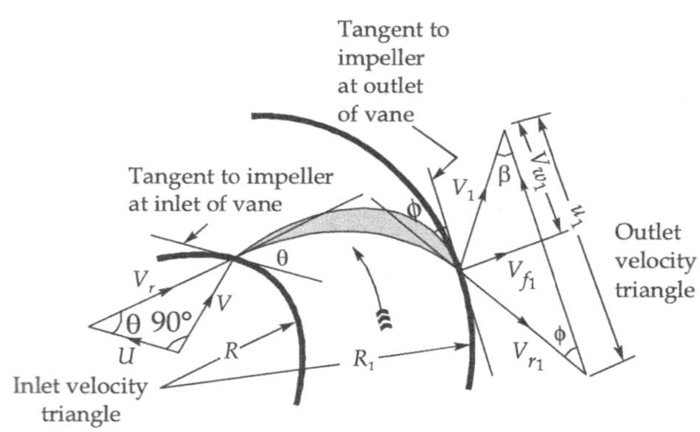

In the figure above.

u_1 = Peripheral or tangential velocity of the impeller at outlet

$$= \frac{\pi D_1 N}{60}$$

Given, D_1 = Dia. of impeller = 300 mm

N = Speed = 1000 rpm

$$\therefore \quad u_1 = \frac{\pi\times 0.3\times 1000}{60} = 15.7\ \text{m/s}$$

15.63 (a)

Given: Piston area,

$A = 0.1\ m^2$

Stroke length, $L = 0.3$ m

Actual discharge,

$N = 45$ rpm

Height, $(h_s + h_d) = 10$ m

$Q_{act} = 2.4\ m^3/min$

$= \frac{2.4}{60}\ m^3/s = 0.04\ m^3/s$

For double acting reciprocating pump theoretical discharge

$$Q_{theo} = \frac{2ALN}{60}$$

$$Q_{theo} = \frac{2(0.1)(0.3)(45)}{60}$$

$= 0.045\ m^3/s$

Slip $= Q_{theo} - Q_{act}$

$= 0.045 - 0.04 = 0.005\ m^3/s$

15.64 (a)

$$\text{Percentage slip} = \left(\frac{Q_{theo} - Q_{act}}{Q_{theo}}\right) \times 100$$

Given, Slip < 0

$\Rightarrow$ $Q_{theo} < Q_{act}$

If $Q_{act} > Q_{theo}$

$$\left\{C_d = \frac{Q_{act}}{Q_{theo}}\right\}$$

$\Rightarrow$ $C_d > 1$

15.65 (d)

$$Q_{theo} = \frac{ALN}{60} = \frac{\frac{\pi}{4}(0.25)^2 \times (0.35) \times (60)}{60}$$

$\Rightarrow$ $Q_{theo} = 0.017180\ m^3/s = 17.18$ litre/s

$\therefore$ Coefficient of discharge,

$$C_d = \frac{Q_{act}}{Q_{theo}} = \frac{16.5}{17.18} = 0.96$$

15.66 (a)

The purpose of air-vessel is to minimize the flow fluctuation and smooth out pressure.

In a multi-cylinder reciprocating pump typically number of cylinder work together to reduce the flow pulsations and vibration that can be occured in a single acting reciprocating pump. Therefore if the fluctuation of velocity is already minimum in multi cylinder system, then air vessel may not be required.

15.68 (d)

Manometric head

■■■■

16 Open Channel Flow

16.1 **Assertion (A):** Surges can be positive or negative.

Reason (R): Negative surges may occur when a gate at the head of a channel is suddenly opened or when a gate at tail end of a channel is suddenly closed. [ESE : 2000]

16.2 Consider the following statements:

1. In an open channel flow, energy grade line is obtained by adding datum head, pressure head and velocity head.
2. In an open channel, hydraulic grade line is the free surface itself.
3. For a pipe and an open channel of source dimension, the hydraulic gradient line is located at the same height above datum.
4. Energy gradient line of an open channel is always horizontal.

Which of these statements are correct?

(a) 1, 2 and 3 (b) 1 and 4
(c) 1, 3 and 4 (d) 2, 3 and 4

[ESE : 2000]

16.3 For a wide rectangular channel using Manning's formula, the differential equation of gradually varied flow (with the usual notations) is given by

(a) $\dfrac{dy}{dx} = S_0 \dfrac{\left[1-\left(\dfrac{y_c}{y}\right)^{10/3}\right]}{\left[1-\left(\dfrac{y_0}{y}\right)^{3}\right]}$

(b) $\dfrac{dy}{dx} = S_0 \dfrac{\left[1-\left(\dfrac{y_0}{y}\right)^{10/3}\right]}{\left[1-\left(\dfrac{y_c}{y}\right)^{3}\right]}$

(c) $\dfrac{dy}{dx} = S_0 \dfrac{\left[1-\left(\dfrac{y_0}{y}\right)^{3}\right]}{\left[1-\left(\dfrac{y_c}{y}\right)^{3}\right]}$

(d) $\dfrac{dy}{dx} = S_0 \dfrac{\left[1-\left(\dfrac{y_c}{y}\right)^{3}\right]}{\left[1-\left(\dfrac{y_0}{y}\right)^{3}\right]}$

[ESE : 2000]

16.4 A sluice gate opening in a canal is shown in the given figure. The shapes of water surface profiles at X, Y and Z will be respectively (NDL = Normal Depth Line, CDL = Critical Depth Line, HJ = Hydraulic Jump)

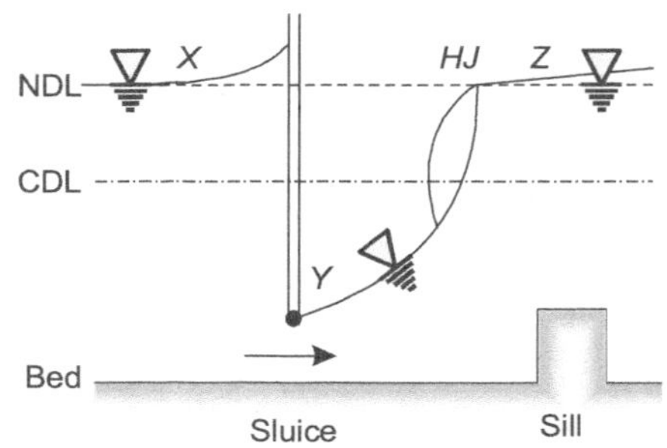

(a) M_1, M_3 and M_1 (b) M_2, M_3 and M_2
(c) S_1, S_3 and S_2 (d) H_2, S_3 and S_1

[ESE : 2000]

16.5 The stage in a river is 4.8 m, the water surface slope is 1 in 10,000 and the discharge in the stream is 600 m^3/s. If the stage remains the same and the water surface slope is 1 in 14,400, then the discharge in the stream will be

(a) 300 m^3/s (b) 400 m^3/s
(c) 600 m^3/s (d) 500 m^3/s

[ESE : 2000]

16.6 Water surface profiles that are asymptotic at one end and terminated at the other end would include

(a) H_2 and S_2 (b) H_3 and S_2
(c) M_2 and H_2 (d) M_2 and H_3

[ESE : 2001]

16.7 A hydraulic jump is always needed in case of

(a) an A_2 profile followed by an A_3 profile
(b) an A_3 profile followed by an A_2 profile
(c) an H_2 profile followed by an M_2 profile
(d) an M_1 profile followed by an M_3 profile

[ESE : 2001]

16.8 For a hydraulically efficient rectangular section, the ratio of width to normal depth is

(a) 0.5 (b) 1.0
(c) 2.0 (d) $2\sqrt{3}$

[ESE : 2002]

16.9 A rigid boundary rectangular channel having a bed slope of $\frac{1}{800}$ has its width and depth of flow equal to 2 m and 1 m respectively. If the flow is uniform and the value of Chezy's constant is 60, the discharge through the channel is

(a) 1.0 m^3/s (b) 1.5 m^3/s
(c) 2.0 m^3/s (d) 3.0 m^3/s

[ESE : 2002]

16.10 Consider the following statements in regard to the critical flow:

1. Specific energy is maximum for a given discharge.
2. Specific force is maximum for a given discharge
3. Discharge is maximum for a given specific force.
4. Discharge is maximum for a given specific energy.

Which of these statements are correct?

(a) 1, 2, 3 and 4 (b) 1 and 2
(c) 2 and 3 (d) 3 and 4

[ESE : 2002]

16.11 Match **List-I** (Slope) with **List-II** (Description) and select the correct answer using the codes given below the lists:

List-I

A. Mild
B. Adverse
C. Limit
D. Critical

List-II

1. is when the flow is at Froude number one along the channel
2. is an example of non-sustaining slope
3. always sustains uniform subcritical flow
4. is one which has smallest critical slope for a given channel shape and roughness

Codes:

	A	B	C	D
(a)	3	2	4	1
(b)	1	4	2	3
(c)	3	4	2	1
(d)	1	2	4	3

[ESE : 2002]

16.12 If F_1 and F_2 are the Froude numbers of flow before and after the hydraulic jump occurring in a rectangular channel, then

(a) $F_2^2 = \dfrac{F_1^2}{\left(-1+\sqrt{1+8F_1^2}\right)^3}$

(b) $F_2^2 = \dfrac{8F_1^2}{\left(-1+\sqrt{1+8F_1^2}\right)^3}$

(c) $F_2^2 = \dfrac{F_1^2}{\left(-0.5+\sqrt{1+8F_1^2}\right)^3}$

(d) $F_2^2 = \dfrac{8F_1^2}{\left(-0.5+\sqrt{1+F_1^2}\right)^3}$

[ESE : 2002]

16.13 Consider the following statements about a hydraulic ram

1. Hydraulic ram does not need external power.
2. It works on the fundamental principle of water hammer.
3. The efficiency of a hydraulic ram is only of the order of 8 to 10%.
4. It can be termed as a low, intermittent discharge low-head pumping installation.

Which of these statements are correct?

(a) 2 and 3 (b) 1, 2, 3 and 4
(c) 1 and 2 (d) 1 and 3

[ESE : 2002]

16.14 Consider the following statements related to centrifugal pumps:

1. Centrifugal pumps with blades facing backward have fast runners.
2. Multistage pumps have two or more impellers installed in series so that the discharge is increased.
3. Diffusion type centrifugal pumps are called turbine pumps.
4. A centrifugal pump with rising characteristics is used when actual lift is small and the amount of flow is constant.

Which of these statements are correct?

(a) 1, 2 and 3 (b) 1, 2 and 4
(c) 1, 3 and 4 (d) 2, 3 and 4

[ESE : 2002]

16.15 The critical depth of water flowing through a rectangular channel of width 5 m when discharge is 12.5 m^3/s is

(a) $(2.25)^{1/2}$ m (b) $(16)^{1/2}$ m
(c) $(0.46)^{1/3}$ m (d) $(0.64)^{1/3}$ m

[ESE : 2003]

Directions: The following items consists of two statements; one labelled as **'Assertion (A)'** and the other as **'Reason (R)'**. You are to examine these two statements carefully and select the answers to these items using the codes given below:

Codes:

(a) both A and R are true and R is the correct explanation of A
(b) both A and R are true but R is not a correct explanation of A
(c) A is true but R is false
(d) A is false but R is true

16.16 Assertion (A): At the critical state of flow, the specific force is a minimum for the given discharge.

Reason (R): For a minimum value of specific force, the first derivative of force with respect to depth should be unity.

[ESE : 2003]

16.17 Assertion (A): The vertical co-ordinate of the apex of the Q-curve is $\frac{2}{3}E$.

Reason (R): The horizontal co-ordinate of the apex of E-curve is 1.5 times $\left(\frac{q^2}{g}\right)^{1/3}$.

[ESE : 2003]

16.18 The steps involved in the procedure of computing length of backwater curves are given below with usual notations:

1. Insert all DH values and average value of slope in the differential equation.
2. Assume a new depth of flow y for the other end of the reach and insert corresponding values of A, P, v, R, $\frac{v^2}{2g}$, H and S.
3. Determine or assume the initial conditions of depth, channel characteristics and discharge.
4. Compute average S value and the change in specific head ΔH.
5. Repeat the process for each reach adding resulting ΔL values to obtain total required distance ΔL until the depth or distance desired is reached.

The correct sequence of these steps is

(a) 1-2-3-4-5 (b) 3-1-5-2-4
(c) 3-2-4-1-5 (d) 1-4-5-2-3

[ESE : 2003]

16.19 Consider the following statements which relate to different types of water surface profiles. Curve types conform to usual classifications y_c and y_0 (critical and normal depths):

1. Type-3 curves lie between y_c and y_0
2. All curves where $y_c < y_0$ are unaffected by upstream disturbances.
3. All curves where $y_c > y_0$ are influenced by downstream disturbances.
4. All curves approaching the 'y' line approach it asymptotically except for C curve where $y_0 = y_c$.

Which of these statements are correct?

(a) 1, 2, 3 and 4 (b) 1, 3 and 4
(c) 2, 3 and 4 (d) 1 and 2

[ESE : 2003]

16.20 Flow duration curve is a plot of
(a) flow against its time of occurrence in chronological order
(b) flow in ascending order against percentage time in chronological order
(c) flow that equalled or exceeded against percentage time
(d) flow against duration of time for which it is sustained

[ESE : 2003]

16.21 The momentum correction factor for a flow through open channel is given by

(a) $\frac{1}{AV^2}\int_A v^2 dA$ (b) $\frac{1}{AV}\int_A v^2 dA$

(c) $\frac{1}{A}\int_A v^2 dA$ (d) $\frac{1}{AV^2}\int_A v dA$

[ESE : 2003]

16.22 Consider the following statements regarding the conditions to be satisfied for the maximum discharge through a trapezoidal channel section with side slope 1: n, bed width b, flow depth d and having a fixed bed slope:
1. Sloping sides should have an angle of 30° with vertical.
2. Hydraulic mean depth equals half the flow depth.
3. Length of sloping sides should be equal to twice the bottom width.

Which of these statements are correct?
(a) 1, 2 and 3 (b) 1 and 2
(c) 2 and 3 (d) 1 and 3

[ESE : 2004]

16.23 Water can flow with 1 m depth in alternatively four channels of different sections as shown below:

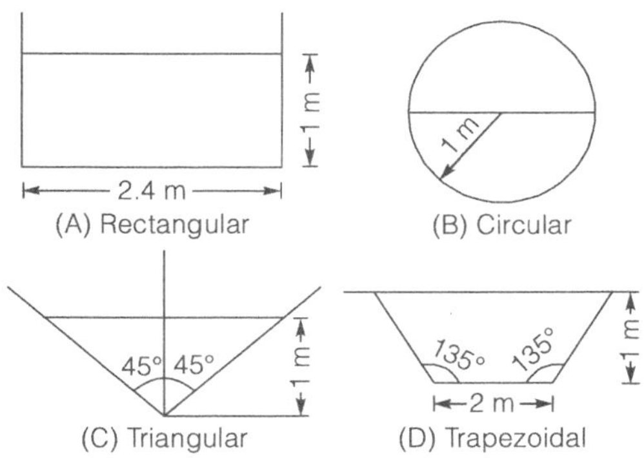

Which one of the following sequences shows their hydraulic radii, arranged in descending order?
(a) D-C-B-A (b) D-A-B-C
(c) A-B-C-D (d) A-B-D-C

[ESE : 2004]

16.24 In a reach of open channel flow, with $q = 2$ m²/s, depth of flow at section A is 1.5 m and at section B, 1.6 m. The difference in specific energies at two sections can be correspondingly taken as 0.09 m. The bed slope is 1 in 2000. The mean energy slope between the sections is given as 0.003. What is the length of reach AB?
(a) 54 m with A on the upstream
(b) 54 m with B on the upstream
(c) 36 m with A on the upstream
(d) 36 m with B on the upstream

[ESE : 2004]

Directions: The following items consists of two statements; one labelled as **'Assertion (A)'** and the other as **'Reason (R)'**. You are to examine these two statements carefully and select the answers to these items using the codes given below:

Codes:
(a) both A and R are true and R is the correct explanation of A
(b) both A and R are true but R is not a correct explanation of A
(c) A is true but R is false
(d) A is false but R is true

16.25 Assertion (A): When water flows through a long stretch of a narrow rectangular channel, the water surface at the walls is higher than that at the central axis.

Reason (R): Resistive effort of atmospheric air is felt more at the walls than away from the walls.

[ESE : 2004]

16.26 Assertion (A): At the critical state of flow, the specific force is a minimum for the given discharge.

Reason (R): For a minimum value of specific force, the first derivative of force with respect to depth should be unity.

[ESE : 2005]

16.27 Match List-I (Type of curve) with List-II (Flow condition) and select the correct answer using the codes given below the lists:

	List-I		List-II
A.	M_1	1.	Slope upward in the direction of flow
B.	H_3	2.	Back water profile
C.	S_2	3.	Hydraulic jump occurs
D.	A_2	4.	Hydraulic drop occurs

Codes:

	A	B	C	D
(a)	1	3	4	2
(b)	2	4	3	1
(c)	1	4	3	2
(d)	2	3	4	1

[ESE : 2005]

16.28 The Froude number of a hydraulic jump is 5.5. The jump can be classified as a/an

(a) undular jump (b) oscillating jump
(c) weak jump (d) steady jump

[ESE : 2005]

16.29 Consider the following statements in respect of specific energy of flow in an open channel of fixed width:

1. There is only one specific energy curve for a given channel.
2. Alternate depths are the depths of flow at which the specific energy is the same.
3. Critical flow occurs when the specific energy is minimum.

Which of these statements are correct?

(a) 1 and 2 (b) 1 and 3
(c) 2 and 3 (d) 1, 2 and 3

[ESE : 2006]

16.30 If the Froude number of flow in a rectangular channel at a depth of flow of y_0 is F_0, then what is y_c / y_0 equal to?

(a) $F_0^{1/3}$ (b) $F_0^{2/3}$
(c) $F_0^{3/2}$ (d) $F_0^{-1/2}$

[ESE : 2006]

16.31 A hydraulically efficient trapezoidal section of open channel flow carries water at the optimal depth of 0.6 m. Chezy coefficient is 75 and bed slope is 1 in 250. What is the discharge through the channel?

(a) 1.44 m^3/s (b) 1.62 m^3/s
(c) 1.92 m^3/s (d) 2.24 m^3/s

[ESE : 2006]

16.32 In the step methods (both direct and standard), the computations must

(a) proceed downstream in subcritical flow
(b) proceed upstream in subcritical flow
(c) always proceed upstream
(d) always start at a control section

[ESE : 2006]

16.33 Which of the following equations are used for the derivation of the differential equation for water surface profile in open channel flow?

1. Continuity Equation
2. Energy Equation
3. Momentum Equation

Select the correct answer using the codes given below:

(a) 1, 2 and 3 (b) 1 and 3 only
(c) 1 and 2 only (d) 2 and 3 only

[ESE : 2006]

16.34 In a wide rectangular channel if the normal depth is increased by 20%, then what is the approximate increase in discharge?

(a) 25% (b) 30%
(c) 35% (d) 40% [ESE : 2006]

16.35 A right-angled triangular channel symmetrical in section about the vertical carries a discharge of 5 m^3/s with a velocity of 1.25 m/s. What is the approximate value of the Froude number of the flow?

(a) 0.3 (b) 0.4
(c) 0.5 (d) 0.6

[ESE : 2007]

16.36 Flow depths across a sluice gate are 2.0 m and 0.5 m. What is the discharge (per metre width)?

(a) 1.0 m^2/s (b) 1.4 m^2/s
(c) 2.0 m^2/s (d) 2.8 m^2/s

[ESE : 2007]

16.37 A rectangular channel is 6 m wide and discharges 30 m^3s^{-1}. The upstream depth is 2.0 m, acceleration due to gravity is 10 ms^{-2}. Then, what is the specific energy (approximate)?

(a) 2.5 (b) 0.3
(c) 2.3 (d) None of these

[ESE : 2007]

16.38 For a smooth hump in a sub-critical flow to function as a broad crested weir, the height ΔZ of the hump above the bed must satisfy which one of the following?

(a) $\Delta Z \geq (E_1 - y_c)$ (b) $\Delta Z \geq (E_1 - E_c)$
(c) $\Delta Z \leq (E_1 - y_c)$ (d) $\Delta Z \leq (E_1 - E_c)$

(E_1 = Specific head upstream of the hump,
E_c = Specific head at the critical depth y_c)
(Neglect friction effects)

[ESE : 2007]

16.39 Match **List-I** (Flow section type) with **List-II** (Critical discharge is proportional to) where y is the depth of flow and select the correct answer using the codes given below the lists:

List-I	List-II
A. Shallow parabolic	1. $y(z^{3/2})$
B. Triangular	2. $y^{3/2}$
C. Rectangular	3. $y^{5/2}$
D. Trapezoidal	4. y^2

Codes:

	A	B	C	D
(a)	2	3	4	1
(b)	4	1	2	3
(c)	2	1	4	3
(d)	4	3	2	1

[ESE : 2007]

16.40 Which one of the following statements is not correct?

A control section in an open channel is the site
(a) where the flow quantiy can be controlled
(b) at which flow is known to be critical
(c) where the discharge can be measured
(d) where the specific energy is determined

[ESE : 2007]

16.41 What is the normal depth in a wide rectangular channel carrying 0.5 m^2/s discharge at a bed slope of 0.0004 and Manning's $n = 0.01$?

(a) 0.13 m (b) 0.32 m
(c) 0.43 m (d) 0.50 m

[ESE : 2007]

16.42 Flow happens at a critical depth of 0.5 m in a rectangular channel of 4 m width. What is the value of discharge?

(a) 5.4 m^3/s (b) 5.1 m^3/s
(c) 4.9 m^3/s (d) 4.4 m^3/s

[ESE : 2007]

16.43 A hydraulic jump occurs at the top of a spillway. The depth before jump is 0.2 m. The sequent depth is 3.2 m. What is the energy dissipated in m (approximate)?

(a) 27 (b) 10.5
(c) 15 (d) 42

[ESE : 2007]

16.44 A rectangular channel 3 m wide is laid on a slope of 0.0002. When the depth of flow in the channel is 1.5 m what is the average boundary shear stress (nearly)?

(a) 0.3 N/m^2 (b) 0.15 N/m^2
(c) 3.0 N/m^2 (d) 1.5 N/m^2

[ESE : 2007]

16.45 In connection with a gradually varied flow with notations y_0 = normal depth, y_c = critical depth and y = depth in the gradually varied flow, match **List-I** with **List-II** and select the correct answer using the codes given below the lists:

List-I	List-II
A. $y_c > y_C > y$	1. M_1
B. $y_0 > y > y_c$	2. S_3
C. $y > y_c > y_0$	3. M_2
D. $y > y_0 > y_c$	4. S_1

Codes:

	A	B	C	D
(a)	4	1	2	3
(b)	2	3	4	1
(c)	4	3	2	1
(d)	2	1	4	3

[ESE : 2007]

16.46 In a wide rectangular channel, an increase in the normal depth by 20% corresponds to how much (approximate) increase in discharge?

(a) 12% (b) 20%
(c) 36% (d) 48%

[ESE : 2007]

16.47 The sequent depth ratio of a hydraulic jump in a rectangular channel is 16.48. What is the Froude number (approximate) at the beginning of the jump?

(a) 9.0 (b) 12.0
(c) 5.0 (d) 8.0

[ESE : 2007]

16.48 When discussing most efficient section of flow into open channels, what is the perimeter *P* as a proportion of depth of flow *h* (i.e., *P*/*h*) for (i) a triangular section, and (ii) a trapezoidal section, respectively?

(a) 2.25, 2.83
(b) 2.25, 3.15
(c) 2.83, 3.15
(d) 2.83, 3.46

[ESE : 2008]

16.49 Consider the following statements:

1. Hydraulically most efficient channel section for an open channel flow will carry maximum discharge for a given area of cross section.
2. For a given cross sectional area hydraulic radius is maximum when the wetted perimeter is minimum.

Which of these statements is/are correct?

(a) 1 only
(b) 2 only
(c) Both 1 and 2
(d) Neither 1 nor 2

[ESE : 2008]

16.50 The type of jump that forms when initial Froude number lies between 2.5 and 4.5 is

(a) weak jump
(b) steady jump
(c) undular jump
(d) oscillating jump

[ESE : 2008]

16.51 Consider the following factors:

1. The change in the shape and size of the channel cross-section
2. The change in the slope of the channel
3. The presence of obstruction
4. The change in the frictional forces at the boundaries

Which of the above factors would cause a gradually varied flow?

(a) 1, 2, 3 and 4
(b) 1, 2 and 3 only
(c) 2 and 4 only
(d) 3 and 4 only

[ESE : 2009]

Directions: The following items consists of two statements; one labelled as **'Assertion (A)'** and the other as **'Reason (R)'**. You are to examine these two statements carefully and select the answers to these items using the codes given below:

Codes:

(a) both A and R are true and R is the correct explanation of A
(b) both A and R are true but R is not a correct explanation of A
(c) A is true but R is false
(d) A is false but R is true

16.52 Assertion (A): Total energy of flow decreases in the direction of flow.

Reason (R): The specific energy may decrease, increase or remain constant.

[ESE : 2009]

16.53 Assertion (A): For a hydraulically efficient channel, the hydraulic radius is equal to half the depth of flow.

Reason (R): A hydraulically efficient channel has the minimum perimeter for a given area of flow.

[ESE : 2009]

16.54 Assertion (A): Energy dissipation in a hydraulic jump is mainly caused by the large eddies in turbulence.

Reason (R): Large eddies transport the fluid over large distances, thus causing the mixing effect of turbulence.

[ESE : 2010]

16.55 Assertion (A): An economical channel section gives maximum discharge for a given cross-sectional area.

Reason (R): An economical channel section has smooth surface for reduced friction.

[ESE : 2010]

16.56 A partially open sluice gate is suddenly raised to its full opening. The resulting surge waves at the gate are

(a) a positive wave travels towards the gate from the upstream side while a negative wave travels downstream from the gate
(b) a positive wave travels from the gate onto the upstream side while a positive wave travels downstream from the gate
(c) a negative wave travels from the gate onto the upstream side while a positive wave travels downstream from the gate
(d) a negative wave travels towards the gate from the upstream side whereas a positive wave travels downstream from the gate

[ESE : 2010]

16.57 Consider the following statements regarding specific energy of the flow in an open channel:

1. There is only one specific energy curve for a given channel.
2. Alternate depths are the depths of flow at which the specific energy is the same.
3. Critical flow occurs when the specific energy is minimum for the flow rate.

Which of these statements is/are correct?

(a) 1 only (b) 1 and 2 only
(c) 2 and 3 only (d) 1, 2 and 3

[ESE : 2010]

16.58 Match **List-I** with **List-II** and select the correct answer using the code given below the lists:

List-I

A. Sudden closure of a sluice gate
B. Hydraulic jump in a stilling basin
C. Spreading of irrigation water in a field
D. Flow in a main irrigation canal

List-II

1. Uniform flow
2. Rapidly varied flow
3. Unsteady flow
4 Spatially varied flow

Code:

	A	B	C	D
(a)	1	2	4	3
(b)	3	2	4	1
(c)	1	4	2	3
(d)	3	4	2	1

[ESE : 2011]

16.59 The water surface profile in the flow situation as shown in the figure below is

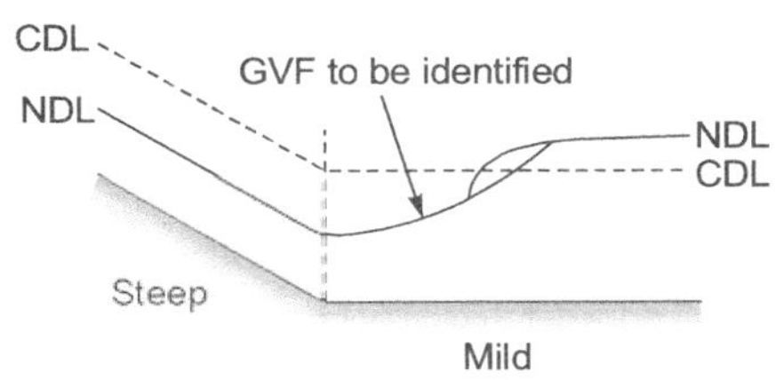

(a) S3 (b) M3
(c) S2 (d) M1

[ESE : 2011]

16.60 Best side slope for most economical trapezoidal section in open channel flow, wherein side slopes are defined by: X horizontal to 1 vertical, is when X equals

(a) 0.404 (b) 0.500
(c) 0.577 (d) 0.673

[ESE : 2011]

16.61 Consider the following statements in respect of critical flow in a wide rectangular channel:

1. The specific energy is minimum for a given discharge.
2. The discharge is maximum for a given specific energy.
3. The specific force is minimum for a given discharge.
4. The Froude number is equal to unity.

Which of these statements are correct?

(a) 1, 2 and 3 only (b) 1, 2, 3 and 4
(c) 1, 2 and 4 only (d) 2, 3 and 4 only

[ESE : 2012]

16.62 For a hydraulically efficient rectangular channel of bed width 4.0 m, the depth of flow is

(a) 4 m (b) 0.5 m
(c) 1 m (d) 2 m [ESE : 2012]

16.63 The velocity with which an elementary surge wave can travel upstream in a channel with depth y = 1.6 m and velocity V = 2.4 m/s is (Take g = 10 m/s^2)

(a) 16 m/s (b) 13.6 m/s
(c) 2.4 m/s (d) 1.6 m/s

[ESE : 2012]

16.64 Consider the following statements related to water surface profile in gradually varied flow in an open channel:

1. M_1 and S_1 curves approach Y_0 line asymptotically; and tend to be horizontal as y tends to ∞.
2. M_2 and S_2 curves meet Y_0 line horizontally, and Y_0 line asymptotically.
3. M_3 and S_3 curves meet Y_0 line normally, and also meet the channel bed normally
4. C_1 and C_2 curves will be slightly curved if Chezy's equation is used; otherwise, they may tend to be straight lines.

Which of these statements are correct?

(a) 1, 2, 3 and 4 (b) 1 and 4 only
(c) 2 and 3 only (d) 3 and 4 only

[ESE : 2012]

Directions: Each of the next items consists of two statements, one labeled as the **'Statements (I)'** and the other as **'Statement (II)'**. You are to examine these two statements carefully and select the answers to these items using the codes given below :

Codes:

(a) Both Statement (I) and Statement (II) are individually true and Statement (II) is the correct explanation of Statement (I)

(b) Both Statement (I) and Statement (II) are individually true but Statement (II) is NOT the correct explanation of Statement (I)

(c) Statement (I) is true but Statement (II) is false

(d) Statement (I) is false but Statement (II) is true

16.65 Statement (I): The best hydraulic section always has the minimum excavation.

Statement (II): The best hydraulic section gives the minimum area for a given discharge.

[ESE : 2012]

16.66 Statement (I): A given channel may be classifiable as mild for one discharge, critical for another discharge, and steep for yet another discharge.

Statement (II): Normal depth and critical depth are independent functions of the discharge along with, or without, other appropriate parameters.

[ESE : 2012]

16.67 Statement (I): An economical channel section gives maximum discharge for a given cross-sectional area.

Statement (II): An economical channel section has smooth surface with reduced friction.

[ESE : 2013]

16.68 Statement (I): In the case of an open channel flow, for a hydraulically efficient section, the hydraulic radius is equal to 50% of depth of flow in the channel.

Statement (II): Wetted perimeter is minimum for a hydraulically efficient section of an open channel.

[ESE : 2013]

16.69 A rectangular channel carries a uniform flow with a Froude number of 2.83. The ratio of critical depth to normal depth of this flow is:

(a) 1.68 (b) 2.83
(c) 2.00 (d) 4.75 [ESE : 2013]

16.70 Which of the following equations are used for the derivation of the differential equation for water surface profile in an open channel?

1. Continuity equation
2. Energy equation
3. Momentum equation

(a) 1, 2 and 3 (b) 1 and 3 only
(c) 2 and 3 only (d) 1 and 2 only

[ESE : 2013]

16.71 When flow is critical through open channels, which of the following statements are correct?

1. The specific energy is minimum for a given discharge.
2. The discharge is maximum for a given specific energy.
3. The specific force is minimum for a given discharge.
4. Froude number of the flow is equal to unity.

(a) 1, 2 and 3 only (b) 1, 2 and 4 only
(c) 1, 2, 3 and 4 (d) 2, 3 and 4 only

[ESE : 2013]

16.72 For a hydraulically efficient rectangular channel of bed width 5 m, the hydraulic radius is equal to

(a) 2.5 m (b) 1.25 m
(c) 5 m (d) 2 m

[ESE : 2014]

16.73 A triangular open channel has a vortex angle of 90° and carries flow at a critical depth of 0.3 m. The discharge in the channel is

(a) 0.41 m^3/s (b) 0.11 m^3/s
(c) 0.21 m^3/s (d) 0.31 m^3/s

[ESE : 2014]

16.74 At a hydraulic jump, the depths at the two sides are 0.3 m and 1.4. The head loss in the jump is nearly

(a) 1.0 m (b) 0.95 m
(c) 0.79 m (d) 0.45 m [ESE : 2014]

16.75 It is expected that due to extreme cold weather, the entire top surface of canal carrying water will be covered with ice sheet for some days. If the discharge in the canal were to remain unaltered, this condition would lead to

(a) No change in the depth of flow
(b) Decrease in the depth of flow
(c) Gradually decreasing flow
(d) Increase in the depth of flow **[ESE : 2014]**

16.76 Consider the following statements regarding a gradually varied flow in a prismatic open channel.
1. Total energy line remains parallel to the water surface.
2. The rate of energy loss at the section is inversely proportional to d^x where d is flow depth with index x.
3. In analyzing the flow, the energy equation of flow is not applicable.
4. Pressure distribution at every section can be assumed to be hydrostatic.

Which of the above statements are correct?
(a) 1 and 4 (b) 2 and 4
(c) 1 and 2 (d) 2 and 3
[ESE : 2014]

16.77 At a hydraulic jump, the depths at its two sides are 0.3 m and 1.2 m. The head loss in the jump is
(a) 1.0 m (b) 0.8 m
(c) 0.5 m (d) 0.45 m **[ESE : 2015]**

16.78 Field observations are carried out to assess the discharge of a river. Measurements are taken in a 2000 m straight reach. Slope is approximately 1 in 4000. Bed slope is determinably to a possible accuracy of 0.4 cm; wetted perimeter is determinable within 4% of possible error; and sectional area within 6% of possible error. Using Chezy's equation, the assessed discharge will be accurate to within
(a) 9.6% (b) 10.8%
(c) 11.4% (d) 12.7% **[ESE : 2015]**

Directions: The following items consists of two statements; one labelled as **'Assertion (A)'** and the other as **'Reason (R)'**. You are to examine these two statements carefully and select the answers to these items using the codes given below:

Codes:
(a) Both A and R are true and R is the correct explanation of A
(b) Both A and R are true but R is not a correct explanation of A
(c) A is true but R is false
(d) A is false but R is true

16.79 Assertion (A): Open channel flow in a channel is said to be critical when the specific force is maximum for a given discharge.
Reason (R): Direct integrating for steady non-uniform flow by Bresse's method was developed for very wide rectangular channels.
[ESE : 2015]

16.80 Assertion (A): A moving hydraulic jump is called a surge.
Reason (R): The travel of a wave is faster in the upper portion than in the lower portion in case of positive surges.
[ESE : 2015]

16.81 In a trapezoidal channel with bed width of 2 m, and side slopes of 2 v on 1 h, critical flow occurs at a depth of 1 m. What will be the quantity of flow and the flow velocity? Take g as 10 m/s^2.
(a) 7.22 m^3/s and 3.10 m/s
(b) 6.82 m^3/s and 2.89 m/s
(c) 7.22 m^3/s and 2.89 m/s
(d) 6.82 m^3/s and 3.10 m/s
[ESE : 2016]

16.82 A 7.5 m wide rectangular channel conveys 12 m^3/s of water with a velocity of 1.5 m/s. The specific energy head of the flow is
(a) 1.18 m (b) 1.78 m
(c) 2.18 m (d) 2.78 m
[ESE : 2016]

16.83 The sequent depth ratio in a hydraulic jump formed in a rectangular horizontal channel is 10. The Froude number of the supercritical flow is
(a) 12.2 (b) 10.4
(c) 7.42 (d) 4.21
[ESE : 2016]

16.84 A partially open sluice gate discharges water at 6 m/s with a depth of 40 cm in a rectangular horizontal channel of width 5 m. What would be the post-jump depth of flow on the downstream of the gate by taking g as 10 m/s^2?

(a) 1.51 m (b) 1.70 m
(c) 1.85 m (d) 1.95 m

[ESE : 2016]

16.85 For a hydraulically efficient rectangular channel of bed width 5 m, the hydraulic radius is

(a) 0.5 m (b) 1.25 m
(c) 2.75 m (d) 4.25 m

[ESE : 2016]

16.86 At a hydraulic jump, the flow depths are 0.4 m and 5 m at the upstream and downstream, respectively. The channel is wide rectangular. The discharge per unit width is nearly

(a) 5.8 m^2/s (b) 6.4 m^2/s
(c) 7.3 m^2/s (d) 8.3 m^2/s

[ESE : 2016]

Directions: Each of the next items consists of two statements, one labelled as the **'Statement (I)'** and the other as **'Statement (II)'**. Examine these two statements carefully and select the answers to these items using the codes given below :

Codes:

(a) Both Statement (I) and Statement (II) are individually true and Statement (II) is the correct explanation of Statement (I)
(b) Both Statement (I) and Statement (II) are individually true but Statement (II) is *not* the correct explanation of Statement (I)
(c) Statement (I) is true but Statement (II) is false
(d) Statement (I) is false but Statement (II) is true

16.87 **Statement (I):** In open channel flow, maximum velocity does not occur on the surface.

Statement (II): Three is wind drag on the free surface of an open channel.

[ESE : 2016]

16.88 **Statement (I):** Flow over sharp-crested weirs, standing wave flumes and abrupt free overfalls at ends of long straight channels are examples of rapidly varied flow.

Statement (II): The above-listed flows are all essentially local phenomena and can be utilized for flow measurement in open channels.

[ESE : 2016]

16.89 Hydraulic jump forms in a horizontal rectangular channel carrying a unit discharge of 1.019 $m^3/sec/m$ at a depth of 10.19 mm. This jump is classified as

(a) Weak jump (b) Oscillating jump
(c) Steady jump (d) Strong jump

[ESE : 2017]

16.90 In wide rectangle channel, the normal depth is increased by 20%. This would mean an increase in the discharge of the channel nearly by

(a) 20% (b) 26%
(c) 36% (d) 56%

[ESE : 2017]

16.91 At a sluice gate across a rectangle channel, the upstream flow conditions are depth of 2.0 m, velocity of flow of 1.25 m/sec. The flow conditions at the vena contracta just downstream of the gate can be taken as: depth of 0.44 m; velocity of flow of 5.68 m/sec. What is the total thrust on the gate on its upstream face (to the nearest 10 units)?

(a) 770 kgf (b) 800 kgf
(c) 825 kgf (d) 870 kgf

[ESE : 2017]

16.92 In a rectangular open channel, 2.0 m wide, water flows at a depth of 0.8 m. It discharges over an aerated sharp-crested weir over the full width, with depth over weir crest being 0.25 m. $C_c = 0.61$. Adjusting for velocity head of approach, what would be the discharge through the channel? $\sqrt{2g} = 4.43$ units

(a) 0.439 m^3/sec (b) 0.445 m^3/sec
(c) 0.453 m^3/sec (d) 0.461 m^3/sec

[ESE : 2018]

16.93 An open channel is of isosceles triangle shape, with side slopes *I* vertical and *n* horizontal. The ratio of the critical depth to specific energy at critical depth will be

(a) $\frac{2}{3}$ (b) $\frac{3}{4}$
(c) $\frac{4}{5}$ (d) $\frac{5}{6}$

[ESE : 2018]

16.94 A 2 m wide rectangular channel carries a discharge of 10 m^3/s. What would be the depth of flow if the Froude number of the flow is 2.0?

(a) 1.72 m (b) 1.36 m
(c) 0.86 m (d) 0.68 m

[ESE : 2018]

16.95 M_3 profile is indicated by which of the following conditions?

(a) $y_0 > y_c > y$ (b) $y_0 > y_0 > y_c$
(c) $y_c > y_0 > y$ (d) $y > y_c > y_0$

[ESE : 2018]

16.96 The sequent depth ratio in a rectangular channel is 14. The Froude number of the supercritical flow will be

(a) 6.62 (b) 7.55
(c) 8.45 (d) 10.25 [ESE : 2018]

16.97 In a hydraulic jump, the depths on the two sides are 0.4 m and 1.4 m. The head loss in the jump is nearly

(a) 0.45 m (b) 0.65 m
(c) 0.80 m (d) 0.90 m

[ESE : 2018]

16.98

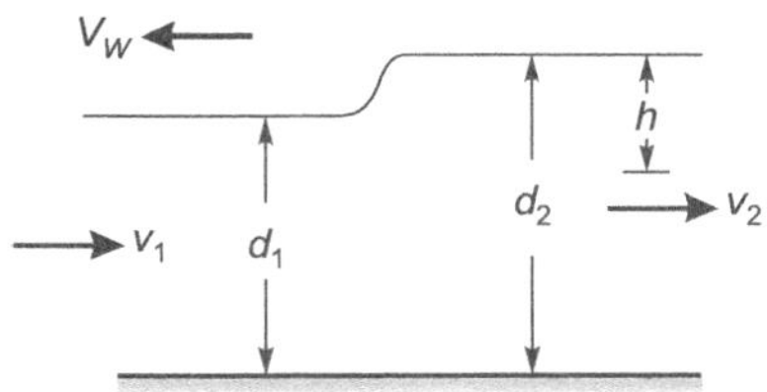

Consider the occurrence of a surge at the water surface of a wide rectangular channel flow, as in the figure, where the one-dimensionally considered velocities are v_1 and v_2 and the depths are d_1 and d_2, with the surge height h, whereby $d_2 - d_1 = h$, moving at a speed of V_W over depth d_1. Joint application of continuity and momentum principles will indicate the surge front speed V_W, to be

(a) $V_W = \sqrt{gd_1}\left(1+\frac{3}{2}\frac{h}{d_1}\right)^{1/2}$

(b) $V_W = \sqrt{gd_1}\left(1+\frac{3}{2}\frac{h}{d_1}\frac{1}{2}\left(\frac{h}{d_1}\right)^2\right)^{1/2}$

(c) $V_W = \sqrt{gd_1}\left(1+\frac{h}{d_1}\right)^{1/2}$

(d) $V_W = \sqrt{gd_1}\left(1+\left(\frac{h}{d_1}\right)^2\right)^{1/2}$

[ESE : 2018]

16.99 **Statement (I)** : A nomogram is a ready reckoner to compute any two hydraulic parameters like discharge, pipe diameter, pipe slope and flow velocity in the pipe if the other two are known.

Statement (II) : Hydraulic parameters can be determined by using Manning's or Chezy's formulae; and a Nomogram is an organized compilation of a number of such, varied computations.

(a) Both Statement (I) and Statement (II) are individually true and Statement (II) is the correct explanation of Statement (I)
(b) Both Statement (I) and Statement (II) are individually true but Statement (II) is NOT the correct explanation of Statement (I)
(c) Statement (I) is true but Statement (II) is false
(d) Statement ( ) is false but Statement (II) is true

[ESE : 2018]

16.100 Floating logs of wood tend to move to the mid-river reach on the water surface. This is due to

(a) Least obstruction from the banks.
(b) 2-cell transverse circulation in the flow.
(c) Fastest velocity along the mid-river reach.
(d) Near-symmetry of the isovels across the section is conducive to principle of least work.

[ESE : 2018]

16.101 Which of the following assumptions is/are made in the analysis of hydraulic jump?

1. It is assumed that before and after jump formation the flow is essentially two-dimensional and that the pressure distribution is hydrostatic.
2. The length of the jump is small so that the losses due to friction on the channel floor are small and hence neglected.

3. The channel floor is horizontal or the slope is so gentle that the weight component of the water mass comprising the jump is very high.

(a) 1 only (b) 2 only
(c) 3 only (d) 1, 2 and 3

[ESE : 2019]

16.102 Which one of the following statements is correct regarding flow in open channel?

(a) The curve for kinetic energy is a parabola
(b) The curve for potential energy is a parabola
(c) Specific energy is asymptotic to the vertical axis
(d) At critical depth the specific energy is maximum

[ESE : 2020]

16.103 Which one of the following statement is correct regarding critical state of flow through a channel section?

(a) Specific energy is a minimum for a given discharge
(b) Specific energy is a maximum for a given discharge
(c) The Froude number is greater than two
(d) The discharge is a minimum for a given specific force

[ESE : 2020]

16.104 The depth of flow of a channel section at which the specific energy is minimum, is called

(a) Critical velocity (b) Critical depth
(c) Critical energy (d) Subcritical flow

[ESE : 2022]

16.105 Which one of the following statements is correct for flow in open channels?

(a) When the depth of flow in a channel is greater than the critical depth, the flow is said to be supercritical flow
(b) When the depth of flow in a channel is equal to critical depth, the flow is said to be super-critical flow
(c) When the depth of flow in a channel is less than the critical depth, the flow is said to be subcritical flow
(d) When the depth of flow in a channel is greater than the critical depth, the flow is said to be subcritical flow

[ESE : 2023]

16.106 If Q = discharge into the channel, a = Cross sectional area at the entrance, and D_n is depth at the channel entrance, what is the equation for depth of the reservoir (D_r) when the flow in the channel is subcritical?

(a) $D_r = D_n - \frac{Q}{2ga^2}$

(b) $D_r = D_n - \frac{Q^2}{2ga^2}$

(c) $D_r = D_n + \frac{Q}{2ga^2}$

(d) $D_r = D_n + \frac{Q^2}{2ga^2}$

[ESE : 2024]

16.107 Consider the following statements regarding uniform flow in channel :

1. The depth of flow and wetted area are constant at every section along the channel reach.
2. The velocity of flow and discharge are varying along the channel reach.
3. The total energy line, water surface and the channel bottom are all parallel.

Which of the above statements is/are correct?

(a) 1 and 2 (b) 2 and 3
(c) 3 only (d) 1 and 3

[ESE : 2024]

■■■■

Answers Open Channel Flow

16.1	(b)	16.2	(a)	16.3	(b)	16.4	(a)	16.5	(d)	16.6	(c)	16.7	(b)	16.8	(c)
16.9	(d)	16.10	(d)	16.11	(a)	16.12	(b)	16.13	(c)	16.14	(c)	16.15	(d)	16.16	(c)
16.17	(a)	16.18	(c)	16.19	(c)	16.20	(c)	16.21	(a)	16.22	(b)	16.23	(b)	16.24	(d)
16.25	(c)	16.26	(c)	16.27	(d)	16.28	(d)	16.29	(c)	16.30	(b)	16.31	(b)	16.32	(d)
16.33	(c)	16.34	(c)	16.35	(b)	16.36	(d)	16.37	(c)	16.38	(b)	16.39	(d)	16.40	(b)
16.41	(c)	16.42	(d)	16.43	(b)	16.44	(d)	16.45	(b)	16.46	(c)	16.47	(b)	16.48	(d)
16.49	(c)	16.50	(d)	16.51	(a)	16.52	(b)	16.53	(d)	16.54	(c)	16.55	(c)	16.56	(c)
16.57	(c)	16.58	(b)	16.59	(b)	16.60	(c)	16.61	(b)	16.62	(d)	16.63	(d)	16.64	(*)
16.65	(d)	16.66	(c)	16.67	(c)	16.68	(d)	16.69	(c)	16.70	(d)	16.71	(c)	16.72	(b)
16.73	(b)	16.74	(c)	16.75	(d)	16.76	(b)	16.77	(c)	16.78	(c)	16.79	(d)	16.80	(b)
16.81	(c)	16.82	(a)	16.83	(c)	16.84	(a)	16.85	(b)	16.86	(c)	16.87	(b)	16.88	(b)
16.89	(d)	16.90	(c)	16.91	(a)	16.92	(c)	16.93	(c)	16.94	(c)	16.95	(a)	16.96	(d)
16.97	(a)	16.98	(b)	16.99	(a)	16.100	(b)	16.101	(b)	16.1042	(a)	16.103	(a)	16.104	(b)
16.105	(d)	16.106	(d)	16.107	(d)										

Explanations Open Channel Flow

16.1 (b)

(i) When the gate is partly closed suddenly, positive surge moves upstream and negative surge moves downstream.

(ii) When the gate is opened suddenly, positive surge moves downstream and negative surge moves upstream.

16.3 (b)

Since, $H = z + E$ (H = total energy head, E specific energy)

differentiating above equation, we get

$$\frac{dH}{dx} = \frac{dz}{dx} + \frac{dE}{dx}$$

or,
$$\frac{dE}{dx} = S_0 - S_f$$

For wide rectangular channel

Using Manning's equation

$$Q = \frac{1}{n} A R^{2/3} S_0^{1/2}$$

$\because$
$$R = \frac{By}{B+2y} \approx y \qquad \{\because B >> y\}$$

Hence,
$$y_n^{5/3} \propto \frac{1}{\sqrt{S_0}}$$

and
$$y^{5/3} \propto \frac{1}{\sqrt{S_f}}$$

$$\frac{S_f}{S_0} = \left(\frac{y_n}{y}\right)^{10/3}$$

and
$$\frac{Q^2T}{gA^3} = \frac{Q^2B}{g(By)^3} = \frac{q^2}{gy^3} = \left(\frac{y_c}{y}\right)^3$$

$$\frac{dy}{dx} = S_0\left[\frac{1-\left(\frac{y_n}{y}\right)^{10/3}}{1-\left(\frac{y_c}{y}\right)^3}\right]$$

Using Chezy's formula

Since,
$$Q = AC\sqrt{R}\sqrt{S_0} \quad \{A = By\,;\, R = y\}$$

Then,
$$y_n^{3/2} \propto \frac{1}{\sqrt{S_0}}$$

and
$$y^{3/2} \propto \frac{1}{\sqrt{S_f}}$$

Hence,
$$\frac{S_f}{S_0} = \left(\frac{y_n}{y}\right)^3$$

and
$$\frac{Q^2T}{gA^3} = \left(\frac{y_c}{y}\right)^3$$

$$\frac{dy}{dx} = S_0\left[1-\left(\frac{y_n}{y}\right)^3 \Big/ 1-\left(\frac{y_c}{y}\right)^3\right]$$

$$\frac{dy}{dx} = S_0\frac{1-(y_0/y)^N}{1-(y_c/y)^M}$$

For Chezy's formula $N = 3$; $M = 3$

For Manning's formula $N = 10/3$; $M = 3$

N is related to conveyance $\left(\frac{1}{n}AR^{2/3}\right)$.

16.4 (a)

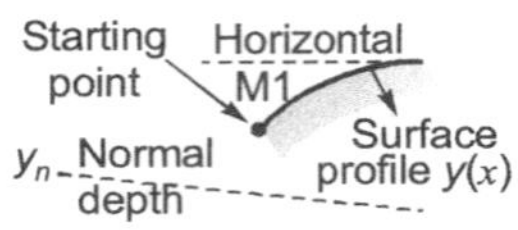

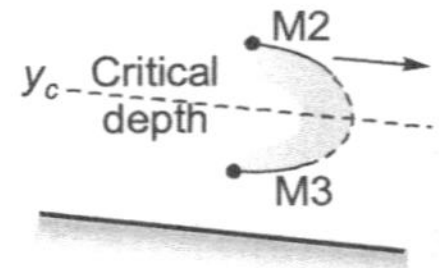

The surface profile is: M_1, M_3, *HJ* and M_1 Since NDL is above CDL so the bed slope is mild.

16.5 (d)

$$\frac{Q_2}{Q_1} = \sqrt{\frac{S_2}{S_1}}$$

$$\therefore \quad Q_2 = 6000\times\sqrt{\frac{10{,}000}{14{,}400}} = 500 \text{ m}^3\text{/s}$$

16.6 (c)

Channel Slope	Profile Notation	Flow Depth	Froude Number	Profile Slope	Surface Profile
Mild (M) $y_c < y_n$, $S_0 < S_c$	M_1	$y > y_n$	Fr < 1	$\frac{dy}{dx} > 0$	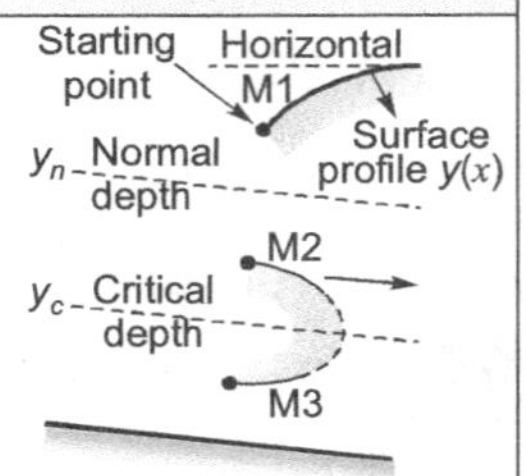
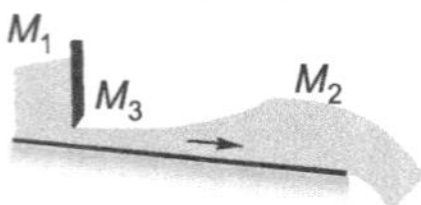	M_2	$y_c < y < y_n$	Fr < 1	$\frac{dy}{dx} < 0$	
	M_3	$y < y_c$	Fr > 1	$\frac{dy}{dx} > 0$	Channel bottom, $S_0 < S_c$
Steep (S) $y_c > y_n$, $S_0 > S_c$	S_1	$y > y_c$	Fr < 1	$\frac{dy}{dx} > 0$	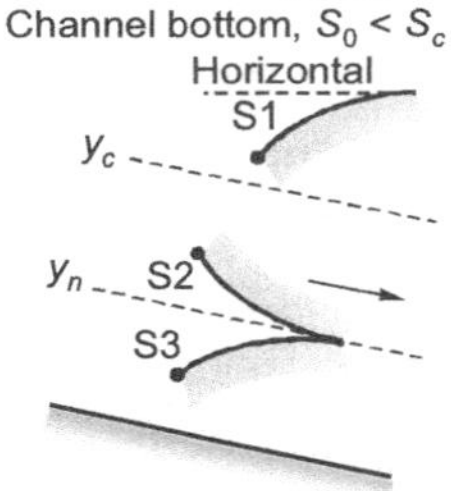
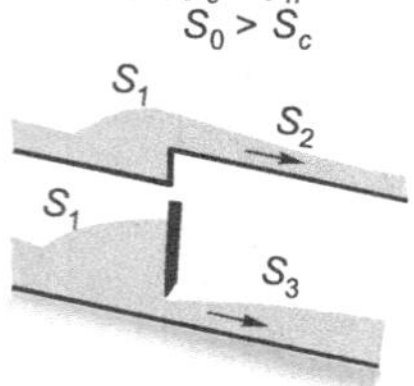	S_2	$y_n < y < y_c$	Fr > 1	$\frac{dy}{dx} < 0$	
	S_3	$y < y_n$	Fr > 1	$\frac{dy}{dx} > 0$	Channel bottom, $S_0 > S_c$
Critical (C) $y_c = y_n$, $S_0 = S_c$	C_1	$y > y_c$	Fr < 1	$\frac{dy}{dx} > 0$	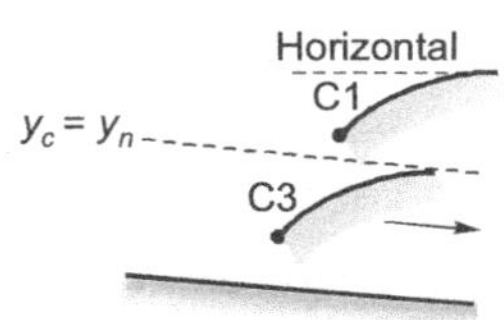
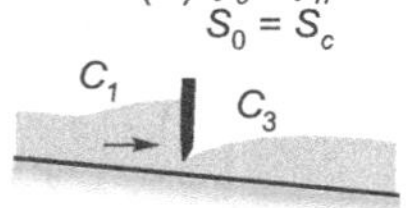	C_3	$y < y_c$	Fr > 1	$\frac{dy}{dx} > 0$	Channel bottom, $S_0 = S_c$
Horizontal (H) $y_n \to \infty$, $S_0 = 0$	H_2	$y > y_c$	Fr < 1	$\frac{dy}{dx} < 0$	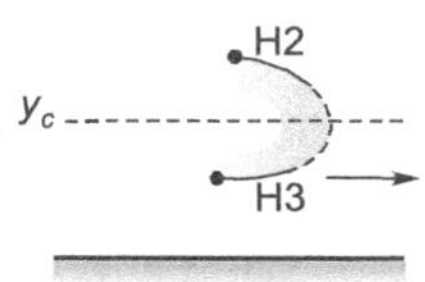
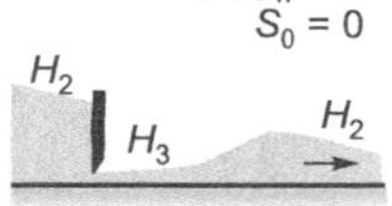	H_3	$y < y_c$	Fr > 1	$\frac{dy}{dx} > 0$	Channel bottom, $S_0 = 0$
Adverse (A) $S_0 < 0$, y_n: does not exist	A_2	$y > y_c$	Fr < 1	$\frac{dy}{dx} < 0$	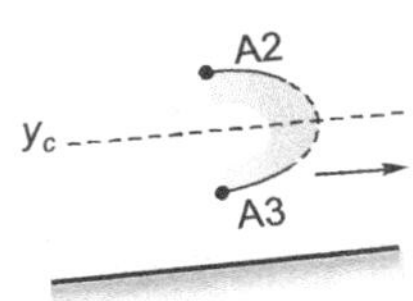
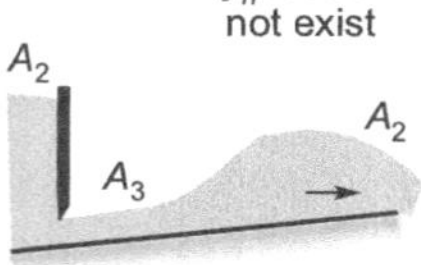	A_3	$y < y_c$	Fr > 1	$\frac{dy}{dx} > 0$	Channel bottom, $S_0 < 0$

16.7 (b)

A hydraulic jump is needed when supercritical flow changes ino subcritical flow or when critical depth line is to be crossed.

16.8 (c)

For hydraulically efficient rectangular section.

$$\frac{B_{em}}{y} = 2.0$$

Note:

Sl. No.	Channel shape	A	P_{em}	B_{em}	R_{em}	T_{em}
1.	Rectangle (half square)	$2y_e^2$	$4y_{em}$	$2y_{em}$	$\frac{y_{em}}{2}$	$2y_{em}$
2.	Trapezoidal (half regular hexagon)	$\sqrt{3}y_e^2$	$2\sqrt{3}y_{em}$	$\frac{2}{\sqrt{3}}y_e$	$\frac{y_{em}}{2}$	$\frac{4}{\sqrt{3}}y_e$
3.	Circular (semicircle)	$\frac{p}{2}y_e^2$	πy_e	–	$\frac{y_{em}}{2}$	$2y_{em}$
4.	Triangle (vertex angle = 90°)	y_e^2	$2\sqrt{2}y_{em}$	–	$\frac{y_{em}}{2\sqrt{2}}$	$2y_{em}$

16.9 (d)

Hydraulic depth,

$$R = \frac{A}{P} = \frac{2\times 1}{2\times 2} = 0.5\text{m}$$

$$Q = C\sqrt{RS} = 60\times\sqrt{0.5\times\frac{1}{800}}\times 2\times 1$$

$$= 3.0\ \text{m}^3/\text{s}$$

16.10 (d)

At critical flow, specific energy and specific force is minimum for a given discharge. In other words for a given specific energy or specific force, discharge is maximum at critical flow.

16.11 (a)

In mild slope critical depth line (CDL) is below the normal depth line (NDL). The flow is subcritical and uniform.

For adverse slope, uniform flow can not be sustained. The CDL is defined by NDL does not exist.

The critical slope has unit Froude number and NDL coincides with CDL.

For a channel of given shape and roughness, critical slope will have a least value under conditions corresponding to a limit value.

The least value of S_c is called the limit slope.

(i) For a trapezoidal section of given geometry and roughness, a given depth of critical flow can be maintained by one and only one critical slope. However, for a given critical slope there can be more than one critical depth.

(ii) For channels of the second kind (m is negative) and for rectangular channels ($m = 0$), only one limit slope exists, Slopes flatter than this cannot be critical and the slopes steeper than this can be critical at two different depths. For a rectangular channel, the limt value of S_{*c} is 8/3 at $\eta_c = 1/6$.

16.12 (b)

$$F_{r_1}^2 = \frac{q^2}{gy_2^3\left(\frac{y_1}{y_2}\right)^3}$$

$$F_{r_1}^2 = \frac{8F_{r_2}^2}{\left(-1+\sqrt{1+8F_r^2}\right)^3} \quad \ldots(i)$$

16.13 (c)

The efficiency of hydraulic ram is about 75%. It is used when large quantity of water available at small height and small quantity of water is to be raised to greater heights.

16.14 (c)

Multistage pumps with impellers in series, increase the head while discharge is constant.

16.15 (d)

Discharge per unit width

$$q = 12.5/5 = 2.5\ \text{m}^3/\text{s/m}$$

Critical depth,

$$y_c = (q^2/g)^{1/3} = (0.64)^{1/3}$$

16.16 (c)

At critical state of flow, specific force is minimum so first derivative of force with respect to depth should be zero.

16.17 (a)

Q-curve is the graph between discharge on horizontal axis and depth of flow on vertical axis at constant specific energy Q. The apex corresponds to the critical depth.

$$h_c = \frac{2}{3}E = \left(\frac{q^2}{g}\right)^{1/3}$$

So vertical co-ordinate of Q curve is $\frac{2}{3}E$.

E-curve is the graph between specific energy on horizontal axis and depth of flow on vertical axis. The apex again corresponds to critical depth of flow.

$$E_c = \frac{3}{2}h_c$$

16.19 (c)

For $y_c < y_0$ the curves are M_1, M_2 and M_3. The control section for mild profiles is downstream so upstream disturbance does not affect them.

Similarly, for $y_c > y_0$ the curves are S_1, S_2, and S_3. The control section for steep profiles is upstream and downstream disturbance does not affect them.

Type-2 curve lie between y_c and y_0.

Type-3 curves are above y_c or y_0 whichever is large.

Type-1 curves are below y_c or y_0 whichever is smaller.

16.20 (c)

The flow duration curve or discharge frequency curve is a plot that shows the percentage of time that flow in a stream is likely to equal or exceed some specified value of interest. A flow duration curve characterizes the ability of the basin to provide flows of various magnitudes. It is extremely useful for the design of structures on a stream.

16.21 (a)

Channel boundary resistances alter the velocity distribution along the cross-section, with varying influence that reduces from the wall of the channel to the water free surface. This non-uniform velocity distribution needs to be taken into consideration for appropriate design of the related structures. The energy and momentum coefficients are used to correct these effects.

$$\text{Momentum correction factor} = \frac{1}{AV^2}\int_0^A v^2 dA$$

$$\text{Kinetic Energy correction factor} = \frac{1}{AV^3}\int_0^A v^3 dA$$

16.22 (b)

Hydraulically efficient trapezoidal channel is half hexagon. Length of sloping sides will be equal to the bottom width. The sides will make 60° angle from horizontal and 30° with vertical.

16.23 (b)

Figure	Area (A)	Perimeter (P)	Hydraulic Radius
A.	2.4	4.4	0.55
B.	$\pi/2$	π	0.5
C.	1	$2\sqrt{2}$	0.35
D.	3	$2+2\sqrt{2}$	0.62

So the arrangement in descending order of hydraulic radii is $D-A-B-C$

16.24 (d)

Critical depth,

$$y_c = \left(\frac{q^2}{g}\right)^{1/3} = \left(\frac{2^2}{9.81}\right)^{1/3} = 0.74\text{m}$$

As $y_A = 1.5$ m and $y_B = 1.6$ m are more than y_c so the flow is subcritical.

$\therefore\ E_A < E_B$

Length of reach AB assuming B on the upstream will be

$$L_{AB} = \frac{E_B - E_A}{S_0 - \bar{S}_f} = \frac{0.09}{\frac{1}{2000} - 0.003} = -36\text{m}$$

Since L_{AB} is negative so B is on the upstream.

16.25 (c)

The velocity is higher in the middle so the water surface level will be lower in the middle and higher at middle.

16.26 (c)

For minimum value of specific force first derivative of force with respect to depth should be zero.

16.27 (d)

When a S_2 profile ends with a sudden drop of bed, the surface drops suddenly and it is called hydraulic drop. This can happen with any profile except A_2.

There can be no hydraulic jump in S_2 profile.

Note: A adverse slope the datum rises along the direciton of flow.

16.28 (d)

The hydraulic jumps in horizontal rectangular channels are classified into five categories based on Froude number F_1 of the super-critical flow as:

Range	Type
$1.0 < F_1 \leq 1.7$	Undular jump
$1.7 < F_1 \leq 2.5$	Weak jump
$2.5 < F_1 \leq 4.5$	Oscillating jump
$4.5 < F_1 \leq 9.0$	Steady jump
$F_1 > 9.0$	Strong or choppy jump

16.29 (c)

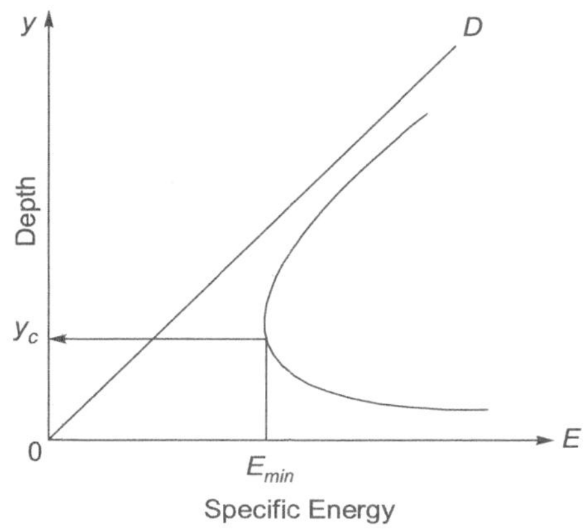

16.30 (b)

$$F_0 = \frac{V}{\sqrt{gy_0}} = \frac{q}{\sqrt{gy_0^3}} \text{ and } y_c^3 = \frac{q^2}{g}$$

$$\therefore \frac{y_c}{y_0} = F_0^{2/3}$$

16.31 (b)

$Q = CA\sqrt{RS}$

For hydraulically efficient trapezoidal section,

$$A = \sqrt{3}y_e^2 = \sqrt{3} \times (0.6)^2 = 0.624\text{m}$$

$$R = \frac{y_e}{2} = \frac{0.6}{2} = 0.3 \text{ m}$$

$$\therefore Q = 75 \times 0.624 \times \sqrt{0.3 \times \frac{1}{250}} = 1.62 \text{ m}^3/\text{s}$$

16.32 (d)

Flow competition will always start from control section.

The calculations must proceed upstream in subcritical flows and downstream in supercritical flows to keep the errors minimum.

16.33 (c)

Momentum equation is not used for derivation of differential equat on for water surface profile in open channels.

16.34 (c)

For very wide rectangular channel,

$$P = y$$

As per Chezy's equation

$$Q \propto y_0^{3/2}$$

$$\frac{\delta Q}{Q} \times 100 = \frac{3}{2} \times 20\% = 30\%$$

As per Manning's equation

$$Q \propto y_0^{5/3}$$

$$\frac{\delta Q}{Q} \times 100 = \frac{5}{3} \times 20\% = 33.33\%$$

Close answer is 36%.

16.35 (b)

We know that

$$Q = AV$$

$$\Rightarrow \quad 5 = A \times 1.25$$

$$\Rightarrow \quad A = 4 \text{ m}^2$$

For a symmetrical right angled triangular channel,

$$m = 1$$

$$\therefore \quad A = my^2 = 1 \times y^2 = y^2$$

$$T = 2my = 2 \times 1 \times y = 2y$$

But $\quad A = 4 \text{ m}^2$

$$\Rightarrow \quad y^2 = 4$$

$$\Rightarrow \quad y = 2 \text{ m}$$

Hydraulic depth, $D = \frac{A}{T} = \frac{4}{2 \times 2} = 1$

Froude Number $= \frac{V}{\sqrt{gD}} = \frac{1.25}{\sqrt{9.81 \times 1}} = 0.4$

16.36 (d)

If the loss of energy across the sluice gate is assumed to be negligible, then by energy equation, we have

$$y_1 + \frac{V_1^2}{2g} = y_2 + \frac{V_2^2}{2g}$$

But discharge per unit width,

$$q = V_1 y_1 = V_2 y_2$$

$$\therefore \quad y_1 + \frac{q^2}{2gy_1^2} = y_2 + \frac{q^2}{2gy_2^2}$$

$$\Rightarrow \quad \frac{q^2}{g} = \frac{2y_1^2 y_2^2}{y_1 + y_2}$$

$$\Rightarrow \quad \frac{q^2}{9.81} = \frac{2\times(2)^2\times(0.5)^2}{2+0.5}$$

$$\Rightarrow \quad q^2 = 0.8\times 9.81$$

$$\Rightarrow \quad q = 2.8 \text{ m}^2/\text{s}$$

16.37 (c)

We know that

$$Q = AV$$

$$\Rightarrow \quad 30 = 6\times 2\times V$$

$$\Rightarrow \quad V = 2.5 \text{ m/s}$$

The specific energy may be given as

$$E = y+\frac{V_1^2}{2g} = 2+\frac{(2.5)^2}{2\times 9.81} = 2.3 \text{ m}$$

16.38 (b)

As the flow over broad crested weir will be critical, which is possible only the chocking.

$$E_c \le E_1 - \Delta Z$$

$$\Rightarrow \quad \Delta Z \ge E_1 - E_c$$

16.39 (d)

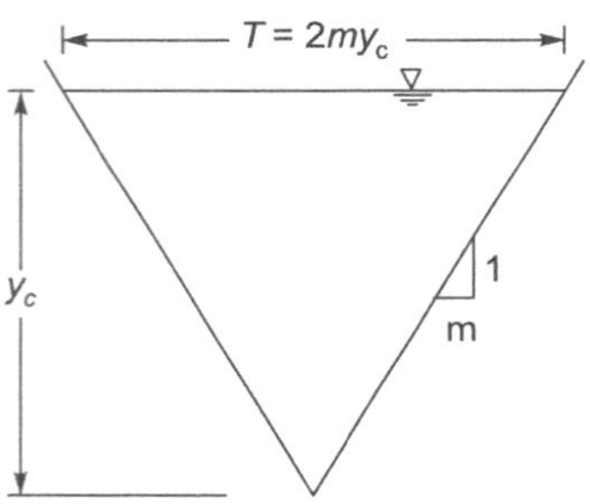

$$\frac{Q^2T}{gA^3} = 1$$

$$\Rightarrow \quad \frac{Q^2(2my_c)}{g(my_c^2)^3} = 1$$

$$A = \frac{Q^2(2my_c)}{g(my_c^2)^3} = my_c^2$$

$$\therefore \quad Q \propto y_c^{5/2}$$

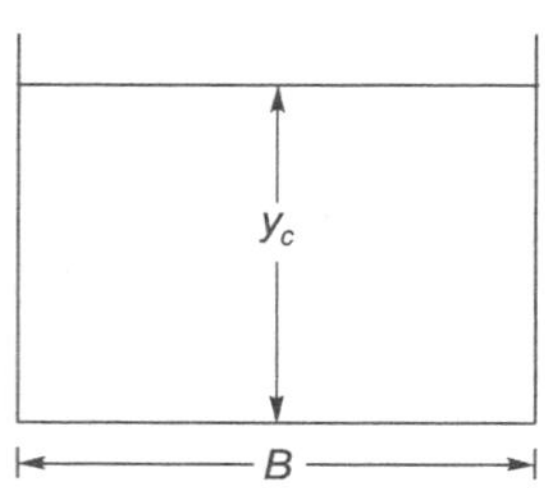

$$\frac{Q^2T}{gA^3} = 1$$

$$\frac{Q^2\times B}{gB^3y_c^3} = 1$$

$$A = By_c$$

$$\Rightarrow \quad Q \propto y_c^{3/2}$$

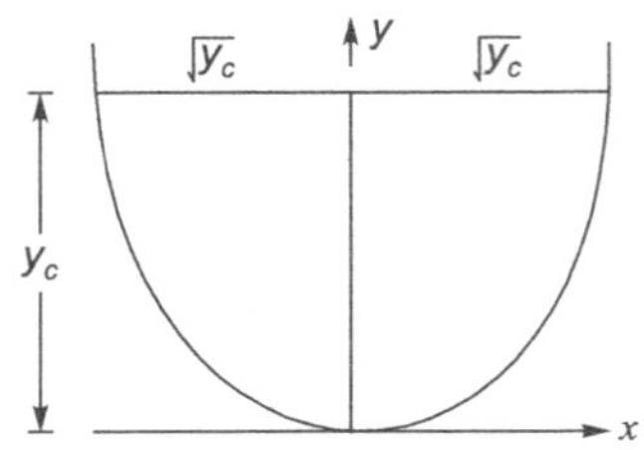

Equation $\quad y = x^2$

$$\therefore \quad y_c = kx^2$$

$$\Rightarrow \quad x = k\sqrt{y_c}$$

$$T = 2k\sqrt{y_c}$$

$$A = \frac{2}{3}\times\left(2k\sqrt{y_c}\right)\times y_c$$

$$= \frac{4}{3}ky_c^{3/2}$$

$$\frac{Q^2\times T}{gA^3} = 1$$

$$Q^2 \propto y_c^{\frac{9}{2}-\frac{1}{2}} \propto y_c^4$$

$$\Rightarrow \quad \frac{Q^2\times 2ky_c^{1/2}}{g\left(\frac{4k^2}{3}\right)^3(y_c)^{a/2}} = 1 \qquad (Q \propto y_c^2)$$

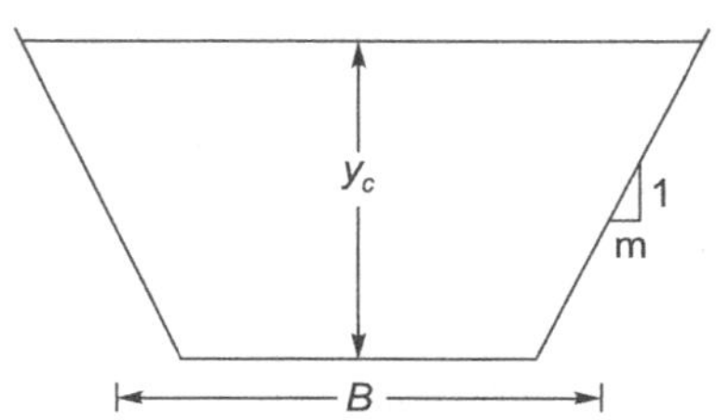

$$\frac{Q^2T}{gA^3} = 1$$

$$\Rightarrow \quad \frac{Q^2(B+2my_c)}{g(B+y_c)^3y_c^3} = 1$$

$$T = B + 2my_c$$

$$A = \frac{1}{2}(2B + 2my_c)y_c$$

$$= (B + my_c)y_c$$

$$Z^2 = \frac{A^3}{T}$$

16.40 (b)

A control section is defined as a section in which a fixed relationship exists between the discharge and depth of flow. Weirs, spillways, sluice gates are some typical examples of structures which give rise to control sections. Any GVF profile will have at least one control section. In the synthesis of GVF profiles occurring in serially connected channel elements, the control sections provide a key to the identification of proper profile shapes.

16.41 (c)

For a wide rectangular channel, the hydraulic radius (R) is approximately equal to the depth of flow. By Manning's equation, we have

$$Q = \frac{1}{n} A R^{2/3} S_0^{1/2}$$

Here $A = By$

$R = y$

$$\therefore \quad Q = \frac{1}{n} \times By \times y^{2/3} S_0^{1/2}$$

$$\Rightarrow \quad \frac{Q}{B} = \frac{1}{n} \times y^{5/3} S_0^{1/2}$$

$$\Rightarrow \quad q = \frac{1}{n} \times y^{5/3} S_0^{1/2}$$

$$\Rightarrow \quad \frac{0.5 \times 0.01}{(0.0004)^{1/2}} = y^{5/3}$$

$$\Rightarrow \quad y = 0.43 \text{ m}$$

16.42 (d)

For a rectangular channel, critical depth is given by

$$y_c^3 = \frac{q^2}{g}$$

$$\Rightarrow \quad q^2 = y_c^3 g$$

$$\Rightarrow \quad q^2 = (0.5)^3 \times 9.81$$

$$\Rightarrow \quad q = 1.107 \text{ m}^2/\text{s}$$

But discharge,

$$Q = q \times B = 1.107 \times 4 = 4.4 \text{ m}^3/\text{s}$$

16.43 (b)

For a rectangular channel in horizontal bed.

The energy dissipation is given by

$$E_L = \frac{(y_2 - y_1)^3}{4y_1 y_2} = \frac{(3.2 - 0.2)^3}{4 \times 0.2 \times 3.2}$$

$$= 10.55 \text{ m}$$

16.44 (d)

The average shear stress (τ_0) at the boundary of a channel in uniform flow is given by

$$\tau_0 = \gamma R S_0$$

But $$R = \frac{A}{P} = \frac{By}{B + 2y}$$

$$= \frac{3 \times 1.5}{3 + (2 \times 1.5)} = 0.75 \text{ m}$$

$$\therefore \quad \tau_0 = 9810 \times 0.75 \times 0.0002$$

$$= 1.47 = 1.5 \text{ N/m}^2$$

16.46 (c)

The Manning's equation for a wide rectangular channel is given as

$$Q = \frac{1}{n} \times B \times y^{5/3} \times S_0^{1/2}$$

Therefore for same value of n, B and S_0, we get

$$Q \propto y^{5/3}$$

$$\Rightarrow \quad \frac{Q_1}{Q_2} = \left(\frac{y_1}{y_2}\right)^{5/3}$$

$$\Rightarrow \quad \frac{100}{Q_2} = \left(\frac{100}{120}\right)^{5/3}$$

$$\Rightarrow \quad Q_2 = \frac{100}{0.738}$$

$$\Rightarrow \quad Q_2 = 136$$

$\therefore$ Increase in discharge

$$= Q_2 - Q_1 = 136 - 100 = 36\%$$

16.47 (b)

The sequent depth ratio for a rectangular channel is given by

$$\frac{y_2}{y_1} = \frac{1}{2}\left(-1 + \sqrt{1 + 8F_1^2}\right)$$

$$\Rightarrow \quad 16.48 = \frac{1}{2}\left(-1 + \sqrt{1 + 8F_1^2}\right)$$

$$\Rightarrow 1 + 8F_1^2 = 1153.28$$

$$\Rightarrow \quad F_1 = \sqrt{144}$$

$$\Rightarrow \quad F_1 = 12.0$$

16.48 (d)

(i) For most efficient triangular section, $m = 1$

$$\therefore \quad P = 2y\sqrt{m^2+1} \qquad [\because y = h]$$

$$\Rightarrow \quad \frac{P}{h} = 2\sqrt{2}$$

$$\Rightarrow \quad \frac{P}{h} = 2.83$$

(ii) For most efficient trapezoidal section,

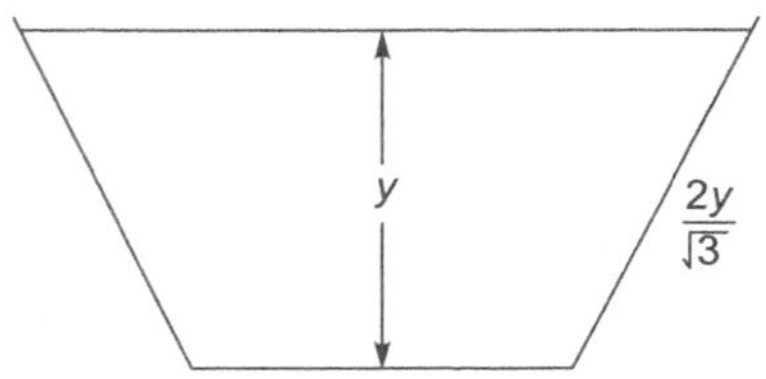

$$P = \frac{6y}{\sqrt{3}}$$

$$\therefore \quad \frac{P}{y} = 2\sqrt{3}$$

16.49 (c)

Most economical channel carries maximum discharge for given area of cross section.

$$\text{Hydraulic radius} = \frac{\text{Cross sectional area}}{\text{Wetted parameter}}$$

Hydraulic radius is maximum when wetted parameter is minimum for a given cross-section area.

16.50 (d)

Jump	Froude number
Undular jump	$1 < Fr_1 < 1.7$
Weak jump	$1.7 < Fr_1 < 2.5$
Oscillating jump	$2.5 < Fr_1 < 4.5$
Steady jump	$4.5 < Fr_1 < 9$
Strong jump	$Fr_1 > 9$

16.51 (a)

A steady nonuniform flow in a prismatic channel with gradual changes in its water surface is termed as gradually varied flow (GVF).

16.52 (b)

Total energy of flow in real fluid always decreases in the direction of flow. The specific energy depends on elevation of bed of channel relative to elevation of total energy line so in varied flow it can decrease or increase. For uniform flow specific energy is constant.

Both statements are correct but reason does not explain assertion.

16.53 (d)

Assertion is incorrect as it is not a necessary condition.

In case of triangular channel,

$$R = \frac{y}{2\sqrt{2}}$$

16.54 (c)

Energy y dissipation in a hydraulic jump is due to large scale eddies caused due to high turbulence. These eddies can seen only over short distance.

16.55 (c)

It is not necessary that surface is smooth in economical section for reduced friction. However if surface is smooth then it allows better discharge flow and increases efficiency. Economic section must have minimum wetted perimeter.

16.56 (c)

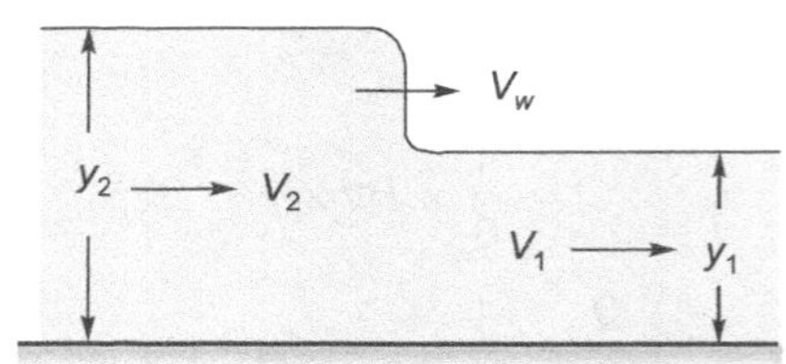

Positive wave travels downstream from gate as it results in increase in depth of flow and negative wave travels from gate onto upstream as it results in decrease in depth of flow.

16.57 (c)

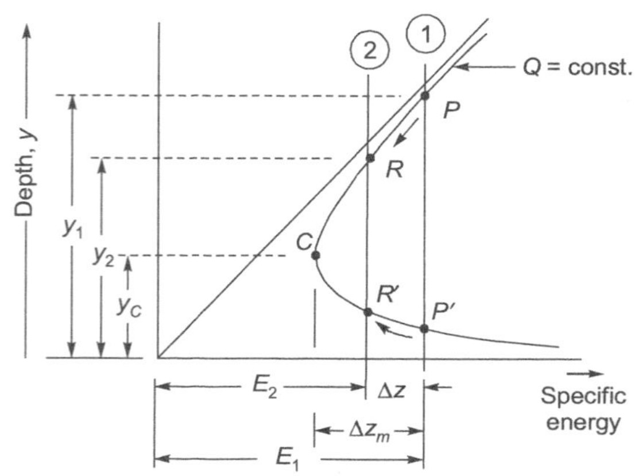

Critical flow occurs when specific energy and specific force is minimum and discharge is maximum.

Alternative depths are depth with same specific energy.

There can be many specific energy curve becasue it changes with change in 'Q' and geometry.

16.58 (b)

If the flow properties, say the depth of flow, in a channel remains constant along the length of the channel it is called uniform flow. A unobstructed flow over the significant length of a irrigation canal will be a uniform flow.

If the change of depth in a varied flow is rapid and appreciable it is called a rapidly varied flow. In a stilling basin hydraulic jump is confined either partly or entirely for energy dissipation. Hydraulic jump is associated with rapid and significant change in depth and other flow properties.

An unsteady flow occurs when the flow properties, such as the depth or discharge at a section changes with time. Sudden closure of a sluice gate creates such phenomena on the upstream of the gate.

If some flow is added to or abstracted from the varied flow it is know as a spatially varied flow. When irrigation water spreads over a field some water percolates down into the soil creating a spatially varied flow.

16.59 (b)

On a mild slope channel the flow below critical depth line is termed as M_3.

16.60 (c)

A most economical trapezoidal channel is half-hexagon such that its internal angle is 120°.

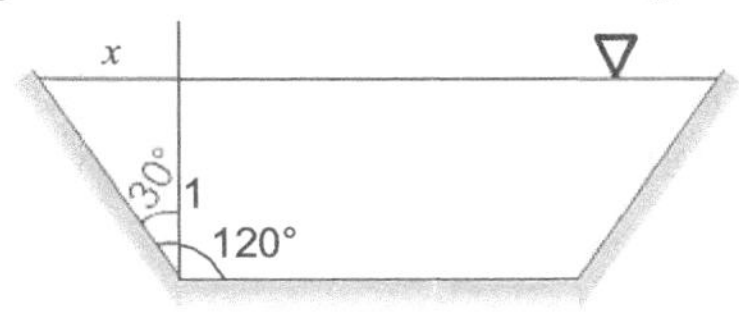

$$\tan 30° = \frac{x}{1}$$

or $$\frac{1}{\sqrt{3}} = \frac{x}{1}$$

or $$x = 0.577$$

16.61 (b)

At critical flow, specific energy and specific force is minimum for a given discharge. In other words for a given specific energy or specific force discharge is maximum at critical flow and Froude number is equal to unity.

16.62 (d)

For a most economical rectangular channel

$$y = \frac{b}{2} = \frac{4}{2} = 2\text{m}$$

16.63 (d)

$$C = \sqrt{gy} = \sqrt{10 \times 1.6} = 4 \text{ m/s}$$

∴ Velocity of surge wave travelling

Upwards = C – V

$= 4 - 2.4 = 1.6$ m/s

16.65 (c)

While carrying a discahrge 'Q' through the channel, if the permissible velocity is not be exceeded, then the area $A = \frac{Q}{V_P}$ represents the min area required for the best hydraulic section. However the area required to be excavated to achieve the flow are will depend upon the removal of overburdne. Hence the excavation may not be the min excavation even though area is min.

16.66 (c)

It depends on discharge whether a given channel is mild, steep or critical. Hence Statement (I) is correct.

16.69 (c)

$$F_r = 2.83$$

$$= \frac{v}{\sqrt{gy_n}} = \frac{q}{y_n\sqrt{gy_n}} \quad [\because q = vy_n]$$

$$\Rightarrow \quad (2.83)^2 = \frac{(q^2/g)}{y_n^3} = \frac{y_c^3}{y_n^3}$$

$$\Rightarrow \quad \frac{y_c}{y_n} = (2.83)^{2/3} = 2$$

16.70 (d)

Only energy and continuity equations are sufficient

16.71 (c)

The characteristics of critical flow are

(i) The specific energy is minimum for the given discharge.

(ii) The Froude number is equal to unity.

(iii) For a given specific energy the discharge is maximum at the critical flow.

(iv) The velocity head is equal to half the hydraulic depth in a channel of small slope.

(v) Specific force is minimum for the given discharge.

(vi) For a given discharge specific force for the flow weill be min is always half of a square.

16.72 (b)

For most efficient rectangular channel,

Hydraulic radius = $\frac{y}{2}$

and depth, $y = \frac{B}{2}$

$$\therefore \quad R = \frac{y}{2} = \left(\frac{B/2}{2}\right) = \frac{B}{4}$$

$$= \frac{5}{4} = 1.25 \text{ m}$$

16.73 (b)

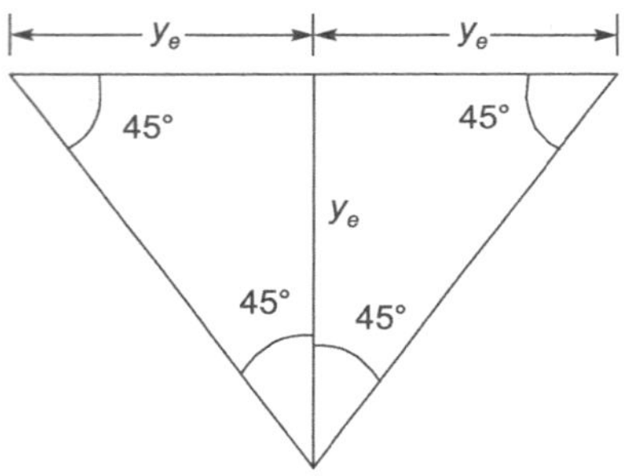

$$\therefore \quad \frac{Q^2(2y_c)}{gy_c^6} = 1$$

$$\Rightarrow \quad y_c^5 = \left(\frac{2Q^2}{g}\right)$$

$$T = 2y_e$$

$$A = y_e^2$$

16.74 (c)

Assuming channel to be rectangular horizontal and frictionless.

$$E_L = \frac{(y_2 - y_1)^3}{4y_1 \cdot y_2} = \frac{(1.4 - 0.3)^3}{4 \times 1.4 \times 0.3} = 0.79 \text{ m}$$

16.75 (d)

As per the question, we have to keep the discharge constant. With the formation of ice resistance to flow (↑) hence velocity will (↓). So the keep the discharge constant, area of flow must is (↑).

Note: Practically Q will fall due to formation of ice, but in these question we have keep Q constant and change the other variables accordingly.

16.76 (b)

From Manning's equation

$$Q = \frac{1}{n} AR^{2/3} S^{1/2}$$

For a wide rectangular channel R = y

$$\therefore \quad Q = \frac{1}{n}(By)y^{2/3} S_E^{1/2}$$

$$\Rightarrow \quad S_E = \frac{q^2 n^2}{y^{10/3}}$$

$$\therefore \text{Slope of TEL (SE)} = \frac{dh}{dx} = \frac{q^2 n^2}{y^{10/3}}$$

$$\therefore \quad \frac{dh}{dx} \propto \frac{1}{y^{10/3}}$$

As the bed slope of the channel are very small, so the water surface well have very small curvature hence the normal acc can be neglected. Therefore force GVF the pressure distributor is assumed to be hydrostatic.

16.77 (c)

Assuming channel to be horizontal, frictionless, rectangular channel.

Head loss,

$$\therefore \quad h_L = \frac{(y_2 - y_1)^3}{4y_1 y_2} = \frac{(1.2 - 0.3)^3}{4 \times 1.2 \times 0.3} = 0.50 \text{ m}$$

Background: Discharge over a aerated shart-crested were considering velocity of approach is given by

$$q = \frac{2}{3} C_d \sqrt{2g}\, h^{3/2}$$

Where h is the head causing the flow, given by

$$h = y + \frac{V^2}{2g}$$

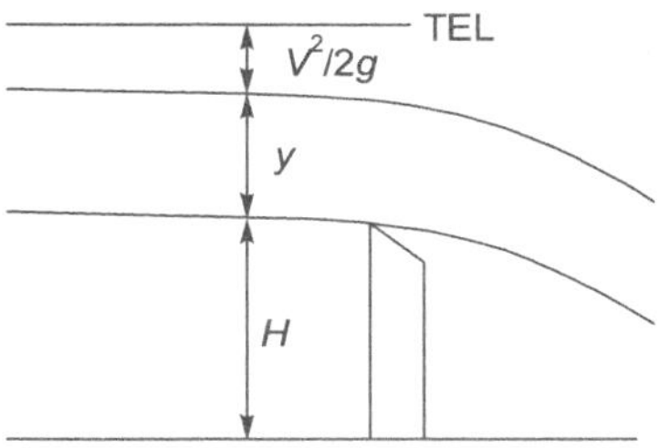

Considering the velocity of appraoch.

16.78 (c)

Given:

Length = 2000 m, Slope = 1/4000

$$\text{Percentage error in slope} = \frac{0.4}{\left(\frac{2000}{4000}\right)\times 100}\times 100$$

$$\Rightarrow \frac{dS}{S}\times 100 = 0.8\%$$

From chezy's equation

$$Q = AC\sqrt{RS} = AC\sqrt{\frac{A}{P}S}$$

$$= CA^{3/2}P^{-1/2}S^{1/2}$$

$$\Rightarrow \frac{dQ}{Q}\times 100 = \frac{3}{2}\frac{dA}{A}\times 100 + \frac{1}{2}\frac{dP}{P}\times 100 + \frac{1}{2}\frac{dS}{S}\times 100$$

$$\frac{3}{2}\frac{dA}{A}\times 100 + \frac{1}{2}\frac{dP}{P}\times 100 + \frac{1}{2}\frac{dS}{S}\times 100$$

$$= \frac{3}{2}\times 6 + \frac{1}{2}\times 4 + \frac{1}{2}\times 0.8$$

$$= 9 + 2 + 0.4 = 11.4\%$$

16.79 (d)

For critical flow, specific energy and specific force is minimum for a given discharge.

16.81 (c)

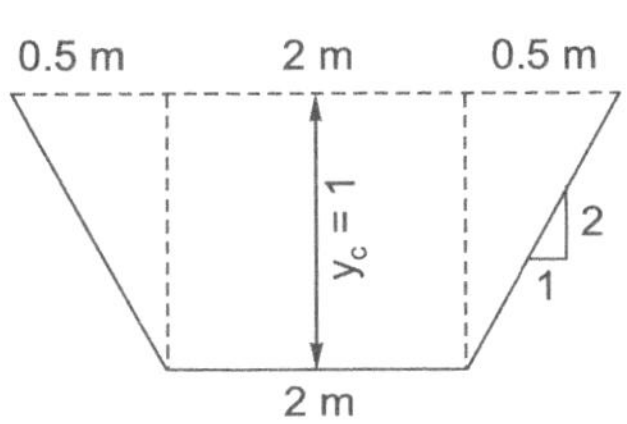

$$A = \frac{1}{2}(2+3)\times 1 = 2.5\text{ m}^2$$

$$T = 3.0\text{ m}$$

For critical flow

$$\frac{Q^2T}{gA^3} = 1$$

$$\Rightarrow \frac{Q^2\times 3.0}{10\times 2.5^3} = 1$$

$$\Rightarrow Q = 7.216\text{ m}^3/\text{s}$$

Critical velocity,

$$V_c = \frac{Q}{A} = \frac{7.216}{2.5}$$

$$= 2.886\text{ m}^3/\text{s}$$

16.82 (a)

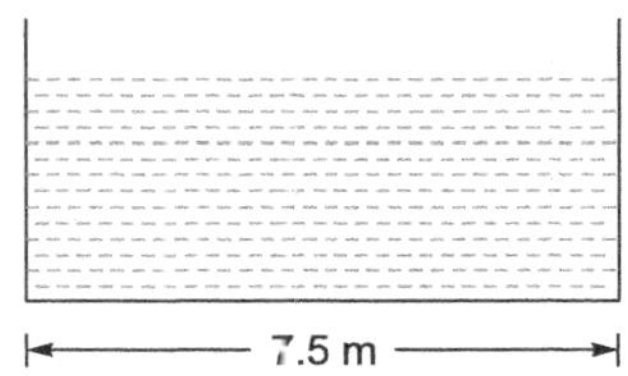

$$Q = 12\text{ m}^3/\text{s}$$

$$V = 1.5\text{ m/s}$$

$$V = \frac{Q}{A}$$

$$1.5 = \frac{12}{7.5\times y}$$

$$y = 1.067\text{ m}$$

$$E = y + \frac{v^2}{2g}$$

$$= 1.067 + \frac{1.5^2}{2\times 9.8} = 1.18\text{ m}$$

16.83 (c)

$$\frac{y_2}{y_1} = 10$$

$$\frac{y_2}{y_1} = \frac{1}{2}\left[-1+\sqrt{1+8F_1^2}\right]$$

$$10 = \frac{1}{2}\left[-1+\sqrt{1+8F_1^2}\right]$$

$$20 = -1+\sqrt{1+8F_1^2}$$

$$21 = \sqrt{1+8F_1^2}$$

$$441 = 1+8F_1^2$$

$$8F_1^2 = 440$$

$$F_1 = \sqrt{\frac{440}{8}} = 7.42$$

16.84 (a)

Assuming the jump to be free jump.

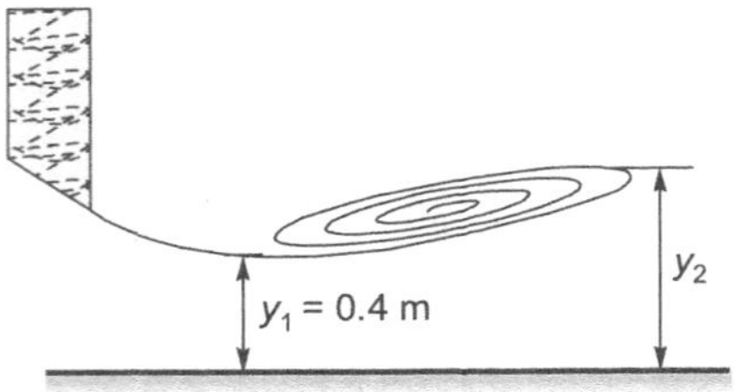

y_1 = Pre jump depth = Depth at vena contracta
y_2 = Post jump depth

$$q = v_1 y_1 = 6 \times 0.4 = 2.4 \text{ m}^3/\text{s}$$

$$F_1^2 = \frac{q^2}{g y_1^3} = \frac{2.4^2}{10 \times 0.4^3} = 9$$

$$\frac{y_2}{y_1} = \frac{1}{2}\left[-1 + \sqrt{1 + 8F_1^2}\right]$$

$$= \frac{1}{2}\left[-1 + \sqrt{73}\right] = 1.51 \text{ m}$$

16.85 (b)

Most efficient rectangular channel is always half of a square.

$$R = \frac{A}{P} = \frac{B \cdot y}{B + 2y}$$

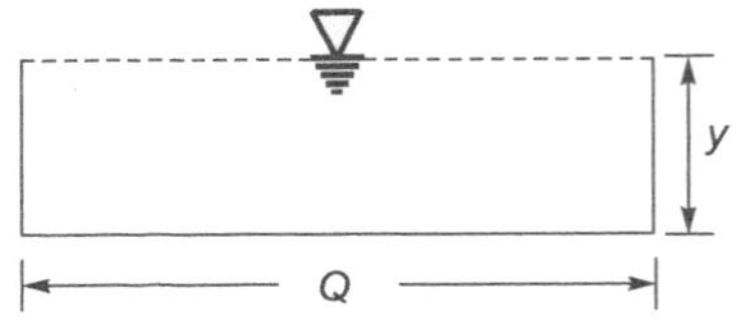

For hydraulic efficient rectangular channel.

$$B = 2y$$

$$\therefore \quad R = \frac{2y^2}{2y + 2y} = \frac{y}{2}$$

$$= \frac{B}{4} = \frac{5}{4} = 1.25$$

16.86 (c)

$$\frac{2q^2}{g} = y_1 y_2 (y_1 + y_2)$$

$$\frac{2q^2}{9.8} = 0.4 \times 5(0.4 + 5) = 2 \times 5.4 = 10.8$$

$$q = \sqrt{\frac{10.8 \times 9.8}{2}} = 7.3 \text{ m}^2/\text{s}$$

16.89 (d)

Upstream Fr_1	Depth Ratio y_2/y_1	Fraction of Energy Dissipation	Discription	Surface Profile
<1	1	0	*Impossible jump.* Would violate the second law of thermodynamics.	
1–1.7	1–2	<5%	*Undular jump (or standing wave).* Small rise in surface level. Low energy dissipation. Surface rollers develop near Fr = 1.7.	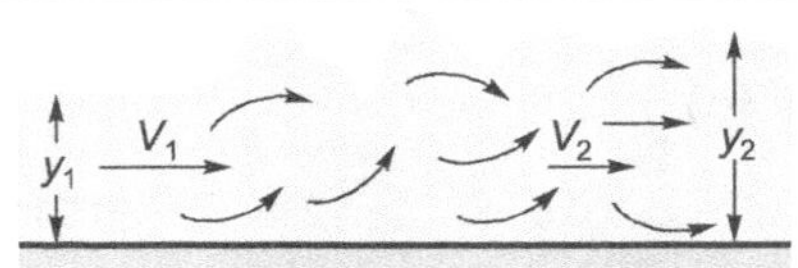
1.7–2.5	2–3.1	5–15%	*Weak jump.* Surface rising smoothly, with small rollers. Low energy dissipation.	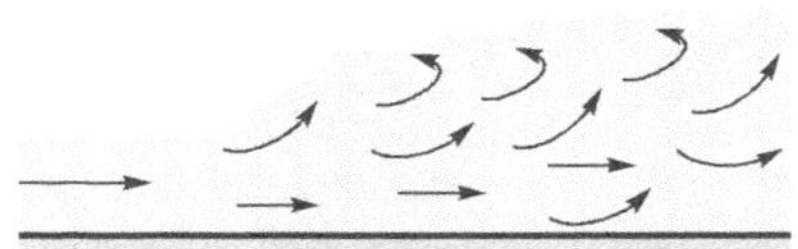
2.5–4.5	3.1–5.9	15–45%	*Oscillating jump.* Pulsations caused by entering jets at the bottom generate larger waves that can travel for miles and damage earth banks. Should be avoided in the design of stilling basins.	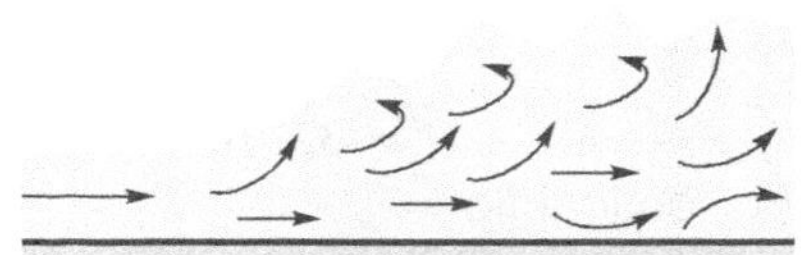

4.5–9	5.9–12	45–70%	*Steady jump.* Stable, well-balanced, and insensitive to downstream conditions. Intense eddy motion and high level of energy dissipation within the jump. Recommended range for design.	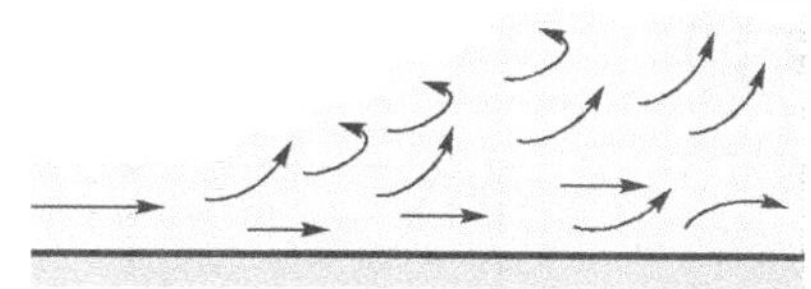
>9	>12	70–85%	*Strong jump.* Rough and intermittent. Very effective energy dissipation but may be uneconomical compared to other designs.	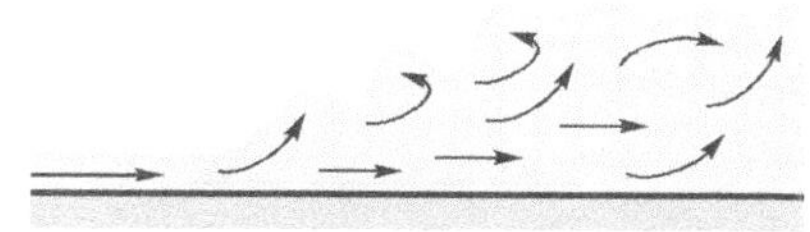

$$F_1^2 = \frac{q^2}{gy^3}$$

$$\therefore \quad F_1^2 = \frac{(1.019)^2}{9.81\times(0.1019)^3}$$

$$F_1 = 10$$

$\therefore$ $F_1 = 10$ is greater than 9 i.e. $F_1 > 9$ [Strong jump]

16.90 (c)

For OCF, if nothing is given then use Manning's equation for calculation of discharge. And force a wide rectangular channel

$$R = y$$

$$Q = \frac{1}{n}AR^{2/3}S^{1/2}$$

$$\Rightarrow \quad Q = \frac{1}{n}(By)y^{2/3}S^{1/2}$$

$$Q \propto y^{5/3}$$

$$Q = ky^{5/3}$$

16.91 (a)

$$F = \frac{\rho(y_1-y_2)^3}{2(y_1+y_2)} = \frac{1000}{2}\frac{(2-0.44)^3}{(2+0.44)}$$

$= 762.39$ kgf $\simeq 770$ kgf

16.92 (c)

Approximate calculation

$$Q = \frac{2}{3}C_d\sqrt{2g}\,BH^{3/2}$$

$$Q = \frac{2}{3}\times 0.61\times 4.43\times 2\times 0.25^{3/2}$$

$= 0.450$ m³/s

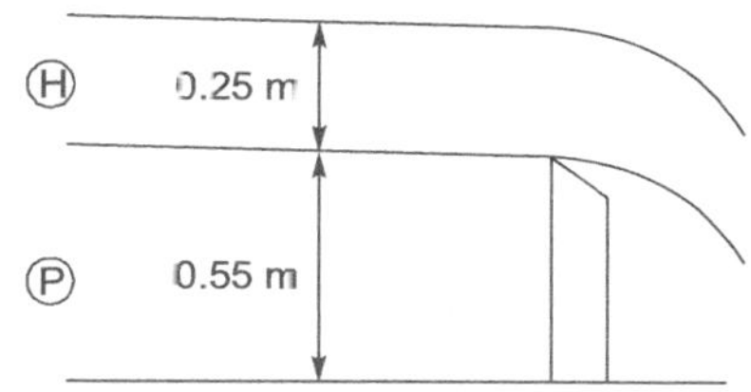

Considering the velocity of approah as (V_a)

$$V_a = \frac{Q}{(P+H)\times B} = \frac{0.450}{0.80\times 2.0} = 0.281$$

$$\therefore \quad h_a = \frac{V_a^2}{2g} = \frac{0.281^2}{2\times 9.81} = 4.038\times 10^{-3}$$

$$Q = \frac{2}{3}\times C_d\times\sqrt{2g}B(H+h_a)^{3/2}$$

$$= \frac{2}{3}\times 0.61\times 4.43\times 2\left[0.25+4.038\times 10^{-3}\right]^{3/2}$$

$= 0.461$ m³/s

Background: Discharge over a aerated shart-crested were considering velocity of approach is given by

$$q = \frac{2}{3}C_d\sqrt{2g}\,h^{3/2}$$

Where h is the head causing the flow, given by

$$h = y + \frac{V^2}{2g}$$

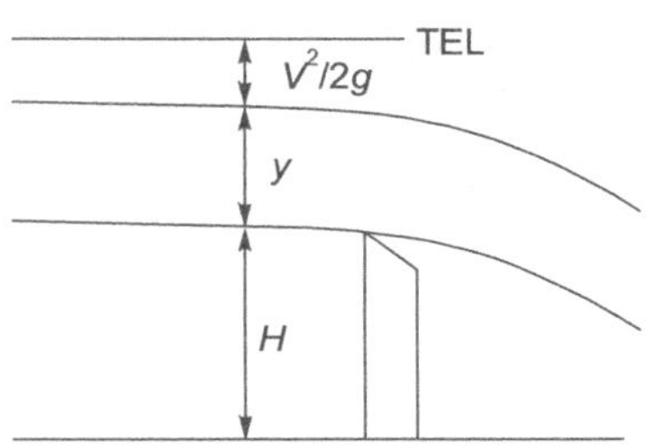

Considering the velocity of appraoch.

16.93 (c)

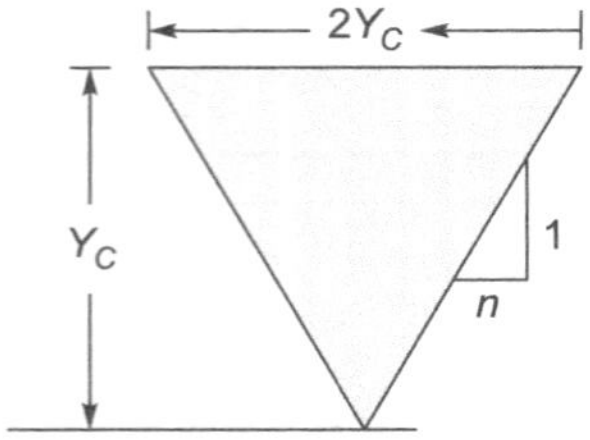

For isosceles triangle the central angle = 90°
∴Side slope is 1 V : 1 H

$$\text{Area} = \frac{1}{2}(2Y_C)(Y_C) = Y_C^2$$

$$\text{Specific energy } (E_C) = Y_C + \frac{V_C^2}{2g}$$

$$\therefore \quad E_C = Y_C + \frac{Q^2}{2gA^2} \quad \text{... (i)}$$

Again for critical flow $\frac{Q^2T}{gA^3} = 1$ ... (ii)

From (i) and (ii)

$$\therefore \quad E_C = Y_C + \frac{A}{2T}$$

$$\Rightarrow \quad E_C = Y_C + \frac{Y_C^2}{2\times(2Y_C)} = \frac{5}{4}Y_C$$

$$\therefore \quad \frac{Y_C}{E_C} = \frac{4}{5}$$

Note: The above formula is a general formula can be used for any shape of channel.

16.94 (c)

Method-I

$B = 2$ m

$Q = 10$ m³/s

$$F = \frac{V}{\sqrt{gy}} = \frac{Q/By}{\sqrt{gy}}$$

$$2 = \frac{10/2y}{\sqrt{9.8y}}; \quad 4 = \frac{25}{9.8y^3}$$

$$y^3 = \frac{25}{4\times 9.8}; \quad y = 0.86 \text{ m}$$

Method-II

$$F_r^2 = \frac{Q^2T}{gA^3}$$

$$\Rightarrow \quad 2^2 = \frac{10^2\times 2}{9.81\times 2^3\times y^3}$$

16.95 (a)

Y = Actual depth of flow at any section of GVF profile.

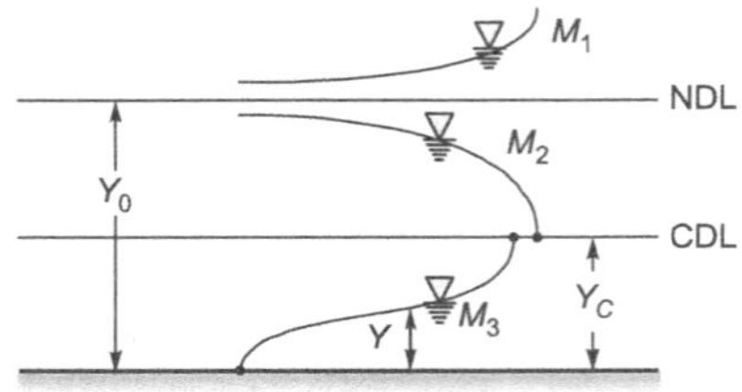

$$Y_0 > Y_C > Y$$

$$\therefore \quad M_3 = \text{Profile}$$

16.96 (d)

$$\frac{y_2}{y_1} = 14$$

Belangere momebntum equation for a horizontal, frictikonless rectangular channel is

$$\frac{y_2}{y_1} = \frac{1}{2}\left[-1+\sqrt{1+8F_1^2}\right]$$

$$14 = \frac{1}{2}\left[-1+\sqrt{1+8F_1^2}\right]$$

$$F_1 = 10.25$$

***Note:** The above equation to be used only for a horizontal frictionless rectangular channel.*

16.97 (a)

Assuming the channel to be a rectangular channel

$$E_L = \frac{(y_2-y_1)^3}{4y_1y_2} = \frac{(1.4-0.4)^3}{4\times 0.4\times 1.4}$$

$$= \frac{1}{1.6\times 1.4} = 0.446 \text{ m}$$

***Note:** General formula for $h_L = E_1 - E_2$*

16.98 (b)

Positive surge moving u/s neglect V and solve

$$\frac{V_w^2}{gy_1} = \frac{1}{2}\frac{y_2}{y_1}\left(\frac{y_2}{y_1}+1\right)$$

$$V_w = \sqrt{gy_1}\sqrt{\frac{1}{2}\left(\frac{y_2}{d_1}\right)+\frac{1}{2}\left(\frac{y_2}{y_1}\right)}$$

$$= \sqrt{gy_1}\left(\frac{1}{2}\left(\frac{d_2}{d_1}\right)^2+\frac{1}{2}\left(\frac{d_2}{d_1}\right)\right)^{1/2}$$

$$= \sqrt{gd_1}\left(\frac{1}{2}\left(\frac{h+d_1}{d_1}\right)^2+\frac{1}{2}\left(\frac{h+d_1}{d_1}\right)\right)^{1/2}$$

$$= \sqrt{gd_1}\left(\frac{1}{2}\left(\frac{h}{d_1}+1\right)^2+\frac{1}{2}\left(\frac{h}{d_1}+1\right)\right)^{1/2}$$

$$= \sqrt{gd_1}\left(\frac{1}{2}\left(\left(\frac{h}{d_1}\right)^2+\frac{2h}{d_1}+1\right)+\frac{1}{2}\left(\frac{h}{d_1}+1\right)\right)^{1/2}$$

$$V_w = \sqrt{gd_1}\left(\frac{1}{2}\left(\frac{h}{d}\right)^2+\frac{3}{2}\frac{h}{d_1}+1\right)^{1/2}$$

16.101 (b)

Before and after the jump formation, the flow is essentially one dimensional.

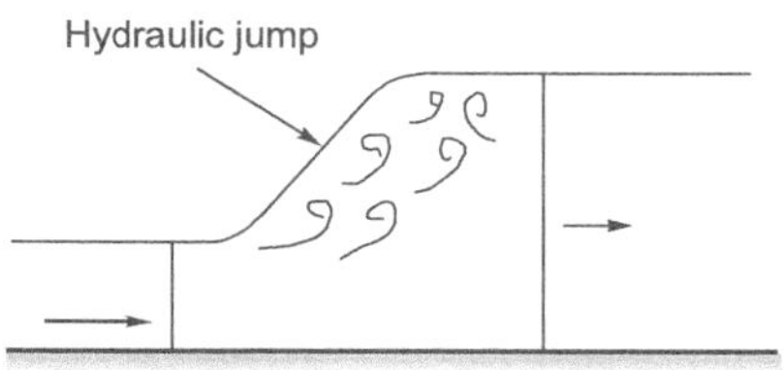

The channel floor is horizontal or the slope is so gentle that the weight component of the water mass comprising the jump is negligibly small.

16.102 (a)

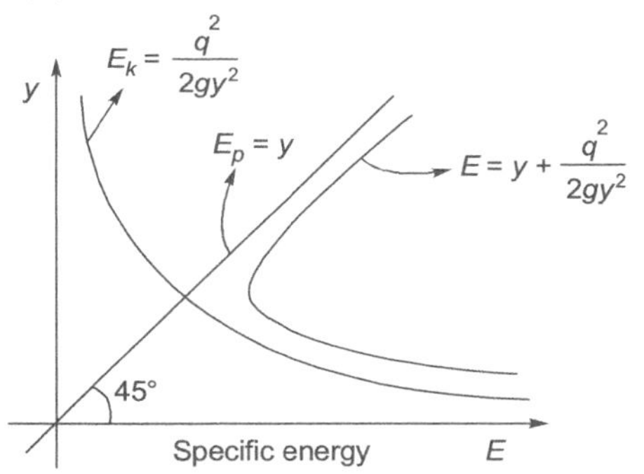

Consider a rectangular channel,

Specific energy,

$$E = y+\frac{q^2}{2gy^2}$$

...q = Discharge per unit width

$$E = E_p + E_k$$

$\therefore$ The curve of kinetic energy $\left(E_k = \frac{q^2}{2gy^2}\right)$ is parabolic while the curve of potential energy ($E_p = y$) is a straight line making an angle 45°. Specific energy curve is asymptotic to $E_p = y$ line and horizontal axis (E-axis).

16.103 (a)

For a given discharge specific energy is minimum when the flow is critical.

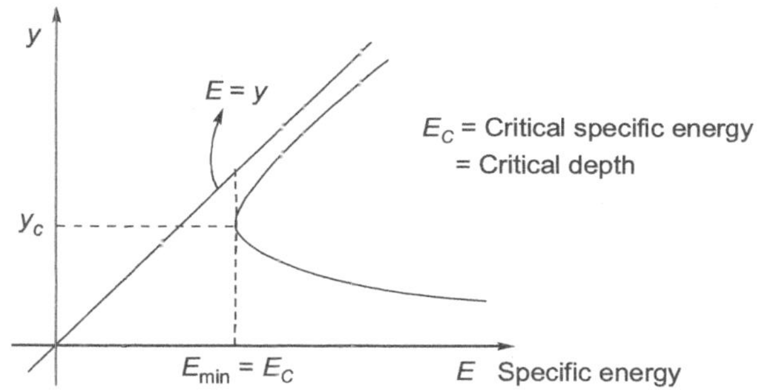

16.104 (b)

The depth at which specific energy is minimum is called critical depth.

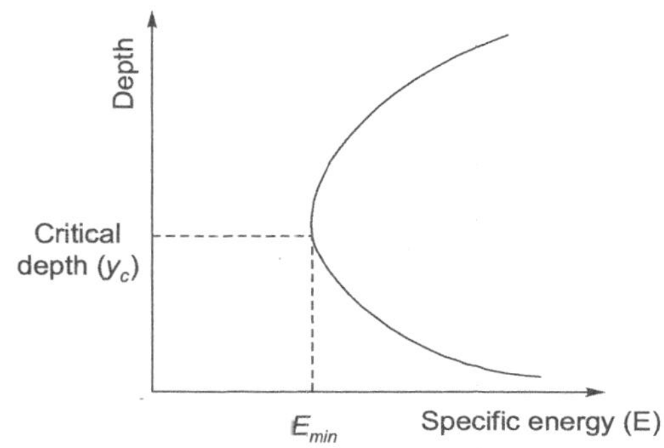

16.105 (d)

Critical flow: $F_r = 1,\ v = v_c,\ y = y_c$

Sub-critical flow: $F_r < 1,\ v < v_c,\ y > y_c$

Super critical flow: $F_r > 1,\ v > v_c,\ y < y_c$

F_r : Froude number

v_c : critical velocity

y_c : critical depth

16.106 (d)

For reservoir,

$$y_1+\frac{v_1^2}{2g} = y_2+\frac{v_2^2}{2g}$$

$$\Rightarrow \quad D_n+\frac{Q^2}{2ga^2} = D_r+0 \qquad [v_2 = 0]$$

$$\Rightarrow \quad D_r = D_n+\frac{Q^2}{2ga^2}$$

16.107 (d)

The velocity and discharge should remain constant along the channel reach.

Statements 1 and 3 are correct.

UNIT

II

Engineering Hydrology

Syllabus

Hydrological cycle, Ground water hydrology, Well hydrology and related data analysis; Streams and their gauging; River morphology; Flood, drought and their management.

Contents

■■■■

1 Precipitation and General Aspects of Hydrology

1.1 In a water-shed, four raingauges I, II, III and IV are installed. The depths of normal annual rainfall at these stations are 60, 75, 80 and 100 cm respectively. The rain gauge at station III went out of order during a particular year. The annual rainfall for that year, recorded at the remaining three stations was 90, 60 and 70 cm. The rainfall at station III can be considered as

(a) 60 cm (b) 70 cm
(c) 80 cm (d) 120 cm

[ESE : 2002]

1.2 Regional hydrological cycle is shown in the figure.

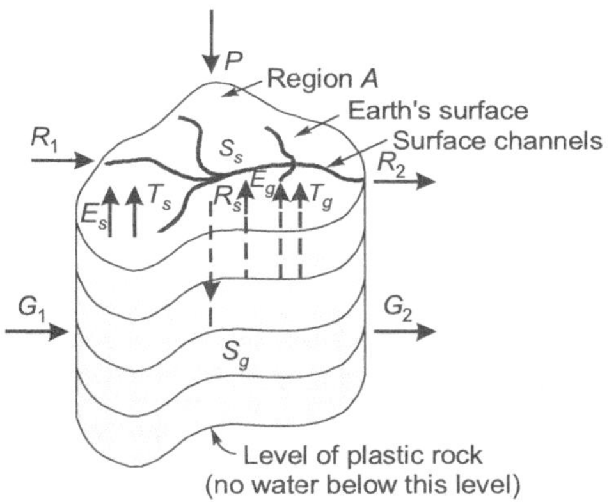

The correct hydrologic budget equations

(a) $P + R_1 - R_2 + R_g - E_s - T_s - I = DS_s$
(b) $I + G_1 - G_2 - R_g - E_g - T_g = DS_g$
(c) $P - (R_2 - R_1) - (E_s + E_g) - (T_s + T_g) - (G_2 - G_1) = D(S_s + S_g)$
(d) $P - R - G - E - T = DS_s$

[ESE : 2002]

1.3 The moving average of annual precipitation record is carried out to determine

(a) Trend
(b) Annual mean
(c) Extreme annual variation
(d) Extreme seasonal variation

[ESE : 2003]

1.4 Match **List-I** (Type of precipitation) with **List-II** (Principal causes) and select the correct answer using the codes given below the lists:

List-I
A. Convective
B. Cyclonic
C. Frontal
D. Orographic

List-II
1. Atmospheric disturbance
2. Mountain barrier
3. Pressure difference
4. Temperature difference
5. Warm and cold air masses

Codes:

	A	B	C	D
(a)	1	4	5	2
(b)	4	3	5	2
(c)	1	4	2	5
(d)	4	3	2	5

[ESE : 2003]

1.5 Match **List-I** (Hydrological terms) with **List-II** (Relationship/Nature of curve) and select the correct answer using the codes given below the lists:

List-I
A. Theissen Polygon
B. Mass Curve
C. Hyetograph
D. DAD curve

List-II
1. Average depth of rainfall over an area
2. Relationship of rainfall intensity and time
3. Relationship of accumulated rainfall and time
4. Relationship of river run-off and time
5. Always a falling curve

Codes:

	A	B	C	D
(a)	1	3	2	5
(b)	1	5	3	2

(c) 4 3 2 5
(d) 4 5 3 2 [ESE : 2003]

1.6 Which one of the following defines Aridity Index (AI)?

(a) $AI = \frac{\text{PET} - \text{AET}}{\text{PET}} \times 100$

(b) $AI = \frac{\text{PET}}{\text{AET}} \times 100$

(c) $AI = \frac{\text{AET}}{\text{PET}} \times 100$

(d) $AI = \frac{\text{AET} - \text{PET}}{\text{AET}} \times 100$

(where AET = Actual Evapotranspiration and PET = Potential Evapotranspiration)

[ESE : 2004]

1.7 The area between the two isohyetes 45 cm and 55 cm is 100 km^2, and that between 55 cm and 65 is 150 km^2. What is the average depth of annual precipitation over the basin of 250 km^2?

(a) 50 cm (b) 52 cm
(c) 56 cm (d) 60 cm

[ESE : 2005]

1.8 For which one of the following purposes is the double mass curve used ?

(a) Checking on the consistency of precipitation records.
(b) Prediction of annual precipitation.
(c) Defining which periods of storm should be analyzed to obtain the maximum useful information from storm rainfall records.
(d) For estimating the capacity of a reservoir.

[ESE : 2005]

1.9 Which one of the following is not a major type of storm precipitation?

(a) Frontal storm (b) Air mass storm
(c) Orographic storm (d) Continental storm

[ESE : 2006]

1.10 Double mass curves are used

(a) to check on the consistency of precipitation records
(b) as basis for storm rainfall analysis
(c) to determine average rainfall over an area
(d) to indicate rainfall distribution [ESE : 2006]

1.11 A drainage coefficient

(a) decides the choice of the method of the drainage
(b) decides the kind of crop that can be grown on the land
(c) is the depth of water that can be removed from the drainage area in unit time.
(d) is the flow of water from the soil into the tile laterals per unit time

[ESE : 2006]

1.12 What is the Probable Maximum Precipitation (PMP)?

(a) Projected precipitation for a 100 year return period
(b) Maximum precipitation for all past recorded storms
(c) Upper limit of rainfall, which is justified climatologically
(d) Effective precipitable water [ESE : 2007]

1.13 Which one of the following is a method of extending the length of record for a frequency curve at a station?

(a) Double Mass Curve method
(b) The Station Year method
(c) Thiessen method
(d) Isohyetal method [ESE : 2007]

1.14 The quantitative statement of the balance between water gains and losses in a certain basin during a specified period of time is known as which one of the following?

1. Water budget
2. Hydrologic budget
3. Groundwater budget

Select the correct answer using the codes given below:

(a) 1 only (b) 2 only
(c) 3 only (d) none of these

[ESE : 2007]

1.15 The rainfall hyetograph shows the variation of which one of the following?

(a) Cumulative depth of rainfall with time
(b) Rainfall depth with area
(c) Rainfall intensity with time
(d) Rainfall intensity with cumulative depth of rainfall [ESE : 2007]

1.16 Inconsistency of rainfall data can be checked by which one of the following?
(a) Normal ratio method
(b) Mass curve method
(c) Double-mass curve method
(d) Depth duration frequency curve
[ESE : 2008]

1.17 What is the chemical symbol for ice as per UNESCO terminology?
(a) H_8O_4 (b) H_2O
(c) H_6O_3 (d) H_4O_2
[ESE : 2008]

1.18 What is 'Hydrological Cycle'?
(a) Processes involved in the transfer of moisture from sea to land
(b) Processes involved in the transfer of moisture from sea back to sea again
(c) Processes involved in the transfer of water from snowmelt in mountains to sea
(d) Processes involved in the transfer of moisture from sea to land and back to sea again
[ESE : 2009]

1.19 Consider the following with respect to 'double-mass curve':
1. Plot of accumulated rainfall with respect to two chronological orders
2. Plot for estimating multiple missing rainfall data
3. Plot for checking the consistency of the rainfall data
4. Plot of accumulated annual rainfall of a station vs. accumulated rainfall of a group of stations

Which of these statements are correct?
(a) 1 and 3 (b) 2 and 3
(c) 3 and 4 (d) 1 and 4
[ESE : 2009]

1.20 Generally to estimate PMP, $P_m = 42.16D^{0.475}$ is used (P_m is maximum depth of precipitation, D = duration). What are the units of P_m and D in the equation?
(a) mm, sec (b) cm, sec
(c) mm, hr (d) cm, hr
[ESE : 2009]

1.21 The coefficient of variation of the rainfall for six rain gauge stations in catchments was found to be 29.54%. The optimum number of stations in the catchments for an admissible 10% error in the estimation of the mean rainfall will be
(a) 3 (b) 6
(c) 9 (d) 12
[ESE : 2010]

1.22 Ombrometer (pluviometer) is used to measure
(a) soil moisture stress of a plant
(b) rainfall depth
(c) leaf area
(d) root zone depth
[ESE : 2010]

1.23 Which of the following are pertinent to the realization of hydrological cycle?
1. Latitudinal differences in solar heating of the Earth's surface.
2. Inclination of the Earth's axis
3. Uneven distribution of land and water
4. Coriolis effect.

(a) 1, 2 and 3 only (b) 1, 2 and 4 only
(c) 2, 3 and 4 only (d) 1, 2, 3 and 4
[ESE : 2012]

1.24 Which of the following statements are correct?
1. A fine sprinkle of precipitation of small and rather uniform water drops with all drop diameters below 0.10 mm is called as drizzle.
2. The precipitation of liquid water with every drop diameter less than 0.5 mm is known as rain.
3. Precipitation in the form of balls or irregular lumps of ice each over 8 mm in diameter is called as hail.
4. Dew is formed directly by condensation on the ground mainly during night when the surface has been cooled by outgoing radiation.

(a) 1 and 2 (b) 2 and 3
(c) 3 and 4 (d) 1 and 4
[ESE : 2013]

1.25 The double mass curve technique is used:
(a) To find the average rainfall over a number of years
(b) To estimate the missing rainfall data
(c) To check the consistency of rain gauge records
(d) To find the minimum number of rain gauges required in a basin
[ESE : 2013]

1.26 Statement (I): Condensation of water vapour into droplets precedes the precipitation process.

Statement (II): Formation of precipitation droplets is predicted on the presence of condensation nuclei.

(a) Both Statement (I) and Statement (II) are individually true and Statement (II) is the correct explanation of Statement (I)
(b) Both Statement (I) and Statement (II) are individually true but Statement (II) is NOT the correct explanation of Statement (I)
(c) Statement (I) is true but Statement (II) is false
(d) Statement (I) is false but Statement (II) is true

[ESE : 2013]

1.27 Consider the following statements:

1. The normal annual rainfall of a station is obtained as the arithmetic average of the successive annual rainfall in the last 30 years.
2. The 'normal rainfall' are updated by deleting the needful number of oldest years data from the record and adding the needful number of most recent years data to the record.
3. The standard deviation computed for the rainfall of the same 30 years of rainfall data is taken as a measure of the variability of the rainfall during the self same set of years.
4. If the observed rainfall in any year is less than the current normal annual rainfall, then that year is called a dry year.

Which of the above statements are correct?

(a) 1, 2, 3 and 4 (b) 1, 2 and 3 only
(c) 2 and 4 only (d) 1, 3 and 4 only

[ESE : 2014]

1.28 The double mass curve analysis is adopted to

(a) Estimate the missing rainfall data
(b) Obtain intensities of rainfall at various durations
(c) Check the consistency of data
(d) Obtain the amount of storage needed to maintain a demand pattern [ESE : 2014]

1.29 The successive annual rainfall magnitudes at a place for a period of 10 years from 2001 to 2010, both inclusive, are 30.3, 41.0, 33.5, 34.00, 33.3, 36.2, 33.6, 30.2, 35.5 and 36.3 cm. The mean and median values of this annual rainfall series are respectively

(a) 33.8 cm and 34.39 cm
(b) 34.39 cm and 33.8 cm
(c) 34.39 cm and 40.2 cm
(d) 33.8 cm and 40.2 cm [ESE : 2014]

1.30 Orographic rain occurs when the air is cooled sufficiently as a result of

(a) lifting due to flow over a mountain barrier
(b) relative movement of two large air masses
(c) violent upthrow of air arising from localized heating
(d) cyclonic conditions [ESE : 2015]

1.31 A Double-Mass-Curve Analysis is useful in

(a) Consistency Analysis
(b) Frequency Analysis
(c) Storage Computation Analysis
(d) Guessing missing data in cases of non-homogeneous terrain

[ESE : 2015]

1.32 Consider the following statements concerning precipitation.

1. The Isohyetal map method of determining the average precipitation is considered to be better than the Thiessen method.
2. There is no possibility of damaging the storms because of cloud seeding activity.
3. Water that percolates through the soil emerges as the dry weather flow in streams.

Which of the above statements are correct?

(a) 1 and 2 (b) 1 and 3
(c) 2 and 3 (d) 1, 2 and 3

[ESE : 2017]

1.33 Annual rainfall values at station A in mm for the years 2001 to 2010 are given in the table below. If simple central 3-year moving mean of this rainfall record is calculated, the maximum and minimum values in the moving mean list would be

Years	2001	2002	2003	2004	2005	2006	2007	2008	2009	2010
Annual Rainfall P at station A (mm)	586	621	618	639	689	610	591	604	621	650

(a) 689 mm and 602 mm
(b) 649 mm and 602 mm
(c) 689 mm and 586 mm
(d) 549 mm and 586 mm [ESE : 2018]

1.34 Which one of the following points should be kept in mind while selecting the site for a rain gauge station?

(a) The site where a rain gauge is set up should be close to a meteorological observatory.
(b) The rain gauge should be on the top. of a hill.
(c) A fence, if erected to protect the rain gauge from cattle etc. should be located within twice the height of the fence.
(d) The distance between the rain gauge and the nearest object should be at least twice the height of the object.

[ESE : 2019]

1.35 A lake has an area of 15 km^2. Observation of hydrological variable during a certain year has shown as follows:
Precipitation = 700 mm/year;
Average inflow Q_{in} = 1.4 m^3/s
Average outflow Q_{out} = 1.6 m^3/s.
Assume that there is not net water exchange between the lake and the groundwater. What is the evaporation during this year?

(a) 480 mm (b) 280 mm
(c) 380 mm (d) 10 mm

[ESE : 2021]

1.36 A catchment has six rain gauge stations, In a year, the annual rainfall recorded by the gauges are as follows:

Station	A	B	C	D	E	F
Rainfall (cm)	82.6	102.9	180.3	110.3	98.8	136.7

For a 10% error in the estimation of the mean rainfall, what is the optimum number of stations in the catchment? (Take $\bar{P}$ = 118.6, σ_{m-1} = 35.04 and ε = 10)

(a) 10 (b) 9
(c) 8 (d) 7

[ESE : 2022]

1.37 Which one of the following precipitations results from ocean air streams passing over land and being deflected upward by coastal mountains, thus cooling below saturation temperature and spilling moisture?

(a) Convective precipitation
(b) Frontal precipitation
(c) Orographic precipitation
(d) Cyclonic precipitation

[ESE : 2024]

Answers Precipitation and General Aspects of Hydrology

1.1 (c)	1.2 (c)	1.3 (a)	1.4 (b)	1.5 (a)	1.6 (a)	1.7 (c)	1.8 (a)	1.9 (d)
1.10 (a)	1.11 (c)	1.12 (c)	1.13 (b)	1.14 (a)	1.15 (c)	1.16 (c)	1.17 (b)	1.18 (d)
1.19 (c)	1.20 (d)	1.21 (c)	1.22 (b)	1.23 (d)	1.24 (c)	1.25 (c)	1.26 (a)	1.27 (d)
1.28 (c)	1.29 (b)	1.30 (a)	1.31 (a)	1.32 (d)	1.33 (b)	1.34 (d)	1.35 (b)	1.36 (b)
1.37 (c)								

Explanations Precipitation and General Aspects of Hydrology

1.1 (c)

Since the normal annual rainfall for stations I, II and IV does not lie within 10% of that of normal annual rainfall at station III. So,

Using Normal Ratio Method

$$P_{III} = \frac{80}{3}\left[\frac{90}{60}+\frac{60}{75}+\frac{70}{100}\right] = 80 \text{ cm}$$

1.2 (c)

The hydrologic budget consists of inflows, outflows, and storage as shown in the following equation: Inflow = Outflow +/- Changes in Storage. Inflows add water to the different parts of the hydrologic system, while outflows remove water. Storage is the retention of water by parts of the system.

In the Hydrologic Budget Equation:

$P - R - E - T - G = \Delta S$

Where, P - Total Precipitation
R – Net Runoff
E – Total Evaporation
T – Total Transpiration
G – Net Ground Water Flow
ΔS – Total change in storage

1.3 (a)
A moving average plot, in which the average value of precipitation of three or five consecutive time intervals is plotted at the mid-value of the time interval is useful in smoothening out the variations and bringing out the trend.

1.4 (b)
Convective precipitation is caused by convection where the surface layer of the atmosphere is heated causing the moisture laden air to rise. As it rises it cools to form clouds. The unequal heating of the earth's surface causes convection.

Cyclonic precipitation is caused by lifting of an air mass due to the pressure difference. Cyclonic precipitation may be either frontal or non-frontal cyclonic precipitation.

Frontal rainfall occurs when warm air is forced to rise over cold air. The moisture in the warm air condenses as it cools which causes clouds and rain.

Orographic or relief rainfall is caused when masses of air pushed by wind are forced up the side of elevated land formations, such as large mountains.

1.5 (a)
Methods used to convert point rainfall values at different raingauge stations into an average value over a catchment include

- Arithmetical Mean Method
- Thiessen Polygon Method
- Isohyetal Method

A Mass curve is the graph of the cumulative values of a hydrologic quantity (such as precipitation or runoff), generally as the Ordinate, plotted against time or date as the Abscissa.

1.6 (a)
The departure of AI from its corresponding normal value is known as AI anomaly, represents moisture shortage. Based on AI anomaly, the intensity of agricultural drought is classified as follows:

AI anomaly	Severity class
$0 < AI \le 25$	mild arid
$25 < AI \le 50$	moderate arid
> 50	severe arid

1.7 (c)
Average depth of annual precipitation

$$= \frac{\left(\frac{45+55}{2}\right)\times 100 + \left(\frac{55+65}{2}\right)\times 150}{100+150} = 56 \text{ cm}$$

1.8 (a)
The double-mass curve can be used to adjust inconsistent precipitation data.
A graph is plotted between accumulated precipitation of problem station and average accumulated precipitation of group of base station. A break in slope of resulting graph indicates change in precipitation of problem station beyond the period of change.

1.9 (d)
Orographic storm: A rain or snow event that results from, or is significantly enhanced by, lifting of an air mass over a mountain barrier.

Air Mass Storm, also called an "ordinary" thunderstorm, is a thunderstorm that is generally weak and usually not severe. The lifting source, which is a crucial factor in thunderstorm development, is usually the result of uneven heating of the surface, though they can be induced by weather fronts and other low-level boundaries associated with wind convergence.

Frontal thunderstorms occur along the boundaries of weather fronts (e.g. cold front).

1.11 (c)
The rate at which water is removed by a drain is called drainage. It is expressed as the depth of water in cm or metres, to be removed in 24 hours from the drainage area.

1.12 (c)
The Probable Maximum Precipitation (PMP) may be defined as the maximum depth of precipitation for a given duration that may possibly occur on a given catchment at any time of the year. Hence such a precipitation would result from the possible severe most storm that may result from the worst possible combinations of hydrological conditions in the area.
Probable Maximum Flood (PMF) is the extreme flood that is physically possible in a region as a result of severe most combinations, including rare combinations of meteorological and hydrological factors. The PMF is used in situations where the

failure of the structure would result in loss of life and catastrophic damage and as such complete security from potential floods is sought.

1.13 (b)

When long records are not available, records at two or more stations are combined to get one long record for the purposes of recurrence interval calculation. This method is known as station year method. In this method all the records which are available from the various stations situated in a meteorologically homogeneous area are combined together, so as to represent the record of an imaginary representative station called the station year.

1.14 (a)

A water budget equation is used to describe the flow of water in and out of a system in a specific time period. The system can be a particular basin or a catchment.

1.15 (c)

A rainfall hyetograph is a plot of intensity of rainfall against the time interval.

1.16 (c)

The checking for inconsistency of a record is done by the double-mass curve technique. This technique is based on the principle that when each recorded data comes from the same parent population, they are consistent.

1.18 (d)

Most of the earth's water sources such as rivers, lakes, oceans, ground water, etc. get their supplies from rain, while the rain water in itself is derived from the evaporation from these sources. Water is infact lost to the atmosphere as vapor from the earth, which is then precipitated back in the form of rain, snow, hail, dew, sleet or frost, etc. This evaporation and precipitation continues forever and thereby, a balance is maintained between the two. This process is known as hydrologic cycle.

1.19 (c)

Double mass curve is a plot for checking the inconsistency of the rainfall data. This technique is based on the principle that when each recorded data comes from the same parent population, they are consistent. In this technique the data of the annual (or monthly mean) rainfall of any station x and also the average rainfall of the group of base stations covering a long period is arranged in reverse chronological order (i.e. the latest record as the first entry and the oldest record as the last entry in the list). The accumulated precipitation of the station $x(\Sigma P_x)$ and the accumulated values of the average of the group of base stations, (ΣP_{av}) are calculated starting from the latest record values of ΣP_x are plotted against ΣP_{av} for various consecutive time periods.

1.20 (d)

Based upon the rainfall records available all over the world, a list of world's greatest recorded rainfalls of various duration can be assembled. When this data is plotted upon a log-log paper, an enveloping straight line drawn to the plotted points obeys the equation

$$P_m = 42.16\, D^{0.475}$$

where P_m = extreme rainfall depth in cm and D = duration in hours. The values obtained from this equation are of use in probable maximum precipitation (PMP) estimations.

1.21 (c)

The number of optimum stations can be given as

$$N = \left(\frac{C_v}{\varepsilon}\right)^2 = \left(\frac{29.54}{10}\right)^2$$

$$= 8.72 \approx 9 \text{ stations}$$

1.22 (b)

Rainfall is expressed in cm or mm which is volume per unit area. A device which is used to measure the rainfall is called raingauge. Term such as pluviometer ombrometer and hyetoemeter are sometimes used to designate a raingauge.

1.24 (c)

Rain: When water droplets are of size 0.5 mm to 6.0 mm then it is classified as rain.

Classification of rain as per intensity of rainfall is as shown below:

Intensity (mm/hr)	Rain
< 2.5 mm/hr	Light rain
2.5 mm/hr to 7.5 mm/hr	Moderate rain
> 7.5 mm/hr	Heavy rain

Drizzle: Rain droplets of size less than 0.5 mm and rain intensity of less than 1mm/h.

Hail: It is a showery precipitation in the form of irregular pellets or lumps of ice of size more than 8 mm.

Dew: Water droplets condensed from the air, usually at night, onto cool surfaces near the ground. Dew forms when the temperature of the surfaces falls below the dew point of the surrounding air usually due to radiational cooling.

1.26 (a)

When water is saturated locally in the atmosphere, the water condenses and precipitates. Sufficient water nuclei must be present to aid condensation and thus cause precipitation.

1.27 (d)

The Normal rainfall is calculated as the average value of rainfall of a particular date, month or year over a specified period of 30 years. The 30 year Normal rainfall is recomputed every decade to account for factors that may affect the amount of rainfall on that area like change in land use, environment etc.

The Standard deviation shows the rainfall variability over a particular period.

1.28 (c)

The trend of the rainfall records at a station may slightly change after some years due to a change in the environment (or exposure) of a station either due to coming of new building, fence, planting of trees or cutting of forest nearby, which affect the catch of the gauge due to change in the wind pattern or exposure. The consistency of records at the station is tested by a double mass curve by plotting the cumulative annual rainfall at a station against the concurrent cumulative values of mean annual rainfall for a group of surrounding stations, for the number of year of record.

1.29 (b)

$$\text{Mean value} = \frac{\sum x}{n}$$

$$= \frac{30.3 + 41.0 + 35.5 + 34 + 33.3 + 36.2 + 33.6 + 30.2 + 35.5 + 36.3}{10}$$

$= 34.39$ cm

If we arrange the various rainfall magnitude in increasing order, then we get,

30.2, 30.3, 33.3, 33.5, 33.6, 34, 35.5, 36.2, 36.3, 41.0

If no. of terms are even, then median will

$$= \left[\frac{\left(\frac{n}{2}\right) + \left(\frac{n}{2}+1\right)}{2}\right] = \frac{33.6 + 34}{2} = 33.8 \text{ cm}$$

1.30 (a)

Orographic precipitation, also known as relief precipitation, is precipitation generated by a forced upward movement of air upon encountering a physiographic upland. This lifting can be caused by two mechanisms:

1. The upward deflection of large scale horizontal flow by the orography.
2. The upward vertical propagation of moist air up an orographic slope caused by daytime heating of the mountain barrier surface.

1.32 (d)

Cloud seeding enhence the precipitation.

1.33 (b)

Maximum mean precipitation,

$$\bar{P}_{max} = \frac{618 + 639 + 689}{3}$$

$$= 648.67 \text{ mm} \simeq 649 \text{ mm}$$

Minimum mean precipitation,

$$\bar{P}_{min} = \frac{610 + 591 + 604}{3} \simeq 602 \text{ mm}$$

1.34 (d)

The rain-gauge station should not be too close to the buildings or trees etc. The nearness of such objects affects entry rainfall in the funnel. Its distance from other objects should not be less than twice the height of the object above the rim of the gauge.

1.35 (b)

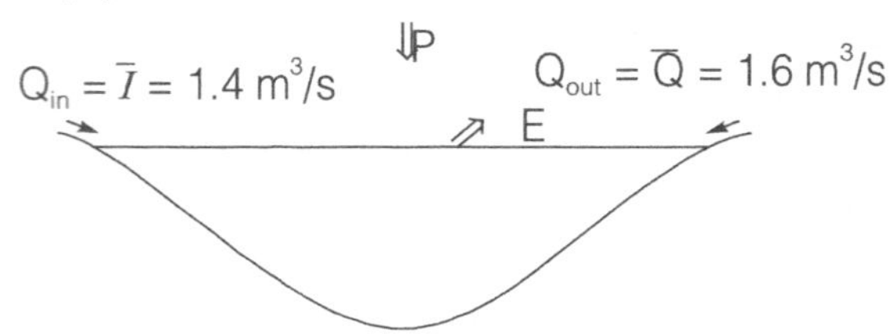

Given: Lake area,

$$A = 15 \text{ km}^2$$

Precipitation,

$$P = 700 \text{ mm/year}$$

Let E (in m) be the evaporation from lake during this year

From hydrologic water budget equation

$$(\bar{I}\Delta t + PA) - (\bar{Q}\Delta t + EA) = \Delta S$$

As there is not net water exchange between the lake and the groundwater,

$$\therefore \quad \Delta S = 0$$

$$(1.4 \times 365 \times 24 \times 3600) + (0.7 \times 15 \times 10^6) = (1.6 \times 365 \times 24 \times 3600) + (E \times 15 \times 10^6)$$

$$\Rightarrow \quad E = \frac{4192800}{15 \times 10^6} = 0.27952 \text{ m}$$

$$= 279.52 \text{ mm} \simeq 280 \text{ mm}$$

1.36 (b)

Optimum number of stations,

$$N = \left(\frac{C_v}{E}\right)^2$$

Where, $C_v = \dfrac{\sigma}{\bar{P}} \times 100$

$$= \frac{35.04}{118.6} \times 100 = 29.545\%$$

$$\Rightarrow \quad N = \left(\frac{29.545}{10}\right)^2 = 8.73 \simeq 9$$

1.37 (c)

Precipitation caused by hills or mountain ranges deflecting the moisture laden air masses upward, causing them to cool and precipitate their moisture is called orographic precipitin.

■■■■

2 Evaporation, Transpiration, Evapotranspiration and Stream Flow Measurement

2.1 Match **List-I** with **List-II** and select the correct answer using the codes given below the lists:

List-I	List-II
A. Anemometer	1. Humidity
B. Rain simulator	2. Evapotranspiration
C. Lysimeter	3. Infiltration
D. Hygrometer	4. Wind speed

Codes:

	A	B	C	D
(a)	4	3	1	2
(b)	3	4	1	2
(c)	4	3	2	1
(d)	3	4	2	1

[ESE : 2001]

2.2 The Penman's evapo-transpiration equation is based on
(a) water budget method
(b) energy balance method
(c) mass transfer method
(d) energy balance and mass transfer approach
[ESE : 2002]

2.3 The slope area method is extensively used in
(a) development of rating curve
(b) estimation of flood discharge based on high-water marks
(c) cases where shifting control exists
(d) cases where back-water effect is present
[ESE : 2002]

2.4 Penman's equation is based on
(a) energy budgeting only
(b) Energy budgeting and water budgeting
(c) Energy budgeting and mass transfer
(d) Water budgeting and mass transfer
[ESE : 2006]

2.5 In the case of large rivers, a number of equidistant vertical sections of the total width of flow are identified, for the purpose of finding by numerical integration, the total discharge on any day. On each section, the mean velocity is taken as the arithmetic average of two typical depths on that section. Then the mean velocity is worked out for that section. Usually, the mean velocity of any section, corresponds to which one of the following? (V represents the point velocity at the given section and the depth such as $0.1d$, $0.2d$.... etc.)

(a) $\dfrac{V_{0.1d} + V_{0.9d}}{2}$ (b) $\dfrac{V_{0.2d} + V_{0.8d}}{2}$

(c) $\dfrac{V_{0.3d} + V_{0.7d}}{2}$ (d) $\dfrac{V_{0.4d} + V_{0.6d}}{2}$

[ESE : 2006]

2.6 Match **List-I** with **List-II** and select the correct answer using the code given below the lists:

List-I	List-II
A. Location	1. Perennial
B. Stability	2. Degrading
C. Variation of discharge	3. Tidal
D. Plan form	4. Braided

Codes:

	A	B	C	D
(a)	4	2	1	3
(b)	3	1	2	4
(c)	4	1	2	3
(d)	3	2	1	4

[ESE : 2006]

2.7 A standard, ground-based evaporation pan, corresponding to Indian Standards, is installed at the banks of a surface reservoir. The water surface area on a particular day is 100 hectares. The recorded evaporation loss from the pan, on a certain day, is nearly 4.0 cm. What is the anticipated evaporation loss from the reservoir for that day?
(a) $(1.8 \text{ to } 2) \times 10^4$ m^3 per day
(b) $(2.5 \text{ to } 2.75) \times 10^4$ m^3 per day
(c) $(3.0 \text{ to } 3.25) \times 10^4$ m^3 per day
(d) $(3.8 \text{ to } 4.05) \times 10^4$ m^3 per day [ESE : 2007]

2.8 How is the average velocity along the vertical in a wide stream obtained?
(a) By averaging the velocities at 0.2 and 0.8 depth from surface.
(b) By measuring velocity at 0.6 depth below the surface.
(c) By measuring velocity at half the depth.
(d) By measuring velocity at 0.1 times the depth below the surface.
[ESE : 2007]

2.9 Maximum possible discharge from a small catchment corresponding to a particular rainfall intensity is independent of which one of the following?
(a) Soil moisture conditions
(b) Drainage characteristics of catchment
(c) Area of the catchment
(d) Duration of the rainstorm [ESE : 2008]

2.10 Consider the following chemical emulsions:
1. Methyl alcohol
2. Cetyl alcohol
3. Stearyl alcohol
4. Kerosene

Which of the above chemical emulsions is/are used to minimize the loss of water through the process of evaporation?
(a) 1 only (b) 1 and 4
(c) 2 and 4 (d) 2 and 3 [ESE : 2009]

2.11 Consider the following with respect to measurement of stream flow during flood:
1. Timing of the travel of floats released in the stream
2. Use of weir formula for spillways provided on a dam
3. Calculation of flow through a contracted opening at a bridge
4. Using a current meter

Which of the above is/are reliable and accurate?
(a) 1 only (b) 4 only
(c) 3 and 4 (d) 2 and 3
[ESE : 2009]

2.12 Calibration of a current meter for use, in channel flow measurement is done in a
(a) wind tunnel (b) water tunnel
(c) towing tank (d) flume
[ESE : 2010]

2.13 A stilling well is required when the stage measurement is made by employing
(a) Bubble gauge
(b) Float gauge recorder
(c) Vertical staff gauge
(d) Inclined staff gauge [ESE : 2011]

2.14 Consider the following statements:
1. Over the oceans there is more evaporation than precipitation.
2. On land it is more precipitation than evapo-transpiration.

Which of these statements are correct?
(a) Both 1 and 2 (b) Neither 1 nor 2
(c) 1 only (d) 2 only
[ESE : 2012]

2.15 Wading technique is used:
(a) To determine velocity of sea waves during Tsunami
(b) To determine thickness of canal lining in alluvial soils
(c) To measure the volume of dredging material in harbours
(d) To determine velocity of flow in shallow streams
[ESE : 2013]

2.16 Steep rise in the flow-mass curve during a certain period indicates:
(a) Very high evaporation losses during that period
(b) Flash floods during that period
(c) Sudden spurt in irrigation demand during that period
(d) Sudden rise in demand for water to meet hydropower generation
[ESE : 2013]

2.17 **Statement (I):** The deeper a lake, the lesser the evaporation in summer and the more in winter.

Statement (II): Heat storage in water bodies affects seasonal evaporation.
(a) Both Statement (I) and Statement (II) are individually true and Statement (II) is the correct explanation of Statement (I)
(b) Both Statement (I) and Statement (II) are individually true but Statement (II) is NOT the correct explanation of Statement (I)

(c) Statement (I) is true but Statement (II) is false
(d) Statement (I) is false but Statement (II) is true

[ESE : 2017]

2.18 The relative humidity h is a measure of air's capacity, at its existing temperature, to absorb further moisture, and is defined by the relation

(a) $\frac{e}{e_s} \times 100$ (b) $\frac{e_s}{e} \times 100$

(c) $\frac{2e_s}{e} \times 100$ (d) $\frac{2e}{e_s} \times 100$

Where: e = Vapour pressure, e_s = Saturation vapour pressure

[ESE : 2020]

2.19 Which one of the following is defined as the process by which the water leaves a living plant during photosynthesis, through its leaves, to enter the atmosphere as water vapour?

(a) Transpiration (b) Evapotranspiration
(c) Stomata (d) Evaporation

[ESE : 2022]

■■■■

Answers Evaporation, Transpiration and Stream Flow Measurement

2.1 (c) 2.2 (d) 2.3 (b) 2.4 (c) 2.5 (b) 2.6 (d) 2.7 (c) 2.8 (b) 2.9 (d)
2.10 (d) 2.11 (b) 2.12 (c) 2.13 (b) 2.14 (a) 2.15 (d) 2.16 (b) 2.17 (a) 2.18 (a)
2.19 (a)

Explanations Evaporation, Transpiration and Stream Flow Measurement

2.1 (c)

An anemometer is a device used for measuring wind speed, and is a common weather station instrument.

A hygrometer is an instrument used for measuring the moisture content in the atmosphere.

A lysimeter is a measuring device which can be used to measure the amount of actual evapotranspiration which is released by plants. Evapotranspiration is estimated in terms of amount of water required to maintain constant moisture conditions within Lysimeter measured either volumetrically or gravimetricall through an arrangement made in the Lysimeter.

In filtrometer and rain simulator is used to measure infiltration.

2.2 (d)

Penman's Equation is based on a combination of both energy balance and mass transfer approach.

2.3 (b)

When it is not possible to make direct measurement especially during floods, slope-area method is a most commonly used indirect method of measurement. In this method discharge is computed on the basis of a uniform flow equation involving channel characteristics, water surface profile and a roughness coefficient. The drop in water surface profile for a uniform reach of channel represents losses caused by bed roughness. Discharge is calculated by using Manning's equation in this method.

2.4 (c)

Penman's Equation is based on a combination of both energy budgeting and mass transfer approach.

2.5 (b)

In shallow stream of depth up to about 3.0 m, the average velocity

$$\bar{V} = V_{0.6d}$$

This is called single point observation.

In moderately deep streams the velocity is observed at two points.

$$\therefore \quad \bar{V} = \frac{V_{0.2d} + V_{0.8d}}{2}$$

In rivers having large flood flows, only surface velocity (V_s) is measured within a depth of about 0.5 m from the surface. The average velocity $\overline{V} = kV_s$ where k is in the range of 0.85 to 0.95.

2.6 (d)

(i) A perennial stream is one which always carries some flow. There is considerable amount of groundwater flow contribution throughout the year. The discharge varies considerably throughout the year.

(ii) If the river bed is constantly getting eroded to reduce and dissipate available excess land slope, then the river is known as a degrading river. Such river is in the process of achieving stability.

(iii) Tidal rivers, undergo periodical rise and fall in its water level, depending upon the nature of the tide.

The distance up to which tidal effect is experienced depends upon various factors such as shape and configuration of the river, the tidal range, etc. Thus location of river and ocean confluence is important.

2.7 (c)

Evaporation from reservoir = $C_p \times$ pan evaporation

For I.S. pan, C_p = 0.65 to 1.1 and average value of C_p = 0.80.

Reservoir evaporation$|_{min}$

$= 0.65 \times 4.0 \times 10^{-2} \times 100 \times 10^4$ m³/day

$= 2.6 \times 10^4$ m³/day

Reservoir evaporation$|_{max}$

$= 1.1 \times 4.0 \times 10^{-2} \times 100 \times 10^4$ m³/day

$= 4.4 \times 10^4$ m³/day

Reservoir evaporation$|_{avg}$

$= 0.8 \times 4.0 \times 10^{-2} \times 100 \times 10^4$ m³/day

$= 3.2 \times 10^4$ m³/day

The closest option is option (d)

2.8 (b)

The velocity profile shows how the average velocity is usually at 0.6 times the total depth from the water surface, or 0.4 times the total depth from the bottom of the channel. This is why, in shallow channels (< 2.5 ft or < 0.75 m), current meter measurements are made at four tenths of the total depth (from the bottom).

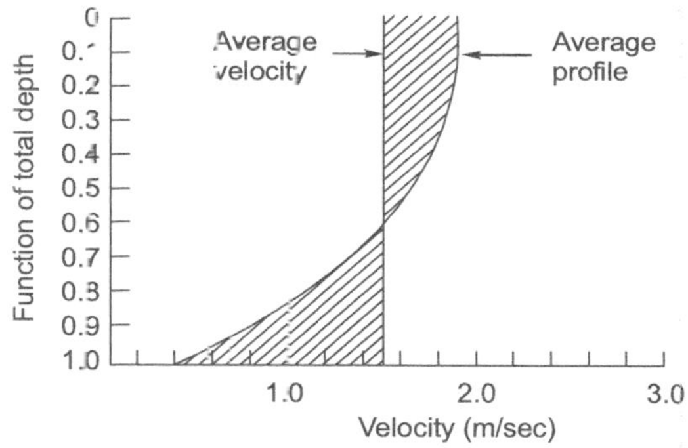

2.10(d)

Certain chemicals such as cetyl alcohol (hexadecanol) and stearyl alcohol (octadecanol) forms monomolecular layers on a water surface. These layers act as evaporation inhibitors by preventing the water molecules to escape past them.

2.11 (b)

Quick stream flow measuring is best done with a water current meter to measure water current velocity. For larger streams, the current meter may be used to measure a vertical profile of water velocity at several points across a stream channel.

Using a weir formula for Spillways provided on a dam & in case of calculation of flow through a contracted opening at a bridge causes losses.

Timing of the travel of floats released in the stream is not reliable.

2.12(c)

The relation between the stream velocity and revolutions per second of the current meter is called the calibration equation. The calibration equation is unique to each instrument and is determined by towing the instrument in a special tank called towing tank. Towing tank is a long channel containing still water with arrangements for moving a carriage longitudinally over its surface at constant speed. The current meter to be calibrated is mounted on the carriage with the rotating element immersed to a specified depth in the water body in the tank. The carriage is then towed at a predetermined constant speed (v) and the corresponding average value of revolutions per second (N_s) of the instrument determined. This experiment is repeated over the complete range of velocities and a best-fit linear relation in the form of the equation $v = aN_s + b$ where a and b are constants of the current meter.

2.13 (b)

The Float-operated stage recorder is the most common type of automatic stage recorder in use. In this a float operating in a stilling well is balanced by means of a counterweight over the pulley of a recorder.

To protect the float from debris and to reduce the water surface wave effects on the recording, a stilling wells are provided in all float-type stage recorder installations.

2.14 (a)

Evaporation is more prevalent over the oceans than precipitation, while over the land, precipitation routinely exceeds evaporation. Most of the water that evaporates from the oceans falls back into the oceans as precipitation. Only about 10 percent of the water evaporated from the oceans is transported over land and falls as precipitation. Once evaporated, a water molecule spends about 10 days in the air. The process of evaporation is so great that without precipitation runoff, and groundwater discharge from aquifers, oceans would become nearly empty.

2.15 (d)

The current meter is held at the requisite depth below the surface of the stream vertically by an observer who stands in the water. This technique is known as Wading and is used to determine the velocity of flow in shallow streams.

2.16 (b)

Flow mass curve is a plot of the cumulative discharge volume against time plotted in chronological order, the slope of the mass at any point represents rate of flow at that instant, so flash flood during a period will show steep rise in the flow mass curve.

2.18 (a)

$$\text{Relatively humidity} = \frac{e_a}{e_s} \times 100 \text{ or } \frac{e}{e_s} \times 100$$

Here, e or e_a = Actual vapour pressure of air in mm of Hg

e_s = Saturation vapour pressure of air in mm of Hg

2.19 (a)

Transpiration is the process by which water leaves the body of a living plant and reaches the atmosphere as water vapour during photosynthesis.

■■■■

3 Infiltration, Runoff and Hydrograph

3.1 If the base period of a 6-hour hydrograph of a basin is 84 hours, then a 12 hours unit hydrograph derived from this 6 hour unit hydrograph will have a base period of

(a) 72 hours (b) 78 hours
(c) 84 hours (d) 90 hours

[ESE : 2000]

3.2 Which of the following principles relate to a Unit Hydrograph?

1. The hydrographs of direct runoff due to effective rainfall of equal duration have the same time base.
2. Effective rainfall is not uniformly distributed within its duration.
3. Effective rainfall is uniformly distributed throughout the whole area of drainage basin.
4. Hydrograph of direct run off from a basin due to a given period of effective rainfall reflects the combination of all the physical characteristics of the basin.

Select the correct answer using the codes given below:

(a) 1, 2 and 3 (b) 1, 2 and 4
(c) 2, 3 and 4 (d) 1, 3 and 4

[ESE : 2001]

3.3 A 3-hour storm on a small drainage basin produced rainfall intensities of 3.5 cm/hr, 4.2 cm/hr and 2.9 cm/hr in successive hours. If the surface runoff due to the storm is 3 cm, then the value of ϕ-index will be

(a) 2.212 cm/hr (b) 2.331 cm/hr
(c) 2.412 cm/hr (d) 2.533 cm/hr

[ESE : 2001]

3.4 Consider the following statements:

1. An ephemeral stream is one which has a base-flow contribution.
2. Flow characteristics of a stream depend upon rainfall and catchment characteristics and also the climatic factors which influence evapo-transpiration.
3. Sequent Peak Algorithm is used for estimating run off from rainfall.

Which of these statements is/are correct?

(a) 1, 2 and 3 (b) 1 and 3
(c) 2 and 3 (d) 2 alone

[ESE : 2002]

3.5 Match **List-I** with **List-II** and select the correct answer using the codes given below the lists:

List-I

A. ϕ-index
B. Lysimeter
C. Dilution technique
D. Snyder's equation

List-II

1. Used for measurement of evapo-transpiration for given vegetation
2. Used for flow measurement
3. Average rainfall above which the rainfall volume is equal to the runoff volume
4. Relates the basin lag to the basin characteristics

Codes:

	A	B	C	D
(a)	3	1	2	4
(b)	4	2	1	3
(c)	3	2	1	4
(d)	4	1	2	3

[ESE : 2002]

3.6 A 6-hour rainstorm with hourly intensities of 7, 18 25, 17, 11 and 3 mm/hour produced a runoff of 39 mm. Then, the ϕ-index is

(a) 3 mm/hour (b) 7 mm/hour
(c) 8 mm/hour (d) 10 mm/hour

[ESE : 2002]

3.7 The Unit-Hydrograph theory is based on the assumption of
(a) nonlinear response and time invariance
(b) linear response and nonlinear time variance
(c) time invariance and linear response
(d) nonlinear response and nonlinear time variance [ESE : 2002]

3.8 A two-hour storm hydrograph has 5 units of direct runoff. The two-hour unit hydrograph for this storm can be obtained by dividing the ordinates of the storm hydrograph by
(a) 2 (b) 2/5
(c) 5 (d) 5/2 [ESE : 2002]

3.9 **Assertion (A):** Unit hydrograph theory is not applicable to catchment areas larger than 5000 sq. km.
Reason (R): Rainfall is not uniformly distributed on large catchment areas.
(a) both A and R are true and R is the correct explanation of A
(b) both A and R are true but R is not a correct explanation of A
(c) A is true but R is false
(d) A is false but R is true [ESE : 2002]

3.10 The following items related to hydrologic requirements in a hydropower system. Match **List-I** with **List-II** and select the correct answer using the codes given below the lists:

List-I
A. Stream flow in order of magnitude is plotted as ordinate and per cent of time as abscissa.
B. Cumulative value of stream flow is plotted against time for whole period of record.
C. Plot of available power against per cent of time.
D. Plot of flow in a stream against time at specific interval.

List-II
1. Mass curve
2. Flow duration curve
3. Hydrograph
4. Power duration curve

Codes:

	A	B	C	D
(a)	2	3	4	1
(b)	4	1	2	3
(c)	2	1	4	3
(d)	4	3	2	1

[ESE : 2002]

3.11 Which one of the following characteristics describes a watershed system in system's parlance?
(a) Linear
(b) Non-linear
(c) Linear and time-invariant
(d) Non-linear and time-variant [ESE : 2003]

3.12 A 6-hour storm with hourly intensities of 7, 18, 25, 12, 10 and 3 mm per hour produced a run-off of 33 mm. Then the ϕ-index is
(a) 7 mm/h (b) 3 mm/h
(c) 10 mm/h (d) 8 mm/h
[ESE : 2003]

3.13 A catchment area of 90 hectares has a run-off coefficient of 0.4. A storm of duration larger than the time of concentration of the catchment and of intensity 4.5 cm/hr creates a peak discharge rate of
(a) 11.3 m^3/s (b) 0.45 m^3/s
(c) 450 m^3/s (d) 4.5 m^3/s
[ESE : 2003]

3.14 In Snyder's method of synthetic unit hydrograph development, basin lag is taken as
(a) the time interval between centroid of the rainfall excess and surface runoff
(b) the time interval from mid point of the unit rainfall excess to the peak of the unit hydrograph
(c) independent of rainfall duration
(d) independent of catchment characteristics
[ESE : 2004]

3.15 A storm with 14 cm precipitation produced a direct runoff of 8 cm. The time distribution of the storm is as shown in the table below:

Time from start (h)	Incremental Rainfall in cm
1	1.0
2	2.0
3	2.8
4	3.3
5	2.5
6	1.8
7	0.6

What is the value of ϕ-index of the storm?
(a) 0.5 cm/h (b) 0.7 cm/h
(c) 0.8 cm/h (d) 0.9 cm/h
[ESE : 2004]

3.16 Match **List-I** with **List-II** and select the correct answer using the codes given below the lists:

List-I
A. Rising limb of a hydrograph
B. Falling limb of a hydrograph
C. Peak rate of flow
D. Drainage density

List-II
1. Depends on intensity of rainfall
2. Function of total channel length
3. Function of catchment slope
4. Function of storage characteristics

Codes:

	A	B	C	D
(a)	3	4	1	2
(b)	1	4	3	2
(c)	3	2	1	4
(d)	1	2	3	4

[ESE : 2004]

3.17 Match **List-I** with **List-II** and select the correct answer using the codes given below the lists:

List-I
A. Unit Hydrograph
B. Synthetic unit Hydrograph
C. Darcy's law
D. Rational method

List-II
1. Design flood
2. Permeability
3. Ungauged basin
4. 1 cm runoff

Codes:

	A	B	C	D
(a)	2	3	4	1
(b)	2	1	4	3
(c)	4	3	2	1
(d)	4	1	2	3

[ESE : 2004]

3.18 A 6 hour storm has 6 cm of rainfall and the resulting runoff was 3 cm. If ϕ-index remains at the same value, which one of the following is the runoff due to 12 cm of rainfall in 9 hours in the catchment?
(a) 4.5 cm (b) 6.0 cm
(c) 7.5 cm (d) 9.0 cm [ESE : 2005]

3.19 A catchment has an area of 150 hectares and a run-off/rainfall ratio of 0.40. If due to 10 cm rainfall over the catchment, a stream flow at the catchment outlet lasts for 10 hours, what is the average stream flow in the period?
(a) 60,000 m^3/hr (b) 100 m^3/minute
(c) 3.5 m^3/s (d) 1.33 m^3/s
[ESE : 2005]

3.20 Viewing watershed as a system, which one of the following assumptions is made in the Unit Hydrograph theory?
(a) Non-linearity
(b) Both linearity and time variance
(c) Both time invariance and non-linearity
(d) Both linearity and time invariance
[ESE : 2005]

3.21 For a given storm, other factors remaining same,
(a) basins with large drainage densities give smaller flood peaks
(b) low drainage density basins give shorter time bases of hydrographs
(c) the flood peak is independent of the drainage density
(d) basins having low drainage density give smaller peaks in flood hydrographs
[ESE : 2005]

3.22 A DRH due to a storm over a basin has a time base of 90 hours with straight line portions of the hydrograph with flow rates of 0, 10, 70, 90, 40 and 0 m^3/s at elapsed durations of 0, 10, 20, 30, 50 and 90 hours as indicated on the above diagram, respectively. The catchment area is 300 km^2. What is the rainfall excess in the storm?

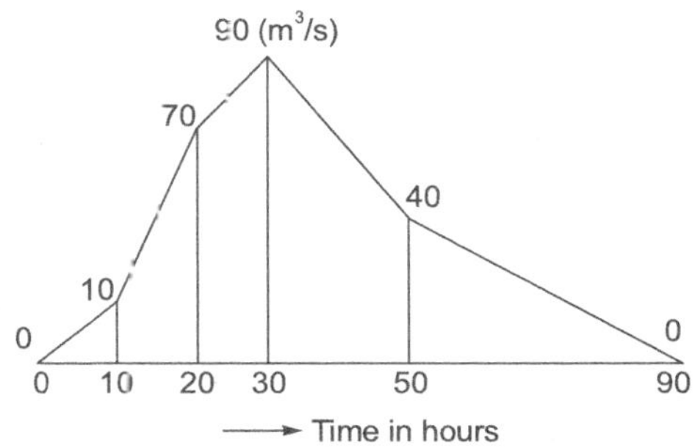

(a) 2.83 cm (b) 3.46 cm
(c) 3.87 cm (d) 4.02 cm
[ESE : 2006]

3.23 The time of concentration at the outlet in an urban area catchment of 1.5 km^2 area with a run off coefficient of 0.42 is 28 minutes. The maximum depth of rainfall with a 50 year return period for this time of concentration is 48 mm. What is the peak flow rate at the outlet for this return period?

(a) 12 m^3/s (b) 14 m^3/s
(c) 16 m^3/s (d) 18 m^3/s

[ESE : 2006]

3.24 Match **List-I** (Technique/Principle) with **List-II** (Purpose) and select the correct answer using the code given below the lists:

List-I
A. ϕ-Index
B. Slope - area method
C. Flow duration curve
D. Dilution technique

List-II
1. Dependable flow
2. Reservoir regulation
3. Steady stream discharge determination
4. Run-off volume
5. Unsteady stream discharge determination

Codes:

	A	B	C	D
(a)	3	5	1	4
(b)	4	1	2	3
(c)	3	1	2	4
(d)	4	5	1	3

[ESE : 2006]

3.25 A 252 km^2 catchment area has a 6 hr UH which is a triangle with time base of 35 hours. What is the peak discharge of the DRH due to 5 cm effective rainfall in 6 hr from that catchment?

(a) 45 cumecs (b) 115 cumecs
(c) 200 cumecs (d) 256 cumecs

[ESE : 2006]

3.26 A 4-hour unit hydrograph of a basin can be approximated as a triangle with a base period of 48 hours and peak ordinate of 300 m^3/s. What is the area of the catchment basin?

(a) 7776 km^2 (b) 5184 km^2
(c) 2592 km^2 (d) 1294 km^2

[ESE : 2006]

3.27 Assertion (A): The unit hydrograph cannot be applied for areas less than 5000 sq. km.

Reason (R): The run-off hydrograph reflects the physiographic factors of a catchment.

(a) both A and R are true and R is the correct explanation of A
(b) both A and R are true but R is not a correct explanation of A
(c) A is true but R is false
(d) A is false but R is true

[ESE : 2006]

3.28 Match **List-I** (Parameter) with **List-II** (Relatable Term) and select the correct answer using the codes given below the lists:

List-I
A. Rainfall Intensity
B. Rainfall Excess
C. Rainfall Averaging
D. Mass Curve

List-II
1. Isohyets
2. Cumulative rainfall
3. Hyetograph
4. Direct Runoff hydrograph

Codes:

	A	B	C	D
(a)	1	3	2	4
(b)	3	4	1	2
(c)	1	3	4	2
(d)	3	4	2	1

[ESE : 2007]

3.29 The total observed runoff volume during a 4 hour storm with a uniform intensity of 2.8 cm/hr is 25.2×10^6 m^3 from a basin of 280 km^2 area. What is the average infiltration rate for the basin?

(a) 3.6 mm/hr (b) 4.8 mm/hr
(c) 5.2 mm/hr (d) 5.5 mm/hr

[ESE : 2007]

3.30 In constructing a 4 hour synthetic unit hygrograph for a basin, the lag time is estimated to be 40 hours. When will the peak discharge in the Synthetic Unit Hydrograph occur from the start of the storm?

(a) 36 hours (b) 40 hours
(c) 42 hours (d) 44 hours

[ESE : 2007]

3.31 Consider the following statements:

1. Only the surface flow constitutes the flood hydrograph due to an isolated storm.
2. For a given storm, the flood peak is dependent on the drainage density.

3. Fan shaped catchments give narrow hydrograph with low peak.

Which of these statements is/are correct?

(a) 1, 2 and 3 (b) 1 and 3
(c) 2 only (d) 3 only [ESE : 2007]

3.32 A unit hydrograph for a watershed is triangular in shape with base period of 20 hours. The area of the watershed is 500 ha. What is the peak discharge in m^3/hour?

(a) 7000 (b) 6000
(c) 5000 (d) 4000 [ESE : 2007]

3.33 Kirpich equation is used to determine which one of the following?

(a) Run-off from a given rainfall
(b) Base time of a unit hydrograph
(c) Time of concentration in run-off hydrograph
(d) None of the above

[ESE : 2008]

3.34 A 4 hr storm had 4 cm of rainfall and the resulting direct runoff was 2.0 cm. If the ϕ-index remains at the same value, the runoff due to 10 cm of rainfall in 8 hr in the catchment is

(a) 6.0 cm (b) 7.5 cm
(c) 2.3 cm (d) 2.8 cm [ESE : 2008]

3.35 A triangular direct runoff hydrograph due to a storm has a time base of 60 hr and a peak flow of 30 m^3/s occurring at 20 hr from the start. If the catchment area is 300 km^2, what is the rainfall excess in the storm?

(a) 50 mm (b) 20 mm
(c) 10.8 mm (d) 8.3 mm

[ESE : 2009]

3.36 A 3 hr unit hydrograph U_1 of a catchment of area 235 km^2 is in the form of a triangle with peak discharge 30 m^3/s. Another 3 hr unit hydrograph U_2 is also triangular in shape and has the same base width as U_1, but has a peak flow of 90 m^3/s. What is the catchment area of U_2?

(a) 117.5 km^2 (b) 235 km^2
(c) 470 km^2 (d) 705 km^2

[ESE : 2009]

3.37 A catchment area of 60 ha has a runoff coefficient of 0.40. If a storm of intensity 3 cm/h and duration longer than the time of concentration occurs in the catchment, then what is the peak discharge?

(a) 2.0 m^3/s (b) 3.5 m^3/s
(c) 4.5 m^3/s (d) 2.5 m^2/s

[ESE : 2009]

3.38 The land use of an area and the corresponding runoff coefficients are as follows:

Sl. No.	Land use	Area (ha)	Runoff coefficient
1.	Roads	10	0.70
2.	Lawn	20	0.10
3.	Residential area	50	0.30
4.	Industrial area	20	0.80

What is the equivalent runoff coefficient?

(a) 0.15 (b) 0.36
(c) 0.40 (d) 0.51

[ESE : 2009]

3.39 A catchment of area 200 ha has a runoff coefficient 0.5. A storm of duration larger than the time of concentration of the catchment and of intensity 3.6 cm/h causes a peak discharge of

(a) 5 m^3/s (b) 10 m^3/s
(c) 100 m^3/s (d) 360 m^3/s

[ESE : 2010]

3.40 A catchment consists of 30% area with runoff coefficient 0.40 with the remaining 70% area with runoff coefficient 0.60. The equivalent runoff coefficient will be

(a) 0.48 (b) 0.54
(c) 0.63 (d) 0.76

[ESE : 2010]

3.41 A direct-runoff hydrograph due to isolated storm was triangular in shape with a base of 80 h and peak of 200 m^3/s. If the catchment area is 1440 km^2, the effective rainfall of the storm is

(a) 20 cm (b) 10 cm
(c) 5 cm (d) 2 cm

[ESE : 2010]

3.42 The shape of the recession limb of a hydrograph depends on

(a) basin as well as storm characteristics
(b) storm characteristics only
(c) basin characteristics only
(d) base flow only [ESE : 2010]

3.43 An S-curve hydrograph has been obtained for catchments of 270 km^2 from a 3 hour unit hydrograph. The equilibrium discharge for the S-curve is

(a) 750 m^3/s (b) 277.8 m^3/s
(c) 250 m^3/s (d) 187 m^3/s

[ESE : 2010]

3.44 The rainfall on five successive days on a catchment was 3, 6, 9, 5 and 1 cm respectively. If the ϕ-index for the storm can be assumed to be 3 cm/day, the total direct runoff from the catchment due to this storm is

(a) 11 cm (b) 24 cm
(c) 9 cm (d) 20 cm [ESE : 2012]

3.45 The excess runoff hydrograph from a catchment area 10 km^2 due to a storm of 6 hrs duration has been observed to be triangular in shape. The peak flow is observed to be 10 m^3/s and the base length is 20 hrs. The rainfall excess in the catchment is

(a) 5.1 cm (b) 3.6 cm
(c) 4.5 cm (d) 2.5 cm [ESE : 2012]

3.46 An urban area is located in plains having "average climatic conditions". The impervious area for which drainage must be provided is 3.6 ha and the design rainfall intensity is 2.0 cm/hr. The drains will be designed for a runoff of

(a) 0.05 m^3/s (b) 0.10 m^3/s
(c) 0.20 m^3/s (d) 0.40 m^3/s

[ESE : 2012]

3.47 The theory of synthetic hydrograph based on flood routing techniques is based on the principle that rainfall impulse (net storm rain) is modified by the factors.

1. The time of travel of the flow volume in channel and overland flow
2. Storage
3. Both translation and storage

(a) 1 and 2 only (b) 1, 2 and 3
(c) 1 and 3 only (d) 2 and 3 only

[ESE : 2013]

3.48 A mean annual runoff of 2 m^3/s from a catchment of area 10 km^2 represents an effective rainfall of nearly:

(a) 530 cm (b) 590 cm
(c) 630 cm (d) 658 cm

[ESE : 2013]

3.49 The peak magnitude of a flood hydrograph during 4 hr study duration over a catchment is 300 m^3/s. The total depth of rainfall is 6 cm, and the infiltration loss during the said 4 hr period is 2 cm. A constant uniform base flow of 20 m^3/s is premised throughout. The peak value of the corresponding 4 hr unit hydrograph is

(a) 75 m^3/s (b) 70 m^3/s
(c) 50 m^3/s (d) 40 m^3/s

[ESE : 2014]

3.50 The rainfall on 5 successive days in a catchment was 2, 7, 8, 4 and 3 cm. If the ϕ index for the storm is 3 cm/day, the total direct run-off volume generated from a 195 km^2 catchment is

(a) 19.5 M m^3 (b) 23.4 M m^3
(c) 15.6 M m^3 (d) 32.5 M m^3

[ESE : 2014]

3.51 The best estimate of runoff represented by 57 mm of runoff depth from a basin area 3300 km^2 is nearly

(a) 2300 cumec-days
(b) 2225 cumec-days
(c) 2175 cumec-days
(d) 2020 cumec-days

[ESE : 2014]

3.52 A direct runoff hydrograph due to a storm idealized into a triangular shape has a peak flow rate of 60 m^3/s occurring at 25 hours from its start. If the base width of this hydrograph is 72 hours, and the catchment area is 777.6 km^2, the run-off from the storm is

(a) 1 cm (b) 2 cm
(c) 5 cm (d) 10 cm

[ESE : 2014]

3.53 If the direction of advance of a storm is the same as the direction of the resulting runoff in the drainage basin, such runoff will generally

(a) Be more than the 'rational' run-off through a short time
(b) Result in increased infiltration through a short time

(c) Be less than the 'rational' runoff through a short time

(d) Result in increased evaporation, as also transpiration, loss [ESE : 2014]

Directions: The following items consists of two statements; one labelled as **'Statement (I)'** and the other as **'Statement (II)'**. You are to examine these two statements carefully and select the answers to these items using the codes given below:

Codes:

(a) Both Statement (I) and Statement (II) are individually true and Statement (II) is the correct explanation of Statement (I).

(b) Both Statement (I) and Statement (II) are individually true but Statement (II) is not the correct explanation of Statement (I).

(c) Statement (I) is true but Statement (II) is false.

(d) Statement (I) is false but Statement (II) is true.

3.54 Statement (I): Instantaneous unit hydrograph (IUH) is used in theoretical analysis of rainfall excess-runoff characteristics of a catchment.

Statement (II): For a given catchment, IUH, being independent of rainfall characteristics, is indicative of the catchment storage characteristics.

[ESE : 2014]

3.55 Statement (I): Theoretically an infinite number of unit graphs are possible for a given basin.

Statement (II): The rainfall duration and its areal distribution affect the hydrograph.

[ESE : 2014]

3.56 A 4-hour rainfall in a catchment of 250 km^2 produces rainfall depths of 6.2 cm and 5 cm in successive 2-hour unit periods. Assuming the ϕ index of the soil to be 1.2 cm/hour, the runoff volume is

(a) 1.6 ha-m (b) 16 ha-m

(c) 160 ha-m (d) 1600 ha-m

[ESE : 2015]

3.57 Consider the following steps which are involved in arriving at a unit hydrograph:

1. Separation of base flow
2. Estimating the surface runoff in volume
3. Estimating the surface runoff in depth
4. Dividing surface runoff ordinate by depth of runoff

Which is the correct sequence of these steps?

(a) 4, 3, 2 and 1 (b) 1, 2, 3 and 4

(c) 4, 2, 3 and 1 (d) 1, 3, 2 and 4

[ESE : 2015]

3.58 S-Curve Hydrograph is the hydrograph

(a) producing 1 cm of runoff over the basin

(b) of flow from a 1 cm intensity rain of infinite duration

(c) having a volume of 1 cm^3

(d) of the total storm duration in any single storm rainfall

[ESE : 2015]

3.59 Surface Runoff represents total water

(a) flowing in surface channels after the rainfall.

(b) obtained after deducting from rainfall water what has infiltrated and/or evaporated, from the total rainfall.

(c) excluding the base flow in surface channels after the rainfall.

(d) flown (or flowing) through all channels over a specified period of time. [ESE : 2015]

3.60 A peak flow of a flood hydrograph due to a six-hour storm is 470 m^3/s. The corresponding average depth of rainfall is 8 cm. Assume infiltration loss of 0.25 cm/hour and a constant base flow of 15 m^3/s. What is the peak discharge of 6 hour unit hydrograph for this catchment?

(a) 60 m^3/s (b) 70 m^3/s

(c) 80 m^3/s (d) 90 m^3/s

[ESE : 2016]

3.61 Over a basin of area 333 km^2, there was a storm for 6h with a uniform intensity of 2 cm/h. The observed runoff was 20×10^6 m^3. The observed rate of infiltration for the basin was

(a) 5 mm/h (b) 10 mm/h

(c) 20 mm/h (d) 40 mm/h

[ESE : 2016]

3.62 In the standard SCS-CN method of modeling runoff due to daily rainfall, if CN = 75, the runoff magnitude for a one-day rainfall of 100 mm is nearly

(a) 17 mm (b) 31 mm

(c) 41 mm (d) 57 mm

[ESE : 2017]

3.63 An 8-hour storm with incremental rainfall during each successive hour is tabulated herewith. What will be the correction to total runoff when, whereas the actual φ index was 0.5 cm/hour, a wrong magnitude of 0.6 cm/hour was adopted in computing the total runoff?

Time from start (hour)	1	2	3	4	5	6	7	8
Incremental rainfall in each hour (cm)	0.4	0.8	1.2	0.9	1.8	1.0	0.3	0.8

(a) +1.0 cm (b) +0.9 cm
(c) +0.8 cm (d) +0.6 cm

[ESE : 2017]

3.64 Statement (I): Collin's method for the determination of the unit hydrograph for a multi-period storm considers residuals as attributable to the unit graph of the period of the largest rain.
Statement (II): Collin's method is *ab initio* premised on the Bernard method where a unit graph is necessarily to be proportional to the net rain.

(a) Both Statement (I) and Statement (II) are individually true and Statement (II) is the correct explanation of Statement (I).
(b) Both Statement (I) and Statement (II) are individually true but Statement (II) is not the correct explanation of Statement (I).
(c) Statement (I) is true but Statement (II) is false.
(d) Statement (I) is false but Statement (II) is true.

[ESE : 2017]

3.65 Horton's infiltration equation was fitted to data from an infiltration test. It was found that the initial infiltration capacity was 20 mm/h, final infiltration capacity was 5 mm/h and the exponential decay constant was 0.5 h^{-1}. If the infiltration was at capacity rates, the total infiltration depth for a uniform storm of 10 h duration would be

(a) 80 mm (b) 50 mm
(c) 30 mm (d) 20 mm

[ESE : 2018]

3.66 The rate of rainfall for the successive 30 min periods of a 3-hour storm are: 1.6, 3.6, 5.0, 2.8, 2.2 and 1.0 cm/hour. The corresponding surface runoff is estimated to be 3.2 cm. Then, the ϕ-index is

(a) 1.5 cm/h (b) 1.8 cm/h
(c) 2.1 cm/h (d) 2.4 cm/h

[ESE : 2018]

3.67 Which of the following will pose difficulties in adopting UHG principles and processes in evaluating flood hydrographs of basins?

1. Non-uniform areal distribution within a storm.
2. Intensity variation within a storm.
3. The centre of the storm varying from storm to storm in case of large catchments.
4. Dividing into a number of sub-basins and routing the individual DRHs through their respective channels to obtain the composite DRH at the basin outlet.
5. Large storages within the catchment.

(a) 1, 3 and 4 only (b) 2, 3 and 4 only
(c) 1, 2 and 5 only (d) 1, 2, 3 and 5 only

[ESE : 2018]

3.68 Rainfall of magnitude 4.3 cm, followed by 3.7 cm, occurred on two consecutive 4h durations on a catchment area of 25 km^2, and there resulted a DRH (after isolation of base flow in the flood flow hydrograph) with the following ordinates starting from the beginning of the rainfall. (Adopt trapezoidal formula)

Time (hours)	0	4	8	12	16	20	24	28	32	36	40	44
DRH (ordinate m^3/sec)	0	9	16	20	20	17.8	13.4	9.4	6.2	3.7	1.8	0

What is the ϕ-index value?

(a) 0.149 cm/h (b) 0.155 cm/h
(c) 0.161 cm/h (d) 0.167 cm/h

[ESE : 2018]

3.69 In intensity-duration analysis by Sherman, the intensity of rainfall i is represented as

(a) $\dfrac{b^n}{(t+a)}$ (b) $\dfrac{a^n}{(t+b)^n}$

(c) $\dfrac{(a+t)^n}{b}$ (d) $\dfrac{a}{(t+b)^n}$

where t is time and a, b, n are constants for the area.

[ESE : 2019]

3.70 Which one of the following terms refers to the time between the end of the net rainfall and the end of the direct runoff hydrograph?
(a) Recession time (b) Time-to-peak
(c) Lag time (d) Concentration time
[ESE : 2024]

3.71 Match the **List I** (Soil and vegetation) and **List-II** (Infiltration rate (mm/hr)) following lists:

List-I	List-II
P. Forested loam	1. 10 – 70
Q. Loam pasture	2. 0 – 4
R. Sand	3. 3 – 15
S. Bare clay	4. 100 – 200

Select the correct answer using the code given below :

	P	Q	R	S
(a)	2	3	1	4
(b)	3	1	2	4
(c)	4	1	3	2
(d)	4	1	2	3

[ESE : 2024]

3.72 Match the following lists :
List I (Type of surface)
P. Wooded areas
Q. Gravel roads and walks
R. Macadamized Roads
S. Watertight roof surfaces
List II (Value of coefficient of runoff)
1. 0.70 – 0.95
2. 0.15 – 0.30
3. 0.01– 0.20
4. 0.25 – 0.60

Select the correct answer using the code given below:

	P	Q	R	S
(a)	2	3	1	4
(b)	3	2	4	1
(c)	3	1	2	4
(d)	4	1	2	3

[ESE : 2024]

■■■■

Answers Infiltration, Runoff and Hydrograph

3.1 (d)	3.2 (d)	3.3 (d)	3.4 (c)	3.5 (a)	3.6 (c)	3.7 (c)	3.8 (c)	3.9 (a)
3.10 (c)	3.11 (c)	3.12 (d)	3.13 (d)	3.14 (b)	3.15 (d)	3.16 (a)	3.17 (c)	3.18 (c)
3.19 (b)	3.20 (d)	3.21 (d)	3.22 (d)	3.23 (d)	3.24 (d)	3.25 (c)	3.26 (c)	3.27 (d)
3.28 (b)	3.29 (d)	3.30 (c)	3.31 (c)	3.32 (c)	3.33 (c)	3.34 (a)	3.35 (c)	3.36 (d)
3.37 (a)	3.38 (c)	3.39 (b)	3.40 (b)	3.41 (d)	3.42 (c)	3.43 (c)	3.44 (a)	3.45 (b)
3.46 (c)	3.47 (c)	3.48 (c)	3.49 (b)	3.50 (a)	3.51 (c)	3.52 (a)	3.53 (a)	3.54 (a)
3.55 (b)	3.56 (d)	3.57 (b)	3.58 (b)	3.59 (b)	3.60 (b)	3.61 (b)	3.62 (c)	3.63 (d)
3.64 (d)	3.65 (a)	3.66 (b)	3.67 (d)	3.68 (b)	3.69 (d)	3.70 (d)	3.71 (c)	3.72 (b)

Explanations Infiltration, Runoff and Hydrograph

3.1 (d)
The 12-hr Unit Hydrograph will be obtained by adding the ordinates of 6-hr and ordinates of 6-hr UH lagged by 6-hr and then dividing the ordinates by 2. So the base period for 12 hr UH will be, 84 + 6 = 90 hr.

Alternate method:
Base period of T hr unit hydrograph is given by

$$t_B' = t_B + (n-1)\,D$$

where, $n = T/D = 12/6 = 2$

$$\therefore \quad t_B' = 84 + (2-1) \times 6 = 90 \text{ hr}$$

3.2 (d)
In a given period of effective rainfall, the hydrograph of direct runoff from a basin reflects the combination of rainfall characteristics and the physical characteristics of the basin.

Effective rainfall is assumed to be constant and uniformly distributed over the basin, i.e., if there are N rain gauges spread uniformly over the basin, then all the gauges should record almost same amount of rainfall during the specified time.

Also the hydrographs of direct runoff due to effective rainfall of equal duration have the same time base.

3.3 (d)

Total rainfall = 3.5 + 4.2 + 2.9 = 10.6 cm

Effective rainfall = 3 cm

Infiltration = 10.6 – 3 = 7.6 cm

$$\text{Now, w-index} = \frac{7.6}{3} = 2.533 \text{ cm/hr}$$

As all rainfall intensities are greater then w-index therefore all will produce rainfall excess. So that

$$\text{w-index} = \phi\text{-index} = 2.533 \text{ cm/hr}$$

3.4 (c)

Ephemeral stream is one which does not have any base flow contribution. It does not have any well defined channel.

3.5 (a)

The rate at which water infiltrates into a ground is called the infiltration capacity. For an initially dry soil subjected to rain, the infiltration capacity curve shows an exponentially decaying trend. Initially the rate of infiltration is high but it soon decreases, as most of the soil gets moist. The rate of infiltration reaches a uniform rate after some time.

For consistency in hydrological calculations, a constant value of infiltration rate for the entire storm duration is adopted. The average infiltration rate is called the Infiltration Index and the two types of indices commonly used are ö – index and W – index.

The ϕ-index is defined as the rate of infiltration above which the rainfall volume equals runoff volume.

A lysimeter is a measuring device which can be used to measure the amount of actual evapotranspiration which is released by plants. By recording the amount of precipitation that an area receives and the amount lost through the soil, the amount of water lost to evapotranspiration can be calculated. Lysimeters are of two types: weighing and non-weighing.

Dilution technique is used for flow measurement of small and turbulent streams.

3.6 (c)

Total rainfall = 7 + 18 + 25 + 17 + 11 +3
= 81 mm

Effective rainfall = 39 mm

Infiltration = 81 – 39 = 42 mm

$$\text{Now w-index} = \frac{42}{6} = 7 \text{ mm/hr}$$

As ϕ-index > *w*-index

∴ 3 mm/hr and 7mm/hr will not produce any run-off.

∴ Net infiltration for 4 hours
= 42 –10 = 32 mm

$$\phi\text{-index} = \frac{32}{4} = 8 \text{ mm/hr}$$

3.8 (c)

The 2 hour unit hydrograph is obtained by dividing the ordinates of storm hydrograph by the direct runoff depth which is 5 units.

3.9 (a)

The size of catchment imposes upper limit and lower limit for application of UH. Above 5000 km^2 the centre of the storm can vary from storm to storm and each can give different DRHs under otherwise identical situations. The rainfall will be nonuniform in these conditions. The lower limit for UH theory is 200 ha. At this level of area, a Number of factors affect the rainfall-runoff relationship and the UH is not accurate enough for the prediction of DRH.

3.10 (c)

A flow duration curve is a plot that shows the percentage of time that flow in a stream is likely to equal or exceed some specified value of interest.

A Mass curve is the graph of the cumulative values of a hydrologic quantity (such as precipitation or runoff), generally as the Ordinate, plotted against time or date as the Abscissa.

A hydrograph is a graph showing the rate of flow (discharge) versus time past a specific point in a river, or other channel or conduit carrying flow.

3.12 (d)

Total rainfall = 7 + 18 + 25 + 12 + 10 + 3 = 75 mm

Effective rainfall = 33 mm

Infiltration = 75 – 33 = 42 mm

$$w\text{-index} = \frac{42}{6} = 7 \text{ mm/hr}$$

As it is greater than 3 mm/hr and 7 mm/hr

So these intensities will not produce rainfall excess.

$$\therefore \phi\text{-index} = \frac{(18+25+12+10)-33}{4} = 8 \text{ mm/hr}$$

3.13 (d)

$Q_P = kiA/36$

where

Q_P is Peak discharge in m^3/s

i is intensity in cm/hour

A is Area in ha

$$\therefore Q_P = \frac{0.4 \times 4.5 \times 90}{36} = 4.5 \text{ m}^3/\text{s}$$

If area is in km^2 and intensity in mm/hr, then

$$Q_p(\text{m}^3/\text{s}) = \frac{kiA}{3.6}$$

3.14 (b)

Snyder was the one to develop a synthetic Unit Hydrograph based on a study of watersheds in the Appalachian Highlands. In basins ranging from 10 – 10,000 mi^2

Snyder relations are $t_p = C_t (L\,L_C)^{0.3}$

where t_p = basin lag (hr)

L = length of the main stream from the outlet to the divide (mi)

L_C = length along the main stream to a point nearest the watershed centroid (mi)

C_t = Coefficient usually ranging from 1.8 to 2.2

3.15 (d)

Total rainfall = 14 cm

Direct runoff = 8 cm

Infiltration = 14 – 8 = 6 cm

$$w\text{-index} = \frac{6}{7} = 0.86 \text{ cm/hr}$$

It is greater than 0.6 cm/hr

$$\therefore \phi\text{-index} = \frac{(1+2+2.8+3.3+2.5+1.8)-8}{6} = 0.9 \text{ cm/hr}$$

3.16 (a)

The rising limb of a hydrograph is also known as concentration curve and it represents the increase in discharge due to the gradual building up of storage in channels and over the catchment surface.

3.17 (c)

In India, only a small number of streams are gauged (i.e., stream flows due to single and multiple storms, are measured). There are many drainage basins (catchments) for which no stream flow records are available and unit hydrographs may be required for such basins. In such cases, hydrographs may be synthesized directly from other catchments, which are hydrologically and meteorologically homogeneous, or indirectly from other catchments through the application of empirical relationship. Unit hydrograph thus derived is called Synthetic Unit Hydrograph.

Darcy's law at constant elevation is a simple proportional relationship between the instantaneous discharge rate through a porous medium, the viscosity of the fluid and the pressure drop over a given distance. This equation for single phase fluid flow is the defining equation for absolute permeability (single phase permeability).

3.18 (c)

$$\phi - \text{index} = \frac{6-3}{6} = 0.5 \text{ cm/hr}$$

Infiltration for 9 hour rainfall

= 9 × 0.5 = 4.5 cm

Runoff due to 12 cm rainfall

= 12 – 4.5 = 7.5 cm

3.19 (b)

Runoff = $150 \times 10^4 \times 0.4 \times 0.1 = 60{,}000 \text{ m}^3$

$$\text{Average stream flow} = \frac{60{,}000}{10 \times 60} = 100 \text{ m}^3/\text{minute}$$

3.20 (d)

The primary assumption of unit hydrograph theory is that the rainfall has uniform distribution, both in space – with minimal variations across the basin – and in time; in other words, the rainfall rate does not vary much during the event.

3.21 (d)

Drainage density is the ratio of total length of stream channels to the area of catchment. Large drainage density creates quick flowing off or runoff down the channel hence gives a faster response (shorter time base) and high peak discharge.

3.22 (d)

$$\text{Area under DRH} = \frac{1}{2}\times 10\times 10 + \frac{(10+70)\times 10}{2} + \left(\frac{70+90}{2}\right)\times 10 + \left(\frac{90+40}{2}\right)\times 20 + \frac{40}{2}\times 40$$

$$= 3350 \text{ m}^3\text{hr/s} = 12.06 \times 10^6 \text{ m}^3$$

$$\therefore \text{ Rainfall excess} = \frac{12.06\times 10^6}{300\times 10^6} = 0.0402 \text{ m} = 4.02 \text{ cm}$$

3.23 (d)

The mean intensity of precipitation for a duration equal to time of concentration and an exceedence probability P,

$$i_{t,cp} = \frac{48}{(28/60)} = \frac{720}{7} \text{mm/h}$$

Peak discharge from Rational formula

$$Q = \frac{C\, i_{t,cp} A}{3.6} = 0.42\times\frac{720}{7}\times\frac{1.5}{3.6} = 18 \text{ m}^3/\text{s}$$

3.24 (d)

Dilution technique is used for stream flow measurement. The two dilution techniques used are (1) the steady feed method and (2) the instantaneous, point - source time indigenous method.

Advantages: They condense in closed conduits, such as penstocks, sewers pipe lines, where current-meter measurements are difficult, and they are fast and accurate.

Disadvantages: Expensive for measuring large stream and the special equipments required for the measurements of concentration.

When it is not possible to make direct measurement especially during floods, slope-area method is a most commonly used indirect method of measurement. In this method discharge is computed on the basis of a uniform flow equation involving channel characteristics, water surface profile and a roughness coefficient. The drop in water surface profile for a uniform reach of channel represents losses caused by bed roughness. Discharge is calculated using Manning's equation is used in this method.

A flow duration curve is a plot that shows the percentage of time that flow in a stream is likely to equal or exceed some specified value of interest.

3.25 (c)

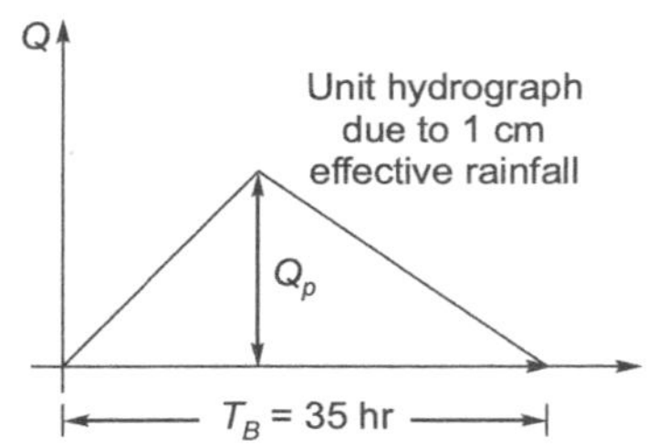

From the unit hydrograph

$$\frac{Q_p T_B}{2} = A\times\left(\frac{1}{100}\right)$$

$$\frac{Q_P\times 35\times 3600}{2} = \frac{252\times 10^6}{100}$$

$$\therefore \quad Q_p = 40 \text{ m}^3/\text{s}$$

The peak discharge due to 5 cm effective rainfall is 5 × 40 = 200 m^3/s

3.26 (c)

$$\frac{Q_P T_B}{2} = A\times\frac{1}{100}$$

$$A = \frac{300\times 48\times 3600\times 100}{2\times 10^6} = \frac{3\times 48\times 36}{2} = 2592 \text{ km}^2$$

3.27 (d)

A unit hydrograph is applicable for an area greater than 2 km^2 and less than 5000 km^2

3.28 (b)

A hyetograph is a graphical representation of the distribution of rainfall over time.

Isohyet is a line on a map connecting points having the same amount of rainfall in a given period.

A Mass curve is the graph of the cumulative values of a hydrologic quantity (such as precipitation or runoff), generally as the Ordinate, plotted against time or date as the Abscissa.

Direct runoff hydrograph (DRH) is effective

rainfall occurring uniformly over a watershed at a uniform rate over a particular period of time.

3.29 (d)

The average infiltration rate is also called *W*-index. Thus

$$W = \frac{P - R - I_a}{t}$$

where, P = Total storm precipitation depth
R = Total storm runoff depth
t = time of rainfall excess
I_a = Initial losses in terms of depth

$$\therefore \quad P = 2.8 \times 10 \times 4 = 112 \text{ mm}$$

$$R = \frac{25.2 \times 10^6 \times 10^3}{280 \times 10^6} = 90 \text{ mm}$$

$$t_c = 4 \text{ hr}$$

$$I_a = 0$$

$$\therefore \quad W = \frac{112 - 90}{4} = 5.5 \text{ mm/hr}$$

Alternate solution:

Total rainfall = 2.8 cm/hr × 4 hr = 11.2 cm

$$\text{Run-off} = \frac{25.2 \times 10^6}{280 \times 10^6} \times 100 \text{ cm} = 9 \text{ cm}$$

$$\therefore \text{Rate of infiltration} = \frac{11.2 - 9}{4} = 0.55 \text{ cm/hr}$$

$$= 5.5 \text{ mm/hr}$$

3.30 (c)

The time when the peak discharge occurs from the start of the storm may be given as

$$t_{pk} = \frac{t_R}{2} + t_P = \frac{4}{2} + 40 = 42 \text{ hours}$$

3.31 (c)

The flood hydrograph consists of flow in all the three phases of runoff, viz. surface runoff, interflow and base flow.

The factors affecting flood hydrograph are shape, size, slope and drainage density. A large drainage density creates situation conducive for quick disposal of runoff down the channels. This fast response is reflected in a pronounced peak discharge.

Fan shaped i.e. nearly semicircular shaped catchments give high peak and narrow hydrographs while elongated catchments give broad and low-peaked hydrographs.

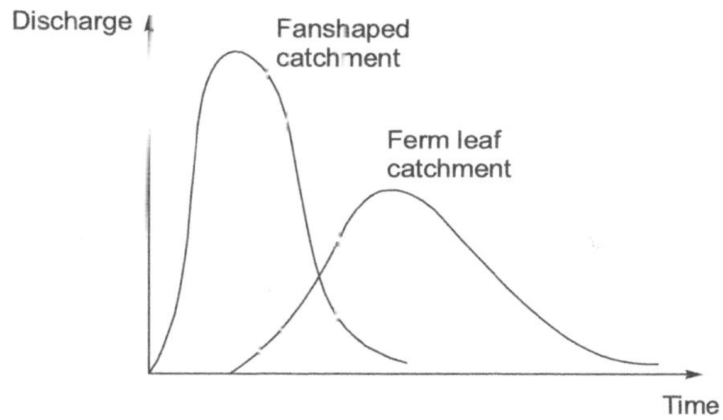

3.32 (c)

The unit hydrograph is shown below:

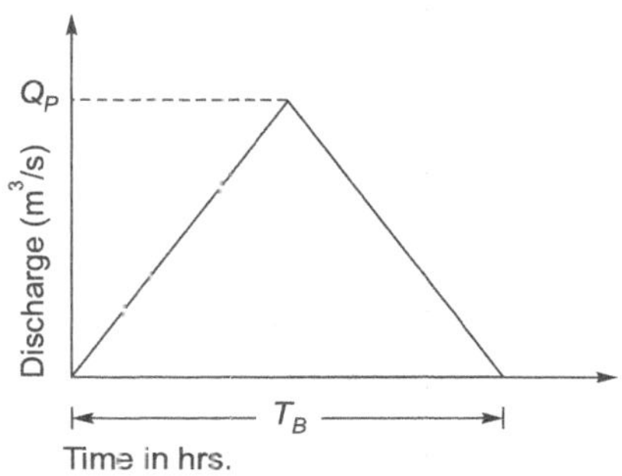

From this Unit Hydrograph

$$\frac{Q_P T_E}{2} = A \times \left(\frac{1}{100}\right)$$

$$\Rightarrow \quad \frac{Q_P \times 20}{2} = 500 \times 10^4 \times \frac{1}{100}$$

$$\Rightarrow \quad Q_P = 5000 \text{ m}^3/\text{s}$$

3.33 (c)

Kirpich equation is an empirical relation used for the estimation of the time concentration. It is given as

$$t_c = 0.01947\, L^{0.77} S^{-0.385}$$

where t_c = time of concentration in minutes
L = maximum length of travel of water (m)
S = slope of the catchment = $\frac{\Delta H}{L}$
ΔH = difference in elevation between the most remote point on the catchment outlet.

3.34 (a)

$$(\phi - \text{index})_4 = \frac{P - R}{t_e} = \frac{4 - 2}{4} = 0.5 \text{ cm/hr}$$

But, $(\phi\text{-index})_4 = (\phi\text{-index})_8$

$$\Rightarrow \quad 0.5 = \frac{10 - R}{8}$$

$$\Rightarrow \quad R = 10 - 4$$

$$\Rightarrow \quad R = 6 \text{ cm}$$

3.35 (c)

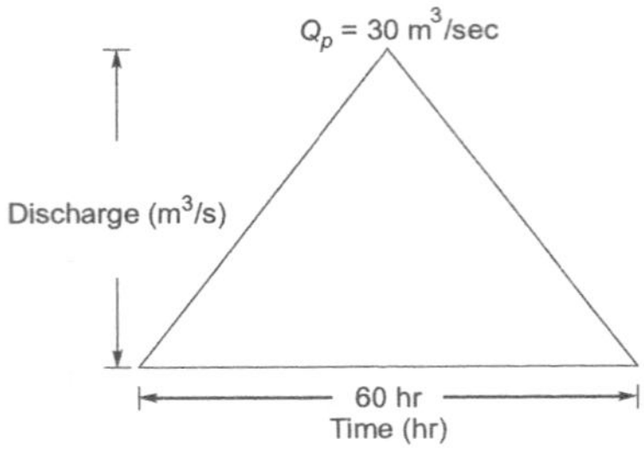

Area of direct runoff hydrograph

= Total volume of excess rainfall (V)

$$\Rightarrow \frac{1}{2}\times 30\times 60\times 60\times 60$$

= Total volume of excess rainfall (V)

$\Rightarrow$ Depth of excess rainfall

$$= \frac{V}{\text{catchment area}}$$

$$= \frac{30\times 60\times 60\times 60\times 10^3}{2\times 300\times 10^6} = 10.8 \text{ mm}$$

3.36 (d)

Time base of both the unit hydrographs is same. Let it be t.

$$\therefore \quad \frac{1}{2}\times 30\times t\times \frac{1}{235} = \frac{1}{2}\times 90\times t\times \frac{1}{A_2}$$

$$\Rightarrow \quad A_2 = 235\times 3$$

$$\Rightarrow \quad A_2 = 705 \text{ km}^2$$

3.37 (a)

The peak discharge is given by

$$Q_p = kiA$$

$$= 0.40\times \frac{3\times 10^{-2}}{60\times 60}\times 60\times 10^4 = 2.0 \text{ m}^3/\text{s}$$

3.38 (c)

$$K_{eq} = \frac{K_1A_1 + K_2A_2 + K_3A_3 + K_4A_4}{A_1 + A_2 + A_3 + A_4}$$

$$= \frac{0.70\times 10 + 0.10\times 20 + 0.30\times 50 + 0.80\times 20}{10 + 20 + 50 + 20}$$

$$= 0.4$$

3.39 (b)

As per rational formula, the peak value of discharge is expressed as

$$Q_p = KAi$$

$$= 0.5\times 200\times 10^4\times \frac{3.6\times 10^{-2}}{3600}$$

$$= 10 \text{ m}^3/\text{s}$$

3.40 (b)

The equivalent runoff coefficient is given by

$$K = \frac{K_1A_1 + K_2A_2}{A_1 + A_2}$$

$$= \frac{0.4\times 0.3 + 0.6\times 0.7}{0.3 + 0.7} = 0.54$$

3.41 (d)

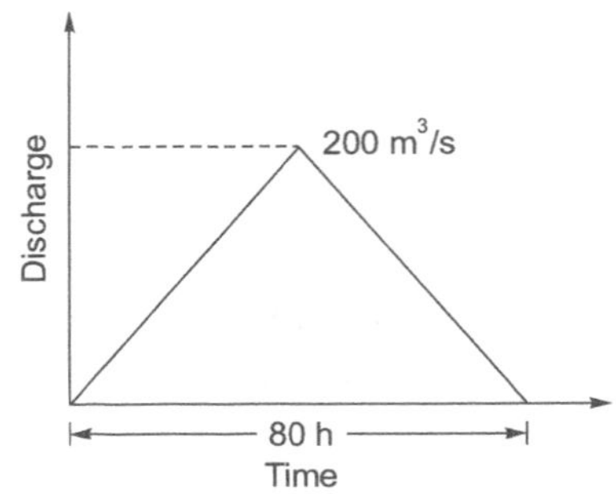

$\therefore$ Total runoff = Area of triangle

$$= \frac{1}{2}\times 80\times 3600\times 200$$

$$= 28.8\times 10^6 \text{ m}^3$$

Area of catchment = 1440 km²

$$\therefore \text{ Effective rainfall} = \frac{\text{Total runoff}}{\text{Area of catchment}}$$

$$= \frac{28.8\times 10^6}{1440\times 10^6} = 0.02 \text{ m} = 2 \text{ cm}$$

3.42 (c)

The climatic factors control the rising limb and the recession limb is independent of storm characteristics, being determined by catchment characteristics only.

3.43 (c)

The equilibrium discharge is expressed as

$$Q_s = \frac{A}{D}\times 10^4 \text{ m}^3/\text{h}$$

where A is area of catchment in km² and D is duration in hours.

$$\therefore \quad Q_s = \frac{270}{3}\times 10^4 = 90\times 10^4 \text{ m}^3/\text{h}$$

$$= \frac{90\times 10^4}{3600} \text{ m}^3/\text{s} = 250 \text{ m}^3/\text{s}$$

3.44 (a)

As ϕ-index = 3 cm/day

$\therefore$ Storm of intensities 3 cm/day and 1 cm/day

will not produce rainfall excess.
So that,
Run-off = (6 + 9 + 5) cm – 3 cm/day × 3 day
= 11 cm

3.45 (b)

$$\text{Total runoff in m}^3 = \left(\frac{1}{2}\times 20\times 60\times 60\times 10\right)$$
$$= 360{,}000 \text{ m}^3$$
$$\therefore \quad \text{Rainfall excess} = \left(\frac{360{,}000}{10\times 10^6}\right)\text{m}$$
$$= 0.036 \text{ m} = 3.6 \text{ cm}$$

3.46 (c)

The runoff, Q is given by

$$Q = \frac{1}{36}\text{k.p.A}$$
$$K = 1 \text{ (impervious)}$$
$$\therefore \quad Q = \frac{1}{36}\times 1\times 2\times 3.6 = 0.2 \text{ m}^3/\text{sec}$$

3.47 (c)

According to Snyder the shape of synthetic unit hydrograph is likely to be affected by basin characteristics like area, topography, shape of the slope, drainage density and channel storage. So, according to the theory synthetic hydrograph is affected by:

1. The time of travel, i.e. translation time of the flow volume in channel and overland flow.
2. Both translation and storage capacity of the basin.

3.48 (c)

$$Q = 2 \text{ m}^3/\text{s}$$
$$= 2 \times 3600 \times 24 \times 365 \text{ m}^3/\text{yr}$$
$$= 63.072 \times 10^6 \text{ m}^3/\text{yr}$$
$$A = 10 \text{ km}^2 = 10 \times 10^6 \text{ m}^2$$

$\therefore$ Annual volume

$$= 10 \times 10^6\, y = 63.072 \times 10^6$$
$$\Rightarrow \quad y = 6.3072 \text{ m} = 630.72 \text{ cm}$$

3.49 (b)

Effective rainfall
$$= 6 - 2 = 4 \text{ cm}$$
Peak of effective hydrograph
$$= 300 - 20 = 280 \text{ m}^3/\text{sec.}$$
The peak value of 4 hr unit hydrograph
$$= \frac{280}{4} = 70 \text{ m}^3/\text{sec.}$$

3.50 (a)

Total direct runoff
$$= (7 - 3) + (3 - 3) + (4 - 3) = 10 \text{ cm}$$
So, Volume generated
$$= \frac{10}{100}\times 195\times 10^6 \text{ m}^3$$
$$= 19.5 \times 10^6 \text{ m}^3 = 19.5 \text{ Mm}^3$$

3.51 (c)

Best estimated runoff
$$= \frac{\text{Total volume generated}}{\text{duration}}$$
$$= \frac{(57\times 10^{-3})\times 3300\times 10^6}{24\times 60\times 60}$$
$$= 2177.083 \simeq 2175 \text{ cumec – days}$$

3.52 (a)

$$\text{Total runoff} = \frac{1}{2}\times 60\times 72\times 3600$$
$$= 7776000 \text{ m}^3$$
Depth of runoff
$$= \frac{\text{Total runoff}}{\text{Catchment area}}$$
$$= \frac{7776000}{777.6\times 10^6}\text{m} = \frac{1}{100}\text{m}$$
$$= 0.01 \text{ m} = 1 \text{ cm}$$

3.54 (a)

In order to obtain the runoff hydrograph resulting from a storm of varying intensities, it is preferable to have a unit hydrograph of very short duration, theoretically, the shortest duration is zero. Accordingly if duration of effective rainfall approach, zero, the unit hydrograph is called as an instantaneous unit hydrograph, generally abbreviated as IUH.

The advantage of IUH over unit hydrograph (Sharman, 1932) is that the IUH overcomes the problem of duration of rainfall and the restriction on uniform distribution of rainfall.

3.55 (b)

The unit hydrograph (UH) of a basin is defined as the hydrograph resulting from an effective rainfall of 1 unit evenly distributed over the basin during the time D. Theoretically an infinite number of unit hydrograph, are possible for a given basin. The following hypothesis are considered while the linear unit hydrograph is used as a transfer function.

1. The unit hydrograph reflects the ensemble of the physical characteristics of the river basin.
2. The effective rainfall is uniformly distributed over the catchment.
3. The shape characteristics of the unit hydrograph are independent of time. Therefore duration of the unit hydrograph is constant regardless the effective rainfall intensity.

3.56 (d)

Net runoff depth

$= (6.2 - 2 \times 1.2) + (5 - 2 \times 1.2)$

$= 6.4$ cm $= 0.064$ m

Runoff volume of catchment

= Runoff depth × Area

$= 0.064 \times 250 \times 10^6 \times 10^{-4}$ ha-m

= 1600 ha-m

3.57 (b)

- The separation of base flow gives Direct Runoff Hydrograph (DRH).
- The area under the DRH gives volume of surface runoff.
- The effective depth of precipitation can be obtained by using index or volume of surface runoff divided by the catchment area.
- Finally the ordinates of UH can be obtained by dividing the ordinates or DRH by effective depth of precipitation.

3.58 (b)

S-curve hydrograph is the hydrograph of flow from a 1 cm intensity rain of infinite duration.

3.59 (b)

According to water balance,

Runoff + Evapotranspiration + Infiltration/ Recharge = Precipitation

3.60 (b)

Peak discharge of 6 hr hydrograph = 470 m^3/s

∴ Peak discharge of D.R.H

= 470 − base flow

= 470 − 15 = 455 m^3/s

Now, Rainfall excess depth

= Total rainfall − Infiltration losses

$= 8$ cm $- 0.25 \times 6$

$= 8 - 1.5 = 6.5$ cm

∴ Peak discharge of UH

$$= \frac{455}{6.5} = 70 \text{ m}^3/\text{s}$$

3.61 (b)

$A = 333$ km^2

Storm duration is 6 hr and intensity 2 cm/hr

Let rate of infiltration be x mm/hr

$$(2 - 0.1x) \times 10^{-2} \times 6 \times 333 \times 10^6 = 20 \times 10^6$$

$x = 9.99$ mm/hr $= 10$ mm/hr

3.62 (c)

According to standard SCS-CN method,

$$S = \left(\frac{1000}{CN}\right) - 10$$

where, S = Potential maximum retention

CN = Curve number

$$S = \frac{1000}{75} - 10 = 3.33$$

Actual runoff,

$$Q = \frac{(P - 0.2S)^2}{(P + 0.8S)}$$

P = Total rainfall in inch

$$= \frac{100}{25.4} = 3.937 \text{ in}$$

$$Q = \frac{(3.937 - 0.2 \times 3.33)^2}{3.937 + 0.8 \times 3.33}$$

= 1.62 inch = 41.17 mm

3.63 (d)

Total runoff

$= (0.8 + 1.2 + 0.9 + 1.8 + 1.0 + 0.8) - \phi\text{-index} \times T_e$

Now, total runoff as per correct value of ϕ-index, is

$\phi = (0.8 + 1.2 + 0.9 + 1.8 + 1.0 + 0.8) - 0.5 \times 6$

$= 6.5 - 3.0 = 3.5$ cm

∴ Correction $= 3.5 - 2.9 = 0.6$ cm

3.64 (d)

Collin's method for determination of unit hydrograph is an iterative method used for a multi-period storm.

Collins introduced a method of approximation to obtain a unique solution of set of equation by taking the assumptions:– assume a unit hydrograph and apply it to all excess rainfall blocks of the hyetograph except the largest.

3.65 (a)

$$f_i = 20 \text{ mm/hr}$$
$$k = 0.5/\text{hr}$$
$$f_f = 5 \text{ mm/hr}$$
$$f_t = f_f + (f_i - f_f)e^{-kt}$$
$$\Rightarrow \quad f_t = 5 + (20 - 5)e^{-0.5t}$$
$$= 5 + 15\, e^{-0.5t}$$
$$\text{Total infiltration} = \int_0^t f_t\, dt = \int_0^{10} \left(5 + 15e^{-0.5t}\right) dt$$
$$= 80 \text{ mm}$$

3.66 (b)

Total rainfall

$$= (1.6 + 3.6 + 5.0 + 2.8 + 2.2 + 1.0) \times \frac{30}{60}$$
$$= 8.1 \text{ cm}$$
$$W_{index} = \frac{(8.1 - 3.2)}{3} = 1.63 \text{ cm/hr}$$

As $\phi_{index} > W_{index}$

hence storm of intensity 1.6 cm/hr and 1.0 cm/hr will not produce any rainfall excess.

$$\phi_{index} = \frac{(3.6 + 5.0 + 2.8 + 2.2) \times \frac{30}{60} - 3.2}{2}$$
$$= 1.8 \text{ cm/hr}$$

3.68 (b)

$$\text{Run-off, } Q = \frac{0.36\, \Sigma o.t}{A} = 6.756 \text{ cm}$$
$$\phi_{index} = W_{index} = \frac{P - Q}{t} = \frac{(4.3 + 3.7) - 6.756}{8}$$
$$= 0.155 \text{ cm/hr}$$

3.69 (d)

Empirical IDF fomula by Sherman is given by:

$$i = \frac{a}{(t + b)^n}$$

Where, i = rainfall intensity (mm/hour)

t = duration in minutes

a, b & n are constants.

3.70 (d)

Concentration time starts from end of rainfall to end of recession limb.

3.72 (b)

$$\text{Coefficient of runoff} = \frac{\text{Runoff}}{\text{Precipition}}$$

Water tight roof surfaces will have lowest infiltration rate, therefore having highest runoff.

■■■■

4 Floods, Flood Routing & Flood Control

4.1 The probability that a 100 year flood is equalled or exceeded, at least once in 100 years is

(a) 99% (b) 64%
(c) 36% (d) 1%

[ESE : 2001]

4.2 As a flood wave passes a given section of a river, the time of occurrence of the maximum stage and that of the maximum discharge will be such that

(a) the maximum discharge passes down before the maximum stage is attained
(b) the maximum stage is attained before the maximum discharge passes down
(c) the two events occur simultaneously
(d) no specific sequence would be universally assignable

[ESE : 2001]

4.3 **Assertion (A):** In routing a flood hydrograph through a reservoir, the peak of the outflow hydrograph will be smaller than the inflow hydrograph and it occurs after the peak of the inflow.

Reason (R): In linear reservoir routing, the storage is a function of both outflow and inflow discharges.

(a) both A and R are true and R is the correct explanation of A
(b) both A and R are true but R is not a correct explanation of A
(c) A is true but R is false
(d) A is false but R is true

[ESE : 2002]

4.4 Match **List-I** (Floods) with **List-II** (Para-meters) and select the correct answer using the codes given below the lists:

List-I

A. Standard Project Flood (SPF)
B. Maximum Probable Flood (MPF)
C. Design Flood
D. Maximum Flood

List-II

1. Includes catastrophic floods
2. Includes floods of severe conditions
3. Peak flow obtained from observed data
4. Flood of desired recurrence interval

Codes:

	A	B	C	D
(a)	2	1	4	3
(b)	1	2	3	4
(c)	2	1	3	4
(d)	1	2	4	3

[ESE : 2003]

4.5 Consider the following statements:

1. Time-area histogram method aims at developing an IUH.
2. Isochrone is a line joining equal rainfall on a map.
3. Linear reservoir is a reservoir having straight boundaries.
4. Linear channel is a fictitious channel in which an inflow hydrograph passes through with only translation and no attenuation.

Which of these statements are correct ?

(a) 1, 2 and 3 (b) 1 and 4
(c) 2, 3 and 4 (d) 1, 2, 3 and 4

[ESE : 2003]

4.6 A bridge has an expected life of 50 years and is designed for a flood magnitude of return period 100 years. What is the risk associated with this hydrologic design?

(a) $1-(0.99)^{50}$ (b) $(0.5)^{50}$
(c) $(0.99)^{50}$ (d) $(0.99)^{100}$

[ESE : 2004]

4.7 Which one of the following statements is correct in respect of the two important aspects of flood forecast – (1) reliability of the forecast, and (2) the time available in between the forecast and the occurrence of flood?

(a) Meteorological forecast is least reliable and time available is also the least
(b) Hydrological forecast is most reliable but the time available is the least
(c) River forecast is least reliable and the time available is the maximum
(d) River forecast is most reliable but the time available is the least

[ESE : 2004]

4.8 Clark's method aims at which one of the following?
(a) Developing an IUH due to a instantaneous rainfall excess over catchment
(b) Developing stage-discharge relationship
(c) Measurement of infiltration
(d) Flood routing through channels

[ESE : 2004]

4.9 Which one of the following flood routing methods involve the concepts of wedge and prism storages?
(a) Coefficient method
(b) Muskingum method
(c) Pul's method
(d) Lag method

[ESE : 2005]

4.10 In a river carrying a discharge of 142 m³/s, the stage at a station A was 3.6 m and the water surface slope was 1 in 6000. If during a flood, the stage at A was 3.6 m and the water surface slope was 1 in 3000, what was the flood discharge (approximately)?
(a) 284 m³/s (b) 200 m³/s
(c) 164 m³/s (d) 96 m³/s

[ESE : 2006]

4.11 Which one of the following statements is not correct?
(a) Storage and pondage can be obtained from flow duration curve.
(b) Primary or firm power corresponds to maximum stream flow condition.
(c) Secondary power is occasionally called surplus power.
(d) Often, flash boards are put on dams to augment the pondage at low flow durations.

[ESE : 2006]

4.12 The basic equation of flood routing through a reservoir can be modified for discrete successive intervals Δt by which one of the following?
(a) $\left[\frac{I_1+I_2}{2}\right]\Delta t+\left[S_1+\frac{Q_1\Delta t}{2}\right]=\left[S_2+\frac{Q_2\Delta t}{2}\right]$
(b) $\left[\frac{I_1+I_2}{2}\right]\Delta t+\left[S_1-\frac{Q_1\Delta t}{2}\right]=\left[S_2+\frac{Q_2\Delta t}{2}\right]$
(c) $\left[\frac{I_1+I_2}{2}\right]\Delta t+\left[S_1+\frac{Q_1\Delta t}{2}\right]=\left[S_2-\frac{Q_2\Delta t}{2}\right]$
(d) $\left[\frac{I_1+I_2}{2}\right]\Delta t+\left[S_1-\frac{Q_1\Delta t}{2}\right]=\left[S_2-\frac{Q_2\Delta t}{2}\right]$

[ESE : 2006]

4.13 By using Gumbel's method, the flood discharge with a return period of 500 years at a particular township neighbourhood was estimated as 18000 m³/s with a probable error of 2000 m³/s. What are the 95% confidence probability limits of the 500 year flood at the location?
(a) 16100 m³/s to 19900 m³/s
(b) 17050 m³/s to 18950 m³/s
(c) 14080 m³/s to 21920 m³/s
(d) 13600 m³/s to 22400 m³/s

[ESE : 2006]

4.14 Recorded annual maximum 24-hour rainfall magnitudes at stations 'X' are as under:

Years	1960	1961	1962	1963	1964	1965	1966	1967	1968	1969
Rainfall (cm)	12.0	6.0	4.8	7.9	12.0	14.2	13.6	6.0	3.7	2.9

What is the return period, in years, for a 6.0 cm annual rainfall according to (i) Hazen formula and (ii) Weibull formula; respectively?
(a) $\frac{10}{7}, \frac{11}{7}$ (b) $\frac{20}{13}, \frac{22}{13}$
(c) $\frac{5}{3}, \frac{11}{7}$ (d) $\frac{20}{11}, \frac{11}{6}$

[ESE : 2006]

4.15 During a flood in a stream of width 200 m (taken as rectangular section), the gauge reading was found to rise by 18 cm per hour. What would be the difference in discharges at two sections, each 250 m on either side of the gauge station?
(a) 10 m³/s less at downstream section
(b) 7.5 m³/s less at downstream section
(c) 5 m³/s less at downstream section
(d) 5 m³/s more at downstream section

[ESE : 2007]

4.16 From the analysis of rainfall data at a particular station, it was found that a rainfall of 400 mm had a return period of 20 years. What is the probability of rainfall equal to or greater than 400 mm occurring at least once in 10 successive years?

(a) $(0.95)^{10}$ (b) $1-(0.95)^{10}$
(c) $1-(0.05)^{10}$ (d) $(0.05)^{10}$

[ESE : 2008]

4.17 A culvert is designed for a flood magnitude of return period 100 years and has expected life of 20 years. The risk in this hydrologic design is

(a) $1-0.99^{20}$ (b) $1-0.01^{20}$
(c) $1-0.09^{20}$ (d) $1-0.10^{20}$

[ESE : 2010]

4.18 Consider the following statements.
In case of flood routing in a river channel by Muskingum method, the coefficient x represents:

1. A dimensionless constant indicating the relative importance of inflow and outflow in determining storage.
2. A storage constant having the dimension of time.
3. In natural channels, x usually varies between 0.1 and 0.3.
4. When the values of x equals 0.5, there exists the influence of both inflow and outflow on storage.

Which of these statements are correct?

(a) 1, 2, 3 and 4 (b) 1, 3 and 4 only
(c) 1, 2 and 3 only (d) 2, 3 and 4 only

[ESE : 2011]

4.19 Consider the following situation in a flow mass curve study when demand line drawn from a ridge in the mass curve does not intersect the mass curve again. This means that

(a) The storage is not adequate
(b) The demand cannot be met by the inflow as the reservoir will not refill
(c) the reservoir was not full at the beginning
(d) The reservoir is wasting water by spill

[ESE : 2011]

4.20 The hydrologic risk of a 100-year flood occurring during the 2-year service life of a project is

(a) 9.8% (b) 9.9%
(c) 19.9% (d) 1.99% [ESE : 2012]

4.21 The design flood commonly adopted in India for barrages and minor dams is

(a) Probable maximum flood
(b) A flood of 50-100 years return period
(c) Peak flood
(d) Standard project flood or a 100-year flood, whichever is higher

[ESE : 2012]

4.22 The Muskingham method of flood routing is a

(a) Form of hydraulic routing of a flood
(b) Form of reservoir routing
(c) Complete numerical solution of St. Venant equations
(d) Hydrological channel routing method

[ESE : 2012]

4.23 It is proposed to design a causeway along a village road. The return period for the annual maximum flood of a given magnitude was found to be 5 years. The probability that this flood magnitude will be exceeded at least once during next 2 years is:

(a) 0.8 (b) 0.5
(c) 0.45 (d) 0.36 [ESE : 2013]

4.24 The probability of a 10 years flood to occur at least once in the next 5 years is

(a) 35% (b) 41%
(c) 60% (d) 65%

[ESE : 2014]

4.25 That extreme flood that is physically possible in a region from a severe-most combination, including rare combination of metrological and hydrological factors, is designated as the

(a) Design flood
(b) Standard project flood
(c) Probable maximum flood
(d) Flash flood

[ESE : 2014]

4.26 Return Period Refers to

(a) The Probability of Exceedance of an event
(b) The Probability of Non-Exceedance of an event
(c) The Inverse of the Probability of Exceedance of an event
(d) The Inverse of the Probability of Non-Exceedance of an event

[ESE : 2015]

4.27 Probability of a 10-year flood to occur at least once in the next 5 years is
(a) 35% (b) 40%
(c) 50% (d) 65%
[ESE : 2015]

4.28 The probability of a 10 year flood to occur at least once in the next 4 years is
(a) 45% (b) 35%
(c) 30% (d) 20%
[ESE : 2017]

Directions: The following items consists of two statements; one labelled as **'Statement (I)'** and the other as **'Statement (II)'**. You are to examine these two statements carefully and select the answers to these items using the codes given below:

Codes:
(a) Both Statement (I) and Statement (II) are individually true and Statement (II) is the correct explanation of Statement (I).
(b) Both Statement (I) and Statement (II) are individually true but Statement (II) is not the correct explanation of Statement (I).
(c) Statement (I) is true but Statement (II) is false.
(d) Statement (I) is false but Statement (II) is true.

4.29 **Statement (I):** The fundamental storage equation through a river reach considers that the total inflow is balanced by total outflow plus the change in storage in the reach over the routing period as has been considered.
Statement (II): To be adaptable for actual computations, the storage equation is recast in the form

$$\frac{1}{2}(I_1 + I_2)t + \left(S_1 - \frac{1}{2}D_1t\right) = \left(S_2 + \frac{1}{2}D_2t\right)$$

where suffixes 1 and 2 denote values at start and end respectively, of the routing interval *t*, the I's and D's denote, respectively, the inflow and outflow at the respective points of time and the S's denote the storage in the reach at the respective point of time.
[ESE : 2017]

4.30 **Statement (I):** When a loop rating curve is prepared at a gauging station when a flood wave passes, the maximum discharge occurs earlier than the maximum stage.
Statement (II): When a flood wave passes through a reach, during rising stages, some water may get into depressions on the floodplains of the river and, may be, only part of this spill returns to the channel runoff when flood levels recede.
[ESE : 2017]

4.31 **Statement (I):** The Muskingum method of routing a flood through a river reach is not a simple recast of the storage equation

$$I = D + \frac{d}{dt}(S)$$

Statement (II): The Muskingum method first develops trial storage loops considering contributions from both inflow and outflow and also the algebraic difference between them.
[ESE : 2017]

4.32 Which of the following statements relates to a retarding reservoir?
1. There are no gates at the outlets and hence the possibility of human error in reservoir operation is eliminated.
2. The high cost of gate installation and also its operation is saved.
3. An automatic regulation may cause coincidence of flood crest farther downstream where two or more channels taking off from retarding reservoirs join together.

(a) 1, 2 and 3 (b) 1 and 2 only
(c) 1 and 3 only (d) 2 and 3 only
[ESE : 2019]

4.33 A bridge has an expected life of 25 years and is designed for a flood magnitude of return period 100 years. What is the risk of this hydrologic design?
(a) $1-\left(\frac{100}{99}\right)^{25}$ (b) $\left(\frac{99}{100}\right)^{25}$
(c) $1-\left(\frac{99}{100}\right)^{25}$ (d) $\left(\frac{100}{99}\right)^{25}$
[ESE : 2021]

4.34 Consider the following statements regarding channel routing:
1. In channel routing, the flood hydrograph at various sections of the reach is predicted by considering a channel reach and an input hydrograph at the upstream end.

2. As the flood wave moves down the river, the shape of the wave does not change.
3. Flood waves passing down a river have their peaks attenuated due to friction.
4. The addition of lateral inflows can cause an increase of attenuation.

Which of the above statements are not correct?

(a) 1 and 3 only (b) 2 and 3 only
(c) 1 and 4 only (d) 2 and 4 only

[ESE : 2021]

4.35 Which of the following are the only two factors which govern the storage capacity of the reservoir?

(a) Inflow to reservoir and the outflow from the reservoir
(b) Inflow and catchment area
(c) Catchment area and outflow
(d) Height of reservoir and catchment area

[ESE : 2022]

Answers Floods, Flood Routing & Flood Control

4.1 (b)	4.2 (a)	4.3 (c)	4.4 (a)	4.5 (b)	4.6 (a)	4.7 (d)	4.8 (a)	4.9 (b)
4.10 (b)	4.11 (b)	4.12 (b)	4.13 (c)	4.14 (d)	4.15 (c)	4.16 (b)	4.17 (a)	4.18 (b)
4.19 (b)	4.20 (d)	4.21 (d)	4.22 (d)	4.23 (d)	4.24 (b)	4.25 (c)	4.26 (c)	4.27 (b)
4.28 (b)	4.29 (a)	4.30 (a)	4.31 (a)	4.32 (a)	4.33 (c)	4.34 (d)	4.35 (c)	

Explanations Floods, Flood Routing & Flood Control

4.1 (b)

The probability corresponding to 100 years return period

$$P = \frac{1}{100} = 0.01$$

The probability of the flood exceeding at least once in 100 years is

$$R = 1 - (1 - P)^{100} = 1 - 0.99^{100}$$
$$= 0.634 = 64\%$$

4.2 (a)

Due to storage effect maximum stage will occur after the maximum inflow discharge passes down.

4.3 (c)

Flood routing is the technique of determining the flood hydrograph at a section of a river by utilizing the data of flood flow at one or more upstream sections. The hydrologic analysis of problems such as flood forecasting, flood protection, reservoir design and spillway design invariably include flood routing.

The peak of the outflow hydrograph will be smaller than of the inflow hydrograph because of the effects of channel/reservoir storage and resistance to flow. This reduction in the peak value is called attenuation.

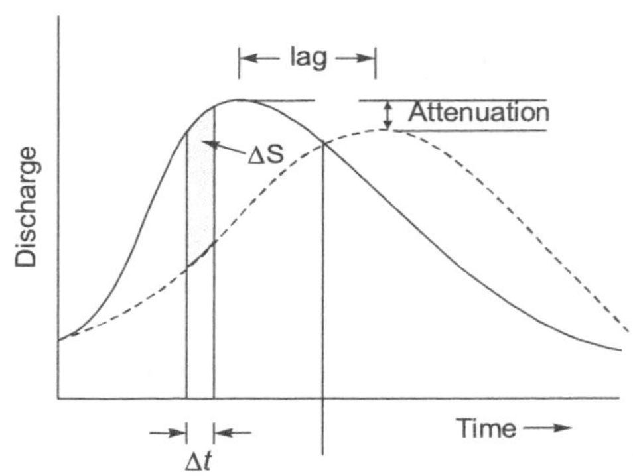

4.4 (a)

Probable Maximum Flood (PMF): This is the flood resulting from the most severe combination of critical meteorological and hydrological conditions that are reasonably possible in the region. The PMF is computed by using the Probable Maximum Storm (PMS) which is an estimate of the physical upper limit to storm rainfall over the catchment. This is obtained from the studies of all the storms that have occurred over the region and maximizing them for the most critical atmospheric conditions.

Standard Project Flood (SPF): This is the flood resulting from the most sever combination of meteorological and hydrological conditions considered reasonably characteristic of the region. The SPF is computed from the Standard Project Storm (SPS) over the watershed considered and may be taken as the largest storm observed in the region of the watershed.

Design flood is the value of the instantaneous peak discharge adopted for the design of a particular project or any of its structures. In addition to the considerations of the flood characteristics, frequencies and potentiality of the contributing drainage area above the structure, social, economic and other non-hydrologic considerations which are likely to have influence are considered in deriving a design flood.

The term "design flood" is used to denote the maximum flood flow that could be passed without damage or serious threat to the stability of engineering structures.

4.5 (b)

Clark's method of IUH as time area histogram method is used to develop an Instantaneous unit hydrograph due to the Instantaneous rainfall excess over a catchment.

Isochrones are imaginary lines across the catchment from where water particles traveling downward take the same time to reach the catchment outlet. The concept of Isochrones is helpful for explaining the effective duration of a uniform rainfall on the shape of hydrograph.

A linear reservoir is characterized by a linear relationship between the storage and the output.

A linear channel is characterized by continuity and momentum equations. It is a fictitious channel in which the time required to translate a discharge Q through a given reach is constant, in such channels inflow hydrograph passes through with only translation and no attenuation.

4.6 (a)

Risk = Probability of flood being exceeded at least once over a period of n years.

$$= 1-\left(1-\frac{1}{100}\right)^{50} = 1-(0.99)^{50}$$

4.7 (d)

Meteorological forecast is long range forecast gives maximum time for advance notice but it is least reliable.

Hydrological forecast is based on rainfall-runoff relationship and gives warning 2 to 5 days in advance. Its reliability is more than meteorological forecast.

River forecast is most reliable but the advance warning is 12 to 40 hours only.

4.8 (a)

Clark's method, also known as Time-area histogram method aims at developing an IUH due to an instantaneous rainfall excess over a catchment. It is assumed that the rainfall excess first undergoes pure translation and then attenuation. The translation is achieved by a travel time-area histogram and the attenuation by routing the results of the above through a linear reservoir at the catchment outlet.

4.9 (b)

Wedge and prism storage concept is used in channel routing using Muskingum method.

4.10 (b)

Discharge is directly proportional to $\sqrt{S}$.

$$\therefore \quad Q_2 = Q_1 \times \sqrt{\frac{S_2}{S_1}}$$

$$= 142 \times \sqrt{\frac{1}{3000} \times 6000}$$

$$= 200.8 \text{ m}^3/\text{s}$$

4.11 (b)

The firm power corresponds to minimum flow condition.

4.12 (b)

The basic continuity equation for reservoir routing is

$$\left(\frac{I_1+I_2}{2}\right)\Delta t - \left(\frac{Q_1+Q_2}{2}\right)\Delta t = S_2 - S_1$$

$$\left(\frac{I_1+I_2}{2}\right)\Delta t + \left(S_1 - \frac{Q_1 \Delta t}{2}\right) = S_2 + \frac{Q_2 \Delta t}{2}$$

4.13 (c)

For confidence probability C, the confidence interval (if variate x_T is bounded by values x_1 and x_2) is given by:

$$x_1 = x_T - f(c)\, S_e$$

$$x_2 = x_T + f(c)\, S_e$$

Given x_T = 18000 m^3/s

S_e = 2000 m^3/s

For 95% confidence probability $f(c) = 1.96$

$$\therefore \quad x_1 = 18000 - 1.96 \times 2000 = 14080 \text{ m}^3\text{/s}$$

$$x_2 = 18000 + 1.96 \times 2000 = 21920 \text{ m}^3\text{/s}$$

4.14 (d)

The data are arranged in descending order

Rainfall	Order (m)
14.2	1
13.6	2
12.0	3
12.0	4
7.9	5
6.0	6
6.0	7
4.8	8
3.7	9
2.9	10

Probability, $P = \left(\dfrac{m-0.5}{N}\right)$ Hazen's formula

$= \dfrac{m}{N+1}$ Weibull's formula

Return period $T = \dfrac{1}{P}$

For 6.0 cm rainfall $m = 6$

(i) $T = \dfrac{10}{5.5} = \dfrac{20}{11}$

(ii) $T = \dfrac{N+1}{m} = \dfrac{11}{6}$

4.15 (c)

$$Q_1 - Q_2 = \frac{18 \times 10^{-2}}{60 \times 60} \times (250 + 250) \times 200$$

$$\Rightarrow Q_1 - Q_2 = 5 \text{ m}^3\text{/s}$$

4.16 (b)

The probability of occurrence of an event at least once over a period of 'n' successive years is called the risk, R. It is given as

$$R = 1 - (1 - P)^n$$

where $P = \dfrac{1}{T} = \dfrac{1}{20} = 0.05$

$$\therefore \quad R = 1 - (1 - 0.05)^{10} = 1 - (0.95)^{10}$$

4.17 (a)

The probability of occurrence of an event at least once over a period of n successive years is called the risk, R. Thus the risk is given by

$$R = 1 - \left(1 - \frac{1}{T}\right)^n$$

$$= 1 - \left(1 - \frac{1}{100}\right)^{20} = 1 - 0.99^{20}$$

4.18 (b)

Muskingum equation relates the storage in a reservoir to the outflow (Q) and inflow (I)

$$S = K[x \cdot I + (1 - x) \cdot Q]$$

In this the parameter x is known as weighting factor and takes a value between 0 and 0.5. The coefficient K is known as storage-time constant and has the dimensions of time. Normally, for natural channels, the value of x lies between 0 to 0.3. For a given reach, the values of x and K are assumed to be constant. When $x = 0$, $S = KQ$, valid for linear reservoir. When $x = 0.5$, $S = K[0.5I + 0.5\,Q]$. There exists the influence of both inflow and outflow on storage.

4.19 (b)

Demand line drawn from the ridge denotes that reservoir is full at the beginning, i.e. at another point on mass curve if demand line intersects the mass curve again, then the reservoir is assumed to be full at that point of inter section. If demand line does not meets the mass curve again, it means reservoir will not refill again as the inflow into the reservoir could not meet the demand. If mass curve meets the demand line and then moves above the demand line, reservoir gets filled and water strarts spilling off from reservoir.

4.20 (d)

The risk is given by,

$$R = 1 - \left(1 - \frac{1}{T}\right)^n = 1 - \left(1 - \frac{1}{100}\right)^2$$

$$= 0.0199 = 1.99\%$$

4.21 (d)

For permanent barrages and minor dams with capacity less than 60 Mm^3 design flood recommended is SPF or flood with a return period of 100 years whichever gives higher value.

4.22 (d)

In reservoir routing, the storage is a unique function of outflow discharge, $S = f(Q)$.

However, in channel routing the storage is a function of both outflow and inflow discharge and hence a different routing method is needed from muskingum method of hydrological channel routing storage is given by

$$S = k\left[xI^m + (1-x)Q^m\right]$$

4.23 (d)

$$p = 1-q^n = 1-\left(1-\frac{1}{T}\right)^n$$

$$\Rightarrow \quad p = 1-\left(1-\frac{1}{5}\right)^2 = 1-0.64 = 0.36$$

4.24 (b)

Favorable probability

$$(p) = \frac{1}{10} = 0.1$$

Unfavorable probability (q)

$$= 1-p = 1-0.1 = 0.9$$

Probability to flood occur atleast once in the next 5 year

$$= 1-q^n = 1-(0.9)^5$$
$$= 0.40951 \simeq 0.41 \text{ or } 41\%$$

4.25 (c)

Maximum Probable Flood (MPF) differs from the Standard Project Flood (SPF) in that it includes the extremely rare and catastrophic floods and is usually confined to spillway design of very high dams. The SPF is generally around 80% of the MPF for the basin.

4.26 (c)

The estimated likelihood of an event to occur is called recurrence interval or Return Period. It is the inverse of the probability that the event will be exceeded in one year, or the inverse of the expected number of occurrences in a year.

4.27 (b)

Probability corresponding to a 10 year return period is

$$P = \frac{1}{10} = 0.1$$

The probability of flood occurring at least once in 5 year is

$$= 1-(1-P)^5 = 1-(1-0.1)^5$$
$$= 1-0.5904 = 0.4095 = 40.95\%$$

4.28 (b)

Probability of an event occurring atleast once in 'n' successive years is called risk:

$$\text{Risk} = 1-(q^n) = 1-\left(1-\frac{1}{T}\right)^n = 1-\left(1-\frac{1}{10}\right)^4$$
$$= 0.35 \text{ or } 35\%$$

4.29 (a)

$$\frac{1}{2}(I_1+I_2)t + \left(S_1 - \frac{1}{2}D_1t\right) = \left(S_2 + \frac{1}{2}D_2t\right)$$ is the fundamental routing equation which says that

Total inflow − Total outflow = Change in storage

4.31 (a)

Any routing method is based on the fundamental equation i.e.,

$$I - D = \frac{d}{dt}(S)$$

(Here symbol D denotes outflow).

4.32 (a)

Flood control reservoirs are adopted to prevent loss of life and valuable property due to floods. There are two types of single purpose flood control reservoirs namely:

(i) Retarding reservoirs
(ii) Detention basins

(i) **Retarding reservoirs:** A retarding reservoir is the one in which the spillway and outlets are not controlled by gates or valves but instead, they moderate the flood peak by passing discharge downstream, depending upon reservoir elevation.

Advantages: (a) There are no gates at the outlets and hence the possibility of human error in reservoir operation is eliminated. (b) The cost of expensive gate installation and operation is saved.

Disadvantages: A serious disadvantage of retarding reservoir is that automatic regulation may cause coincidence of flood crest farther downstream where two or more channels taking off from retarding reservoirs join together.

Note: The most ideal location of such reservoirs is immediately above a city or above the area to be protected from flood furiosity.

4.33 (c)

Given: Design life,

$$n = 25 \text{ years}$$

Return period,

$$T = 100 \text{ years}$$

Probability of occurrence,

$$P = \frac{1}{T}$$

Risk, $\bar{R} = 1-q^n$

(q = probability of non-occurrence of the event in n-years)

$$= 1-(1-P)^n = 1-\left(1-\frac{1}{T}\right)^n$$

$$= 1-\left(1-\frac{1}{100}\right)^{25} = 1-\left(\frac{99}{100}\right)^{25}$$

4.34 (d)

In channel routing the change in shape of hydrograph as it travels down a channel is studied. By considering a channel reach and an input hydrograph of the upstream end, this form of routing aims to predict the flood hydrograph of various section of reach.

Hence statement (1) is correct.

- The shape of the wave will change hence statement 2 is incorrect.
- Flood waves passing down a river have their peaks attenuated due to friction because major losses is due to friction and minor losses are due to eddies formation. Hence statement (3) is correct.
- In routing we find outflow hydrograph with the help of given inflow hydrograph.
- Due to addition of lateral flow, attention decreases but not increases.

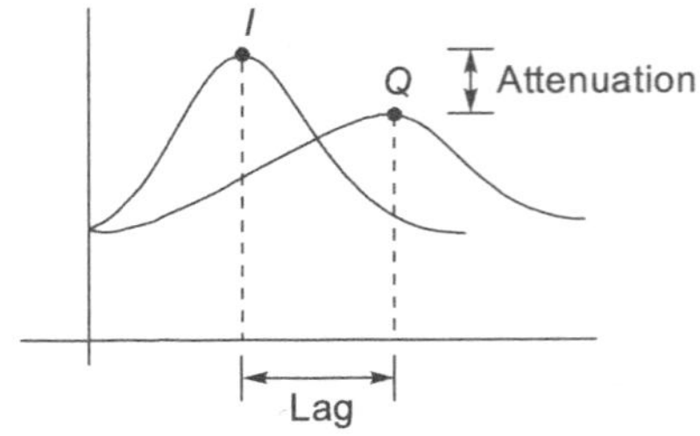

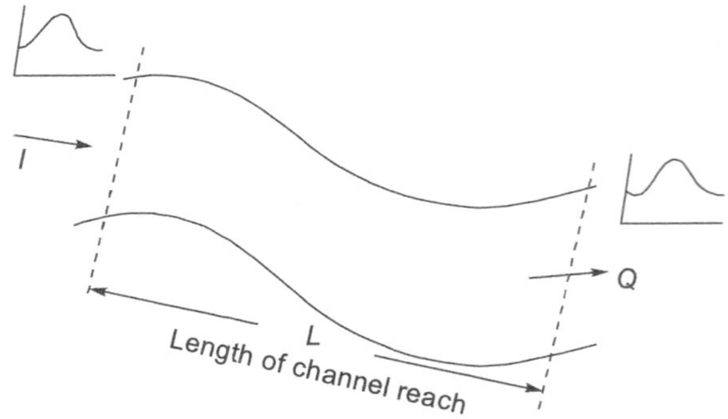

4.35 (c)

Storage capacity of a reservoir is function of outflow discharge only. Where this discharge depends on catchment area. Thus, storage capacity of channel is function of outflow and inflow discharge.

■■■■

UNIT

III

Water Resources Engineering

Syllabus

Multipurpose uses of Water, River basins and their potential; Irrigation systems, water demand assessment; Resources - storages and their yields; Water logging, canal and drainage design. Gravity dams, falls, weirs, Energy dissipators, barrage Distribution works, Cross drainage works and head-works and their design; Concepts in canal design, construction & maintenance; River training, measurement and analysis of rainfall.

Contents

1 Introduction to Irrigation and Methods of Irrigation

1.1 **Assertion (A):** Sprinkler method of irrigation has a higher water application/use efficiency.
Reason (R): Sprinkler system causes less interference in cultivation and other farming operations.
(a) both A and R are true and R is the correct explanation of A
(b) both A and R are true but R is not a correct explanation of A
(c) A is true but R is false
(d) A is false but R is true
[ESE : 2002]

1.2 **Assertion (A):** In the border strip method of irrigation, the size of the strip depends on soil characteristics, slope of the land and discharge.
Reason (R): Border strip method is a controlled type of subsurface irrigation method.
(a) both A and R are true and R is the correct explanation of A
(b) both A and R are true but R is not a correct explanation of A
(c) A is true but R is false
(d) A is false but R is true
[ESE : 2010]

1.3 Depending upon the source from which the water is drawn, flow irrigation can be sub-divided into
1. River canal irrigation
2. Reservoir or tank irrigation
3. Combined storage and lift irrigation
4. Combined storage and diversion irrigations
Which of the above designations are relevant?
(a) 1, 2 and 3 only
(b) 1, 2 and 4 only
(c) 1, 3 and 4 only
(d) 2, 3 and 4 only
[ESE : 2019]

1.4 In drip irrigation system, which one of the following emitters is not based on definitions by American Society of Agricultural Engineers (ASAE)?
(a) Emitter
(b) Pulsating emitter
(c) Long path emitter
(d) Multi-outlet emitter [ESE : 2020]

1.5 Which one of the following is that (low) water content of the soil at which plants can be longer extract sufficient water for this growth?
(a) Wilting point (b) Tail water
(c) Irrigating head (d) Capillary water
[ESE : 2021]

1.6 Which one of the following techniques of water distribution in the farms is sometimes called wild flooding?
(a) Free flooding
(b) Check flooding
(c) Border flooding
(d) Furrow irrigation method [ESE : 2023]

1.7 Which one of the following conditions is favourable for sprinkler irrigation method?
(a) When the land topography is regular
(b) When the soil is not easily erodible
(c) When the water tale is low
(d) When the land soil is excessively permeable
[ESE : 2023]

1.8 In drip irrigation method, the drip nozzles are fixed on laterals, discharging water at a very small rate of the order of
(a) 0.1 to 1 litre per hour
(b) 1 to 2 litre per hour
(c) 2 to 10 litre per hour
(d) 10 to 15 litre per hour
[ESE : 2023]

Answers Introduction to Irrigation and Methods of Irrigation

1.1 (c) 1.2 (c) 1.3 (b) 1.4 (b) 1.5 (a) 1.6 (a) 1.7 (d) 1.8 (c)

Explanations Introduction to Irrigation and Methods of Irrigation

1.1 (c)

Sprinkler system consists of network of pipes running across the field. Thus it will cause interference in cultivation and other operation.

1.2 (c)

Border irrigation method is a type of surface irrigation method.

1.3 (b)

Flow irrigation is that type of irrigation in which the supply of irrigation water available is at such a level that it is conveyed on to the land by gravity flow. Examples are river canal irrigation reservoir or tank irrigation, combined storage and diversion irrigation under gravity flow.

1.5 (a)

Wilting point is that water content at which plant can no longer extract sufficient water for its growth, and wilts up.

Capillary water is that water which is attached to the soil molecules by surface tension against gravitational forces.

1.7 (d)

Sprinkler irrigation is more useful where:

i. The land cannot be prepared for surface methods.
ii. Slopes are excessive.
iii. Topography is irregular.
iv. Soil is erosive.
v. Soil is excessively permeable.
vi. Depth of soil is shallow over gravel or sand.

■■■■

2 Water Logging and Quality of Irrigation Water

2.1 Acidic soils are reclaimed by
(a) leaching of the soil
(b) using limestone as a soil amendment
(c) using gypsum as a soil amendment
(d) provision of drainage
[ESE : 2002]

2.2 Leaching is a process
(a) by which alkali salts present in the soil are dissolved and drained away
(b) by which alkali salts in soil come up with water
(c) of draining excess water of irrigation
(d) which controls water-logging
[ESE : 2003]

2.3 A soil sample has an exchangeable sodium percentage of 16%, its electrical conductivity is 3.2 milli-Mhos/cm and pH of 9.5. How is the soil classified?
(a) Saline soil (b) Saline-alkaline soil
(c) Alkaline soil (d) None of the above
[ESE : 2007]

2.4 An identified source of irrigation water has ion concentrations of Na^+, Ca^{++} and Mg^{++} as 20, 10 and 8 milliequivalent per litre, respectively. The SAR of this water is approximately
(a) 2.06 (b) 6.67
(c) 2.67 (d) zero
[ESE : 2011]

2.5 Consider the following statements regarding water logging:
1. Water logging is the rise of groundwater table leading to possible increase in salinity resulting in a reduction in the yield of crops.
2. Water logging cannot be eliminated in certain areas but can be controlled only if the quantity of water percolating into that soil is checked and reduced.

Which of the above statements is/are CORRECT?
(a) 1 only
(b) 2 only
(c) Both 1 and 2
(d) Neither 1 nor 2 [ESE : 2018]

2.6 Consider the following data for irrigation water

	Concentration	Milli-equivalents per litre
1	Na^+	24
2	Ca^{++}	3.6
3	Mg^{++}	2

The Sodium-Absorption Ratio (SAR) is nearly
(a) 13.1 (b) 14.3
(c) 15.5 (d) 16.7 [ESE : 2019]

2.7 Which one of the following is not a major deterrent in water harvesting through water tanks?
(a) Deforestation mainly due to population pressure in the catchments of tank systems
(b) Siltation
(c) Lack of maintenance and repairs and breaches of tank embankments
(d) Shallow depth of water tanks [ESE : 2020]

2.8 Which one of the following is not the main cause for soil salinity and sodicity?
(a) Irrigation mismanagement
(b) Poor land levelling
(c) Use of heavy machinery, resulting in no soil compaction
(d) Leaching without adequate drainage
[ESE : 2020]

2.9 Consider the following statements related to water logging control:
1. It is evident that water logging can be controlled only if the quantity of water into the soil below is checked and reduced.

2. Attempts should be made to reduce the seepage of water from the canals and water courses.
3. The entire irrigable land should receive canal water in all seasons.

Which of the above statements are correct?

(a) 1 and 2 only (b) 2 and 3 only
(c) 1 and 3 only (d) 1, 2 and 3

[ESE : 2021]

2.10 A tile drainage system draining 12 hectares, flows at a design capacity for two days, following a storm. If the system is designed using a drainage coefficient of 1.25 cm, how much of water will be removed during this period?

(a) 1500 m^3 (b) 4500 m^3
(c) 3000 m^3 (d) 3500 m^3

[ESE : 2021]

2.11 If the sodium ions increase to about 10% or more, the soil becomes

(a) less permeable and of poorer tilth
(b) more permeable
(c) good tilth
(d) more permeable and of better tilth

[ESE : 2023]

2.12 Which one of the following is the cause of water logging?

(a) Adequate natural drainage
(b) Adequate surface drainage
(c) Normal rains
(d) Submergence due to floods [ESE : 2023]

2.13 Which of the following is a very effective method to control water logging?

(a) Lining of canal and water courses
(b) Flat topography
(c) Adequate natural drainage
(d) Adequate surface drainage [ESE : 2023]

■■■■

Answers Water Logging and Quality of Irrigation Water

2.1 (b) 2.2 (a) 2.3 (c) 2.4 (b) 2.5 (c) 2.6 (b) 2.7 (d) 2.8 (c) 2.9 (a)
2.10 (c) 2.11 (a) 2.12 (d) 2.13 (a)

Explanations Water Logging and Quality of Irrigation Water

2.1 (b)

As compared to most of the methods, due to reduced runoff and percolation, water available to the roots is comparatively more. Usually, soil is affected by salinity and alkalinity which may be originally present or may come from irrigation water. However, if soil becomes acidic, it can be reclaimed by basic salt such as lime stone.

Water application efficiency,

$$\eta = \frac{\text{Water stored in root zone}}{\text{Water delivered to the field}}$$

2.2 (a)

Leaching is the process in which land is flooded with adequate depth of water. The alkali salts present in soil, get dissolved with this water, which percolate down to join the water table or are drained away by sub-surface drains.

Leaching requirement, $LR = \frac{D_d}{D_i}$

D_d is Depth of water drained out per unit area

D_i is Depth of irrigation water applied per unit area

$$D_i = C_u + D_d$$

C_u is Consumptive use

$$\therefore \quad LR = \frac{D_i - C_u}{D_i} = \frac{EC_i}{EC_d}$$

EC_i is Electrical conductivity of irrigation water

EC_d is Electrical conductivity of drained water.

2.3 (c)

The classification of soils as per electrical conductivity (EC) in milli mhos/cm, Exchangeable sodium percentage (ESP) and pH values is tabulated below:

Classification	EC	ESP	pH
Saline soil or white alkali	> 4	< 15	≤ 8.5
Alkaline soil or Non-saline soil or sodic soil or Black Alkali	< 4	> 15	8.5 to 10.0
Saline-alkali soil	> 4	> 15	≤ 8.5

2.4 (b)

SAR = Sodium Absorption Ratio

$$= \frac{Na^+}{\sqrt{\frac{Ca^{++} + Mg^{++}}{2}}} = \frac{20}{\sqrt{\frac{10+8}{2}}} = 6.67$$

2.5 (c)

Water-logging can cause the salts (NaCl, Na_2CO_3, Na_2SO_4) to build up in excess in the soil. Due to water logging & also due to salt accumulation, the yield of crops may be reduced significantly. Hence, statement 1 is correct.

Water logging cannot be eliminated in certain low lying areas and it can only be controlled. Percolation of water in the soil is due to excessive irrigation. If excessive irrigation water is applied, it may ultimately lead to water logging. Hence, over-irrigation must be avoided in these areas to reduce percolation. Statement 2 is also correct.

2.6 (b)

$$SAR = \frac{[Na^+]}{\sqrt{\frac{[Ca^{2+}]+[Mg^{2+}]}{2}}} = \frac{24}{\sqrt{\frac{3.6+2}{2}}}$$

$$= \frac{24}{\sqrt{2.8}} = \frac{24}{1.673} = 14.35$$

2.8 (c)

Heavy machinery usage will result only in the change in physical properties of soil but there will not be any change in chemical properties so salt concentration of soil will remain unaffected.

2.9 (a)

Water logging control: It is evident that water logging can be controlled only if the quantity of water into the soil below is checked and reduced. To achieve this, the inflow of water into the under ground reservoir should be reduced and the out flow from this reservoir should be increased, as to keep the highest position of water table at least about 3 m below the ground surface. The various measures adopted for water logging control.

- **Lining of canals and after courses:**

Attempts should be made to reduce the seepage of water from the canals and water courses.

- **Reducing the intensity of irrigation.**
- **By introducing crop rotation.**
- **By providing intercepting drain**
- **By provision of efficient drainage system**

2.10 (c)

Given:

Area, A = 12 ha

Drainage coefficient = 1.25

Since, drainage coefficient is expressed as the depth of water in cm or m, to be removed in 24 hrs from the drainage area.

Hence, for the data given, we may infer that for a given drainage area, 1.25 cm depth of water is removed in 24 hours.

∴ Volume of water drained in 2-days

$$= 2 \times (1.25 \times 10^{-2}) \times 12 \times 10^4$$
$$= 3000 \text{ m}^3$$

2.11 (a)

If the percentage of sodium is more, the aggregation of soil grains breakdown, and the soil becomes less permeable and of poorer tilth.

■■■■

3 Water Requirement of Crops

3.1 If the depth is 8.64 cm on a field over a base period of 10 days, then the duty is
(a) 10 hectares per cum/s
(b) 100 hectares per cum/s
(c) 864 hectares per cum/s
(d) 1000 hectares per cum/s [ESE : 2000]

3.2 10 m^3/s of water is diverted to a 32 hectare field for 4 hours. Soil probing after irrigation showed that 0.3 m of water had been stored in the root zone. Water application efficiency in this case would be
(a) 96% (b) 66.67%
(c) 48% (d) 24% [ESE : 2001]

3.3 A canal was designed to supply the irrigation needs of 1000 ha of land growing rice of 140 days base period and having a delta of 130 cm. If the canal water is used to irrigate wheat of base period 119 days and having a delta of 50 cm, the area that can be irrigated is
(a) 452 ha (b) 904 ha
(c) 1105 ha (d) 2210 ha
[ESE : 2002]

3.4 The delta for a crop having base period 120 days is 70 cm. What is the duty?
(a) 2480 hectare/cumec
(b) 1481 hectare/cumec
(c) 148 hectare/cumec
(d) 1.481 hectare/cumec [ESE : 2004]

3.5 For a culturable command area of 1000 hectare with intensity of irrigation of 50%, the duty on field for a certain crop is 2000 hectare/cumec. What is the discharge required at head of water course with 25% losses of water?
(a) 3/16 cumec (b) 1/4 cumec
(c) 1/3 cumec (d) 1/2 cumec
[ESE : 2004]

3.6 What is the moisture depth available for evapotranspiration in root zone of 1 m depth soil, if dry weight of soil is 1.5 gm/cc, field capacity is 30% and permanent wilting point is 10%?
(a) 450 mm (b) 300 mm
(c) 200 mm (d) 150 mm
[ESE : 2004]

3.7 Consider the following terms relating to irrigation requirements:
1. Consumptive irrigation requirement
2. Net irrigation requirement
3. Field irrigation requirement
4. Gross irrigation requirement

For a given set up, which one of the following is the correct relation?
(a) 1 > 2 > 3 > 4 (b) 1 < 2 < 3 < 4
(c) (1 = 2) < 3 < 4 (d) 1 < (2 = 3)< 4
[ESE : 2006]

3.8 The moisture tension for a soil is 8 atmospheres. The soil is then at
(a) permanent wilting point
(b) field capacity
(c) optimum moisture content
(d) equivalent moisture
[ESE : 2006]

3.9 If the discharge required for different crops is 0.4 cumecs in the field and the capacity factor and time factors are 0.8 and 0.5 respectively, then what is the design discharge of the distributary at its head ?
(a) 0.80 cumecs (b) 0.16 cumecs
(c) 1.0 cumecs (d) 1.24 cumecs
[ESE : 2007]

3.10 During a particular stage of the growth of a crop, consumptive use of water is 2·8 mm/day. If the amount of water available in the soil is 25% of 80 mm depth of water, what is the frequency of irrigation?

(a) 9 days (b) 13 days
(c) 21 days (d) 25 days

[ESE : 2009]

3.11 Consider the following statements:
Irrigation water has to be supplied to the crops when the moisture level falls

1. below field capacity
2. to wilting point
3. below wilting point

Which of these statements is/are correct?

(a) 1 only (b) 2 only
(c) 3 only (d) 2 and 3 only

[ESE : 2009]

3.12 In a canal irrigation project, 76% of the culturable command area (CCA) remained without water during Kharif season; and 58% of CCA remained without water during Rabi season in a particular year. Rest of the areas got irrigated in each crop respectively. What is the intensity of irrigation for the project in the year ?

(a) 134% (b) 76%
(c) 66% (d) 58%

[ESE : 2009]

3.13 The discharge required for Rabi and Kharif crops are 0.4 m^3/s and 0.3 m^3/s respectively. The capacity and time factors are 0.8 and 0.5 respectively at each season. The design discharge of the distributary at its head is

(a) 0.8 m^3/s (b) 0.16 m^3/s
(c) 1.0 m^3/s (d) 1.24 m^3/s

[ESE : 2010]

3.14 Consider the following zones:

1. Saturation zone
2. Capillary zone
3. Intermediate zone
4. Soil water zone

Which of these does not relate to the zone of aeration in the soil profile?

(a) 1 and 2 only (b) 2 and 3 only
(c) 4 only (d) 1 only

[ESE : 2011]

3.15 **Assertion (A):** The duty of water decreases as the point of its measurement moves away from the field of application.
Reason (R): Duty depends on soil characteristics.

(a) both A and R are true and R is the correct explanation of A
(b) both A and R are true but R is not a correct explanation of A
(c) A is true but R is false
(d) A is false but R is true [ESE : 2011]

3.16 A rice crop is to be irrigated in a field covering an area of 2400 ha, the duty and base period of rice are given as 860 ha/cumec and 120 days respectively. The volume of water required in the field is nearly:

(a) 500 ha-m (b) 1400 ha-m
(c) 2000 ha-m (d) 2880 ha-m

[ESE : 2013]

3.17 In an irrigation project, in a certain year, 70% and 46% of the culturable command area in Kharif and Rabi, respectively, remained without water and rest of the area got irrigation water. The intensity of irrigation in that year for the project was:

(a) 116% (b) 84%
(c) 42% (d) 58% [ESE : 2013]

3.18 The gross command area for a distributary is 5000 hectares, 80% of which is culturable irrigable. The intensity of irrigation for Rabi is 50%; and for Kharif is 30%. The average duty at the head of the distributory is 2000 hectares/cumec for Rabi season and 900 hectares/cumec for Kharif season. The discharge required at the head of the distributory from average demand considerations is

(a) 1.00 cumecs (b) 1.33 cumecs
(c) 2.33 cumecs (d) 3.33 cumecs

[ESE : 2013]

3.19 The following data were noted from an irrigation field:

1. Field capacity = 20%
2. Permanent wilting point = 10%
3. Permissible depletion of available soil moisture = 50%
4. Dry unit weight of soil = 15 kN/m^3
5. Effective rainfall = 50 mm

The net irrigation requirement per metre depth of soil will be

(a) 75 mm (b) 125 mm
(c) 50 mm (d) 25 mm

[ESE : 2014]

3.20 For an irrigated field having: Field capacity = 30%, Permanent wilting point = 10%, Permissible depletion of available moisture = 40%, Dry weight of soil = 14.70 kN/m^3, Unit weight of water = 9.8 kN/m^3 and Effective rainfall = 30 mm, what is the net irrigation requirement per meter depth of soil?

(a) 300 mm (b) 150 mm
(c) 120 mm (d) 90 mm

[ESE : 2014]

3.21 A lift irrigation scheme using a discharge of 96 m^3/hr is planned to raise a crop with an average Δ of 0.375 m. Intensity of irrigation is 60%. Assuming 3600 hours of working of the tubewell for a year, the culturable command area is

(a) 96 Ha (b) 48 Ha
(c) 144 Ha (d) 120 Ha

[ESE : 2014]

3.22 A field measures 40 hectares. When 8 cumecs of water was supplied for 6 hours, 30 cm of water was stored in the root zone. The field application efficiency is nearly.

(a) 70% (b) 80%
(c) 85% (d) 90% [ESE : 2014]

3.23 Consumptive Use refers to the loss of water as a result of

(a) Evaporation and Transpiration
(b) Crop Water Requirement
(c) Evaporation and Infiltration
(d) Evaporation and Transpiration from the cropped area

[ESE : 2015]

3.24 The discharge capacity required at the outlet to irrigate 3000 ha of sugarcane having a kor depth of 173 mm and a kor period of 30 days is

(a) 2.0 m^3/s (b) 1.0 m^3/s
(c) 20 m^3/s (d) 0.20 m^3/s

[ESE : 2015]

Directions: The following items consists of two statements; one labelled as **'Assertion (A)'** and the other as **'Reason (R)'**. You are to examine these two statements carefully and select the answers to these items using the codes given below:

Codes:

(a) Both A and R are true and R is the correct explanation of A
(b) Both A and R are true but R is not a correct explanation of A
(c) A is true but R is false
(d) A is false but R is true

3.25 Assertion (A): If the soil moisture is only slightly more than the wilting coefficient, the plant must expend extra energy to obtain the water and hence the plant will not grow healthily.

Reason (R): Excessive water supply retards plant growth.

[ESE : 2015]

3.26 Assertion (A): Duty of drip irrigation is very high.

Reason (R): Losses are least in drip irrigation.

[ESE : 2015]

3.27 Consider the following statements in connection with soil-water-crop relationship:

1. Water utilization by plants is mainly from capillary water.
2. The amount of irrigation water required to meet the evapotranspiration needs of the crop during its full growth duration is its consumptive irrigation requirements.
3. The depth of water required to bring the soil-moisture level of a given soil up to its field capacity is called hygroscopic water.
4. With continuous increase in quantity of water applied the yield of most crops increase up to a certain limit and then is expected to be constant.

Which of the above statements are correct?

(a) 1 and 2 (b) 2 and 3
(c) 3 and 4 (d) 1 and 4

[ESE : 2017]

3.28 Consider the following data

Root zone depth = 2 m

Existing water content = 5%

Dry density of soil = 15 kN/m^3
Water applied to the soil = 500 m^3
Water loss due to evaporation and deep percolation = 10%
Area of plot = 1000 m^2
The field capacity of the soil will be nearly
(a) 16.8% (b) 17.7%
(c) 18.8% (d) 19.7% [ESE : 2019]

3.29 A Persian wheel with an average discharge of 230 litre/minute irrigates 1 hectare wheat crop in 50 hours. The average depth of irrigation will be nearly
(a) 4 cm (b) 5 cm
(c) 6 cm (d) 7 cm [ESE : 2020]

3.30 If wheat requires 7.5 cm of water after every 28 days and the base period for wheat is 140 days, what is the value of delta for wheat?
(a) 7.5 cm (b) 27.5 cm
(c) 37.5 cm (d) 17.5 cm
[ESE : 2021]

3.31 What is the delta for a crop when its duty is 864 hectares/ cumec on the field and the base period of this crop is 120 days?
(a) 120 cm (b) 140 cm
(c) 160 cm (d) 172 cm
[ESE : 2021]

3.32 The monthly consumptive use values for paddy are tabulated below. What is average monthly consumptive use?

Month	Dates	Rice (Loam Soil) C_u in cm
June	1-30 } Nursery	29.69
July	1-12 } Nursery	8.76
July	13-31	14.38
August	1-31	22.73
September	1-30	21.29
October	1-31	25.50
November	1-24	15.06

(a) 7.7 cm (b) 23.1 cm
(c) 26.69 cm (d) 137.41 cm
[ESE : 2022]

3.33 Which one of the following is the total area, bounded within the irrigation boundary of a project, which can be economically irrigated without considering the limitation of the quantity of available water?
(a) Gross command area
(b) Cultivable command area
(c) Net and gross sown area
(d) Net and gross irrigated area
[ESE : 2023]

3.34 If the cultivation method is faulty and less efficient, resulting in the wastage of water, the duty of the water will naturally be
(a) less
(b) more
(c) average
(d) depending upon cultivators
[ESE : 2023]

3.35 Water attached to soil particles through loose chemical bonds is termed as
(a) Capillary water
(b) Gravity water
(c) Hygroscopic water
(d) Field capacity [ESE : 2024]

3.36 The field capacity and permanent wilting point for a given 0.8 m root zone soil are 35 and 10 per cent, respectively. At a given time, the soil moisture in the given soil is 20 per cent when a farmer irrigates the soil with 250 mm depth of water. Assuming bulk specific gravity of the soil as 1.6, what is the amount of water wasted from the consideration of irrigation?
(a) 23.2% (b) 58%
(c) 5.8% (d) 19.2%
[ESE : 2024]

3.37 Match the following lists:

List-I (Soil texture)	List-II (Field capacity(%))
P. Sand	1. 5 – 10
Q. Loam	2. 27 – 35
R. Silty clay	3. 18 – 25
S. Clay	4. 32 – 40

Select the correct answer using the code given below:

	P	Q	R	S
(a)	4	3	2	1
(b)	1	3	2	4
(c)	3	1	2	4
(d)	4	2	1	3

[ESE : 2024]

■■■■

Answers Water Requirement of Crops

3.1 (d)	3.2 (b)	3.3 (d)	3.4 (b)	3.5 (c)	3.6 (b)	3.7 (b)	3.8 (a)	3.9 (c)
3.10 (c)	3.11 (a)	3.12 (c)	3.13 (c)	3.14 (d)	3.15 (a)	3.16 (d)	3.17 (b)	3.18 (b)
3.19 (d)	3.20 (d)	3.21 (*)	3.22 (a)	3.23 (d)	3.24 (a)	3.25 (b)	3.26 (a)	3.27 (a)
3.28 (d)	3.29 (d)	3.30 (c)	3.31 (a)	3.32 (b)	3.33 (a)	3.34 (a)	3.35 (c)	3.36 (a)
3.37 (b)								

Explanations Water Requirement of Crops

3.1 (d)

$$D \times \Delta = 864 \times B$$

$$\therefore \quad D = \frac{864 \times 10}{8.64} = 1000 \text{ ha/m}^3\text{/s}$$

3.2 (b)

The volume of water present in root zone,

$$V_r = 0.3 \times 32 \times 10^4 = 9.6 \times 10^4 \text{ m}^3$$

The volume of water supplied/delivered,

$$V_s = 10 \times 4 \times 3600 = 1.44 \times 10^5 \text{ m}^3$$

Water application efficiency,

$$\eta = \frac{V_r}{V_s} \times 100 = 66.67\%$$

3.3 (d)

The discharge (Q) in canal remains same.

Duty for rice, $D_1 = \dfrac{1000}{Q}$ ha/cumec

Duty for wheat, $D_2 = \dfrac{A_2}{Q}$ ha/cumec

Now $\dfrac{D_1\Delta_1}{B_1} = \dfrac{D_2\Delta_2}{B_2}$

or $\dfrac{A_1\Delta_1}{B_1} = \dfrac{A_2\Delta_2}{B_2}$

$$\therefore \quad A_2 = \frac{119}{140} \times \frac{130}{50} \times 10002 = 2210 \text{ ha}$$

3.4 (b)

$$\text{Duty} = \frac{864 \times 120}{70} = 1481 \text{ ha/cumec}$$

3.5 (c)

The area to be irrigated = 0.5 × 1000 = 500 ha

The discharge needed on field

$$= \frac{500}{2000} = 0.25 \text{ cumec}$$

For 25% loss of water at the head of water course, the discharge required at the head of water course

$$Q = \frac{0.25}{0.75} = \frac{1}{3} \text{ cumec}$$

3.6 (b)

$$\text{Moisture depth} = \frac{\gamma_d}{\gamma_w} \times (FC - PWP)d$$

$$= \frac{1.5}{1} \times (0.3 - 0.1) \times 1$$

$$= 0.3 \text{ m} = 300 \text{ mm}$$

3.7 (b)

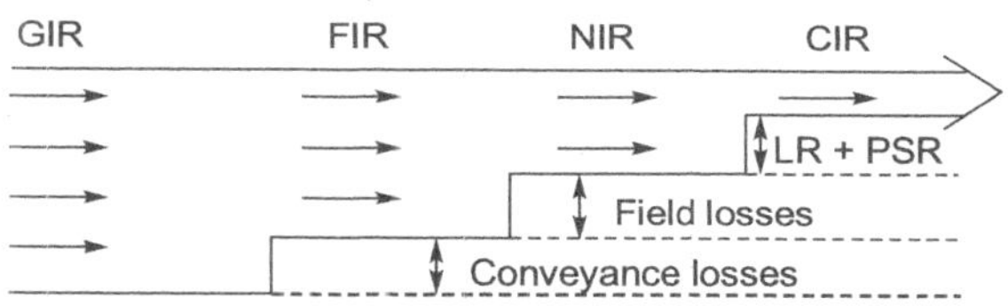

Thus, GIR > FIR > NIR + CIR

3.8 (a)

Soil moisture tension is the pressure which must be exerted on the soil in order to extract water from it.

It is a measure of tenacity with which the water is held by the soil. It is also known as capillary potential.

At field capacity moisture tension is about $\frac{1}{10}$ to $\frac{1}{3}$ atm. At permanent wilting point moisture tension is 7-32 atm.

It is measured by Tensiometer.

3.9 (c)

Design discharge

$$= \frac{\text{Discharge required in the field}}{\text{Time factor} \times \text{capacity factor}}$$

$$= \frac{0.4}{0.5 \times 0.8} = 1.0 \text{ cumecs}$$

3.10 (c)

Water available to the soil $= \frac{25}{100} \times 80 = 20$ mm

Additional water required = 80 – 20 = 60 mm

$\therefore$ Frequency of irrigation $= \frac{60}{2.8} = 21.43$ days

Thus the crop should be irrigated after every 21 days.

3.11 (a)

The soil moisture is not allowed to be depleted up to the wilting point, as it would result in considerable fall in crop yield. The optimum level up to which the soil moisture may be allowed to be depleted in the root zone without fall in crop yield has to be worked out for every crop and soil by experimentation. The irrigation water should be supplied as soon as the moisture falls upto this optimum level and its quantity should be just sufficient to bring the moisture content up to its field capacity.

Note: There is no point in irrigating above FC as any water added above FC will be drained away as gravity water.

3.12 (c)

The annual intensity of irrigation is the sum total of intensities of irrigation of all the seasons of the year.

Intensity of irrigation for kharif season

$= 100 - 76 = 24\%$

Intensity of irrigation for rabi season

$= 100 - 58 = 42\%$

$\therefore$ Annual intensity of irrigation

$= 24 + 42 = 66\%$

3.13 (c)

Design discharge for Rabi

$$= \frac{0.4}{0.8 \times 0.5} = 1\ \text{m}^3/\text{s}$$

Design discharge for Kharif

$$= \frac{0.3}{0.8 \times 0.5} = 0.75\ \text{m}^3/\text{s}$$

Thus design discharge will be maximum out of the two values i.e., 1 m^3/s.

3.14 (d)

(i) *Soil-water zone:* The soil-water zone extends from the ground surface down upto the major root zone. The soil in this zone becomes saturated either during irrigation or when rainfall occurs. However, the soil in this zone remains saturated only for a short duration after irrigation or rainfall, because the excess water drains through the soil under the influence of gravity. This excess water which cannot be retained by the soil is known as gravitational water.

(ii) *Intermediate zone:* The intermediate zone extends from the lower edge of soil-water zone to the upper limit of the capillary zone. This zone usually contains non-moving vadose water which is held by molecular and surface tension forces in the form of hygroscopic and capillary water.

(iii) *Capillary zone:* The capillary zone extends from the water table upto the limit of capillary rise of water. This pore space is saturated just about the water table but above that, it may or may not be saturated.

(iv) *Saturation zone:* In the zone of saturation all the interstices are filled with water under hydrostatic pressure. The zone of saturation is bounded at the top by either a limiting surface of saturation called water table or an overlying impermeable strata and extends down to underlying impermeable strata.

3.15 (a)

As we move away from the field, discharge increases. The losses in conveyance are taken into consideration viz. seepage and evaporation. Seepage depends on soil characteristics. Hence duty is influenced by soil characteristics. Further it is also influenced by evaporation.

3.16 (d)

$$\Delta = 8.64\frac{B}{D} = \frac{8.64 \times 120}{860} = 1.2056 \text{ m}$$

$\therefore$ Vol. $= 2400 \times 1.2056$ ha-m

$= 2893 \approx 2880$ ha-m

3.17 (b)

% of area under irrigation

$= (100 - 70) + (100 - 46)$

$= 30 + 54 = 84\%$

3.18 (b)

Gross Command Area= 5000 hectares

Culturable Command Area = 4000 hectares

Area irrigated by Rabi = 2000 hectares

Area irrigated by Kharif = 1200 hectares

Average duty for Rabi season = 2000 hec/cumecs
Average duty for Kharif season = 900 hec/cumecs
Discharge required at the head of distributory
(Average demand consideration)

$$= \text{Maximum of} \begin{bmatrix} \text{Rabi season} \\ \text{Kharif season} \end{bmatrix}$$

$$= \text{Maximum of} \begin{bmatrix} 2000/2000 = 1 \text{ cumec} \\ 1200/900 = 1.33 \text{ cumecs} \end{bmatrix}$$

= 1.33 cumecs

3.19 (d)

Net irrigation requirement
= Consumption Use – Effective Rainfall

$$= 0.5 \times (0.2 - 0.1) \times \frac{15}{10} \times 1000$$

$$= 0.05 \times \frac{15}{10} \times 1000 = 75 \text{ mm}$$

Effective rainfall = 50 mm
∴ Net irrigation requirement
= 75 mm – 50 mm = 25 mm

3.20 (d)

Total irrigation water requirement

$$= 0.4(0.3 - 0.1) \times \frac{14.7}{9.8} \times 1000$$

$$= 0.4 \times 0.2 \times 1.5 \times 1000$$

$$= 120 \text{ mm}$$

∴ Net irrigation requirement

$$= 120 - 30 - 90 \text{ mm}$$

3.21 (*)

By applying mass conservation,
$0.375 \times 0.6 \times A = 3600 \times 96$
$A = 1536000 \text{ m}^2$ (Where A = Area)
or $A = 153.6$ ha
None of the options match.

3.22 (a)

Total water applied = $8 \times 3600 \times 6 = 172800 \text{ m}^3$
Height of water applied

$$= \frac{\text{Volume of water}}{\text{Area}} = \frac{172800}{40 \times 10^4}$$

= 0.432 m = 43.2 cm

∴ Field application efficiency

$$= \frac{30}{43.2} \times 100 = 70\%$$

3.23 (d)

Crop consumptive water use is the amount of water transpired during plant growth plus what evaporates from the soil surface and fatigue in the crop area.

3.24 (a)

Required discharge capacity

$$= \frac{\text{Total volume of water}}{\text{kor period}}$$

$$= \frac{3000 \times 10^4 \times 173 \times 10^{-3}}{30 \times 24 \times 60 \times 60} = 2 \text{ m}^3/\text{s}$$

3.25 (b)

Due to excessive water supply roots do not grow properly because excess water affects soil aeration and hence plants roots do not grow properly.

3.26 (a)

Drip irrigation, also known as trickle irrigation, is an irrigation method that saves water and fertilizer by allowing water to drip slowly to the roots of plants.

Since the losses are less, more area can come under the influence of irrigation for the same discharge or if the same area is considered, discharge requirement will be less as compared to other methods. Thus, duty of drip irrigation is very high.

3.27 (a)

Water mainly used by plants is from capillary water.

CIR (consumption irrigation requirement) $= C_u - P_{eff}$

C_u = Consumptive use or evapotranspiration.

From definition of CIR, it is clear that, the portion of evapotranspiration to be satisfied by irrigation is called CIR.

Hygroscopic water is water below P.W.P.

Yield of crop increases only upto optimum depth, beyond that yield starts reducing.

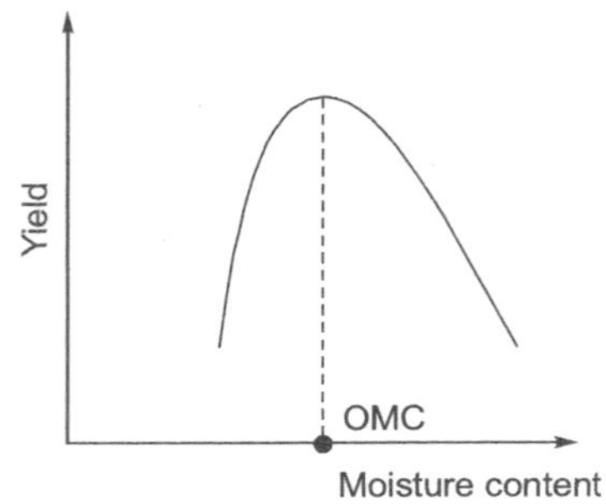

3.28 (d)

Depth of water added into the soil after removing

$$10\% \text{ losses} = \frac{(500 \times 0.9)\,m^3}{1000\ m^2} = 0.45\ m$$

$$d_w = \frac{\gamma_d}{\gamma_w} \times d \times (F.C. - \text{Existing moisture})$$

F.C. = 0.14715 + 0.05 = 0.19715

or, F.C. $\simeq$ 19.715%

Note : If γ_w = 10 kN/m^3 is taken, then it can be solved without calculator and answer will be F.C. = 20% = 19.7%

Alternatively,

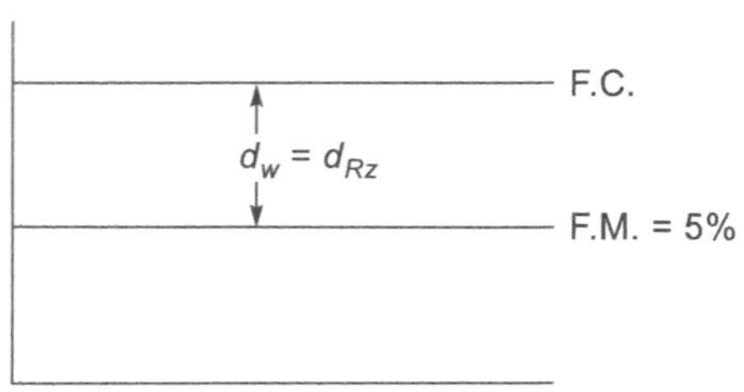

$d_w = d_{Rz}$ = Depth of water getting stored in the R.Z. to ↑ M.C. upto F.C.

$$d_{Rz} = \frac{V_{Rz}}{\text{Area}} = \text{Volume getting stored in the R.Z.}$$

$$V_{Rz} = V_F = \text{Losses}$$

$$V_{Rz} = 500 - \frac{10}{100} \times 50 = 450\ m^3$$

$$d_{Rz} = d_w = \frac{V_{R_z}}{\text{Area}} = \frac{450}{1000} = 0.45\ m$$

Using,

$$d_w = \frac{\gamma_d}{\gamma_w} \times d \times \left(\frac{F.C.}{100} - \frac{E.M.}{100}\right)$$

$$\Rightarrow\ 0.45 = \frac{15}{9.81} \times 2 \times \left(\frac{F.C.}{100} - \frac{5}{100}\right)$$

F.C. = 19.71%

3.29 (d)

Volume supplied = Discharge × Time

$$V = 230 \frac{l}{\text{min}} \times 50\ \text{hr}$$

$$= \frac{230 \times 10^{-3} m^3 \times 50 \times 60\ \text{min}}{\text{min}} = 690\ m^3$$

Area irrigated = 1 ha

$A = 10^4\ m^2$

$$\text{Depth of irrigation} = \frac{\text{Volume}}{\text{Area}} = \frac{690}{10000}\ m$$

$$= 0.069\ m = 6.9\ cm \simeq 7\ cm$$

3.30 (c)

Given: For wheat crop:

Base period, B = 140 days

Watering interval = 28 days

Assuming the base period to be representing the crop period, as per usual practice, we can infer that the water is required at an average interval of 28 days up to a total period of 140 days. This means that $\frac{140}{28} = 5$ No. of waterings are required

The depth of water required each time = 7.5 cm

∴ Total depth of water required in 140 days

= 5 × 7.5 = 37.5 cm

Hence, delta (Δ) for wheat = 37.5 cm.

3.31 (a)

Given: Duty, D = 864 ha/cumec

Base period, B = 120 days

We know that, $D = \frac{8.64B}{\Delta}$

$$\Delta = \frac{8.64 \times 120}{864} = 1.2\ m \simeq 120\ cm$$

3.32 (b)

Total consumptive use of rise = 29.69 + 8.76 + 14.38 + 22.73 + 21.29 + 25.50 + 15.06

= 137.41 cm

Average daily consumptive use

$$= \frac{137.41}{30+31+31+30+31+24} = 0.776$$

∴Average monthly consumptive use

= 0.776 × 30 = 23.28 ≃ 23.1

3.34 (a)

$$\text{Duty} = \frac{A}{Q} \text{ha/cumec}$$

Since the wastage of water is more, so to irrigate a certain field the discharge required will be more. Thus, duty will be less.

3.36 (a)

Field capacity, FC = 35%

Permanent wilting point, PWP = 10%

Depth of root zone, d = 0.8 m

Moisture content, M_0 = 20%

Now, depth of water provided,

$$d_w = \frac{\gamma}{\gamma_w} \times d \times (FC - M_0)$$

$$= 1.6 \times 0.8 \times (0.35 - 0.20)$$

$$= 0.192 \text{ m} = 192 \text{ mm}$$

So, percentage of water wasted

$$= \frac{(250 - 192) \times 100}{250} = 23.2\%$$

■■■■

4 Design of Unlined and Lined Canals

4.1 In order to ensure that no scouring takes place in the bed of a channel of bed slope '*S*' constructed on alluvial soil of particle size '*d*' cm, the flow velocity should be restricted to

(a) $4.85\ d^{1/2}\ S^{-1/6}$ (b) $4.85\ d^{-1/2}\ S^{1/6}$
(c) $0.48\ d^{1/2}\ S^{-1/6}$ (d) $0.48\ d^{-1/2}\ S^{1/6}$

[ESE : 2000]

4.2 The Lacey's silt factor for a particular alluvium is 2.0. This alluvium would comprise

(a) medium sand of size 0.5 mm
(b) coarse sand of size 0.75 mm
(c) medium bajri of size 1.3 mm
(d) coarse bajri of size 2.4 mm [ESE : 2001]

4.3 Match **List-I** (Surface) with **List-II** (Appro-ximate range of Manning's *n*) and select the correct answer using the codes given below the lists:

List-I

A. Gravel river bed with 50 mm diameter bed
B. Badly maintained unlined irrigation channel
C. Concrete lined channel
D. Channel with brick lining

List-II

1. 0.02 to 0.022
2. 0.025 to 0.04
3. 0.013 to 0.017
4. 0.04 to 0.08

Codes:

	A	B	C	D
(a)	4	3	2	1
(b)	4	2	3	1
(c)	1	3	2	4
(d)	1	2	3	4

[ESE : 2001]

4.4 In alluvial channels carrying clear water, the ratio of maximum tractive shear stress on the sides and that on the channel bed is approximately

(a) 0.5 (b) 1.76
(c) 0.76 (d) 1.5 [ESE : 2001]

4.5 Consider the following statements:
Garret's diagram for the design of irrigation channel is based on:

1. Kennedy's theory
2. Lacey's theory
3. Kutter's formula
4. Manning's formula

Which of these statements are correct?

(a) 1 and 3 (b) 1 and 4
(c) 2 and 3 (d) 2 and 4

[ESE : 2002]

4.6 The total number of independent equations that form the Lacey's regime theory is

(a) 2 (b) 3
(c) 4 (d) 6

[ESE : 2002]

4.7 Balanced depth of cutting of canal is

(a) half the total depth of a canal
(b) half of full supply depth
(c) the maximum cut that an excavator can take
(d) where volume of cutting is equal to volume of filling

[ESE : 2002]

4.8 **Assertion (A):** With tile lining of canals, permissible velocity of flow is lower than that with concrete lining.
Reason (R): The surface of tile lining becomes rough due to loss of surface material with high velocity.

(a) both A and R are true and R is the correct explanation of A
(b) both A and R are true but R is not a correct explanation of A
(c) A is true but R is false
(d) A is false but R is true

[ESE : 2003]

4.9 Which of the following pairs are correctly matched?

1. Silt factor : Average size of silt particles
2. Silt load : Volume of suspended sediments flowing with water in unit time.
3. Silt charge : Weight of silt per unit volume of water
4. Silt grade : Gradation between different silt particles

Select the correct answer using the codes given below :

(a) 1 and 4 (b) 3 and 4
(c) 2 and 3 (d) 1 and 2 **[ESE : 2003]**

4.10 The channel section can be designed on the basis of Lacey's Theory. The steps are mentioned below:

1. Finding out the perimeter
2. Finding out the velocity
3. Calculation of the silt factor
4. Finding out the area

What is the correct sequence of the steps?

(a) 4-2-3-1 (b) 3-1-4-2
(c) 4-1-3-2 (d) 3-2-4-1

[ESE : 2005]

4.11 In Lacey's regime theory, the velocity of flow is proportional to

(a) Qf^2 (b) Q/f^2
(c) $(Qf^2)^{1/6}$ (d) $(Q/f^2)^{1/6}$

[ESE : 2005]

4.12 Match **List-I** with **List-II** and select the correct answer using the codes given below the lists:
(S = bed slope, q = discharge intensity, Q = Discharge)

List-I

A. Mean velocity in a Lacey's regime channel
B. Mean velocity in a lined channel
C. Normal scour depth in an alluvial channel
D. Wetted perimeter of a Lacey regime channel

List-II

1. $S^{1/2}$
2. $S^{1/3}$
3. $q^{2/3}$
4. $q^{-2/3}$
5. $Q^{1/2}$

Codes:

	A	B	C	D
(a)	2	5	3	1
(b)	3	1	4	5
(c)	2	1	3	5
(d)	3	5	4	1

[ESE : 2006]

4.13 The permissible tractive force in an erodible channel depends upon which of the following?

1. Angle of the repose of the material
2. Particle size
3. Sediment content of water
4. Wetted perimeter of channel

Select the correct answer using the codes given below:

(a) 1, 2 and 4 (b) 1, 2 and 3
(c) 1 and 3 only (d) 2 and 4 only

[ESE : 2008]

4.14 Consider the following statements:
Lining of irrigation canals has necessarily to be carried out in the reaches where the channel passes through

1. sandy soil
2. coarse aggregate soil
3. clay soil
4. fine silt and clay

Select the correct answer using the codes given below:

(a) 1 and 3 (b) 3 only
(c) 1 and 2 (d) 3 and 4

[ESE : 2008]

4.15 The two sides of a trapezoidal channel are lined with cement plaster, n_w = 0.012, while the earthen bed has n_b = 0.025, the area is divided into three parts as in the figure.

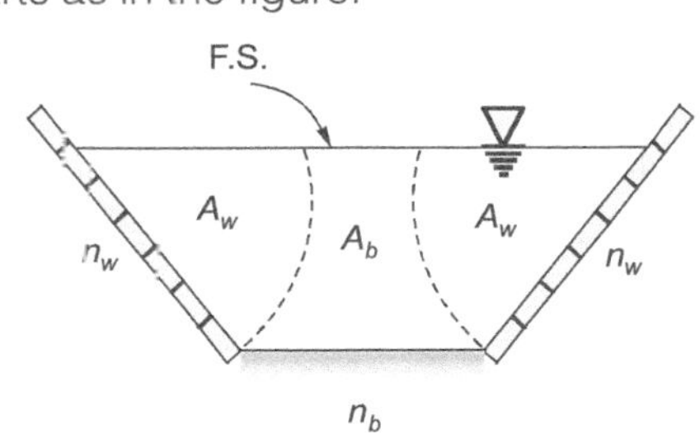

Consider the following assumptions to find the discharge through this section:

1. The average velocity is the same for all the 3 parts.
2. Discharge is the same in all the 3 parts.
3. Bed slope is the same for all the 3 parts.
4. Hydraulic mean radius is the same for all the 3 parts.

Which of the above assumptions are correct?
(a) 1 and 3 (b) 1 and 4
(c) 2 and 3 (d) 2 and 4

[ESE : 2010]

4.16 Garret's diagrams are used to
(a) separate base flow from total runoff
(b) correct inconsistency in rainfall data
(c) determine reservoir capacity
(d) design channels

[ESE : 2010]

4.17 Assertion (A): Lining a canal is always beneficial and economical.

Reason (R): The seepage losses are greatly reduced and extra water is available for irrigation.
(a) both A and R are true and R is the correct explanation of A
(b) both A and R are true but R is not a correct explanation of A
(c) A is true but R is false
(d) A is false but R is true

[ESE : 2010]

4.18 Two different channels, *M* and *N*, in two different sites are designed based on Lacey's theory, to carry same quantum of discharge. But the bed material of *M* is found to be finer than that of *N*.
(a) Channel *M* will have steeper longitudinal slope
(b) Channel *N* will have steeper longitudinal slope
(c) Channels *M* and *N* can have same longitudinal slopes
(d) Silting is more in *M* than in *N*

4.19 The relation between suspended sediment transport Q_s and stream flow Q is often represented by an equation of the form

$$Q_s = K \cdot Q^n$$

where n commonly varies between:
(a) 1/3 and 1/2 (b) 0.2 and 0.3
(c) 1 and 3 (d) 0.6 and 0.8

[ESE : 2011]

4.20 Consider the following statements:
The general depth of scour calculated by Lacey's formula in a river represents the depth below the:
1. Maximum flood level in the river
2. Minimum flow water level in the river
3. Normal flow water level in the river
4. Existing river bed level

Which of these statements is/are correct?
(a) 1, 2, 3 and 4 (b) 1, 2 and 3 only
(c) 2, 3 and 4 only (d) 1 only

[ESE : 2011]

4.21 Consider the following statements:
1. A regime channel will have side slopes of value 0.5 H : 1 V.
2. Lacey's regime formula is applicable to regime channels with sediment concentration more than 5000 ppm
3. For a Lacey regime channel the Manning roughness coefficient is estimated by Strickler's formula
4. The mean velocity in a Lacey channel is proportional to $R^{2/3}$ where R = hydraulic radius.

Which of these statements are correct?
(a) 1 and 2 (b) 1 and 4
(c) 2 and 3 (d) 3 and 4

[ESE : 2013]

4.22 The minimum size of gravel that will not move in the bed of a wide rectangular channel of depth 0.8 m and longitudinal slope 0.0041 is
(a) 11 mm (b) 23 mm
(c) 36 mm (d) 57 mm

[ESE : 2013]

4.23 Consider a wide rigid-boundary canal with bed slope of 0.004, Manning's $n = 0.18$ and depth of flow 0.88 m. The average boundary shear stress will be
(a) 17.2 N/mm^2 (b) 18.2 N/mm^2
(c) 19.2 N/mm^2 (d) 16.2 N/mm^2

[ESE : 2014]

4.24 Statement (I): The sharp corners in the cross-section of a canal are rounded.

Statement (II): The corners may not become zones of stagnation.
(a) Both Statement (I) and Statement (II) are individually true and Statement (II) is the correct explanation of Statement (I).
(b) Both Statement (I) and Statement (II) are individually true but Statement (II) is not the correct explanation of Statement (I).

(c) Statement (I) is true but Statement (II) is false.
(d) Statement (I) is false but Statement (II) is true.

[ESE : 2014]

4.25 In a Lacey regime channel:
1. The bed load is zero.
2. The suspended load is zero.
3. The bed slope is a function of the full supply discharge and the silt size.

Which of the above statements is/are correct?

(a) 1 and 2 (b) 3 only
(c) 2 and 3 (d) 2 only

[ESE : 2014]

4.26 A regime canal carries silt of median size 0.25 mm. The Lacey's silt factor of this silt is

(a) 0.66 (b) 0.88
(c) 0.44 (d) 0.22

[ESE : 2014]

4.27 A channel designed by Lacey's theory has a velocity of 0.88 m/sec. The silt factor is 1.1. Then hydraulic mean depth will be

(a) 1.95 m (b) 1.76 m
(c) 1.63 m (d) 1.50 m

[ESE : 2014]

4.28 Assertion (A): An alluvial channel is defined as a channel in which the flow transports sediment of the same physical characteristics as the material in the wetted surface of the channel.

Reason (R): This ensures that the channel cross-section and the channel slope do not change.

(a) Both A and R are true and R is the correct explanation of A
(b) Both A and R are true but R is not a correct explanation of A
(c) A is true but R is false
(d) A is false but R is true

[ESE : 2015]

4.29 Assertion (A): The excavation of side slope of an irrigation canal for clayey type of soil should be made at 1 : 1 (i.e. 1 horizontal to 1 vertical) which is taken as nearly equal to the angle of internal friction of the soil.

Reason (R): The angle of internal friction represents the stable slope when the excavated soil, or soil in loose conditions, assumes when dumped in situ.

(a) Both A and R are true and R is the correct explanation of A
(b) Both A and R are true but R is not a correct explanation of A
(c) A is true but R is false
(d) A is false but R is true

[ESE : 2015]

4.30 The conditions to be satisfied for a channel in 'Regime' as per-Lacey are
1. Constant discharge
2. Silt grade and silt concentration are constant
3. The channel is flowing in unlimited incoherent alluvium of the same alluvial character as that transported.

Which of the above statements are correct?

(a) 1 and 2 only (b) 1, 2 and 3
(c) 1 and 3 only (d) 2 and 3 only

[ESE : 2016]

4.31 A channel designed by Lacey's theory has a mean velocity of 1.1 m/s. The silt factor is 1.1. Then hydraulic mean radius will be

(a) 1.13 m (b) 2.27 m
(c) 3.13 m (d) 4.27 m

[ESE : 2016]

Directions: Each of the next items consists of two statements, one labelled as the **'Statement (I)'** and the other as **'Statement (II)'**. Examine these two statements carefully and select the answers to these items using the codes given below :

Codes:

(a) Both Statement (I) and Statement (II) are individually true and Statement (II) is the correct explanation of Statement (I)
(b) Both Statement (I) and Statement (II) are individually true but Statement (II) is ***not*** the correct explanation of Statement (I)
(c) Statement (I) is true but Statement (II) is false
(d) Statement (I) is false but Statement (II) is true

4.32 Statement (I) : A channel in alluvium running with constant discharge and constant sediment charge will first form its flow section and then its final longitudinal slope.

Statement (II) : If a channel in alluvium has a section too small for a given discharge and slope steeper than required, degradation and aggradation happen and then the flow section attains final regime.

[ESE : 2016]

4.33 **Statement (I):** The shear stress exerted by the stream flow on the bed is responsible for the movement of bed sediment particles.

Statement (II): The sediment will move when the shear stress crosses a threshold limit designated as a critical shear stress τ_c.

[ESE : 2016]

4.34 Consider the following statements regarding design of channel by Lacey and Kennedy:

1. The theoretical concept of silt transportation is the same in both the theories.
2. Lacey improved upon Kennedy's formula.
3. There are no defects in either the theories of Lacey or of Kennedy.

Which of the above statements are correct?

(a) 1 and 2 (b) 1 and 3
(c) 2 and 3 (d) 1, 2 and 3

[ESE : 2017]

4.35 Which one of the following is not the major factor influencing seepage from a canal?

(a) Characteristics of the soil traversed by the canal system
(b) Area wetted by the canal
(c) Location of the canal
(d) Frequencies of canal usage

[ESE : 2020]

4.36 Which of the following are the principal factors influencing the choice of a particular method of lining?

1. Availability and cost of the material at the site or within easy reach.
2. Velocity of flow in the channel.
3. Cost of maintenance.

(a) 1 and 2 only (b) 1 and 3 only
(c) 1, 2 and 3 (d) 2 and 3 only

[ESE : 2020]

4.37 What is the hydraulic radius of a stable canal carrying a discharge of 27 m^3/s using Lacey's method? (Assume silt factor is 1.0)

(a) 1.44 m (b) 2.67 m
(c) 3.14 m (d) 4.28 m

[ESE : 2021]

4.38 Consider the following statements regarding loss of water in canals:

1. The water lost by evaporation is generally very small, as compared to the water loss by seepage in certain channels.
2. In percolation, there exists a one of continuous saturation from the canal to the water-table and a direct flow is established.
3. In absorption, a small saturated soil zone exists around the canal section and is surrounded by a zone of decreasing saturation.

Which of the above statements are correct?

(a) 1 and 2 only (b) 2 and 3 only
(c) 1 and 3 only (d) 1, 2 and 3

[ESE : 2021]

4.39 Which one of the following conditions is correct for a channel to behave in true regime?

(a) Discharge is non-uniform
(b) Flow is non-uniform
(c) Silt grade is varying
(d) Silt change is constant

[ESE : 2021]

■■■■

Answers Design of Unlined and Lined Canals

4.1 (a)	4.2 (c)	4.3 (b)	4.4 (c)	4.5 (a)	4.6 (b)	4.7 (d)	4.8 (a)	4.9 (c)
4.10 (d)	4.11 (c)	4.12 (c)	4.13 (b)	4.14 (c)	4.15 (*)	4.16 (d)	4.17 (d)	4.18 (b)
4.19 (c)	4.20 (d)	4.21 (b)	4.22 (c)	4.23 (*)	4.24 (a)	4.25 (b)	4.26 (b)	4.27 (b)
4.28 (c)	4.29 (d)	4.30 (b)	4.31 (*)	4.32 (d)	4.33 (a)	4.34 (a)	4.35 (c)	4.36 (c)
4.37 (a)	4.38 (d)	4.39 (d)						

Explanations Design of Unlined and Lined Canals

4.1 (a)

Flow velocity, $V = \frac{1}{n}R^{2/3}S^{1/2}$

For no scouring in the channel

$$d = 11\,RS$$

$$\therefore \quad R = \frac{d}{11S}$$

By Stickler's formula

$$n = \frac{d^{1/6}}{24}$$

$$\therefore \quad V = \left(\frac{24}{d^{1/6}}\right)\left(\frac{d}{11S}\right)^{2/3} S^{1/2}$$

$$= 4.8\,d^{1/2}S^{-1/6}$$

4.2 (c)

Lacey's silt factor

$$f = 1.76\sqrt{d_{mm}}$$

where d_{mm} is dia. in mm

$$\therefore \quad d_{mm} = \left(\frac{f}{1.76}\right)^2 = 1.3$$

4.3 (b)

The gravel bed will be more rough than brick lined channel.

Similarly badly maintained unlined irrigation channel will have higher 'n' value than concrete lined channel.

4.4 (c)

Maximum tractive shear stress on the sides and on the bed of a channel is $0.75\,\gamma RS$ and $0.97\gamma RS$ respectively.

The ratio of of maximum tractive shear stress on the sides and that on the channel bed is

$= 0.75\,\gamma RS / 0.97\gamma RS = 0.7731$

Hence nearest answer is (c), 0.76.

4.5 (a)

Garret's diagram is graphical solution of Kennedy's and Kutter's equations to save enormous mathematical calculations in the design of irrigation channels. These have been drawn for trapezoidal channel with side slopes as $\frac{1}{2}$H : 1V on the assumption that irrigation channels adopt approximately this shape even though they were constructed on different side slopes.

4.6 (b)

A rigid bed canal has one degree of freedom i.e. for a given channel a change in discharge would cause a change in the depth only.

A man-made alluvial channel has three degrees of freedom i.e. for a given channel, change in discharge can cause changes in width, depth and bed slope. Thus three equations independent of each other are needed, as given by Lacey, to represent three degrees of freedom viz. depth, width and gradient. A natural alluvial river has four degrees of freedom as its planiform can also alter.

4.7 (d)

Balanced depth of cutting of canal is where volume of cutting = volume of filling. This reduces the transportation expenses of soil.

4.8 (a)

Limiting velocities in different type of linings

	Type of lining	Safe limiting velocity
1.	Boulder lining	1.5 m/s
2.	Burnt clay tile lining	1.8 m/s
3.	Cement concrete lining	2.7 m/s

4.9 (c)

Silt factor $f = 1.76\sqrt{d_{mm}}$ is related to square root of average size of silt particles (d_{mm}) but not the average size itself.

Silt load or sediment load includes bed load and suspended load.

Silt charge is the weight of silt per unit volume of water.

Silt grade refers to the type and size of silt.

4.10 (d)

Design procedure of Lacey's theory:

(i) Calculate silt factor

$$f = 1.76\sqrt{d_m}$$

(ii) Find out velocity

$$V = \left(\frac{Qf^2}{140}\right)^{1/6} = 10.8\, R^{2/3}R^{1/3}$$

(iii) Find out area $A = Q/V$

(iv) Find out perimeter $P = 4.75\sqrt{Q}$

As per Lacey's theory slope,

$$S = \frac{f^{5/3}}{3340Q^{1/6}}$$

4.12 (c)

Mean velocity in Lacey's regime channel

$$V = 10.68\, R^{2/3}\, S^{1/3}$$

Mean velocity in lined channel given by Manning's

$$V = \frac{1}{n}R^{2/3}S^{1/2}$$

Normal scour depth in alluvial channel

$$= 1.35\left(\frac{q^2}{f}\right)^{1/3}$$

Wetted perimeter,

$$P = 4.75\sqrt{Q}$$

4.13 (b)

- Angle of repose controls the shear stress between channel bed and particle.
- Tractive force exerted by water is proportional to particle size.
- Sediment content controls the value of γ in the Tractive force equation.

Note: The actual shear stress depends on the wetted perimeter.

4.14 (c)

Lining of irrigation canals is necessary in reaches where channel passes through Sandy soils and coarse aggregate soils as percolation in these soils is high.

4.15 (*)

Only bed slope will be same for all the three parts.

$$V_{avg} = \frac{1}{n}R^{2/3}S^{1/2}$$

$\because$ n is variable, therefore V_{avg} will vary.

Hence discharge will also vary. Further relation between areas is not provided as $Q = AV$ at a section.

4.16 (d)

Garret's diagram gives the graphical method of designing a channel. Basic aim is to save time. It is based on Keneddy's theory and Kutter's formula.

4.17 (d)

Lining of a canal is always beneficial. It reduces seepage losses and extra water is available for irrigation. But it may not be economical always as cost of lining can be very expensive.

The economical justification of lining is done as per IS 10430 : 2000 (Annexure A).

4.18 (b)

When discharge is same, channel with finer material will have less silt factor.

$$f = 1.76\sqrt{d_{mm}}$$

$$S = \frac{5/3}{3340Q^{1/6}}$$

If Q is constant

$$S \propto f^{5/3}$$

Thus channel N will have steeper longitudinal slope

Note: The judgement will be taken based on Lacey's equations.

4.19 (c)

$$Q_s = KQ^n$$

From Meyer-Peter and Muller formula n varies between 1 and 3.

4.20 (d)

The general depth of scour calculated by Lacey's formula in a river represents depth below maximum flood level or HFL.

4.21 (b)

Although a channel in true regime attain semi-elliptical cross-section, its equation is not provided by lacey. Thus for design purpose we may assume channel to be trapezoidal with side slopes 0.5H : 1V. Although

4.22 (c)

For a particle on bed to be stationary

$$d \geq 11\ RS$$

Thus, $d_{min} = 11\ RS$

Where, $R = \dfrac{By}{B+2y} \simeq y \qquad [\because B \gg y]$

$$= 11(0.8)0.0041$$
$$= 0.03608 \text{ m} = 36 \text{ mm}$$

4.23 (*)

Average boundary shear stress for rigid bondary canal:

$$\tau_{av} = \gamma_w RS$$
$$= 9.81 \times 0.88 \times 0.004$$
$$= 0.03453 \text{ kN/m}^2$$
$$= 34.53 \times 10^{-6} \text{ N/mm}^2$$

4.25 (b)

In Lacey's regime channel

Silt factor, $f = 1.76\sqrt{d_{mm}}$

Bed slope, $S = \dfrac{f^{5/3}}{3340Q^{1/6}}$

Thus S is function of f and so of d (silt size) also. The bed load and suspended load are a function of d (silt size) and thus they vary, not equal to zero.

4.26 (b)

Lacey's silt factor (f)

$$= 1.76\sqrt{d_{mm}} = 1.76\sqrt{0.25} = 0.88$$

4.27 (b)

Hydraulic mean depth,

$$R = \frac{5}{2}\frac{V^2}{f} = \frac{5}{2}\times\frac{(0.88)^2}{1.1} = 1.76 \text{ m}$$

4.28 (c)

The channel which passes through the region of alluvial soil is known as alluvial channel. The cross-section and slope of an alluvial channel may change due to silting and erosion of wetted surface of the channel.

It is the final regime and true regime in which there is the flow of transports sediment of same physical characteristics which ensures that the channel cross-section and the channel slope do not change.

4.29 (d)

Angle of 1 : 1 is the angle of repose of moist clay which is the gradient of slope at which soil settles naturally when in loose state.

4.30 (b)

Conditions to be satisfied for Regime as per Lacey are:

(i) Channel should flow uniformly in "incoherent unlimited alluvium" of same character as that transported by the water;
(ii) Silt grade should be constant
(iii) Silt charge should be constant;
(iv) Discharge should be constant.
(v) Flow is uniform

4.31 (*)

Hydraulic mean radius $= R = \dfrac{5}{2}\cdot\dfrac{v^2}{f}$

v = mean velocity = 1.1 m/s

f = silt factor = 1.1

$$R = \frac{5}{2}\times\frac{1.1^2}{1.1} = 2.75 \text{ m}$$

None of the options match.

4.32 (d)

The channel in alluvium running with a constant discharge first adjusts its depth of flow and its longitudinal slope. Finally, it attains final regime after its base width has also become content.

4.34 (a)

Concept of silt transportation from both Kennedy's and Lacey's theory is weight of water. Limitations are there in both Kennedy's as well as Lacey's theory.

4.36 (c)

That type of lining is preferred for which material required is easily available within small distance. If alignment of area if such that slope is very high then velocity of canal will also be high hence in such core stronger material for lining such or concrete lining should be used.

To have better benefit - cost ratio that type of lining should be used in which cost of maintenance is less.

4.37 (a)

Given:

Discharge, $Q = 27\ m^3/s$

Silt factor, $f = 1$

As per Lacey's

Velocity of flow,

$$V = \left(\frac{Qf^2}{140}\right)^{1/6} = \left(\frac{27 \times 1^2}{140}\right)^{1/6} = 0.76\ m/s$$

Hydraulic radius,

$$R = \frac{5}{2}\frac{V^2}{f} = \frac{5}{2} \times \frac{(0.76)^2}{1} = 1.444\ m \simeq 1.44\ m$$

4.38 (d)

Evaporation and seepage losses of channels are:

1. **Evaporation:** The water lost by evaporation is generally very small, as compared to the water lost by seepage in certain channels. Evaporation losses are generally of the order of 2 to 3 per cent of the total losses. They depend upon all those factors, on which the evaporation depends, such as temperature, wind velocity, humidity, etc. In summer season, these losses may be more but seldom exceed about 7% of the total water diverted into the main canal.
2. **Seepage:** There may be two different conditions of seepage, i.e., (i) Percolation, (ii) Absorption.

   (i) Percolation: In percolation, there exists a zone of continuous saturation from the canal to the water-table and a direct flow is established. Almost all the water lost from the canal, joins the ground water reservoir.

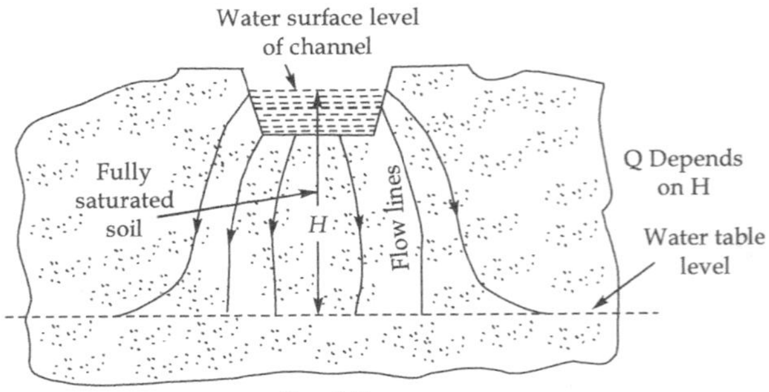

Percolation

The loss of water depends upon the difference upon the difference of the top water surface level of the channel and the level of the water-table (i.e., H) as shown in figure.

   (ii) Absorption: In absorption, a small saturated soil zone exists round the canal section, and is surrounded by zone of decreasing saturation. A certain zone just above the water-table is saturated by capilarity. Thus, there exists an unsaturated soil zone between the two saturated zones, as shown in figure.

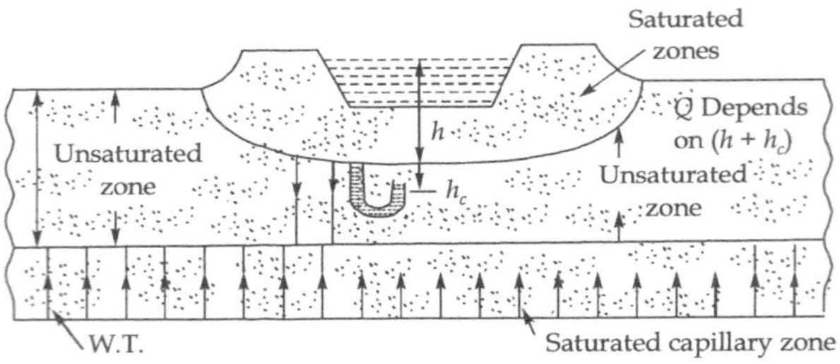

Absorption

4.39 (d)

An artificially constructed channel having a certain fixed section and a certain fixed slope can behave in regime only if the following condition are satisfied:

(i) Discharge is constant
(ii) Flow is uniform
(iii) Silt charge is constant
(iv) Silt grade is constant

A designed channel shall be in true regime if the above conditions are satisfied.

5 Design and Construction of Gravity Dam

5.1 The base width of a solid gravity dam is 25 m. The material of the dam has a specific gravity of 2.56 and the dam is designed as an elementary profile ignoring uplift. What is the approximate allowable height of the dam?

(a) 64 m (b) 40 m
(c) 164 m (d) 80 m

[ESE : 2005]

5.2 What is the critical combination of vertical and horizontal earthquake accelerations to be considered for checking the stability of a gravity dam in reservoir full condition ?

(a) Vertically upward and horizontally downstream
(b) Vertically upward and horizontally upstream
(c) Vertically downward and horizontally upstream
(d) Vertically downward and horizontally downstream

[ESE : 2009]

5.3 What is the height of wave which is likely to be generated by a wind of 80 km/hr in a reservoir having a fetch of 50 km?

(a) 0.5 m (b) 1.0 m
(c) 2.0 m (d) 3.0 m

[ESE : 2009]

5.4 The wave height, in metres, generated on the surface of a reservoir, having a fetch length $F = 30$ km, due to wind blowing on the surface of the reservoir at a velocity of 30 km/hr, is

(a) 0.26 m (b) 0.96 m
(c) 0.52 m (d) 1.2 m

[ESE : 2010]

5.5 The maximum height of a low gravity dam of elementary profile made of concrete of relative density 2.5 and safe allowable stress of foundation material 3.87 MPa without considering uplift force is about

(a) 113 m (b) 217 m
(c) 279 m (d) 325 m

[ESE : 2010]

5.6 At the base of a gravity dam section, the vertical stress at the toe was found to be 2.4 MPa. If the downstream face of the dam has a slope of 0.707 horizontal : 1 vertical, and if there is no tail water, the maximum principal stress at the toe of the dam is

(a) 1.7 MPa (b) 2.4 MPa
(c) 3.6 MPa (d) 4.8 MPa

[ESE : 2014]

5.7 The top width and the bottom width at river thalweg level of a concrete gravity dam are 6 m and 24 m respectively. The vertical section of the dam at the thalweg alignment is 38 m above the thalweg. The upstream face is vertical and the depth of water on the upstream side is 36 m. The moment of hydrostatic force about the toe at that level is

(a) 77760 kN-m (b) 25292 kN-m
(c) 6480 kN-m (d) 51840 kN-m

[ESE : 2014]

5.8 Consider the following statements related to uplift pressure in gravity dams:

1. A drainage gallery reduces the uplift pressure at all levels below the gallery.
2. A drainage gallery below upstream water level reduces the uplift pressure at all levels below the upstream water level.
3. A grout curtain near the heel reduces seepage and uplift pressure everywhere on the gravity dam whatever the upstream water level.

Which of the above statements are correct?

(a) 2 and 3 only (b) 1 and 2 only
(c) 1 and 3 only (d) 1, 2 and 3

[ESE : 2016]

5.9 Consider the following statements in respect of dams:

1. In a gravity, dam, grout curtain is provided near the toe to reduce the exit gradient.
2. A drainage gallery with its drainage pipe system provided in gravity dam reduces the uplift pressure at all levels below the upstream water level.
3. An earthquake acceleration of 0.1 g acting vertically downward causes a decreases of 10% in the unit weight of concrete and of water in a gravity dam.
4. The Tehri dam is a gravity dam.

Which of the above statements are correct?

(a) 1 and 3 (b) 1 and 4
(c) 2 and 3 (d) 2 and 4

[ESE : 2017]

5.10 A masonry dam 8 m high, 1.5 m wide at the top and 5 m wide at the base retains water to a depth of 7.5 m, the water face of the dam being vertical. If the weight of water is 9.81 kN/m^3, weight of masonry is 22 kN/m^3, the maximum intensity of stress developed at the base will be nearly

(a) 196 kN/m^2 (b) 182 kN/m^2
(c) 160 kN/m^2 (d) 148 kN/m^2

[ESE : 2019]

5.11 Consider the following data regarding a theoretical profile of a dam:

Permissible value of compressible stress σ = 350 tonnes/m^2

Specific gravity of concrete, s = 2.4

Uplift coefficient, c = 0.6 m

The value of γ = 1

The height of base width will be nearly

(a) 125 m and 63 m (b) 175 m and 63 m
(c) 125 m and 93 m (d) 175 m and 93 m

[ESE : 2019]

5.12 Which one of the following is the correct assumption made in two dimensional design of gravity dams?

(a) The loads are transferred to the abutments by beam action
(b) The foundation and dam behave as separate units
(c) The materials in the foundation and body of the dam are isotropic and homogeneous
(d) The stresses developed in the foundation and body of the dam are not within elastic limit

[ESE : 2023]

5.13 If V is the wind velocity and F is the fetch, F < 32 km, what is the formula for wave height (h_w) in gravity dams?

(a) $h_w = 1.032\sqrt{VF} + 0.76 + 0.27F^{\frac{1}{4}}$

(b) $h_w = 0.032\sqrt{VF} - 0.76 - 0.27F^{\frac{1}{4}}$

(c) $h_w = 1.032\sqrt{VF} - 0.76 + 0.27F^{\frac{1}{4}}$

(d) $h_w = 0.032\sqrt{VF} + 0.76 - 0.27F^{\frac{1}{4}}$

[ESE : 2024]

5.14 If c_1 = a dimensionless pressure coefficient, α_h = horizontal acceleration factor, ρ= mass density of water, g = acceleration due to gravity, h = depth of the reservoir, what is the formula for the variation of horizontal hydrodynamic earthquake pressure with depth (P_e) in gravity dams?

(a) $P_e = C_1\alpha_h \frac{g}{h}$ (b) $P_e = \frac{C_1\alpha_h}{gh}$

(c) $P_e = C_1\alpha_h gh$ (d) $P_e = C_1 g(h+\alpha_h)$

[ESE : 2024]

■■■■

Answers **Design and Construction of Gravity Dam**

5.1 (b) 5.2 (c) 5.3 (c) 5.4 (b) 5.5 (a) 5.6 (c) 5.7 (a) 5.8 (a) 5.9 (c)
5.10 (b) 5.11 (c) 5.12 (c) 5.13 (d) 5.14 (c)

Explanations Design and Construction of Gravity Dam

5.1 (b)

For elementary profile

$$B = \frac{H}{\sqrt{G_C - C}}$$

When uplift is ignored C = 0

$$\therefore \quad B = \frac{H}{\sqrt{G_C}}$$

Or $$H = 25 \times \sqrt{2.56} = 40 \text{ m}$$

5.2 (c)

When the vertical acceleration is acting downward, it reduces effective weight and stability of the dam. Thus it is the worst case from design point of view. When the reservoir is full, the force due to horizontal upstream acceleration would produce worst results because it will act along with hydrostatic water pressure in downward direction.

5.3 (c)

Waves are generated on the surface of reservoir by the blowing winds, which causes a pressure towards the downstream side. Wave pressure depends upon the wave height. Wave height in metres may be given by the equation.

(i) $h_w = 0.032\sqrt{VF} + 0.763 - 0.271(F)^{1/4}$

for $F < 32$ km

(ii) $h_w = 0.032\sqrt{VF}$ for $F \geq 32$ km

where V is wind velocity in km/hr and F is fetch in km.

$$\therefore \quad h_w = 0.032\sqrt{80 \times 50} = 2.02 \text{ m}$$

5.4 (b)

Wave height generated on the surface of a reservoir is given by

$$h_w = 0.032\sqrt{VF}$$

$$= 0.032 \times \sqrt{30 \times 30}$$

$$= 0.96 \text{ m}$$

However, it may be noted that the above formula is applicable for $F > 32$ km in strict terms.

5.5 (a)

The limiting height of a low concrete gravity dam without considering uplift force is given by

$$H_{max} = \frac{f}{\gamma_w (G+1)}$$

$$= \frac{3.87 \times 10^3}{9.81(2.5+1)}$$

$$= 112.7 \text{ m} \simeq 113 \text{ m}$$

5.6 (c)

$$\tan\alpha = \frac{B}{H} = 0.707$$

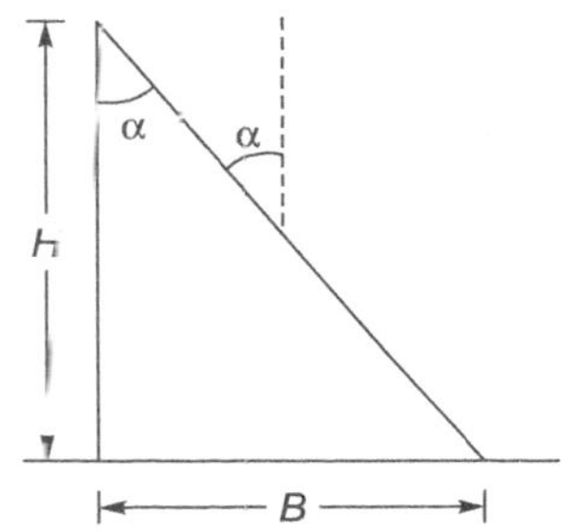

$$p_{max\,toe} = 2.4 \text{ MPa}$$

$$\sigma_{max} = p_{max\,toe} \sec^2\alpha$$

$$= p_{max\,toe}(1 + \tan^2\alpha)$$

$$= 2.4 \text{ MPa} (1 + 0.707^2)$$

$$= 2.4 \times 1.5 \text{ MPa} = 3.6 \text{ MPa}$$

5.7 (a)

$$F_H = \rho g \bar{H} \times \text{Area}$$

$$= 10^3 \times 9.81 \times \frac{36}{2} \times 36 \times 1$$

$$= 6356.88 \text{ kN}$$

Moment of hydrostatic force about toe

$$= F_H \times \frac{H}{3} = 6356.88 \times \frac{36}{3}$$

$$= 76282.56 \text{ kNm}$$

5.8 (a)

A drainage gallery reduces the uplift pressure at all levels below as well as above the gallery. Hence statement 1 is incomplete.

5.9 (c)

Tehri dam is Rock and Earth fill dam.

5.10 (b)

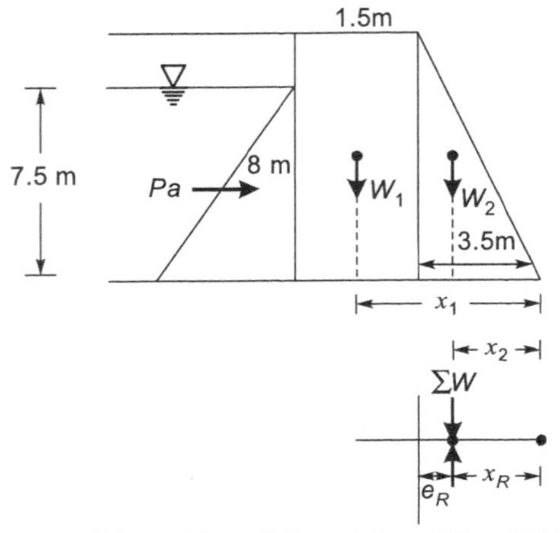

$$W_1 = 1.5 \times 8.0 \times 1.0 \times 22 = 264 \text{ kN}$$

$$x_1 = 5.0 - \frac{1.5}{2} = 4.25 \text{ m}$$

$$W_2 = \frac{1}{2} \times 3.5 \times 8.0 \times 22 = 308 \text{ kN}$$

$$x_2 = 3.5 \times \frac{2}{3} = 2.3333 \text{ m}$$

$$P_a = \frac{1}{2} \times \gamma \cdot H^2$$

$$= \frac{1}{2} \times 9.81 \times 7.5^2 = 275.91 \text{ kN}$$

$$h = \frac{1}{3} \times 7.5 = 2.50 \text{ m}$$

$$\Sigma W \cdot x_R = W_1 x_1 + W_2 x_2 - P_a h$$

$$x_R = \frac{(264 \times 4.25 + 308 \times 2.33333) - 275.91 \times 2.50}{264 + 308}$$

$$= 2.012 \text{ m}$$

$$e_R = \frac{5}{2} - 2.012 = 0.488 \text{ m}$$

Maximum pressure

$$= \frac{\Sigma W}{A}\left(1 + \frac{6e_R}{B}\right)$$

$$= \frac{(264 + 308)}{5.0 \times 1.0}\left(1 + \frac{6 \times 0.488}{5.0}\right)$$

$$= 181.39 \text{ kN/m}^2$$

5.11 (c)

Critical or limiting height of dam,

$$H_{lim} = \frac{F}{\gamma_w (G_c - C + 1)}$$

$$= \frac{350 \text{ tonnes/m}^2}{1(2.4 - 0.6 + 1)}$$

Base width,

$$B_{min} = \frac{H_{lim}}{\sqrt{G_c - C}}$$

$$= \frac{125}{\sqrt{2.4 - 0.6}} = \frac{125}{\sqrt{1.8}}$$

$$= 93.14 \text{ m} \simeq 93 \text{ m}$$

5.12 (c)

The assumptions made in the two dimensional gravity method are:

1. The dam is considered to be composed of a number of cantilevers, each of which is 1 m thick and acts independent of the other
2. No loads are transferred to the abutments by beam action
3. The foundation and the dam behave as a single unit
4. The materials in the dam body and foundation are isotropic and homogeneous; and
5. The stresses developed in the dam and foundation are within elastic limits and no movement of the foundation is caused due to transfer of loads.

5.13 (d)

When F < 32 km,
height of wave,

$$h_w = 0.032\sqrt{VF} + 0.76 - 0.27F^{\frac{1}{4}}$$

When $F > 32$ km

$$h_w = 0.032\sqrt{VF}$$

5.14 (c)

As per Zanger method for calculating hydrodynamic pressure, P_e

$$P_e = C_1 \times k_h \times \rho_w g \times h$$

$$= C_1 \times \frac{\alpha_h}{g} \times \rho_w g \times h$$

$$= C_1 \times \alpha_h \times \rho_w \times h$$

NOTE: No option is correct but (c) is closest.

6 Conveyance Structures for Canals & Outlets/Modules

6.1 Gibb's module is a type of outlet which ensures
(a) constant discharge even if the water levels in the supply channel and water course fluctuate
(b) variable discharge as per the need
(c) constant discharge into the water course when the water levels in the supply channel vary
(d) constant discharge for varying water levels in the water course for a given water level in the supply channel

[ESE : 2001]

6.2 The following data pertain to a natural drain crossing an irrigation canal:

Item	Canal data	Drainage data
Flow (m^3/s)	5	500
Bed level (m)	120	116
Depth of flow (m)	0.8	10

Which one of the following types of cross-drainage should be recommended in this case?
(a) Aqueduct
(b) Syphon aqueduct
(c) Syphon
(d) Super-passage

[ESE : 2001]

6.3 Modular limit of a canal outlet is the ratio of
(a) rate of change of discharge of outlet to that of distributary.
(b) water depth above outlet crest to the full supply depth of the channel
(c) water depth above the crest on downstream to that on upstream of outlet
(d) rate of change of discharge of an outlet to the rate of change of water level of the channel.

[ESE : 2002]

6.4 A canal fall is a control structure
(a) located at a place where the country slope is flatter than the canal bed slope
(b) located most economically where the depth of cutting is less than the balancing depth
(c) the location of which is independent of the command to be served
(d) designed to secure raising of water surface on its upstream

[ESE : 2002]

6.5 **Assertion (A):** Canal escape serves as a safety valve for a canal.
Reason (R): Canal escape discharges the excess water in the parent canal due to sudden closure of outlets by the farmers.
(a) both A and R are true and R is the correct explanation of A
(b) both A and R are true but R is not a correct explanation of A
(c) A is true but R is false
(d) A is false but R is true

[ESE : 2003]

6.6 The worst condition of uplift on the floor of a syphon aqueduct occurs when there is
(a) high flood flow in the drainage with canal dry
(b) full supply flow in the canal with drainage dry
(c) high flood flow in the drainage with canal running full
(d) Water is at drainage bed and canal is dry

[ESE : 2003]

6.7 Match **List-I** (Relative position of canal and drainage channel) with **List-II** (Type of cross drainage work) and select the correct answer using the code given below the lists:

List-I
A. Canal taken above the drainage channel at its grade
B. Drainage channel taken above the canal at its bed slope
C. Canal taken below the drainage channel
D. Drainage channel taken below the canal

List-II
1. Canal siphon

2. Drainage siphon
3. Aqueduct
4. Super passage

Codes:

	A	B	C	D
(a)	2	4	1	3
(b)	3	1	4	2
(c)	2	1	4	3
(d)	3	4	1	2

[ESE : 2005]

6.8 What type of cross drainage work is provided when the canal runs below the drain, with FSL of canal well below the bed of the drain?

(a) Aqueduct (b) Super passage
(c) Level crossing (d) Siphon aqueduct

[ESE : 2006]

6.9 Which type of fall can be generally used for a moderate discharge of 40–60 cumecs and low fall heights of 1 to 1.5 m?

(a) Vertical drop fall (b) Ogee fall
(c) Glacis fall (d) Baffle fall

[ESE : 2006]

6.10 If the sensitivity of an irrigation module is 0.5, then what percent variation in outlet discharge will be caused by 50 per cent variation in canal water depth?

(a) 100% (b) 50%
(c) 25% (d) 12.5% [ESE : 2006]

6.11 Consider the following statements:
An aqueduct is a cross drainage work in which
1. a canal is carried over the drainage channel.
2. a drainage channel is carried over the canal.
3. both drainage channel and canal are at the same level.

Which of these statements is/are correct?

(a) 1 only (b) 1 and 2 only
(c) 2 and 3 only (d) 1, 2 and 3

[ESE : 2007]

6.12 What are the recommended setting options of an adjustable proportional module worked with an open flume type outlet?

1. 3/10 2. 9/10
3. 1/2 4. 5/3

Select the correct answer using the codes given below:

(a) 1 and 2 (b) 1 and 3
(c) 3 and 4 (d) 2 and 4

[ESE : 2007]

6.13 Semi-module outlets are those outlets in which

(a) discharge gets affected by the change in water level of field channel
(b) discharge gets affected by the change in water level of the distributing channel but not with the change in water level of field channel
(c) discharge is independent of water levels in the distributing channel and the field channel
(d) None of the above

[ESE : 2008]

6.14 A submerged pipe outlet is an example of

(a) semi-modular outlet
(b) non-modular outlet
(c) rigid module
(d) adjustable proportional module

[ESE : 2010]

6.15 Consider the following statements:

1. In a super-passage, the drain runs over the canal.
2. In a siphon, the drain runs below the canal.
3. In a siphon aqueduct, type-II, the canal banks are made of RCC walls.

Which of these statements are correct?

(a) 1 only (b) 1 and 2 only
(c) 2 and 3 only (d) 1 and 3 only

[ESE : 2011]

6.16 Flexibility of an outlet may be defined as the ratio of the rate of change of

(a) Outlet discharge to the rate of change of water level of the parent channel
(b) Outlet discharge to the rate of change of the discharge of the parent channel
(c) Parent channel discharge to the rate of change of the outlet discharge
(d) Parent channel water level to the rate of change of the outlet discharge

[ESE : 2011]

6.17 Consider the following statements:
Mitra's hyperbolic transition design is based on the principle that

1. Flow depth in the canal, as well as the discharge, is constant
2. Width of the canal varies along with the discharge
3. Rate of change of velocity per unit length of transition is constant throughout the length of the transition

Which of these statements are correct?

(a) 1, 2 and 3 (b) 1 and 2 only
(c) 2 and 3 only (d) 1 and 3 only

[ESE : 2011]

6.18 If the channel index at an irrigation outlet is 5/3, "setting" of an orifice type outlet in order to have proportionality is

(a) 0.90 (b) 0.67
(c) 0.30 (d) 0.15 [ESE : 2014]

6.19 Which of the following outlets are of semi-modular type?

1. Khanna's module
2. Adjustable proportional module
3. Submerged pipe outlet
4. Open flume outlet
5. Kennedy's gauge outlet

(a) 2, 4 and 5 (b) 1, 2 and 4
(c) 2, 3 and 5 (d) 3, 4 and 5

[ESE : 2014]

6.20 By considering the channel index as 5/3, the setting of an orifice type irrigation outlet to have proportionality is

(a) 0.90 (b) 0.67
(c) 0.30 (d) 0.15 [ESE : 2015]

6.21 In a siphon aqueduct, the worst condition of uplift on the floor occurs when

(a) The canal is full and the drainage is empty, with water table at drainage bed level.
(b) The canal is empty and the drainage is full, with water table at drainage bed level.
(c) Both the canal and the drainage are full.
(d) The canal is empty and the drainage is full, with water table below the floor.

[ESE : 2018]

6.22 Consider the following data for a drain:
$L = 50$ m, $a = 10$ m, $b = 10.3$ m and $k = 1 \times 10^{-5}$ m/s
If the drains carry 1% of average annual rainfall in 24 hrs, the average annual rainfall for which this system has been designed will be

(a) 78 cm (b) 84 cm
(c) 90 cm (d) 96 cm [ESE : 2019]

6.23 **Statement I :** Cross drainage structure, aqueduct is constructed to negotiate an aligned channel over, below or at the same level of a stream.
Statement II : Aqueduct is constructed, when full supply level (FSL) of canal is much higher than high flood level (HFL) of a stream.

[ESE : 2023]

■■■■

Answers Conveyance Structures for Canals & Outlets/Modules

6.1 (a)	6.2 (b)	6.3 (*)	6.4 (c)	6.5 (a)	6.6 (b)	6.7 (d)	6.8 (b)	6.9 (c)
6.10 (c)	6.11 (a)	6.12 (a)	6.13 (b)	6.14 (b)	6.15 (d)	6.16 (b)	6.17 (d)	6.18 (c)
6.19 (a)	6.20 (c)	6.21 (a)	6.22 (b)	6.23 (d)				

Explanations Conveyance Structures for Canals & Outlets/Modules

6.1 (a)

Gibb's module is a rigid module or modular outlet and the discharge through it is constant within certain limits, irrespective of the fluctuations of water level in distributary or water course or both.

6.2 (b)

The difference in bed level of the canal and drainage is 4 m while the flow depth in drainage is 10 m. Thus HFL of drain at 126 m is higher than the canal bed at 120 m. Therefore syphon aqueduct is most suitable.

6.3 (*)

The ratio of depth of water level over crest on the downstream of the module (H_2) to the depth of water level over crest on the upstream of the module (H_1) is called drowning ratio or submergence ratio (H_2/H_1).

The rate of change of discharge through the outlet to the rate of change of water level of the distributary or channel is called sensitivity.

The rate of change of discharge of outlet to that of distributary is called flexibility.

Modular limits are extreme values of any one or more variables, beyond which an outlet becomes incapable of acting as a module or semi-module. The range between the lowest and highest limiting values of various such factors is known as modular range. Modular limit is not a ratio.

6.4 (c)

When the available natural ground slope is steeper than the designed bed slope of the channel, the difference is adjusted by constructing canal falls or drops. The location of fall depends upon the topography, command area and the possibility of combining it with a bridge, regulator, or some other masonry work, since such combinations often result in economy and better regulation but it is independent of the command area to be served. It must also be located most economically where the depth of cutting is equal to the depth of filling.

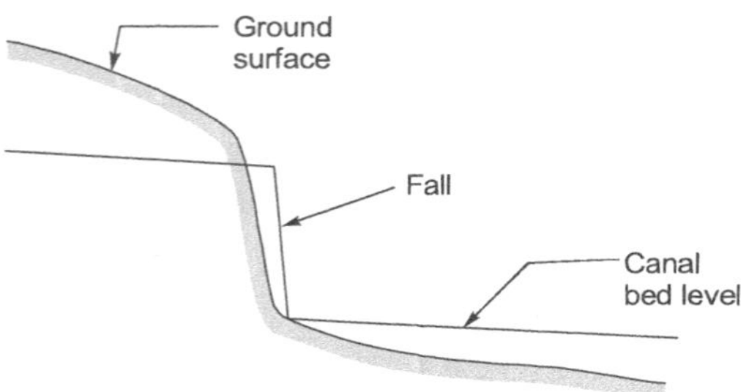

6.5 (a)

Apart from sudden closure by the farmers, excess water can be drained off during the times of flood or excess discharges to avoid submergence.

6.6 (b)

The floor of the aqueduct or syphon-aqueduct is subjected to uplift due to two causes:

(a) *Uplift due to water-table:* This force acts where the bottom floor is depressed below the drainage bed, especially in syphon aqueducts. The maximum uplift under the worst condition would occur when there is no water flowing in the drain and the water-table has risen up to the drainage bed.

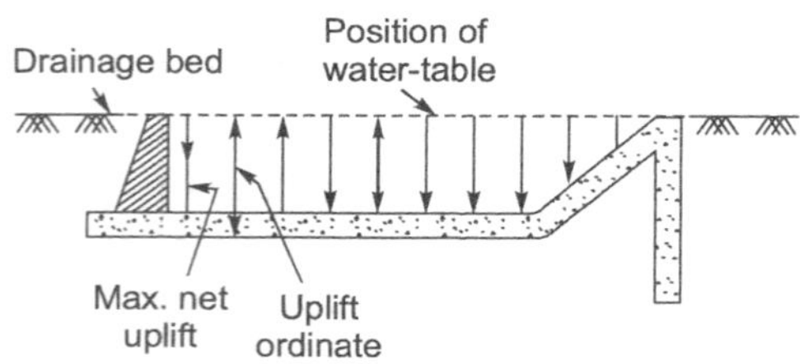

(b) *Uplift due to seepage of water from the canal to the drainage:* The maximum uplift due to this seepage ocurs when the canal is running full and there is no water in the drain.

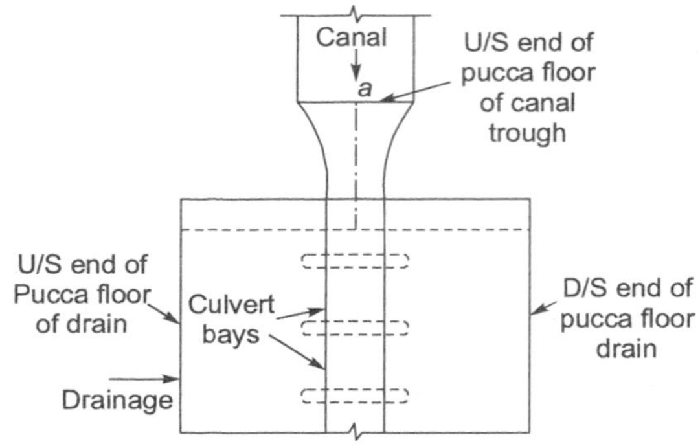

6.7 (d)

An aqueduct is a watercourse constructed to convey water, which carries canal above the drainage channel at its grade. Drainage siphon is similar to aqueduct except that stream bed is depressed locally below the canal. Super passage is reverse of aqueduct drainage channel is carried above the canal at its bed slope. Canal siphon is similar to super passage but canal is taken below drainage channel.

6.8 (b)

If bed level of canal is sufficiently above the HFL of the drain, an aqueduct is the right choice. Sometimes bed level of canal may be little below the HFL of drain so that water flows under syphonic action. This structure is known as aqueduct syphon. If bed level of the drain is sufficiently above the canal FSL, a super passage may be constructed. If the canal FSL is little above bed level of drain to allow syphonic action, the structure is called Canal Syphon. Aqueduct or super-passage type of drainage works are used when high flood discharge is large and continues for a sufficient time.

A level crossing is used when the canal water and drain water are allowed to intermingle with each other. This is adopted when high flood drainage discharge is short lived.

6.9 (c)

Ogee fall has convex and concave curves. Due to smooth transition, the kinetic energy of the flow is not at all dissipated. This causes erosion of downstream bed and banks. It is not used these days.

Well (Cylindrical or syphon well) fall is necessary in the case of falls greater than 1.8 m and for discharges greater than 0.29 m^3/s. Thus it is used for larger drop and smaller discharges.

Simple vertical drop type and Sarda type falls are used for small discharges which vary over wide range. They are suitable for discharges upto 15 cumecs and drops upto 1.5 m.

Straight Glacis falls are designed such that hydraulic jump occurs on the glacis, causing sufficient energy dissipation. They can be adopted suitably for discharges upto 60 cumecs and drops upto 1.5 m and can be even flumed.

Baffle fall (English fall) may be used for all discharges when drop is more than 1.5 m.

Note: Glacis type fall is suitable as a metering device. Baffle fall can also be flumed to measure the discharge.

6.10 (c)

$$\text{Sensitivity, } S = \frac{(dq/q)}{(dy/y)}$$

where q is discharge through outlet
y is depth in distributary canal.

Given $S = 0.5$

$dq/q = ?$

$$\therefore \quad \frac{dq}{q} = \frac{dy}{y} \times S = 50 \times 0.5 = 25\%$$

6.11 (a)

An aqueduct and syphon aqueduct are cross drainage works in which the canal is taken over the natural drain, such that the drainage water runs below the canal either freely or under syphoning pressure. Super passage and syphon are cross drainage works in which the drain is taken over the canal such that the canal water runs below the drain either freely or under syphoning pressure.

Level crossing is a cross drainage work in which the canal water and drain water are allowed to intermingle with each other. A level crossing is generally provided when a large canal and a huge drainage (such as a stream or a river) approach each other practically at the same level.

6.12 (a)

The ratio of the depth of sill level of the outlet below the FSL of the distributary to the full supply depth of the distributary, is known is setting.

$$\text{Setting} = \frac{\text{Outlet index (m)}}{\text{Channel index (n)}}$$

For a wide trapezoidal channel, the discharge is proportional to $y^{5/3}$. Therefore for such channels, $n = 5/3$. The discharge through an orifice type outlet is proportional to $H^{1/2}$ and thus for such an outlet,

$$m = \frac{1}{2}$$

$$\therefore \quad \text{Setting} = \frac{1/2}{5/3} = \frac{3}{10}$$

Similarly for a weir type outlet,

$$m = \frac{3}{2}$$

$$\therefore \quad \text{Setting} = \frac{3/2}{5/3} = \frac{9}{10}$$

The outlet is said to be proportional when the rate of change of outlet discharge equals the rate of change of channel discharge.

6.13 (b)

The various available types of outlets (modules) are:

(i) Non-modular outlets are those through which the discharge depends upon the difference of head between distributary and water course. Common examples are open sluice outlet, drowned pipe outlet and submerged pipe outlet.

(ii) Semi modules or flexible modules are those through which the discharge is independent of the water level of the water course but depends only upon the water level of the distributary so long as a minimum working head is available. Common examples are pipe outlet, venturi flume, open flume and orifice semi-module.

(iii) Rigid modules or modular outlets are those through which the discharge is constant and fixed within limits, irrespective of the fluctuations of the water levels of either the distributary or of the water course or of both. Common example is Gibb's module.

6.14 (b)

Refer solution of 6.13.

6.15 (d)

A super passage is just like a bridge in which the natural drain is carried over the canal. A super passage is constructed where the bed of the drain is well above the canal F.S.L.

A syphon serves the same purpose as superpassage and is constructed where the F.S.L. of the canal is higher than the bed of the drain.

Aqueducts are of following three types

Type I: In this type the sides of the aqueduct (or syphon aqueduct) are earthen banks with complete earthern slopes.

Type II: In this type also the canal continues in its earthen section over the drain, but the outer slopes of the earthen banks are replaced by retaining walls.

Type III: In this type the earthen banks are not provided through the aqueduct (or syphon aqueduct) and the canal water is carried in a masonry or concrete trough.

6.16 (b)

An outlet is a sort of head regulator through which water is released from a distributing channel into a water course or field channel.

Flexibility: It is defined as the ratio of the rate of change of discharge of an outlet to the rate of change of discharge of the distributing channel.

Proportionality: If the flexibility is equal to one then the outlet is termed as proportional outlet.

Sensitivity: It is defined as the ratio of the rate of change of discharge of an outlet to the rate of change in the level of the water surface in the distributing channel.

Efficiency: It is defined as the ratio of the head recovered to the head put in. Efficiency is a measure of conservation of head by an outlet.

Setting: It is the ratio of the depth below full supply level of the crest of an outlet to the full supply depth of the channel at that point.

Adjustability: Adjustability means capacity for modification of the outlet to take into account changed condition such as silting or scouring of the distributing channel.

6.17 (d)

For water depth to remain constant in the transition and the trough, A.C. Mitra proposed hyperbolic transitions. The transition is designed on the basis that the rate of change of velocity per unit length of the transition is constant throughout the transition length. It assumes that the width of the canal varies along with the velocity.

6.18 (c)

For orifice type outlet,

$$m = \frac{1}{2}$$

$$\therefore \quad \text{Setting} = \frac{H}{D} = \frac{1}{2}\times\frac{3}{5} = \frac{3}{10} = 0.3$$

6.19 (a)

Common types of semi modulus are:-

(i) Open fume outlet

(ii) Adjustable orifice semi-module

(iii) Pipe outlet discharging freely into air

(iv) Kennedy's gauge outlet or Venturi-flume outlet.

6.20 (c)

Outlet index for orifice type outlet = 1/2

Channel index = 5/3

$$\text{Setting} = \frac{\text{Outlet index}}{\text{Channel index}} = \frac{1/2}{5/3} = 0.3$$

6.21 (a)

Refer answer of 6.8.

6.22 (b)

As per the standard notations:

$a \rightarrow$ distance between impervious strata from the centre of the drain

$b \rightarrow$ maximum height of drained water level above the impervious floor

$l \rightarrow$ length of the drain

$s \rightarrow$ centre to centre spacing between the drain

The question cannot be solved since spacing 's' is not given.

Note: If in place of L = 50m, it would have been s = 50m, then it would be solved like this:

$$s = \frac{4k}{q}(b^2 - a^2)$$

$$50 = \frac{4\times10^{-5}\times(10.3^2 - 10^2)}{q}$$

$$\therefore \quad q = 4.872\times10^{-6}\ \text{m}^2/\text{s}$$

$$q = \frac{\frac{1}{100}\times\bar{p}\times(s\times1)}{8.64\times10^4}$$

where, $\rightarrow$ annual average rainfall

$$\bar{p} = 0.841\ \text{m} \simeq 84\ \text{cm}$$

6.23 (d)

Aqueduct is a cross drainage work in which high flood level (HFL) of drain is much below the bed level of the canal. So that drainage water flows freely under gravity.

Therefore most appropriate answer is (d).

7 Theories of Seepage

7.1 Consider the following statements:

1. In designing a hydraulic structure in permeable foundation by Khosla's theory, the slope correction is applicable to piles located at the ends of the sloped floor only in a structure with one or more sloped floors.
2. Khosla's theory can be correctly applied in alluvium of finite depth.
3. Length of flow has lesser effect on exit gradient than the depth of piles.
4. Intercepts between hydraulic gradient line and free water surface on the glacis and horizontal floor downstream is the unbalanced uplift in a structure.

Which of these statements related to the design of hydraulic structure are correct?

(a) 1, 2 and 3 (b) 1, 3 and 4
(c) 2, 3 and 4 (d) 1, 2 and 4

[ESE : 2000]

7.2 Consider the following statements:

The downstream impervious floor of concrete for a barrage has ruptured. This can be due to

1. insufficient length of upstream impervious floor
2. insufficient length of downstream impervious floor
3. insufficient depth of downstream pile
4. choking of inverted filter

Which of these statements are correct?

(a) 1 and 2 (b) 1 and 4
(c) 2, 3 and 4 (d) 1, 2, 3 and 4

[ESE : 2001]

7.3 In a barrage on pervious foundation, sheet piles are provided both upstream and downstream of the barrage to reduce uplift pressure and to prevent piping. Which one of the following statements is true in this regard?

(a) Compared to downstream sheet pile, the upstream sheet pile is more effective in reducing uplift and piping
(b) Compared to upstream sheet pile, the downstream sheet pile is more effective in reducing uplift and piping
(c) Downstream sheet pile is more effective in reducing uplift while the upstream sheet pile is more effective in reducing piping
(d) Upstream sheet pile is more effective in reducing uplift while the downstream sheet pile is more effective in reducing piping

[ESE : 2001]

7.4 At a certain point in the floor of weir, the uplift pressure head due to seepage is 4.5 m. If the relative density of concrete is 2.5, the minimum thickness of floor required at this point to counteract the uplift pressure is

(a) 1 m (b) 2 m
(c) 3 m (d) 4 m

[ESE : 2002]

7.5 Match **List-I** (Theory) with **List-II** (Propounded by) and select the correct answer using the codes given below the lists:

List-I	List-II
A. Exit gradient	1. G. Lacey
B. Alluvial canal	2. L.K. Sherman
C. Unit hydrograph	3. A.N. Khosla
D. Boundary layer	4. C. Inglis
	5. T.V. Karman
	6. L. Prandtl

Codes:

	A	B	C	D
(a)	1	3	2	6
(b)	6	2	3	5
(c)	3	1	2	6
(d)	3	1	4	2

[ESE : 2003]

7.6 Which one of the following is the purpose of providing the downstream sheet pile in a barrage?
(a) To control failure due to piping by high value of exit gradient
(b) To control failure due to scour
(c) To stop failure due to sliding
(d) To stop failure due to uplift pressure

[ESE : 2007]

7.7 Consider the following statements:
1. Giving equal weightages to horizontal and vertical creeps for design of weir foundations is one of the drawbacks of Kennedy's theory.
2. Khosla's theory of design of foundations for weirs is based on potential theory.
3. Piping problem can be reduced by increasing the length of floor.
4. In Lane's weighted creep theory, horizontal creep is given less weightage compared to vertical creep.

Which of these statements is/are correct?
(a) 1 (b) 2, 3 and 4
(c) 2 and 4 only (d) 3 and 4 only

[ESE : 2009]

7.8 A road passes through a cut with impervious strata at a depth of 1.25 m from the sub-grade level. To intercept seepage flow, a longitudinal pipe drain in trench filled with filter material and clay seal is constructed. What will be the required depth of the trench?
(a) 2.5 m (b) 1.5 m
(c) 1.25 m (d) 0.5 m

[ESE : 2013]

7.9 A 20 m long horizontal concrete floor under a barrage on a permeable foundation retains a 5 m head of water and has a 5 m deep downstream end pile. The exit gradient is
(a) 1 in 4 (b) 1 in 5
(c) 1 in 6 (d) 1 in 8

[ESE : 2016]

7.10 Khosla's formulae for assessing pressure distribution under weir floors are based on
(a) Potential flow in permeable layers just beneath the floors
(b) Boundary layer flow with pressure drop longitudinally
(c) Conformal transformation of potential flow into the w plane
(d) Simplification of 3-D flow

[ESE : 2018]

7.11 Consider the following statements with respect to weir under discussion:
1. Its design corresponds to soft sandy foundation.
2. The difference in weir crest and downstream river bed may not exceed 3 m.
3. When water passes over it, the longitudinal location of the formation of a hydraulic jump is variable.

This weir is of the type
(a) Vertical drop weir
(b) Masonry or concrete sloping weir
(c) Dry stone slope weir
(d) Parabolic weir

[ESE : 2019]

7.12 Consider the following data while designing an expansion transition for a canal by Mitra's method:
Length of flume = 16 m
Width of thorat = 9 m
Width of canal = 15 m
If B_x is the width at any distance x from the flumed section, the values of B_x at x = 8 m and at x = 16 m are nearly
(a) 10.8 m and 15 m (b) 11.3 m and 15 m
(c) 10.8 m and 13 m (d) 11.3 m and 13 m

[ESE : 2019]

7.13 Which of the following are the causes of failure of weirs?
1. Rupture of floor due to uplift.
2. Rupture of floor due to suction caused by standing wave.
3. Scour on the upstream and downstream of the weir.

(a) 1 and 2 only
(b) 1 and 3 only
(c) 1, 2 and 3
(d) 2 and 3 only

[ESE : 2020]

■■■■

Answers Theories of Seepage

7.1 (d) 7.2 (a) 7.3 (d) 7.4 (c) 7.5 (c) 7.6 (a) 7.7 (b) 7.8 (c) 7.9 (b)
7.10 (c) 7.11 (b) 7.12 (b) 7.13 (c)

Explanations Theories of Seepage

7.1 (d)

The exit gradient, $G_E = \dfrac{H}{d}\dfrac{1}{\pi\sqrt{\lambda}}$

where $\lambda = \dfrac{1+\sqrt{1+\alpha^2}}{2}$

and $\alpha = \dfrac{b}{d}$

b is length of floor
d is depth of vertical cut off

With increase in *b*/*d* the factor $\dfrac{1}{\pi\sqrt{\lambda}}$ decreases and *H*/*d* reduces with increase in depth of vertical cutoff. Thus both '*b*' and '*d*' affect equally on the exit gradient.

7.2 (a)
The rupturing of downstream impervious floor means that uplift pressure is very high and weight of concrete or masonry floor is not able to counter balance the same. It means that the length of upstream and downstream floor is not sufficient to reduce the uplift pressure.

7.3 (d)
Upstream sheet pile

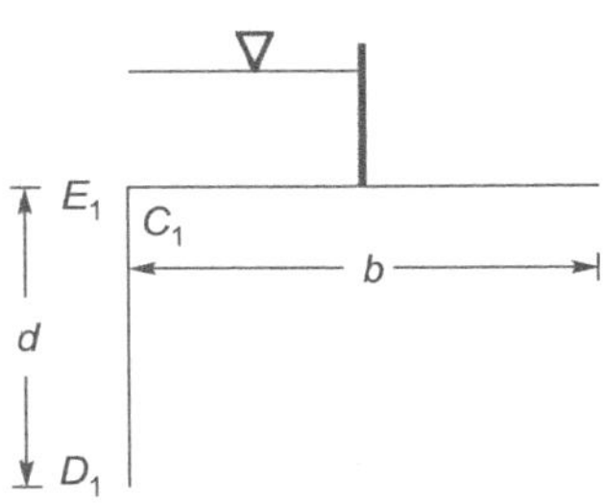

$\phi_{C_1} = 100 - \phi_E$
$\phi_{D_1} = 100 - \phi_D$
Downstream sheet pile

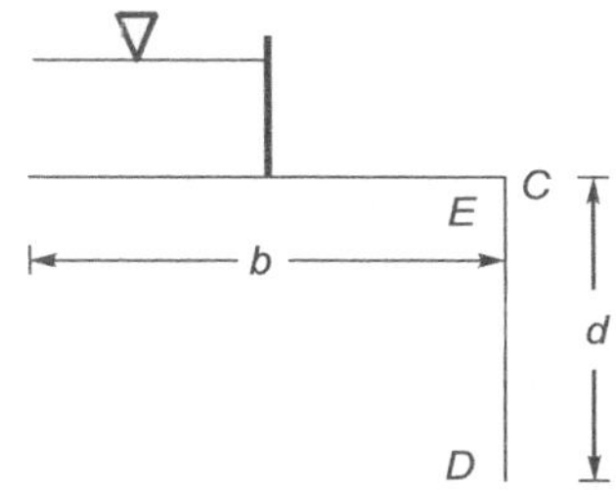

$$\phi_E = \frac{1}{\pi}\cos^{-1}\left(\frac{\lambda-2}{\lambda}\right)$$

$$\phi_D = \frac{1}{\pi}\cos^{-1}\left(\frac{\lambda-1}{\lambda}\right)$$

Where $\lambda = \dfrac{1+\sqrt{1+\alpha^2}}{2}$ and $\alpha = \dfrac{b}{d}$

$\phi_D < 50\%$
So $\phi_{D_1} > 50\%$ for all values of α.
For normal range of $0.4 < \alpha < 0.8$
$\phi_E = 50\%$
So $\phi_{C_1} = 50\%$
Thus upstream piles are more effective in reducing uplift pressure. While down stream piles are more effective in reducing exit gradient.

7.4 (c)
To counter the uplift pressure at this point,
Buoyant weight of concrete=seepage force on area
$(2.5 - 1) \times t \times 1 = 4.5 \times 1$
$t = 3$ m

7.6 (a)
Downstream sheet pile is provided to reduce high value of exit gradient. It increases the seepage path and reduces head at the end.

7.7 (b)
Giving equal weightage to horizontal and vertical creeps for design of weir foundation is one of the drawbacks of Bligh's theory.

7.8 (c)

The maximum depth of trench can be equal to depth impervious strata below the subgrade level of the pavement which is equal to 1.25 m here. To intercept the seepage flow, longitudinal pipe drain in trench filled with filler material and clay seal is constructed.

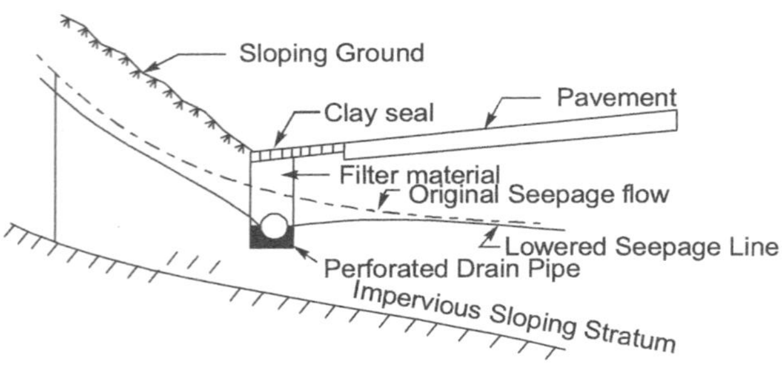

Interception of Sewage Flow by Surface drains

7.9 (b)

$$G_E = \frac{H}{d} \times \frac{1}{\pi\sqrt{\lambda}}$$

$$\lambda = \frac{1+\sqrt{1+\alpha^2}}{2} = \frac{1+\sqrt{1+4^2}}{2} = 2.56$$

$$\left(\because \alpha = \frac{b}{d} = \frac{20}{5} = 4\right)$$

$$G_E = \frac{5}{5} \times \frac{1}{3.15 \times \sqrt{2.56}} = \frac{1}{5}$$

7.10 (c)

Khosla's theory is also referred as theory of conformal transformation as it uses conformal mapping (preserving the original angles) to depict the nature of flow below an impervious floor. The theory is based on the flow-net approach.

7.11 (b)

Weirs constructed of masonry or concrete with sloping glacis are usually used now a days and their design in based upon modern concepts of sub-surface flow (Khosla's theory). Sheet piles of sufficient lengths are driven at the ends of upstream and downstream floor. The hydraulic jump is formed on the downstream sloping glacis so as to dissipate the energy of the flowing water.

This type of weir is exclusively used on permeable foundations (sandy soil). These are provided with a low crest with the differences in weir crest and d/s river not exceeding 3 m usually.

7.12 (b)

B_x is given by

$$B_x = \frac{B_n \times B_f \times L_f}{B_n L_f - x(B_n - B_f)}$$

B_n - normal width of canal

B_f - width of throat of canal

L - length of flume = 16 m

$$B_x(x = 8\text{m}) = \frac{15 \times 9 \times 16}{15 \times 16 - 8 \times (15 - 9)}$$

$$= 11.25 \text{ m}$$

$$B_x(x = 16\text{m}) = \frac{15 \times 9 \times 16}{15 \times 16 - 16 \times (15 - 9)}$$

$$= 15 \text{ m}$$

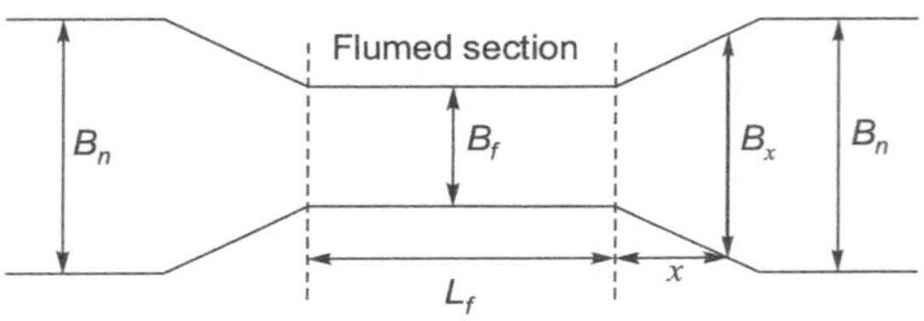

7.13 (c)

Rupture of floor due to uplift:

If the weight of floor is insufficient to resist the uplift pressure, the floor may burst and fail due to rupture.

Rupture of floor due to suction caused by standard wave.

The standing wave or hydraulic jump formed at the d/s of the weir causes suction which also acts in the direction of uplift pressure. If floor thickness is insufficient then it may fail due to rupture.

Scour on the upstream and downstream of the weir.

If the natural water way of a river is contracted, the water may scour the bed both at upstream and downstream of the structure. The scour holes so formed may progress towards the structure causing its failure.

8 River Training Works & Diversion Headworks

8.1 In curved reach of a meandering river, both deposition of sediments and erosion of bank occur. Which one of the following statements is true in this regard?

(a) Deposition of sediments occurs in the inner bank while the outer bank is subjected to erosion.
(b) Deposition of sediments occurs in the outer bank while the inner bank is subjected to erosion.
(c) In the direction of flow, the outer bank undergoes erosion first and the sediment is later deposited on the outer bank downstream.
(d) In the direction of flow, the inner bank undergoes erosion first and the sediments are subsequently deposited on the inner bank downstream.

[ESE : 2001]

8.2 In a river, silt excluder and silt ejector are constructed

(a) at a location after the head regulator and at the head of the canal, respectively
(b) at the head of the canal and at a location after the head regulator, respectively
(c) at the same location
(d) at specific locations depending upon diverse factors and their locations do not follow a set pattern

[ESE : 2001]

8.3 **Assertion (A):** Cut-offs are provided on the downstream base of hydraulic structures.
Reason (R): Hydraulic structures fail by overturning about the downstream base due to excessive erosion at the base caused by overflowing water.

(a) both A and R are true and R is the correct explanation of A
(b) both A and R are true but R is not a correct explanation of A
(c) A is true but R is false
(d) A is false but R is true

[ESE : 2002]

8.4 Which one of the following is the correct sequence in the increasing order of the Froude number of flow assumed by the bed form of an alluvial stream with movable bed material?

(a) Ripple-Plane bed-Dune-Plane bed-Antidune
(b) Dune-Ripple-Plane bed-Antidune-Plane bed
(c) Plane bed-Ripple-Dune-Plane bed-Antidune
(d) Plane bed-Ripple-Antidune-Dune-Plane bed

[ESE : 2004]

8.5 What is the regime scour depth for a channel in soil with silt factor of unity and carrying 8 m^2/s of discharge intensity in accordance with Lacey's regime theory?

(a) 3.6 m (b) 4 m
(c) 5.4 m (d) 25. 6 m

[ESE : 2004]

8.6 Denehy's groyne is a special type of groyne which is

(a) pointing upstream (b) pointing downstream
(c) Hockey type (d) T-headed

[ESE : 2005]

8.7 Which of the following are the purpose of a groyne as a river training structure?

1. It contracts a river channel to improve its depth.
2. It protects the river bank.
3. It does not allow silt to deposit in the vicinity.
4. It trains the flow along a certain course.

Select the correct answer using the codes given below:

(a) 1 and 2 (b) 1, 2 and 4
(c) 1 and 3 (d) 2, 3 and 4

[ESE : 2005]

8.8 Groynes are adopted for river bank protection works. When it is placed inclined downstream in the direction of flow in the river, it is designated as which one of the following?

(a) Repelling groyne
(b) Attracting groyne
(c) Neither repelling nor attracting groyne
(d) Fixed groyne

[ESE : 2007]

8.9 The flood plain of a river carries a discharge of 2000 m^3/s. What are the values of the meander length and dominant flow width?

(a) 160 m, 48 m (b) 180 m, 42 m
(c) 200 m, 38 m (d) 220 m, 36 m

[ESE : 2007]

8.10 Under which one of the following categories is the river Ganga classified in the reach through UP and Bihar?

(a) Straight river (b) Meandering river
(c) Braided river (d) Deltaic river

[ESE : 2009]

8.11 Consider the following statements related to undersluices provided in diversion weirs on permeable foundations :

1. They are fully gate-controlled and have crest at the same level as the weir crest when no silt excluders are provided.
2. They scour the silt deposited on the river bed in the pockets upstream of the canal head regulator.
3. It is not necessary to provide end pile line on the downstream end of the undersluice floor.
4. The discharge capacity of the undersluice is 10-15% of the maximum flood or two times the maximum discharge of the offtaking canal or maximum winter discharge, whichever is the highest.

Which of these statements is/are correct?

(a) 1 (b) 2 and 4 only
(c) 2, 3 and 4 (d) 3 and 4 only

[ESE : 2009]

8.12 Consider the following statements:

1. High water training is undertaken to protect against damage due to floods.
2. Low water training is undertaken to provide sufficient depth for navigation.
3. Mean water training is undertaken to provide efficient disposal of sediment load.

Which of these statements is/are correct?

(a) 1 and 2 only (b) 1, 2 and 3
(c) 2 and 3 only (d) 2 only

[ESE : 2010]

8.13 Heavy scour at the head and shank of guide banks can lead to undermining of the stone pitching and consequent failure of the guide bank. This situation is avoided by providing

(a) spurs (b) vertical cutoffs
(c) marginal bunds (d) launching apron

[ESE : 2010]

8.14 Consider the following statements Morphological characteristics of a river are represented by:

1. Changes in the river form.
2. Changes in the characteristics of the river bed as a result of variation of discharge in the river.
3. No changes in the plan bed form.
4. No changes in the river bed form.

Which of these statements are correct?

(a) 1 and 2 (b) 1 and 3
(c) 2 and 3 (d) 3 and 4

[ESE : 2011]

8.15 Match **List-I** with **List-II** and select the correct answer using the code given below the lists:

List-I

A. Deflector
B. Biff wall
C. Cistern
D. Baffle wall

List-II

1. Low-height wall across canal down- stream of the fall
2. Vertical end wall with horizontal projection
3. Short wall at down-stream end of the fall
4. Pond level below the fall

Code:

	A	B	C	D
(a)	1	2	4	3
(b)	3	2	4	1
(c)	1	4	2	3
(d)	3	4	2	1

[ESE : 2011]

8.16 Which one of the groyne arrangements represents an 'attracting groyne'?

(a)

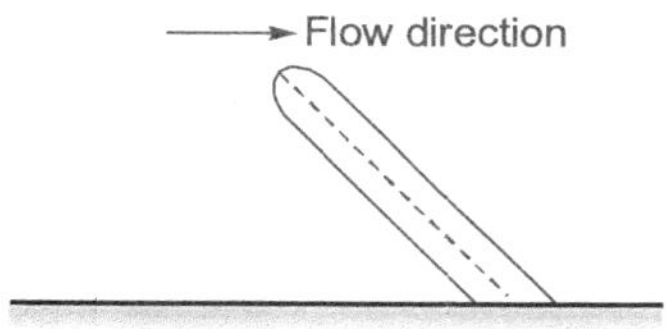

(b)

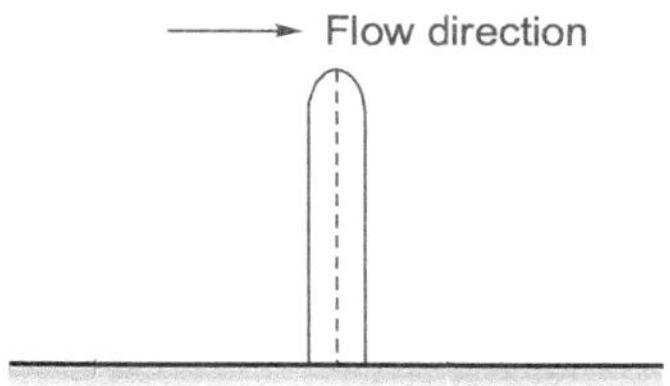

(c)

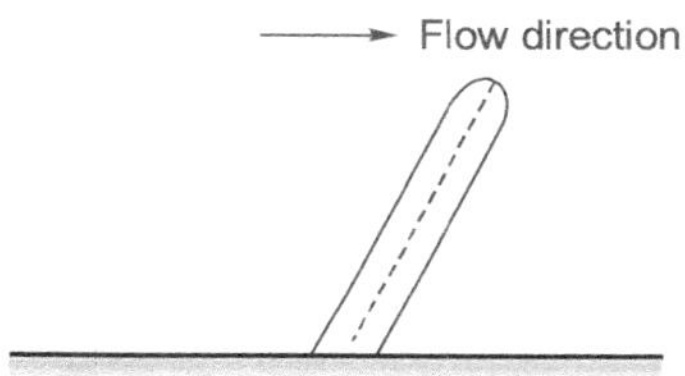

(d)

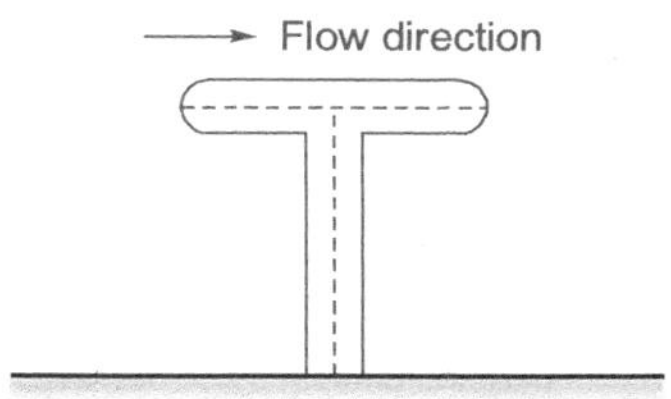

[ESE : 2011]

8.17 Consider the following statements:
The function of a launching apron at the end of downstream impervious apron of a weir is
1. to protect the downstream pile from the scour holes progressing in the upstream direction.
2. when the scour holes are formed, to provide for the stones of the falling apron to settle down in the holes and cover them.
3. to provide relief of the uplift pressure at the downstream end.
4. to provide extra length for proper formation of hydraulic jump.

Which of these statements is/are correct?
(a) 1, 2, 3 and 4 (b) 3 only
(c) 2 only (d) 1 and 2 only

[ESE : 2010]

8.18 The local scour depth in front of a semicircular shaped rectangular pier having width equal to W aligned parallel to the flow below the surrounding bed is
(a) 2.0 W (b) 1.5 W
(c) 1.2 W (d) 1.0 W

[ESE : 2012]

8.19 Consider the following statements
1. Groynes are constructed projecting from the bank into the river at the bank to be protected from flood.
2. Repelling groyne projects downstream into the river-flow from the point of its origin from the bank.
3. Attracting groyne projects upstream into the river from the point of its origin from the bank.
4. Perpendicular groyne projects normal to the river flow from the point of its origin from the bank.

Which of these statements are correct?
(a) 1 and 4 (b) 1 and 2
(c) 2 and 3 (d) 3 and 4

[ESE : 2013]

8.20 A barrage on a major river in the Gangetic plains has been design for a flood discharge 7000 m^3/s. It has been provided with a waterway of 360 m length. The looseness factor of this barrage is
(a) 1.7 (b) 1.1
(c) 0.7 (d) 0.1

[ESE : 2016]

8.21 Tortuosity of a meandering river is the ratio of
(a) Meander length to width of the meander
(b) Meander length to width of the river
(c) Curved length along the river to the direct axial length of the river
(d) Direct axial length of the river to the curved length along the river

[ESE : 2017]

8.22 Objective for river training are
1. High flood discharge may pass safely through the reach.
2. Sediment load (including bed and suspended load) may be transported efficiently.
3. By making the river course unstable whereby to increase bank erosion.

Select the correct answer using the codes given below:

(a) 1, 2 and 3 (b) 1 and 2
(c) 2 and 3 (d) 1 and 3

[ESE : 2017]

8.23 The purpose of constructing a 'Groyne' is to

(a) Expand a river channel to improve its depth
(b) Encourage meandering
(c) Train the flow along a certain course
(d) Reduce the silting in the river bed

[ESE : 2019]

8.24 Which of the following are the objectives for river training?

1. High flood discharge may pass safely and quickly through the reach.
2. To make the river course stable and reduce bank erosion to minimum.
3. To check flow through canal.
4. To provide a sufficient draft for navigation as well as good course for it.

(a) 1, 2 and 3 only (b) 1, 3 and 4 only
(c) 1, 2 and 4 only (d) 2, 3 and 4 only

[ESE : 2020]

8.25 The chief aim of river training is

(a) to protect water from loss
(b) bed scouring
(c) to achieve ultimate stability of river with the aid of river training measures.
(d) pitching of banks and provision of launching aprons.

[ESE : 2021]

8.26 Statement (I): Spur length is kept longer than 1.5 to 2 times the depth of flow.

Statement (II): Shorter spur length in deeper rivers induces swirling motion on both the upstream and downstream sides of the spur.

(a) Both Statement (I) and Statement (II) are individually true; and Statement (II) is the correct explanation of Statement (I)
(b) Both Statement (I) and Statement (II) are individually tru'e; but Statement (II) is NOT the correct explanation of Statement (I)
(c) Statement (I) is true; but Statement (II) is false
(d) Statement (I) is false; but Statement (II) is true

[ESE : 2021]

8.27 Which one of the following is a gap developed in the canal bank due to erosion of some portion of the bank?

(a) Canal breach
(b) Sub canal
(c) Temporary outlet used for irrigation
(d) Fault

[ESE : 2022]

8.28 Which one of the following stages does the river bed consist of a mixture of boulders, gravels, shingles and alluvial sand deposit created by itself?

(a) Rocky stage
(b) Incised river stage
(c) Boulder river stage
(d) Rivers in alluvial flood plains stage

[ESE : 2022]

8.29 Formation of successive bends of reverse order may lead to the formation of a complete S curve called

(a) Concave or outer edge
(b) Scouring
(c) Meander
(d) Convex or inner edge

[ESE : 2022]

8.30 The permeable groynes made from timber stakes or wooden piles, are called

(a) Balli spurs (b) Tree spurs
(c) Balli crates (d) Wire crates

[ESE : 2022]

8.31 The distance between the outer edges of clockwise and anti-clockwise loops of the meander

(a) Meander length (b) Meander belt
(c) Meander ratio (d) Cross-overs

[ESE : 2022]

■■■■

Answers River Training Works & Diversion Headworks

8.1 (a)	8.2 (b)	8.3 (b)	8.4 (c)	8.5 (c)	8.6 (d)	8.7 (b)	8.8 (b)	
8.9 (*)	8.10 (b)	8.11 (b)	8.12 (b)	8.13 (d)	8.14 (a)	8.15 (a)	8.16 (c)	8.17 (c)
8.18 (a)	8.19 (a)	8.20 (*)	8.21 (c)	8.22 (b)	8.23 (c)	8.24 (c)	8.25 (c)	8.26 (a)
8.27 (a)	8.28 (c)	8.29 (c)	8.30 (a)	8.31 (b)				

Explanations River Training Works & Diversion Headworks

8.1 (a)

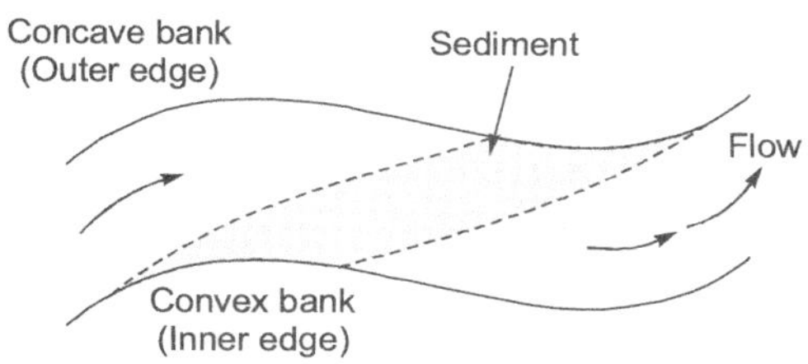

The sediment scouring occurs at outer edge and deposition occurs on the inner bank at same place and at outer edge downstream.

8.2 (b)

Silt excluders are constructed on the bed of the river, upstream of the head regulator. While silt ejectors are constructed on the bed of the canal and a little distance downstream from the head regulator.

Silt excluders remove silt from the water before it enters canal. Silt ejectors extract the silt from the canal water after the silted water has travelled a certain distance in the off-take canal.

8.3 (b)

In order to dissipate residual energy and to keep scour away from the pucca floor, loose stone talus is provided on the downstream of D/S pucca floor and upstream of U/S pucca floor. The erosion at the downstream will cause the structure to fail by overturning about downstream base.

Cut-offs on downstream prevent undermining of foundation by reducing exit gradient.

8.4 (c)

At low flow velocities, the bed does not move at all and its form is plain. With the increase in velocity (Froude number) the bed forms change from ripples to dunes with Froude number less than one. When the Froude number is further increased the dunes are washed away and a flat surface is left. As the Froude number approaches unity, the surface waves breaks and the whole wave system moves upstream leading to formation of antidunes.

Plane bed $(\tau_0 - \tau_c)$ – Ripples – Dunes –Plane bed $(\tau_0 > \tau_c)$ – Antidunes

8.5 (c)

Regime scour depth,

$$R = 1.35\left(\frac{q^2}{f}\right)^{1/3}$$

$$= 1.35\times\left(\frac{8^2}{1}\right)^{1/3} = 5.4 \text{ m}$$

8.6 (d)

Denehy's T-shaped groyne is a special type of groyne developed in India for protection of bank. It comprises of ordinary groyne with an extra cross groyne at the head giving it T-shape.

Hockey shaped groyne is an attracting type of groyne.

8.7 (b)

Groynes are embankment type structures, constructed transverse to the river flow, extending from the bank into the river. They are constructed to protect the bank from which they are extended, by deflecting the current away from the bank. These are called repelling groynes. On the upstream side of a repelling groyne, a still water pocket is formed, where the suspended sediments carried by the river, gets deposited. For low water training or training for depth groynes can be used. Herein they contract the width of the channel to provide sufficient depth for navigation during low water periods.

In order to train the flow along a certain course guide banks are most commonly used. However groynes can also be used to deflect the flow or attract the flow.

8.8 (b)

Groynes or spurs are the structures constructed transverse to the river flow extending from the bank into the river. Groynes may be aligned either perpendicular to the bank line or at an angle pointing upstream or downstream. These are also known as spur dikes or transverse dikes, and constitute the most widely used river training works.

According to function seved by the groynes these may be classified as

(i) *Attracting groynes:* A groyne pointing downstream tends to attract the river flow towards the bank on which it is provided and hence it is called an attracting groyne. The attracting groynes safeguard the opposite bank against the attack of the current as they attract the current towards the bank adjacent to them.

(ii) *Repelling groynes:* A groyne pointing upstream tends to repel the river flow away from the bank on which it is provided and hence it is called a repelling groyne.

As compared to attracting groynes, repelling groynes are more effective and do not cause any trouble. As such repelling groynes are commonly used for the purpose of river training and bank protection.

(iii) *Deflecting groynes:* A groyne either perpendicular to the bank or pointing slightly upstream and having a relatively short length tends to only deflects the flow without repelling it and hence it is called a deflecting groyne. A deflecting groyne, however gives only local protection.

(iv) *Sedimenting groynes:* A groyne which dampens the velocity of flow and thus causes deposition of sediment carried by the river without repelling or deflecting the flow, is called a sedimenting groyne.

8.9 (*)

Meander Parameters:

River flood plains	Incided river
$M_b = 153.42\sqrt{Q}$	$M_b = 102.16\sqrt{Q}$
$M_l = 53.61\sqrt{Q}$	$M_l = 46.01\sqrt{Q}$
$W = 8.84\sqrt{Q}$	$W = 4.53\sqrt{Q}$

Here

Q – Discharge in cumecs

M_b – Meander width (m)

M_l – Meander Length (m)

W – effective width (m)

8.10 (b)

Formation of successive bends of reverse order in a river may lead to the formation of a complete *S* curve called meander. Such a river is called meandering river.

When a river flows in two or more channels around alluvial islands, it is known as a braided river.

Deltaic river is a river which gets divided into branches before joining the sea, thus forming a Δ shaped delta.

In plain regions, the tendency of a river of meander is predominant as one can observe in regions of UP and Bihar.

8.11 (b)

The under sluices are the openings provided at the base of the weir or barrage with their crest at low level. It is necessary to provide end pile on downstream end of undersluice to prevent undermining due to head difference.

8.12 (b)

Training type:

(i) Low water- navigational depth

(ii) Mean water- sediment disposal

(iii) High water- flood control

8.13 (d)

To ensure the safety against undermining of stone pitching and consequent failure of guide bank and damage to the toe of well protected bank from scouring by the action of floods launching apron is provided at the toe of the bank.

8.15 (a)

Cistern: The cistern or cistern element is that portion of the fall on the downstream of the crest wall in which the surplus energy of water leaving the crest is dissipated and the subsequent flow stilled before it passes on to the lower level chanel. The object of providing a cistern are

(i) To reduce the intensity of impact of the dropping jet of water against the downstream floor.

(ii) To provide water cushion to dissipate the energy of the falling jet, and

(iii) To produce reverse flow by providing a suitable end wall to ensure an impact in the cistern.

Deflector or Deflector wall: A deflector or deflector wall may also be provided at the end of the cistern if high velocity jet persists upto the end of the cistern.It causes a reverse roller to form which causes a limited scour away from the toe where it does not endanger the structure and piling up of material against the toe.

Biff wall: A biff wall which may also be provided at the end of the cistern if high velocity jet persists upto the end of the cistern. Its function is also similar to that of a deflector.

Baffle wall: A baffle wall is a low wall which may be provided at the end of the cistern. It provides a deep pool of water behind itself in the cistern and thus help to cause dissipation of surplus energy. However, to be effective for this purpose the baffle walls should be sufficiently high.

8.17 (c)

A launching apron is an apron of loosely packed stones. The function of a launching apron is to protect the impervious floor and the pile from the scour holes progressing towards the floor and the pile. The protection is provided by a launching apron by forming a protective covering of stones over a certain slope below the bed of the river at which the apron is originally laid to the bottom of the deepest scour likely to occur. When scour occurs, the position attained by this apron is called launched position and the apron is the known as launched apron.

8.18 (a)

As per the data, the local scour depth below the surrounding bed is $2W$. This data needs to be memorized.

8.19 (a)

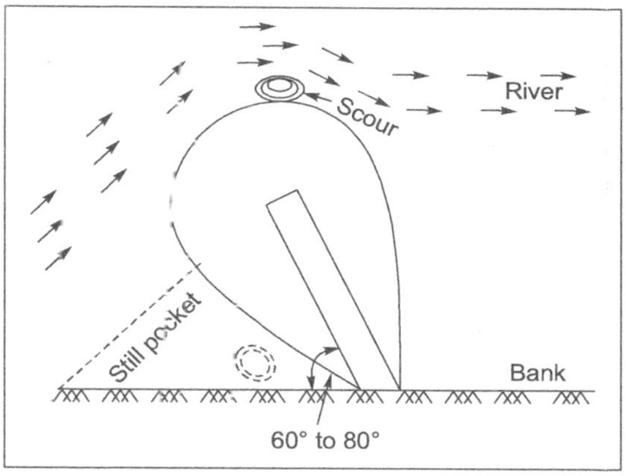

Repelling Groyne

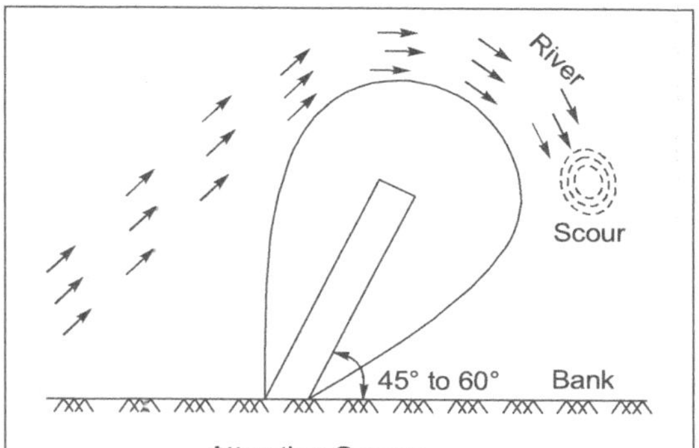

Attracting Groyne

8.20 (*)

$$\text{Looseness factor} = \frac{\text{Actual width}}{\text{Regime width}}$$

$$\text{Actual width} = 360 \text{ m (given)}$$

$$\text{Regime width} = 4.75\sqrt{Q}$$

Given, Q = discharge for which barrage has been designed

$$\text{Regime width} = 4.75\sqrt{7000} = 397.41$$

$$\text{Looseness factor} = \frac{360}{397.41} = 0.9$$

8.21 (c)

Tortuosity means the sharpness of any curve. For a meandering river it is defined as

$$\text{Tortuosity} = \frac{\text{Curved length along the river}}{\text{Direct axial length of the river}}$$

8.22 (b)

River training's purpose is to stabilise the river course and to reduce bank erosion.

8.23 (c)

Groynes are the structures constructed transverse to the river flow and extend from the bank into river upto a limit

They perform one or more of following functions depending on their type:

1. Training the river along the desired course to reduce the concentration of flow at a point of attack.
2. Creating a low flow for silting up in the vicinity.
3. Protecting the bank by keeping the flow away from it.

In general, all the types of groynes train the flow along a certain course.

8.24 (c)

Objectives of river training are high water training: So that high flood discharge may pass safely.

Mean water training : To have efficient disposal of sediments.

Low water training : To provide sufficient depth of navigation during low weather season.

To reduce bank erosion and make the river course stable.

8.25 (c)

River training, in its wider aspects, covers all those engineering works which are constructed on a river, so as to guide and confine the flow to the river channel, and to control and regulate the river bed configuration, thus ensuring safe and effective disposal of floods and sediment load.

The chief aim of river training is to achieve ultimate stability of the river with the aid of river-training measures.

8.27 (a)

Canal Breach or Canal Bank Breach can happen due to erosion of soil from the sides. It is mostly observed in unlined canals.

8.28 (c)

In rocky stage, river mostly contains rocks or very large boulders.

In incised river stage, river flows sufficiently below the natural ground surface as opposed to the one flowing in flood plains, both of which carry mostly sand and shingle.

Boulder river stage is characterised by steep slopes and their beds consists of a mixture of boulders, gravel, shingle and sand.

8.30 (a)

Permeable groynes do permit restricted flow through them. The groynes are more or less temporary structures & are susceptible to damage by floating debris, etc. The common materials used as permeable groynes, are:

1. Trees used as groynes, called **Tree spur.**
2. Timber stakes or wooden piles called **Balli spurs**.
3. Stone filled in **balli crates**.
4. Stone filled in **wire crates.**

8.31 (b)

The distance between outer banks of two adjacent clockwise and anticlockwise meanders is known as meander belt width.

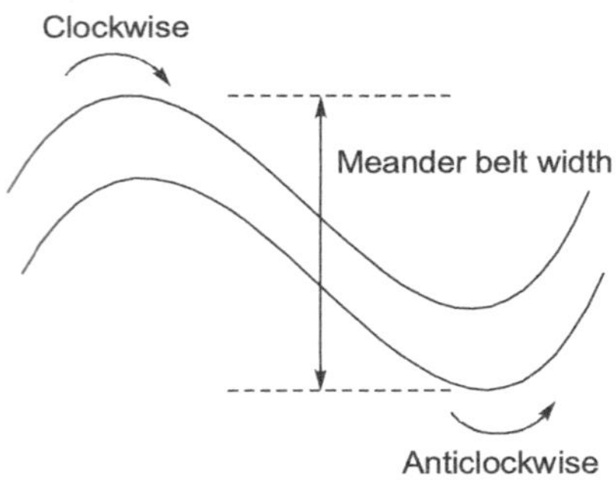

■■■■

9 Dams (general), Spillways & Energy Dissipators

9.1 Match **List-I** (Energy dissipation) with **List-II** (Water level and slope condition) and select the correct answer using codes given below the lists:

List-I

A. Roller bucket
B. Ski-jump bucket
C. Standing wave basin with depressed floor
D. Standing wave basin with raised floor

List-II

1. TWL (tail channel water level) is slightly above JWL (jump height water level) and the slope of the channel is mild
2. TWL is considerably above JWL and the slope of the channel is mild
3. TWL is slightly below JWL and the slope of the channel is mild
4. TWL is considerably less than JWL and the slope of the channel is steep

Codes:

	A	B	C	D
(a)	4	2	1	3
(b)	2	4	1	3
(c)	2	4	3	1
(d)	4	2	3	1

[ESE : 2000]

9.2 Which one of the following sets is used to control the seepage through the foundations of an earth dam?

(a) Chimney drain, upstream blanket and cut-off trench
(b) Cut-off sheet piles, upstream blanket and cut-off trench
(c) Upstream blanket, cut-off sheet piles and chimney drain
(d) Relief wells, upstream blanket and chimney drain

[ESE : 2000]

9.3 The life of a reservoir is determined by its capacity (C), volume of annual inflow into the reservoir (I) and concentration of sediment in the incoming flow (C_s). Life will be more if

(a) C, I and C_s are high
(b) C and I are high but C_s is low
(c) C is high but I and C_s are low
(d) C, I and C_s are low [ESE : 2001]

9.4 An effective storage of a flood control reservoir is

(a) the amount of water which can be supplied from it in a particular interval of time
(b) the storage between the minimum and maximum reservoir levels under ordinary operating conditions
(c) the useful storage plus the surcharge storage less the valley storage
(d) the storage volume of flood water above maximum reservoir level

[ESE : 2002]

9.5 A constant angle arch dam when compared to a constant radius arch dam utilizes concrete quantity of about

(a) 33% (b) 43%
(c) 73% (d) 143% [ESE : 2002]

9.6 A check dam is a

(a) flood control structure
(b) soil conservation structure
(c) river training structure
(d) water storage structure [ESE : 2003]

9.7 The ideal condition for energy dissipation in the design of spillways is the one when the tail water rating curve

(a) lies above jump rating curve at all discharges
(b) coincides with the jump rating curve at all discharges
(c) lies below jump rating curve at all discharges
(d) lies either above or below the jump rating curve depending upon discharge

[ESE : 2003]

9.8 Consider the following statements:

The function of a cut-off in an earth dam is

1. to reduce uplift pressures on the dam
2. to prevent undermining of foundation
3. to reduce loss of stored water
4. to support the dam

Which of these statements are correct?

(a) 1 and 3 (b) 2 and 4
(c) 2 and 3 (d) 3 and 4 **[ESE : 2003]**

9.9 Which one of the following equations represents the downstream curve of the "Ogee" spillway (where x and y are the co-ordinates of the crest profile measured from the apex of the crest, and H is the design head)?

(a) $x^{1.85} = 2H^{0.85}y$ (b) $x = 2H^{1.85}y^{0.85}$
(c) $x^{0.85} = 2H^{1.85}y$ (d) $x = 2H^{0.85}y^{1.85}$

[ESE : 2004]

9.10 For a saddle siphon, the maximum operative head is 6.25 m. The width and height of the throat of the siphon are 4 m and 2 m respectively. The coefficient of discharge is 0.90. How many units are required to pass a flood of 300 cumec? (Take g = 10 m/s^2)

(a) One (b) Two
(c) Three (d) Four **[ESE : 2004]**

9.11 Match **List-I** (Factor) with **List-II** (Ratio) and select the correct answer using the codes given below the lists:

List-I

A. Load factor
B. Capacity factor
C. Diversity factor
D. Plant use factor

List-II

1. $\dfrac{\text{Sum of individual maximum demands}}{\text{Simultaneous maximum demands}}$

2. $\dfrac{\text{Maximum demands}}{\text{Station capacity}}$

3. $\dfrac{\text{Average load}}{\text{Maximum load}}$

4. $\dfrac{\text{Energy produced}}{\text{Installed capacity} \times \text{time in hours}}$

Codes:

	A	B	C	D
(a)	1	2	3	4
(b)	3	4	1	2
(c)	1	4	3	2
(d)	3	2	1	4

[ESE : 2005]

9.12 Match **List-I** with **List-II** and select the correct answer using the codes given below the lists:

List-I

A. Storage
B. Pondage
C. Pondage factor
D. Daily load factor

List-II

1. Small storage
2. Ratio of total inflow to the total number of days power plant operates in a week
3. Large reservoirs
4. Average plant load × peak load
5. Ratio of average load to peak load

Codes:

	A	B	C	D
(a)	3	2	1	5
(b)	5	1	2	4
(c)	3	1	2	5
(d)	5	2	1	4

[ESE : 2005]

9.13 **Assertion (A):** The USBR type-II stilling basin length requirement is less than that in type-III basin for similar design conditions.

Reason (R): Energy dissipation is primarily accomplished by hydraulic jump in USBR type-II stilling basin.

(a) both A and R are true and R is the correct explanation of A
(b) both A and R are true but R is not a correct explanation of A
(c) A is true but R is false
(d) A is false but R is true **[ESE : 2006]**

9.14 During a certain week a power plant turns out 8,400,000 kW-hr and the peak load during the week is 100,000 kW. What is the load factor during the week?

(a) 40% (b) 45%
(c) 50% (d) 60% **[ESE : 2007]**

9.15 For calculating the maximum flood discharge in an alluvial stream, which is the best suited relation?

(a) $V \propto R^{2/3} S^{1/3}$ (b) $V \propto R^{2/3} S^{1/2}$
(c) $V \propto R^{1/2} S^{1/2}$ (d) $V \propto D^{0.64}$

[ESE : 2008]

9.16 Which of the following categories best describes the Hirakud reservoir?
(a) Reservoir for irrigation and power
(b) Reservoir for flood control, power and irrigation
(c) Reservoir for irrigation and water supply
(d) Reservoir for recreation and fishery
[ESE : 2009]

9.17 A ski-jump bucket is generally used as an energy dissipator when the tail water
(a) is greater than 1.1 times the required conjugate depth for the formation of hydraulic jump; and the river bed rock is 'good'
(b) depth is lesser than the depth required for the jump formation; and the bed of the river channel is composed of 'sound' rock
(c) depth is equal to the depth required for the jump formation, and the river bed rock is 'good'
(d) depth is 1.3 times the required for the jump formation and the river bed is composed of 'weak' rock
[ESE : 2010]

9.18 **Assertion (A):** A seepage passing through the body of an earth dam affects the weight of dam.
Reason (R): The specific weight of submerged soil is not dependent on the porosity of soil.
(a) both A and R are true and R is the correct explanation of A
(b) both A and R are true but R is not a correct explanation of A
(c) A is true but R is false
(d) A is false but R is true [ESE : 2010]

9.19 A constant centre (in plan-view) arch dam is best suited for
(a) A V-shaped gorge
(b) A U-shaped gorge
(c) Both U-shaped and V-shaped gorges
(d) Multi-peaked gorges
[ESE : 2011]

9.20 A discharge of 72 m^3/s is to be allowed through siphon spillways of 2 m width and 75 cm depth with working head of 8 m. The number of spillways to be provided will be (Take coefficient of discharge for the spillways = 0.64)
(a) 2 (b) 4
(c) 6 (d) 8
[ESE : 2012]

9.21 The flip bucket energy dissipator for a spillway is suitable where:
1. The tail water depth is low.
2. The rock on the downstream is fragile and is erodible.
3. The rock on the downstream is good and non-erodib e.
(a) 1, 2 and 3 (b) 1 and 2 only
(c) 2 and 3 only (d) 1 and 3 only
[ESE : 2013]

9.22 In a ski-jump bucket provided in an overflow spillway, the lip angle is 30°, and the actual velocity of flow entering the bucket is 30 m/s. The maximum vertical height attained by the trajectory of the jet, measured above the lip of the bucket, is nearly
(a) 45 m (b) 35 m
(c) 22 m (d) 11 m [ESE : 2015]

9.23 A new reservoir has a capacity of 12 Mm^3 and its catchment area is 400 km^2. The annual sediment yield from this catchment is 0.1 ha.m/km^2 and the trap efficiency can be assumed constant at 90%. The number of years it takes for the reservoir to lose 50% of its initial capacity is, nearly
(a) 177 years (b) 77 years
(c) 17 years (d) 7 years
[ESE : 2016]

9.24 **Statement (I):** The ogee spillway is a control weir having an S-shaped crest profile which provides a high discharge coefficient without causing cavitations.

Statement (II): The crest profile of ogee spillway conforms to the lower nappe of flow over a ventilated sharp-crested weir and ensures a constant discharge coefficient for all heads.
(a) Both Statement (I) and Statement (II) are individually true and Statement (II) is the correct explanation of Statement (I)
(b) Both Statement (I) and Statement (II) are individually true but Statement (II) is ***not*** the correct explanation of Statement (I)
(c) Statement (I) is true but Statement (II) is false
(d) Statement (I) is false but Statement (II) is true
[ESE : 2016]

9.25 Statement (I) : The trap efficiency of a reservoir increases with age as the reservoir capacity is reduced by sediment accumulation.

Statement (II): The trap efficiency is a function of the ratio of reservoir capacity to the total inflow. A small reservoir on a large stream has a low trap efficiency.

(a) Both Statement (I) and Statement (II) are individually true and Statement (II) is the correct explanation of Statement (I)
(b) Both Statement (I) and Statement (II) are individually true but Statement (II) is ***not*** the correct explanation of Statement (I)
(c) Statement (I) is true but Statement (II) is false
(d) Statement (I) is false but Statement (II) is true

[ESE : 2016]

9.26 Consider the following statements concerning engineering geology:

1. Dams on sedimentary impervious strata dipping upstream will be unsafe.
2. In general, igneous rocks are hard and are suitable for construction of large hydraulic structures.
3. Construction of a major dam must be preceded by thorough geotechnical investigations.
4. Construction of dams is never indicated in any active zone of seismicity.

Which of the above statements are correct?

(a) 1 and 2 only (b) 1 and 4 only
(c) 2 and 3 only (d) 3 and 4 only

[ESE : 2017]

9.27 Statement (I) : A sloping glacis is always preferred over a horizontal bed for locating a hydraulic jump.

Statement (II) : The hydraulic jump is the best dissipator of energy of the flowing water.

(a) Both Statement (I) and Statement (II) are individually true and Statement (II) is the correct explanation of Statement (I)
(b) Both Statement (I) and Statement (II) are individually true but Statement (II) is NOT the correct explanation of Statement (I)
(c) Statement (I) is true but Statement (II) is false
(d) Statement (I) is false but Statement (II) is true

[ESE : 2018]

9.28 The volume of water below the minimum pool level in a reservoir is known as

(a) Useful storage (b) Surcharge storage
(c) Dead storage (d) Bank storage

[ESE : 2019]

9.29 A reservoir with uncontrolled and ungated outlets is known as

(a) Retarding basin
(b) Storage reservoir
(c) Controlled reservoir
(d) Detention basin [ESE : 2022]

9.30 In which one of the following types of falls, the energy dissipation on a straight glacis remain incomplete due to vertical component of velocity remaining unaffected?

(a) Montage type falls (b) Gravity falls
(c) Inglis falls (d) Baffle falls

[ESE : 2023]

9.31 What should be done to dissipate the residual energy, if the tail water deficiency is small?

(a) Construction of baffle wall
(b) Sloping apron
(c) Stilling basin
(d) Ski jump bucket

[ESE : 2023]

■■■■

Answers Dams (general), Spillways & Energy Dissipators

9.1 (c) 9.2 (b) 9.3 (d) 9.4 (c) 9.5 (b) 9.6 (b) 9.7 (b) 9.8 (c) 9.9 (a)
9.10 (d) 9.11 (b) 9.12 (c) 9.13 (c) 9.14 (c) 9.15 (b) 9.16 (b) 9.17 (b) 9.18 (c)
9.19 (b) 9.20 (c) 9.21 (d) 9.22 (d) 9.23 (c) 9.24 (c) 9.25 (d) 9.26 (c) 9.27 (b)
9.28 (c) 9.29 (a) 9.30 (a) 9.31 (a)

Explanations Dams (general), Spillways & Energy Dissipators

9.1 (c)

When TWC is slightly above JWC and slope of channel is mild, standing wave basin with raised floor is constructed to match them.

When TWC is considerably above JWL and the slope of channel is mild, roller bucket is chosen for desired energy dissipation.

When TWC is slightly below JWL and slope of channel is mild, standing wave basin with depressed floor is preferred to level them.

When TWC is considerably below JWL and slope of channel is steep, ski jump bucket is used to reduce scourging potential energy of water.

9.2 (b)

For seepage control through embankments/dams following drainage filters can be provided:

(i) Rock toe filter
(ii) Horizontal blanket or filter on downstream
(iii) Chimney drain

For seepage control through foundations the following can be provided :

(i) Impervious cut-off sheet piles
(ii) Relief wells or drain trenches or cut-off trench
(iii) Upstream blanket

9.3 (d)

In determining the life of a reservoir the following factors are considered and investigated :

- Water tightness of the basins
- Stability of the reservoir rim
- Availability of construction material in the reservoir area
- Silting
- Direct and indirect submergence of economic mineral wealth
- Seismo-tectonics

Trap efficiency (TE) is one of the most important factors affecting life of a reservoir. TE of reservoir, over a period, is the ratio of total deposited sediment to the total sediment inflow.

9.4 (c)

Effective storage = Useful storage + Surcharge storage – Valley storage corresponding to the rate of inflow in the reservoir.

9.5 (b)

Arch dams can be divided into four types:

(i) Constant radius arch dams
(ii) Variable radius arch dams
(iii) Constant angle arch dams
(iv) Double curvature arch dams

Constant radius arch dam is simplest in design and construction, but uses the maximum concrete. A constant angle arch dam uses about 43% of concrete used by constant radius arch dam. A variable radius arch dam is an intermediate choice, using around 58% of the concrete used by constant radius arch dam.

9.6 (b)

A check dam is a small dam (weir) constructed across to countered erosion by reducing flow velocity of water thereby conserving soil.

9.7 (b)

When tail water curve coincides with jump rating curve no extra construction cost is required to dissipate energy or to protect the basin as it is ideal condition for energy dissipation in design of spillways. When the tail water-rating curve coincides with jump curve for all values of discharge then the requirement for the sequent, depths are always satisfied and the jump forms at the same location.

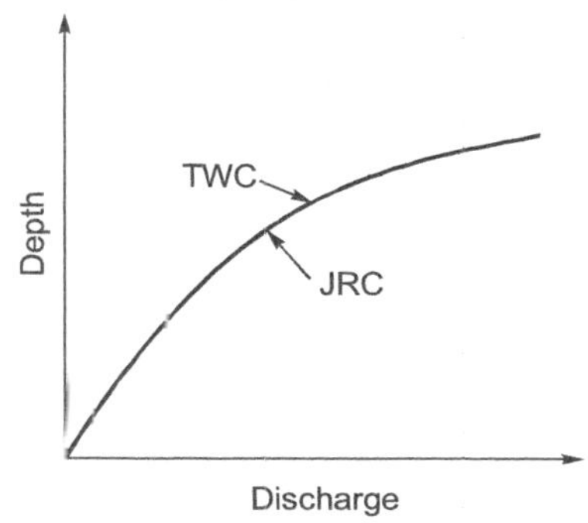

9.8 (c)

Impervious vertical cutoff made of concrete or sheet pile can be provided at the upstream end of the earthen dam. It can control seepage. The downstream cut-offs may reduce exit-gradient and thereby prevent undermining of foundation.

9.9 (a)

The downstream profile has following general equation.

$x^n = KH^{n-1}y$

Slope of upstream face of spillway	K	n
(i) Vertical	2.0	1.85
(ii) 1H : 3V	1.936	1.836
(iii) 1H : 1.5V	1.939	1.810

9.10(d)

The discharge through one unit,

$$Q = C_d A\sqrt{2gh} = 0.90\times(4\times2)\times\sqrt{2\times10\times6.25}$$
$$= 80.5 \text{ m}^3/\text{s}$$

Thus number of units needed $= \dfrac{300}{80.5} = 4$

9.13(c)

USBR Stilling Basins

(i) **Type-I** (with froude number (F_1) between 2.5 to 4.5) has length of stilling basin from $4.3y_2$ to $6y_2$, where y_2 is post jump depth.

(ii) **Type-II** ($F_1 > 4.5$ and $V_1 < 15$ m/s). It is provided with chute blocks, baffle block and end sills. The length of stilling basin is considerably less than Type-I, because the dissipation of energy occurs by the impinging action of the incoming flow against the baffle blocks. However the baffle blocks are subjected to large impact force. They should be designed for extra force due to dynamic action against the upstream face

$F = Z\gamma_w A(y_1 + h_v)$, where $(y_1 + h_v)$ is the specific energy of the flow entering the basin.

(iii) **Type-III** ($F_1 > 4.5$, and $V_1 > 15$ m/s)

In this case baffle blocks are not provided, because

(i) the blocks will be subjected to very high impact forces

(ii) there is possibility of cavitation on the downstream faces of the blocks.

The stilling basin therefore consists of chute blocks and a denoted sill. As the dissipation of energy occurs mainly due to hydraulic jump, the length of Type-III basin is greater than that in Type-II basin.

9.14(c)

Load factor is defined as the ratio of the average load during a certain period to the maximum or peak load during that period. The load factor is thus related to a certain period and therefore, there will be daily load factors, weakly load factors, load for a given month and yearly load factors.

$$\therefore \text{ Load factor} = \frac{8{,}400{,}000}{24\times7\times100{,}000}\times100 = 50\%$$

9.15 (b)

$V = 0.55\text{ m } D^{0.64}$ (Kennedy's)

$V = (1/n)\,R^{2/3}\,S^{1/2}$ (Manning's)

$V = 10.8\,R^{2/3}\,S^{1/3}$ (Lacey's)

Out of above equations, Manning's equation is popularly used for computing actual velocity in OCF.

9.16(b)

Hirakud Dam is built across the Mahanadi River in the state of Odisha in India. Behind the dam extends a lake, Hirakud Reservoir. It is one of the first major multipurpose river valley projects started after India's independence.

9.17(b)

The ski jump bucket type of an energy dissipator can be used for energy dissipation when the tail water depth is insufficient or low at all discharges. The ski jump bucket type of an energy dissipator requires sound and rocky river bed, because a part of the energy dissipation takes place by impact, although some of the energy is dissipated in air by diffusion and aeration.

9.18 (c)

- When seepage passes through the body of an earthen dam, it affects the weight of the dam.
  Effective weight = Actual weight + Seepage Force.
- Specific weight of submerged soil is given by

$$= \gamma_{sub}\times\text{Volume}$$
$$= \left(\frac{G-1}{1+e}\right)\gamma_w\times\text{Volume}$$

As e depends on porosity, hence specific weight of submerged soil also depends on porosity.

9.19 (b)

Constant Center Arch Dam (variable angle) have the following features-

- Good for U-shaped valleys
- Easy construction
- Vertical upstream face
- Appropriate for middle-high dams

9.20(c)

Discharge through one syphon spillway is given by

$Q = C_d . A\times\sqrt{2gH}$

$$= 0.64 \times 2 \times 0.75 \times \sqrt{2 \times 9.81 \times 8} = 12.02 \text{ m}^3/\text{s}$$

$$\therefore \text{ Number of spillways} = \frac{72}{12.02} = 5.98 \cong 6$$

9.21 (d)

The flip bucket, also called ski-jump dissipator, is provided when depth is lesser than the depth required for the jump formation and river bed rock is good.

9.22 (d)

Vertical component of velocity = $v\sin\theta = v\sin30°$

vertical velocity at top is zero, so

$v\sin30° - gt = 0$

$$t = \frac{v\sin30°}{g}$$

$$h = v\sin30° t - \frac{1}{2}gt^2$$

$$h = \frac{1}{2g}v^2\sin^2 30°$$

$$= \frac{30^2 \times \left(\frac{1}{2}\right)^2}{2 \times 9.81} = 11.47 \text{ m}$$

9.23 (c)

Trap efficiency means 90% of sediment will be stored.

Let number of years are x to till reservoir to 50%

$$400 \times 0.1 \times 10^4 \times 0.90 \times x$$
$$= 12 \times 10^6 \times 0.5$$
$$x = 16.67 = 17 \text{ years}$$

9.24 (c)

If the discharge is higher than the designed discharge then negative pressure will develop on the underside of the lower nappe. Hence cavitation will occur.

To get rid of this problem, the ventilated pipe is provided at possible point of cavitation. This will avoid the spawling of concrete due to negative pressure.

9.25 (d)

The sediment trap efficiency is the difference in the sediment yield of the reservoir inflow minus the yield of the outflow, divided by the yield of the inflow. The greater the trap efficiency, the greater the sedimentation.

The trap efficiency of reservoirs typically decreases with increasing age, as the capacity of the reservoir decreases. The trap efficiency is impacted by the type of sediments, the hydraulic characteristics of the reservoir, and how the reservoir is operated.

9.26 (c)

If the faults occur in the upstream side, the downstream dipping faults are dangerous because of the risk of uplift pressure and leakage. But if the faults dip in the upstream side are sealed properly avoiding the leak, there could be benefit of dam stability.

9.27 (b)

The main advantage of a sloping glacis weir is that hydraulic jump is formed on the downstream sloping glacis to dissipate the energy of overflowing water. Statement is explaning the purpose of hydraulic jump.

9.28 (c)

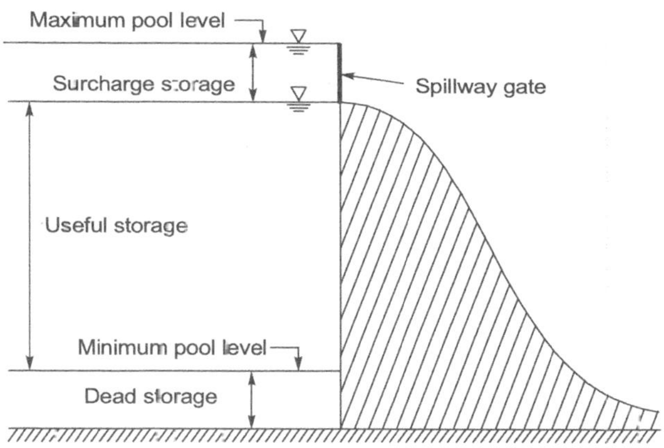

9.29 (a)

A reservoir having gates and valves installed at the spillway and at its sluice outlets is known as a storage reservoir or detention reservoir on the other hand, a reservoir with uncontrolled and ungates outlets is known as retarding reservoir.

9.31 (a)

If the tailwater deficiency is small, baffle wall can be provided to increase its depth. subsidiary dams or rows of friction blocks can also be provided.

■■■■

UNIT

IV

Environmental Engineering

Syllabus

Water Supply Engineering: Sources, Estimation, quality standards and testing of water and their treatment; Rural, Institutional and industrial water supply; Physical, chemical and biological characteristics and sources of water, Pollutants in water and its effects, Estimation of water demand; Drinking water Standards, Water Treatment Plants, Water distribution networks.

Waste Water Engineering: Planning & design of domestic waste water, sewage collection and disposal; Plumbing Systems. Components and layout of sewerage system; Planning & design of Domestic Waste-water disposal system; Sludge management including treatment, disposal and re-use of treated effluents; Industrial waste waters and Effluent Treatment Plants including institutional and industrial sewage management.

Solid Waste Management: Sources & classification of solid wastes along with planning & design of its management system; Disposal system, Beneficial aspects of wastes and Utilization by Civil Engineers.

Air, Noise pollution and Ecology: Concepts & general methodology.

Contents

■■■■

1 Water Demand

1.1 **Assertion (A):** The future population is predicted on the basis of knowledge of the city and its environment.

Reason (R): The future population depends on the trade and expansion of the city, discovery of mineral deposits, power generation etc.

(a) both A and R are true and R is the correct explanation of A
(b) both A and R are true but R is not a correct explanation of A
(c) A is true but R is false
(d) A is false but R is true [ESE : 2000]

1.2 The population figures in a growing town are as follows:

Year	Population
1970	40,000
1980	46,000
1990	53,000
2000	58,000

The predicted population in 2010 by Arithmetic Regression method is

(a) 62,000 (b) 63,000
(c) 64,000 (d) 65,000 [ESE : 2002]

1.3 On which of the following factors, does the population growth in a town normally depend?

1. Birth and death rates 2. Migrations
3. Probabilistic growth 4. Logistic growth

Select the correct statements:

(a) 1 and 4 (b) 1 and 2
(c) 1, 2 and 3 (d) 2 and 3

[ESE : 2003]

1.4 Which one of the following Acts/Rules has a provision for "No right to appeal"?

(a) Environment (Protection) Act, 1986
(b) The Hazardous Waste (Management and Handling) Rules, 1989
(c) Manufacture, Storage and Import of Hazardous Chemicals Rules, 1989
(d) Environment (Protection) Rules, 1992

[ESE : 2004]

1.5 The present population of a community is 28000 with an average water demand of 150 Lpcd. The existing water treatment plant has a design capacity of 6000 m^3/d. It is expected that the population will increase to 48000 during the next 20 years. What is the number of years from now when the plant will reach its design capacity assuming an arithmetic rate of population growth?

(a) 8.6 years (b) 12.0 years
(c) 15.0 years (d) 16.5 years

[ESE : 2006]

1.6 For water supply to a medium town, what is the daily variation factor?

(a) 1.5 (b) 2.5
(c) 3 (d) 3.5 [ESE : 2008]

1.7 Which one of the following factors has the maximum effect on figure of per capita demand of water supply of a given town?

(a) Method of charging of the consumption
(b) Quality of water
(c) System of supply-intermittent or continuous
(d) Industrial demand

[ESE : 2008]

1.8 Population levels over 5 decades of a small town are given below.

Year	1950	1970	1980	1990	2000
Population	2,50,00	2,80,000	3,40,000	4,20,000	4,90,000

The population of the town in the year 2020 estimated by Arithmetic Increase method will be

(a) 5,10,000 (b) 5,90,000
(c) 6,10,000 (d) 6,90,000

[ESE : 2014]

1.9 The population of a city in the year 2000 was 82,300. If average percent increase in population per decade in 35%, the population of the city in the year 2020 estimated by geometrical increase method will nearly be

(a) 100000 (b) 125000
(c) 150000 (d) 175000 [ESE : 2015]

1.10 A water supply distribution system for an averagely-populated township to be designed for
(a) Maximum daily demand
(b) Maximum hourly demand and fire demand
(c) Average demand
(d) Maximum daily demand and fire demand, or maximum hourly demand, whichever is higher
[ESE : 2016]

1.11 In which one of the following industries, the water requirement in kilo litres per unit of production is very high?
(a) Paper industry (b) Steel industry
(c) Sugar industry (d) Fertilizer industry
[ESE : 2020]

1.12 **Statement (I):** Both the Empirical formulae given by American Insurance Association and Buston for the determination of fire demand of water are not suitable for Indian conditions.
Statement (II): Kuchling's formula estimates lesser value fire water demand.
(a) Both Statement (I) and Statement (II) are individually true and Statement (II) is the correct explanation of Statement (I)
(b) Both Statement (I) and Statement (II) are individually true but Statement (II) is NOT the correct explanation of Statement (I)
(c) Statement (I) is true but Statement (II) is false
(d) Statement (I) is false but Statement (II) is true
[ESE : 2021]

1.13 Which one of the following is NOT a factor affecting losses and wastes in water supply scheme?
(a) Metering
(b) Unauthorised connections
(c) Water demand
(d) Pressure in the distribution system
[ESE : 2022]

1.14 Which one of the following forecasting methods for population is also known as uniform increase method?
(a) Arithmetic increase method
(b) Decreasing rate method
(c) Geometric increase method
(d) Simple geographical method [ESE : 2022]

1.15 In a big city having population of 50 lakhs, if 6 moderate fires each of 3 hours break out in a day, what is the approximate amount of water required per person per day? (Assume 3 jet streams simultaneously throwing water from a hydrant with discharge of 1100 litres/minute/stream)
(a) less than 1 litre (b) between 1-3 litre
(c) between 3-5 litres (d) 20 litres
[ESE : 2022]

1.16 The v arious types of water demand, which a city may have, may be broken down into which of the following classes?
1. Domestic water demand
2. Industrial water demand
3. Demand for public uses
Select the correct statements:
(a) 1 and 2 only (b) 2 and 3 only
(c) 1 and 3 only (d) 1, 2 and 3
[ESE : 2022]

1.17 For water supply scheme design of a town or a city, the suitable method of estimating future population by the end of the design period is
(a) increasing rate method
(b) decreasing rate method
(c) exponential curve method
(d) incremental decrease method
[ESE : 2022]

1.18 Match the following lists:
List I (Type of commercial building/institution)
P. Restaurants
Q. Nurses homes and medical quarters
R. Cinemas, concert halls and theatre
S. Factories
List II (Water demand in Litres per day)
1. 45 per capita
2. 70 per seat
3. 135 per capita.
4. 15 per seat

Select the correct answer using the code given below:

	P	Q	R	S
(a)	4	1	2	3
(b)	2	1	4	3
(c)	4	3	2	1
(d)	2	3	4	1

[ESE : 2024]

1.19 If P is population of town/city, Q is total quantity of water required during one year for a town/city; what is the formula for maximum daily per capita demand for a town/city?

(a) $1.8 \times \frac{Q}{P \times 365}$ (b) $1.8 \times \frac{Q}{P \times 24}$

(c) $1.5 \times \frac{Q}{P \times 365}$ (d) $1.5 \times \frac{Q}{P \times 24}$

[ESE : 2024]

1.20 If P_n = Population at the end of n decade, P_o = Present population, P_{av} = Average arithmetic increase in the population (decadal), P_I = Average incremental increase in population, n = Number of decades; what is the mathematical representation of Incremental Increase Method?

(a) $P_n = P_I \times n + (P_{av} + P_o)$

(b) $P_n = P_I + (P_{av} + P_o) \times n$

(c) $P_n = P_o + (P_{av} + P_I) \times n$

(d) $P_n = P_o \times n + (P_{av} + P_I)$

[ESE : 2024]

1.21 If Q is the water requirement in L/min and P is population in thousands, what is Buston's formula to determine fire demand?

(a) $Q = 5663\sqrt{P}$

(b) $Q = 3182\sqrt{P}$

(c) $Q = 4637\left(1 - 0.01\sqrt{P}\right)$

(d) $Q = 100\sqrt{P}$

[ESE : 2024]

■■■■

Answers Water Demand

1.1 (a) 1.2 (c) 1.3 (b) 1.4 (a) 1.5 (b) 1.6 (b) 1.7 (d) 1.8 (c) 1.9 (c)

1.10 (d) 1.11 (a) 1.12 (b) 1.13 (c) 1.14 (a) 1.15 (a) 1.16 (d) 1.17 (b) 1.18 (d)

1.19 (a) 1.20 (c) 1.21 (a)

Explanations Water Demand

1.1 (a)

The trade and expansion of a city, discovery of mineral deposits, power generation etc. lead to increase in number of opportunities thus resulting in migration to that city and increase in future population. Thus the future population of a city is predicted on the basis of knowledge of the city and environment.

1.2 (c)

Year	Population	Increase in population (x)
1970	40000	
1980	46000	6000
1990	53000	7000
2000	58000	5000
Total		18000

∴ Average increase per decade

$$\bar{x} = \frac{18000}{3} = 6000$$

Ther future populations are now computed by using

$$P_n = P_0 + n\bar{x}$$

Thus, opulation after 1 decade bdyond 2000 i.e.,

$$P_{2010} = P_{2000} + 1 + (6000)$$
$$= 58000 + 6000 = 64000$$

1.3 (b)

Birth Rate: The number of live births per thousand of population per year.

Death Rate: The ratio of deaths to the population of a particular area or during a particular period of time, usually calculated as the number of deaths per one thousand people per year.

Migration: Migration also affects the no. of deaths and births in an area. Migrations take into account emigration from and immigration to the city.

1.4 (a)

Section-22 of Environment (Protection) Act, 1986 has a provision for 'No right to appeal'.

1.5 (b)

Method-1

Population after n years,

$$P_n = P_0 + n\bar{x}$$

$\bar{x}$ is arithmetic design growth rate

Given P_0 = 28000

For n = 20; P_{20} = 48000

$$\therefore \quad \bar{x} = \frac{48000 - 28000}{20} = 1000 \text{ per year}$$

The population when design capacity will be required,

$$P_n = \frac{6000 \times 1000}{150} = 40{,}000$$

∴ Number of years to reach the plant at design capacity,

$$n = \frac{P_n - P_0}{\bar{x}} = \frac{40000 - 28000}{1000}$$

$$= 12.0 \text{ years}$$

Note: *Since design capacity of treatment plant is done on the basis of 2 times average daily demand not maximum daily demand, but in the given question solution is done on the basis of simply average daily demand otherwise answer will not match from any of the given option (data will be improper).*

Method-2

Given present population,

$$P_0 = 28000$$

AVerage water demand,

$$q = 150 \text{ lpcd}$$

Design capacityof existing water treatment plant

$$= 6000 \text{ m}^3/\text{day}$$

Total demand at present

$$= 28000 \times 150 \times 10^{-3}$$

$$= 4200 \text{ m}^3/\text{day}$$

After (20 years i.e., 2 decades) population (P_2) increased to 48000

∴ By Arithmetic increase method

$$P_2 = P_0 + 2 \times \bar{x}$$

$$48000 = 28000 + 2 \times \bar{x}$$

$$\bar{x} = 10{,}000$$

∴ Average increase in population per decade = 10,000

Also design capacity Population

4200 m^3/day 28000

6000 m^3/day $\frac{28000}{4200} \times 6000 = 40000$

$$\therefore \quad P_n = P_0 + n\bar{x}$$

$$48000 = 28000 + n \times (10000)$$

$$n = 1.2 \text{ decades}$$

$$\approx 12 \text{ years}$$

∴ After 12 years plant will reach its design capacity.

1.6 (b)

The GOI manual on water supply has recommended the following values of peak factor depending upon the population.

S.No.	Population	Peak factors
1.	Upto 50,000 (small town)	3.0
2	50,001 - 2,00,000 (medium town)	2.5
3.	Above 2 lakh (Large town)	2.0
4.	For Rural water supply scheme, (where supply is effected through standpost for only 6 hr)	3.0

(Based on old GOI manual data as it is an old question)

1.7 (d)

The factors affecting per capita demand are:

(i) Size of the city
(ii) Climatic conditions
(iii) Types of gentry and habits of people
(iv) Industrial and commercial activities
(v) Quality of water supplies
(vi) Pressure in the distribution system
(vii) Development of sewerage facilities
(viii) System of supply
(ix) Cost of water
(x) Policy of metering and method of charging

Among the above factors industrial and commercial activities have the maximum effect on per capita demand. The pressure of industrial and commercial activities at a particular place increases the water consumption by large amounts. Many industries require really huge amounts of water and as such increase the water demand considerably.

1.8 (c)

Year	Population	Increase in population (x)
1960	250000	
		30,000
1970	280000	
		60,000
1980	340000	
		80,000
1990	420000	
		70,000
2000	490000	
Total		240,000

Average increase per decade

$$(\bar{x}) = \frac{240000}{4} = 60{,}000$$

The future population are now computd by using

$$P_n = P_0 + n\bar{x}$$

Thus population after 2-decades beyond 2000

i.e., $P_{2020} = P_{2000} + 2 \times \bar{x}$
$= 490000 + 2 \times (60,000)$
$= 6,10,000$

1.9 (c)

Estimation of population by geometrical increase method

$$P_n = P_0\left(1+\frac{r}{100}\right)^n$$

$$P_{2020} = P_{2000}\left(1+\frac{35}{100}\right)^2$$

Where, $P_0 = 82300$, $n = 2$, $r = 35\%$

$$= 82300\left(1+\frac{35}{100}\right)^2$$

$$= 149991.75 \simeq 1,50,000$$

1.10 (d)

Distribution system is to be designed for greater of maximum hourly demand or maximum daily demand and fire demand.

Units	Design parameters
Water treatment units	Maximum daily demand
Main supply pipes (water mains)	Maximum daily demand
Wells and Tubewells	Maximum daily demand
Demand Reservoir	Average annual demand
Distribution System	Maximum hourly demand } Max. coincident + draft

Note: Coincident draft = Max. daily demand+ Fire demand

1.13 (c)

Water demand doesn't affect losses and wastes in water supply scheme.

1.14 (a)

Arithmetic increase method is based on the assumption that population increase at constant rate i.e., $\frac{dP}{dt} = \text{Constant}$ whereas geometric increase method or uniform percentage growth method is based on assumption that percentage rate growth is constant.

1.15 (a)

Population = 50 lacs

Total duration of fire fighting = 6 × 3 × 60 min

Total volume of water for fire fighting

= 3 × 1100 lit/min × 6 × 3 × 60 min

Per capita fire demand

$$= \frac{3 \times 1100 \times 6 \times 3 \times 60}{5000000} = 0.7 \text{ lit/day}$$

1.17 (b)

As we reach the end of design period, the city passes through its initial exponential growth phase and reaches a phase where the growth happens, but at a decreasing rate.

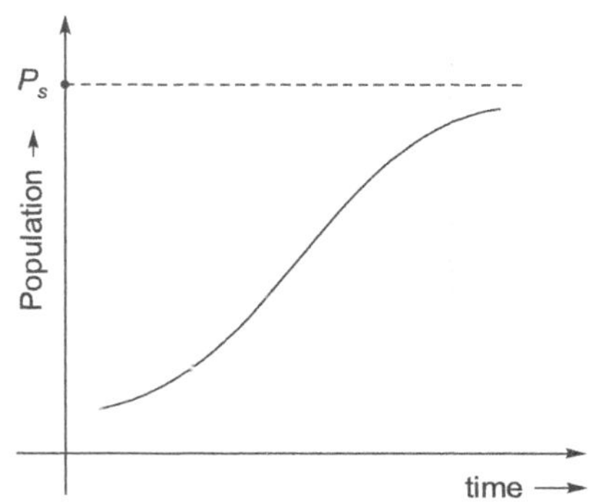

2 Sources of Water and Well

2.1 Specific capacity of a well is the
(a) volume of water that can be extracted by the force of gravity from a unit volume of aquifer
(b) discharge per unit drawdown of the well
(c) drawdown per unit discharge of the well
(d) rate of flow through a unit width and entire thickness of aquifer [ESE : 2000]

2.2 **Assertion (A):** The available yield of a tubewell can be doubled by doubling the diameter of the well.
Reason (R): The yield of a tubewell varies inversely with the logarithm of the reciprocal of the diameter of the well.
(a) both A and R are true and R is the correct explanation of A
(b) both A and R are true but R is not a correct explanation of A
(c) A is true but R is false
(d) A is false but R is true [ESE : 2001]

2.3 Match **List-I** (Equation) with **List-II** (Applicability or principle of equation) and select the correct answer using the codes given below the lists:
List-I
A. Theim's equation
B. Dupit's assumption
C. Bernoulli's equation
D. Continuity equation
List-II
1. is based on energy conservation principle
2. is based on mass conservation principle
3. is applicable to steady flow towards a well in a confined aquifer
4. is applicable to steady flow in an unconfined aquifer

Codes:

	A	B	C	D
(a)	4	3	2	1
(b)	3	4	2	1
(c)	4	3	1	2
(d)	3	4	1	2

[ESE : 2002]

2.4 The performance of a well is measured by its
(a) Specific capacity
(b) Specific yield
(c) Storage coefficient
(d) Permeability coefficient [ESE : 2002]

2.5 Match **List-I** (Tests) with **List-II** (Features) and select the correct answer using the codes given below the lists:
List-I
A. Pumping test
B. Recuperation Test
C. Pressure Test
D. Jar Test
List-II
1. The gradual rise of water level in well is observed as time progresses
2. Rate of pumping is adjusted to constant level of water in well
3. Vigorous mixing of the chemical followed by slow mixing
4. Pipeline is filled up with water, allowed to stand for sometime and then at least double the maximum pressure is applied

Codes:

	A	B	C	D
(a)	1	2	3	4
(b)	2	1	4	3
(c)	1	2	4	3
(d)	2	1	3	4

[ESE : 2003]

2.6 Consider the following statements:
A well development
1. involves reversal of flow through the well screen
2. increases permeability towards the well
3. decreases permeability towards the well
4. is continued till sand/silt free water is pumped out

Which of these statements is/are correct?

(a) 1, 3 and 4 (b) 1, 2 and 4
(c) 3 only (d) 1 and 4

[ESE : 2003]

2.7 Match **List-I** (Well hydraulics parameters) with **List-II** (Definition) and select the correct answer using the codes given below the lists:

List-I

A. Specific yield
B. Safe yield
C. Specific capacity
D. Field capacity

List-II

1. Discharge per unit drawdown of well
2. Same as specific retention
3. Measure of water that can be removed by pumping
4. Limit of withdrawal from well without depletion of the aquifer
5. Water-bearing capacity of aquifer

Codes:

	A	B	C	D
(a)	4	3	2	5
(b)	3	4	1	2
(c)	4	3	1	2
(d)	3	4	2	5

[ESE : 2003]

2.8 **Assertion (A):** The yield of a well varied from 10 m^3/day to 20 m^3/day when the aquifer area changed from 50 m^2 to 75 m^2.

Reason (R): The yield is found to be directly proportional to the area of an aquifer opening into a well.

(a) both A and R are true and R is the correct explanation of A
(b) both A and R are true but R is not a correct explanation of A
(c) A is true but R is false
(d) A is false but R is true [ESE : 2004]

2.9 Which one of the following correctly defines aquiclude?

(a) A saturated formation of earth material which not only stores water but also yields it in sufficient quantity.
(b) A formation through which only seepage is possible and thus the yield is insignificant compared to an aquifer.
(c) A geological formation which is neither porous nor permeable.
(d) A geological formation which is essentially impermeable to the flow of water.

[ESE : 2004]

2.10 The discharge per unit drawdown at the well is known as

(a) Specific yield (b) Specific storage
(c) Specific retention (d) Specific capacity

[ESE : 2005]

2.11 Consider the following statements:

1. Infiltration galleries are placed along the river beds at a depth 4 to 6 m.
2. The draw-down for the infiltration galleries is more than that for radial wells.
3. Clogging of pipe pores in infiltration galleries is less than that for radial wells.
4. The cost of extracting unit volume of water is more in case of infiltration galleries as compared to radial wells.

Which of these statements is/are correct?

(a) 1 and 2 (b) 1, 2 and 3
(c) 2, 3 and 4 (d) 4 only

[ESE : 2005]

2.12 How is the determination of aquifer parameters *S* (Storage Coefficient) and *T* (Transmissivity Coefficient) done?

1. By recording the drawdown in a pumped well at different time intervals.
2. By recording the drawdown in installed observation wells at different time intervals.

Select the correct answer using the codes given below:

(a) 1 only (b) 2 only
(c) Both 1 and 2 (d) Neither 1 nor 2

[ESE : 2007]

2.13 A groundwater basin consists of 10 km^2 area of plains. The maximum groundwater table fluctuation has been observed to be 1.5 m. Consider specific yield of the basin as 10%. What is the available groundwater storage in million cubic metres?

(a) 1.0 (b) 1.5
(c) 2.5 (d) 2.0 [ESE : 2009]

2.14 An unconfined aquifer of porosity 30%, permeability 35 m/day and specific yield of 0.20 has an area of 100 km^2. If the water table falls by 0.25 m during a drought, the volume of water lost from storage, in million cubic metres, is
(a) 2.0 (b) 5.0
(c) 1.0 (d) 4.0 [ESE : 2010]

2.15 **Assertion (A):** In a strainer type tube well, strainer pipes are surrounded by wire mesh.
Reason (R): This prevents the fine particles from entering the well pipe.
(a) both A and R are true and R is the correct explanation of A
(b) both A and R are true but R is not a correct explanation of A
(c) A is true but R is false
(d) A is false but R is true [ESE : 2010]

2.16 **Statement-I** : Rainwater is collected and harvested using storage structures like underground tanks for future use and also for recharging the aquifer.
Statement-II : Rainwater harvesting pits allow the rainwater to percolate and recharge the aquifer.
(a) Both Statement (I) and Statement (II) are individually true and Statement (II) is the correct explanation of Statement (I)
(b) Both Statement (I) and Statement (II) are individually true but Statement (II) is NOT the correct explanation of Statement (I)
(c) Statement (I) is true but Statement (II) is false
(d) Statement (I) is false but Statement (II) is true
[ESE : 2012]

2.17 What would be the volume of water stored in a saturated column with a porosity of 0.35 with a cross-sectional area of 1 m^2 and depth of 3 m?
(a) 2.0 m^3 (b) 0.105 m^3
(c) 1.05 m^3 (d) 3.0 m^3 [ESE : 2012]

2.18 The surface joining the static levels in several non-pumping wells penetrating a continuous confined aquifer represents
(a) Water table surface
(b) Capillary fringe
(c) Piezometric surface of the aquifer
(d) Physical top surface of the aquifer
[ESE : 2012]

2.19 Coefficient of permeability of an underground stratum is 0.001 m/s. Discharge obtained from a well of area 20 m^2 dug into this stratum (with drawdown of 2 m) will be
(a) 2400 lpm (b) 2000 lpm
(c) 1200 lpm (d) 1000 lpm
[ESE : 2012]

2.20 **Statement-I** : When a tube well penetrates into a homogenous aquifer and is then pumped, there will occur lowering of water surface. The resultant surface is designated as 'Drawdown curve'.
Statement-II : Since the pressure on the surface of the 'Drawdown curve' is always at atmospheric level, it is called by this name.
(a) Both Statement (I) and Statement (II) are individually true and Statement (II) is the correct explanation of Statement (I)
(b) Both Statement (I) and Statement (II) are individually true but Statement (II) is NOT the correct explanation of Statement (I)
(c) Statement (I) is true but Statement (II) is false
(d) Statement (I) is false but Statement (II) is true
[ESE : 2012]

2.21 Due to a continuous pumping for 7 hrs from a catchment of 25 ha, having porosity of 30% and specific retention of 10%, groundwater level dropped by 1 m. The corresponding change in storage is:
(a) 7.5 ha-m (b) 0.75 ha-m
(c) 1.87 ha-m (d) 0.187 ha-m
[ESE : 2013]

2.22 As recommended by Sichardt, the radius of influence is
(a) Inversely proportional to drawndown
(b) Linearly proportional to drawdown
(c) Independent of drawdown
(d) Proportional to square root of drawdown
[ESE : 2013]

2.23 When 3.5 million m^3 of water was pumped out from an unconfined aquifer of 6.3 km^2 areal extent, the water table was observed to go down by 2.5 m. The specific yield of the aquifer is best approximated as
(a) 32% (b) 28%
(c) 25% (d) 22% [ESE : 2014]

2.24 An unconfined aquifer of porosity 35%, permeability 40 m/day and specific yield of 0.15 has an area of 100 km^2. If the water table falls uniformly throughout the aquifer area by 0.2 m during a drought, the volume of water lost from storage is

(a) 1.5 Mm^3 (b) 3.0 Mm^3
(c) 7.0 Mm^3 (d) 8.0 Mm^3

[ESE : 2014]

2.25 Ground water flows through an aquifer with a cross-sectional area of 1.0×10^4 m and a length of 1500 m. Hydraulic heads are 300 m and 250 m at the ground water entry and exit points in the aquifer respectively. Ground water discharge into stream at the rate of 1500 m^3/day. The corresponding hydraulic conductivity of this aquifer will be

(a) 3.5 m/day (b) 4.5 m/day
(c) 5.0 m/day (d) 5.5 m/day

[ESE : 2014]

2.26 A tubewell of 30 cm diameter penetrates fully into an artesian aquifer. The artesian length is 15 m. The yield from the well under a drawdown of 3 m through the aquifer consisting of sand of effective size of 0.2 mm and coefficient of permeability of 50 m/day, with radius of drawdown of 150 m, is nearly

(a) 240 *l*/s (b) 120 *l*/s
(c) 24 *l*/s (d) 12 *l*/s **[ESE : 2014]**

2.27 In a uniform, semi-infinite aquifer, the dependable discharge of a lone circular open well is increased most easily by

(a) increasing the diameter
(b) making it into one with a square kerb
(c) deepening the well
(d) providing coarser screening filter

[ESE : 2015]

2.28 What is the strainer length required for a deep tube well giving a discharge of 8 litres per second? Assume permissible entrance velocity of 2 cm/second. It is desired to have the strainer of slot sizes 20 mm × 0.2 mm with number of slots per cm length of the strainer as 100.

(a) 8 m (b) 1 m
(c) 12 m (d) 10 m

[ESE : 2015]

2.29 Which of the following statement are correct as regards aquifer characteristics?

1. The storage coefficient is the volume of water released from storage from the entire aquifer due to unit depression of piezometric head.
2. The storage coefficient is the same as the specific yield for water table aquifer.
3. Both the aquifer constant, viz, storage coefficient *S* and Transmissivity *T* are dimensionless numbers.

(a) 1 only (b) 2 only
(c) 3 only (d) 1, 2 and 3

[ESE : 2016]

2.30 Consider the following assumptions as regards field permeability test:

1. The flow is laminar and Darcy's law is valid.
2. The flow is horizontal and uniform at all points in the vertical section.
3. The velocity of flow is proportional to the 'tangent magnitude' of the hydraulic gradient.

Which of the above assumptions are correct?

(a) 1 and 2 only (b) 1 and 3 only
(c) 2 and 3 only (d) 1, 2 and 3

[ESE : 2016]

2.31 Consider the following statements in respect of ground water aquifers:

1. Specific storages is specific capacity per unit depth of the aquifer.
2. Specific capacity is storage coefficient per unit aquifer depth.
3. Specific capacity is a constant for a given well.
4. For one-dimensional flow in a confined aquifer between two water bodies, the piezometric head line in the aquifer is a straight line.

Which of the above statements are correct?

(a) 2 and 3 (b) 2 and 4
(c) 1 and 3 (d) 1 and 4

[ESE : 2017]

2.32 For an unconfined aquifer, the specific yield is 20%, specific retention is 15%, and permeability is 35 m/day. Consider the following statements:

1. The porosity of the aquifer is 35%.
2. The transmissibility is 35 m^2/day.

3. The volume of water lost from storage per meter drop in the water table per 100 km^2 area of the aquifer is 20 million m^3.

Which of the above statements are correct?

(a) 1 and 3 only (b) 1 and 2 only
(c) 2 and 3 only (d) 1, 2 and 3

[ESE : 2017]

2.33 Which of the following situations can be attributed to sustained excessive groundwater pumping in a basin?

1. Drying up of small lakes and streams over a period in spite of normal rainfall.
2. Deterioration of groundwater quality in certain aquifers.
3. Land subsidence in the basin.
4. Increase in seismic activity.
5. Increased cost of groundwater extraction.

(a) 2 and 4 only (b) 1, 2, 3 and 5 only
(c) 3 and 4 only (d) 1 and 5 only

[ESE : 2018]

2.34 Groundwater flows through an aquifer with a cross-sectional area of 1.0×10^4 m^2 and a length of 1500 m. Hydraulic heads are 300 m and 250 m at the groundwater entry and exit points in the aquifer, respectively. Groundwater discharges into a stream at the rate of 750 m^3/day. Then the hydraulic conductivity of the aquifer is

(a) 1.50 m/day (b) 2.25 m/day
(c) 3.50 m/day (d) 4.25 m/day

[ESE : 2018]

2.35 The coefficient of transmissibility T for a confined aquifer can be determined by a pumping-out test together with other relevant observations. The applicable formula is (where Q = discharge, and ΔS = Difference in drawdowns in two wells)

(a) $\frac{Q}{2.72\Delta S}$ (b) $\frac{Q}{1.72\sqrt{\Delta S}}$

(c) $\frac{Q}{2.72}\Delta S$ (d) $\frac{Q}{2.72}\sqrt{\Delta S}$

[ESE : 2019]

2.36 The transition region between unsaturated zone and saturated zone is called

(a) Capillary fringe
(b) Water table
(c) Yadose water zone
(d) Confining bed

[ESE : 2020]

2.37 In a groundwater field test, a tracer took 8 hours to travel between two observation wells which are 56 m apart. The difference in water table elevations in these wells was 0.70 m. The volume of the void of the aquifer is 30% of the total volume of the aquifer. What is the hydraulic conductivity of the aquifer, if the dynamic viscosity of water is 0.995×10^{-3} Ns/m^2?

(a) 4.664 cm/s (b) 3.664 cm/hr
(c) 2.664 mm/s (d) 1.664 cm/hr

[ESE : 2021]

2.38 A fully penetrating well in a confined sand aquifer has a maximum discharge capacity of 1200 l/min. The aquifer is overlain and underlain by impervious formations. The thickness of the aquifer is 20 m. Assume the percentage of the open area of an available strainer to be 15% and bore hole diameter as 15 cm. What is the length of the well screen? (Take safe entrance velocity as 0.02 m/s)

(a) 10.5 m (b) 12.3 m
(c) 13.1 m (d) 14.1 m

[ESE : 2023]

2.39 Which one of the following is the ability of a material to absorb moisture from air and thus to dissolve and become liquid?

(a) Deliquescence (b) Hygroscopicity
(c) Wilting (d) Capillarity

[ESE : 2023]

2.40 Consider the following statements regarding quality difference of ground water and surface water:

1. The turbidity of ground water is high, whereas surface water is little.
2. Total dissolved solids of ground water is lower than surface water.
3. Concentration of inorganic compounds or ions of ground water is higher than surface water.

Which of the above statements is/are correct?

(a) 1 and 2 (b) 2 only
(c) 3 only (d) 1 and 3

[ESE : 2024]

2.41 Consider the following statements regarding ground water :

1. Ground water is exhaustible and is evenly available.
2. Natural replenishment of the ground water resource is a very fast process.
3. Ground water is generally better than surface water in respect of biological characteristics.

Which of the above statements is/are correct?

(a) 1 and 3 only (b) 2 and 3 only
(c) 3 only (d) 1, 2 and 3

[ESE : 2024]

2.42 Match the following lists:

List I (Material)	List II (Specific yield %)
P. Clay	1. 5 – 15
Q. Sand and gravel	2. 15 – 25
R. Sandstone	3. 0.5 – 5
S. Limestone	4. 1 – 10

Select the correct answer using the code given below :

	P	Q	R	S
(a)	2	3	1	4
(b)	1	2	4	3
(c)	3	1	2	4
(d)	4	2	1	3

[ESE : 2024]

2.43 A fully penetrating artesian well is pumped at a rate Q = 1500 m^3/ day from an aquifer whose storage coefficient and transmissivity are 4×10^{-4} and 0.145 m^2/min, respectively. Considering $\omega(u)$ = 8.62, what is the drawdowns at a distance 3 m from the production well after one hour of pumping?

(a) 8.62 m (b) 14.53 m
(c) 4.93 m (d) 2.38 m

[ESE : 2024]

■■■■

Answers Sources of Water & Well

2.1	(b)	2.2	(d)	2.3	(d)	2.4	(a)	2.5	(b)	2.6	(b)	2.7	(b)	2.8	(d)	2.9	(d)
2.10	(d)	2.11	(b)	2.12	(c)	2.13	(b)	2.14	(b)	2.15	(a)	2.16	(a)	2.17	(c)	2.18	(c)
2.19	(a)	2.20	(c)	2.21	(*)	2.22	(b)	2.23	(d)	2.24	(b)	2.25	(b)	2.26	(c)	2.27	(c)
2.28	(d)	2.29	(b)	2.30	(d)	2.31	(b)	2.32	(d)	2.33	(b)	2.34	(b)	2.35	(a)	2.36	(a)
2.37	(a)	2.38	(d)	2.39	(a)	2.40	(c)	2.41	(c)	2.42	(c)	2.43	(c)				

Explanations Sources of Water & Well

2.1 (b)

- Discharge per unit drawdown in the well is called specific capacity.
- It is a measure of performance of well.

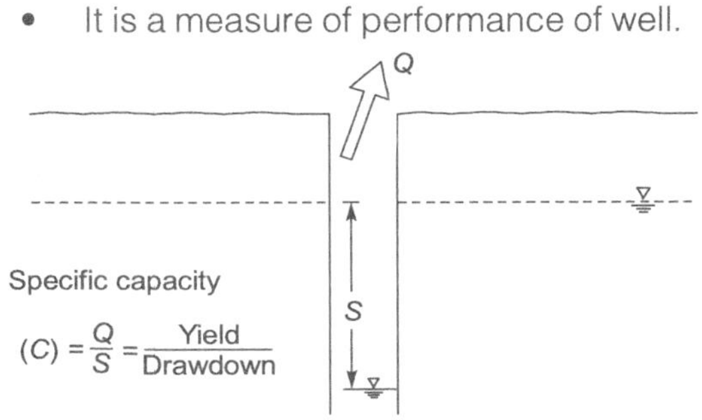

2.2 (d)

Yield of a well in unconfined aquifer,

$$Q = \frac{\pi K\left(d^2 - h_w^2\right)}{2.3\log_{10}\left(R / r_w\right)}$$

Yield of a well in confined aquifer,

$$Q = \frac{2\pi KH\left(D - h_w\right)}{2.3\log_{10}\left(R / r_w\right)}$$

As the yield of well is inversely proportional to the logarithm of the reciprocal of radius (diameter) of the well, so doubling the diameter will not double the yield.

2.3 (d)

- Theim's Equation – Equation of discharge for the steady flow in a confined aquifer.
- Dupit's Equation - Equation for steady flow in an unconfined aquifer.
- Bernoulli's Equation – Based on Energy Conservation principle.
- Continuity Equation – Mass Conservation principle.

2.4 (a)

- Discharge per unit drawdown in the well is called specific capacity.
- It is a measure of performance of well.

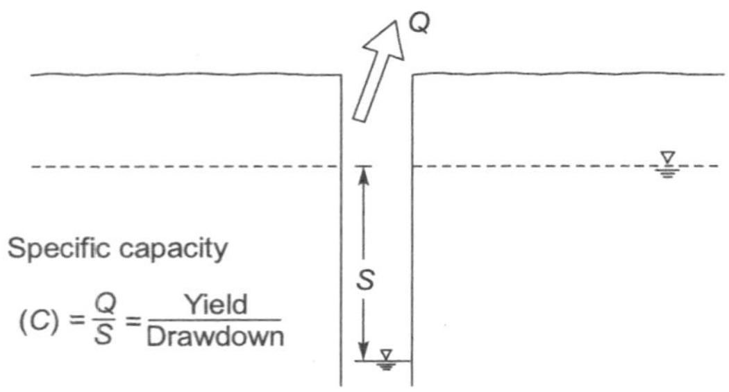

2.5 (b)

Pumping Test: A pump is first of all, installed so as to draw sufficient supplies of water from the open well, and to cause heavy drawdown in its water level. The rate of pumping is then changed and so adjusted that the water level in the well become constant.

Pressure Test: The purpose of pressure test is to investigate the various limits of the pipeline such as reliability, maximum capacity, leaks, fittings and pressure. It is a test in which a test pipeline is filled with water and is allowed to stand for some time and then pressure is applied.

Recuperation test: Although the pumping test gives accurate value of safe yield, it becomes very difficult to adjust the rate of pumping. So as to keep the well water level constant. In such circumstances recuperation test is adopted.

In this method, the water is first of all drained from the well at a fast rate so as to cause sufficient drawdown.

The pumping is then stopped the water level in the well will start rising the time taken by the other measured level is taken noted.

Jar test: The common test which is performed to determine the optimum dosage of coagulent is known as jar test.

2.6 (b)

Well development: Development means the stabilisation of the wells of a well adjacent to the screen, by a process which removes the fine particles from the formation immediately surrounding the well screen, leaving coarser particles to contact and surround the screen.

Objectives of well development are:

(i) to unclog the water bearing formation.

(ii) to increase the porosity and permeability of the water bearing formation in the vicinity of the well.

(iii) to stabilise the sand formation around a screened well, so that the well may yield send free water.

The basic principle in well development is to cause reversals of flow through the screen openings, that will rearrange the aquifer particles.

2.7 (b)

The quantity of water retained by the material against the pull of gravity is called specific retention or field capacity.

Specific yield + Specific retention = Porosity.

So the soil is not fully saturated at field capacity.

Safe yield: Maximum discharge at which withdrawal of groundwater can be done without causing undesirable effects like permanent lowering of water table, dislodging of soil particles, salt water encroachment in coastal areas etc. is called safe yield.

2.8 (d)

Yield of a well $Q = KiA$

Where i is the hydraulic gradient at the surface of well. If i remains constant then discharge is proportional to area of an aquifer opening into the well. So the discharge should change from 10 m^3/day to 15 m^3/day when aquifer area is changed from 50 m^2 to 75 m^2.

2.9 (d)

- Aquifer is a saturated formation of earth material which not only stores water but also yields it in sufficient quantity. e.g. unconsolidated deposits of sand and gravel.

- Aquitard is a formation through which only seepage is possible and thus the yield is insignificant compared to an aquifer e.g. sandy clay unit.
- Aquifuge is a geological formation which is neither porous nor permeable e.g. massive compact rock without any fracture.
- Aquiclude is geological formation which is essentially impermeable to the flow of water although it may contain large amounts of water due to its high porosity e.g. clay.
- So in terms of yield, Aquifer > Aquitard > Aquiclude > Aquifuge.

2.10 (d)

Specific capacity is the discharge per unit drawdown of a well and it measures efficiency of a well.

2.11 (b)

An **Infiltration Gallery** is a structure including perforated conduits in gravel to expedite transfer of water to or from a soil aquifer.

Infiltration galleries may be used to supplement a storm sewer, by directing storm runoff from non-road areas.

Primarily, upright plastic pipes capped with simple grates are placed every 5-8 metres along the low point of a slope, to handle heavy surface runoff. The pipes proceed straight down, about two metres, to a horizontal cross-pipe; this pipe is the secondary system.

The horizontal pipe is then perforated slightly (about 10% of its surface area) and surrounded by gravel. Initially, runoff will exit the pipe and infiltrate the gravel to the soil beyond, dissipating naturally. As flow increases, the water will eventually fill the pipe and need to be dissipated more quickly. Thus, a catch basin is placed at the lowest point of the sloping ground, which is connected to the storm sewer system at large.

2.12 (c)

Storativity or the storage coefficient is the volume of water released from storage per unit decline in hydraulic head in the aquifer, per unit area of the aquifer. Storativity is a dimensionless quantity, and ranges between 0 and the effective porosity of the aquifer

Transmissivity is the rate of flow under a unit hydraulic gradient through a unit width of aquifer of given saturated thickness. The transmissivity of an aquifer is related to its hydraulic conductivity as follows:

$$T = Kb$$

where T is transmissivity $[L^2/T]$, K is Transmissivity Coefficient or permeability and b is aquifer thickness $[L]$.

2.13 (b)

Available ground water storage

$$= 10 \times (10^3)^2 \times 1.5 \times \frac{10}{100}$$
$$= 1.5 \times 10^6 \text{ m}^3 = 1.5 \text{ m}^3$$

2.14 (b)

The actual volume of water that can be extracted by the force of gravity from a unit volume of aquifer material is known as the specific yield (S_y).

Volume of water lost during drought

$$= S_y \times \text{Area} \times \text{drop in water table}$$
$$= 0.2 \times 100 \times 10^6 \times 0.25$$
$$= 5 \times 10^6 \text{ m}^3 = 5 \text{ Mm}^3$$

2.15 (a)

A strainer essentially consist of a perforated or a slotted pipe with a wire mesh wrapped around the pipe with a small annular space between the two.

The wire screen prevents the sand particles from entering the well. The water, therefore, enters the well pipe through the fine mesh, and the sand particles of size greater then the size of the mesh are prevented from entering the well. This reduces the danger of sand removal, and hence large flow velocities can be permitted.

Note: A strainer type tubewell is generally unsuitable for fine sandy strata, because in that case, the size of the mesh openings will have to be considerably reduced, which may result in choking of the strainer.

2.16 (a)

Rainwater harvesting can be done in a lot of ways. The rainwater can be collected in the roof and then directed to storage tanks, or it can be stored in pits in ground to recharge the aquifer and increase ground water level of the area.

2.17 (c)

Volume of water stored

$= 0.35 \times 3 \times 1 = 1.05\ m^3$

2.19 (a)

Discharge is given by

$q = kiA = 0.001 \times 2 \times 20$

(Assuming $L = 1$ m)

$= 0.04\ m^3/sec$

$= 0.04 \times 1000 \times 60$

$= 2400$ lpm

2.20 (c)

Any point on drawdown curve will have pressure equal to atmospheric pressure but slope of the tangent at this point being > 0, the curve exponentially gets lowered towards the well from where water is being drawn and thus it is called as DRAWDOWN CURVE.

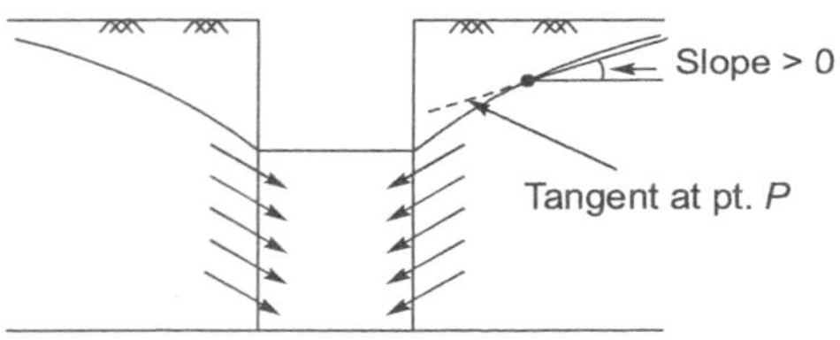

2.21 (*)

$A = 25$ ha

Porosity, $n = 0.3$

Sp. retention $S_r = 0.1$

$\therefore \quad S_y = n - S_r = 0.2$

$\therefore$ Storage change $= 25 \times 0.2 \times 1$

$= 5$ ha-m

2.22 (b)

The radius of influence (R) is commonly known as radius of drawdown, is the radius measured from the centre of the wall to the point at which drawdown is measured.

It is given by $R = 3000\ S\sqrt{k}$

where,

R = radius of drawdown

S = drawdown

k = permeability of aquifer soil

2.23 (d)

By considering volumetric approach

Water pumped $= S_y \times \text{Area} \times \text{drawdown}$

$$S_y = \frac{\text{water pumpdown}}{\text{area} \times \text{drawdown}}$$

$$= \frac{3.5 \times 10^6}{6.3 \times 10^6 \times 2.5} = 0.2222 \simeq 22\%$$

2.24 (b)

Volume lost $= S_y \times \text{Drawdown} \times \text{Area}$

$= 0.15 \times 0.2 \times 100 \times 10^6\ m^3$

$= 3 \times 10^6\ m^3 = 3\ Mm^3$

2.25 (b)

Discharge, $Q = kiA$

$$1500 = k \times \frac{(300 - 250)}{1500} \times 10^4$$

$\therefore \quad k = 4.5$ m/day

2.26 (c)

$$\text{Yield of aquifer } (q) = \frac{2\pi kb(H - h_w)}{ln\left(\frac{R}{r_w}\right)}$$

$(H - h_w = s_w = 3\ m)$

$$q = \frac{2\pi \times \left(\frac{50}{24 \times 60 \times 60}\right) \times 15 \times 3}{ln\left(\frac{150}{0.15}\right)}$$

$$q = \frac{2\pi \times 5.787 \times 10^{-4} \times 15 \times 3}{6.90775}$$

or $\quad q = 0.02368\ m^3/sec.$

$q = 23.68\ l/sec. \simeq 24\ l/sec.$

2.28 (d)

Discharge of tube well

$= 8\ l/s = 8 \times 10^{-3}\ m^3/s$

Permissible entrance velocity,

$= 2\ cm/s = 2 \times 10^{-2}\ m/s$

Let length of strainer is 'l'

$l \times (20 \times 0.2 \times 10^{-6}) \times (2 \times 10^{-2}) \times (100 \times 100)$

$= 8 \times 10^{-3}$

$l = 10$ m

2.29 (b)

Storage Coefficient:

- In general, storage coefficient (S) is defined as the volume of water that aquifer releases or stores per unit surface area per unit decline or rise of water table.

- For an unconfined aquifer, storage coefficient is generally taken as equal to the specific yield.
- For a confined aquifer, it is defined as equal to the volume of water released from the aquifer of full height and unit cross-sectional area when the piezometric surface declines by unity.

2.32 (d)

Porosity, $n = S_y + S_R$

S_y = Specific yield

S_R = Specific retention

$\therefore \quad n = 0.20 + 0.15 = 0.35$

Transmissibility of an aquifer is defined as the ability of an aquifer to pass water through unit width.

$$T = KH$$
$$= 35 \text{ m/d} \times 1 \text{ m} = 35 \text{ m}^2\text{/d}$$

Volume of water lost from storage

$$= S_y \times 100 \times 10^6 \text{ m}^2 \times 1 \text{ m}$$
$$= 0.20 \times 100 \times 10^6 \times 1 = 20 \text{ Mm}^3$$

2.34 (b)

$$Q = kiA$$
$$750 = K\frac{(300-250)\times 10^4}{1500}$$
$$K = 2.25 \text{ m/day}$$

2.35 (a)

Discharge from confined aquifer using Thiem's theory

$$Q = \frac{2\pi kH(s_1 - s_2)}{2.303 \log_{10}(r_2/r_1)}$$

Assuming, $r_2 = 10\, r_1$

$$Q = \frac{2\pi T\Delta S}{2.303 \log_{10} 10} = \frac{2\pi T\Delta S}{2.303}$$
$$T = \frac{Q}{\frac{2.303\,\Delta S}{2\pi}}$$
$$\Rightarrow \quad T = \frac{Q}{2.72\Delta S}$$

2.36 (a)

Between the unsaturated zone and the saturated zone (water table), transition zone is present called as the capillary fringe.

In this zone the voids are saturated or almost saturated with water that is held in place by capillary forces.

2.37 (a)

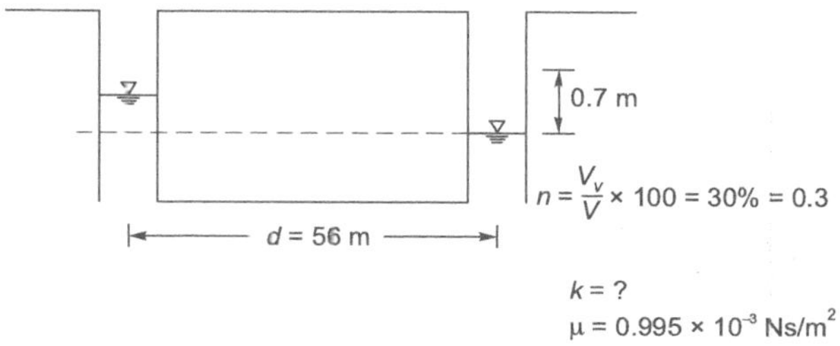

$$t = 8\ h$$

Seepage velocity,

$$V_s = \frac{d}{t} = \frac{56 \text{ m}}{8 \text{ h}} = 7 \text{ m/hr}$$

$\therefore$ Discharge velocity,

$$V = V_s \times n$$
$$= 7 \text{ m/h} \times 0.3$$
$$= 2.1 \text{ m/hr}$$

From Darcy's law:

$$V = ki$$
$$\therefore \quad k = \frac{V}{i} = \frac{2.1 \text{ m/hr}}{\left(\frac{0.7}{56}\right)}$$
$$= 3 \times 56 \text{ m/hr} \quad \left(\because i = \frac{\Delta h}{L} = \frac{0.7}{56}\right)$$
$$= 168 \text{ m/hr}$$
$$= \frac{168 \times 10^2}{3600} \text{ cm/s}$$
$$= 4.664 \text{ cm/s}$$

2.38 (d)

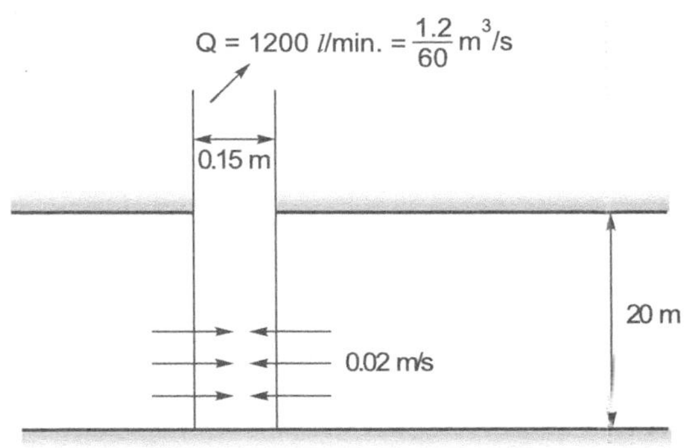

$$\therefore \quad Q = AV$$

$$\Rightarrow 1200 \text{ l/min} = \frac{1.2}{60} \text{m}^3/\text{s}$$

$$= \pi \times 0.15 \times L \times 0.15 \times 0.02 \text{ m/s}$$

$$\Rightarrow \quad L = \left(\frac{1.2}{60 \times 0.02}\right) \times \frac{1}{\pi \times 0.15 \times 0.15}$$

$$\Rightarrow \quad L = \frac{1}{\pi \times 0.15^2} = 14.147 \text{m}$$

Hence, length of well screen = 14.1 m.

2.40 (c)

- Turbidity of ground water is lower in comparison to surface water.
- TDS of ground water is high than surface water due to presence of minerals.

2.41 (c)

- Ground water is not evenly available.
- Natural repensishment of ground water resource is a slow process.

2.42 (c)

Clay will have lowest yield among given materials.

2.43 (c)

Drawdown, $S_w = \dfrac{Q}{4\pi T} \times \omega(u)$

where $\omega(u)$ is the well function,

$$\Rightarrow S_w = \frac{1500 \times \text{m}^3/\text{d} \times 8.62}{4\pi \times 0.145 \text{m}^2/\text{min} \times 24 \times 60 \text{min/d}}$$

$$= 4.928 \simeq 4.93 \text{ m}$$

■■■■

3 Water Quality Parameters

3.1 Which of the following statements are correct?

1. Groundwater is generally free from suspended and dissolved impurities.
2. Suspended matters often contain pathogenic bacteria.
3. Rain water is soft and tasteless.
4. Lake water may contain microscopic organisms.

Select the correct answer using the codes given below:

(a) 1, 2, 3 and 4 (b) 1 and 2
(c) 2, 3 and 4 (d) 1, 3 and 4

[ESE : 2001]

3.2 Match **List-I** (Impurities) with **List-II** (Effects) and select the correct answer using the codes given below the lists:

List-I

A. Dissolved sulphates and chlorides of Ca and Mg
B. Dissolved bicarbonates of Ca and Mg
C. Dissolved fluorides of Na
D. Dissolved organic matter

List-II

1. Hardness and corrosion
2. Bacterial infection
3. Alkalinity and softness
4. Impairment of dental health

Codes:

	A	B	C	D
(a)	2	3	4	1
(b)	1	4	3	2
(c)	2	4	3	1
(d)	1	3	4	2

[ESE : 2001]

3.3 Match **List-I** (Different forms of nitrogen in water) with **List-II** (Inferences) and select the correct answer using the codes given below the lists:

List-I

A. Nitrate nitrogen
B. Total nitrogen
C. Nitrite nitrogen
D. Ammonia nitrogen

List-II

1. Unsatisfactory microbial activity
2. Satisfactory microbial activity
3. Eutrophication may result
4. Recent organic pollution

Codes:

	A	B	C	D
(a)	3	2	1	4
(b)	1	4	3	2
(c)	3	4	1	2
(d)	1	2	3	4

[ESE : 2002]

3.4 Match **List-I** with **List-II** and select the correct answer using the codes given below the lists:

List-I	List-II
A. Protozoa	**1.** Methaemog-lobinemia
B. Bacteria	**2.** Poliomyelitis
C. Presence of nitrate> 45 mg/L	**3.** Dysentry
D. Virus	**4.** Typhoid fever

Codes:

	A	B	C	D
(a)	3	2	1	4
(b)	1	4	3	2
(c)	3	4	1	2
(d)	1	2	3	4

[ESE : 2004]

3.5 Which one of the following organisms is responsible for enteric fever?

(a) ECHO
(b) Salmonella typhi
(c) Entamoeba histolytica
(d) Echinococcus

[ESE : 2004]

3.6 Match **List-I** (Type of impurity) with **List-II** (Effect) and select the correct answer using the codes given below the lists:

List-I
A. Carbonates and bicarbonates of calcium and magnesium
B. Carbonates and bicarbonates of sodium
C. Sulphates and chlorides of calcium and magnesium
D. Oxides of iron and manganese

List-II
1. Permanent hardness
2. Temporary hardness
3. Alkalinity and softness
4. Colour and taste

Codes:

	A	B	C	D
(a)	1	3	2	4
(b)	2	4	1	3
(c)	1	4	2	3
(d)	2	3	1	4

[ESE : 2005]

3.7 A sample of ground water at a pH of 7.0 contains 122 mg/L of bicarbonates. What is the alkalinity of this water (in terms of $CaCO_3$)?
(a) 120 mg/L (b) 60 mg/L
(c) 100 mg/L (d) 200 mg/L
[ESE : 2005]

3.8 What is the equivalent calcium carbonate concentration of 110 mg/L of $CaCl_2$?
(a) 50 mg/L (b) 58.5 mg/L
(c) 100 mg/L (d) 117 mg/L
[ESE : 2006]

3.9 Which of the following cations impart(s) pseudo-hardness to water?
(a) Calcium only
(b) Magnesium only
(c) Calcium and magnesium
(d) Sodium [ESE : 2006]

3.10 What is the most common cause of acidity in water?
(a) Carbon monoxide (b) Nitrogen
(c) Hydrogen (d) Carbon dioxide
[ESE : 2006]

3.11 If total hardness and alkalinity of a water sample are 200 mg/L as $CaCO_3$ and 260 mg/L as $CaCO_3$ respectively, what are the values of carbonate hardness and non-carbonate hardness?
(a) 200 mg/L and zero
(b) Zero and 60 mg/L
(c) Zero and 200 mg/L
(d) 60 mg/L and zero [ESE : 2006]

3.12 Hardness to water is caused by the presence of calcium (Ca^{2+}) and magnesium (Mg^{2+}) ions. Which are the least soluble forms of calcium and magnesium at normal water temperature?
(a) $CaCl_2$ and $MgCO_3$
(b) $Ca(HCO_3)_2$ and $MgCl_2$
(c) $Ca(OH)_2$ and $Mg(HCO_3)_2$
(d) $CaCO_3$ and $Mg(OH)_2$ [ESE : 2006]

3.13 A standard multiple-tube fermentation test was conducted on a sample of water. The results of the analysis for the confirmed test are given below:

Sample Size (mL)	No. of positive results out of 5 tubes	No. of negative results out of 5 tubes
10	4	1
1	2	3
0.1	1	4
0.01	0	5

MPN Index for combinations of positive results when 5 tubes used per dilutions (10 mL, 1.0 mL, 0.1 mL)

Combination of positives	MPN Index per 100 mL
5-4-3	280
4-3-1	33
4-2-1	26
2-1-0	7

Using the above MPN Index table, what is the most probable number (MPN) of the sample?
(a) 280 (b) 33
(c) 26 (d) 70 [ESE : 2006]

3.14 Match **List-I** (Pathogen) with **List-II** (Epidemic) and select the correct answer using the codes given below the lists:

List-I	List-II
A. Bacteria	1. Gastroenteritis
B. Virus	2. Cholera
C. Protozoa	3. Worms
D. Helminth	4. Polio

Codes:

	A	B	C	D
(a)	2	4	1	3
(b)	3	1	4	2
(c)	2	1	4	3
(d)	3	4	1	2

[ESE : 2007]

3.15 Match **List-I** (Parameter) with **List-II** (Impact) and select the correct answer using the codes given below the lists:

List-I

A. Excess sulphates
B. Lack of iodide
C. Excess hardness
D. Excess dissolved oxygen

List-II

1. Greater soap consumption
2. Laxative effect
3. Goitre
4. Corrosion of pipes

Codes:

	A	B	C	D
(a)	2	1	3	4
(b)	4	3	1	2
(c)	2	3	1	4
(d)	4	1	3	2

[ESE : 2007]

3.16 The concentration of hardness producing cations may be estimated using which one of the following?

(a) Conductivity meter
(b) pH meter
(c) Spectrophotometer
(d) Flame photometer [ESE : 2007]

3.17 What does the presence of excess nitrates in river water indicate?

(a) Recent pollution of water with sewage
(b) Past pollution of water with sewage
(c) Intermittent pollution of water with sewage
(d) No pollution of water with sewage

[ESE : 2008]

3.18 Match **List-I** (Type of impurity) with **List-II** (Harm caused) and select the correct answer using the codes given below the lists:

List-I

A. Excess of nitrates
B. Excess of fluorides
C. Lack of iodides
D. Excess of chlorides

List-II

1. Brackish water
2. Goiter
3. Fragile bones
4. Blue babies

Codes:

	A	B	C	D
(a)	4	2	3	1
(b)	1	2	3	4
(c)	4	3	2	1
(d)	1	3	2	4

[ESE : 2009]

3.19 The concentration of OH^- ion in a water sample is measured as 17 mg/L at 25°C. What is the pH of the water sample?

(a) 10 (b) 11
(c) 12 (d) 13 [ESE : 2009]

3.20 A 12.5 mL sample of treated waste-water requires 187.5 mL of odor-free distilled water to reduce the odor to a level that is just perceptible. What is the threshold odor number (TON) for the wastewater sample?

(a) 0.07 (b) 1.07
(c) 15 (d) 16 [ESE : 2009]

3.21 Which one of the following tests of water/wastewater employs Erichrome Black T as an indicator ?

(a) Harcness (b) COD
(c) Residual chlorine (d) DO [ESE : 2009]

3.22 Match **List-I** (Parameters) with **List-II** (Units) and select the correct answer using the code given below the lists:

List-I	List-II
A. Turbidity	1. TON
B. Pathogen	2. TCU
C. Odour	3. JTU
D. Colour	4. MPN

Code:

	A	B	C	D
(a)	2	1	4	3
(b)	3	1	4	2
(c)	2	4	1	3
(d)	3	4	1	2

[ESE : 2011]

3.23 Assertion (A): Fluorides should always be present in drinking water upto a value 1.5 mg/*l*.

Reason (R): Such a water helps clean the teeth well.

(a) both A and R are true and R is the correct explanation of A
(b) both A and R are true but R is not a correct explanation of A

(c) A is true but R is false
(d) A is false but R is true [ESE : 2011]

3.24 The maximum safe permissible limit of sulphates in domestic water supply is
(a) 100 mg/L (b) 200 mg/L
(c) 500 mg/L (d) 600 mg/L
[ESE : 2012]

3.25 One Nephelometry Turbidity Unit (NTU) is equal to the turbidity produced by
(a) 1 mg SiO_2 dissolved in 1*l* of distilled water with the test being run according to absorption principle .
(b) 1 mg SiO_2 dissolved in 1*l* of distilled water with the test being run according to scattering principle.
(c) 1 mg Formazin dissolved in 1*l* of distilled water with the test being run according to absorption principle.
(d) 1 mg Formazin dissolved in 1*l* of distilled water with the test being run according to scattering principle.
[ESE : 2012]

Directions: Each of the next items consists of two statements, one labeled as the **'Statements (I)'** and the other as **'Statement (II)'**. You are to examine these two statements carefully and select the answers to these items using the codes given below :

Codes:
(a) Both Statement (I) and Statement (II) are individually true and Statement (II) is the correct explanation of Statement (I)
(b) Both Statement (I) and Statement (II) are individually true but Statement (II) is NOT the correct explanation of Statement (I)
(c) Statement (I) is true but Statement (II) is false
(d) Statement (I) is false but Statement (II) is true

3.26 **Statement-I** : Fluoride concentrations of approximately 1.0 mg/*l* in drinking water help to prevent dental cavities in children.
Statement-II : During formation of permanent teeth, fluoride combines chemically with tooth enamel resulting in softer and weaker teeth that are less resistant to decay.
[ESE : 2012]

3.27 **Statement-I** : Virus is living organisms in a natural environment including soil.
Statement-II : Virus comes to life after entering the host tissue through contamination.
[ESE : 2012]

3.28 **Statement-I** : The ability of water to conduct electricity, known and measured as the specific conductance, and concentration of total dissolved solids are not relatable on a one-to-one basis.
Statement-II : Many organic molecules and compounds dissolve in water without ionizing and hence are not taken into account while measuring specific conductance.
[ESE : 2012]

3.29 **Statement-I:** Water with heavy algal growth often has pH values as high as 9 to 10.
Statement-II: Non-utilization of the bicarbonate ion as a carbon source by algae can result in substantial accumulation of OH^- ions.
[ESE : 2012]

3.30 EDTA titration method of hardness determination of water sample uses an indicator which combines with hardness-causing divalent cations and forms a coloured complex. The name of the indicator and the colour of the formed complex respectively are
(a) Ferroin and dark blue
(b) Ferroin and wine red
(c) Eriochrome Black T and dark blue
(d) Eriochrome Black T and wine red
[ESE : 2012]

3.31 The most common constituents of alkalinity in natural water are measured by titrating the water sample with 0.02 N H_2SO_4 using.
(a) Eriochrome Black T and Ferroin indicators
(b) Ferroin and Phenolphthalein indicators
(c) Phenolphthalein and Methyl Orange indicators
(d) Methyl Orange and Eriochrome Black T indicators.
[ESE : 2012]

3.32 An approximate estimation of total dissolved solids of a given water sample is often made by measuring

(a) Electrical conductivity of the water sample
(b) Electro-magnetic conductivity of the water sample
(c) Sound conductivity of the water sample
(d) Thermal conductivity of the water sample

[ESE : 2012]

3.33 Which one of the following tests employs Ethylene Diamine Tetra Acetic Acid as a titrating agent?

(a) Chlorides (b) Dissolved oxygen
(c) Hardness (d) Residual chlorine

[ESE : 2012]

3.34 Consider the following statements:

1. Typhoid fever is caused by viral infection.
2. Infectious hepatitis is caused by bacterial infection.
3. Cholera is caused by bacteria.
4. Amoebic dysentery is caused by protozoal infection.

Which of these statements are correct?

(a) 1, 2, 3 and 4 (b) 1, 2 and 3 only
(c) 3 and 4 only (d) 2 and 3 only

[ESE : 2013]

Directions: Each of the next items consists of two statements, one labeled as the **'Statements (I)'** and the other as **'Statement (II)'**. You are to examine these two statements carefully and select the answers to these items using the codes given below :

Codes:

(a) Both Statement (I) and Statement (II) are individually true and Statement (II) is the correct explanation of Statement (I)
(b) Both Statement (I) and Statement (II) are individually true but Statement (II) is NOT the correct explanation of Statement (I)
(c) Statement (I) is true but Statement (II) is false
(d) Statement (I) is false but Statement (II) is true

3.35 Statement (I): Water with heavy algal growths often has pH values as high as 9 to 10.

Statement (II): Non-utilization of the bicarbonate ion as a carbon source by algae can result in substantial accumulation of OH ions.

[ESE : 2013]

3.36 Statement (I): 'Environment' includes abiotic and biotic parameters.

Statement (II): Abiotic parameters include algae, bacteria, animals; and biotic parameters are air, water and soil.

[ESE : 2013]

3.37 The total hardness value obtained from the complete analysis of a water sample was found to be 120 mg/*l*. If the value of carbonate hardness is 50 mg/*l*, the non-carbonate hardness and alkalinity are, respectively

(a) 170 mg/*l* and 70 mg/*l*
(b) 170 mg/*l* and 50 mg/*l*
(c) 70 mg/*l* and 50 mg/*l*
(d) 50 mg/*l* and 70 mg/*l*

[ESE : 2014]

3.38 Assertion (A): Dracontiasis is transmitted by drinking contaminated water.

Reason (R): Dracontiasis can be controlled by filtration of the drinking water.

(a) Both A and R are true and R is the correct explanation of A
(b) Both A and R are true but R is not a correct explanation of A
(c) A is true but R is false
(d) A is false but R is true

[ESE : 2015]

3.39 Electrical conductivity (EC) of water and total dissolved solids (TDS) are relatable as:

The value of EC will

(a) Decrease with increase in TDS
(b) Increase with increase in TDS
(c) Decrease initially and then increase with increase in TDS
(d) Increase nitially and then decrease with increase in TDS

[ESE : 2016]

3.40 The waste water of a certain large colony contains $10^{-5.6}$ mmol/*l* of OH^- ions at 25°C. The pH of this sample is

(a) 8.6 (b) 7.9
(c) 5.4 (d) 4.5

[ESE : 2016]

3.41 Consider the following characteristics of E.coli bacteria:

1. Gram negative
2. Spore-forming
3. Facultative anaerobic
4. Bacillus

Which of the above are correct?

(a) 1, 2 and 3 only (b) 2 and 4 only
(c) 1, 3 and 4 only (d) 1, 2, 3 and 4

[ESE : 2016]

3.42 The concentration of chloride ions in a water sample is estimated by titration with
(a) Sodium thiosulphate reagent using ferrion as an indicator
(b) Ferrous ammonium sulphate reagent using soluble starch as in indicator
(c) Silver nitrate reagent using potassium chromate as an indicator
(d) Silver nitrate reagent using potassium dichromate as in indicator
[ESE : 2017]

3.43 Consider the following statements regarding groundwater pollutants:
1. Most of the groundwaters are generally non-alkaline.
2. A moderate amount of fluoride, about 0.6 mg/*l* to 1.5 mg/*l*, in drinking water would help in good development of teeth.
3. Natural waters do not have dissolved mineral matter in them.

Which of the above statements is/are CORRECT?
(a) 1 only (b) 2 only
(c) 3 only (d) 1, 2 and 3
[ESE : 2018]

3.44 Which one of the following statements related to testing of water for municipal use is correctly applicable?
(a) Pseudo-hardness is due to presence of fluoride in water.
(b) When alkalinity ≥ total hardness, Carbonate hardness in mg/*l* = Total hardness in mg/*l*.
(c) Bicarbonate alkalinity = Total alkalinity – (carbonate alkalinity – hydroxide alkalinity).
(d) Hydroxide alkalinity = Carbonate alkalinity + Bicarbonate alkalinity.
[ESE : 2018]

3.45 Which one of the following compounds of nitrogen, when in excessive amounts in water, contributes to the illness known as infant methemoglobinemia.
(a) Ammoniacal nitrogen
(b) Albuminoid nitrogen
(c) Nitrite
(d) Nitrate [ESE : 2019]

3.46 Reactive substances are
(a) Unstable under normal conditions. They can cause explosions and/or liberate toxic fumes, gases, and vapors when mixed with water.
(b) Easily ignited and burn vigorously and persistently.
(c) Liquids with pH less than 2 or greater than 12.5, the those that are capable of corroding metal containers.
(d) Harmful or fatal when ingested or absorbed.
[ESE : 2020]

3.47 Water bearing stratum, having no confined impermeable over burden lying over it, is known as
(a) An unconfined aquifer
(b) An artesian aquifer
(c) Confined aquifer
(d) Controlled aquifer [ESE : 2022]

3.48 Which one of the following is the one which rests in a pervious stratum and draws its supply from the surrounding material?
(a) Sidetrack well (b) Horizontal well
(c) Deep well (d) Shallow well
[ESE : 2022]

3.49 Storage capacity of a reservoir can be estimated by using
(a) Cuboidal formula
(b) Cylindrical formula
(c) Prismoidal formula
(d) Conical formula
[ESE : 2022]

3.50 Modern commercial turbidimeter which works on the principle of scattering of light at right angles to the incident light, is called
(a) Spectrometer (b) Nephelometer
(c) Optimeter (d) Lightmeter
[ESE : 2022]

3.51 Which one of the following is a disease caused by protozoal infections under water-borne disease?
(a) Infectious hepatitis
(b) Amoebic dysentery
(c) Infectious jaundice
(d) Poliomyelitis [ESE : 2022]

3.52 Which one of the following is an essential requirement of swimming pool water?
(a) It should not have disinfectant dissolved in it

(b) Minimum amount of chemical should be added in treating water of the swimming pools

(c) The water of the pool need not be visible at maximum depth

(d) The pool need not have an efficient surface water removal system

[ESE : 2023]

3.53 The impurities in water which are extremely small size particles and cannot be removed by settling and filtration are known as

(a) Suspended impurities

(b) Colloidal impurities

(c) Dissolved impurities

(d) Picked up impurities

[ESE : 2024]

■■■■

Answers Water Quality Parameters

3.1 (c)	3.2 (d)	3.3 (a)	3.4 (c)	3.5 (b)	3.6 (d)	3.7 (c)	3.8 (c)	3.9 (d)
3.10 (d)	3.11 (a)	3.12 (d)	3.13 (d)	3.14 (a)	3.15 (c)	3.16 (c)	3.17 (b)	3.18 (c)
3.19 (b)	3.20 (d)	3.21 (a)	3.22 (d)	3.23 (c)	3.24 (b)	3.25 (d)	3.26 (c)	3.27 (d)
3.28 (a)	3.29 (c)	3.30 (d)	3.31 (c)	3.32 (a)	3.33 (c)	3.34 (c)	3.35 (c)	3.36 (c)
3.37 (c)	3.38 (b)	3.39 (b)	3.40 (c)	3.41 (c)	3.42 (c)	3.43 (b)	3.44 (b)	3.45 (d)
3.46 (a)	3.47 (a)	3.48 (d)	3.49 (c)	3.50 (b)	3.51 (b)	3.52 (b)	3.53 (c)	

Explanations Water Quality Parameters

3.2 (d)

Note: In given List-II point 3 should be Alkalinity and Hardness in place of Alkalinity and Softness.

3.3 (a)

Eutrophication is caused due to excessive nutrients in the water (nitrates, phosphates etc). The presence of total nitrogen indicates satisfactory microbial activity. Nitrite nitrogen indicates partly decomposed condition i.e. unsatisfactory microbial activity. Ammonia nitrogen presence indicates recent organic pollution.

3.4 (c)

Typhoid fever is caused by bacteria (Salmonella Typhi). Presence of excess nitrates > 45mg/*l* causes blue baby disease called Methemoglobinemia. Dysentry is caused by Protozoan (Entamoeba Histolytic). Poliomyelitis is caused by virus.

3.5 (b)

Salmonella typhi and Salmonella paratyphi (bacteria) cause enteric fever.

Enteric Cytopathogenic Human Orphan ECHO (virus) cause aseptic meningitis, epidemic exanthem, infantile diarrhoea.

Entamoeba histolytica (protozoa) cause dimebiasis (amebic dysentry, amebic enteritis, amebic colitis).

Echmococcus (helminth) cause enchinococcosis (hydlatidosis, granulosis, dog tapeworm).

3.6 (d)

Temporary Hardness – Carbonates and bicarbonates of Ca and Mg

Permanent Hardness – Sulphates and Chlorides of Ca and Mg

Alkalinity and Softness – Carbonates and bicarbonates of Na

Oxides of Iron and Manganese impart colour and taste to water.

3.7 (c)

Alkalinity expressed in equivalents

$$= HCO_3^- + OH^- - H^+$$

Milliequivalent mass of $[HCO_3^-]$

$$= \frac{122}{61} = 2 \text{ meq/L}$$

Milliequivalent mass of $[OH^-]$

$$= \frac{10^{-7}}{17} \text{meq/L}$$

Milliequivalent mass of $[H^+]$ $= \frac{10^{-7}}{1} \text{meq/L}$

The milliequivalent of $[OH^-]$ and $[H^+]$ are negligible. So Alkalinity as $CaCO_3$

$$= 2 \times 50 = 100 \text{ mg/L}$$

3.8 (c)

Molecular weight of $CaCl_2$

$$= 40 + 2 \times 35.0 = 110 \text{ g/mole}$$

Equivalent weight of $CaCl_2 = \frac{110}{2} = 55$

Milliequivalents of $CaCl_2$ in 110 mg/L

$$= \frac{110}{55} = 2 \text{ meq/L}$$

So $CaCl_2$ as $CaCO_3 = 2 \times 50 = 100$ mg/L

3.9 (d)

Monovalent cations impart pseudo-hardness. Hardness is the concentration of multivalent cations.

3.10 (d)

CO_2 reacts with water to form carbonic acid.

3.11 (a)

Since alkalinity is more than total hardness so only carbonate hardness is present.

Carbonate hardness = 200 mg/L

Non-carbonate hardness = zero

3.12 (d)

Carbonates are less soluble than bicarbonates. Bicarbonates begin to convert to less soluble carbonates at pH values above 9.0. The least soluble forms are calcium carbonate and magnesium hydroxide. Chemical precipitation is accomplished by converting calcium hardness to calcium carbonate and magnesium hardness to magnesium hydroxide.

3.13 (d)

In this problem we have more than three dilutions. Here 4-2-1 will give better representation. Corresponding MPN per 100 m*l* = 26.

3.14 (a)

Cholera	:	Bacteria
Worms	:	Helminths
Polio	:	Virus
Gastroenteritis	:	Protozoan

3.16 (c)

Hardness is defined as the concentration of multivalent metallic cations in solution. The multivalent metallic ions most abundant in natural waters are calcium and magnesium. Hardness can be measured by using spectrophotometric techniques or chemical titration to determine the quantity of calcium and magnesium ions in a given sample. Hardness can be measured directly by titration with ethylenediamine tetracetic acid (EDTA) using Eriochrome Black *T* (EBT) as an indicator.

3.17 (b)

The presence of nitrogen in sewage indicates the presence of organic matter and may occur in form of free ammonia, albuminoid nitrogen, nitrites and nitrates. The free ammonia indicated the very first stage of decomposition of organic matter (thrust indicating recently staled sewage); albuminoid nitrogen indicates quantity of nitrogen present in sewage before the decomposition of organic matter is started; the nitrites indicate the presence of partly decomposed (not fully oxidized) organic matter; and nitrates indicate the presence of fully oxidized matter.

3.18 (c)

Concentrations of fluorine upto 1.0 mg/L in drinking water help to prevent dental cavities in children. During formation of permanent teeth, fluoride combines chemically with tooth animal resulting in harder, stronger teeth that are more resistant to decay. When concentration of fluorides exceed 2.2 mg/L, it results in discoloration of teeth called mottling. Excess fluorides also results in bone fluorosis and other skeleton abnormalities.

3.19 (b)

$pOH = -\log[OH^-]$

where $[OH^-]$ is concentration of OH^- ions in moles per litre

$$\text{Moles} = \frac{\text{Weight in gm}}{\text{Molecular mass}} = \frac{17 \times 10^{-3}}{17}$$

$\therefore$ Moles of $OH^- = 10^{-3}$

$\therefore$ $pOH = -\log[10^{-3}]$

$\Rightarrow$ $pOH = 3$

But $pH + pOH = 14$

$\Rightarrow$ $pH = 14 - 3$

$\Rightarrow$ $pH = 11$

3.20 (d)

Threshold odour number is given as

$$TON = \frac{A+B}{A}$$

where A is the volume of odorous water (mL) and B is the volume of odor-free water required to produce a 200 mL mixture.

$$\therefore \quad TON = \frac{12.5+187.5}{12.5} = 16$$

3.21 (a)

The ions involved in water hardness, i.e. Ca^{2+}(aq) and Mg^{2+}(aq), can be determined by titration with a **chelating** agent, ethylene diamine tetra acetic acid (EDTA), usually in the form of disodium salt (H_2Y^{2-}). The titration reaction is:

$$Ca^{2+}(aq) + H_2Y^{2-}(aq) \longrightarrow CaY^{2-}(aq) + 2H^{+}(aq)$$

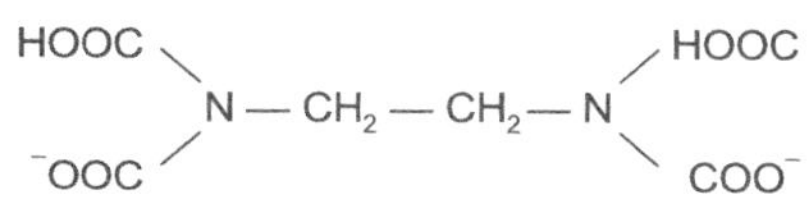

EDTA (anionic form)

Eriochrome Black T is commonly used as indicator for the above titration. The end point of titration is indicated by a sharp colour change from wine red to blue. Titration using Eriochrome Black T as indicator determines total hardness due to Ca^{2+}(aq) and Mg^{2+}(aq) ions. Hardness due to Ca^{2+}(aq) ion is determined by a separate titration at a higher pH, by adding NaOH solution to precipitate $Mg(OH)_2$(s), using hydroxynaphthol blue as indicator.

3.22 (d)

JTU: Jackson turbidity unit is a unit of measure turbidity on silica scale. Other turbidity units are NTU (Nephelometric turbidity unit) and FTU (Formazin Turbidity unit.

The extent of taste or odour present in a particular sample of water is measured by a term called odour intensity, which is related with the threshold odour. The threshold odour or threshold odour number, as it is generally called, represents the dilution ratio at which the odour is hardly detectable. MPN (Most Probable Number) is an index to measure the coliforms in the total water sample.

3.23 (c)

Fluorine upto (1 mg/*l* to 1.5 mg/*l*) is required for the growth of permenant teeth and to prevent dental cavity. If it is more than 1.5 mg/*l* it causes decolourisation of the teeth and result in its mottling (a d sease called flourisis).

3.24 (b)

The safe maximum limits in domestic water supply are:

Sulphates <=200 mg/*l*

Chlorides <=250 mg/*l*

Fluorides <=1.5 mg/*l*

Nitrates <=45 mg/*l*

3.25 (d)

Jackson's Turbidity Unit (JTU) is equal to the turbidity produced by 1 mg SiO_2 in 1 L of distilled water. Formazin, a chemical compound, provides more reproducible standards than SiO_2 and has replaced it as a reference. The term Nephelometre Turbidity Units (NTU) is often used to indicate that the test was run according to the scattering principle.

3.26 (c)

Fluoride concentrations of approximately 1.0 mg/*l* in drinking water help to prevent dental cavities in children. During formation of permanent teeth, fluoride combines chemically with tooth enamel resulting in harder and stronter teeth that are more resistant to decay.

3.27 (d)

Virusus are the smallest biological structure, containing all information necessary for its reproduction. They are obligate parasites, and as such, requ re a host in which to live.

3.28 (a)

Specific conductance and concentration of TDS are not related on one-to-one basis, only ionized substances contribute to specific conductance. Organic molecules and compounds that dissolve without ionizing are not measured. Additionally, the magnitude of specific conductance is influenced by the valancy of the ions in solution, there mobility and relative numbers.

3.29 (c)

The alkalinity present in water may be of mineral origin. It may be present due to atmospheric carbon dioxide mixed in water or due to microbial decomposition of organic matter.

$$CO_2 + H_2O \rightleftharpoons H_2CO_3$$

$$H_2CO_3 \rightleftharpoons H^+ + HCO_3$$

$$HCO_3^- \rightleftharpoons H^+ + CO_3^{2-}$$

$$CO_3^{2-} + H_2O \rightleftharpoons HCO_3^- + OH^-$$

Last reaction is a weak reaction but utilization of bicarbonate by algae drives the reaction to the right and hence sufficient accumulation of $[OH^-]$ takes place. Hence if algae is present in water, it becomes alkaline.

3.30 (d)

In EDTA method, water is titrated with ethylene diamene tetra acetic acid using EBT (Erichrome black T) as indicator. EBT forms red colour with water and titration changes the colour to blue.

3.31 (c)

Alkalinity in natural water is measured by titrating the water sample with 0.02 N H_2SO_4 using phenolphthalein and methyl orange indicators.

3.32 (a)

An approximate analysis of TDS is after made by determining the electrial conductivity of water. The ability of water to conduct electricity is known as specific conductance is a function of its ionic strength.

3.34 (c)

Typhoid fever is caused by bacteria not by viral infection where as infectious hapatitis is caused by virus not by bacteria.

3.35 (c)

Carbonate ions (CO_3^{2-}) react with water (H_2O)to form bicarbonate ions (HCO_3^-) & Hydroxide ions (OH^-).

In case of algae presence in water, it utilizes bicarbonate ions (HCO_3^-). On depletion of bicarbonate ions (HCO_3^-), more Hydroxide ions (OH^-) will be produced thereby increasing the alkalinity of water.

3.36 (c)

Environment includes the following two factors:

1. **Abiotic** factors refer to non-living physical and chemical elements in the ecosystem. Abiotic resources are usually obtained from the lithosphere, atmosphere, and hydrosphere. Examples of abiotic factors are water, air, soil, sunlight, and minerals.
2. **Biotic** factors are living or once-living organisms in the ecosystem. These are obtained from the biosphere and are capable of reproduction. Examples of biotic factors are animals, birds, plants, fungi, and other similar organisms.

3.37 (c)

Carbonate hardness

$$= \text{Minimum of} \begin{cases} \text{Total hardness} \\ \text{Alkalinity} \end{cases}$$

Carbonate hardness is 50 mg/l and total hardness is 120 mg/l

$\therefore$ Alkalinity is 50 mg/l

and, Non-carbonate hardness

= Total hardness – carbonate hardness

= (120 – 50) mg/l = 70 mg/l

3.38 (b)

- Dracontiasis, more commonly known as Guinea worm disease, is a preventable infection caused by the parasite Dracunculus medinensis. Infection affects poor communities in remote parts of Africa that do not have safe water to drink.
- Filtering drinking water using a fine mesh cloth of filter like nylon, can remove guinea worm from water. Hence statement II is not the explanation of statement I.

3.39 (b)

The EC of a solution is a measure of its ability to carry an electric current; the more dissolved ionic solutes in a water, the greater its electrical conductivity.

$$\left(EC\left(p\frac{\text{mho}}{\text{cm}}\right) \text{at } 25°\text{C}\right) \times 0.65 = \text{TDS}(\text{mg}/l).$$

3.40 (c)

$$[OH^-] = 10^{-5.6} \text{ mmole/lt.}$$
$$= 10^{-5.6} \times 10^{-3}$$
$$= 10^{-8.6} \text{ moles/lt.}$$
$$p[OH^-] = -\log[OH^-] = -\log[10^{-8.6}] = 8.6$$
$$pH + pOH = 14$$
$$\Rightarrow \quad pH = 14 - 8.6 = 5.4$$

3.41 (c)

E.coli is:

- Gram negative bacteria i.e. it leave a red strain in a staining technique which is performed to identify cell components.

- Non spore forming bacteria
- Facultative anaerobic i.e. can work both in presence or absence of oxygen.
- Bacillus i.e. they have straight rod shape with square or rounded ends.

3.42 (c)

The Mohr method uses silver nitrate for titration. This corresponds to 1 m*l* of 0.0141 equals to 1 mg chloride in solution. The silver nitrate solution is standardized against standard chloride solution, prepared from sodium chloride (NaC*l*). During the titration, chloride ion is precipitated as white silver chloride.

$$Ag^+ + Cl^- \rightleftharpoons AgCl$$

The potassium chromate as indicator is added to visualize the endpoint, demonstrating presence of excess silver ions. In the presence of excess silver ions, solubility product of silver chromate exceeded and it forms a reddish-brown precipitate. This stage is taken as evidence that all chloride ions have been consumed and only excess silver ions have reacted with chromate ions.

$$2Ag^+ + CrO_4^{2-} \rightleftharpoons Ag_2CrO_4$$

3.44 (b)

$(\text{Total hardness, Alkalinity})_{min}$ = Carbonate hardness

3.45 (d)

Nitrate in water in excess of 45 mg/*l* cause methemoglobinemia, as there are certain types of enzymes found in the intestine of infants which reduces nitrate into nitrite that has very high affinity for hemoglobin hence combines with it and leads to deficiency of O_2 and causes methemoglobinemia.

3.46 (a)

Reactive substance are solids, liquids or gaseous substances, mixture of substances, as well as substances that occur in combinations of such state, which upon contact with water, at their pressure, temperature or other chemical properties represent a risk of dangerous reaction explosion or emission of dangerous gas, vapour, dust or mist.

3.47 (a)

An unconfined aquifer is an aquifer whose upper water surface (water table) is at atmospheric pressure and thus is able to rise and fall.

3.49 (c)

Storage capacity of a reservoir can be estimated by mass curve, prismoidal and trapezoidal method.

3.50 (b)

Baylis turbidmeter works on the principle of scattering of light in the direction of light whereas nephelometer at right angles to the incident light.

3.53 (c)

- Colloidal impurities can be removed by filtration.
- Suspenced impurities and picked up impurities can be removed by sedimentation.

■■■■

4 Water Treatment

4.1 A rural water supply scheme serves a population of 10,000 at the rate of 50 litres per capita per day. For the chlorine dose of 2 ppm, the required amount of bleaching powder with 20% available chlorine will be

(a) 0.5 kg (b) 5 kg
(c) 10 kg (d) 15 kg [ESE : 2000]

4.2 The raw water entering an ideal horizontal settling tank contains following two types of particles:

Particle type	Settling velocity (m/h)	Concentration (mg/L)
I	3	200
II	1	300

When the surface overflow rate of the settling tank is 3 $m^3/m^2/h$, the concentration of the particles in the settled water will be

(a) 100 mg/L (b) 200 mg/L
(c) 300 mg/L (d) 400 mg/L

[ESE : 2000]

4.3 Which one of the following filters will produce water of higher bacteriological quality?

(a) Slow sand filter (b) Rapid sand filter
(c) Pressure filter (d) Dual media filter

[ESE : 2000]

Directions: The following items consists of two statements; one labelled as **'Assertion (A)'** and the other as **'Reason (R)'**. You are to examine these two statements carefully and select the answers to these items using the codes given below:

Codes:

(a) both A and R are true and R is the correct explanation of A
(b) both A and R are true but R is not a correct explanation of A
(c) A is true but R is false
(d) A is false but R is true

4.4 **Assertion (A):** Alum is the most commonly used coagulant in water treatment.

Reason (R): Alum is very effective in killing pathogens present in water.

[ESE : 2000]

4.5 **Assertion (A):** Break-point chlorination is required to ensure free chlorine residuals for better disinfection.

Reason (R): Free chlorine residuals have good disinfecting power and they are usually dissipated quickly in the distribution system.

[ESE : 2001]

4.6 Match **List-I** (Unit in water treatment plant) with **List-II** (Impurities removed) and select the correct answer using the codes given below the lists:

List-I
A. Aerator
B. Rapid sand filter
C. Slow sand filter
D. Sedimentation tank (after coagulation and flocculation)

List-II
1. Excess CO_2 and H_2S
2. Settleable and colloidal mater
3. Suspended matter
4. Suspended, colloidal and bacterio-logical matter

Codes:

	A	B	C	D
(a)	1	3	2	4
(b)	3	1	2	4
(c)	3	1	4	2
(d)	1	3	4	2

[ESE : 2001]

4.7 Consider the following impurities:

1. CO_2 and H_2S
2. Finely-divided suspended matter
3. Disease causing bacteria

4. Excess alkalinity

The correct sequence of the removal of these impurities in a water treatment plant is

(a) 1, 2, 3, 4 (b) 1, 4, 3, 2
(c) 1, 4, 2, 3 (d) 4, 1, 3, 2

[ESE : 2001]

4.8 Which of the following are the common problems associated with the operation of rapid sand-filter?

1. Air binding
2. Cracking of sand beds
3. Bumping of filter beds
4. Mud balls

Select the correct answer using the codes given below:

(a) 1 and 2 (b) 2 and 3
(c) 2, 3 and 4 (d) 1, 2, 3 and 4

[ESE : 2001]

4.9 Which of the following statements regarding industrial water supply is/are correct?

1. Industrial water supplies need not be disinfected.
2. Water for industrial use requires chemical treatment.
3. Standards of purity and methods of treatment of water or industrial use are often different from those for domestic water supplies.
4. It is more economical to use water from surface sources than groundwater sources for industrial water supplies.

Select the correct answer using the codes given below:

(a) 1, 2 and 4 (b) 2, 3 and 4
(c) 3 and 4 (d) 3 only

[ESE : 2001]

4.10 The amount of bleaching powder containing 20% available chlorine needed to chlorinate a rural water supply covering a population of 10000 at 50 Lpcd at the rate of 2 ppm is

(a) 1 kg (b) 5 kg
(c) 0.2 kg (d) 20 kg [ESE : 2001]

4.11 'Air binding' may occur in

(a) Sewers (b) Artesian well
(c) Aerator (d) Filter

[ESE : 2002]

4.12 The purpose of re-carbonation after lime-soda process of water softening is the

(a) removal of excess soda from water
(b) removal of non-carbonate hardness
(c) recovery of lime
(d) conversion of precipitates to soluble form

[ESE : 2003]

4.13 Which of the following treatment processes are necessary for removing suspended solids from water?

1. Coagulation 2. Flocculation
3. Sedimentation 4. Disinfection

Select the correct answer using the codes given below:

(a) 1 and 2 (b) 1, 2 and 3
(c) 2 and 4 (d) 1 and 4

[ESE : 2003]

4.14 Match **List-I** (Filter operating problems) with **List-II** (Effects) and select the correct answer using the codes given below the lists:

List-I

A. Air binding
B. Mud deposition
C. Cracking of bed
D. Sand incrustation

List-II

1. Changes effective size of sand
2. Mud penetrates deeper inside the bed
3. Mounds and balls of mud are formed in the bed
4. Air and gases get locked in the bed

Codes:

	A	B	C	D
(a)	4	3	2	1
(b)	3	4	1	2
(c)	4	3	1	2
(d)	3	4	2	1

[ESE : 2003]

4.15 Which of the following are removed by rapid sand filter from water?

1. Dissolved solids
2. Suspended solids
3. Bacteria
4. Helminths

Select the correct answer using the codes given below:

(a) 1 and 2 (b) 2 and 3
(c) 1 and 3 (d) 2, 3 and 4

[ESE : 2003]

4.16 If the specific gravity of a suspended particle is increased from 2 to 3, the settling velocity will
(a) not change
(b) get doubled
(c) get increased by 1.5 times
(d) get increased by 2.25 times [ESE : 2004]

4.17 Which one of the following filters should be recommended for protected rural water supply project?
(a) pressure filter
(b) slow sand filter
(c) diatomaceous earth filter
(d) rapid sand filter [ESE : 2004]

Directions: The following items consists of two statements; one labelled as **'Assertion (A)'** and the other as **'Reason (R)'**. You are to examine these two statements carefully and select the answers to these items using the codes given below:

Codes:
(a) both A and R are true and R is the correct explanation of A
(b) both A and R are true but R is not a correct explanation of A
(c) A is true but R is false
(d) A is false but R is true

4.18 **Assertion (A):** The fluoride content in drinking water should neither be too low nor too high.
Reason (R): Deficiency of fluoride content causes mottling of teeth and its excess causes dental cavities in children.

[ESE : 2004]

4.19 **Assertion (A):** The settling velocity of a discrete particle will become five times when its diameter doubles.
Reason (R): The settling velocity of a discrete particle is almost proportional to the square of the particle diameter.

[ESE : 2004]

4.20 Which one of the following is a not a specific criterion for calculating surface overflow rate in sedimentation tank design?
(a) Total quantity of water to be treated
(b) Total surface area available in the tank
(c) Total length of the tank
(d) Total depth of the tank [ESE : 2004]

4.21 Which of the following remedial measures are taken to avoid negative head and air binding in a rapid sand filter?
1. Avoiding the occurrence of excessive negative head
2. Pumping in air
3. Avoiding increase in water temperature
4. Control of algae growth

Select the correct answer using the codes given below:
(a) 1, 3 and 4 (b) 2, 3 and 4
(c) 1 and 2 (d) 1, 2, 3 and 4

[ESE : 2004]

4.22 Consider the following statements:
Aeration in water treatment helps in
1. killing pathogens
2. correcting pH
3. precipitating dissolved iron and manganese
4. expelling excess CO_2 and H_2S
5. expelling volatile oils

Which of these statements are correct?
(a) 2, 4 and 5 (b) 3, 4 and 5
(c) 1, 2, 3 and 5 (d) 1, 2, 3 and 4

[ESE : 2004]

4.23 Match **List-I** (Water/Waste water parameter) with **List-II** (Test) and select the correct answer using the codes given below the lists:

List-I	List-II
A. Potability of water	**1.** Mohr's method
B. Chloride	**2.** Orthotolidine test
C. Residual chlorine	**3.** EDTA method
D. Hardness of water	**4.** MF technique

Codes:

	A	B	C	D
(a)	4	3	2	1
(b)	2	1	4	3
(c)	2	3	4	1
(d)	4	1	2	3

[ESE : 2004]

4.24 **Assertion (A):** A small quantity of ammonia is added to water before carrying out disinfection using chlorine.

Reason (R): Chloramines are persistent disinfectants, which provide continued protection against regrowth of micro-organisms in a water distribution system.

(a) both A and R are true and R is the correct explanation of A
(b) both A and R are true but R is not a correct explanation of A
(c) A is true but R is false
(d) A is false but R is true

[ESE : 2005]

4.25 A flash mixer of 2.0 m^3, with a velocity gradient of mixing mechanism equal to 600 s^{-1}, and fluid absolute viscosity of 1.0×10^{-3} N-s/m^2 is continuously operated. What is the power input per unit volume?

(a) 360 W (b) 720 W
(c) 1440 W (d) 300 W

[ESE : 2005]

4.26 Which one of the following types of settling phenomenon can be analyzed by the classic sedimentation laws of Newton and Stokes?

(a) Discrete settling
(b) Flocculent settling
(c) Hindered settling
(d) Compression settling

[ESE : 2005]

4.27 Match **List-I** (Treatment process) with **List-II** (Removed matter) and select the correct answer using the codes given below the lists:

List-I

A. Plain sedimentation
B. Chemical precipitation
C. Slow Sand Filtration
D. Aeration

List-II

1. Dissolved gases
2. Dissolved solids
3. Suspended solids with specific gravity more than 1.0.
4. Floating solids
5. Bacterial cells

Codes:

	A	B	C	D
(a)	5	1	4	2
(b)	3	2	5	1
(c)	5	2	4	1
(d)	3	1	5	2

[ESE : 2005]

4.28 According to the theory of filtration in water treatment, which of the following mechanisms come into play when water is filtered through a filter bed?

1. Mechanical straining
2. Capillary action
3. Centrifugal force
4. Electro-kinetic phenomenon
5. Osmotic force
6. Bacteriological action

Select the correct answer using the codes given below:

(a) 1, 2, 4 and 6 (b) 2, 3 and 5
(c) 3, 4, 5 and 6 (d) 1, 3, 4 and 6

[ESE : 2005]

4.29 Which of the following are associated with alum coagulation?

1. A decrease in alkalinity of treated water
2. Formation of hydroxide flocs of aluminium
3. A slight decrease in pH of treated water
4. An increase in permanent hardness

Select the correct answer using the codes given below:

(a) 1, 2 and 3 (b) 1, 3 and 4
(c) 1, 2, 3 and 4 (d) 2 and 4

[ESE : 2005]

4.30 Chlorides from water are removed by

(a) Lime soda process
(b) Reverse osmosis
(c) Cation exchange process
(d) Chemical coagulation [ESE : 2006]

4.31 In which treatment unit is *Schmutzdecke* formed?

(a) Sedimentation tank
(b) Rapid sand filter
(c) Coagulation tank
(d) Slow sand filter [ESE : 2006]

4.32 Which one of the following chemicals is employed for dechlorination of water?

(a) Sodium sulphite
(b) Sodium bicarbonate
(c) Calcium carbonate
(d) Hydrogen peroxide [ESE : 2006]

4.33 Which one of the following is the correct sequence of slow sand filter (SSF), rapid sand filter (RSF), dual media filter (DMF) and mixed media filter (MMF) in the decreasing order of their filtration rates?

(a) MMF ≈ DMF > RSF > SSF
(b) DMF > RSF > SSF > MMF
(c) RSF > SSF > MMF ≈ DMF
(d) SSF > MMF ≈ DMF > RSF [ESE : 2006]

Directions: The following items consists of two statements; one labelled as **'Assertion (A)'** and the other as **'Reason (R)'**. You are to examine these two statements carefully and select the answers to these items using the codes given below:

Codes:
(a) both A and R are true and R is the correct explanation of A
(b) both A and R are true but R is not a correct explanation of A
(c) A is true but R is false
(d) A is false but R is true

4.34 **Assertion (A):** Flocculating particles settle in a curved path in a long rectangular sedim-entation tank designed for a constant flow rate.

Reason (R): The downward settling velocities of particles as well as their respective horizontal velocities remain constant during their stay in the tank.

[ESE : 2006]

4.35 **Assertion (A):** Large weir overflow rates will result in excessive velocities at the outlet of a settling basin.

Reason (R): These excessive velocities will extend backward into the settling zone, causing particles and flocs which would otherwise be removed as sludge to be drawn into the outlet.

[ESE : 2006]

4.36 Chlorination with Cl_2 produces hypoch-lorous acid (HOCl), which may further dissociate as hypochlorite ion (OCl^-) depending upon pH of the water. The reaction is represented as:

$$HOCl \underset{}{\overset{k}{\rightleftharpoons}} H^+ + OCl^-$$

($k = 2.5 \times 10^{-8}$ moles/L at 20°C)

What is the fraction of HOCl in the water at pH 7.0?

(a) 0.95 (b) 0.80
(c) 0.20 (d) 0.05

[ESE : 2006]

4.37 Which one of the following treatments is economically effective in the control of guinea worm disease?

(a) Chlorination (b) Filtration
(c) Ozonation (d) Sedimentation

[ESE : 2007]

4.38 Match **List-I** (Disinfectant) with **List-II** (Property) and select the correct answer using the codes given below the lists:

List-I
A. Chlorine
B. Ozone
C. Iodine
D. Ultra-violet rays

List-II
1. No carcinogenics results
2. Ineffective in the presence of suspended solids
3. Not affected by the Ammonium ion
4. Feasible residual oxygen

Codes:

	A	B	C	D
(a)	4	3	1	2
(b)	1	2	4	3
(c)	4	2	1	3
(d)	1	3	4	2

[ESE : 2007]

4.39 What is the predominating coagulation mechanism for raw water having high turbidity and high alkalinity?

(a) Ionic layer compression
(b) Adsorption and charge neutralization
(c) Sweep coagulation
(d) Inter particle bridging [ESE : 2007]

4.40 **Assertion (A):** A discrete particle (of diameter d_0) settling in a circular sedimentation tank follows a parabolic path.

Reason (R): The downward settling velocity (v_0) of the discrete particle (of diameter d_0) in a circular sedimentation tank does not change with time.

(a) both A and R are true and R is the correct explanation of A
(b) both A and R are true but R is not a correct explanation of A

(c) A is true but R is false
(d) A is false but R is true [ESE : 2007]

4.41 Match **List-I** (Predominance of compounds) with **List-II** (pH range) and select the correct answer using the codes given below the lists:

List-I	List-II
A. Monochloramine	**1.** Below pH 4.4
B. Dichloramine	**2.** Over pH 7.5
C. Nitrogen trichloride	**3.** Between pH 5 to 6.5

Codes:

	A	B	C
(a)	1	2	3
(b)	2	3	1
(c)	3	1	2
(d)	3	2	1

[ESE : 2008]

4.42 What is the most important design parameter used in designing a continuous flow rectangular sedimentation tank for removal of discrete particles?
(a) Length of the tank
(b) Surface overflow rate
(c) Depth of the tank
(d) Temperature of the water to be treated
[ESE : 2009]

4.43 Which combination of surface water quality parameters will indicate sweep coagulation as the preferred mechanism of coagulation?
(a) High turbidity–low alkalinity
(b) High turbidity–high alkalinity
(c) Low turbidity–high alkalinity
(d) Low turbidity–low alkalinity
[ESE : 2009]

4.44 Which one of the following processes of water softening requires recarbonation?
(a) Lime-soda ash process
(b) Hydrogen-cation exchanger process
(c) Sodium-cation exchanger process
(d) Demineralization
[ESE : 2009]

4.45 Consider the following treatment process units in a water treatment plant:
1. Coagulation 2. Disinfection
3. Sedimentation 4. Filtration
Which is the correct sequence of the process units in the water treatment plant?
(a) 2–4–3–1 (b) 1–4–3–2
(c) 2–3–4–1 (d) 1–3–4–2
[ESE : 2009]

4.46 The correct sequence of treatment of typical turbid surface water is
(a) flocculation, coagulation, sedimentation, filtration
(b) flocculation, coagulation, filtration, sedimentation
(c) coagulation, flocculation, filtration, sedimentation
(d) coagulation, flocculation, sedimentation, filtration [ESE : 2010]

4.47 Consider the following statements:
The appropriate method(s) for removal of fluorides from water comprise
1. addition of alum and lime followed by clarification.
2. passing through beds of activated alumina.
Which of these statements is/are correct?
(a) Neither 1 nor 2 (b) Both 1 and 2
(c) 1 only (d) 2 only
[ESE : 2010]

4.48 Consider the following statements:
In water treatment, the addition of chlorine inactivates the cells of pathogenic bacteria through
1. penetration of the chlorine species through the cell walls.
2. suffocation of the cells.
3. reaction of the chlorine species with the enzyme system of the cells.
Which of these statements are correct?
(a) 1, 2 and 3 (b) 1 and 2 only
(c) 2 and 3 only (d) 1 and 3 only
[ESE : 2010]

4.49 In a water treatment, the optimum time of flocculation is usually given as 30 minutes. In case the time of flocculation is increased beyond this value, the flocs will
(a) become heavy and settle down in flocculation itself
(b) entrap air and will float in the sedimentation tank
(c) break up and defeat the purpose of flocculation
(d) stick to the paddles [ESE : 2010]

4.50 **Assertion (A):** Break point chlorination ensures a residual of free available chlorine.

Reason (R): A super high chlorine dose inactivates the pathogens in a very short time.

(a) both A and R are true and R is the correct explanation of A
(b) both A and R are true but R is not a correct explanation of A
(c) A is true but R is false
(d) A is false but R is true [ESE : 2010]

4.51 If organic sources of carcinogenic compounds in water persist even after chlorination, then what is the correct sequence among treatment processes listed below if all these are considered compulsory?

1. Coagulation
2. Sedimentation
3. Filtration in general
4. Activated carbon bed filtration
5. Flocculation
6. Chlorination

(a) 4-5-3-2-6 and 1
(b) 1-2-3-4-5 and 6
(c) 4-2-3-1-5 and 6
(d) 1-5-2-3-4 and 6 [ESE : 2011]

4.52 Consider the following statements:

The role of the gravel bed in a rapid sand filter is

1. To filter out large suspended matter.
2. To support the sand bed above it.
3. To prevent the escape of sand particles.
4. To uniformly distribute the backwash water.
5. To prevent algae growth.

Which of these statements are correct?

(a) 1, 2, 3, 4 and 5 (b) 2, 3 and 4 only
(c) 3, 4 and 5 only (d) 1, 2 and 3 only

[ESE : 2011]

4.53 Which of the following methods are employed for determination of free and combined chlorine residuals in water?

1. Starch-iodide method
2. Orthotolidine method
3. Amperometric titration method
4. SNORT method
5. DPD method

(a) 1, 2, 4 and 5 (b) 1, 2 and 3 only
(c) 3, 4 and 5 only (d) 2, 3 and 4 only

[ESE : 2011]

4.54 Consider the following statements:

The following factors relate to the process of coagulation

1. Percentage removal is higher when turbidity is more.
2. Addition of activated silica aids in the process of coagulation.
3. pH of water is an important consideration for selecting a coagulant.

Which of these statements are correct?

(a) 1, 2 and 3 (b) 1 and 2 only
(c) 2 and 3 only (d) 1 and 3 only

[ESE : 2011]

4.55 Match **List-I** with **List-II** and select the correct answer using the code given below the lists:

List-I

A. Primary sedimentation
B. Coagulation
C. Flocculation
D. Secondary sedimentation

List-II

1. Differential settling
2. Hindered settling
3. Charge neutralization
4. Growth of flocs
5. Flow-through velocity

Code:

	A	B	C	D
(a)	2	5	4	1
(b)	1	3	4	2
(c)	2	4	3	1
(d)	5	4	3	2

[ESE : 2011]

4.56 Which of the following operational problems relate to the functioning of rapid gravity filter?

1. Inadequate media comprising filter bed
2. Sludge bulking
3. Mud balls
4. Negative head
5. Incrustation of media

(a) 1, 3, 4 and 5 only (b) 1, 2, 3 and 4 only
(c) 2, 3, 4 and 5 only (d) 1, 3, 4 and 5

[ESE : 2011]

Directions: The following items consists of two statements; one labelled as **'Assertion (A)'** and the other as **'Reason (R)'**. You are to examine these two statements carefully and select the answers to these items using the codes given below:

Codes:
(a) both A and R are true and R is the correct explanation of A
(b) both A and R are true but R is not a correct explanation of A
(c) A is true but R is false
(d) A is false but R is true

4.57 **Assertion (A):** Disinfection is the last treatment given to water before it is supplied to consumers.
Reason (R): Any other treatment after disinfection may incidentally also contaminate the water.
[ESE : 2011]

4.58 **Assertion (A):** The bottom layers of water in a deep reservoir are usually not acceptable as raw water in a water supply system.
Reason (R): The bottom water layers may contain products of anaerobic degradation.
[ESE : 2011]

4.59 Conside r the following statements:
1. Carbonate hardness is due to bicarbonates.
2. Non-carbonate hardness is due to sulphates and chlorides of Ca and Mg.
3. Both the hardnesses can be removed by lime-soda method.
4. Both the hardnesses can be removed by ion-exchange method.

Which of these statements are correct.
(a) 1, 2 and 3 only (b) 1, 2 and 4 only
(c) 2, 3 and 4 only (d) 1, 2, 3 and 4
[ESE : 2012]

4.60 Which of the following reasons are responsible for adoption of post-chlorination of water?
1. Chlorine demand is reduced.
2. Possibility of taste and odour formation is reduced.
3. Possibility of carcinogenic compounds is reduced.
4. Chloramines are formed.

(a) 1, 2, 3 and 4 (b) 1, 2 and 3 only
(c) 1 and 4 only (d) 2, 3 and 4 only
[ESE : 2012]

Directions: Each of the next items consists of two statements, one labeled as the **'Statements (I)'** and the other as **'Statement (II)'**. You are to examine these two statements carefully and select the answers to these items using the codes given below :

Codes:
(a) Both Statement (I) and Statement (II) are individually true and Statement (II) is the correct explanation of Statement (I)
(b) Both Statement (I) and Statement (II) are individually true but Statement (II) is NOT the correct explanation of Statement (I)
(c) Statement (I) is true but Statement (II) is false
(c) Statement (I) is false but Statement (II) is true

4.61 **Statement-I :** Chlorides are added to kill pathogens as a disinfection process in the treatment of water.
Statement-II : It forms hypochlorous acid to oxidize the organic compounds including bacteria.
[ESE : 2012]

4.62 **Statement-I :** An epidemic of infective hepatitis is transmitted by drinking contaminated water.
Statement-II : Since infective hepatitis is transmitted by bacteria, it can be controlled by filtration and disinfection of water.
[ESE : 2012]

4.63 **Statement (I):** Rapid sand filters consisting of carefully sieved sand on top of a bed of gravel will only filter particles larger than its pore size.
Statement (II): In addition to staining, adsorption, continued flocculation and sedimentation in the pore spaces help in particles removal mechanism.
[ESE : 2013]

4.64 **Statement (I):** Suspended as well as colloidal particulates are removed from water by addition of 25% alum dosage.
Statement (II): The optimum dosage of alum is determined by performing the 'Jar test'.
[ESE : 2013]

4.65 **Statement (I):** In a rectangular baffled channel, it is easy to alter the G value.
Statement (II): In a rectangular baffled channel, the baffles are rigidity built and so cannot undergo any change.
[ESE : 2013]

4.66 Which of the following statements are the important characteristics of a slow sand filter?

1. Cleaning of filter is done by scraping and sand removal.
2. Lack of pretreatment.
3. Greater efficiency of bacterial removal as compared to rapid sand filter.
4. Efficient in colour, taste and odour removal.

(a) 1, 2, 3 and 4 (b) 1, 2 and 4 only
(c) 2, 3 and 4 only (d) 1, 2 and 3 only

[ESE : 2013]

4.67 A water treatment plant treats 6000 m^3 of water per day. If it consumes 20 kg chlorine per day, then the chlorine dosage would be

(a) 3.00 mg/l (b) 3.75 mg/l
(c) 4.25 mg/l (d) 3.33 mg/l

[ESE : 2014]

4.68 When a turbid stream flows into the ocean, the ionic content of the water increases drastically. This causes coagulation and settling which leads to formation of deposits (deltas). Such coagulation mechanism is called

(a) Adsorption and charge neutralization
(b) Interparticle bridging
(c) Ionic layer compression
(d) Sweep coagulation [ESE : 2014]

4.69 A rapid sand filter for a town with water requirement of 1 MLD is to be provided with rate of filtration at 4000 l/hr/m^2, with backwash system. What would be the size of filter and its underdrainage system to the nearest approximation?

(a) 10 and 0.032 sq.m
(b) 11 and 0.022 sq.m
(b) 12 and 0.045 sq.m
(d) 15 and 0.062 sq.m

[ESE : 2014]

4.70 Consider the following statements in respect of injecting chlorine between rapid sand filter and storage tank:

1. More contact time is available.
2. Chlorine demand is reduced.
3. $NHCl_2$ is formed which is more effective than residual chlorine.
4. Break point chlorination takes place.

Which of the above statements are correct?

(a) 1, 2, 3 and 4 (b) 1, 2 and 3 only
(c) 1 and 2 only (d) 3 and 4 only

[ESE : 2014]

4.71 What is the settling velocity of a discrete particle in a wide body of water when the relevant Reynold's number is less than 0.5? The diameter and specific gravity of the particle are 2×10^{-3} cm and 2.65, respectively. Water temperature is 20°C. (Kinematic viscosity = 2×10^{-2} cm^2/sec)

(a) 0.018 cm/sec (b) 0.025 cm/sec
(c) 0.18 cm/sec (d) 0.25 cm/sec

[ESE : 2014]

4.72 If the length dimension of a square filter bed increase to two times (while the rate of filtration remains unchanged), the amount of water filtered would become

(a) 4 times (b) 2 times
(c) 1 time (d) 16 times

[ESE : 2014]

4.73 **Statement (I):** During an epidemic of infective hepatitis, the supplied water is super chlorinated.
Statement (II): Spore-forming bacteria which cause this disease are effectively removed from water by super chlorination.

(a) Both Statement (I) and Statement (II) are individually true and Statement (II) is the correct explanation of Statement (I)
(b) Both Statement (I) and Statement (II) are individually true but Statement (II) is ***not*** the correct explanation of Statement (I)
(c) Statement (I) is true but Statement (II) is false
(d) Statement (I) is false but Statement (II) is true

[ESE : 2014]

4.74 A certain water needs alum treatment to the extent of 10 p.p.m. How much alum, in quintals per day, would be needed to treat 10 MLD of water?

(a) 10 (b) 1.0
(c) 100 (d) 1000 [ESE : 2015]

4.75 10 MLD water is to be chlorinated in a clear water reservoir (CWR) with 0.8 mg/l chlorine dose with providing contact time of 40 minutes. The required CWR capacity is nearly

(a) 220 m^3 (b) 280 m^3
(c) 28 m^3 (d) 22 m^3

[ESE : 2015]

4.76 What is the rapid sand filter surface area required for filtering of 10 MLD water assuming a filtration rate of 100000 $l/m^2/day$?
(a) 100 m^2 (b) 10 m^2
(c) 1 m^2 (d) 1000 m^2
[ESE : 2015]

4.77 Consider the following statements in respect of slow sand filter and rapid sand filter:
1. The two filters differ in respect of the standards regarding non-uniformity of the sand used in their filtering media.
2. The two filters do not differ in respect of the effective size of the sand used in them.
3. The two filters differ in respect of their respective under-drainage system.
4. The two filters differ in respect of their respective rate of filtration.

Which of the above statements are correct?
(a) 1, 2 and 3 (b) 1, 2 and 4
(c) 2, 3 and 4 (d) 1, 3 and 4
[ESE : 2015]

4.78 How many kg of bleaching powder is needed per day to chlorinate 4 MLD of water of so that, after 40 minutes of contact, there remains residual chlorine of 0.25 mg/l. The input water has a chlorine demand of 1.25 mg/l, and that the bleaching powder has only 25% available chlorine.
(a) 8 kg (b) 20 kg
(c) 24 kg (d) 6.6 kg
[ESE : 2015]

4.79 **Assertion (A):** Coagulation is the process of charge neutralization on colloids.
Reason (R): Flocculation is the process to grow the chargeless colloids into settleable flocs.
(a) Both A and R are true and R is the correct explanation of A
(b) Both A and R are true but R is not a correct explanation of A
(c) A is true but R is false
(d) A is false but R is true
[ESE : 2015]

4.80 Consider the following statements regarding the overflow rate of a sedimentation tank:
1. Temperature of water affects the overflow rate.
2. Size of particle intended to be removed does not affect the overflow rate.
3. Density of particle intended to be removed affects the overflow rate.

Which of the above statements are correct?
(a) 1 and 3 only (b) 1 and 2 only
(c) 2 and 3 only (d) 1, 2 and 3
[ESE : 2015]

4.81 The following are certain operating problems of a rapid sand filter:
1. Sand depth should never be depleted by more than 10 cm.
2. Air binding results due to development of negative head and formation of air bubbles in the filter sand.
3. Water softening with lime-soda leads to incrustation of sand.
4. Bumping of filter bed is caused due to negative head.

Which of the above statements are correct?
(a) 1, 2 and 4 only (b) 1, 2 and 3 only
(c) 3 and 4 only (d) 1, 2, 3 and 4
[ESE : 2016]

4.82 How much bleaching powder is needed to chlorinate 5000 l of water whose chlorine demand is 2 mg/l, assuming that the bleaching powder has 25% available chlorine?
(a) 4 g (b) 40 g
(c) 140 g (d) 400 g [ESE : 2016]

4.83 The design overflow rate of a sedimentation tank is chosen considering
(a) Flow rate through the tank
(b) Diameter of the particle intended to be removed
(c) Volume of the sedimentation tank
(d) Detention time in the tank
[ESE : 2016]

4.84 **Statement (I):** Recarbonation of water softened by, lime-soda process results in increased hardness of the water.

Statement (II): Suspended solids, like $CaSO_4$ and $MgSO_4$, which have not settled in the sedimentation tank get dissolved due to passage of CO_2.
(a) Both Statement (I) and Statement (II) are individually true and Statement (II) is the correct explanation of Statement (I)

(b) Both Statement (I) and Statement (II) are individually true but Statement (II) is ***not*** the correct explanation of Statement (I)
(c) Statement (I) is true but Statement (II) is false
(d) Statement (I) is false but Statement (II) is true

[ESE : 2016]

4.85 Consider the following statements regarding coagulant aids:

1. They are employed when temperature is low as they then accelerate the process of coagulation.
2. They are employed when flocs are small and water is coloured.
3. With their use, water purification capacity of the plant is increased.
4. Most widely used materials are oxidants, absorbent, weighing agents, activated carbon and polyelectrolytes.

Which of the above statements are correct?
(a) 1, 2, 3 and 4 (b) 1, 2 and 3
(c) 1 and 4 (d) 2, 3 and 4

[ESE : 2017]

4.86 A good disinfectant

1. Should be persistent enough to prevent regrowth of organisms in the distribution system.
2. Must be toxic to micro-organism at concentrations well above the toxic thresholds of humans and higher animals.
3. Should have a fast rate of kill of micro-organisms.

Which of the above statements are correct?
(a) 1, 2 and 3 (b) 1 and 2
(c) 1 and 3 (d) 2 and 3

[ESE : 2017]

4.87 How much bleaching powder (having 20% available chlorine) is needed to disinfect 10000 litres of water whose chlorine demand is 1.0 mg/*l*? It should be ensured that, after about 1 hour contact time 0.2 mg/*l* chlorine is available.
(a) 30 g (b) 40 g
(c) 48 g (d) 68 g [ESE : 2017]

4.88 Consider the following statements regarding removal of impurities from water:

1. Settleable solids are removed by filtration.
2. Volatile solids are removed through sedimentation.
3. Dissolved solids are removed through reverse osmosis.
4. Colloidal solids are removed by coagulation.

Which of the above statements are correct?
(a) 1 and 2 (b) 3 and 4
(c) 2 and 3 (d) 1 and 4

[ESE : 2017]

Directions: Each of the next items consists of two statements, one labeled as the **'Statements (I)'** and the other as **'Statement (II)'**. You are to examine these two statements carefully and select the answers to these items using the codes given below :

Codes:
(a) Both Statement (I) and Statement (II) are individually true and Statement (II) is the correct explanation of Statement (I)
(b) Both Statement (I) and Statement (II) are individually true but Statement (II) is NOT the correct explanation of Statement (I)
(c) Statement (I) is true but Statement (II) is false
(d) Statement (I) is false but Statement (II) is true

4.89 **Statement (I):** Softening of clear groundwater should be carried out immediately after collection by pumping out, or from springs.
Statement (II): Iron and manganese precipitates can foul the exchange medium surface if oxidation occurs in, or prior to, the ion-exchange phase.

[ESE : 2017]

4.90 **Statement (I):** Incrustation of sand occurs in rapid sand filter when water softened by lime-soda is fed in for filtration.
Statement (II): The sand gets coated with precipitates of $Ca(OH)_2$ and $Mg(OH)_2$ generated during softening of water by lime-soda process which leads to incrustation of sand.

[ESE : 2017]

4.91 Zero hardness of water is achieved by
(a) Lime-soda process
(b) Ion exchange treatment
(c) Excess lime treatment
(d) Excess alum dosage

[ESE : 2018]

4.92 The capacity of a service reservoir in a campus should cater to
(a) Sum total of balancing storage, breakdown storage and fire reserve.

(b) Sum total of balancing storage and fire reserve.
(c) Sum total of breakdown storage and fire reserve.
(d) Balancing storage only.

[ESE : 2018]

4.93 The purpose of re-carbonation after water softening by the lime soda process is the
(a) Removal of excess soda from the water.
(b) Removal of non-carbonate hardness in the water.
(c) Recovery of lime from the water.
(d) Conversion of precipitates to soluble forms in the water.

[ESE : 2018]

4.94 The demand of water is 150 litres/head/day in a city of one lakh population. The factor of safety is taken as 1.5, detention time as 4 h and overflow rate as 20,000 litres/day/m^2. The area of 3 m deep plain sedimentation tank as per surface loading considerations will be
(a) 1025 m^2 (b) 1075 m^2
(c) 1125 m^2 (d) 1175 m^2

[ESE : 2019]

4.95 Chlorine usage in the treatment of 25,000 m^3/day of water has been 9 kg/day. The residual chlorine after 10 minutes contact is 0.2 mg/l. The chlorine demand of water would be nearly
(a) 0.28 mg/l (b) 0.22 mg/l
(c) 0.16 mg/l (d) 0.12 mg/l

[ESE : 2019]

4.96 **Statement (I) :** Alum works in slightly alkaline range.

Statement (II) : At higher temperatures, viscosity of water (resistance to settling) decreases and flocs settle better.
(a) Both Statement (I) and Statement (II) are individually true, and Statement (II) is the correct explanation of Statement (I)
(b) Both Statement (I) and Statement (II) are individually true, but Statement (II) is NOT the correct explanation of Statement (I)
(c) Statement (I) is true, but Statement (II) is false
(d) Statement (I) is false, but Statement (II) is true

[ESE : 2019]

4.97 Flocculation is the process of
(a) Gently mixing the water and coagulant allowing the formation of large particles of floc.
(b) Removing relatively large floating and suspended debris.
(c) Flow, which is slowed enough so that gravity will cause the floc to settle.
(d) Mixture of solids and liquids collected from the settling tank are dewatered and disposed of.

[ESE : 2020]

4.98 When chlorine is dissolved in water, it reacts to form hypochlorous acid and hypochlorite ions. At pH < 5, chlorine exists in water as
(a) Elemental or molecular chlorine
(b) Remains in the form of hypochlorous acid
(c) Remains in the form of hypochlorite ions
(d) Remains in the form of both hypochlorous acid and hypochlorite ions

[ESE : 2020]

4.99 Which one of the following is the process whereby chemicals are added to a wastewater resulting in a reduction of the forces tending to keep suspended particles apart?
(a) Coagulation (b) Flocculation
(c) Clarification (d) Sedimentation

[ESE : 2021]

4.100 Which on of the following is the advantages of using activated carbon for water treatment?
(a) When used in powdered from after coagulation, it does not aid is coagulation.
(b) It increases the chlorine demand of treated water.
(c) It removes organic matter present in water.
(d) Its overdose is harmful.

[ESE : 2021]

4.101 Which of the following methods adopted for purifying the public water supplies?
1. Screening
2. Sedimentation aided with coagulation
3. Disinfection

Select the correct answer using the codes given below:
(a) 1 and 2 only (b) 2 and 3 only
(c) 1 and 3 only (d) 1, 2 and 3

[ESE : 2022]

4.102 According to Stoke's law, if the diameter (d) is less than 0.1 mm, then the settling velocity of spherical particles is directly proportional to

(a) d^2 (b) d^3
(c) d (d) $d^{1/2}$

[ESE : 2022]

4.103 What is the settling velocity of a discrete particles in water under conditions when Reynold's number is less than 0.5? (Take the diameter and specific gravity of the particles are 5×10^{-3} cm and 2.65 respectively and Kinematic viscosity of water at 20°C is 1.01×10^{-2} cm²/s and g = 9.81 m/s²)

(a) 0.22 cm/s (b) 0.35 cm/s
(c) 0.14 cm/s (d) 0.46 cm/s

[ESE : 2022]

4.104 The rate of filtration of pressure filter as compared to rapid gravity filter is about

(a) 10 times (b) 15 times
(c) 2 to 5 times (d) 6 to 8 times

[ESE : 2022]

4.105 Which one of the following is NOT the common method of disinfection of water?

(a) By boiling of water
(b) By ultra-violet rays
(c) By use of ozone
(d) By use of sodium chloride

[ESE : 2023]

4.106 What is the function of flocculator in the typical water treatment scheme for surface water?

(a) To rapid dispersion of chemical coagulant(s) to encourage destabilization of colloids
(b) To permit the settlement of chemical 'flocs' along with colloidal particles
(c) To provide for gentle mixing of the destabilized colloids to promote agglomeration of colloids into large easily settleable flocs
(d) To remove the flocs and colloids which escape from the settlement in the secondary sedimentation tank

[ESE : 2024]

4.107 What is the function of rapid mixing in the typical water treatment flow scheme for ground water with high hardness?

(a) To drive out the objectionable dissolved gas such as H_2S and CO_2
(b) To permit the settlement of chemical precipitates under gravity
(c) To reform the calcium and magnesium bicarbonates to prevent settlement of $CaCO_3$ under $Mg(OH)_2$ precipitates in the distribution lines
(d) To disperse the lime and soda ash to form the chemical precipitates

[ESE : 2024]

■■■■

Answers Water Treatment

4.1 (b)	4.2 (b)	4.3 (a)	4.4 (c)	4.5 (c)	4.6 (d)	4.7 (c)	4.8 (d)	4.9 (b)
4.10 (b)	4.11 (d)	4.12 (d)	4.13 (b)	4.14 (a)	4.15 (d)	4.16 (b)	4.17 (b)	4.18 (c)
4.19 (d)	4.20 (d)	4.21 (d)	4.22 (b)	4.23 (d)	4.24 (a)	4.25 (a)	4.26 (a)	4.27 (b)
4.28 (d)	4.29 (c)	4.30 (b)	4.31 (d)	4.32 (a)	4.33 (a)	4.34 (c)	4.35 (b)	4.36 (b)
4.37 (b)	4.38 (a)	4.39 (c)	4.40 (b)	4.41 (b)	4.42 (b)	4.43 (b)	4.44 (a)	4.45 (d)
4.46 (d)	4.47 (b)	4.48 (d)	4.49 (b)	4.50 (b)	4.51 (d)	4.52 (b)	4.53 (a)	4.54 (a)
4.55 (b)	4.56 (a)	4.57 (a)	4.58 (a)	4.59 (c)	4.60 (d)	4.61 (d)	4.62 (c)	4.63 (d)
4.64 (d)	4.65 (d)	4.66 (d)	4.67 (d)	4.68 (c)	4.69 (b)	4.70 (c)	4.71 (a)	4.72 (a)
4.73 (c)	4.74 (b)	4.75 (b)	4.76 (a)	4.77 (d)	4.78 (c)	4.79 (b)	4.80 (a)	4.81 (b)
4.82 (b)	4.83 (b)	4.84 (c)	4.85 (a)	4.86 (c)	4.87 (*)	4.88 (b)	4.89 (d)	4.90 (c)
4.91 (b)	4.92 (a)	4.93 (d)	4.94 (c)	4.95 (c)	4.96 (b)	4.97 (a)	4.98 (a)	4.99 (a)
4.100 (c)	4.101 (d)	4.102 (a)	4.103 (a)	4.104 (c)	4.105 (d)	4.106 (c)	4.107 (d)	

Explanations Water Treatment

4.1 (b)

Chlorine dose required

$= 2 \times 10^{-6} \times 10{,}000 \times 50 = 1$ kg/day

Bleaching powder required $= \frac{1}{0.2} = 5$ kg/day

4.2 (b)

All particles having settling velocity equal to or greater than surface overflow rate will be completely removed while particles having settling velocity lower than SOR will be partially removed.

Total removal $= 200 + \frac{1}{3} \times 300 = 300$ mg/L

The concentration of particles in settled water $= (200 + 300) - 300 = 200$ mg/L

4.3 (a)

Slow sand filter can remove upto 98-99% of bacteria present in water and thus has maximum bacteria removal efficiency. In case the water is pre-chlorinated, the efficiency can go upto 99.5% - 99.9%.

4.4 (c)

Alum is very effective coagulant for water treatment and is most commonly used. It is quite cheap, forms excellent stable floc and does not require any skilled supervision.

Destruction of bacteria by different treatment processes.

Process	Percent removal
Coarse screens	0 – 5
Fine screens	10 – 25
Grit chambers	10 – 25
Plain sedimentation	25 – 27
Chemical coagulation	40 – 80
Trickling filters	90 – 95
Activated sludge	90 – 98
Chlorination	98 – 99.9

Alum is not very effective in killing pathogens.

4.5 (c)

The addition of chlorine beyond break point is called break point chlorination. It ensures a residual of 0.2 to 0.3 mg/L of free chlorine. This residual of free chlorine, appearing after break point is not easily removed except by sunlight and therefore it takes care of the future contamination of water in the distribution system. The chloramines formed by reaction of chlorine and ammonia before break point are called combined residual chlorine. These chloramines are less effective disinfectants than free residuals. Chloramines are persistent and provide continued protection against regrowth of pathogens in distribution system.

4.6 (d)

Aerator is used to remove gases like CO_2 and H_2S. Sedimentation tank (after coagulation and flocculation) can remove settleable as well as colloidal particles. Slow sand filter and rapid sand filter both remove suspended matter but slow sand filter is more effective in removing bacteriological matter.

4.7 (c)

First aeration is done to remove gases CO_2 and H_2S and pathogens (disease causing bacteria) will be removed in the last.

4.8 (d)

Common problems associated with operation of rapid sand filter:

(i) Air binding
(ii) Turbidity break through cracking of sand bed
(iii) Mudball formation
(iv) Buildup of emulsified grease
(v) Loss of filter medium due to backwashing
(vi) Gravel moulding or bumping of filter beds.

4.9 (b)

The disinfection of industrial water supplies is necessary in food processing, distillery (alcohol), etc.

4.10 (b)

Chlorine dose required

$= 2 \times 10^{-6} \times 10{,}000 \times 50 = 1$ kg/day

Bleaching powder required $= \frac{1}{0.2} = 5$ kg/day

4.11 (d)

The filtration process causes blocking of filter media due to impurities arrested in them. The downward flow of water is resisted by sand grains and impurities arrested in them. A stage may reach when the frictional resistance exceeds the static head of water above the sand bed most of this resistance is caused by the top 10 to 15 cm sand layer. The bottom sand acts like a vacuum and the water is sucked through the filter media without cleaning. The dissolved air and gases from water are released and they stick to sand grains. This phenomena is called air binding and seriously affects the efficiency of filter.

4.12 (d)

Lime-soda process converts all types of hardness into precipitate of $CaCO_3$ and $Mg(OH)_2$. These precipitate may not be removed completely and will continue settling slowly resulting in deposits in water lines and storage facilities. It is therefore necessary to 'stabilize' the water by converting the supersaturated $CaCO_3$ back to the soluble form. This can be done by recarbonation.

4.13 (b)

In **sedimentation** process the coarser suspended particles are removed by letting the waste water settle in sedimentation basins.

The effectiveness of sedimentation may however be increased by mixing certain chemicals with the water. So as to form flocculent precipitate, the process is known as **flocculation.**

These flocculent precipitate carries the suspended particles as it settles. The process is called chemical coagulation or simply **coagulation**.

But **disinfection** is used for process of killing of germs, by addition of certain chemicals.

4.15 (d)

Dissolved solids cannot be removed in rapid sand filter. Helminth eggs, which range in size from about 10 mm to more than 100 mm, can be removed by many commonly used waste water-treatment processes such as sedimentation, filtration and stabilization ponds. However, some helminth eggs are extremely resistant to environmental stresses and may survive usual waste water and sludge disinfection procedure. Chlorine disinfection and mesophilic anaerobic digestion, for example, are not effective at inactivating many helminth eggs.

4.16 (b)

$$\text{Settling velocity, } v = \frac{g}{18} \times \frac{(S_s - 1)}{\nu} d^2$$

As specific gravity changes from 2 to 3, the settling velocity will change from v to $2v$.

4.17 (b)

Slow sand filter will be able to remove suspended particles and up to 99% pathogenic organisms. So small amount of supplies for rural areas can be ensured by slow sand filters.

4.18 (c)

Concentrations of fluoride approximately 1.0 mg/L in drinking water help to prevent dental cavities in children. Excessive intake of fluoride (> 2 mg/L) may be noticeable discoloration of teeth called mottling.

4.19 (d)

Settling velocity of discrete particles is given by Newton or Stoke's equation. It is propor-tional to square of the particle diameter.

4.20 (d)

$$\text{Surface overflow rate} = \frac{Q}{A_s} = \frac{Q}{LB}$$

A_s is surface area

SOR does not depend upon total depth of the tank.

4.21 (d)

To avoid negative head and air binding, filter is cleaned when the head loss is more than the static head of water over sand bed. Another remedy is to pump in air to reduce the suction effect created due to negative head. Growth of algae can be reduced by avoiding increasing in water temperature. Growth of algae fills the pores of filter media thereby creating negative head.

4.22 (b)

Aeration can not kill pathogens and correct pH of water. Although volatile liquids such as humic acids and phenols can be removed from water by aeration. The removal rates are too low for the process to be practical except in extreme cases where excessive quantities must be reduced to more manageable levels.

4.23 (d)

Potability means absence of pathogenic organisms and they can be detected by Membrane Filtration (MF) technique.

4.24 (a)

The chloramines prevent the formation of compounds formed by reaction of free chlorine with the organics or phenol. The chloramines are predominant species above pH 6.5. Chloramines are persistent and provide continued protection against regrowth of micro-organisms in distribution system.

4.25 (a)

Velocity gradient, $G = \sqrt{\frac{P}{\mu V}}$

Power input per unit volume,

$$P = \mu G^2 V = 1 \times 10^{-3} \times (600)^2 \times 1$$
$$= 360 \text{ W}$$

4.26 (a)

In **Discrete particle settling** (Type I), the particles settle individually without interaction with neighboring particles. This is in accordance to the classic sedimentation laws of Newton and Stokes.

Key features:

- Size, shape and specific gravity of the particles do not change with time.
- Settling velocity remains constant.

4.27 (b)

In plain sedimentation tank, suspended solids with specific gravity more than1 are removed. Chemical precipitation removes dissolved solids. Slow sand filtration removes 98-99% of the bacteria. Aeration removes dissolved gases (CO_2 and H_2S)

4.28 (d)

Principal mechanisms and phenomenon contributing to removal of material within a granular medium depth filter are: (i) Straining (mechanical and chance contact), (ii) Sedimentation, (iii) Impaction, (iv) Interception, (v) Adhesion, (vi) Flocculation, (vii) Chemical adsorption (bonding and chemical interaction), (viii) Physical adsorption (electrostatic forces, electrokinetic forces and vaander Walls forces), (ix) Biological growth.

(x) Centrifugal force in disc filter.

4.29 (c)

The purpose of coagulation is to alter the colloids so that they can adhere to each other. During coagulation a positive ion is added to water to reduce the surface charge to the point where the colloids are not repelled from each other.

Aluminum can be purchased as either dry or liquid alum [$Al_2(SO_4)_3 . 14H_2O$]. When alum is added to a water containing alkalinity, the following reaction occurs :

$$Al_2(SO_4)_3\ 14\ H_2O + 6HCO_3 \rightleftharpoons$$
$$2Al(OH)_3 + 6CO_2 + 14H_2O + 3SO_4^{2-}$$

Such that each mole of alum added uses six moles of alkalinity and produces six moles of carbon dioxide. The above reaction shifts the carbonate equilibrium and decreases the pH. However, as long as sufficient alkalinity is present and CO_2 is allowed to evolve, the pH is not drastically reduced and is generally not an operational problem. When sufficient alkalinity is not present to neutralize the sulfuric acid production, the pH may be greatly reduced :

$$Al_2(SO_4)_3 .\ 14\ H_2O \rightleftharpoons$$
$$2Al(OH)_3 + 8H_2O + 3H_2SO_4$$

If the second reaction occurs, lime or sodium carbonate may be added to neutralize the sulfuric acid. Two important factors in coagulant addition are pH and dose. The optimum dose and pH must be determined from laboratory tests. The optimal pH range for alum is approximately 5.5 to 6.5 with adequate coagulation possible between pH 5 to pH 8 under some conditions.

4.30 (b)

Chlorides are generally present in water in the form of sodium chloride and may be due to leaching of marine sedimentary deposits, pollution from sea water, brine, or industrial and domestic wastes, etc. Their concentrations above 250 mg/L produce brackish taste which is objectionable. It is measured by potentiometric method using titration with silver nitrate solution. In Argentometric method, of chloride measurement, silver nitrate titration in the presence of potassium chromate indicator is used. The red precipitate of silver chromate indicates end of titration.

Brackish taste of chlorides can be removed by:

(i) Evaporation and distillation
(ii) Electrodialysis method
(iii) Reverse osmosis method
(iv) Freezing process
(v) Solar distillation method

Lime soda process and cation exchange process are used for hardness removal. Chemical coagulation is used for colloidal particle removal.

4.31 (d)

The biological layer formed by microbial decomposition of harmless compounds is called *schmutzdecke* or dirty skin. This helps in absorbing and straining out impurities. This layer is formed in slow sand filter.

4.32 (a)

Dechlorination means removing the chlorine from water. This is generally required when super-chlorination has been practised. The common dechlorinating agents are:

(i) SO_2 gas
(ii) Sodium thiosulphate ($Na_2S_2O_3$)
(iii) Activated carbon
(iv) Sodium metabisulphate ($Na_2S_2O_5$)
(v) Ammonia as NH_4OH
(vi) Sodium sulphite (Na_2SO_3)
(vi) Sodium bisulphite ($NaHSO_3$)

4.33 (a)

<table>
<tr><th>S.L.</th><th>Filter</th><th colspan="4">Medium</th><th>Rate of filtration</th></tr>
<tr><td>(i)</td><td>Slow sand filter</td><td colspan="4">Fine sand effective size 0.2mm</td><td>0.12 to 0.32 m/h</td></tr>
<tr><td>(ii)</td><td>Rapid sand filter</td><td colspan="4">Coarse sand effective size 0.45 to 0.55 mm and Coefficient of uniformity 1.65</td><td>2.5 to 5 m/h</td></tr>
<tr><td>(iii)</td><td>Dual media filter</td><td>Effective size (mm)
Uniformity coefficient</td><td colspan="2">Sand
0.5-0.55
< 1.65</td><td>Coal
0.9-1.0
< 1.8</td><td>10 to 20 m/h</td></tr>
<tr><td>(iv)</td><td>Mixed media filter</td><td>Percentage
Effective size (mm)</td><td>Silica sand
30%</td><td>Garnet sand
10%
> 0.15mm</td><td>Anthracite
60%
< 1.0mm</td><td>10 to 20 m/h</td></tr>
</table>

The gradation of mixed media filter is such that the depth to pore size graph is close to ideal filter as shown in figure below. Dual and mixed media filters make possible the direct filtration of water of low turbidity without settling operation.

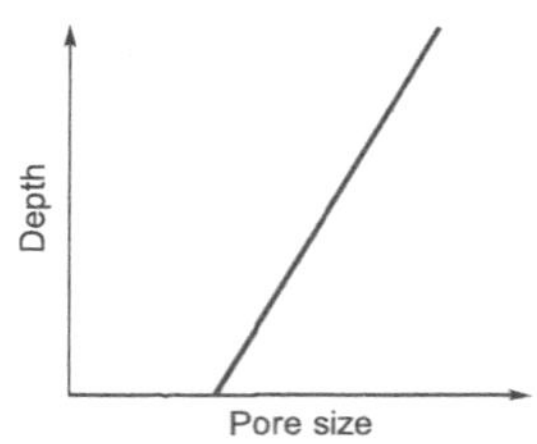

Ideal Filter Gradation

4.34 (c)

Flocculating particles have type-2 settling. The Stoke's equation cannot be used because flocculating particle are continuously changing in shape, size and, if a large aggregate of particles collect, specific gravity because also change of entrapped water in interstitial space. The batch analysis of type-2 settling is represented by plotting isoremoval line.

The slope at any point on isoremoval line is the instantaneous velocity of the fraction of particles represented by that line. The velocity of setting increases with depth due to increase in particle size by continued collision and aggregation with other particle. The horizontal flow velocity remains constant.

4.35 (b)

The large weir overflow rates cause short circuiting and this causes particles to get removed without setting.

4.36 (b)

For pH = 7

Concentration of $[H^+] = 10^{-pH}$ moles/L

Ionization constant

$$k = \frac{[H^+][OCl^-]}{[HOCl]} = 2.5 \times 10^{-8} \text{ moles/L}$$

The total quantity of HOCl and OCl^- present in water is called free available chlorine. The relative distribution of these two species is very important because the killing efficiency of HOCl is about 40 to 80 times that of OCl^-. The percent distribution of HOCl

$$\frac{[HOCl]}{[HOCl]+[OCl^-]} = \frac{1}{1+k\,10^{pH}}$$

$$= \frac{1}{1+2.5\times10^{-8}\times10^{7}} = \frac{1}{1.25} = 0.8$$

4.37 (b)

Most of the standing pond water is contaminated with **Guinea Worm Disease** and anyone who drinks this water stands at a risk of getting infected from it. GWD can be prevented by filtration of the water or by using water from underground sources which is free from contamination of larvae.

4.38 (a)

Chlorine residual is necessary for being chlorine as effective disinfectant.

Iodine has no carcinogenic results ultraviolet rays are ineffective in the presence of suspended solids because they finders the rays ozone on other hand is unaffected by the ammonium ions.

4.39 (c)

The coagulation of metallic salts releases hydrogen ions as well as coagulant species. These hydrogen ions neutralize alkalinity. In water with high turbidity and high alkalinity the pH will be relatively unaffected by coagulant addition. Because of the high alkalinity, absorption and charge neutralization will be less effective than in waters of low alkalinity. Higher coagulant dosage should be used to ensure sweep coagulation.

4.40 (b)

When the flow in circular sedimentation tank enters at the centre and is baffled to flow radially towards the perimeter, the horizontal velocity of the water is continually decreasing as the distance from the centre increases. Thus, a discrete particle with a settling velocity (v_0) is continually undergoing a change in its absolute velocity due to the decrease in horizontal velocity. Thus, the particle path in a circular basin is a parabola as opposed to the straight particle path line in the long rectangular tank.

4.41 (b)

During chlorination, the chlorine (in the form of HOCl) immediately reacts with ammonia present in water to form various chloramines as given below:

$$NH_3 + HOCl \rightarrow \underset{\text{Monochloramine}}{NH_2Cl} + H_2O$$

$$NH_2Cl + HOCl \rightarrow \underset{\text{Dichloramine}}{NHCl_2} + H_2O$$

$$NHCl_2 + HOCl \rightarrow \underset{\text{Nitrogen trichloramine}}{NCl_3} + H_2O$$

The monochloramine predominates at pH value of over 7.5; dichloramine at pH ranges of 5 to 6.5 and nitrogen trichloramine at pH below 4.4. These chloramines so formed are stable and are found to possess disinfecting properties.

4.42 (b)

Surface overflow rate is the most important design parameter in designing a continuous flow rectangular sedimentation tank. If the discharge through the tank is Q, width of tank is B and length of the tank is L, then surface overflow rate is given by

$$v_0 = \frac{Q}{BL}$$

The surface overflow rate can be said to be representing the setting velocity of the slowest setting particles which are 100% removed. Thus, particles having setting velocity greater than or equal to surface overflow rate are 100% removed.

Particles having setting velocity less than surface overflow rate are not 100% removed. If a particle enters the tank at a height '*h*' with a setting velocity v which is less than v_0, then it will be removed if

$$v = \frac{h}{H} \times v_0$$

where *H* is the height of the tank.

4.43 (b)

Higher alkalinity provides higher concentration of OH^- ions which helps in forming insoluble compounds which can be removed easily. High turbidity and high alkalinity in water makes it unaffected by the addition of coagulant. Higher coagulant dosage should be used to ensure sweep coagulation.

High Turbidity and Low Alkalinity: A Low doses of coagulants.

Low Turbidity and Low Alkalinity: A Addition of both turbidity and alkalinity.

Low Turbidity and High Alkalinity: A High doses of coagulants.

4.44 (a)

When hard water is treated with lime [$Ca(OH)_2$] and soda ash [Na_2CO_3], calcium and magnesium salts present in it react with lime and soda ash to form insoluble precipitates of calcium carbonate and magnesium hydroxide. These precipitates can be sedimented out in a sedimentation tank. However, a little quantity may remain as finely divided particles and may cause troubles by getting deposited on the filter to cause enlargement of the sand grains called incrustation of the filter media or in the pipe of distribution system. To prevent this, it is generally necessary that the water be recarbonated by passing carbondioxide gas through it, as it leaves the sedimentation tank. In the recarbonation process, the insoluble carbonates combine with the CO_2 to again form the soluble bicarbonates, as given below:

(i) $CaCO_3 + CO_2 + H_2O \longrightarrow Ca(HCO_3)_2$

(ii) $Mg(OH)_2 + CO_2 \longrightarrow MgCO_3 + H_2O$

(iii) $MgCO_3 + CO_2 + H_2O \longrightarrow Mg(HCO_3)_2$

By the recarbonation process, even though the water regains some of its hardness yet recarbonation is advisable.

4.47 (b)

The appropriate method for removal of fluorides from water is de-fluoridation. It can be achieved by the addition of any of the following $Ca_3(PO_4)_2$, Bone charcoal, Lime, Activated carbon and Reagents.

4.48 (d)

At low concentrations, chlorine probably kills micro-organisms by penetrating the cell and reacting with the enzymes and protoplasm. At higher concentrations, oxidation of the cell wall will destroy the organism.

4.49 (b)

Excessive time of flocculation than its optimum lime of 30 min leads to entrapment of air and floc will float in sedimentation tank.

4.50 (b)

Once chlorine is utilized in disinfection additional Cl_2 applied to the water creates an equal residual. Beyond this all Cl_2 added appears as free Cl_2 residuals.

With increase in does of Cl_2 time required to achieve disinfection decreases as tC^n = constant.

4.51 (d)

Filtration is employed after coagulation, flocculation and sedimentation to remove the finer particles in suspension, which may avoid settling in sedimentation basins even after using chemical coagulation.

Activated carbon is a specially treated carbon, which possesses the property of absorbing and attracting impurities, such as gases, liquid, and finely divided solids. Because of its excellent property of absorbing impurities, it is widely used for removing tastes and odours from public supplies.

4.52 (b)

The bed of a rapid sand filter consists of stratified layer with smallest size in the top most layer. The sand used being bigger in sizes, the penetration of impurities will be deeper and the greater depth of bed is effective.

The gravel below sand supports the filtering media. Gravel helps in uniform distribution of wash water up through the filter media. It also prevents sand from entering into drains and then causing them to choke. The depth of gravel varies from 45 to 60 cm. The size of the gravel will vary from 5 to 25 mm the smallest size layers being at the top.

4.53 (a)

The amount of residual chlorine left in the

chlorinated water after the required contact period, can be experimentally determined by using any of the following tests:

1. Orthotolidine test
2. D.P.D. test
3. Chlorotex test
4. Starch-iodide test

Amperometric titration method is used for the determination of total chlorine concentrations in natural water.

4.54 (a)

The amount of a coagulant required for coagulation depends on the turbidity and colour of the raw water. High turbidity aids in formation of thick and dense flocs leading to effective removal.

Selection of type of coagulant is dependent on pH value of raw water. Alum is used for pH range of 6.5 to 8.5. For pH range of 8.5 and above copperas is used. Ferric sulphate is quite effective in the pH range of 4 to 7 and above 9, whereas Ferric chloride is quite effective in the pH range of 3.5 to 6.5 and above 8.5.

4.55 (b)

Coagulation is a chemical technique which is directed towards the destabilisation of the charged colloidal particles. Flocculation, on the other hand, is the slow mixing technique which promotes the agglomeration of the stabilised particles to form flocs.

4.56 (a)

The operational troubles which are generally encountered during the operation of rapid gravity filters are.

Air binding: Development of negative pressure in the bottom sand layer of rapid sand filter leads to air binding. The negative pressure so developed, tends to release the air dissolved in water. It causes the formation of bubbles, which stick to the sand grains, and affects the working of filter.

Mud balls: The mud from the atmosphere usually accumulates on the sand surface, so as to form a dense mat. During inadequate washing of the filter, this mud may sink down into the sand bed. This mud then sticks to the sand grains and other arrested impurities, thereby forming mud balls. The fine sand contained in the top layers of the filter bed, shrinks and causes the development of shrinkage cracks in the sand bed.

4.57 (a)

The disinfection not only removes the existing bacteria from the water at the plant, but also ensures their immediate killing even afterwards, in the distribution system. The contamination of water during its transit from the treatment plant to the place of its consumption is also thus, prevented by the disinfectants. The chemical which is used as a disinfectant must, therefore, be able to give this residual sterilizing effect for a long period, thus affording some protection against recontamination.

4.59 (c)

Carbonate hardness is only due to carbonate and bicarbonate of multivalent cation.

4.60 (d)

Post chlorination means chlorination of effluent of filters which results in

(i) reduction in Cl_2 demand
(ii) reduction in possibility of taste and odour.
(iii) reduction in possibility of carcinogenic compounds

But chloramines will be formed only when NH_3 is present in water.

4.61 (d)

Chlorine in its various forms is invariably and almost universally used for disinfecting public water supplies. When chlorine is added to water, it forms hypochlorous acid which has an immediate and disastrous effect on most forms of microscopic organism.

4.62 (c)

Hepatitis is transmitted by virus.

4.63 (d)

Filters are also able to remove even particles of size smaller than the size of voids present in filter however these smaller particles are retained in mat layer.

4.64 (d)

- The amount of alum required for coagulation depends on the turbidity and colour of the raw water.

- The dose of alum may very from 5 mg/*l* for relatively clear waters to 85 mg/*l* for highly turbid water. The average normal dose is about 17 mg/*l*.

4.65 (d)

In baffled chanel as they are fixed turbulence can't be altered, hence G remains filxed.

4.66 (d)

Advantages of Slow Sand Filter

- Very effective removal of bacteria, viruses, protozoa, turbidity and heavy metals in contaminated fresh water
- Simplicity of design and high self-help compatibility: construction, operation and maintenance only require basic skills and knowledge and minimal effort
- If constructed with gravity flow only, no (electrical) pumps required
- Local materials can be used for construction
- High reliability and ability to withstand fluctuations in water quality
- No necessity for the application of chemicals
- Easy to install in rural, semi-urban and remote areas, Simplicity of design and operation
- Long lifespan (estimated >10 years)

4.67 (d)

Chlorine dosage

$$= \frac{20\times10^6}{6000\times10^3}\text{mg}/l = 3.33 \text{ mg}/l$$

4.68 (c)

Large numbers of ions are present in Sea water/ Salty water. So, when a turbid stream flows into an ocean, then turbidity and the ions together lead to coagulation by ionic layer compression.

4.69 (b)

$$\text{Size of filter} = \frac{Q}{\text{Rate of filtration}}$$

$$= \frac{1\text{MLD}}{4000\, l/\text{hr/m}^2} = \frac{10^6}{4000\times24}\text{m}^2 = 10.42 \text{ m}^2$$

Use 11 m^2 filter

Size of under drainage system

= 0.2% of filter area = 0.022 m^2

4.71 (a)

For Re < 0.5, we apply Stoke's law equation

$$V_s = \frac{g}{18}(G-1)\frac{d^2}{\nu}$$

$$= \frac{9.81}{18}(2.65-1)\times\frac{\left(2\times10^{-3}\right)^2}{2\times10^{-2}}\frac{\text{m}}{\text{sec}^2}\times\frac{\text{cm}^2/\text{cm}^2}{\text{sec}}$$

$$= \frac{9.81}{18}(1.65)\times\frac{4\times10^{-6}}{2\times10^{-2}} = 0.018 \text{ cm/sec.}$$

4.72 (a)

Assume rate of filtration = q *l*/d/m^2

Given as a square filter, if length (l) dimension of filter becomes two times, the area will become ($l'\times l' = 2l\times2l) = 4l^2$) four times.

So, amount of water filtered = Rate of filteration × Area

= $q\times4A = 4\times(q\times A)$

= 4 times the initial amount of water filtered.

4.73 (c)

During an epidemic of infective hepatitis, the supplied water is super chlorinated because of avoidance of future contamination but the spore forming bacteria, do not cause hepatitis, it is caused by virus.

4.74 (b)

Alum needed

= 10 MLD × 10 ppm =10 × 10^6 L/day × 10 mg/L

= 100 × 10^6 mg/day = 100 kg/day = 1 quintal/ day

4.75 (b)

Required CWR capacity

$$= \left(10\times10^6\right)\times\left(\frac{40}{24\times60}\right)\times10^{-3}\text{ m}^3 = 280 \text{ m}^3$$

4.76 (a)

Required filtering water = 10 MLD = 10 × 10^6 *l*/d

Filtration rate = 100,000 *l*/m^2/day

Now, surface area required

$$= \frac{\text{Required filtering water}}{\text{Filteration rate}} = \frac{10\times10^6}{100{,}000} = 100 \text{ m}^2$$

4.77 (d)

Difference between Slow Sand and Rapid Sand filters

S. No	Item	Slow Sand Filters	Rapid Sand Filters
1.	Rate of filtration	$3 m^3/m^2/d$	$125 m^3/m^2/d$
2.	Size of bed	Large, 200 m^2	Small, 200-400 m^2
3.	Depth of bed	0.3 m of gravel, 1.2 m of sand usually reduced to not less than 0.6 m by scraping.	0.5 m of gravel, 0.7 m of sand or less not, reduced by washing.
4.	Size of sand	Effective size 0.25 to 0.3 to 0.35 mm, uniformity coefficient 2 to 2.5 to 3.	0.45 mm and higher uniformity coefficient 1.5 and lower depending on under-drainage system.
5.	Grain size distribution of sand in filter	Unstratified	Stratified with smallest or lightest grains at top and coarsest or heaviest at bottom.
6.	Under-drainage system	Split tile laterals laid in coarse stone and discharging into tile of concrete main drain	1. Perforated pipe laterals discharging into pipe mains. 2. Porous plates above inlet box 3. Porous blocks with included channels
7.	Loss of head	0.15 m initial to 1 m final	0.3 m initial to 2.5 m final
8.	Length of run between cleanings	20 to 60 days	12 to 72 hours
9.	Penetration of suspended matter	Superficial	Deep
10.	Method of cleaning	1. Scraping off surface layer of sand and washing and storing cleaned sand for periodic re-sanding of bed 2. Washing surface sand in Place by washer traveling over sand bed.	Dislodging and removing suspended matter by upward flow or backwashing which fluidises the bed. Possible use of water or air jets or mechanical rakers to improve scour.
11.	Amount of wash water used in cleaning sand	0.2 to 6% of water filtered	1 to 4-6% of water filtered
12.	Preparatory treatment of water	Generally none	Coagulation, flocculation and sedimentation

4.78 (c)

Required chlorine dose = Chlorine demand + Residual chlorine = 1.25 + 0.25 = 1.5 mg/l

$\therefore$ Total required chlorine per day = $4 \times 10^6 \times 1.5 \times 10^{-6}$ kg = 6 kg

Available chlorine in bleaching powder = 25%

$$\therefore \text{Bleaching powder required per day} = \frac{\text{Total required chlorine per day}}{\text{Available chlorine in beaching per day}} = \frac{6}{0.25} = 24 \text{ kg}$$

4.80 (a)

Stoke's law states that denser and larger particles have a higher settling velocity, thus a higher overflow rate. Hence size and density of particle affects the overflow rate.When the temperature decreases, the rate of settling becomes slower. When the water is colder, the flow in the plant is at its lowest and the detention time in the plant is increased so the floc has time to settle out in the sedimentation basins.

4.81 (b)

- Expansion of sand bed during back washing should be kept within limits to avoid carry over of sand to wash water through. Sand bed should never be depleted more than 10 cm from original thickness for this purpose.
- Air binding is caused due to negative head and formation of air bubbles.
- In lime-soda method, hardness is removed in the form of precipitate which get coated over sand particles and leads to incrustation.
- Sometimes careless or indifferent operation of the filter, bumping or lifting of the filter beds takes plate when switching on the back wash cycle is adopted.

4.82 (b)

Quantity of 100% pure bleaching powder required $= 2 \times 5000 \times 10^{-3} = 10$ gm

Quantity of 25% pure bleaching powder required

$= \frac{10}{0.25} = 40$ gm

4.83 (b)

Design overflow rate of a sedimentation tank is the setting velocity of the particle intended to be removed, which is chosen considering the diameter of the particle intended to be removed.

4.84 (c)

Recarbonation of water softened by lime-soda. Process results in increased hardness of water because in this method precipitate are again made soluble in water. $CaSO_4$ and $MgSO_4$ are dissolved solids not suspended solids.

4.85 (a)

- Coagulation aid is a chemical which when used along with main coagulant improves or accelerates the process of coagulation and flocculation by producing quick forming dense and rapid settling flocs.
- Finely divided clay, bentonites and activated powdered carbon are the most commonly used materials as nuclei to floc formation.
- Activated silica, polyelectrolytes are other coagulation aids.

4.86 (c)

A good disinfectant must be toxic to micro-organism at concentrations well below the toxic threshold of humans and higher animals.

4.87 (*)

Total dosage of chlorine (mg/l) = Total demand (mg/l) + Residual (mg/l)

Total dosage = (1 + 0.2) mg/l = 1.2 mg/l

For 10000 l of water, chlorine required

= 10000 l × 1.2 mg/l = 12000 mg = 12g

This chlorine has to come from bleaching powder which has 20% available chlorine

Hence, bleaching powder required$= \frac{12}{0.2} g = 60 g$

Note: *Since no answer is matching. We can take nearest as correct answer i.e., 68g.*

4.88 (b)

Settleable solids are removed by sedimentation volatile solids which get segregated from fixed solids by heating at 600°C in muffle furnace.

4.89 (d)

Softening of ground water may or may not be done, depending upon the quality of ground water.

4.90 (c)

Incrustation of sand occurs in rapids sand filter when water softened by lime soda is fed in for filtration. During lime soda process, ppt of calcium carbonate (and not calcium hydroxide) is formed along with magnesium hydroxide which leads to incrustation.

4.91 (b)

Lime soda process is a process of reduction of hardness to a permissible value (75-115 mg/l of $CaCO_3$). Excess lime only removes carbonate hardness.

Excess alum dosage will itself induce hardness.

Zero hardness is achieved by the following methods:

1. Demineralization method
2. Ion-exchange/Zeolite method

4.92 (a)

Total capacity of service reservoir = balancing storage + fire storage + emergency storage.

4.93 (d)

Recarbonation is done to make 40 mg/*l* of $CaCO_3$ an d 10 mg/*l* of $Mg(OH)_2$ ppt back soluble in water which are formed very slowly else it would lead to incrustation of pipes in distribution system.

4.94 (c)

$$\text{Design discharge} = 1.5 \times 150 \times 10^5$$
$$= 2.25 \times 10^7 \; l/d$$
$$\text{Overflow rate} = 20000 \; l/d/m^2$$
$$\text{Surface area} = \frac{Q_0}{OFR} = \frac{2.25\times10^7}{2\times10^4}$$
$$= 1125 \; m^2$$

or
$$\text{Volume} = Q_0 \times t_d = 2.25\times10^7 \times \frac{4}{24}$$
$$= 3.75 \times 10^6 \; l$$
$$\text{Surface area} = \frac{V}{H} = \frac{3.7\times10^6\times10^{-3}}{3}$$
$$= 1250 \; m^2$$

As per loading consideration $S_A = 1125 \; m^2$

4.95 (c)

$$\text{Chlorine usage} = 9 \text{ kg/d}$$
$$Q_0 = 25000 \; m^3/d$$
$$\text{Chlorine usage} = \frac{9\times10^6}{25000\times10^3} = 0.36 \text{ mg}/l$$
$$\text{Chlorine usages} = \text{chlorine demand} + \text{residual chlorine}$$
$$0.36 = \text{chlorine demand} + 0.2$$
$$\text{Chlorine demand} = 0.36 - 0.2 = 0.16 \text{ mg}/l$$

4.96 (b)

Statement-II alum works in pH range of 6.5 – 8.5, hence works in slightly alkaline range.

Statement-I with increase in temperature viscosity of water decrease hence resistance to seltting decreasing there by flocs settle better.

4.97 (a)

Flocculation in a process of contact and cohesion whereby the particles of a dispersion form larger size clusters in the pressure of coagulants.

4.99 (a)

- Coagulation is defined as the addition of a chemical to a colloidal dispersion which results in particle destablisation by the reduction in force, which tend to keep particle apart.
- Flocculation is basically slow mixing or agitation process in which destabilized colloidal particles are brought into intimate contact in order to promote agglomeration.
- Sedimentation is a natural process by which solids with higher density than the fluid in which they are suspended, settles under the action of gravity.

4.100 (c)

Treatment with Activated Carbon

- This is a specially treated carbon, which possesses the property of absorbing and attracting impurities.
- Due to absorption property it removes taste, colour and odour and also removes phenol type impurity.
- The activated carbon is mostly used in the powered form, and may be added to water either before or after the coagulation but before filtration.
- In split method, a part of activated carbon is mixed in mixing tank and remaining is mixed before it enters the filter.
- The usual dose of activated carbon varies from 5 to 20 mg/*l*.
- When used in powered from before coagulation, it aids in coagulation.
- It reduces the chlorine demand of treated water.
- It removes the organic matter present in water.
- It removes the tastes, odours-and colour caused by the presence of iron, manganese, phenols, excess chlorine, hydrogen sulphide etc.
- Its overdoes is not harmful.

4.102 (a)

For $d \le 0.1$ mm (i.e., Laminar flow) and discrete settling, stoke's law is valid. According to which, settling velocity of particles,

$$v_s = \frac{g}{18}(G-1)\frac{d^2}{\nu} \Rightarrow v_s \propto d^2$$

4.103 (a)

For discrete particle settling and Reynold's number less than 0.5, stoke's law is valid.

Settling velocity of particles,

$$\therefore \quad v_s = \frac{g}{18}(G-1)\frac{d^2}{\nu}$$

$$v_s = \frac{981\ \text{cm/s}^2}{18} \times \frac{(2.65-1)\times\left(5\times10^{-3}\right)^2\ \text{cm}^2}{1.01\times10^{-2}\ \text{cm}^2/\text{s}}$$

$$\Rightarrow \quad v_s = 0.22\ \text{cm/s}$$

4.104 (c)

Filtration rate of RGF = 3000 to 6000 l/h/m^2

Filtration rate of pressure filter = 6000 to 18000 l/h/m^2

i.e., rate of pressure filter as compared to rapid gravity filter is about 2 to 5 times.

4.105 (d)

Minor methods of disinfection.

i. By boiling of water.

ii. U-V rays.

iii. By use of ozone.

Major method of disinfection.

i. Chlorination.

4.106 (c)

After coagulation, filtration is adopted which induces slow mixing in water and destabilizes the colloids.

4.107 (d)

Lime and soda process is used to remove hardness of water. These chemicals are added in coagulation tank.

■■■■

5 Conveyance and Distribution of Water

5.1 If a tangent drawn parallel to the demand line from a ridge point of a mass curve does not intersect the mass curve again it can be inferred that the

(a) frequency of the flood entering into the reservoir is less
(b) inflow into the reservoir cannot meet the demand
(c) reservoir is overflowing resulting in wastage
(d) reservoir can meet higher demand

[ESE : 2000]

5.2 **Assertion (A):** The leakage losses are less when the water supply is intermittent.
Reason (R): Pressure is less in intermittent water supply.

(a) both A and R are true and R is the correct explanation of A
(b) both A and R are true but R is not a correct explanation of A
(c) A is true but R is false
(d) A is false but R is true [ESE : 2000]

5.3 The purpose of providing a balancing reservoir in a water supply distribution system is to

(a) equalize pressures in the distribution system
(b) store adequate quantity of water to meet requirements in case of break-down of inflow
(c) store adequate fire fighting reserve
(d) take care of fluctuations in the rate of consumption.

[ESE : 2001]

5.4 In the figure shown below, x_2x_4 represents

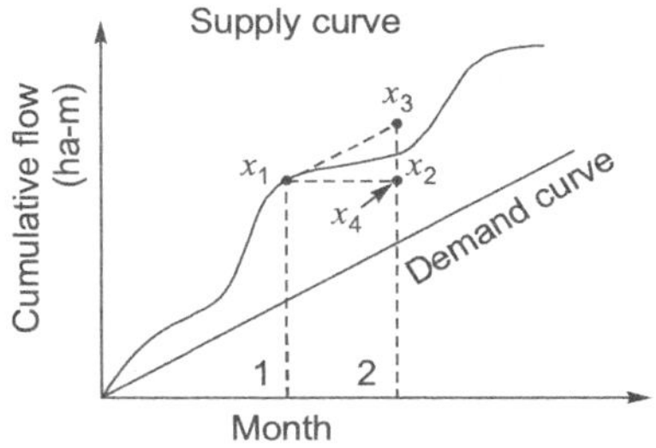

Mass Diagram

(a) accumulated supply
(b) storage requirement
(c) accumulated demand for the period 1 to 2
(d) surplus in the reservoir [ESE : 2001]

5.5 The usual size of residential ferrule bore varies from

(a) 1 mm to 5 mm
(b) 10 mm to 50 mm
(c) 100 mm to 500 mm
(d) 1000 mm to 5000 mm [ESE : 2000]

5.6 The following sketch shows a water supply main from a storage reservoir provided with a sluice valve:

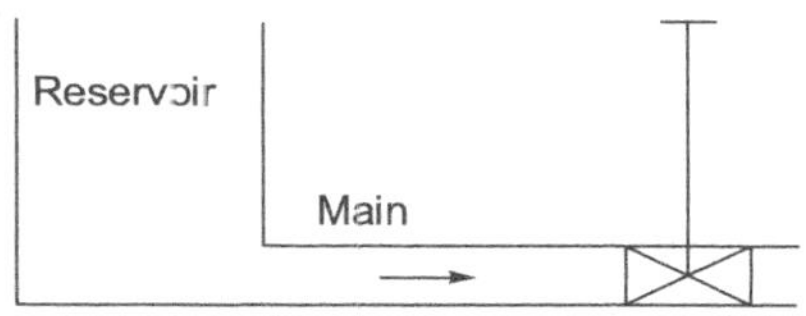

The type of valve most suitable for this pipeline is

(a) air valve
(b) scour valve
(c) pressure relief valve
(d) None of the above [ESE : 2001]

5.7 Match **List-I** with **List-II** and select the correct answer using the codes given below the lists:

List-I (Valves)

A. Sluice valve
B. Check valve
C. Air inlet valve
D. Ball valve

List-II (Uses)

1. Used where gravity flow is required through pipe line
2. Used to maintain constant level of water
3. Used for reversal of flow
4. Used for isolating

Codes:

	A	B	C	D
(a)	1	3	4	2
(b)	4	2	1	3
(c)	1	2	4	3
(d)	4	3	1	2

[ESE : 2002]

5.8 Consider the following statements:
While deciding to locate an intake structure for a city situated on a river bank, intake for water supply should be located
1. in deep waters
2. sufficiently away from shore lines
3. upstream of the populated city
4. near navigational channel

Which of these statements are correct?
(a) 1, 2 and 4 (b) 1, 2 and 3
(c) 2, 3 and 4 (d) 1, 3 and 4

[ESE : 2002]

5.9 Consider the following valves in a water distribution system:
1. Check valve
2. Pressure-reducing valve
3. Air relief valve
4. Scour valve
5. Sluice valve

Which of these work automatically?
(a) 1, 3 and 4 (b) 2, 4 and 5
(c) 3, 4 and 5 (d) 1, 2 and 3

[ESE : 2003]

5.10 Match **List-I** with **List-II** and select the correct answer using the codes given below the lists:

List-I (Fixture)
A. Surge arrester
B. Butterfly valve
C. Scour valve
D. Check valve

List-II (Purpose)
1. Prevention of reversal of flow in a pipeline
2. Regulating or stopping the flow especially in large size conduits
3. Control of water hammer
4. Draining or emptying the pipeline section

Codes:

	A	B	C	D
(a)	1	2	4	3
(b)	3	2	4	1
(c)	1	4	2	3
(d)	3	4	2	1

[ESE : 2004]

5.11 Match **List-I** (Type of Pipe) with **List-II** (Purpose) and select the correct answer using the codes given below the lists:

List-I	**List-II**
A. Steel pipe	1. House plumbing
B. Cast iron pipe	2. Hot water carrying
C. GI Pipe	3. Distribution main
D. PVC pipe	4. Pumping main

Codes:

	A	B	C	D
(a)	4	1	2	3
(b)	4	3	2	1
(c)	2	1	4	3
(d)	2	3	4	1

[ESE : 2004]

5.12 Match **List-I** (Equation/Law) with **List-II** (Related Application) and select the correct answer using the codes given below the lists:

List-I
A. Chick's law
B. Darcy-Weisbach equation
C. Stoke's equation
D. Carmen-kozeny equation

List-II
1. Discrete particle settling
2. Head loss in a pipe
3. Head loss in filters
4. Rate of bacterial kill

Codes:

	A	B	C	D
(a)	4	2	1	3
(b)	3	1	2	4
(c)	4	1	2	3
(d)	3	2	1	4

[ESE : 2005]

5.13 Consider the following statements:
In water supply distribution network,
1. the grid-iron system requires more length of pipe lines and larger number of cut-off valves.
2. the design of the grid-iron system is difficult but economical.
3. employing a grid iron system, the dead ends are completely eliminated.
4. employing a grid-iron system permits more water to be diverted towards the affected point from various directions.

Which of these statement are correct?
(a) 1, 2 and 4 (b) 1, 3 and 4
(c) 2, 3 and 4 (d) 1, 2 and 3

[ESE : 2005]

5.14 Match **List-I** with **List-II** and select the correct answer using the codes given below the lists:

List-I (Equation/Method)
A. Manning's Equation
B. Darcy-Weisbach

C. Hardy Cross Method
D. Rational Method

List-II (Application)

1. Frictional head loss estimation in pipe flow
2. Sanitary sewer design
3. Storm sewer design
4. Water distribution system design

Codes:

	A	B	C	D
(a)	2	1	4	3
(b)	1	4	3	2
(c)	2	1	3	4
(d)	1	4	2	3

[ESE : 2007]

5.15 Which of the following are the advantages of cast-iron pipe for its use in water supply?

1. Resistant to corrosion to a reasonable extent
2. Very easy to join the pipes
3. Easy to transport
4. Longer life

Select the correct answer using the codes given :

(a) 1, 3 and 4 (b) 1, 2 and 3
(c) 1 and 4 only (d) 2, 3 and 4

[ESE : 2008]

5.16 Which one of the following valves is seldom used in water distribution systems because of high head loss characteristics?

(a) Butterfly (b) Globe
(c) Plug (d) Sluice [ESE : 2008]

5.17 Sonoscope is used for which one of the following?

(a) Checking the accuracy of water meters
(b) Regulating the fire hydrants
(c) As a replacement of venturimeter for discharge measurement
(d) Detection of leakage in underground water mains

[ESE : 2008]

5.18 Consider the following statements:

In a water supply system,

1. drain valves are provided at elevated or higher, points to remove accumulated air.
2. reflux valve allows flow in one direction only.
3. drain valves are provided at low points to remove silt and other deposits.

Which of these statements is/are correct?

(a) 1, 2 and 3 (b) 2 only
(c) 2 and 3 only (d) 3 only [ESE : 2010]

5.19 Consider the following statements:

The disadvantages of employing steel pipes in conveyance and distribution of water are, they

1. cannot withstand high negative pressures or vacuums that may be created in them, especially the combined effects of vacuum and external loads of backfill and traffic.
2. are easily affected by acidic or alkaline waters and even atmospheric agencies may produce adverse effects on them.
3. cannot be used for high pressures. (Generally not used for pressures above 7 kg/cm^2)

Which of these statements is/are correct?

(a) 1, 2 and 3 (b) 1 and 2 only
(c) 2 and 3 only (d) 1 only

[ESE : 2010]

5.20 Match **List-I** with **List-II** and select the correct answer using the code given below the lists:

List-I

A. Steel pipe
B. Concrete pipe
C. AC pipe
D. Vitrified clay

List-II

1. Highly resistant to corrosion but can break easily
2. Virtually corrosion resistant
3. Sulphide corrosion
4. Electrolyte corrosion

Code:

	A	B	C	D
(a)	2	3	1	4
(b)	4	3	1	2
(c)	2	1	3	4
(d)	4	[illegible]	3	2

[ESE : 2011]

5.21 Which of the reasons given below favour the use of cast iron pipe for city water supply?

1. Strong in tension; and pipe of large diameter (up to 6 m) can be made.
2. Easy to make joints
3. Durable
4. Corrosion-resistant

(a) 1, 2, 3 and 4 (b) 2, 3 and 4 only
(c) 1, 2, and 4 only (d) 1, 3 and 4 only

[ESE : 2013]

5.22 Consider, the following statements in respect of cast iron pipes employed for water supply

1. Easy to make joints
2. Strong and durable
3. Corrosion resistant
4. Long life

Which of the above statements are correct?
(a) 1, 2 and 3 only (b) 1, 3 and 4 only
(c) 2, 3 and 4 only (d) 1, 2, 3 and 4

[ESE : 2015]

5.23 Consider the following statements regarding valves in a pipe line:
1. In long pipe lines, air will accumulate in the low point of the line and will interfere with the flow.
2. Pressure relief valves are used in pipe lines where pressure may increase beyond the maximum permissible pressure.
3. Non-return valves prevent water flowing back, i.e. in the opposite direction.

Which of the above statements are correct?
(a) 1 and 2 only (b) 2 and 3 only
(c) 1 and 3 only (d) 1, 2 and 3

[ESE : 2015]

5.24 **Assertion (A):** The flow in water distribution pipes takes place due to gravity.
Reason (R): The flow in sewers takes place due to gravity.
(a) Both A and R are true and R is the correct explanation of A
(b) Both A and R are true but R is not a correct explanation of A
(c) A is true but R is false
(d) A is false but R is true [ESE : 2015]

5.25 Consider the following statements regarding anchorage of pipelines conveying water:
1. At bends, pipes tends to pull apart.
2. At bends, forces exerted on the joints due to longitudinal shearing stresses are enormous and the joints may get loosened.
3. To avoid problems by hydrodynamic effects, pipes are anchored using concrete blocks which absorb side thrusts at bends.
4. Pipes are also anchored on steep slopes.

Which of the above statements are CORRECT?
(a) 1, 2 and 3 only (b) 1, 2 and 4 only
(c) 1, 3 and 4 only (d) 1, 2, 3 and 4

[ESE : 2018]

5.26 Which one of the following is not a basic requirement for any well screen?
(a) Resistance to corrosion, incrustation and deterioration
(b) Enough structural strength to prevent collapse
(c) Suitability for excessive movement of sand into the well
(d) Minimum resistance to flow of water into the well

[ESE : 2020]

5.27 Which one of the following is the process of driving, pouring or forcing lead oakum plastic or other material into a joint to make it leak proof?
(a) Caulking (b) Bedding
(c) Benching (d) Haunching

[ESE : 2023]

5.28 What is meant by soffit?
(a) The system of pipes which conveys discharges in separate pipes to the drainage system
(b) A pipe or conduit which is owned by a local authority for conveyance of the sewage
(c) The highest point of the interior of a sewer pipe at any cross section
(d) The horizontal pipe lay below the floor level or below basement to receive the discharge of soil or waste water

[ESE : 2024]

5.29 Consider the following statements regarding a good building/house drainage system:
1. The pipes should be of non- absorbent materials.
2. The system should have traps at all necessary points.
3. The branch drains should be as long as possible.

Which of the above statements are correct?
(a) 1 and 2 only (b) 2 and 3 only
(c) 1 and 3 only (d) 1, 2 and 3

[ESE : 2024]

■■■■

Answers Conveyance and Distribution of Water

5.1 (b) 5.2 (c) 5.3 (d) 5.4 (a) 5.5 (b) 5.6 (c) 5.7 (d) 5.8 (b) 5.9 (d)
5.10 (b) 5.11 (b) 5.12 (a) 5.13 (b) 5.14 (a) 5.15 (c) 5.16 (c) 5.17 (d) 5.18 (c)
5.19 (b) 5.20 (b) 5.21 (b) 5.22 (d) 5.23 (b) 5.24 (d) 5.25 (d) 5.26 (c) 5.27 (a)
5.28 (c) 5.29 (a)

Explanations Conveyance and Distribution of Water

5.1 (b)

In case the demand line drawn from a ridge does not intersect the mass curve again, it means the rate of inflow is less than or equal to rate of demand. So, the reservoir will not be refilled and the demand cannot be met by the inflow.

5.2 (c)

In a continuous system, if leakages exist, great volume of water is wasted because of long duration of flow. As compared to continuous supply system, intermittent supply system has more pressure.

5.3 (d)

Total storage = Balancing storage + Breakdown storage + Fire storage

5.4 (a)

Line x_1x_3 is parallel to demand curve. Between 1 and 2 demand is more than supply.

Accumulated demand between 1 and 2 is x_3x_4.

Accumulated supply between 1 and 2 is x_2x_4.

Depletion of storage between 1 and 2 is x_2x_3.

5.5 (b)

The usual size of residential ferrule varies from 10 mm to 50 mm. For connections of more than 50 mm diameter, a tee branch connection is used.

5.6 (c)

S.No.	Types of valves	Remarks
(i)	Check valve/Reflux valve/ Non-returning valve	It allows the water to flow in one direction only.
(ii)	Sluice valve/Gate valve/ Shut off valves	These valves are provided to regulate the flow of water through the pipe and are essential to divide the main line into several sections. These valves are usually placed at the summits of the pressure conduits
(iii)	Air valves/Air relief valves	It is provided at the summits to release the air pressure. At every summit of a pipeline and U/S of sluice vale, an air valve should be provided.
(iv)	Scour valve/Blow off and Valve/ Drain valve	Similar to sluice valves, but their function is different. It is provided at the dead end of the pipeline.
(v)	Butterfly valve	Butterfly valve are used to regulate and stop the flow especially in large sized conduits. Butterfly valves involve slightly higher head loss than sluice valves and
(vi)	Relief valve/Safety valve/Pressure relief valves/Cutoff valves	These valves protect the pipeline from bursting. When the pressure of the water suddenly exceeds the permissible pressure due to water hammer phenomenon, then the valve is opened
(vii)	Ball valves/Bell float valves	It is used to maintain a constant level in a service reservoir or elevated
(viii)	Needle & Cone valve	It is similar to sluice valve but are more expensive then sluice and butterfly valves.

Keypoints: The valve which works automatically are as follows:

1. Check valve/Reflux valve
2. Pressure relief valve
3. Air relief valve

5.7 (d)

Refer Solution 5.5

5.8 (b)

Factors governing the location of an intake:

- The site should be near the treatment plant.
- The intake must never be located at the downstream or in the vicinity of the point of disposal of wastewater.
- The intake should never be located near the indicational channel.
- There should be sufficient scope for future additions and expansions.
- It must be located in deep water, sufficiently away from the shoreline.
- The intake site should remain easily accessible during floods and should not get flooded.
- In meandering rivers, the intakes should not be located on curves or at least on sharp curves.

5.9 (d)

Refer Solution 5.5

5.10 (b)

Refer Solution 5.5

NOTE: Water hammer pressure are kept under control by providing surge tanks, pressure relief valves etc.

5.11 (b)

Depending upon the construction material, the pressure pipes are of the following types:

S.No.	Types of Pipes	Remarks
(i)	Cast iron pipes Most widely used in water supply for trunk as well as distributory mains.	**Advantages:** • Resistant to corrosion • Longer life • Moderate cost • Easy to join • Strong and durable • Service connection can be made easily. **Disadvantages:** • Water carrying capacity decreases with time. • Cannot be used for high pressure (generally P□ 700 kN/m^2) • When large, they are very heavy and uneconomical.
(ii)	Steel Pipe	• It is frequently used for raw water trunk mains, inverted syphons or pumping mains, where pressures areas high and sizes larger. • Life is generally taken around 40 years under ordinary conditions.
(iii)	GI Pipes (Galvenised iron pipes)	• Wrought iron pipes galvenised with zinc coating are known as GI pipes. • GI pipes are liable to be affected by acidic and alkaline water. • Carrying hot water in water supply lines.
(iv)	PVC pipes (Polyvinyl chloride)	• Used for plumbing waste lines. • It can withstand water pressure upto 100 m.
(v)	Vitrified clay pipes	• Not used for pressure pipe • Extensively used for carrying sewage. • These pipes are free from corrosion.
(vi)	Asbestos pipes	• Silica and cement are converted under pressure, to a stiff material called asbestos. **Advantages:** • The are light and hence easy to transport. • Highly resistant to corrosion. • Require less joint in their layout. • They are very suitable to be used as small size distribution system. **Disadvantages:** • They are costly
(vii)	RCC (Reinforced Cement Concrete) Pipes	• Mostly used as water mains. • They are available in diameters from 200 mm to 1800 m.
(viii)	PSC (Prestressed Concrete Pipe)	• They can be made to withstand higher pressure. • Corrosion resistant.

5.12 (a)

One of the most important variables in the disinfection is contact time. Harriet Chick observed that for a given concentration of disinfectant, the longer the contact time, the greater the kill. In differential form, Chick's law is

$$\frac{dN_t}{dt} = -kN_t$$

$\frac{dN_t}{dt}$ is rate of kill of pathogens with time

k is inactivation rate constant

N_t is number of pathogens at time t.

Carmen-Kozeny equation for head loss in filter is

$$h_f = \frac{f' L(1-e)V_s^2}{e^3 g d_p}$$

h_f is frictional head loss through bed of particles of uniform size d_p in metre.

e is porosity of bed

V_s is filtering velocity $= \frac{Q}{A_s}$ m/s

d_p is diameter of filter media grains, m

Friction factor, $f' = 150\frac{(1-e)}{\text{Re}} + 1.75$

$\text{Re} = \frac{\phi \rho_w V_s d_p}{\mu}$ where ϕ is the shape factor

5.13 (b)

Grid Iron System: It is suitable for cities with rectangular layout, where the water mains and branches are laid in rectangles.

Advantages:

1. Water is kept in good circulation due to the absence of dead ends.
2. In the cases of a breakdown in some section, water is available from some other direction.

Disadvantages:

1. Exact calculation of sizes of pipes is not possible due to provision of valves on all branches.

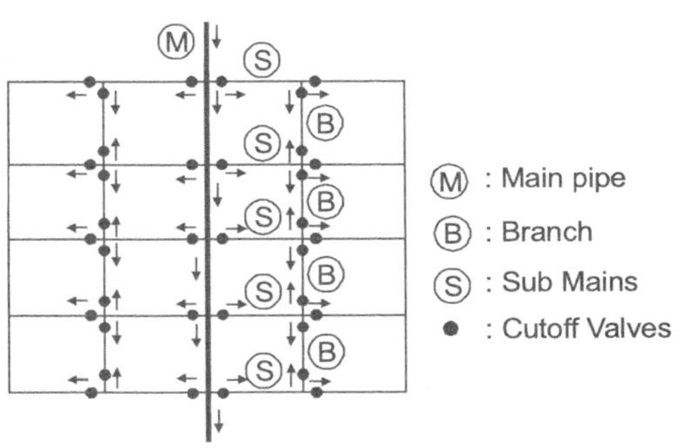

Grid-iron System

5.14 (a)

The **Mannings equation** is an empirical equation that applies to uniform flow in open channels and is a function of the channel velocity, flow area and channel slope. It is extensively used in sanitary sewer design.'

$$V = \frac{1}{n}R^{2/3}S^{1/2}$$

In fluid dynamics, the **Darcy–Weisbach equation** is a phenomenological equation, which relates the head loss, or pressure loss, due to friction along a given length of pipe to the average velocity of the fluid flow for an incompressible fluid.

$$h_f = \frac{fLV^2}{2gD}$$

The **Hardy Cross method** is an iterative method for determining the flow in pipe network systems where the inputs and outputs are known, but the flow inside the network is unknown. The introduction of the Hardy Cross method for analyzing pipe flow network revolutionized municipal water supply design.

The **Rational Method** is a simple technique for estimating a design discharge from a small watershed. It was developed by Kuichling (1889) for small drainage basins in urban areas. It is used extensively in Storm sewer design.

5.15 (c)

Refer Solution 5.11

5.16 (c)

Plug valves are simple and economical shut off valves which are generally not used in plumbing operations because of high head loss and throttling operational characteristics and hence seldom used in water distribution system.

5.17 (d)

The sonoscope/water phone is recommended by the AWWA (American Water Work Association) standards committe in its inspection and maintainace procedure for hydrants. Its regular use promots leak awareness and identifies costly leaks.

5.19 (b)

Steel pipes are comparatively expensive, but they are the strongest and most durable of all water supply systems. They can withstand high water pressure, come in convenient (longer) lengths than most other pipes and thus incur lower installation/ transportation costs. They can also be easily welded.

5.20 (b)

Refer Solution 5.11

5.21 (b)

Advantages of Cast iron pipes

- The main advantage of cast iron pipe fittings are that they are indeed highly resistant to abrasion making them the ideal choice for anyone wanting to drain materials such as sand, dishwasher water and garbage disposal.
- They are highly durable and practically indestructible.
- They do not react with most chemicals and won't melt or burn even in extreme conditions.
- They have a long lifespan which minimizes the need for replacements and repairs.
- Cast iron pipe fittings are very well constructed and as such have fantastic sound-proofing qualities when compared to other materials. So, there is minimal noise and vibration between the fittings and the pipes.
- Their versatility is another great advantage of choosing cast iron pipe fittings. There is a wide range of joining methods available for the fittings and that also allows for simple and efficient installation in all situations.
- Cast iron pipe fittings are cheap to install in comparison to other materials. They are cheaper to buy than many other pipe fittings and they are also one of the easiest types to use and install with minimal fuss.

Disadvantages of Cast iron pipes

- These pipes are very heavy and it requires specific means of transportations.
- It is also not advisable to use cast iron pipes in inaccessible places.
- Because of their heavy load, these pipes are made in shorter lengths. This equates to more jointing and a more complicated layout.

5.22 (d)

Refer Solution 5.11

5.23 (b)

In long pipes, air accumulates at the top most point of the line. **(Also refer solution of 5.8).**

5.24 (d)

The flow in water distribution pipes takes place due to water pressure and in sewer due to gravity.

5.25 (d)

- The pipes try to bull apart and get out of the alignment at bends and other points of unbalanced pressures.
- At such places, the forces exerted on the joints due to longitudinal shearing stresses caused by these imbalanced pressures are enormous and the joint may get loosened, ultimately leading to excessive leakage or failure of the pipe.
- In all such circumstances, in order to prevent the pipes from pulling apart, pipes are anchored by firmly embedding these portions in massive blocks of concrete or masonry, which absorb the side thrust.
- Similarly when pipes are laid on steep slopes, they try to slip and thereby pull apart, pipes are therefore, anchored under such circumstances also.

5.29 (a)

Branch drains should not be very long.

■■■■

6 Biochemical Reactions in Waste Water

6.1 Organisms that mineralize organic matter in an ecosystem are called
(a) producers (b) consumers
(c) decomposers (d) carnivorous
[ESE : 2000]

6.2 In aerobic conditions, the microbial decomposition of organics results in the formation of which one of the following?
(a) Stable and objectionable end products
(b) Unstable and objectionable end products
(c) Unstable and acceptable end products
(d) Stable and unobjectionable end products
[ESE : 2004]

6.3 Presence of nitrogen in a waste water sample is due to the decomposition of
(a) carbohydrates (b) proteins
(c) fats (d) vitamins
[ESE : 2005]

6.4 A heterotroph is an organism that obtains
(a) its cell carbon from an inorganic source
(b) its energy from the oxidation of simple inorganic compounds
(c) its cell carbon as well as its energy from organic matter
(d) its energy from a natural ecosystem
[ESE : 2007]

6.5 **Assertion (A):** The rate of biomass production will be always lower than the rate of food utilization in a biological system having a mixed culture of micro-organism.
Reason (R): Catabolism converts part of the food into waste products.
(a) both A and R are true and R is the correct explanation of A
(b) both A and R are true but R is not a correct explanation of A
(c) A is true but R is false
(d) A is false but R is true
[ESE : 2009]

6.6 In an aerobic attached-culture system, the biomass at the biofilm-medium surface interface experiences
(a) aerobic and endogenous metabolism
(b) anaerobic and endogenous metabolism
(c) anaerobic and exogenous metabolism
(d) aerobic and exogenous metabolism
[ESE : 2010]

6.7 The group of micro-organisms involved in production of methane from acetic acid (or acetate) in anaerobic waste water treatment processes is:
(a) Methanothrix and Methanobacterium
(b) Methanobacterium and Methanosarcina
(c) Methanosarcina and Methanospirillum
(d) Methanothrix and Methanosarcina
[ESE : 2011]

Directions: Each of the next items consists of two statements, one labeled as the **'Statements (I)'** and the other as **'Statement (II)'**. You are to examine these two statements carefully and select the answers to these items using the codes given below :

Codes:
(a) Both Statement (I) and Statement (II) are individually true and Statement (II) is the correct explanation of Statement (I)
(b) Both Statement (I) and Statement (II) are individually true but Statement (II) is NOT the correct explanation of Statement (I)
(c) Statement (I) is true but Statement (II) is false
(d) Statement (I) is false but Statement (II) is true

6.8 **Statement (I):** The rate of biomass production will always be lower than the rate of food utilization in a biological system having a mixed culture of microorganisms.
Statement (II): Catabolism converts part of the food into waste products.
[ESE : 2014]

6.9 **Statement (I):** The rate of biomass production will be always lower than the rate of food utilization in a biological system having a mixed culture of micro-organisms.
Statement (II): Catabolism converts part of the food into waste products. [ESE : 2017]

6.10 Which one of the following is the process in which ammonia is oxidized to nitrites and then to nitrates by aerobic bacteria?
(a) Nitrification (b) Denitrification
(c) Adsorption (d) Regeneration
[ESE : 2021]

6.11 The quantity of nitrogen present in wastewater before the decomposition of organic matter has started, is indicated by
(a) Albuminoid Nitrogen
(b) Free Ammonia
(c) Organic Nitrogen
(d) Nitrate Nitrogen [ESE : 2021]

6.12 The first stage of sewage purification by the action of anaerobic bacteria is termed as
(a) Putrefaction (b) Liquefaction
(c) Clarification (d) Oxidation
[ESE : 2023]

■■■■

Answers Biochemical Reactions in Waste Water

6.1 (c) 6.2 (d) 6.3 (b) 6.4 (c) 6.5 (a) 6.6 (b) 6.7 (d) 6.8 (a) 6.9 (a)
6.10 (a) 6.11 (*) 6.12 (a)

Explanations Biochemical Reactions in Waste Water

6.1 (c)
Decomposer class includes saprophytic organisms that help in nutrients and element recycling process. These are bacteria, fungi and protozoans. They reduce organic matter to simple compounds and elements.

6.2 (d)
In aerobic conditions, the microbial activity results in stable and unobjectionable end products. In anaerobic conditions, methane and carbondioxide (CO_2) are produced which are objectionable.

6.3 (b)
The nitrogen compounds from organisms are eliminated in the form of urea, uric acid and ammonia. All of these substances are produced from protein metabolism.

6.4 (c)
Organisms that derive both energy and material from inorganic sources are called autotrophs, while that obtain both energy and material from organic compounds are called heterotrophs. Phototrophs are organisms which utilize sunlight for an energy source and inorganic substances for a material source.

6.5 (a)
If all the food were converted to biomass, then the rate of food utilization would equal the rate of biomass production. But, since catabolism converts part of the food into waste products, the rate of food utilization will be greater than the rate of biomass production.

6.6 (b)
Attached-culture systems employ reactors in which waste water is contacted with microbial films attached to surfaces. Surface area for biofilm growth is increased by placing a porous medium in the reactor. The organisms attach themselves to the

medium and grow into dense films of a viscous, jellylike nature. Waste water passes over this film in thin sheets with dissolved organics passing into the biofilm due to concentration gradients within the film. Suspended particles and colloids may be retained on the sticky surfaces where they are decomposed into soluble products. Oxygen from the wastewater and from air in the void spaces of the medium provide oxygen for aerobic reactions at the biofilm surface. Waste products from the metabolic processes diffuse outward and are carried away by the water or air currents moving through the voids of the medium. Growth of the biofilm is restricted to one direction-outward from the solid surface. As the film grows thicker, concentration gradients of both oxygen and food develop. Eventually, both anaerobic and endogenous metabolism occur at the biofilm-medium surface interface. The attachment mechanism is a weakened and the shearing action of the wastewater flowing across the film pulls it from its mooring and washes it away. This process known as sloughing, is function of both the hydraulic and organic loading rate. Biofilm is quickly reestablished in places cleared by sloughing.

6.7 (d)

For the biomass converting acetal to methane, there are two classes of methanogens Methanothrix and Methanosarcina. The first class has a low specific activity and a low half velocity constant which causes it to predominate in systems with low steady state acetate concentrations. This is true for most operating systems because they are lowly loaded. The second class-Methanosarcina have relatively high half velocity constants and specific activities. In addition the nutritional requirements of methanosarcina is more than Methanothrix.

6.8 (a)

The rate of food utilization would be equal to rate of biomass production if all food were converted to biomass. Catabolism provides the chemical energy necessary for the maintenance and growth of cells. Since energy is obtained from food, therefore a part of food gets used up in catabolism and the rest of the food gets converted into biomass.

6.9 (a)

Refer Solution 6.8

6.10 (a)

There are two group of chemoautotrophic bacteria that can be associated with the process of nitrification One group (nitrosomonas) derive its energy through the oxidation of ammonium to nitrite, whereas the other group (nitrobacter) obtains energy through the oxidation of nitrite to nitrate.

Nitrification of ammonia to nitrate is a two step process.

$$NH_3 \rightarrow NH_4 \xrightarrow{\text{Nitro-somonas}} NO_2 \xrightarrow{\text{Nitro-bacter}} NO_3$$

6.11 (*)

There are two possible ways in which the question can be interpreted:

- *The nitrogen which we measure are in the forms of NH_3, NO_2, NO_3 and N_2 in the measurement process. If we take the quantity of nitrogen before the decomposition of organic matter is started, it will be free ammonia which will be measured.*
- *The other scenario can be quantity of nitrogen after ammonia leaves the system. In that case, organic or albuminoid nitrogen will be present in the system. So the other answer can be either (a) or (c).*

6.12 (a)

The meaning of putrefaction is the anerobic decomposition of organic matter by bacteria and fungi that results in obnoxiously odorous product.

■■■■

7 Waste Water Quality Characteristics

7.1 A 2% solution of a sewage sample is kept at an incubation temperature of 20°C. If initial DO (Dissolved Oxygen) and final DO values after 5 days' incubation period are 8.5 mg/L and 5.5 mg/L respectively, then the BOD will be
(a) 50 mg/L (b) 150 mg/L
(c) 250 mg/L (d) 350 mg/L
[ESE : 2000]

7.2 For a waste, the 5-day BOD at 20°C is found to be 200 mg/L. For the same waste, 5-day BOD at 30°C will be
(a) less than 200 mg/L
(b) more than 200 mg/L
(c) 200 mg/L
(d) zero, as the bacteria cannot withstand such a high temperature
[ESE : 2000]

7.3 The ultimate BOD value of a waste
(a) increases with temperature
(b) decreases with temperature
(c) remains the same at all temperatures
(d) doubles with every 10°C rise in temperature
[ESE : 2000]

7.4 In which one of the following tests is the organic matter in the waste water used as food by micro-organisms?
(a) BOD
(b) Most probable number
(c) COD
(d) Chlorine demand
[ESE : 2000]

7.5 Match **List-I** (Water quality) with **List-II** (Method of determination) and select the correct answer using the codes given below the lists:

List-I	List-II
A. Hardness	1. Winkler's method
B. Chlorine	2. EDTA method
C. DO	3. Orthotolidine test
D. Chloride	4. Mohr method

Codes:

	A	B	C	D
(a)	2	3	1	4
(b)	2	4	1	3
(c)	1	3	2	4
(d)	1	4	2	3

[ESE : 2001]

7.6 A certain waste has a BOD of 162 mg/L and its flow is 1000 cubic metres per day. If the domestic sewage has a BOD of 80 gram per capita, then the population equivalent of the waste would be
(a) 20.25 (b) 1296
(c) 2025 (d) 12960
[ESE : 2001]

7.7 The correct statement of comparison of ultimate BOD, COD, Theoretical Oxygen Demand (ThOD) and 5-day BOD (BOD_5) is
(a) $BOD_u > COD > ThOD > BOD_5$
(b) $COD > ThOD > BOD_u > BOD_5$
(c) $ThOD > COD > BOD_u > BOD_5$
(d) $COD > BOD_u > BOD_5 > ThOD$
[ESE : 2002]

7.8 **Assertion (A):** U.K. Royal Commission on Sewage Disposal of 1898-1915 classified British rivers on the basis of 65° F, 5-day BOD.
Reason (R): British rivers do not have flow time to open sea more than 5 days.
(a) both A and R are true and R is the correct explanation of A
(b) both A and R are true but R is not a correct explanation of A
(c) A is true but R is false
(d) A is false but R is true
[ESE : 2002]

7.9 Which one of the following pairs is not correctly matched?

(a) BOD/COD = 0 : Waste-water is toxic

(b) BOD/COD $\leq$ 0.2 : Acclimatization of seed is necessary

(c) BOD/COD $\geq$ 0.6 : Waste-water is non-biodegradable

(d) BOD = COD = 0 : Waste-water is devoid of organic matter

[ESE : 2004]

7.10 A waste water sample of 2 mL is made upto 300 mL in BOD bottle with distilled water. Initial DO of the sample is 8 mg/L and after 5 days it is 2 mg/L. What is its BOD?

(a) 894 mg/L (b) 900 mg/L

(c) 300 mg/L (d) 1200 mg/L

[ESE : 2004]

7.11 What is 5 days 20°C BOD equal to?

(a) 3 days 27°C BOD (b) 4 days 30°C BOD

(c) 6 days 32°C BOD (d) 7 days 35°C BOD

[ESE : 2005]

7.12 Match **List-I** (Equipment) with **List-II** (Parameter) and select the correct answer using the codes given below the lists:

List-I	List-II
A. Tintometer	1. Temperature
B. Nephelometer	2. Colour
C. Imhoff cone	3. Turbidity
D. Muffle furnace	4. Settleable solids
	5. Volatile solids

Codes:

	A	B	C	D
(a)	4	3	1	5
(b)	2	5	4	3
(c)	4	5	1	3
(d)	2	3	4	5

[ESE : 2006]

7.13 In context of water polluted with sewage, what does BOD signify?

(a) Biological oxygen demand

(b) Bacteriological oxygen demand

(c) Biochemical oxygen demand

(d) Biology of degradation

[ESE : 2006]

7.14 Match **List-I** (Process/Bacteria) with **List-II** (Energy/Material) and select the correct answer using the codes given below the lists:

List-I

A. Anabolism

B. Autotrophs

C. Catabolism

D. Heterotrophs

List-II

1. Providing energy for the synthesis of new cells and maintenance of other cell functions
2. Obtaining energy and material for growth from organic sources
3. Providing the material necessary for cell growth
4. Obtaining energy and material for growth from inorganic sources

Codes:

	A	B	C	D
(a)	1	4	3	2
(b)	3	2	1	4
(c)	1	2	3	4
(d)	3	4	1	2

[ESE : 2006]

7.15 What is the theoretical oxygen demand of 300 mg/L glucose solution?

(a) 300 mg/L (b) 320 mg/L

(c) 350 mg/L (d) 400 mg/L

[ESE : 2009]

7.16 The figure below shows, BOD curve when the experiment was conducted at 20°C. If the experiment is conducted at 30°C, then the portion AB of the curve

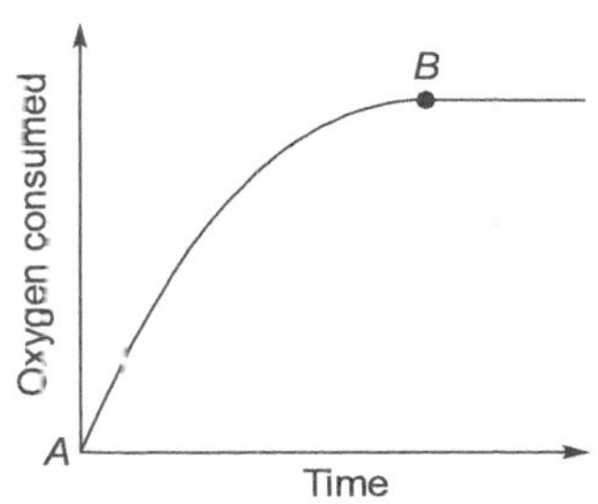

(a) shifts to the left

(b) shifts to the right

(c) remains unchanged

(d) shrinks

[ESE : 2010]

7.17 Which one of the following tests employs ferroin indicator?

(a) Chemical oxygen demand
(b) Ammonia nitrogen
(c) Nitrate nitrogen
(d) Fluoride [ESE : 2011]

7.18 Match **List-I** with **List-II** and select the correct answer using the code given below the lists:

List-I
A. Viruses in water
B. Depletion of oxygen
C. Excess nitrates in water
D. Excess fluorides in water

List-II
1. Parasite-based disease
2. Fish extinction
3. Methemeglobinemia
4. Mottling of teeth

Code:

	A	B	C	D
(a)	1	2	3	4
(b)	4	2	3	1
(c)	1	3	2	4
(d)	4	3	2	1

[ESE : 2011]

7.19 A sample of sewage is estimated to have a 5 day 20° C BOD of 250 mg/*l*. If the test temperature be 30°C, in how many days will the same value of BOD be obtained?

(a) 1.5 days (b) 2.5 days
(c) 3.3 days (d) 7.5 days

[ESE : 2012]

7.20 **Statement-I** : The BOD test is conducted for 5 days at 20°C.

Statement -II : The amount of oxygen utilized by microorganisms anaerobically is called BOD.

(a) Both Statement (I) and Statement (II) are individually true and Statement (II) is the correct explanation of Statement (I)
(b) Both Statement (I) and Statement (II) are individually true but Statement (II) is NOT the correct explanation of Statement (I)
(c) Statement (I) is true but Statement (II) is false
(d) Statement (I) is false but Statement (II) is true

[ESE : 2012]

7.21 The dissolved oxygen in a water sample is generally estimated by modified Winkler method. The water sample is taken in a 300 m*l* BOD bottle and the dissolved oxygen is fixed by adding

(a) 1 ml of conc. H_2SO_4 followed by addition of 1 mnl of $MnSO_4$ reagent and contents mixed thoroughly, thereafter waiting for 5-10 minutes, and then 1 ml of alkali-iodide-azide reagent is added.
(b) 1 ml of alkali-iodide-azide reagent followed by addition of 1 ml of conc. H_2SO_4 and then the contents mixed thoroughly, thereafter waiting for 5-10 minutes, 1 ml of $MnSO_4$ regent is added.
(c) 1 ml of $MnSO_4$ followed by addition of 1 ml of alkali-iodide-azide reagents and then the contents mixed thoroughly, thereafter waiting for 5-10 minutes and then 1 ml of conc. H_2SO_4 is added.
(d) 1 ml of conc. of H_2SO_4 followed by addition of 1 ml of alkali-iodide-azide reagent and then the contents mixed thoroughly, thereafter waiting for 5-10 minutes and then 1 ml of $MnSO_4$ reagent is added.

[ESE : 2013]

7.22 Consider the following statements:

1. The quality of waste water determined on the basis of DO.
2. The BOD test is based on DO.
3. Determination of DO helps in controlling corrosion.

Which of the above statements are correct?

(a) 1, 2 and 3 (b) 1 and 2 only
(c) 1 and 3 only (d) 2 and 3 only

[ESE : 2014]

7.23 The dissolved oxygen in a water sample is generally estimated by modified Winkler method. Accordingly, approximately 200 ml volume of dissolved-oxygen fixed solution shall be titrated with

(a) Sodium thiosulphate reagent using soluble starch as an indicator
(b) Sodium thiosulphate reagent using ferroin as an indicator
(c) Ferrous ammonium sulphate reagent using soluble starch as an indicator

(d) Ferrous ammonium sulphate reagent using ferroin as an indicator

[ESE : 2014]

7.24 If the average sewage from a city is 95×10^6 *l*/day and the average 5 day BOD is 300 mg/*l* against a standard of 75 mg/day per capita at 20°C, then the population equivalent of the city will be

(a) 28,500 (b) 285×10^6
(c) 3,80,000 (d) 380×10^6

[ESE : 2014]

7.25 **Statement (I):** Dilution of the wastewater sample with organic-free, oxygen-saturated water is necessary to measure 5-day 20°C BOD values greater than 7 mg/*l*.

Statement (II): The saturation for oxygen in water at 20°C is approximately 9 mg/*l*.

(a) Both Statement (I) and Statement (II) are individually true and Statement (II) is the correct explanation of Statement (I).
(b) Both Statement (I) and Statement (II) are individually true but Statement (II) is not the correct explanation of Statement (I).
(c) Statement (I) is true but Statement (II) is false.
(d) Statement (I) is false but Statement (II) is true.

[ESE : 2014]

7.26 Consider the following statements regarding biochemical oxygen demand (BOD) of river water.

1. The BOD rate constant varies with river water temperature.
2. The BOD rate constant does not depend on the BOD of the river water.
3. The BOD rate constant is often different for different river waters.
4. The BOD rate constant cannot be determined in a laboratory.

Which of the above statements are correct?

(a) 1 and 4 (b) 1 and 3
(c) 2 and 3 (d) 2 and 4

[ESE : 2015]

7.27 **Statement (I) :** Dilution of the wastewater sample with organic-free, oxygen-saturated water is necessary to measure 5-day 20°C BOD values greater than 7 mg/*l*.

Statement (II): Saturation concentration of oxygen in water at 20°C is approximately 9 mg/*l*.

(a) Both Statement (I) and Statement (II) are individually true and Statement (II) is the correct explanation of Statement (I).
(b) Both Statement (I) and Statement (II) are individually true but Statement (II) is not the correct explanation of Statement (I).
(c) Statement (I) is true but Statement (II) is false.
(d) Statement (I) is false but Statement (II) is true.

[ESE : 2017]

7.28 Five-days BOD of a 10% diluted sample having D_C = 6.7 mg/*l*, D_s = 2 mg/*l* and consumption of oxygen in blank = 0.5 mg/*l*, will be

(a) 22 mg/*l* (b) 42 mg/*l*
(c) 62 mg/*l* (d) 82 mg/*l*

[ESE : 2018]

7.29 Consider the following statements with reference to be mixing of industrial waste water with domestic waste water:

1. The industrial waste water can be mixed with domestic water when it has higher BOD.
2. The industrial waste water can be mixed with domestic water when the pH value of industrial waste water in highly alkaline.

Which of the above statements is/are correct?

(a) 1 only (b) 2 only
(c) Both 1 and 2 (d) Neither 1 nor 2

[ESE : 2019]

7.30 The waste water from a factory having a pH of 10, contains KOH only. For waste water discharge is 80 m^3/day. The total quantity of KOH per day will be nearly

(a) 4.5 kg/day (b) 5.4 kg/day
(c) 6.3 kg/day (d) 7.2 kg/day

[ESE : 2019]

7.31 Which of the following chemical parameters are associated with the organic content of water?

1. Biological Oxygen Demand (BOD).
2. Chemical Oxygen Demand (COD).
3. Total Organic Carbon (TOC) and Total Oxygen Demand (TOD).

(a) 1 and 2 only (b) 1 and 3 only
(c) 2 and 3 only (d) 1, 2 and 3

[ESE : 2020]

7.32 The domestic sewage of a town was tested for total solids and the following results were obtained:
Weight of sample of sewage = 1000 gm
Weight of solids after evaporation of liquid = 0.952 gm
Weight of dry residue after ignition = 0.516 cm
What is the value of volatile solids?
(a) 952 ppm (b) 516 ppm
(c) 436 ppm (d) 694 ppm
[ESE : 2021]

7.33 Which one of the following is a measure of light-emitting properties of waste water?
(a) Turbidity (b) pH
(c) Alkalinity (d) Volatility
[ESE : 2023]

7.34 Which one of the following is defined as the oxygen required for the microorganisms to carry out biological decomposition of dissolved solids or organic matter in the water under aerobic conditions at standard temperature?
(a) Biochemical oxygen demand
(b) Chemical oxygen demand
(c) Total oxygen demand
(d) Theoretical oxygen demand [ESE : 2023]

7.35 What needs to be maintained to avoid the odour problem in tannery effluent treatment?
(a) Low pH
(b) pH between 7 – 8
(c) pH between 9.5 – 10
(d) High pH [ESE : 2024]

Answers Waste Water Quality Characteristics

7.1 (b)	7.2 (b)	7.3 (c)	7.4 (a)	7.5 (a)	7.6 (c)	7.7 (c)	7.8 (a)	7.9 (c)
7.10 (b)	7.11 (a)	7.12 (d)	7.13 (c)	7.14 (a)	7.15 (b)	7.16 (a)	7.17 (a)	7.18 (a)
7.19 (c)	7.20 (c)	7.21 (c)	7.22 (a)	7.23 (a)	7.24 (d)	7.25 (b)	7.26 (b)	7.27 (b)
7.28 (b)	7.29 (d)	7.30 (*)	7.31 (d)	7.32 (c)	7.33 (a)	7.34 (a)	7.35 (d)	

Explanations Waste Water Quality Characteristics

7.1 (b)

BOD = (Initial DO – Final DO) × Dilution Factor

$$= (8.5 - 5.5) \times \left(\frac{100}{2}\right) = 150 \text{ mg/L}$$

7.2 (b)

The effect of temperature on BOD can be approximately given by the von't Hoff-Arrhenius models :

$$K_T = K_{20}(1.047)^{T-20}$$

At $T = 30°C,\ K_T > K_{20}$

$$BOD_{5,30°C} = L_0\left(1 - e^{-k_T t}\right)$$

$$\therefore \quad BOD_{5,30°C} > BOD_{5,20°C} = 200 \text{ mg/L}$$

7.3 (c)

- The total oxygen required for complete oxidation of biodegradable organic matter is called as ultimate BOD.
- Ultimate BOD is independent of temperature.

7.4 (a)

Biochemical oxygen demand is the amount of oxygen required for microbial metabolism of organic compounds in water. This demand occurs over some variable period of time depending on temperature, nutrient concentrations, and the enzymes available to indigenous microbial populations. The amount of oxygen required to completely oxidize the organic compounds to carbon dioxide and water through generations of microbial growth, death, decay, and cannibalism is total biochemical oxygen demand (total BOD).

7.5 (a)

- Chlorine residuals can be tested by orthotolidine test, DPD test, Chlorotex test and Starch-Iodide test.
- The dissolved oxygen of sewage is generally determined by Winkler's method which is an oxidation - reduction process carried out chemically to liberate iodine in amount equivalent to the quantity of DO originally present.

7.6 (c)

Daily BOD contributed by waste

$$= 162 \times \frac{1000 \times 1000}{10^3} = 162{,}000 \text{ g/day}$$

Population equivalent

$$= \frac{162{,}000}{80} = 2025$$

7.7 (c)

$$\text{ThOD} > \text{COD}$$

As during the performance of COD test entire organic matter is not decomposed even by the action of strongest oxidising reagent. But theoretically entire organic matter is considered to be decomposed

7.8 (a)

British rivers were classified by Royal Commission on Sewage Disposal of 1898-1915 on basis of 65°F, 5 day BOD because at that time their rivers did not have flow time to open sea more than 5 days.

7.9 (c)

$BOD_5/COD \geq 0.63$ means waste water is biodegradable

7.10 (b)

BOD_5 = [initial DO – Final D O] × dilution factor

$$= (8 - 2) \times \frac{300}{2} = 900 \text{ mg/L}$$

7.11 (a)

$$BOD_t = L_0 (1 - e^{-kt})$$

and $k_T = k_{20} (1.047)^{T-20}$

BOD_5 at 20°C will be equal to a BOD_t for given temperature when $k_T t$ is equal to $k_{20} \times 5$.

$$k_T t = k_{20} \times 5$$

or
$$\frac{k_T}{k_{20}} = \frac{5}{t} = (1.047)^{T-20}$$

$$t = \frac{5}{(1.047)^{T-20}}$$

T	t(days)
27°C	3.63
30°C	3.16
32°C	2.88
35°C	2.51

There is no correct answer but (a) is closest

7.12 (d)

Nephelometer is based on scattering principle for measurement of turbidity. It uses formazin, a chemical compound, for standard. Formazin is more reproducible standard than SiO_2.

Tintometer is a color measuring instrument which compares colour of water in Nessler tubes which contain solutions of platinum Cobalt dissolved in water.

7.14 (a)

Metabolic processes are of two types:

(i) Catabolism provides energy for synthesis of new cells as well as for maintenance of other cell functions.

(ii) Anabolism provides the material necessary for cell growth. When an external food source is interrupted the organism will use stored food for maintenance energy in a process called endogenous catabolism. Enzymes play major role in biochemical reactions.

The chemical formula for bacterial cell is assumed to be $C_5H_7O_2N$. Organisms that derive energy and material from inorganic sources are called autotrophs, while bacteria that obtain both energy and material from organic compounds are called heterotrophs. Phototrophs, bacteria which utilize sunlight for an energy source and inorganic substances for a material source, play important role in natural water purification processes.

7.15 (b)

The equation for the total oxidation of glucose is as follows:

$$\underset{180\text{ g}}{C_6H_{12}O_6} + \underset{192\text{ g}}{6O_2} \longrightarrow \underset{264\text{g}}{6CO_2} + \underset{108\text{g}}{6H_2O}$$

∵ 180 g of glucose require 192 g of oxygen for complete oxidation

∴ Requirement of oxygen for 300 mg of glucose

$$= 300 \times \frac{192}{180} = 320 \text{ mg}$$

Thus the theoretical oxygen demand of 300 mg/L of glucose solution in 320 mg/L.

7.16 (a)

The BOD of water at any time t is expressed as

$$y_t = L_o (1 - e^{-kt})$$

The value of k is temperature dependent. Because micro-organisms are more active at higher temperatures, the value of k increases with increasing temperatures. The value of y_t approaches L_0 asymptotically indicating that the total, or ultimate, BOD (y_u) is equal to the initial oxygen equivalent of the water L_0. Thus, at higher temperatures the value of y_t approaches to ultimate BOD earlier, shifting point B on the curve towards the left.

7.17 (a)

Ferroin sulfate or ferroin indicator is prepared by combining Phenanthraline to Ferrous sulphate in water. If is used for detecting COD.

7.18 (a)

The presence of too much of nitrate in water may adversely affect the health of infants, causing a disease technically called "methemoglobinemia" (commonly called "blue baby disease". The nitrate concentration in domestic water supplies is generally limited to 45 mg/l.

A fluoride concentration of less than 0.8-1.0 ppm may be harmful and may cause dental caries (tooth decay) due to the formation of excessive cavities in the teeth of young children during calcination of their permanent teeth. Higher fluoride concentrations, greater than 1.5 ppm or so, may again be harmful, causing spotting and discoloration of teeth, (a disease called fluorosis), which with continued excessive consumption of fluorides, may even cause deformation of bones. Depletion of oxygen from water is related to fish extinction.

7.19 (c)

$$BOD_{5,\,20°C} = L\left(1-10^{-K_D \times t}\right)$$

The effect of temperature on BOD can be given by

$$K_{D_{T°C}} = K_{D_{20°C}}(1.047)^{T-20}$$

$$\therefore \quad K_{D_{30°C}} = K_{D_{20°C}}(1.047)^{10}$$

$$K_{D_{30°C}} = 1.583 K_{D_{20°C}}$$

$$\therefore \quad L\left(1-10^{-K_D \times 5}\right) = L\left(1-10^{-1.583 \times K_D \times t}\right)$$

$$\Rightarrow \quad 10^{-K_D \times 5} = 10^{-1.583 \times K_D \times t}$$

$$\therefore \quad t = \frac{5}{1.583} = 3.16 \text{ days}$$

7.20 (c)

If sufficient oxygen is present in water the useful aerobic bacteria will flourish and cause the biological decomposition of waste and organic matter, thereby reducing the carbonaceous material from water. The amount of oygen required in the process until oxidation gets completed is known as BOD. Polluted waters will continue to absorb oxygen for many months and it is not practically possible to determine this ultimate oxygen demand. Hence, the BOD of water during the first five days at 20°C is generally taken as standard demand.

7.21 (c)

The Winkler Method is a technique used to measure dissolved oxygen in fresh water systems. It uses titration to determine dissolved oxygen in the water sample.

Reagent List:

- 2 ml Manganese sulphate
- 2 ml alkali-iodide-azide
- 2 ml concentrated sulfuric acid
- 2 ml starch solution
- Sodium thiosulfate

Procedure:

1. A 300-mL glass Biological Oxygen Demand (BOD) stoppered bottle is filled till brim with sample water.
2. Manganese sulphate is added to the collection bottle by inserting the calibrated pipette just below the surface of the liquid.
3. Alkali-iodide-azide reagent is also added in the same manner.
4. The sample is mixed by inverting it several times. If oxygen is present, a brownish-orange cloud of precipitate or floc will appear.
5. Then concentrated sulfuric acid is added. At this point, the sample is "fixed" and can be stored for up to 8 hours if kept in a cool, dark place.
6. This sample is then titrated with sodium thiosulfate to a pale straw color.

7. Then starch solution is added so a blue color forms.
8. Titration is done until the sample turns clear.
9. The concentration of dissolved oxygen in the sample is equivalent to the number of milliliters of titrant used. Each mL of sodium thiosulphate added in steps 6 and 8 equals 1 mg/L dissolved oxygen.

7.23 (a)

The D.O. content of sewage is generally determined by the Winkler's method which is an oxidation-reduction process carried out chemically to liberate iodine in amount equivalent to the quantity of dissolved oxygen originally present. In this test, sodium thiosunphate is using as reagent and soluble starch as an indicator.

7.24 (d)

Population equivalent of city

$$= \frac{\text{BOD produced per day in city}}{\text{Standard BOD per capital}}$$

$$= \frac{95 \times 10^6 \times 300}{75} = 380 \times 10^6$$

7.26 (b)

BOD rate constant (k_D) varies with temperature (T) as,

$$k_{D_{T°C}} = k_{D_{20°C}} [1.047]^{T-20}$$

k_D is computed from BOD values measured at various times as it depends upon the nature of the organic matter present in sewage.

7.28 (b)

[IS 3025 : Part 44]

$$\text{BOD(mg/}l) = \frac{D_1 - D_2 - (B_1 - B_2)F}{P} \times 100$$

D_1 – Initial DO sample (mg/l)

D_2 – DO of sample after incubation (mg/l)

B_1 – DO of seed before incubation (mg/l)

B_2 – DO of seed after incubation (mg/l)

F – Ratio of seed in diluted sample to seed in control

P – % dilution of sample

$$\text{BOD} = \frac{6.7 - 2 - 0.5}{10} \times 100 = 42 \text{ mg/}l$$

(f is to be used only when correction factor is given)

7.29 (d)

As per **IS: 3306 1974** maximum permissible limits for industrial effluent discharges for BOD and pH intc public sewers is 350 mg/l and 5.5 – 9 respectively. Hence both options are correct.

7.30 (*)

$$\text{pH} = 10,\ \text{pH} + \text{pOH} = 14$$

$$\text{pOH} = 14 - 10 = 4$$

$$-\log_{10}[\text{OH}] = 4$$

$$\Rightarrow [\text{OH}^-] = 10^{-4} \text{ moles/}l$$

$$\text{K}^+ + \text{OH} = \text{KOH}$$

1 mole of OH^- ion forms 1 moles of KOH

Moles of KOH formed = 10^{-4} moles/l

Total quantity of KOH formed

$= 10^{-4} \times 80 \times 10^3 \times 56 \times 10^{-3}$

$= 0.448$ kg/d $= 0.45$ kg/d

7.31 (d)

BOD stands for biochemical oxygen demand not biologica oxygen demand.

7.32 (c)

Weight of sewage sample

= 1000 g

After evaporation of water, weight

= 0.952 g

∴Weight of water

= 1000 – 0.952 g = 999.048 g

After further ignition weight

= 0.516 g

∴ Weight of volatile solids which are loss during ignition

= 0.952 – 0.516

= 0.436 g = 436 mg

436 mg of volatile solids in 999.048 g of water. Taking sewage density as roughly 1 kg/lit. or 1000 g/lit.

∴436 mg in 1 lit. = 436 ppm

7.33 (a)

Turbidity is the measure of extent to which light is either absorbed or scattered by suspended solids present in water.

■■■■

8 Disposal of Sewage Effluent

8.1 When a sewage is disposed off in a river, the rate of depletion of dissolved oxygen of the river mainly depends on
(a) biochemical oxygen demand of the sewage
(b) chemical oxygen demand of the sewage
(c) total organic carbon present in the sewage
(d) dissolved oxygen present in the sewage
[ESE : 2000]

8.2 From ecological considerations, the minimum level of Dissolved Oxygen (DO) necessary in the rivers and streams is
(a) 1 mg/L (b) 2 mg/L
(c) 4 mg/L (d) 8 mg/L
[ESE : 2000]

8.3 Which one of the following sets of processes is a part of self-purification of streams?
(a) Settling, bio-degradation and desalination
(b) Settling, bio-degradation and aeration
(c) Floatation, ion exchange and desalination
(d) Desalination, ion exchange and reverse osmosis
[ESE : 2000]

8.4 When sewage enters a flowing river, the rapid depletion of dissolved oxygen is due to
(a) change in temperature in river water
(b) the suspended particles in river and waste
(c) respiratory activity of aquatic plants in the river
(d) microbial activity [ESE : 2000]

8.5 Sewage may be disposed of without treatment into a water body if the available dilution is
(a) less than 150 (b) more than 150
(c) more than 300 (d) more than 500
[ESE : 2001]

8.6 When wastewater is disposed of into a running stream, four zones are formed. In which one of the following zones will the minimum level of dissolved oxygen be found?
(a) Zone of degradation
(b) Zone of active decomposition
(c) Zone of recovery
(d) Zone of clear water [ESE : 2001]

8.7 Sewage sickness signifies
(a) diseases caused by sewage
(b) soil pores getting clogged and preventing free circulation of air when sewage is continuously applied on land
(c) raw sewage is applied and used for irrigating vegetables which are eaten raw
(d) disposal of septic sewage on land
[ESE : 2003]

8.8 In dissolved oxygen sag curve, the sag curve results because
(a) it is a function of rate of addition of oxygen to the stream
(b) it is a function of rate of addition of oxygen from the stream
(c) it is a function of both addition and depletion of oxygen from the stream
(d) the rate of addition of oxygen is linear but the rate of depletion is non-linear
[ESE : 2004]

8.9 A municipal sewage has BOD_5 of 200 mg/L. It is proposed to treat it and dispose off into a marine environment. For what minimum efficiency should the sewage treatment plant be designed?
(a) 85% (b) 60%
(c) 50% (d) 33.67%
[ESE : 2005]

8.10 In which type of lakes, does a perfect ecological equilibrium among the producers, decomposers and consumer groups of organisms exist?
(a) Senescent lakes
(b) Mesotrophic lakes
(c) Oligotrophic lakes
(d) Eutrophic lakes [ESE : 2005]

8.11 What is eutrophication of lakes primarily due to?
(a) Multiplication of bacteria
(b) Excessive inflow of nutrients
(c) Increase in benthic organisms
(d) Thermal and density currents [ESE : 2007]

8.12 Which one of the following can fix atmospheric nitrogen?
(a) Green algae (b) Blue green algae
(c) Red algae (d) Brown algae
[ESE : 2008]

8.13 A drain carrying sewage of BOD = 200 mg/L and flow rate of 50 m^3/s joins a river whose upstream BOD is 8 mg/L and flow rate is 500 m^3/s. Assume immediate and complete mixing of drain with the river. What is the estimated downstream BOD of the river flow?
(a) 20.4 mg/L (b) 25.4 mg/L
(c) 104.4 mg/L (d) 70.4 mg/L
[ESE : 2009]

8.14 Effluent from a wastewater treatment plant (flow rate = 8640 m^3/d, temperature = 25°C) is discharged to a surface stream (flow rate = 1.2 m^3/s, temperature = 15°C). What is the temperature of the stream after mixing?
(a) 10°C (b) 15.77°C
(c) 20°C (d) 24.99°C
[ESE : 2009]

8.15 **Assertion (A):** The BOD gets removed at a very fast rate immediately after sewage is discharged into a river.

Reason (R): A part of the BOD in the sewage is due to settleable organic matter therein.
(a) both A and R are true and R is the correct explanation of A
(b) both A and R are true but R is not a correct explanation of A
(c) A is true but R is false
(d) A is false but R is true [ESE : 2011]

8.16 Consider the following statements:

The time of BOD assimilation in a stream can be affected by
1. Ratio of stream depth to flow width.
2. Stream BOD value.
3. BOD rate constant.

Which of these statements are correct?
(a) 1, 2 and 3 (b) 1 and 2 only
(c) 2 and 3 only (d) 1 and 3 only
[ESE : 2012]

8.17 Sewage sickness relates to
(a) Toxicity of sewage interfering with 'response' to treatment
(b) Destruction of aquatic flora and fauna due to gross pollution of receiving bodies of water by the sewage
(c) Reduction in the waste purifying capacity of the soil
(d) Clogging of pores in soil due to excessive application of sewage leading to obstruction of land aeration thereby leading to septic conditions in land.
[ESE : 2016]

8.18 A drain carrying 5 m^3/s of wastewater with its BOD of 100 mg/l joins a stream carrying 50 m^3/s flow with it BOD of 5 mg/l. What will be the value of the BOD of the combined flow after complete mixing?
(a) 3.6 mg/l (b) 13.6 mg/l
(c) 33.6 mg/l (d) 53.6 mg/l
[ESE : 2016]

8.19 Environmental flow of a river refers to the quantity, quality and timing of the flow
(a) Required in the river to sustain the river ecosystem.
(b) Required to maintain healthy ecological conditions in the command area of a river development project.
(c) Generated by the ecosystem of the catchment of the river.
(d) As the minimum requirement to support the cultural practices of the community living on the banks of the river.
[ESE : 2018]

8.20 Critical dissolved oxygen (D.O.) deficit occurs in which one of the following zones of pollution of 'oxygen sag curve' in case of self-purification of natural stream?

(a) Zone of recovery
(b) Zone of active decomposition
(c) Zone of degradation
(d) Zone of clear water

[ESE : 2019]

8.21 Which one of the following type of ecology is dealt with autecology?

(a) Synecology
(b) Community ecology
(c) Ecosystem ecology
(d) Individual species ecology

[ESE : 2020]

■■■■

Answers Disposal of Sewage Effluent

8.1 (a)	8.2 (c)	8.3 (b)	8.4 (d)	8.5 (d)	8.6 (b)	8.7 (d)	8.8 (c)	8.9 (c)
8.10 (c)	8.11 (b)	8.12 (b)	8.13 (b)	8.14 (b)	8.15 (d)	8.16 (c)	8.17 (d)	8.18 (b)
8.19 (a)	8.20 (b)	8.21 (d)						

Explanations Disposal of Sewage Effluent

8.1 (a)

Rate of depletion of dissolved oxygen, coincides with the rate of BOD exertion.

$$\therefore \quad \frac{dy}{dt} = -\frac{dC}{dt}$$

where C is the concentration of DO

$$y_t = L_0 - L_t$$

$$\frac{dC}{dt} = \frac{dL_t}{dt} = -kL_0e^{-kt}$$

y_t is BOD at any time t

L_0 is ultimate BOD

L_t is oxygen equivalent of the organics at time t

8.2 (c)

The generally accepted minimum amount of DO that will support a large population of various fishes is from 4 to 5 mg/l. When the DO drops below 3 mg/l, even the hardy fish die.

8.3 (b)

Self purification of streams include physical, chemical and biological process.

A. **Physical processes:**
Dilution; sedimentation and resuspension; filtration; gas transfer; and heat transfer.

B. **Chemical Processes:**
Chemical conversion (oxidation and reduction)

C. **Biological process:**
Metabolic processes in micro-organisms.

8.4 (d)

The biodegradable matter present in sewage is consumed by microorganisms through aerobic respiration. This decreases the dissolved oxygen in river.

8.5 (d)

Dilution factor	Purification needed
(i) > 500	No treatment is required
(ii) Between 300-500	Primary treatment such as plain sedimentation. The effluent SS $\not>$150 ppm
(iii) Between 150-300	Screening, sedimentation and chemical precipitation. The effluent SS $\not>$ 600 ppm
(iv) Less than 150	Complete treatment The effluent SS $\not>$ 30 ppm BOD_5 at 18.3°C $\not>$ 20 ppm

8.6 (b)

When a single large charge of sewage is poured into a clean stream, the water becomes turbid, sunlight is shut out of the depths, and green plants, which by photosynthesis remove carbon dioxide from the water and release oxygen to it, die off. Depending on the stream velocity, the water soon turns nearly black. Odorous sulfur compounds are formed and solids settle to the bottom, forming sludge. The settled solids soon decompose, forming gases such as ammonia, carbon dioxide, and methane or marsh gas. Scavenging organisms increase in number until they match the food supply. The oxygen resources are drawn upon heavily and, when overloaded, become exhausted. Life in such waters is confined to anaerobic bacteria (which exist when no oxygen is available), larvae of certain insects such as mosquitoes, and a few worms. There are no fish; turtles are generally the only forms of higher life present. This condition is known as the **zone of degradation.**

In a second zone, or **zone of decomposition**, more solids settle out, the water becomes somewhat clearer, and sunlight penetrates the surface. Oxygen is absorbed from the atmosphere at the air-water interface permitting the establishment of aerobic (oxygen available) conditions. The aerobic bacteria continue the conversion of organic matter into nitrates, sulfates. and carbonates. These, together with the carbon dioxide produced by decomposition as well as by bacteria and plant life, are food sources. With sunlight now penetrating the water, and with abundant food, algae begin to flourish and form a green scum over the surface.

In the third zone, or **zone of recovery**, algae become more numerous and self-purification proceeds more rapidly. Green plants utilizing carbon dioxide and oxygen will liberate in the say time more oxygen than is consumed, thus hastening the recovery of the stream. Simultaneously, the fish that require little oxygen such as catfish are also found. As the dissolved oxygen increases, more types of fish appear.

After recovery, in the **zone of clearer water**, fish find the stream highly favorable, as the algae support various aquatic insects and other organisms on which fish feed. The water is clear or turbid according to concentration of algae, and may have odour for the same reason.

Particular	Zones of Pollution				
	Clear Water	Zone of Degradation	Zone of active decomposition	Zone of Recovery	Zone of Clearer water
1. DO SAG Curve	Saturation Level; 40% DO; Zero 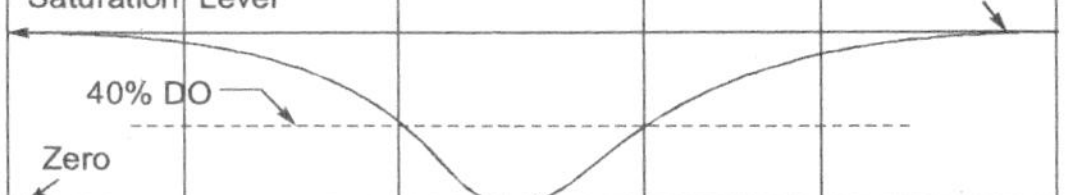				100%
2. Physical Indices	Clear Water No Bottom Sludges	Floating Solids, Bottom Sludge Colour, Turbidity	Grenish and Darker colour, CH_4, H_2S, CO_2 N_2 Gases Evolution, Black Scum at Top	Turbid with Bottom Sludge	Clearer water with no bottom sludge

8.7 (d)

When sewage is applied continuously on a piece of land, the soil pores or voids may get filled up and clogged with sewage matter retained in them. Thus free circulation of air will be prevented and anaerobic conditions will develop within the pores. Sewage sickness is the condition when soil pores get filled up and clogged with sewage matter due to continuous application of waste water effluents. This develops anaerobic conditions and foul gases like methane, carbon-dioxide and hydrogen sulphide are evolved.

In order to prevent sewage sickness:

(i) Sewage should be given primary treatment
(ii) The soil chosen for effluent irrigation/sewage farming should be sandy or loamy.
(iii) A proper under drainage system (open jointed drains) should be designed.
(iv) Land should be given rest for some time and ploughed thoroughly.
(v) Rotation of crops to be followed.
(vi) Shallow depths of water should be applied.

8.8 (c)

The rate of change of deficit

$$\frac{dD}{dt} = k_1L_t - k_2D$$

k_1, k_2 are reaction rate constants for oxygen removal and addition respectively.

L_t is BOD at time t

D is oxygen deficit

8.9 (c)

The tolerance limit for BOD_5 in marine environment disposal is 100 mg/L

Minimum efficiency needed,

$$\eta_{min} = \frac{200-100}{100} \times 100 = 50\%$$

8.10 (c)

A perfect ecological system means that there are producers, consumers, and decomposers present in the system in such a manner that food and energy flows from one level to the other. The pollutants (nitrogen and phosphorous) disturb this equilibrium by increasing productivity of a lake. As the productivity of a lake increases, its water quality reduces and equilibrium is disturbed. Oligotrophic lakes have low level of productivity so they can be considered to be perfect ecological system.

8.11 (b)

Eutrophication is a natural process under which lakes get infested with algae and silt up gradually to become shallower and more productive through the entry and cycling of nutrients like carbon, nitrogen and phosphorous. This natural process of eutrophication in fact can always get its carbon and nitrogen requirements from the atmospheric gases like CO_2 and NO_2; while the requirement of phosphorous is met by its presence in natural run off due to disintegration of rocks, which produce phosphorous. The increased phosphorous in lake water, will cause accelerated eutrophication of lakes and is called cultural eutrophication.

8.12 (b)

Blue Green algae is also called cyanobacteria. Cyanobacteria are oxygenic photosynthetic bacteria that are widespread in marine, freshwater and terrestrial environments, and many of them are capable of fixing atmospheric nitrogen.

8.13 (b)

$$BOD_{mix} = \frac{Q_S BOD_S + Q_R BOD_R}{Q_S + Q_R}$$

$$= \frac{50 \times 200 + 500 \times 8}{50 + 500} = 25.45 \text{ mg/L}$$

8.14 (b)

Temperature of the mix

$$T = \frac{Q_w T_w + Q_s T_s}{Q_w + Q_s}$$

$$= \frac{8640 \times 25 + 1.2 \times 24 \times 60 \times 60 \times 15}{8640 + 1.2 \times 24 \times 60 \times 60}$$

$$= 15.77°C$$

8.15 (d)

Below the point where sewage is discharged into the river-stream zone of degradation or zone of pollution is found. This zone is characterised by water becoming dark and turbid with formation of sludge deposits at the bottom. DO is reduced to about 40% of the saturation value. There is an increase in carbon dioxide content; reoxygenation (i.e., reaeration occurs but is slower than deoxygenation.

8.17 (d)

Sewage sickness relates to clogging of pore in soil due to excessive application of sewage leading to obstruction of land aeration thereby leading to septic condition in land.

8.18 (b)

$$\text{BOD mix} = \frac{Q_S BOD_S + Q_R BOD_R}{Q_S + Q_R}$$

$$= \frac{5 \times 100 + 50 \times 5}{5 + 50} = \frac{750}{55}$$

$$= 13.6 \text{ mg}/l$$

8.19 (a)

Environmental flows describe the quantity, timing, and quality of water flows required to sustain freshwater and estuarine ecosystems and the human livelihoods and well being that depend on these ecosystems. In the Indian context river flows required for cultural and spiritual needs assumes significance.

8.20 (b)

Critical dissolved oxygen (DO) deficit occurs in zone of active decomposition

Table: Showing zone of pollution along a river stream

	← Zones of pollution →				
	Clear water	Zone of degradation	Zone of active decomposition	Zone of recovery	Zone of clearer water
Dissolved oxygen sag curve (100% – 40%)	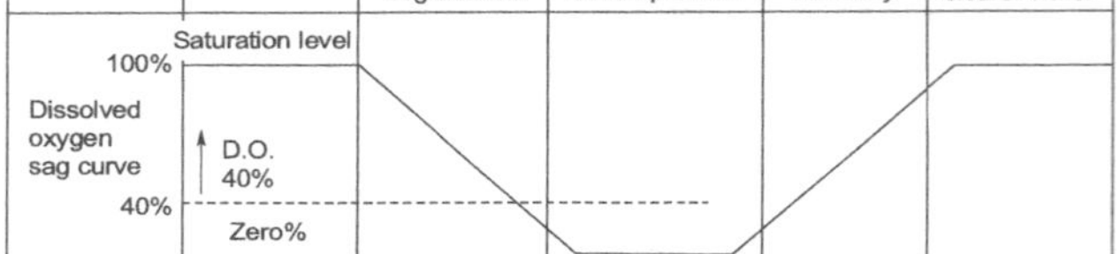				
Physical Indices	Clear, water, no bottom sludge, no colour	Floating solids; bottom sludge present, colour getting turbid	Darker and greyish colour, evolution of gases like CH_4, CO_2, H_2S etc. lot of sludge coming to the surface forming an ugly scum layer at top	Turbid with bottom sludge	Clear water with no bottom sludge
Fish presence	Ordinary fish like game, pan, food & forage etc. present.	Tolerant fishes like carp, buffalo, gary, etc. present	No fish present	Tolerant fish like carp, buffalo, etc. are present	Ordinary fish like game, pan, food, and forage, etc. present
Bottom Animals	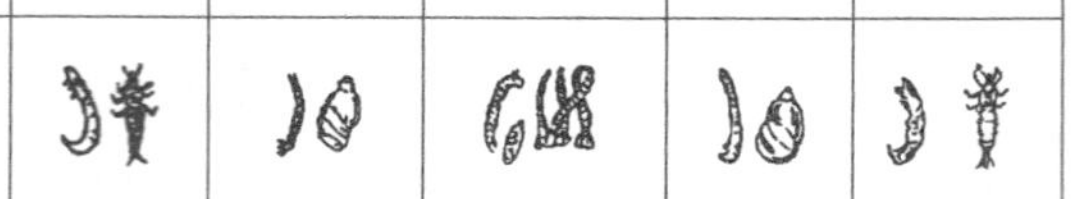				
Algae & Protozoa etc. called plankton	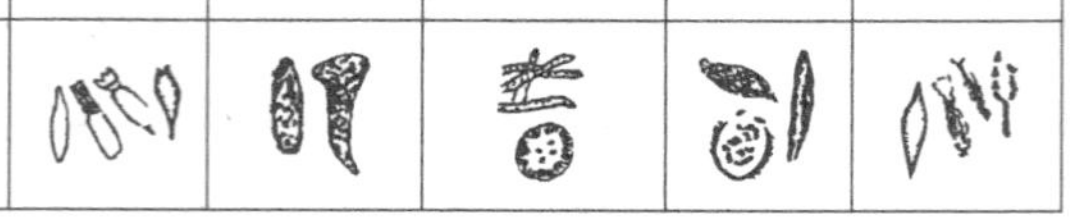				

Here,

Zone-I : Zone of degradation

Zone-II : Zone of active decomposition

Zone-III : Zone of recovery

Zone-IV : Zone of clear water

9 Treatment of Waste Water

9.1 The function of algae in an oxidation pond is to
(a) provide a mat over the surface of the oxidation pond so as to prevent evaporation of water
(b) provide oxygen for bacteria to degrade organic matter
(c) provide a greenish appearance to the pond
(d) prevent the odour nuisance

[ESE : 2000]

9.2 Consider the following pairs of treatment units and impurities removed:
1. Grit chamber : Sand and silt
2. Detritus tank : Organic matter
3. Primary sedimentation tank : Suspended impurities
4. Aeration tank of activated sludge process plant : Oil and grease

Which of these pairs are correctly matched?
(a) 1 and 2 (b) 1, 2, 3 and 4
(c) 2, 3 and 4 (d) 1 and 3

[ESE : 2000]

9.3 Consider the following advantages:
1. Lower BOD concentration in supernatant liquor.
2. Production of a sludge with excellent dewatering capacity.
3. Recovery of methane gas.
4. Lower operation cost.

Which of these are the advantages of anaerobic digestion over aerobic digestion?
(a) 1 and 4 (b) 1 and 2
(c) 2, 3 and 4 (d) 1, 2 and 3

[ESE : 2000]

9.4 The MLSS concentration in the aeration tank of extended aeration activated sludge process is 4000 mg/L. If one litre of sample settled in 30 minutes and the measuring cylinder showed a sludge volume of 200 mL, then the sludge volume index would be nearly
(a) 200 (b) 150
(c) 100 (d) 50 [ESE : 2000]

9.5 Match **List-I** (Treatment units) with **List-II** (Type of processes) and select the correct answer using the codes given below the lists:

	List-I		List-II
A.	Trickling filter	**1.**	Symbiotic
B.	Activated sludge process	**2.**	Extended aeration
C.	Oxidation ditch	**3.**	Suspended growth
D.	Oxidaton pond	**4.**	Attached growth

Codes:

	A	B	C	D
(a)	3	4	2	1
(b)	4	3	1	2
(c)	3	4	1	2
(d)	4	3	2	1

[ESE : 2001]

9.6 If the moisture content of a sludge is reduced from 98% to 96%, the volume of sludge will decrease by
(a) 2% (b) 20%
(c) 25% (d) 50% [ESE : 2001]

9.7 In a high rate trickling filter, the problem of ponding can be solved by
(a) flooding and raking
(b) chlorination and supply of air
(c) raking and chlorination
(d) flooding and supply of air [ESE : 2001]

9.8 The least expensive and most suitable excreta disposal unit for rural areas would be the
(a) soak pit
(b) pit privy
(c) leaching cesspool
(d) septic tank [ESE : 2001]

9.9 Consider the following statements:
The process of activated sludge can be explained as

1. a physical action whereby the finer suspended particles of sewage form a sublayer for a bacterial film at the surface
2. a chemical action whereby the finer suspended particles and colloidal solids are combined into masses of large bulk
3. a biochemical action whereby the sludge flocs so formed act as vehicles for aerobic bacteria oxidizing the organic matter

Which of these statements are correct?

(a) 1, 2 and 3 (b) 1 and 2
(c) 2 and 3 (d) 1 and 3

[ESE : 2001]

9.10 **Assertion (A):** Settled sewage is subjected to biological treatment.

Reason (R): Any biological treatment is preceded by a settling tank.

(a) both A and R are true and R is the correct explanation of A
(b) both A and R are true but R is not a correct explanation of A
(c) A is true but R is false
(d) A is false but R is true

[ESE : 2002]

9.11 At a sewage treatment plant for a flow of 3 m^3/s, the cross-sectional area of grit chamber will be about

(a) 3 m^2 (b) 10 m^2
(c) 25 m^2 (d) 30 m^2

[ESE : 2002]

9.12 In the preliminary treatment of wastewater skimming tanks are often included in the treatment scheme. Various features of skimming tanks are that these

1. can remove general floating matter
2. can remove oily and greasy matter
3. have detention time of 30 min
4. employ compressed air blown through diffusers

Which of these statements are correct?

(a) 1 and 3 (b) 2 and 4
(c) 2 and 3 (d) 1 and 2

[ESE : 2002]

9.13 Consider the following data in the design of grit chamber:

1. Specific gravity of grit = 2.7
2. Size of grit particle = 0.21 mm
3. Viscosity of water = 1.0×10^{-2} cm^2/s

The settling velocity (cm/s) of the grit particle will be

(a) 1 to 2.5 (b) 2.6 to 5.0
(c) 5.1 to 7.8 (d) > 7.8

[ESE : 2003]

9.14 Amongst the various sewage treatment methods, for the same discharge, the largest area is needed for

(a) trickling filter (b) anaerobic pond
(c) oxidation ditch (d) oxidation pond

[ESE : 2003]

9.15 Consider the following treatment steps in a conventional wastewater treatment plant:

1. Primary sedimentation
2. Grit removal
3. Disinfection
4. Secondary sedimentation
5. Screening
6. Secondary treatment unit

The correct sequence of these steps is

(a) 5-2-1-6-4-3 (b) 1-2-4-5-3-6
(c) 2-3-4-5-6-1 (d) 6-5-4-3-2-1

[ESE : 2003]

9.16 In an activated sludge process, the sludge volume index can be controlled by

(a) aeration
(b) adding chlorine
(c) reducing recycling ratio
(d) increasing the depth of aeration tank

[ESE : 2003]

9.17 A primary sedimentation tank is not required for

(a) activated sludge system
(b) extended aeration system
(c) trickling filtration system
(d) tapered activated sludge process using pure oxygen for aeration [ESE : 2003]

9.18 Match **List-I** (Parameter) with **List-II** (Treatment unit or process) and select the correct answer using the codes given below the lists:

List-I

A. F/M ratio
B. Solar energy

C. Effluent recirculation
D. Volatile fatty acids

List-II

1. Anaerobic digester
2. Detritus tank
3. Waste stabilization pond
4. Trickling filter
5. Activated sludge process

Codes:

	A	B	C	D
(a)	2	3	5	4
(b)	5	3	4	1
(c)	5	1	4	3
(d)	4	1	2	3

[ESE : 2004]

9.19 An industry manufacturing urea produces waste water, which largely contains urea and ammonia. The treatment plant consists following units for effective control of nitrogen:

1. Nitrification followed by denitrification
2. Hydrolysis unit
3. Ammonia stripping by air
4. Lime dosing and mixing unit

Which one of the following is the correct sequence of above unit operations in the treatment plant?

(a) 3-2-1-4 (b) 2-3-1-4
(c) 2-4-3-1 (d) 4-2-3-1

[ESE : 2004]

9.20 Which one of the following reactors can create a problem of short circuiting due to density currents?

(a) Batch (b) Complete mix
(c) Plug flow (d) Fluidized bed

[ESE : 2004]

9.21 Match **List-I** (Treatment process) with **List-II** (Related terms) and select the correct answer using the codes given below the lists:

List-I

A. Lagoons
B. Trickling filter
C. Oxidation ponds
D. Activated sludge process

List-II

1. Attached growth system
2. Algae-bacteria symbiotic relationship
3. Extended aeration
4. Low cost treatment method

Codes:

	A	B	C	D
(a)	4	3	2	1
(b)	2	1	4	3
(c)	4	1	2	3
(d)	2	3	4	1

[ESE : 2005]

9.22 Which one of the following statements is not correct?

(a) Settling and sludge digestion occurs in septic tanks in one compartment.
(b) Settling and sludge digestion occurs in imhoff tank in different compartments.
(c) Septic tank is a low-rate anaerobic unit whereas an imhoff tank is a high rate anaerobic unit.
(d) The rate of sludge accumulation in septic tank is approximately 40-70 litres/ capita/year.

[ESE : 2005]

9.23 Consider the following statements in regard to aerobic and anaerobic treatment processes:

1. Biomass production in the aerobic treatment process is more as compared to the anaerobic treatment process.
2. Start-up period is more in the aerobic treatment process as compared to the anaerobic treatment process.
3. Energy consumption and production is more in the aerobic treatment process as compared to the anaerobic treatment process.

Which of these statements is/are correct?

(a) 1 and 2 (b) 2 and 3
(c) Only 2 (d) Only 1

[ESE : 2005]

9.24 When the recirculation ratio in a high rate trickling filter is unity, then what is the value of the recirculation factor?

(a) 1 (b) > 1
(c) < 1 (d) Zero

[ESE : 2005]

9.25 Among the following, which is/are not pre-treatment unit(s)?

(a) Bar-screen and grit chamber

(b) Flow equalization and proportioning tank
(c) Neutralization for pH adjustment tank
(d) Nutrient removal tank [ESE : 2005]

9.26 After which of the following treatment units, the turbidity is maximum?
(a) Chlorination
(b) Primary sedimentation
(c) Flocculation basin
(d) Secondary sedimentation [ESE : 2006]

9.27 Match **List-I** (Physical properties of filtering material for trickling filters) with **List-II** (Limiting value) and select the correct answer using the codes given below the lists:

List-I	List-II
A. Crushing strength N/mm^2	1. 12.0
B. Hardness	2. 100.0
C. Per cent wear	3. 4.0
D. Specific gravity	4. 2.6

Codes:

	A	B	C	D
(a)	2	1	2	4
(b)	2	4	3	1
(c)	3	4	2	1
(d)	2	1	3	4

[ESE : 2007]

9.28 Match **List-I** with **List-II** and select the correct answer using the codes given below the lists:

List-I
A. Suspended solids
B. Nutrients
C. Heavy metals
D. Dissolved inorganic solids

List-II
1. May cause eutrophication
2. Toxic, may interfere with effluent reuse
3. May interfere with effluent reuse
4. Cause sludge deposits

Codes:

	A	B	C	D
(a)	4	1	2	3
(b)	2	3	4	1
(c)	4	3	2	1
(d)	2	1	4	3

[ESE : 2007]

9.29 In conventional activated sludge process, MLSS is generally kept in which range?
(a) < 100 mg/L
(b) 1000 – 2000 mg/L
(c) 2000 – 3000 mg/L
(d) 3000 – 5000 mg/L [ESE : 2007]

9.30 Which one of the following is considered as the thermophilic range of sludge digestion?
(a) 60°C to 70°C (b) 50°C to 57°C
(c) 29°C to 40°C (d) 20°C to 30°C
[ESE : 2008]

9.31 Consider the following statements:
In the context of sludge volume index (SVI),
1. the SVI is expressed as the volume (in cc) of the activated sludge for one gram of dry weight of the sludge.
2. a higher value of SVI indicates a light and fluffy sludge which is not easily settleable.

Which of these statements is/are correct?
(a) 1 only (b) 2 only
(c) Both 1 and 2 (d) Neither 1 nor 2
[ESE : 2008]

9.32 An industry has a sewage treatment plant which produces sludge with a moisture content of 98%. With the solid content remaining the same, the sludge is thickened so that the moisture content now is 96%. If the original quantity of sludge is P, what is the quantity of the thickened sludge?
(a) 0.98P (b) 3P/4
(c) 2P/3 (d) P/2 [ESE : 2008]

9.33 In a well operating activated sludge process unit, what is the value of sludge volume index?
(a) < 50 (b) 100-150
(c) 200-300 (d) > 300
[ESE : 2008]

9.34 Which of the following pollutants are generally not removed in a sewage treatment plant?
(a) Inorganic suspended solids
(b) Dissolved organic solids
(c) Oil and grease
(d) Dissolved inorganic solids [ESE : 2009]

9.35 Match **List-I** (Water/Wastewater treatment) with **List-II** (Operating problem) and select the correct answer using the codes given below the lists:

List-I
A. Trickling filter
B. Activated sludge process

C. Rapid gravity filter
D. Anaerobic sludge digester

List-II
1. Negative head
2. Fly-breeding
3. Sludge bulking
4. pH reduction

Codes:

	A	B	C	D
(a)	4	3	1	2
(b)	2	3	1	4
(c)	4	1	3	2
(d)	2	1	3	4

[ESE : 2009]

9.36 Which one of the following types of samples is relevantly employed for the design of wastewater treatment plant?
(a) Grab sample
(b) Composite sample
(c) Integrated sample
(d) Any sample [ESE : 2009]

9.37 Secondary clarifier of an activated sludge process must be designed for effluent clarification and solids thickening, both of which relate directly to the
(a) surface area
(b) transport velocity due to sludge withdrawal
(c) gravity settling of solids relative to the water
(d) underflow solids concentration

[ESE : 2010]

9.38 Consider the following statements:
In a sewage treatment process
1. detention time required for anaerobic decomposition is more as compared to that for aerobic decomposition.
2. Anaerobic decomposition is more energy-consuming as compared to aerobic decomposition.
3. Anaerobic decomposition is a more sensitive process as compared to aerobic decomposition.

Which of these statements is/are correct?
(a) 1, 2 and 3 (b) 2 only
(c) 3 only (d) 1 and 3 only

[ESE : 2010]

9.39 Shallow ponds in which dissolved oxygen is present at all depths are called
(a) aerobic lagoons
(b) aerobic ponds
(c) facultative lagoons
(d) facultative ponds [ESE : 2010]

9.40 In the facultative pond systems, the aerobic zone may get extended downwards due to
(a) calm waters along with weak sunlight
(b) mixing by wind action along with weak sunlight
(c) mixing by wind action along with penetration by sunlight
(d) calm waters along with penetration by sunlight

[ESE : 2010]

9.41 Deep ponds, in which oxygen is absent except, perhaps, across a relatively thin surface layer, are called
(a) aerobic ponds
(b) anaerobic ponds
(c) facultative ponds
(d) polishing ponds [ESE : 2011]

9.42 Consider the following statements:
Activated sludge process can be said to comprise:
1. Conversion of dissolved organic matter into biological flocs
2. Removal of dissolved BOD of the waste water
3. Digestion of the sludge

Which of these statements are correct?
(a) 1, 2 and 3 (b) 1 and 2 only
(c) 2 and 3 only (d) 1 and 3 only

[ESE : 2011]

9.43 During sewage treatment, effluent from which one of the following treatment units has minimum wt/vol amount of suspended solids?
(a) Detritus channel
(b) Primary sedimentation tank
(c) Secondary sedimentation tank
(d) Activated sludge process aeration tank

[ESE : 2011]

9.44 When sufficient energy through mechanical mixing is supplied to keep the entire contents, including the sewage solids, mixed and aerated, the reactor is termed

(a) an aerobic lagoon
(b) an aerobic pond
(c) a facultative lagoon
(d) a facultative pond [ESE : 2011]

9.45 The manufacturer of aeration devices reports the oxygen transfer rate of the device obtained through laboratory tests carried under standard conditions. Such standard conditions are

(a) wastewater at zero DO, 25°C and 760 mm Hg
(b) Tapwater at zero DO, 0°C and 700 mm Hg
(c) Tapwater at zero DO, 20°C and 760 mm Hg
(d) Wastewater at zero DO, 0°C and 700 mm Hg

[ESE : 2011]

Directions: The following items consists of two statements; one labelled as **'Assertion (A)'** and the other as **'Reason (R)'**. You are to examine these two statements carefully and select the answers to these items using the codes given below:

Codes:

(a) both A and R are true and R is the correct explanation of A
(b) both A and R are true but R is not a correct explanation of A
(c) A is true but R is false
(d) A is false but R is true

9.46 **Assertion (A):** In the secondary sedimentation tank of a sewage treatment plant the settling particles form a blanket which descends and captures more particles.

Reason (R): The particles are flocculant and in very high concentration whenever there is an efficient working activated sludge, or trickling filter, process. [ESE : 2011]

9.47 **Statement-I:** The sludge processing utilizes the aerobic digestion in which it is converted to CO_2 and methane.

Statement-II: The acid-forming bacteria convert the complex organics such as fats, proteins and carbohydrates into organic fatty acids. The methane-forming bacteria convert organic acids to CO_2 and CH_4.

[ESE : 2012]

9.48 **Statement-I** : Chlorophyll-bearing plants take water and carbondioxide to synthesize carbohydrates.

Statement-II : Wasted food ultimately leads to production of various natural resources like water and sunlight energy.

[ESE : 2012]

9.49 What is the correct sequence of the following treatment units employed in a conventional sewage treatment plant?

1. Screen chamber
2. Primary settling tank
3. Trickling filter
4. Grit chamber
5. Secondary settling tank

(a) 5, 4, 2, 3 and 1 (b) 1, 4, 2, 3 and 5
(c) 5, 2, 3, 4 and 1 (d) 1, 4, 3, 2 and 5

[ESE : 2013]

9.50 Which of the following parameters are employed in the design of trickling filter?

1. Hydraulic loading rate
2. Organic loading rate
3. Detention time
4. Weir loading rate

(a) 1, 2, 3 and 4 (b) 1 and 2 only
(c) 2 and 3 only (d) 3 and 4 only

[ESE : 2013]

9.51 In activated sludge process, 1 MLD (by volume) of secondary sludge has to be returned to keep the required MLSS concentration in the aeration tank. This sludge has a water content of 99%. If the sludge water content is reduced to 98%, what volume of the sludge will be needed to be recycled?

(a) 0.25 MLD (b) 0.5 MLD
(c) 0.75 MLD (d) 1.00 MLD

[ESE : 2013]

Directions: Each of the next items consists of two statements, one labeled as the **'Statements (I)'** and the other as **'Statement (II)'**. You are to examine these two statements carefully and select the answers to these items using the codes given below :

Codes:

(a) Both Statement (I) and Statement (II) are individually true and Statement (II) is the correct explanation of Statement (I)
(b) Both Statement (I) and Statement (II) are individually true but Statement (II) is NOT the correct explanation of Statement (I)

(c) Statement (I) is true but Statement (II) is false
(d) Statement (I) is false but Statement (II) is true

9.52 Statement (I): Plastic media of 25 to 100 mm size are employed in super-rate trickling filters for treatment of waste-water having a very high BOD.

Statement (II): Plastic media have high specific surface area, high void space and low weight. Hydraulic loading rate and organic loading rate are 30 to 90 $m^3/d/m^2$ and 1000 to 2000 $g/d/m^3$ respectively, which are much higher than in stone media trickling filter. [ESE : 2013]

9.53 Statement (I): The cross-section of grit channel is designed for a uniform velocity at all flow rates.

Statement (II): A U-type of cross-section maintains a uniform velocity at all flow rates. [ESE : 2013]

9.54 Consider the following statements:

1. Grease data is a measure of effectiveness of primary settling tanks.
2. Volatile acid data provide information concerning anaerobic degradation of organic matter.
3. Total and volatile solids data are indispensable in the design and operation of sludge digestion, vacuum filter and incineration units.
4. Ammonia and organic nitrogen data are helpful in controlling aerobic treatment costs to a minimum.

Which of these statements are correct?

(a) 1, 2, 3 and 4 (b) 1, 2 and 3 only
(c) 1, 2 and 4 only (d) 2, 3 and 4 only
[ESE : 2013]

9.55 In an aerobic attached culture system, the biomass present at the biofilm-medium surface interface experiences

(a) Aerobic and endogenous metabolism
(b) Anaerobic and exogenous metabolism
(c) Anaerobic and endogenous metabolism
(d) Aerobic and exogenous metabolism
[ESE : 2014]

9.56 A sewage sludge has a water content of 99%. What will be the concentration of suspended solids in the sludge?

(a) 10 mg/*l* (b) 100 mg/*l*
(c) 1000 mg/*l* (d) 10,000 mg/*l*
[ESE : 2014]

9.57 Statement (I): When waste water is treated at an oxidation ditch, anaerobic sludge digester is not required.

Statement (II): Oxidation ditch admits an extended aeration process whereby sludge gets mineralized.

(a) Both Statement (I) and Statement (II) are individually true and Statement (II) is the correct explanation of Statement (I).
(b) Both Statement (I) and Statement (II) are individually true but Statement (II) is not the correct explanation of Statement (I).
(c) Statement (I) is true but Statement (II) is false.
(c) Statement (I) is false but Statement (II) is true.
[ESE : 2014]

9.58 A sludge had 100 m^3 volume when its moisture content was 95%. What would be its volume if its moisture content changed to 90%?

(a) 200 m^3 (b) 50 m^3
(c) 94.7 m^3 (d) 105.5 m^3
[ESE : 2015]

9.59 The Sludge Volume Index for mixed liquor having suspended solids concentration of 2000 mg/*l* and showing a settled volume of 200 m*l* from a one litre sample would be

(a) 0.1 (b) 1000
(c) 100 (d) 10 [ESE : 2015]

9.60 The different actions that take place in anaerobic decomposition process are

1. Alkaline fermentation
2. Acid fermentation
3. Acid regression
4. Methane formation

What is the correct sequence of these actions (from earlier to later)?

(a) 4, 3, 1 and 2 (b) 2, 3, 1 and 4
(c) 4, 1, 3 and 2 (d) 2, 1, 3 and 4
[ESE : 2015]

9.61 Consider the following statements:

1. Ammonia nitrogen is a measure of nitrogen present as ammonium hydroxide and ammonium salts. It will progressively decrease as sewage gets treated.

2. Organic nitrogen is the total nitrogenous matter in sewage excepting that present as ammonia nitrogen, nitrites and nitrates. It becomes ammonia in anaerobic decomposition and nitrites or nitrates in aerobic decomposition.

Which of the above statements is/are correct?

(a) 1 only (b) Both 1 and 2
(c) 2 only (d) Neither 1 nor 2

[ESE : 2015]

9.62 What is the required plan size of a square sedimentation tank (as the primary sedimentation tank in sewage treatment), given that its effect depth is 3 m, and the flow rate is 40 MLD with admissible surface loading of 1000,000 $l/m^2/day$?

(a) 23.5 m × 23.5 m
(b) 30 m × 30 m
(c) 20 m × 20 m
(d) 15 m × 15 m [ESE : 2015]

9.63 **Assertion (A):** Anaerobic digestion of sewage is unsuitable in the vicinity of a crowded locality.
Reason (R): Aerobic digestion of sewage is costly but is suitable at a crowded locality.

(a) Both A and R are true and R is the correct explanation of A
(b) Both A and R are true but R is not a correct explanation of A
(c) A is true but R is false
(d) A is false but R is true [ESE : 2015]

9.64 100 m^3 of sludge holds a moisture content of 95%. If its moisture content changes to 90%, the volume of this sludge will then be

(a) 40.5 m^3 (b) 50 m^3
(c) 75 m^3 (d) 94.7 m^3

[ESE : 2016]

9.65 Consider the following statements in respect of flow equalization in a waste-water treatment plant:

1. Biological treatment is enhanced because shock loadings are eliminated or minimized.
2. Flow equalization is an attractive option for upgrading the performance of overloaded treatment plants.
3. Inhibiting substances can be diluted and pH can be stabilized.
4. Thickening performance of primary sedimentation tanks following grit removal is improved.

Which of the above statements are correct?

(a) 1, 2, 3 and 4
(b) 1, 2 and 3 only
(c) 1, 2 and 4 only
(d) 3 and 4 only [ESE : 2016]

9.66 Consider the following statements in respect of aerated grit chamber:

1. The grit accumulates at the bottom in advancing spiral-flow aeration tanks, locationally preceded by grit chambers, led to the eventual development of aerated grit chambers.
2. The excessive wear on grit handling equipment is a major factor for the popularity of aerated grit chambers.
3. Aerated grit chambers are designed to provide detention period of 1 minute at maximum rate of flow.
4. Diffusers are located at 0.45 m to 0.6 m above the bottom of the chamber.

Which of the above statements are correct?

(a) 1, 2 and 3 only (b) 3 and 4 only
(c) 1, 2 and 4 only (d) 1, 2, 3 and 4

[ESE : 2016]

9.67 Consider the following statements in respect of anaerobic sludge digester:

1. It is less expensive compared to several other methods available.
2. Processing of separable solids impacts the environment to a minimum.
3. Quantity of separated solids requiring disposal is minimal.
4. Digested sludge is very readily dewaterable.

Which of the above statements are correct?

(a) 1, 2, 3 and 4 (b) 1, 2 and 3 only
(c) 3 and 4 only (d) 1, 2 and 4 only

[ESE : 2016]

9.68 Consider the following statements as regards a Septic Tank:

1. The size required is large and uneconomical when serving more than roughly 100 persons.
2. It can remove around 90% of BOD and 80% of suspended solids.

3. As compared to the sludge holding of a plain sedimentation tank, a septic tank can hold about 90% less of sludge volume.
4. Frequent removal of sludge is not required.

Which of the above statements are correct?
(a) 1 and 2 (b) 2 and 3
(c) 3 and 4 (d) 1 and 4

[ESE : 2016]

9.69 Consider the following statements regarding contact stabilization process:
1. Primary settling tank is not required in some cases.
2. BOD removal occurs in two stages.
3. Aeration volume requirements are approximately 50% of those of a conventional - or tapered - aeration plant.
4. Returned sludge is aerated for 30 min to 90 min in sludge aeration tank.

Which of the above statements are correct?
(a) 1, 2, 3 and 4 (b) 1 and 4
(c) 1, 2 and 3 (d) 2, 3 and 4

[ESE : 2017]

9.70 Consider the following statements regarding waste stabilization ponds:
1. The pond has a symbiotic process of waste stabilization through algae on one hand and bacteria on the other.
2. The oxygen in the pond is provided by algae through photosynthesis.
3. The detention period is of the order of two to three days.
4. The bacteria which develop in the pond are aerobic bacteria.

Which of the above statements are CORRECT?
(a) 1 and 2 only (b) 2 and 3 only
(c) 3 and 4 only (d) 1 and 4 only

[ESE : 2018]

9.71 **Statement (I)** : Anaerobic sludge digester, by itself, is considered to be the better method than other methods of sludge treatment.

Statement (II) : During Anaerobic sludge digestion, CH_4 is produced; also rodents and other pests are attracted when digester sludge is dried.

(a) Both Statement (I) and Statement (II) are individually true and Statement (II) is the correct explanation of Statement (I)
(b) Both Statement (I) and Statement (II) are individually true but Statement (II) is NOT the correct explanation of Statement (I)
(c) Statement (I) is true but Statement (II) is false
(d) Statement (I) is false but Statement (II) is true

[ESE : 2018]

9.72 Consider the following statements with reference to bioenergy as a renewable energy source:
1. Plants ensure continuous supply of gas due to their continuous growth.
2. Cost of obtaining energy from biogas is less than that from fossil fuels.
3. Digestion of sludge may produce H_2S and NO_x which are injurious to human health.
4. 'Floating dome' installation is the preferred option as it supplies gas at constant pressure irrespective of quantity of gas produced.

Which of the above statements are CORRECT?
(a) 1, 2 and 3 only (b) 1, 2 and 4 only
(c) 2, 3 and 4 only (d) 1, 3 and 4 only

[ESE : 2018]

9.73 The MLSS concentration in an aeration tank is 2000 mg/l and the sludge volume after 30 minutes of settling in a 1000 ml graduated cylinder is 176 ml. The value of sludge density index (SDI) will be nearly
(a) 3.34 g/ml (b) 2.22 g/ml
(c) 1.14 g/ml (d) 0.26 g/ml

[ESE : 2019]

9.74 Which one of the following gases is the principal by-product of anaerobic decomposition of the organic content in waste water?
(a) Carbon monoxide
(b) Ammonia
(c) Hydrogen sulphide
(d) Methane

[ESE : 2019]

9.75 Which one of the following type of treatments will be used for neutralization of alkaline effluent?
(a) Lime stone treatment
(b) Caustic lime treatment
(c) Carbon dioxide treatment
(d) Hydrochloric acid treatment

[ESE : 2020]

9.76 A tank into which raw or partly treated sewage is collected, left to stay, and discharged at such a rate as may be necessary for subsequent treatment, is called

(a) Dosing tank (b) Sedimentation tank
(c) Skimming tank (d) Settling tank

[ESE : 2021]

9.77 In the context of sludge conditioning, Elutriation is synonymous to

(a) Washing (b) Heating
(c) Compacting (d) Filtering

[ESE : 2021]

9.78 Sludge thickening is commonly achieved by the following method:

1. Gravity thickening
2. Air flotation
3. Centrifugation

Which of the above methods are correct?

(a) 2 and 3 only
(b) 1 and 2 only
(c) 1 and 3 only
(d) 1, 2 and 3 [ESE : 2021]

9.79 Which one of the following is a grit-removal unit which also removes silt as well as some organic matter along with grit?

(a) Detritus tank
(b) Skimming tank
(c) Detention tank
(d) Suspension tank [ESE : 2021]

9.80 A sewage has a suspended solid contents as 250 mg/litre. The sedimentation tank removes 55% of the suspended solids and the water content of the sludge is 95%. What is the quantity of sludge produced after treating 6.5 $\times 10^6$ litres of sewage? (Assume the specific gravity of sludge as 1.0)

(a) 11.875 cu.m (b) 14.275 cu.m
(c) 17.875 cu.m (d) 20.275 cu.m

[ESE : 2023]

9.81 With standard notations, the total subsurface soil area required for the soak pits or dispersion trenches is given by

(a) $Q = 100/\sqrt{t}$ (b) $Q = 130/\sqrt{t}$
(c) $Q = 200/\sqrt{t}$ (d) $Q = 260/\sqrt{t}$

[ESE : 2023]

9.82 Chlorination, liming and recarbonation are the part of which one of the following in the context of unit operations for waste water treatment?

(a) Ion transfer (b) Solute stabilization
(c) Solid transfer (d) Solid concentration

[ESE : 2023]

9.83 The theoretical time taken by a particle of water to pass between entry and exit of a settling tank is known as

(a) Weir loading (b) Detention period
(c) Velocity of flow (d) Overflow rate

[ESE : 2023]

9.84 The most commonly adopted detention period of grit chambers may vary from

(a) 10 to 40 seconds
(b) 45 to 90 seconds
(c) 90 to 120 seconds
(d) 120 to 135 seconds [ESE : 2023]

9.85 What is the logical extension of the pit privy?

(a) Septic tank and tile field
(b) Composting toilet
(c) Seepage pits
(d) Plastic pipe with holes

[ESE : 2024]

9.86 What is a drawback of aerobically digested sludges?

(a) Odor reduction
(b) Enhanced putrescible
(c) Degraded of organic acids
(d) Difficulty in dewatering

[ESE : 2024]

■■■■

Answers Treatment of Waste Water

9.1 (b)	9.2 (d)	9.3 (c)	9.4 (d)	9.5 (d)	9.6 (d)	9.7 (c)	9.8 (d)	9.9 (a)
9.10 (c)	9.11 (b)	9.12 (b)	9.13 (b)	9.14 (d)	9.15 (a)	9.16 (c)	9.17 (b)	9.18 (b)
9.19 (d)	9.20 (c)	9.21 (c)	9.22 (c)	9.23 (d)	9.24 (b)	9.25 (d)	9.26 (c)	9.27 (d)
9.28 (a)	9.29 (c)	9.30 (b)	9.31 (c)	9.32 (d)	9.33 (b)	9.34 (d)	9.35 (b)	9.36 (a)
9.37 (a)	9.38 (d)	9.39 (b)	9.40 (c)	9.41 (b)	9.42 (b)	9.43 (c)	9.44 (a)	9.45 (c)
9.46 (a)	9.47 (d)	9.48 (c)	9.49 (b)	9.50 (b)	9.51 (b)	9.52 (a)	9.53 (c)	9.54 (a)
9.55 (c)	9.56 (d)	9.57 (a)	9.58 (b)	9.59 (c)	9.60 (b)	9.61 (b)	9.62 (c)	9.63 (b)
9.64 (b)	9.65 (a)	9.66 (c)	9.67 (a)	9.68 (c)	9.69 (d)	9.70 (a)	9.71 (b)	9.72 (c)
9.73 (c)	9.74 (d)	9.75 (d)	9.76 (a)	9.77 (a)	9.78 (d)	9.79 (a)	9.80 (c)	9.81 (b)
9.82 (a)	9.83 (b)	9.84 (a)	9.85 (a)	9.86 (d)				

Explanations Treatment of Waste Water

9.1 (b)

Algae are autotrophic, photosynthetic organism and, even though they do not utilize organic compounds directly, yet play significant role in the natural purification process. In the presence of sunlight, algae metabolize the waste products of hetero-trophic bacteria (CO_2, NO_3^-, PO_4^{3-}, etc.) while obtaining energy from sunlight.

9.2 (d)

Detritus tanks are grit chambers designed for flow with smaller flow velocity and longer detention periods so as not only the grit, etc. is separated but also very fine sand particles settle. The organic matter settled in this tank is separated from the grit by currents through baffles.

9.3 (c)

The advantages of anaerobic digestion over aerobic digestion are:

- Recovery of energy in the form of methane. This leads to lower operation cost.
- Anaerobically digested sludge contains nutrients and organic matter which improves the fertility of soil. It also has excellent dewatering capacity.

9.4 (d)

Sludge volume index (SVI)

$$= \frac{V_{0b}(\text{mL/L})}{X_{0b}(\text{mg/L})} \times 1000 = \frac{200}{4000} \times 1000 = 50 \text{ mL/gm}$$

9.5 (d)

Suspended culture systems are:

(i) Activated sludge process and its variants like extended aeration.
(ii) Aerated lagoons and oxidation ponds
(iii) Aerobic digestion with aeration.

Attached culture systems are:

(i) Trickling filters
(ii) Rotating biological contractors
(iii) Packed bed reactors
(iv) Anaerobic packed and fluidized beds

9.6 (d)

$$(100 - p_1)\, V_1 = (100 - p_2)\, V_2$$

$$\Rightarrow \quad \frac{V_2}{V_1} = \left(\frac{100 - p_1}{100 - p_2}\right)$$

$$\Rightarrow \quad \frac{V_2}{V_1} = \frac{100 - 98}{100 - 96} = \frac{2}{4} = 0.5$$

∴ Decrease in volume is 50%

9.7 (c)

Poncing is caused by clogging of filter media due to heavy growth of fungi and algae. This can be controlled by raking and chlorinating. Chlorination kills the algae. Other algicide like copper sulphate can be added to the sewage and bed should be left resting for some time.

Facultative bacteria are the predominating organisms in trickling filters. In the lower reaches of the filter nitrifying bacteria will also be present. The fungi are also responsible for waste stabilization under low pH-conditions or with certain industrial wastes. At time fungi growth can be so rapid that the filter clogs and ventilation becomes restricted.

Algae can grow only in the upper reaches of the filter where sunlight is available. Generally algae do not take part in waste degradation. The algae may cause clogging of filter surface which may cause odours.

9.8 (d)

A septic tank may be defined as the primary sedimentation tank directly admitting raw sewage and removing 60-70% of dissolved matter from it. It is most suitable excreta disposal unit for rural areas.

9.9 (a)

- In ASP, activated sludge is mixed with settled sewage in suspension that provides the necessary contact surface for the growth of aerobic film of bacteria at the surface (physical action).
- After mixing suspended particles and colloidal solids in the mixed liquor combine to form masses of large bulk i.e., biomass (chemical action) which is settled in secondary sedimentation tank.
- The settled sludge consists of active microorganisms which them oxidizes organic matter (Biological action).

9.10 (c)

Almost all biological treatments are preceded by a settling tank except in case of oxidation ponds, septic and Imhoff tank and UASBR.

So reason is wrong.

9.11 (b)

The flow velocity in grit chamber is about 0.25 to 0.3 m/s. Therefore cross-sectional area

$$= \frac{3}{0.3} = 10 \text{ m}^2$$

9.12 (b)

Skimming tanks are used for removal of oily and greasy matter. A detention time of 3 to 5 minutes is usually sufficient. The rise of oil and greasy matter is brought about by allowing compressed air into the sewage from diffusers. The oily and greasy matter should be removed because they interfere in secondary treatment process.

9.13 (b)

Particle size (0.21 mm) is between 0.1 mm and 1 mm so the settling velocity in transition region will be given by Newton's law

$$v^2 = \frac{4}{3}g\frac{(\rho_p - \rho_w)d}{C_D\rho_w}$$

where $C_D = \frac{24}{\text{Re}}$ (laminar) when Re < 1

$$C_D = \frac{24}{\text{Re}} + \frac{3}{\sqrt{\text{Re}}} + 0.34$$

(transition) when $1 < \text{Re} < 10^4$

$C_D = 0.4$ (turbulent) when $\text{Re} > 10^4$

Finding the settling velocity in transition region is a trial and error process. However, the velocity of particle can be calculated by assuming laminar flow and the velocity in transition region will be less than the calculated value.

$$v = \frac{g}{18}(S_s - 1)\frac{d^2}{\nu}$$

$$= \frac{9.81}{18} \times (2.7 - 1) \times \frac{(0.21 \times 10^{-3})^2}{10^{-2} \times 10^{-4}}$$

$$= 0.0408 \text{ m/s} = 4.08 \text{ cm/s}$$

Therefore settling velocity will line in range 2.6 to 5.0 cm/s.

9.14 (d)

Oxidation Ponds are also known as stabilization ponds or lagoons. They are used for simple secondary treatment of sewage effluents. Within an oxidation pond heterotrophic bacteria degrade organic matter in the sewage which results in production of cellular material and minerals.

9.15 (a)

Screening – Grit removal – Primary sedimentation – Secondary treatment unit – Secondary sedimentation – Disinfection will be the correct sequence.

If flow meter is also given then it will come after grit removal.

9.16 (c)

Sludge volume index is used to indicate the physical state of the sludge produced in a biological aeration system. It is defined as the volume occupied in ml by one gram of solids in the mixed liquor after settling for 30 minutes. The usual adopted range of SVI is between 50-150 ml/gm and such a value indicates good sludge. It can be controlled by reducing recycling rate.

9.17 (b)

In extended aeration system long detention times are provided with low F/M ratio and high MLSS concentration to maintain culture in endogenous phase. Primary sedimentation tank is frequently avoided in this process, but grit chamber or communitor is often provided for screening.

9.18 (b)

The F/M ratio is a process control number that helps to determine the proper number of microorganisms for the system. To do this calculation, the following information will be needed:

- Influent Flow into your activated sludge system.
- Influent BOD (mg/*l*) concentration into your aeration tank.
- Mixed Liquor Volatile Suspended Solids Concentration (mg/*l*)
- Volume (in m^3) of your aeration system

The term Food to Microorganism Ratio (F/M) is actually a measurement of the amount of incoming food (kg of Influent BOD) divided by the kg of Microorganisms in the system. If case of an activated sludge system, F/M ratio is determined regularly.

9.19 (d)

The waste from fertilizer manufacturing unit consists of:

(i) Very high nitrogen content as Ammonia and Urea
(ii) High Phosphates and Fluorides
(iii) High Arsenic
(iv) Oils

Removal of wastes:

(i) Oils are removed in a gravity separator.
(ii) Arsenic is segregated by hydrolysis. (water removal and concentration)
(iii) Phosphates and fluorides are segregated and chemically coagulated by lime and settled down in a settling tank.
(iv) Ammonia is removed by Lagooning or air stripped tower packed with red wool stakes.
(v) Nitrification and denitrification can be used for urea removal.

9.20 (c)

In Batch reactor flow is neither entering nor leaving the reactor i.e. flow enters is treated and then is discharged and the cycle repeats. The liquid content of the reactor are mixed completely. For example, the BOD test in BOD bottle is a batch reactor. These reactors are used for activated-sludge process with sequence batch reactors and mixing of concentrated solutions into working solutions.

In complete-mix reactor, it is assumed that complete mixing occurs instantaneously and uniformly throughout the reactor as fluid particles enter the reactor. Fluid particles leave the reactor in proportion to their statistical population. They are used in aerated lagoons, aerobic sludge digestion. Complete-mix reactor with recycle are used in activated sludge treatment.

In plug flow reactors fluid particles passes through the reactor in the same sequence in which they entered. The particles retain their identity and they remain in the reactor for a time equal to the theoretical retention time.

Fluidized-bed reactor is similar to the packed-bed reactor in many respects, but the packing material is expanded by the upward movement of fluid through the bed.

Non ideal flow is defined as short circuiting that occurs when a portion of arrives at the outlet before the bulk of the flow that entered the reactor during the same period arrives. This condition can occur incompletely mix or plug flow reactor. Mixing will break the layer of different density in completely mixed reactor. So density current short circuiting will occur in plug flow reactor.

9.21 (c)

A **trickling filter** is a fixed-bed biological reactor that operates under aerobic conditions. Pre-settled wastewater is continuously 'trickled' or sprayed over the filter. As the water migrates through the pores of the filter, organics are aerobically degraded by the biofilm covering the filter material.

An **activated sludge process** refers to a multi-chamber reactor unit that makes use of highly concentrated microorganisms to degrade organics and remove nutrients from wastewater to produce a high-quality effluent. To maintain aerobic conditions and to keep the activated sludge suspended, a continuous and well-timed supply of oxygen is required.

9.22 (c)

Both septic tank and imhoff tank are anaerobic units having same rate of anaerobic process.

9.23 (d)

Aerobic processes are more efficient than anaerobic processes with respect to biomass conversion and thus have a greater value of yield. Upon the addition of biomass, the organisms require time to acclimatize to their new environment before significant cell division and biomass production. The anaerobic bacteria are very sensitive to pH, salinity and temperature. So their acclimatization period is more in anaerobic treatment process.

Although energy consumption for aeration is more in aerobic process but energy production is more in anaerobic process.

9.24 (b)

Recirculation factor,

$$F = \frac{1+\left(\frac{R}{I}\right)}{\left(1+0.1\frac{R}{I}\right)^2} = \frac{2}{(1.1)^2} > 1$$

9.25 (d)

Nutrient Removal Tank removes the nitrogen and phosphorus present as nutrients in waste water. This is not a part of Preliminary treatment.

9.26 (c)

Flocculation basin is used for formation of flocs through coagulation. After this process turbidity is highest. Primary sedimentation causes settling of particles and turbidity reduces.

9.27 (d)

The ideal filter material is low-cost and durable, has a high surface to volume ratio, is light, and allows air to circulate. Whenever it is available, crushed rock or gravel is the cheapest option. The particles should be uniform and 95% of them should have a diameter between 7 and 10 cm. A material with a specific surface area between 45 and 60 m^2/m^3 for rocks and 90 to 150 m^2/m^3 for plastic packing is normally used. Larger pores are less prone to clogging and provide for good air circulation. Primary treatment is also essential to prevent clogging and to ensure efficient treatment.

9.28 (a)

Eutrophication is the depletion of oxygen in a water body, which kills aquatic animals. It is a response to the addition of excess nutrients, mainly phosphates, which induces explosive growth of plants and algae, the decaying of which consumes oxygen from the water. One example is the "bloom" or great increase of phytoplankton in a water body as a response to increased levels of nutrients. Eutrophication is almost always induced by the discharge of phosphate-containing detergents, fertilizers, or sewage, into an aquatic system.

9.29 (c)

Process	MLSS Range in mg/L
Conventional ASP	1500-3000
Completely mixed ASP	3000-6000
Extended Aeration ASP	3000-6000

9.30 (b)

The process of sludge digestion is greatly influenced by temperature, i.e. rate of sludge digestion is more at higher temperatures and vice versa. There are two distinct temperature zones:

(i) **Thermophilic digestion:** In this zone of high temperature, digestion is brought about by heat loving thermophilic organisms. The temperature in this zone ranges between 40° to 60°C with 54°C being optimum.

(ii) **Mesophilic digestion:** In this zone of moderate temperature, digestion is brought about by common mesophilic organisms. The temperature in this zone ranges between 25° to 40°C with 29°C being optimum.

9.31 (c)

Sludge Volume Index (SVI) is defined as the volume occupied in mL by one gram of solids in the mixed liquor after settling for 30 minutes. The usual adopted range of SVI is 50-150 mL/gm and such a value indicates good settling sludge.

If SVI is high, it denotes sludge with poor settling characteristics.

9.32 (d)

If the sludge with volume V_1 and moisture content p_1 reduces to volume V and moisture content p, then

$$V = V_1\left[\frac{100 - p_1}{100 - p}\right] = P\left[\frac{100-98}{100-96}\right] = \frac{P}{2}$$

9.33 (b)

Sludge Volume Index is a calculation that indicates the tendency of activated sludge solids (aerated solids) to thicken or to become concentrated during the sedimentation/thickening process. SVI is calculated in the following manner: (1) allow a mixed liquor sample from the aeration basin to settle for 30 minutes; (2) determine the suspended solids concentration for a sample of the same mixed liquor; (3) calculate SVI by dividing the measured (or observed) wet volume (m*l*/*l*) of the settled sludge by the dry weight concentration of MLSS in grams/*l*.

As per GOI manual SVI lies between 80-150 m*l*/gm.

9.34 (d)

Pollutants removed:

- Inorganic suspended solids
- Dissolved organic solids
- Oil and grease

Process of treatment:

- Primary sedimentation tank
- Trickling filter, Activated Sludge Process
- Skimming Tank

Dissolved inorganic solids are not removed in water treatment process.

9.36 (a)

The physical and chemical characteristics of sewage vary from top to bottom of sewage depth as well as with time as from morning to evening. Samples are taken at a point beneath the surface where the turbulence is thoroughly mixing up the sewage particles. This is called a grab sample. Such grab samples are collected at regular intervals during a day.

9.37 (a)

Secondary clarifiers for activated sludge must accomplish two objectives. First, they must produce an effluent sufficiently clarified to meet discharge standards. Secondly, they must concentrate the biological solids to minimize the quantity of sludge that must be handled. Thus, secondary clarifiers must be designed for effluent clarification and solids thickening, both of which relate directly to the surface area.

9.38 (d)

An anaerobic digestion, supernatant liquor produced is of high BOD concentration. Hence it should be retracted along with the raw sewage. This is the disadvantage of anaerobic sludge digestion.

Advantages are:

(i) Sludge produced has excellent dewatering capacity.

(ii) Greater production of methane.

(iii) Lesser operating cost, because energy is recovered in the form of methane.

(iv) Lesser capital cost.

9.39 (b)

Shallow ponds in which dissolved oxygen is present at all depths are called aerobic ponds. Although some oxygen is provided by diffusion from the air, the bulk of the oxygen in ponds is provided by photosynthesis. Lagoons are distinguished from ponds in that oxygen for lagoons is provided by artificial aeration.

When sufficient energy is supplied to keep the entire contents, including the sewage solids, mixed and aerated, the reactor is called an aerobic lagoon. When only enough energy is supplied to mix the liquid portion of the lagoon, solids settle to the bottom in areas of low velocity gradients and proceed to degrade anaerobically. This facility is called a facultative lagoon, and the process differs from that in the facultative pond only in the method by which oxygen is supplied.

9.40 (c)

Aerobic conditions are maintained in the upper portions of the facultative pond by oxygen generated by algae, and, to a lesser extent by penetration of atmospheric oxygen. Stagnant conditions in the sludge along the bottom prevent oxygen transfer to that region and anaerobic conditions prevail there. The boundary between the aerobic and anaerobic zones is not stationary. Mixing by wind action and penetration by sunlight may extend the aerobic area downward. Conversely calm waters and weak lighting result in the anaerobic layer rising toward the surface.

9.41 (b)

Stabilisation ponds may be classified as aerobic, facultative or anaerobic, depending upon the mechanism of purification. In a totally aerobic pond, the stabilisation of water is brought about by aerobic bacteria, which flourish in the presence of oxygen. In an anaerobic pond, however, the stabilisation of waste is mainly brought about by the usual anaerobic conversion of organic wastes to carbon dioxide, methane, and gaseous end products, with eruption of foul odours and pungent smells.

In a facultative pond, the upper layers work under aerobic conditions, while the anaerobic conditions prevail in the bottom layers. The upper aerobic layer of the pond acts as a good check against the evolution of the foul odours from such a pond.

Aerobic ponds are usually of depth of 0.5 m

Facultative ponds are usually of depth of 1 to 1.5 m

Anaerobic ponds are usually of depth of 2.5 to 4 m

9.42 (b)

Activated sludge plant involves:

1. wastewater aeration in the presence of a microbial suspension,
2. solid-liquid separation following aeration,
3. discharge of clarified effluent,
4. wasting of excess biomass, and
5. return of remaining biomass to the aeration tank.

In activated sludge process wastewater containing organic matter is aerated in an aeration basin in which micro-organisms metabolize the suspended and soluble organic matter. Part of organic matter is synthesized into new cells and part is oxidized to CO_2 and water to derive energy. In activated sludge systems the new cells formed in the reaction are removed from the liquid stream in the form of a flocculent sludge in settling tanks. A part of this settled biomass, described as activated sludge is returned to the aeration tank and the remaining forms waste or excess sludge.

9.43 (c)

Effluent from SST has the minimum weight/volume ratio of suspended solids because most of the solids are removed in PST & SST.

9.44 (a)

Aerated lagoons are common in small communities. These systems use aerators to mix the contents of the pond and add oxygen to the wastewater. They are sometimes referred to as partial-mix or complete-mix lagoons depending on the extent of aeration.

9.45 (c)

760 mm of Hg is standard atmospheric pressure.

9.46 (a)

The secondary sedimentation facility is provided after the biological reactor to facilitate the sedimentation of the cells produced during biological oxidation of organic matter. If these cells produced are not removed, complete treatment will not be achieved as these cells will represent about 30 to 60% of the organic matter present in untreated wastewater in aerobic treatment. Depending on the type of reactor used fraction of these settled cells is returned back to the reactor and remaining cells are wasted as excess sludge for further treatment.

9.47 (d)

- The sludge processing utilizes anaerobic digestion in which it is converted to CO_2 and CH_4.
- The acid formers produces acid and alcohols from complex organics methane formers converts organic acids to CO_2 and CH_4.

9.48 (c)

Sunlight is not produced from wasted food.

$$\text{Carbon Dioxide + Water} \xrightarrow{\text{Light}} \text{Glucose + Oxygen}$$

$$6CO_2 + 6H_2O \xrightarrow{\text{Chlorophyll}} C_6H_{12}O_6 + 6O_2$$

9.49 (b)

Process Flow-sheet of Conventional Domestic Sewage Treatment Plant.

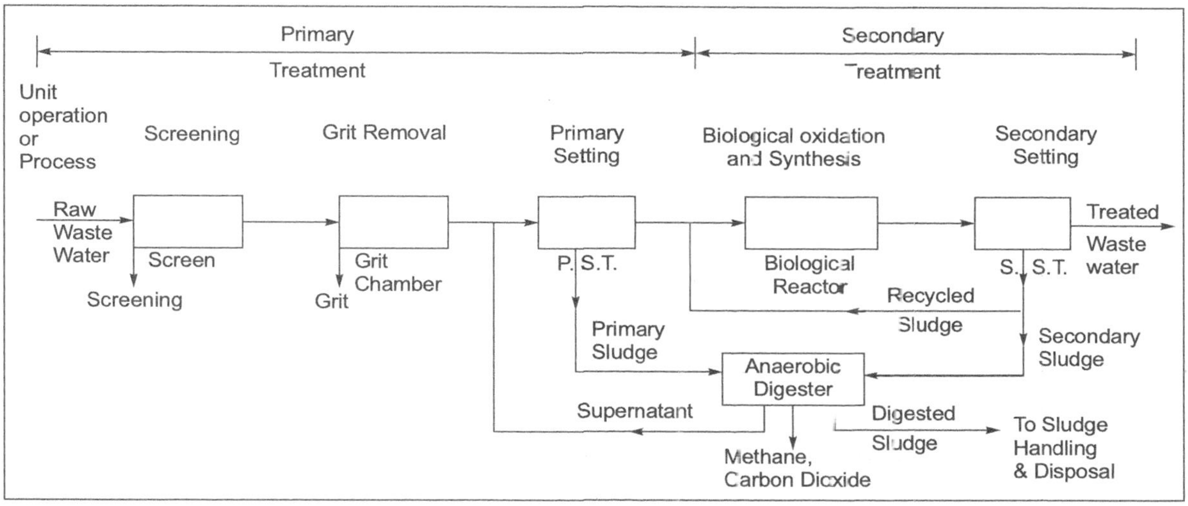

9.50 (b)

The design of trickling filter primarily involves finding out diameters and depth of the tank, which is based on the values of the filter loadings that are:

- Hydraulic Loading Rate
- Organic Loading Rate

9.51 (b)

$$V_1(100-p_1) = V_2(100-p_2)$$

$$\Rightarrow \quad V_2 = (1\,\text{MLD})\left(\frac{100-99}{100-98}\right)$$

$$= 1\left(\frac{1}{2}\right) = 0.5\ \text{MLD}$$

9.52 (a)

In super rate trickling filters plastic sheets of size 600 × 600 × 1200 mm are used. Plsssic media have high specific areas, high void space and low unit weight.

These are used for strong industrial waste having hydraulic loading rate in the range of 30-90 $m^3/d/m^2$ and organic loading rates of 1000-4000 $g/d/m^3$.

9.53 (c)

Uniform velocity section is 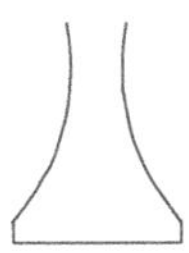 and not the U-type section.

9.54 (a)

Grease data is a measure of effectiveness of primary settling tanks.

Volatile acid test is an important digester control test much better than in monitoring because volatile acid test gives more warning of digester problem.

9.55 (c)

Attached-culture systems employ reactors in which waste water is contacted with microbial films attached to surfaces. Surface area for biofilm growth is increased by placing a porous medium in the reactor. The organisms attach themselves to the medium and grow into dense films of a viscous, jellylike nature. Waste water passes over this film in thin sheets with dissolved organics passing into the biofilm due to concentration gradients within the film. Suspended particles and colloids may be retained on the sticky surfaces where they are decomposed into soluble products. Oxygen from the wastewater and from air in the void spaces of the medium provide oxygen for aerobic reactions at the biofilm surface. Waste products from the metabolic processes diffuse outward and are carried away by the water or air currents moving through the voids of the medium. Growth of the biofilm is restricted to one direction-outward from the solid surface. As the film grows thicker, concentration gradients of both oxygen and food develop. Eventually, both anaerobic and endogenous metabolism occur at the biofilm-medium surface interface. The attachment mechanism is a weakened and the shearing action of the wastewater flowing across the film pulls it from its mooring and washes it away. This process known as sloughing, is function of both the hydraulic and organic loading rate. Biofilm is quickly reestablished in places cleared by sloughing.

9.56 (d)

Moisture content = 99 %

$$= \frac{\text{weight of water}}{\text{weight of solid}}$$

Hence, 100 kg wet sludge makes

$$\frac{100\,\text{kg}}{1000\,\text{kg/m}^3} = 0.1\text{ m}^3$$

∴ Concentration of suspended solid

$$= \frac{1\text{kg}}{0.1\text{m}^3} = \frac{1\times10^6\text{ mg}}{0.1\times10^3 l} = 1\times10^4\text{ mg}/l$$

9.58 (b)

$$V_1(100-p_1) = V_2(100-p_2)$$

$$\Rightarrow 100(100-95) = V_2(100-90)$$

$$\therefore V_2 = 50\text{ m}^3$$

9.59 (c)

Sludge volume index is the volume occupied in mL by one gm of solids in the mixed liquor after settling for 30 minutes.

$$\therefore \text{SVI} = \frac{200\,mL}{2000\times10^{-2}\text{ gm}} = 100\text{ mL/gm}$$

9.60 (b)

The factors that determine the removal efficiency of biodegradable organic matter are:

- The nature and composition of the organic matter to be removed.
- Suitability of environmental factors.
- Sludge retention time in the reactor.
- The intensity of mixing, hence contact between bacterial biomass and organic matter.
- Specific loading of organic matter with respect to bacterial sludge mass, and retention time.

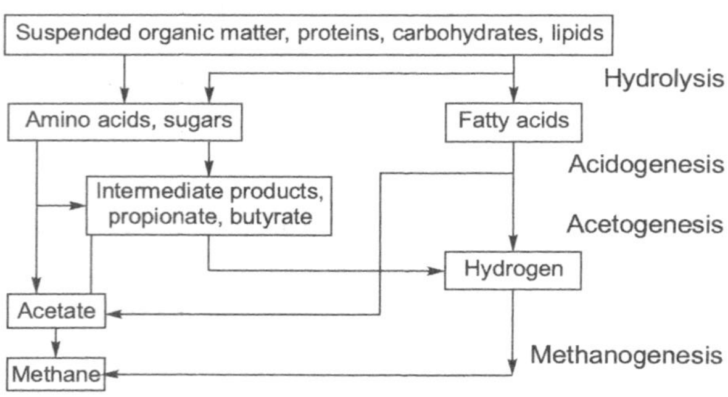

Different steps are necessary for the anaerobic digestion of proteins, carbohydrates, and lipids. Four different phases can be distinguished in the overall conversion process of organic matter to biogas as (1) Hydrolysis, (2) Acidogenesis, (3) Acetogenesis and (4) Methanogenesis.

9.62 (c)

$$\text{Required surface area} = \frac{40\times10^6}{10{,}000} = 400\text{ m}^2$$

(Square sedimentation tank)

$$\therefore B^2 = 400$$

$$\Rightarrow B = L = 20\text{ m}$$

9.63 (b)

Aerobic digestion of sewage is costly but suitable for big cities because in aerobic digestion system end products are an odourless, humus like, biological stable end product.

9.64 (b)

$$V_1(100-P_1) = V_2(100-P_1)$$

$$V_1 = 100\text{ m}^3$$

$$P_1 = 95\%$$

$$P_2 = 90\%$$

$$100(100-95) = V_2(100-90)$$

$$100\times5 = V_2(10)$$

$$\Rightarrow V_2 = 50\text{ m}^3$$

9.65 (a)

Flow equalization basin is located after screening and grit removal and before primary sedimentation.

9.66 (c)

An aerated grit chamber consists of a standard spiral flow aeration tank provided with air diffusion tubes placed on one side of the tank. The grit particles tend to settle down to the bottom of the tank at rates dependent upon the particle size and the bottom velocity of roll of the spiral flow, which in turn depends on the rate of air diffusion through diffuser tubes and shape of aeration tank. The heavier particles settle down whereas the lighter organic particles are carried with roll of the spiral motion. It is designed for a typical detention time of 3 minutes at maximum flow.

9.67 (a)

An anaerobic digestion, supernatant liquor

produced is of high BOD concentration. Hence it should be retracted along with the raw sewage. This is the disadvantage of anaerobic sludge digestion.

Advantages are:

(i) Sludge produced has excellent dewatering capacity.

(ii) Greater production of methane.

(iii) Lesser operating cost, because energy is recovered in the form of methane.

(iv) Lesser capital cost.

9.68 (c)

A septic tank is a watertight chamber made of brick work, concrete, fiberglass, PVC or plastic, through which blackwater from cistern or pour-flush toilets and greywater through a pipe from inside a building or on outside toilet flows for primary treatment. Settling and anaerobic processes reduce solids and organics, but the treatment is only moderate. Effluent is infiltrated into the ground or transported via a sewer to a centralized treatment plant. Accumulating faecal sludge needs to be dug out of the chamber regularly and correctly disposed off.

9.69 (d)

Contact Stabilisaton Process or Biosorption Process:

In this process, the sewage and recycled or returned sludge are mixed and aerated for a comparatively shorter period of 0.5 to 1.5 hr in a special mixing tank, called contact tank.

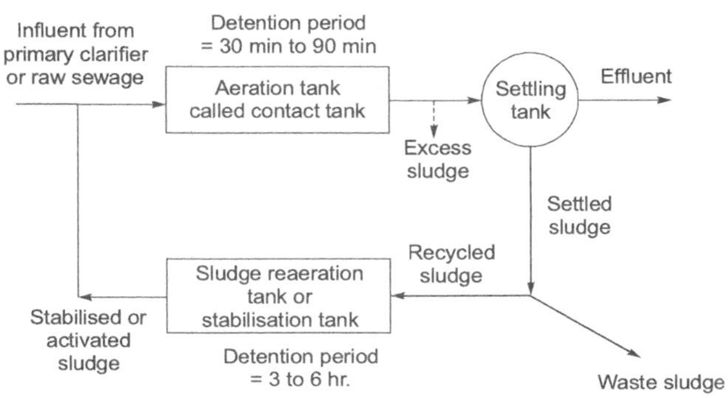

9.70 (a)

In stabilisation pond symbiotic relation exist between algae and aerobic micro-organisms. O_2 is provided by algae during photosynthesis and from atmosphere.

Detention period is 2-6 weeks. Bacteria in pond is both aerobic and anaerobic.

The following stages are involved in the biological action involved in the process of sludge digestion.

Hydrolysis → Acid fermentation → Acid regression → Alkaline fermentation → Methane formation

9.71 (b)

The anaerobic digester plant is a better method for handling of sludge as it produces large amount of energy in the form of methane. In this rodent and pests are attracted when sludge is dried.

9.73 (c)

$$SVI = \frac{V_s}{X_s} \times 10^3 \text{ (ml/gm)}$$

V_s = volume of solids settled in 30 min = 175 ml

X_s = concentration of solids = 2000 mg/l

$$SVI = \frac{175}{2000} \times 10^3 = 87.5 \text{ (ml/gm)}$$

$$SDI = \frac{100}{SVI} = \frac{100}{87.5} = 1.14 \text{ (g/}l\text{)}$$

9.74 (d)

During anaerobic decomposition of organic content in waste water major gases produced are CH_4 (65-70%) and CO_2 (30%).

9.76 (a)

Dosing tanks were used previously when the discharge was not huge and it was possible to collect and let the sewage (untreated or partially treated) be left to stay even for several hours. Now a days, various other techniques are used. Now a days, dosing tanks are used only for low rate reactors.

9.77 (a)

Elutriation is a process of sludge conditioning in which sludge is washed either by fresh water or plant effluent to reduce the sludge alkalinity and fine particles, thus decreasing the amount of coagulant required in sludge dewatering.

9.78 (d)

Centrifugation

- Centrifuges are used to thicken and dewater sludge and their application is limited to waste activated sludge.
- Thickening by centrifugation involves the settling of sludge particles under the influence of centrifugal force.
- Where there is a space restriction, this method is adopted.

9.79 (a)

- Detritus tanks are nothing but rectangular grit chambers, designed to flow with smaller flow velocity (of about 0.09 m/sec) and longer detention periods (about 3 to 4 minutes) so as to separate out not only the larger grit, etc., but also to separate out the very fine sand particles, etc.
- Due to this, a large amount of organic matter will also settle out along with the inorganic grit, sand, etc.
- This organic material is then separated from the grit by control of currents in the tank through baffles, or by controlled aeration of the flow through the tank.

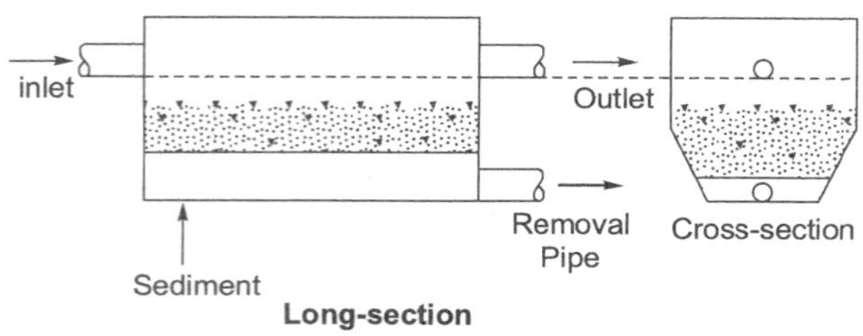

- The rising air bubbles will then separate the lighter organic matter from the descending grit. The grit is removed continuously by means of scraper mechanism.

9.80 (c)

Suspended solids, $S_0 = 250$ mg/l

Efficiency of sedimentation tank,

$$\eta = 0.55$$

So, concentration of suspendd solids in sludge,

$$= 137.5 \text{ mg/L}$$

Therefore, suspended solids in 6.5×10^6 litres

$= 137.5 \times 10^{-6} \times 6.5 \times 10^6$

$$= 893.75 \text{ kg}$$

Now, water content of sludge = 95%

$\therefore$ Total weight of water in sludge

$$= \frac{893.75 \times 0.95}{0.05} = 16981.25 \text{ kg}$$

So, total weight of sludge

$= (16981.25 + 893.75)$ kg $= 17875$ kg

Hence,volume of sludge produces

$$= \frac{17875 \text{ kg}}{1 \times 1000 \text{ kg/m}^3} = 17.875 \text{ m}^3$$

9.81 (b)

As per Cl: 21.2.4 of CPHEEO (1993)

For a dispersion trench, $Q = \dfrac{130}{\sqrt{t}}$

where, Q is in l/m^2/day.

t is percolation rate (min/cm).

■■■■

10 Design of Sewers & Sewerage System

10.1 For the design of a storm sewer in a drainage area, if the time of concentration is 20 minutes, then the duration of rainfall will be taken as

(a) 10 min (b) 20 min
(c) 30 min (d) 40 min

[ESE : 2000]

Directions: The following items consists of two statements; one labelled as **'Assertion (A)'** and the other as **'Reason (R)'**. You are to examine these two statements carefully and select the answers to these items using the codes given below:

Codes:

(a) both A and R are true and R is the correct explanation of A
(b) both A and R are true but R is not a correct explanation of A
(c) A is true but R is false
(d) A is false but R is true

10.2 **Assertion (A):** Sewers are not allowed to flow full.

Reason (R): Reserve space in the sewer takes care of fluctuations in the sewage flow.

[ESE : 2000]

10.3 **Assertion (A):** The design of all non-circular sections is based upon getting a "hydraulically equivalent section".

Reason (R): The chart of hydraulic elements is very useful in sewer design.

[ESE : 2001]

10.4 Match **List-I** (Unit) with **List-II** (Purpose) and select the correct answer using the codes given below the lists:

List-I

A. Leaping weir
B. Gutter inlet
C. Inverted syphon
D. Catch basin

List-II

1. To prevent grit, sand, debris, etc. from entering the storm sewer.
2. To carry the sewer below a stream or railway line
3. To drain rain water from roads to the storm sewer
4. To separate storm water and the sanitary sewage

Codes:

	A	B	C	D
(a)	4	3	1	2
(b)	4	3	2	1
(c)	3	4	2	1
(d)	3	4	1	2

[ESE : 2001]

10.5 In sanitary plumbing of buildings, a two pipe system signifies

(a) separate soil pipes and waste pipe without vent pipes
(b) a soil-cum-waste pipe and a ventilating pipe
(c) separate soil and waste pipe and a common ventilating pipe
(d) separate soil pipe and waste pipe, each with its own vent pipe

[ESE : 2001]

10.6 Which of the following are storm water regulators?

1. Side weir
2. Leaping weir
3. Syphonic spillway
4. Float actuated gates or valves
5. Inverted syphon

Select the correct answer using the codes given below:

(a) 1, 2, 3 and 4 (b) 1, 3 and 5
(c) 2, 3, 4 and 5 (d) 1, 2, 4 and 5

[ESE : 2001]

10.7 **Assertion (A):** Most important activity in sewer line construction is to start constructing it from the tail end and to check levels with a boning rod.

Reason (R): Construction of sewer line from tail end is recommended because required number of pumping stations may be incorporated in sewer network design.

(a) both A and R are true and R is the correct explanation of A
(b) both A and R are true but R is not a correct explanation of A
(c) A is true but R is false
(d) A is false but R is true

[ESE : 2003]

10.8 A sewer is commonly designed to attain self-cleansing velocity at

(a) peak hourly rate of flow
(b) average hourly rate of flow
(c) minimum hourly rate of flow
(d) sewer running half full

[ESE : 2003]

10.9 Match **List-I** with **List-II** and select the correct answer using the codes given below the lists:

List-I

A. Test with sound waves in the audible frequency range
B. Fire flow tests
C. Hydraulic gradient tests
D. Coefficient tests

List-II

1. To determine the ability of a distribution system to transmit water with adequate residual pressure
2. Location and isolation of leaks
3. To determine the efficiency and adequacy of a distribution system during days of high demand
4. To determine the internal condition of pipeline with respect to friction loss

Codes:

	A	B	C	D
(a)	2	1	3	4
(b)	2	3	1	4
(c)	4	1	3	2
(d)	4	3	1	2

[ESE : 2004]

10.10 Consider the following statements:

1. The velocity of flow in the rising main should be less than 0.8 m/s at any time.
2. Maximum velocity of flow is generally limited to 1.8 m/s and never allowed to exceed 3.0 m/s.

In the design of large sewage pumping stations, which of the above conditions must be satisfied?

(a) 1 only (b) 2 only
(c) Both 1 and 2 (d) Neither 1 nor 2

[ESE : 2004]

10.11 If a sewer *X* is to be designed to generate equivalent self-cleansing action as in sewer *Y*, then

(a) velocity in sewer *X* must be equal to velocity in sewer *Y*
(b) slope of sewer *X* must be equal to slope of sewer *Y*
(c) tractive force intensity generated in sewer *X* must be same as that in sewer *Y*
(d) the roughness coefficient of *X* sewer material should be same as that of *Y* sewer material

[ESE : 2004]

10.12 Consider the following statement regarding building manholes:

1. They must be provided at every change of alignment, gradient or diameter.
2. They must be provided at the head of all sewers.
3. They must be provided at every junction of two or more sewers.
4. They must be provided at every 100 m along straight runs of sewers.

Which of these statements are correct?

(a) 1, 2, 3 and 4 (b) 1, 3 and 4
(c) 1, 2 and 3 (d) 2 and 4

[ESE : 2004]

10.13 Which of the following statements is correct?

A combined sewer is one which transports domestic sewage and

(a) storm water
(b) industrial waste
(c) overhead flow
(d) industrial wastes and storm water

[ESE : 2004]

10.14 Consider the following statements:

The basic difference between water pipes and sewer pipes is

1. in the material used for the pipes
2. in the pressure of the liquid flow
3. in the suspended solids they carry

Which of these statements is/are correct?

(a) 1 and 3 (b) 1 only
(c) 2 and 3 (d) 1, 2 and 3

[ESE : 2004]

10.15 In the design consideration of sewerage system, the sewers must have which one of the following?

(a) Maximum velocity of flow
(b) Only 50 per cent of maximum velocity of flow
(c) Minimum velocity of not less than cleansing velocity of flow
(d) High pressure at all times [ESE : 2008]

10.16 What is the depth of waterseal in the traps?

(a) < 2.5 cm (b) 2.5 – 7.5 cm
(c) 7.5 – 12.5 cm (d) Not less than 15 cm

[ESE : 2008]

10.17 Assertion (A): Laterals of minimum specified diameter in sewerage system have to be laid at slopes designed for self-cleaning velocity.

Reason (R): For the specified minimum lateral diameter at specified slopes, a minimum flow rate is not essential to maintain self-cleansing velocity.

(a) both A and R are true and R is the correct explanation of A
(b) both A and R are true but R is not a correct explanation of A
(c) A is true but R is false
(d) A is false but R is true

[ESE : 2011]

10.18 A sewer of 400 mm diameter and slope 1 in 400, running half-full, has a flow velocity of 0.82 m/sec. What velocity of flow will be obtained if the slope is made 1 in 100?

(a) 3.28 m/s (b) 1.64 m/s
(c) 0.82 m/s (d) 0.41 m/s

[ESE : 2012]

10.19 A sewer has a diameter of 300 mm and slope of 1 in 400. While running full it has a mean velocity of 0.7 m/s. If both the diameter and slope are doubled (to respectively be 600 mm and 1 in 200), what will be the changed mean velocity when running half-full ? Use Manning's formula.

(a) 1.59 m/s (b) 2.80 m/s
(c) 0.90 m/s (d) 1.00 m/s

[ESE : 2012]

10.20 At the same mean velocity of flow, the ratio of head loss per unit length for a sewer pipe running full to that for the same pipe flowing half-full is:

(a) 2.0 (b) 1.67
(c) 1.0 (d) 0.67

[ESE : 2013]

10.21 The self-cleansing velocity in a sewer depends on:

1. BOD (soluble)
2. Slope of the sewer
3. Ratio of depth of flow sewage to sewer diameter ratio

(a) 1, 2 and 3 (b) 1 and 2 only
(c) 2 and 3 only (d) 1 and 3 only

[ESE : 2013]

10.22 One sewer has a dia. of 300 mm and another one has a dia. of 600 mm. When both run half-full, what will be the ratio of velocities in the two pipes if the slope of both pipes is the same?

(a) 1 (b) $\frac{1}{2}$
(c) $\left(\frac{1}{2}\right)^{2/3}$ (d) $\left(\frac{1}{2}\right)^{3/2}$

[ESE : 2013]

10.23 If the slope of sewer A is 1/100 and that of sewer B is 1/400, the velocity of flow in the two sewers will have a ratio of (size of both the sewers being same)

(a) $\frac{1}{2}$ (b) 1
(c) $(2)^{2/3}$ (d) 2

[ESE : 2013]

10.24 As per Manual on Sewerage and Sewage Treatment, Central Public Health and Environmental Engineering Organization, Ministry of Urban Development, Government of India, when a sewer connects with another sewer having a difference of 600 mm in level between the water lines (peak flow levels) of main line and the invert level of branch line, the connection may be provided using

(a) Side entrance manhole
(b) Junction manhole
(c) Flushing manhole
(d) Drop manhole [ESE : 2014]

10.25 A combined sewage system is more appropriate for developed areas where
(a) Rainfall occurs for a very few days in a year
(b) Rainfall occurs almost uniformly throughout the year
(c) Air temperatures are nearly uniform throughout the year
(d) Air temperatures through the year include extremes during certain runs of days

[ESE : 2016]

10.26 **Statement (I):** The invert of the lower (outgoing) larger size sewer is depressed suitably so as to match with the crowns of the upper (incoming) and the smaller size sewers.
Statement (II): It ensures that the smaller sewer runs full by backwater effect from the larger sewer even if the larger sewer does not run full.
(a) Both Statement (I) and Statement (II) are individually true and Statement (II) is the correct explanation of Statement (I)
(b) Both Statement (I) and Statement (II) are individually true but Statement (II) is ***not*** the correct explanation of Statement (I)
(c) Statement (I) is true but Statement (II) is false
(d) Statement (I) is false but Statement (II) is true

[ESE : 2017]

10.27 The rain intensity over 54 hectares of land is 50 mm/h, 30% of area consists of roof surfaces with runoff rate as 0.9. 30% is open field with runoff rate of 0.2 and remaining 40% is road network with runoff rate 0.4. The storm water flow will be nearly,
(a) 2.6 m^3/s (b) 3.7 m^3/s
(c) 4.8 m^3/s (d) 4.9 m^3/s

[ESE : 2019]

10.28 Which one of the following the merit of combined sewer system?
(a) Rain water dilutes the sewage, therefore, it can be easily and economically treated.
(b) Initial cost is high as compared with separate system.
(c) If the whole sewage is to be disposed off by pumping, it is uneconomical.
(d) During heavy rains, the overflowing of sewers will endanger the public health.

[ESE : 2021]

■■■■

Answers Design of Sewers & Sewerage System

10.1 (b)	10.2 (a)	10.3 (b)	10.4 (b)	10.5 (d)	10.6 (a)	10.7 (c)	10.8 (c)	10.9 (a)
10.10 (b)	10.11 (c)	10.12 (a)	10.13 (d)	10.14 (c)	10.15 (c)	10.16 (b)	10.17 (b)	10.18 (b)
10.19 (a)	10.20 (c)	10.21 (c)	10.22 (c)	10.23 (d)	10.24 (d)	10.25 (b)	10.26 (c)	10.27 (b)
10.28 (a)								

Explanations Design of Sewers & Sewerage System

10.1 (b)

Maximum runoff will be obtained from the rain having a duration equal to the time of concentration, and this is called the critical rainfall duration. Therefore duration of rainfall should be 20 min.

10.2 (a)

Generally sewers run half to three-fourth full at maximum discharge. The extra space thus left above is provided:

- To account for fluctuations in flow.
- Large scale infiltration of storm water
- Unforeseen increase in population or water consumption and consequent increase in sewage production.

10.3 (b)

Hydraulically equivalent section means that two sections can carry same discharge, while flowing full, on the same grade.

10.4 (b)

- Gutter inlet is used to drain rain water from roads to storm sewer.
- Leaping Weir is used to separate storm water and the sanitary sewage.
- Inverted siphon is used to carry the sewer below a stream or railway line.
- Grit, sand, debris etc. do settle in catch basins and their entry into the sewer is thus prevented.

10.5 (d)

The two-pipe system is the best and most improved system of plumbing. There are two sets of vertical pipes, one for excreta (night soil) as soil pipe and another for sullage as waste pipe. The soil pipes as well as waste pipes are separately ventilated, by providing, separate vent pipe or anti-siphonage pipe. This system has four vertical pipes.

10.6 (a)

The storm water regulators are :
(i) Leaping weir
(ii) Overflow weir
(iii) Siphon spillway

10.7 (c)

The construction of sewer line is started from the tail end so that sewers of tail length may be utilized even during the initial periods of its construction, thus ensuring that the functioning of sewerage scheme need not to wait till the completion of entire scheme.

10.8 (c)

- The flow velocity in a sewer should be such that neither the suspended solids in the sewage gets silt up nor the sewer pipe material gets scoured.
- First condition, limits the minimum velocity and second condition, limits the maximum velocity.
- The minimum velocity also called as self cleansing velocity is the velocity such that suspended solid cannot settle down.
- It must be developed at least once a day and a sewer is commonly designed to attain the self cleansing velocity at minimum hourly rate of flow.

10.9 (a)

Fire flow test: A test to determine how much after a fire hydrant can produce to fight fire.

A fire flow test will tell the fire department exactly how many liters per minute can be pumped and not drop the main pressure below the valve. This ensures minimum residual pressure.
Testing with sound waves is done to detect leakages in the distribution system.
Other methods to detect leakage are:
(i) direct observations
(ii) using sounding rods which work in audible frequency range.
(iii) plotting hydraulic gradient line
(iv) using waste detection meters like Deacon's water waste detection meter.

10.10 (b)

Rising main may be of cast iron or asbestos cement pressure pipe. The velocity of flow in the rising main should not be less than 0.75 m/s at any time of flow to avoid settling of particles. The maximum velocity limitation is due to continuous abrasion caused by the suspended particles.

10.11 (c)

Velocity and discharge are functions of tractive force intensity which depends upon friction coefficient as well as flow velocity. For equal self cleansing tractive force intensity should be same in both sewers.

10.12 (a)

- A manholes is an opening constructed on the alignment of a sewer for facilitating a person access to the sewer for the purpose of inspection, testing, cleaning and removal of obstruction from the sewer line.
- Manholes should be built at every change of alignment, gradient or diameter, at the head of all sewers and branches and at every junction of two or more sewer.
- The spacing of manholes above 90 to 150 m may be allowed on straight runs for sewers of diameter 0.9 to 1.5 m spacing of manholes at 150 to 200 m may be allowed on straight runs for sewers of 1.5 to 2 m diameter.

10.14 (c)

Cast iron pipes are generally used for sewer pipes which can also be used as distribution pipes. The basic difference is pressure flow in water pipe and suspended solids carried in sewage.

10.15 (c)

The velocity which will even scour the deposited particles of a given size, must be developed in the sewers, at least once a day, so as not to allow any deposition in the sewers is called self-cleansing velocity. The generation of such a minimum self-cleansing velocity in the sewer, at least once a day, is important, because if certain deposition takes place and is not removed, it will obstruct free flow causing further deposition and finally leading to the complete blocking of the sewer.

10.16 (b)

Traps are the fittings placed at the ends of the soil pipes or the sullage pipes to prevent the passage of foul gases from the pipes to the outside. This is possible because traps do enclose or maintain water seal between the pipe and outside. This water depth does not allow gases to escape to the outside of pipe. The efficiency and effectiveness of a trap will depend upon the depth of water seal. Greater is this depth more effective the trap will be. This water seal generally varies from 25 mm to 75 mm; 50 mm being quite common in most of the traps.

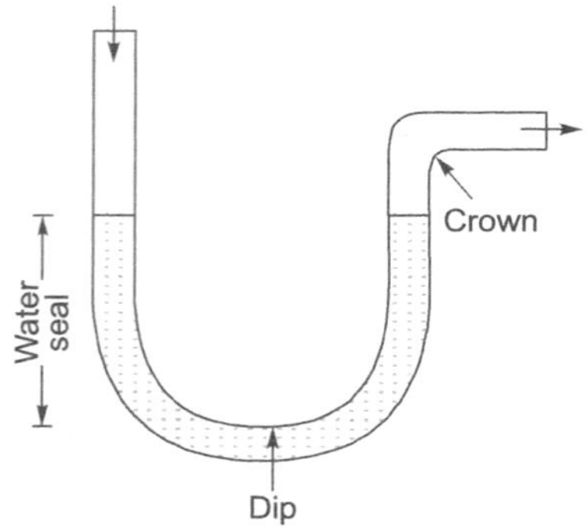

10.17 (b)

In sewage system, cleaning becomes easier if laterals are laid at a slopes designed for self cleansing velocity.

- However, at minimum discharge, self cleansing velocity may not be maintained. Hence, we provide a flushing system to flush out the deposited solids.

10.18 (b)

Velocity of flow is given by Manning's formula

$$V = \frac{1}{n}R^{2/3}S^{1/2}$$

$$\therefore \quad V \propto S^{1/2}$$

$$\therefore \quad V_1 = \frac{1}{n}(R)^{2/3}\left(\frac{1}{400}\right)^{1/2}$$

$$V_2 = \frac{1}{n}(R)^{2/3} \times \left(\frac{1}{100}\right)^{1/2}$$

$$\therefore \quad \frac{V_2}{V_1} = \left(\frac{400}{100}\right)^{1/2}$$

$$\therefore \quad V_2 = 2 \times 0.82$$

$$= 1.64 \text{ m/s}$$

10.19 (a)

According to Manning's formula

$$V_1 = \frac{1}{n}(R_1)^{2/3}(S_1)^{1/2} = 0.7$$

$$\Rightarrow \quad V_1 = \frac{1}{n}\left(\frac{D_1}{4}\right)^{2/3}(S_1)^{1/2} = 0.7$$

Now, $D_2 = 2D_1;\ S_2 = 2S_1$

$$\therefore \quad V_2 = \frac{1}{n}\left(\frac{2D_1}{4}\right)^{2/3}(2S_1)^{1/2}$$

$$\therefore \quad V_2 = 2.24\, V_1$$

$$= 1.57 \text{ m/s} \cong 1.59 \text{ m/s}$$

10.20 (c)

Head loss per unit length (i.e., slope of the sewer pipe)

Case-I: Flowing full

Hydraulic $R = \frac{D}{4}$

Velocity of flow V_1

Given, $V_1 = V_2$

$$\frac{1}{\eta}\left(\frac{D}{4}\right)^{2/3}(S_1)^{1/2} = \frac{1}{\eta}\left(\frac{D}{4}\right)^{2/3}(S_2)^{1/2}$$

$$S_1 = S_2$$

$$\therefore \quad \frac{S_1}{S_2} = 1.0$$

Case-II: Flowing half full

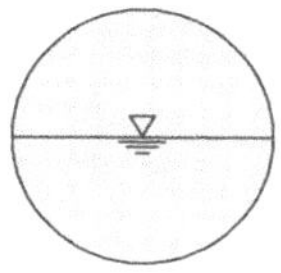

$$r = \frac{\dfrac{\left(\dfrac{\pi}{4}D^2\right)}{2}}{\dfrac{\pi D}{2}} = \frac{D}{4}$$

Velocity of flow V_2

Here $S = i$

S = hydraulic gradient

10.21 (c)

Self-cleaning velocity does not depend on BOD.

10.22 (c)

$$V = \frac{1}{n}R^{2/3}\sqrt{S}$$

$$\frac{V_1}{V_2} = \left(\frac{R_1}{R_2}\right)^{2/3}$$

$$= \frac{(A_1/P_1)}{(A_2/P_2)}^{2/3} = \left[\frac{\dfrac{\pi}{8}D_1^2 \Big/ \dfrac{\pi D_1}{2}}{\dfrac{\pi}{8}D_2^2 \Big/ \dfrac{\pi D_2}{2}}\right]^{2/3}$$

$$= \left(\frac{D_1}{D_2}\right)^{2/3} = \left(\frac{1}{2}\right)^{2/3} \quad [\because D_2 = 2D_1]$$

10.23 (d)

$$\frac{V_1}{V_2} = \sqrt{\frac{S_1}{S_2}} = \sqrt{\frac{1}{100} \times \frac{400}{1}} = 2$$

10.24 (d)

A manhole which is constructed to connect the high level branch sewer (> 0.6 m) to the low level main sewer by vertical dropping pipe is known as drop manhole.

10.25 (b)

If rainfall occurs almost uniformly throughout the year, single sewer would be appropriate as its advantage will be available for entire year.

10.26 (c)

The invert of the lower larger size sewer is depressed suitably so as to match with the crowns of the upper and the smaller size sewers as it prevents backing up of sewage.

10.27 (b)

Given data:

$A = 54 \times 10^4 \text{ m}^2$

$i = 50 \text{ mm/hr}$

$$k_{eq} = \frac{k_1A_1 + k_2A_2 + k_3A_3}{A_1 + A_2 + A_3}$$

$$= \frac{(0.3A \times 0.9) + (0.3A \times 0.2) + (0.4A \times 0.4)}{0.3A + 0.3A + 0.4A}$$

$$= 0.27 + 0.06 + 0.16 = 0.49$$

$\therefore$ Discharge,

$$Q = KiA$$

$$= \frac{0.49 \times 50 \times 10^{-3} \times 54 \times 10^4}{3600 \text{ sec}} \text{m}^3$$

$$= 3.675 \simeq 3.7 \text{ m}^3\text{/sec}$$

10.28 (a)

When the drainage is taken along with sewage, it is called combined sewer system.

Advantages:

- The storm water dilutes the sewage and hence its strength is reduced.
- The cleaning of sewers is easy as they are large in size.
- The maintenance cost is less.
- The self-cleansing velocity is easily achieved.
- This system requires only one set of the sewer, thus it becomes economical.

Disadvantages:

- The load on the treatment plant becomes high.
- The storm water is unnecessarily polluted.
- The sewers are large in diameter.
- This system proves to be uneconomical when pumping is required for the lifting of sewage.
- During a heavy storm, the combined sewer may be overflow which may create trouble for the people.

11 Municipal Solid Waste Management

11.1 Aerobic method of composting practiced in India is called
(a) Bangalore method (b) Nagpur method
(c) Delhi method (d) Indore method
[ESE : 2000]

11.2 Which of the following statements related to C/N (Carbon/Nitrogen) ratio is not correct?
(a) Lower initial C/N ratio leads to loss of nitrogen and slows down the rate of decomposition
(b) Higher initial C/N ratio leads to cell destruction to obtain nutrition
(c) Higher initial C/N ratio leads to lower conservation of nitrogen in the finished compost
(d) An initial C/N ratio of 30 to 50 is optimal for composting
[ESE : 2000]

11.3 Which one of the following methods would be best suited for disposal of plastic and rubber waste?
(a) Composting (b) Incineration
(c) Pyrolysis (d) Sanitary landfill
[ESE : 2001]

11.4 Which one of the following comprehensive classifications is used for different types of solid wastes?
(a) Residential, commercial and treatment plant wastes
(b) Food, demolition and construction wastes
(c) Municipal, industrial and hazardous wastes
(d) Rubbish, special wastes and wastes from open areas
[ESE : 2001]

11.5 Which one of the following terms, correctly describes 'Biomagnification'?
(a) Reproduction of micro-organisms
(b) Observation of micro-organisms under a microscope
(c) Ability of micro-organisms to form zo-ological film
(d) Concentration of toxic materials in the food chain [ESE : 2001]

11.6 The description of solid waste collected is as follows:

Night soil	:	35 t
Rubbish	:	40 t
Debris	:	25 t
Garbage	:	40 t

The organic solids in the above composition is
(a) 35 t (b) 60 t
(c) 100 t (d) 75 t
[ESE : 2002]

11.7 In a sanitary landfill, decomposition and chemical changes within organic content of the solid waste goes on. Consequential changes within landfill can be
1. temperature changes within landfill
2. production of gases like H_2S, CO, CO_2 and CH_4
3. destruction of pathogens
4. production of other gases like SO_2 and NO_2

Which of these statements are correct?
(a) 1, 2, 3 and 4 (b) 1, 2 and 3
(c) 1 and 4 (d) 2 and 3
[ESE : 2002]

11.8 Bangalore method and Indore method of disposing solid wastes are
(a) identical
(b) different as Bangalore method is an anaerobic method
(c) different as Bangalore method does not contain human excreta

(d) different as Indore method is an incineration method

[ESE : 2003]

11.9 The term 'Refuse' generally does not include
(a) putrescible solid waste
(b) excreta
(c) non-putrescible solid waste
(d) ashes

[ESE : 2003]

11.10 Which one of the following statements explains the term pyrolysis?
(a) Solid waste is heated in closed containers in oxygen-free atmosphere
(b) Solid waste is incinerated in presence of oxygen
(c) Wastewater is treated with oxygen
(d) Dissolved solids from water are removed by glass distillation

[ESE : 2003]

11.11 **Assertion (A):** Composting is the most commonly used process for the decomposition of the organic components of municipal solid wastes.
Reason (R): Bangalore method is a common anaerobic method used for biological conversion of organic components of municipal solid wastes.
(a) both A and R are true and R is the correct explanation of A
(b) both A and R are true but R is not a correct explanation of A
(c) A is true but R is false
(d) A is false but R is true [ESE : 2005]

11.12 Consider the following statements:
In solid waste management
1. density separation of solid wastes can be accomplished by air classifiers.
2. iron recovery from solid wastes can be done by magnetic separators.
3. aluminium separation from solid wastes can be accomplished by eddy current separators.

Which of these statements are correct?
(a) 1 and 2 (b) 2 and 3
(c) 1 and 3 (d) 1, 2 and 3

[ESE : 2005]

11.13 Which among the following brings about the Hazardous Waste Management and Handling Rules in India?
(a) Central Pollution Control Board
(b) Ministry of Environment and Forests
(c) Ministry of Urban Development
(d) Ministry of Rural Development

[ESE : 2007]

11.14 The term 'biological magnification' indicates which one of the following?
(a) Likelihood of increasing size of animals during evolution
(b) Magnification pertaining to microscopy
(c) Accumulation of pollutants in soil
(d) Accumulation of pollutants in successive consumers

[ESE : 2007]

11.15 The daily cover of MSW landfills consists of which one of the following?
(a) Compacted soil (b) Geomembrane
(c) Geotextile (d) Geocomposite

[ESE : 2007]

11.16 Which one of the following methods of solid waste management conserves energy most efficiently in the form of gas or oil?
(a) Incineration with heat recovery
(b) Combusting
(c) Fluidized-bed incineration
(d) Pyrolysis

[ESE : 2007]

11.17 A solid waste sample has been segregated and one of the components has been subjected to elemental analysis. The result of analysis in per cent by mass revealed C (40%), H (6.0%) O (44%), N (0.3%). What is the likely waste component?
(a) Food waste (b) Paper and cardboard
(c) Plastic waste (d) Leather waste

[ESE : 2008]

11.18 Biological magnification of pesticides takes place through which of the following?
(a) Population pyramids
(b) Hydrologic cycle
(c) Food chains
(d) Air cycle

[ESE : 2008]

11.19 Which of the following parameters is not included in the routine characterization of solid waste for its physical composition?
(a) Moisture content
(b) Density
(c) Particle size analysis
(d) Energy value [ESE : 2008]

11.20 Bangalore and Indore process of composting are which of the following?
(a) Both anaerobic processes
(b) Both aerobic processes
(c) Anaerobic process and aerobic process, respectively
(d) Aerobic process and anaerobic process, respectively
[ESE : 2008]

11.21 Which one of the following parameters is not included in the routine characterization of solid waste for its physical composition?
(a) Moisture content
(b) Density
(c) Particle size analysis
(d) Energy value [ESE : 2009]

Directions: The following items consists of two statements; one labelled as **'Statement (I)'** and the other as **'Statement (II)'**. You are to examine these two statements carefully and select the answers to these items using the codes given below:

Codes:
(a) Both Statement (I) and Statement (II) are individually true and Statement (II) is the correct explanation of Statement (I)
(b) Both Statement (I) and Statement (II) are individually true but Statement (II) is NOT the correct explanation of Statement (I)
(c) Statement (I) is true but Statement (II) is false
(d) Statement (I) is false but Statement (II) is true

11.22 **Assertion (A):** Municipal Solid Waste is disposed off in the Transport Safe Disposal Facility (TSDF) to convert it into organic compost.
Reason (R): The organic Municipal Solid Waste is converted into compost by worms; and the process is called 'Vermicomposting'.
[ESE : 2012]

11.23 **Assertion (A):** For aerobic composting of refuse, it is mandatory to maintain the C/N ratio between 50 and 60.
Reason (R): C/N ratio is one of the important factors in composting. It is an important constituent as loss of nitrogen is to be prevented and maximum proportion of the nitrogen in compost is to be preserved.
[ESE : 2013]

11.24 Which of the following help to prevent water pollution due to land-disposal of waste?
1. Proper consolidation of waste to reduce pore space and permeability.
2. Disposal over impervious strata.
3. Layer of impervious soil of the top and the sides of the deposited solid waste.
(a) 1 and 2 only (b) 1 and 3 only
(c) 1, 2 and 3 (d) 2 and 3 only
[ESE : 2015]

11.25 Consider the following statements regarding excreta disposal without water carriage system:
1. Pit-Privy is a pit in the ground with the toilet seat located directly over it.
2. Bore-Hole Latrines do not cause nuisance due to files and odour.
3. Aqua-Privy works on the same principle as a septic tank.
4. In the context of a Bore-Hole Latrine, a pit of about 30 cm to 40 cm diameter is dug to a depth of 4 m to 8 m.
Which of the above statements are correct?
(a) 1, 2 and 3 only (b) 2 and 4 only
(c) 1, 3 and 4 only (d) 1, 2, 3 and 4
[ESE : 2015]

Directions: The following items consists of two statements; one labelled as **'Statement (I)'** and the other as **'Statement (II)'**. You are to examine these two statements carefully and select the answers to these items using the codes given below:

Codes:
(a) Both Statement (I) and Statement (II) are individually true and Statement (II) is the correct explanation of Statement (I)
(b) Both Statement (I) and Statement (II) are individually true but Statement (II) is NOT the correct explanation of Statement (I)

(c) Statement (I) is true but Statement (II) is false
(d) Statement (I) is false but Statement (II) is true

11.26 Assertion (A): Composting is basically a treatment method for inorganic waste from a community.

Reason (R): In the incineration method of refuse disposal, the refuse is burnt off and the volume is much reduced.

[ESE : 2015]

11.27 Statement (I) : Aerobic condition in composting of refuse can be confirmed by temperature measurements.

Statement (II): Aerobic reactions are exothermic. [ESE : 2016]

11.28 Consider the following statements regarding pyrolysis:

1. It is an irreversible chemical change brought about by the action of heat in oxygen free atmosphere.
2. Range of temperature is 500°C to 1000°C.
3. Internal heating causes organic matter to decompose physically and chemically rather than burn.
4. It is highly exothermic process.

Which of the above statements are correct?

(a) 1, 2, 3 and 4 (b) 1, 3 and 4
(c) 2 and 4 (d) 1, 2 and 3

[ESE : 2017]

11.29 The moisture content of a certain Municipal Solid Waste with the following composition will be

	Wet, %weight	Dry, %weight
Food waste	10	03
Paper	35	30
Yard waste	20	10
Others	35	20

(a) 100% (b) 63%
(c) 37% (d) 13%

[ESE : 2018]

Directions: The following items consists of two statements; one labelled as **'Statement (I)'** and the other as **'Statement (II)'**. You are to examine these two statements carefully and select the answers to these items using the codes given below:

Codes:

(a) Both Statement (I) and Statement (II) are individually true and Statement (II) is the correct explanation of Statement (I)
(b) Both Statement (I) and Statement (II) are individually true but Statement (II) is NOT the correct explanation of Statement (I)
(c) Statement (I) is true but Statement (II) is false
(d) Statement (I) is false but Statement (II) is true

11.30 Statement (I) : The field capacity of Municipal solid waste is the total moisture that can be retained in a waste sample against gravity.

Statement (II) : The field capacity of Municipal solid waste is of critical importance in determining the volume of leachate in landfills.

[ESE : 2018]

11.31 Statement (I) : Proximate analysis of MSW is carried out to determine moisture content, volatile matter, and fixed carbon.

Statement (II) : Ultimate analysis of MSW is carried out to determine the full range of chemical composition and the energy value.

[ESE : 2018]

11.32 In solid waste management, waste utilization is achieved by

(a) Recover Reclamation and Reproduce
(b) Reuse, Reclamation and Recycling
(c) Recover, Recycling and Reproduce
(d) Reuse, Reproduce and Recycling

[ESE : 2020]

Answers Municipal Solid Waste Management

11.1 (d)	11.2 (a)	11.3 (c)	11.4 (c)	11.5 (d)	11.6 (d)	11.7 (b)	11.8 (b)	11.9 (b)
11.10 (a)	11.11 (b)	11.12 (d)	11.13 (b)	11.14 (d)	11.15 (a)	11.16 (d)	11.17 (a)	11.18 (c)
11.19 (d)	11.20 (c)	11.21 (d)	11.22 (b)	11.23 (d)	11.24 (c)	11.25 (c)	11.26 (d)	11.27 (a)
11.28 (d)	11.29 (c)	11.30 (b)	11.31 (b)	11.32 (b)				

Explanations Municipal Solid Waste Management

11.1 (d)

Indore method is aerobic while Bangalore method is anaerobic for composting.

11.2 (a)

Low C/N ratio leads to loss of carbon, leaving nitrogen in the system which will combines with hydrogen to form ammonia (NH_3). This can kill or inhibit the growth of bacteria especially the methane producers.

11.3 (c)

Pyrolysis is a form of treatment that chemically decomposes organic materials by heat in the absence of oxygen. Pyrolysis typically occurs under pressure and at operating temperatures above 430°C (800°F). In practice, it is not possible to achieve a completely oxygen-free atmosphere. Because some oxygen is present in any pyrolysis system, a small amount of oxidation occurs. If volatile or semi-volatile materials are present in the waste, thermal desorption will also occur.

Organic materials are transformed into gases, small quantities of liquid, and a solid residue containing carbon and ash. The off-gases may also be treated in a secondary thermal oxidation unit. Particulate removal equipment is also required. Several types of pyrolysis units are available, including the rotary kiln, rotary hearth furnace, and fluidized bed furnace. These units are similar to incinerators except that they operate at lower temperatures and with less air supply.

11.4 (c)

Three general categories of solid waste and their sub classification are as follows:

1. Municipal wastes:
   (i) Food wastes
   (ii) Rubbish
   (iii) Ashes and residues
   (iv) Demolition and construction wastes
   (v) Special wastes
   (vi) Treatment-plant wastes
2. Industrial wastes :
   (i) Rubbish, ashes, demolition and construction wastes, special wastes
3. Hazardous wastes :
   (i) Radioactive substances
   (ii) Chemicals (corrosive, reactive, toxic)
   (iii) Biological wastes
   (iv) Flammable wastes
   (v) Explosives

11.5 (d)

Biomagnification is the increasing concentration of a substance, such as a toxic chemical, in the tissues of organisms at successively higher levels in a food chain. This increase can occur as a result of:

- Persistence – where the substance cannot be broken down by environmental processes
- Food chain energetics – where the substance concentration increases progressively as it moves up a food chain
- Low or non-existent rate of internal degradation or excretion of the substance – often due to water-insolubility

Biological magnification often refers to the process whereby certain substances such as pesticides or heavy metals move up the food chain, work their way into rivers or lakes, and are eaten by aquatic organisms such as fish, which in turn are eaten by large birds, animals or humans. The substances become concentrated in tissues or internal organs as they move up the chain. Bioaccumulants are substances that increase in concentration in living organisms as they take in contaminated air, water, or food because the substances are very slowly metabolized or excreted.

11.6 (d)

Night soil, and garbage are organic wastes

Debris is waste from demolition and construction.

Rubbish includes all non-putrescible wastes except ashes.

Thus organic solids = 35 + 40 = 75 t.

11.7 (b)

Gases found in landfills include air, ammonia, carbon dioxide, carbon monoxide, hydrogen, sulfide, methane, nitrogen and oxygen. Carbon dioxide and methane are the principal gases produced from the anaerobic decomposition of the organic solid waste components.

The landfilling operation is biological method of waste treatment, since the waste is stabilized by aerobic as well as anaerobic processes. During initial phase aerobic bacteria and fungi deplete the available oxygen. As a result of aerobic respiration, the temperature in the fill increases.

11.8 (b)

Bangalore method: This method of composting is an anaerobic method and was developed at Bangalore in India in 1939. It is recommended where night soil and refuse are used for preparing the compost. The method overcomes many of the disadvantages of the Indore method, such as the problem of heap protection from adverse weather, nutrient losses from high winds and strong sun, frequent turning requirements, and fly nuisance. However, the time required for the production of finished compost is much longer. The method is suitable for areas with scanty rainfall.

Indore method: This method of composting is an aerobic method through passive aeration and was made at Indore in India in mid 1920s.

11.9 (b)

The solid waste generated through domestic and commercial activities is classified as Municipal Solid Wastes (MSW).

The animal excreta (gobar) and human excreta also does not stand included in the Municipal Solid Waste.

Refuse density is 300 to 600 kg/m^3.

Density of rubbish < garbage < ashes

11.10 (a)

Upon heating in an oxygen free atmosphere during pyrolysis, most organic substances can be split through a combination of thermal cracking & condensation reactions in gaseous liquid and solid fractions.

11.11 (b)

Indore method is aerobic while Bangalore method is anaerobic and is hence more clean than the Indore method.

11.12 (d)

Processing techniques used to recover materials and to prepare wastes for further processing

Processing technique	Function	Representative equipment and/or facilities and applications
Mechanical size & shape alteration	Alteration of the size and shape of solid-waste components	Equipment used to reduce the size of solid waste includes hammermills, shredders, roll crushers, grinders, chippers, jaw crushers, rasp mills, and hydropulpers.
Mechanical component separation	Separation of recoverable materials, usually at a processing facility	Trommels and vibrating screens are used for unprocessed and processed wastes; disk screens for processed wastes; zigzag, vibrating air, rotary air, and air knife classifiers for processed wastes. Jig, pneumatic, sink/float, inertial, inclined or shaking table flotation, and optical sorting are used to separate the light and heavy materials in solid wastes.
Magnetic and electro-mechanical separation	Separation of ferrous & non-ferrous materials from processed solid wastes	Magnetic separation is used for ferrous materials; eddy current separation for aluminum; electrostatic separation for glass in wastes free of ferrous and aluminum scrap; magnetic fluid separation for nonferrous materials from processed wastes.
Drying and dewatering	Removal of moisture from solid wastes	Convection, conduction, and radiation dryers have been used for solid wastes and sludge. Centrifugation and filtration are used to dewater treatment plant sludge.

11.15 (a)

The municipal solid waste (MSW) is simply brought and dumped on the ground, spread in layers of about 0.5 m thickness, and compacted. Another layer of 0.5 m thickness is then placed on top of the previous layer and also compacted. Layering and compacting are repeated until a height of about 1.5 m is reached. At this point and at the end of a working day, a cover of earth of about 0.15 to 0.3 m thickness is compacted on the top and side slopes of the compacted heap, which is called a cell. This cover is called the daily cover.

11.16 (d)

Upon heating in an oxygen free atmosphere, most organic substances can be split through a combination of thermal cracking and condensation reactions into gaseous, liquid and solid fractions. This process is called pyrolysis. In contrast to the combustion process, which is highly exothermic the pyrolytic process is highly endothermic. Hence pyrolysis is also called destructive distillation. The converted products derived from solid wastes through pyrolysis like gas and oil possess a high energy content. Thus, these products can be stored and used for energy production by an additional conversion step.

11.17 (a)

Higher percentage of carbon and oxygen indicate that it is food waste.

11.19 (d)

Energy content, chemical content are parameters used in the chemical composition.

The parameters which are included in the routine characterization of solid waste for its physical composition are:

(i) Particle size analysis
(ii) Moisture content
(iii) Density

11.21 (d)

The parameters which are included in the routine characterization of solid waste for its physical composition are particle size analysis, moisture content and density.

Energy content, chemical content are the parameters used in chemical composition.

11.22 (b)

Vermicompost is the end product of the breakdown of organic matter by earthworms and is generally used in Municipal Solid Waste Management.

11.23 d)

Optimum C/N ratio is 30-50 and it is one of the most important factors affecting composting process. Hence assertion is false.

11.25 (c)

- A Pit Privy is a small structure, separate from a main building, which covers a pit in the ground with a toilet seat located directly over it.
- A Bore-Hole Latrine is a hole of 30 cm to 40 cm diameter drilled into the ground to a depth of 4m to 8m to receive and store the excreta. The disadvantage of this type of latrine is that it causes odour, attracts flies and may lead to groundwater contamination.
- An aqua privy is like a single chamber septic tank, except the toilet is located directly over the tank.

11.26 (d)

The organic material, excluding plastics, rubber and leather are separated from municipal solids wastes and are subjected to bacterial decomposition, the end product remaining after dissimilatory and assimilatory bacterial activity is humus or compost. The entire process involving both separation and bacterial conversion of the organic solid wastes is known as composting.

11.28 (d)

Pyrolysis is a highly endothermic process.

11.29 (c)

Total weight be 100 kg

	Wet wt.	Dry wt.	Moisture
Food water	10 kg	3 kg	7 kg
Paper	35 kg	30 kg	5 kg
Yard waste	20 kg	10 kg	10 kg
Others	35 kg	20 kg	15 kg
			Total = 37 kg

$$M.C. = \frac{\text{weight of water}}{\text{total weight}} \times 100$$

$$= \frac{37}{100} \times 100 = 37\%$$

11.31 (b)

Approximate analysis of municipal solid waste is carried out to determine.

(i) Moisture
(ii) Volatile matter
(iii) Ash
(iv) Fixed carbon

Ultimate analysis of solid waste is used to characterize the chemical composition of organic matter. They are also used to define the proper mix of solid waste materials to achieve suitable C/N ratios for bioconversion process.

11.32 (b)

Reuse, Reclamation and Recycling

12 Air Pollution

12.1 **Assertion (A):** In recent years, there has been a progressive increase in the percentage of solar ultraviolet rays reaching the earth's surface.

Reason (R): In the last few years, there has been a progressive increase in the percentage of carbon dioxide in the earth's atmosphere.

(a) both A and R are true and R is the correct explanation of A
(b) both A and R are true but R is not a correct explanation of A
(c) A is true but R is false
(d) A is false but R is true

[ESE : 2000]

12.2 Which one of the following plume behaviours occurs when atmospheric inversion begins from the ground level and continues?

(a) Looping (b) Fumigation
(c) Coning (d) Fanning

[ESE : 2000]

12.3 Which one of the following pollutants or pairs of pollutants is formed due to photochemical reactions?

(a) CO alone (b) O_3 and PAN
(c) PAN and NH_3 (d) NH_3 and CO

[ESE : 2000]

12.4 In the context of basic concept of an ecological system, the most appropriate definition of ecology is that it is a study of the

(a) inter-relationship between organisms and the environment
(b) relationship of human species with the industry
(c) relationship of human species with natural resources
(d) relationship of human species with air

[ESE : 2000]

12.5 The role of the bed material in a "packed tower" used for removing particulate matter from gaseous emissions is to

(a) act as a filter bed to capture the particulates within the pores
(b) provide a large surface area on which the particulate matter can be collected
(c) reduce the flow of gas
(c) uniformly distribute the spray of water

[ESE : 2000]

12.6 Which of the following air pollutants are responsible for the greenhouse effect?

1. Methane
2. Carbon dioxide
3. Chlorofluorocarbons
4. Nitrogen oxides

Select the correct answer using the codes given below:

(a) 2 and 3
(b) 1 and 2
(c) 1, 3 and 4
(d) 1, 2, 3 and 4 [ESE : 2001]

12.7 Which of the following pairs are correctly matched?

1. Ringelmann chart : To grade density of smoke
2. Pneumoconiosis : Disease caused due to coaldust
3. PAN : Secondary air pollutant

Select the correct answer using the codes given below:

(a) 2 and 3 (b) 1 and 2
(c) 1 and 3 (d) 1, 2 and 3

[ESE : 2001]

Directions: The following items consists of two statements; one labelled as **'Assertion (A)'** and the other as **'Reason (R)'**. You are to examine these two statements carefully and select the answers to these items using the codes given below:

Codes:

(a) both A and R are true and R is the correct explanation of A
(b) both A and R are true but R is not a correct explanation of A
(c) A is true but R is false
(d) A is false but R is true

12.8 **Assertion (A):** The surface of sea water is on the rise.

Reason (R): A thick layer of gases enveloping the earth does not allow heat to pass into space from the earth at a rate which is as much as the rate at which the heat coming from space penetrates it towards the earth.

[ESE : 2002]

12.9 **Assertion (A):** Wet scrubber removes particulates from a gaseous stream.

Reason (R): In the wet scrubber water droplets come into contact with the particulates.

[ESE : 2002]

12.10 The following is a well known formula for estimating the plume rise:

$$\Delta h = \frac{V_s d}{u}\left(1.5 + 0.0096\frac{Q_h}{V_s d}\right)$$

where the letters have their usual meaning.

The estimated plume rise (by the above formula) with a stack gas having heat emission rate 2000 kJ/s, the wind speed 4 m/s, stack gas speed 8 m/s inside a stack diameter of 1 m at the top is

(a) 7.8 m
(b) 8.7 m
(c) 3.15 m
(d) $\frac{8000}{\pi}(1.5 + 0.0024\pi)$

[ESE : 2002]

12.11 In sampling of stack gases for measurement of concentration of Suspended Particulate Matter (SPM), the flue gases are sucked in the instrument at

(a) any rate of flow from mid diameter of chimney
(b) any point of chimney cross-section and at any rate of flow
(c) a constant rate of flow but at four equidistant points along the diameter
(d) controlled positions and controlled velocities along the chimney diameter to get isokinetic conditions

[ESE : 2002]

12.12 Match **List-I** (Air pollutants) with **List-II** (Harmful effects) and select the correct answer using the codes given below the lists:

List-I	**List-II**
A. SPM	1. Blood haemoglobin
B. NO	2. Vegetation
C. CO	3. Respiratory system
D. SO_2	4. Building materials

Codes:

	A	B	C	D
(a)	3	4	1	2
(b)	1	2	3	4
(c)	3	2	1	4
(d)	1	4	3	2

[ESE : 2002]

12.13 Match **List-I** (Air pollutants) with **List-II** (Emitted mainly by) and select the correct answer using the codes given below the lists:

List-I

A. Hydrocarbons
B. Particulates and gases
C. Sulphur dioxide
D. Carbon monoxide

List-II

1. Coal burning
2. Gasoline fuel
3. Tyres
4. Carburettor

Codes:

	A	B	C	D
(a)	3	4	2	1
(b)	4	3	2	1
(c)	3	4	1	2
(d)	4	3	1	2

[ESE : 2002]

12.14 In urban air pollution, the most poisonous gas is supposed to be carbon monoxide. It is hazardous because it

(a) affects our sense of smell
(b) is carcinogenic in nature
(c) combines with haemoglobin
(d) causes blindness

[ESE : 2003]

12.15 The graph shows the relationship of ambient lapse rates to the dry adiabatic lapse rate under different conditions of stability. Match stability situations **A, B, C** and **D** (as given in the graph) with the classes of stability as follows:

1. Super adiabatic
2. Dry adiabatic
3. Sub adiabatic
4. Inversion

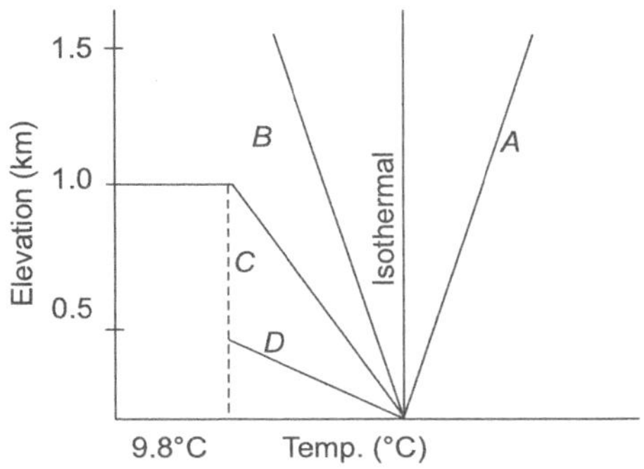

Select the correct answer using the codes given below:

	A	B	C	D
(a)	3	4	1	2
(b)	4	3	2	1
(c)	3	4	2	1
(d)	4	3	1	2

[ESE : 2003]

12.16 Which one of the following procedures is used for sampling of the flue gas in a chimney for SPM?
(a) Isothermal sampling
(b) Isokinetic sampling
(c) Adiabatic condition
(d) Variable rate of sampling

[ESE : 2004]

12.17 Which one of the following units is employed for the removal of particulate matter above 50 μm in size?
(a) Gravity settling chamber
(b) Cyclone
(c) Fabric filter
(d) Electrostatic precipitator

[ESE : 2004]

12.18 Which type of light energy is effectively absorbed by CO_2 in the lower boundary of the troposphere?
(a) X-rays (b) UV-rays
(c) Visible light (d) Infra-red rays

[ESE : 2005]

12.19 Which one of the following pairs is not correctly matched?

Plume Behaviour	**Atmospheric Condition**
(a) Looping	: Stable
(b) Fumigation	: Inversion above and lapse below the stack
(c) Fanning	: Inversion
(d) Trapping	: Inversion above and below the stack with lapse in between

[ESE : 2005]

12.20 Which is the major pollutant present in photochemical smog?
(a) PAN (b) SO_2
(c) HC (d) NO_2

[ESE : 2005]

12.21 Which one of the following is the range of ozonosphere in atmosphere?
(a) Troposphere to Stratosphere
(b) Tropopause to Stratopause
(c) Tropopause to Mesosphere
(d) Stratosphere to Mesosphere

[ESE : 2006]

12.22 Which one of the following pairs is not correctly matched?
(a) Coriolis effect : The effect of earth's rotation on wind direc-tion and velocity
(b) PAN : Found during photo-chemical smog
(c) Cyclone : Employed for parti-culate matter removal
(d) Wind rose : Employed in forecast of pollutant dispersion in ambient air

[ESE : 2006]

12.23 Match **List-I** (Air pollutant) with **List-II** (Impact on human health) and select the correct answer using the codes given below the lists:

List-I
A. Particulates
B. Carbon monoxide
C. Sulphur oxides

D. Photochemical oxidants

List-II

1. Impairs transport of O_2 in bloodstream
2. Irritation of mucous membranes of respiratory tract
3. Causes coughing, shortness of breath, headache, etc.
4. Causes respiratory illness

Codes:

	A	B	C	D
(a)	2	3	4	1
(b)	4	1	2	3
(c)	2	1	4	3
(d)	4	3	2	1

[ESE : 2007]

12.24 Which type of plume may occur during winter nights?

(a) Looping (b) Inversion
(c) Coning (d) Lofting

[ESE : 2007]

12.25 Pneumoconiosis is caused due to inhalation of which one of the following?

(a) Silica (b) NO_x
(c) Lead (d) Cadmium

[ESE : 2008]

12.26 Which one of the following conditions of automobile gives maximum unburned hydrocarbons?

(a) Idling (b) Cruise
(c) Acceleration (d) Deceleration

[ESE : 2008]

12.27 Electrostatic precipitator is most useful for which one of the following industries?

(a) Tannery
(b) Hydroelectric power generation
(c) Thermal power generation
(d) Textile factory

[ESE : 2008]

12.28 Match **List-I** (Air pollutant) with **List-II** (Effect) and select the correct answer using the codes given below the lists:

List-I	List-II
A. CO	1. Acid rain
B. CO_2	2. Acute toxicity
C. SO_2	3. Ozone liberation
D. NO_x	4. Greenhouse effect

Codes:

	A	B	C	D
(a)	4	3	1	2
(b)	4	3	2	1
(c)	2	4	1	3
(d)	3	4	1	1

[ESE : 2008]

12.29 Which one of the following toxic gases has physiological action as asphyxiant?

(a) SO_2 (b) NO_2
(c) Cl_2 (d) CO

[ESE : 2009]

12.30 Assuming annual travel for each vehicle to be 20000 km, what is the quantity of NO_x produced from 50000 vehicles with emission rate of 2.0 g/km/vehicle?

(a) 1800 tonnes (b) 1900 tonnes
(c) 2000 tonnes (d) 2100 tonnes

[ESE : 2009]

12.31 What are the air pollutants responsible for acid rain within and downwind areas of major industrial emissions?

(a) Hydrogen sulfide and oxides of nitrogen
(b) Sulfur dioxide and oxides of nitrogen
(c) Carbon dioxide and hydrogen sulfide
(d) Methane and hydrogen sulfide

[ESE : 2009]

12.32 Consider the following air pollutants:

1. NO_x 2. PAN
3. CO_2 4. CO

Which of the above air pollutants is/are present in an auto exhaust gas?

(a) 1 only (b) 1 and 2
(c) 2 and 3 (d) 1, 3 and 4

[ESE : 2009]

12.33 Assertion (A): Traffic 'Smog' is likely to occur in regions where vehicle mileage is considerable and there is a low incidence of sunlight.

Reason (R): Traffic 'Smog' is caused by the reaction of oxides of nitrogen and some of the hydrocarbons in presence of bright sunlight.

(a) both A and R are true and R is the correct explanation of A

(b) both A and R are true but R is not a correct explanation of A
(c) A is true but R is false
(d) A is false but R is true

[ESE : 2010]

12.34 Which of the following are responsible for the formation of photochemical smog?
1. Light intensity
2. Ratio of hydrocarbons to nitric oxide
3. CO_2
4. Hydrocarbon reactivity
5. SO_2

(a) 1, 2, 3, 4 and 5 (b) 1, 2 and 4 only
(c) 2, 3 and 4 only (d) 2, 3 and 5 only

[ESE : 2011]

12.35 Which of the following factors contribute to formation of photochemical smog?
1. Stable atmospheres
2. NO_x
3. Solar insolation
4. CO

(a) 1, 2, 3 and 4 (b) 2, 3 and 4 only
(c) 1 and 4 only (d) 1, 2 and 3 only

[ESE : 2012]

12.36 Environmental impact assessment includes
(a) Environmental statement
(b) Environmental management plan
(c) Risk and hazard assessment and mitigation
(d) All of the above

[ESE : 2012]

12.37 Statement-I : Air pollutant concentration and time of retention increase due to inversion.
Statement-II : During winter, the heavy cold layer in the atmosphere retains the hot toxic pollutants for a longer period in the atmosphere.
(a) Both Statement (I) and Statement (II) are individually true and Statement (II) is the correct explanation of Statement (I)
(b) Both Statement (I) and Statement (II) are individually true but Statement (II) is NOT the correct explanation of Statement (I)
(c) Statement (I) is true but Statement (II) is false
(d) Statement (I) is false but Statement (II) is true

[ESE : 2012]

12.38 Consider the following statements:
1. Particulates have irregular shapes.
2. Size can be determined by an equivalent aerodynamic diameter by comparing with a perfect sphere.
3. Particulates larger than 10 μ are said to settle relatively quickly since their settling velocity is not less than 10 cm/min.
4. The particles roughly the size of bacteria have aerodynamic diameter of 0.1 μm to 10 μm.

Which of the statements is/are correct?
(a) 1, 2, 3 and 4 (b) 1 and 3 only
(c) 1, 2 and 4 only (d) 2, 3 and 4 only

[ESE : 2012]

12.39 Consider the following statements in regard to an electrostatic precipitator:
1. Power requirement is small compared to other devices.
2. Immune to variable particulate loadings and flow rates.
3. 99+ percent efficiency is obtainable. Very small particles can be collected, wet or dry.
4. Can handle both gases and mists at high volume of flow.

Which of these statements are correct?
(a) 1, 2, 3 and 4 (b) 1, 3 and 4 only
(c) 1, 2 and 3 only (d) 2, 3 and 4 only

[ESE : 2013]

12.40 The major source of 'Carbon monoxide' in the urban atmosphere is due to
(a) Decomposition of organics
(b) Chemical reaction between VOC and NO_x
(c) Incomplete combustion of fuel
(d) Incomplete combustion in the presence of sunlight

[ESE : 2014]

12.41 Consider the following statements relating zones of atmosphere to altitude:
1. Temperature decreases with altitude in troposphere.
2. Speed of sound decreases with altitude in troposphere.
3. Temperature increases with altitude to stratosphere.
4. Ozone is present in stratosphere which protects people from harmful effects of solar radiation.

Which of the above statements are correct?

(a) 1, 2 and 3 only (b) 1, 2 and 4 only
(c) 3 and 4 only (d) 1, 2, 3 and 4

[ESE : 2014]

12.42 The principle involved in collection and sampling of particulate matter in which the particles are drawn through a device by deflecting them from their original paths is called
(a) Filtration
(b) Electrostatic precipitation
(c) Impaction
(d) Gravitational settling

[ESE : 2014]

12.43 **Statement (I):** Electrostatic precipitator's (ESP) can remove particles of sub-micron size.

Statement (II): Due to high voltage supply, ESPs attract any charged particle.
(a) Both Statement (I) and Statement (II) are individually true and Statement (II) is the correct explanation of Statement (I).
(b) Both Statement (I) and Statement (II) are individually true but Statement (II) is not the correct explanation of Statement (I).
(c) Statement (I) is true but Statement (II) is false.
(d) Statement (I) is false but Statement (II) is true.

[ESE : 2014]

12.44 Consider the following statements in respect of electrostatic precipitators:
1. Power requirement is very small compared to other air pollution control devices and so they are cheaper to perform than other devices.
2. Can handle both gases and mists for high volume flow.
3. Very small particles can be collected, either wet or dry.

Which of the above statements are correct?
(a) 1 and 2 only (b) 2 and 3 only
(c) 1 and 3 only (d) 1, 2 and 3

[ESE : 2015]

12.45 Consider the following statements related to ozone:
1. Tropospheric ozone is harmful
2. Stratospheric ozone is beneficial
3. During prevalence of photochemical smog, O_3 is formed

Which of the above statements are correct?
(a) 1, 2 and 3 (b) 1 and 2 only
(c) 1 and 3 only (d) 2 and 3 only

[ESE : 2015]

12.46 Consider the following statements in respect of effect of air pollutants on vegetation:
1. Necrosis refers to killing to tissue.
2. Chlorosis refers to loss or reduction of green plant pigment.
3. Leaf abscission refers to the dropping of leaves.
4. Leaf epinasty refers to a downward curvature of a leaf due to a higher rate of growth on the upper surface.

Which of the above statements are correct?
(a) 1, 2 and 3 only (b) 1, 2, 3 and 4
(c) 2, 3 and 4 only (d) 1, 2 and 4 only

[ESE : 2015]

12.47 Consider the following statements regarding ecology:
1. Climax ecosystem is a stage in the evolution of an ecosystem, at which all the species are in dynamic equilibrium among themselves as also with the environment.
2. Ecological niche means all the physical, chemical and biological factors that a species needs in order to live and reproduce exist.
3. Edge effect refers to the presence of rich and unique biological diversity found in an ecotone.

Which of the above statements are correct?
(a) 1, 2 and 3 (b) 1 and 2 only
(c) 1 and 3 only (d) 2 and 3 only

[ESE : 2015]

12.48 **Assertion (A):** Gases are normally formless fluids and can be changed to liquid or solid states by change of temperature and pressure.
Reason (R): Smog refers to the occurrence of a heavy, cloudy, hazy floating layer in the atmosphere formed by a mixture of smoke, dust, fog and mist.
(a) Both A and R are true and R is the correct explanation of A

(b) Both A and R are true but R is not a correct explanation of A
(c) A is true but R is false
(d) A is false but R is true [ESE : 2015]

12.49 Consider the following properties of fluorine:
1. It is a member of the halogen family.
2. It is a greenish yellow diatomic gas.
3. Chlorine, iodine, bromine and helium are members of the halogen family.
4. Even fireproof asbestos burns in the ambience of fluorine.

Which of the above are correct?
(a) 1, 2 and 3 only (b) 3 and 4 only
(c) 1, 2 and 4 only (d) 1, 2, 3 and 4
[ESE : 2016]

12.50 Consider the relevance of the following features for causing photochemical smog
1. Air stagnation.
2. Abundant sunlight.
3. High concentration of NO_x in atmosphere.
4. High concentration of SO_2 in atmosphere.

Which of the above features are correct?
(a) 1, 2 and 4 only (b) 3 and 4 only
(c) 1, 2 and 3 only (d) 1, 2, 3 and 4
[ESE : 2016]

12.51 Consider the following statements related to ecology
1. All the physical, chemical and biological factors that a species needs in order to live and reproduce is called ecological niche.
2. The boundary zone between two ecosystems is known as ecotone.
3. The forests in the Arctics are known as tundra.
4. A biome usually has a distinct climate and life forms adapted to that climate. Biome is more extensive and complex than an ecosystem.

Which of the above statements are correct?
(a) 1, 2 and 3 only (b) 1, 2 and 4 only
(c) 3 and 4 only (d) 1, 2, 3 and 4
[ESE : 2016]

12.52 Consider the following laws of ecology suggested by Barry Commoner:
1. Everything is connected to every thing else.
2. Everything must go somewhere.
3. Nature knows best.
4. There is no such thing as a free lunch.

Which of the above statements are correct?
(a) 1, 2 and 3 only (b) 1, 2 and 4 only
(c) 3 and 4 only (d) 1, 2, 3 and 4
[ESE : 2016]

12.53 Consider the following statements regarding air pollution:
1. The pollutant caused by incomplete combustion or organic matter is carbon monoxide.
2. Depletion of ozone in outer atmosphere may trigger skin cancer.
3. Acid rains are caused by SO_2 and NO_x.
4. The permissible standard for SO_2 in air for residential areas in India is 80 $\mu g/m^3$.

Which of the above statements are correct?
(a) 1, 2, 3 and 4 (b) 1, 3 and 4
(c) 2 and 4 (d) 1, 2 and 3
[ESE : 2017]

Directions: The following items consists of two statements; one labelled as **'Statement (I)'** and the other as **'Statement (II)'**. You are to examine these two statements carefully and select the answers to these items using the codes given below:

Codes:
(a) Both Statement (I) and Statement (II) are individually true and Statement (II) is the correct explanation of Statement (I)
(b) Both Statement (I) and Statement (II) are individually true but Statement (II) is NOT the correct explanation of Statement (I)
(c) Statement (I) is true but Statement (II) is false
(d) Statement (I) is false but Statement (II) is true

12.54 **Statement (I):** Tiny quantities of over 30 rare gases would warm the atmosphere over the Earth more rapidly than CO_2.
Statement (II): A single molecule of some CFCs, methane and nitrous oxide absorbs as much heat as 15,000 molecules, 25 molecules and 230 molecules of CO_2, respectively.
[ESE : 2017]

12.55 **Statement (I):** The impact of Green House Gas emission on the environment may comprise accelerated increase in global warming as well as a significant rise in mean sea levels.

Statement (II): Green House Gas emission is responsible for decreased land masses, increased population densities and food shortages.

[ESE : 2018]

12.56 Consider the following statements in respect of the troposphere:

1. The gaseous content constantly churns by turbulence and mixing.
2. Its behaviour makes the weather.
3. The ultimate energy source for producing any weather change is the sun.
4. The height of the troposphere is nearly 11 km at the equatorial belt and is 5 km at the poles.

Which of these are true of the troposphere?

(a) 1, 2 and 3 only (b) 1, 2 and 4 only
(c) 1, 3 and 4 only (d) 2, 3 and 4 only

[ESE : 2018]

12.57 Fanning type of plume behaviour takes place when

(a) Super-adiabatic lapse rate prevails with light to moderate wind speed
(b) Extreme inversion conditions exist in the presence of light wind
(c) There exists a strong super-adiabatic lapse rate above a surface of inversion
(d) Plume is caught between two inversion layers

[ESE : 2019]

12.58 A thermal power plant burns coal at the rate of 8 t/h. The coal has sulphur content of 4.5%. The rate of emission of SO_2 Will be

(a) 180 g/s (b) 200 g/s
(c) 220 g/s (d) 240 g/s

[ESE : 2019]

12.59 Electrostatic precipitators are used for removal of

1. Gaseous contaminants
2. Liquid contaminants
3. Particulate contaminants

(a) 1 only (b) 2 only
(c) 3 only (d) 1, 2 and 3

[ESE : 2020]

12.60 Which one of the following is a method of wet mechanical analysis of a fine-grained material?

(a) Partial sedimentation
(b) Sedimentation into dirty water
(c) Observation of partial sedimented soil
(d) Elutriation

[ESE : 2022]

Answers Air Pollution

12.1 (b)	12.2 (d)	12.3 (b)	12.4 (a)	12.5 (b)	12.6 (d)	12.7 (d)	12.8 (a)	12.9 (a)
12.10 (a)	12.11 (d)	12.12 (c)	12.13 (d)	12.14 (c)	12.15 (b)	12.16 (b)	12.17 (a)	12.18 (d)
12.19 (a)	12.20 (a)	12.21 (d)	12.22 (d)	12.23 (b)	12.24 (b)	12.25 (a)	12.26 (a)	12.27 (c)
12.28 (c)	12.29 (b)	12.30 (c)	12.31 (b)	12.32 (d)	12.33 (d)	12.34 (b)	12.35 (d)	12.36 (d)
12.37 (a)	12.38 (c)	12.39 (b)	12.40 (c)	12.41 (d)	12.42 (b)	12.43 (a)	12.44 (b)	12.45 (a)
12.46 (b)	12.47 (a)	12.48 (b)	12.49 (c)	12.50 (c)	12.51 (d)	12.52 (d)	12.53 (a)	12.54 (a)
12.55 (b)	12.56 (a)	12.57 (b)	12.58 (b)	12.59 (c)	12.60 (d)			

Explanations Air Pollution

12.1 (b)

Depletion of ozone layer (in stratosphere and mesosphere) has led to increased exposure of earth's surface to ultraviolet radiation. The ozone depleting substances are chlorofluoro carbons (CFCs).

Increased percentage of carbon dioxide has led to global warming (greenhouse effect).

12.2 (d)

When an inversion layer occurs at a short distance above the top of the stack and super adiabatic conditions prevail below the stack, then plume is said to be **fumigating (fig)**. In such a case, the pollutants cannot escape above the top of the stack because of inversion layer, and they will be brought down near the ground due to turbulence in the region above the ground and below the inversion , caused by strong lapse rate. This represents quite a bad case of atmospheric conditions for dispersion.

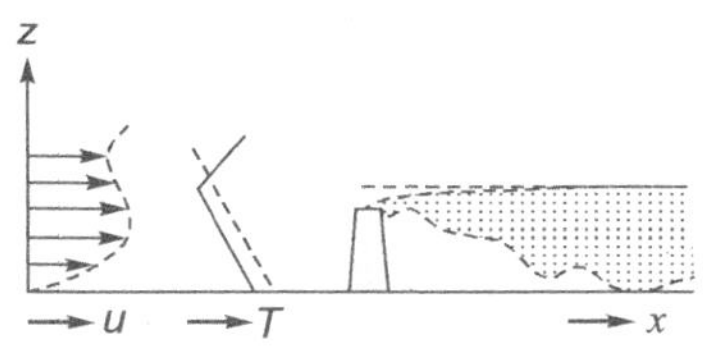

SYMBOLS:

- - - - - - - - - - - Adiabatic lapse rate

———————— Environmental lapse rate

12.3 (b)

Photochemical reactions form O_3, PAN, formaldehyde, etc.

12.4 (a)

The scientific study of interaction among organisms and their environment is called ecology.

12.5 (b)

In the packed tower, packing material (bed material) is used to increase the contact time between vapour and liquid. The material chosen for packing has a large surface to volume ratio and a large void ratio that offers minimum resistance to gas flow.

12.6 (d)

Carbon dioxide is considered to be major greenhouse gas, as it is responsible for about 60% of the total greenhouse effect caused by all the greenhouse gases. Other green-house gases are CH_4, NO_x and chlorofluoro carbons.

12.7 (d)

The **Ringelmann scale** is a scale for measuring the apparent density of Smoke. It has 5 levels of density inferred from a grid of black lines on a white surface which, if viewed from a distance, merge into known shades of grey.

Pneumoconiosis is an occupational lung disease and a restrictive lung disease caused by the inhalation of dust, often in mines and from agriculture.

Peroxyacetyl nitrate (**PAN**) is a component of photochemical smog, which is a mixture of air pollutants that includes both gases and particulates, some of which react with sunlight.

12.8 (a)

Entrapping of infrared radiation emitted from earth's surface by greenhouse gases results in increase of temperature. This causes melting of polar ice caps and high mountains. The melted ice flows to oceans and the surface of sea water is on rise due to this.

12.9 (a)

Wet scrubber removes gaseous pollutant and particulate matter. The particulate matter by incorporating them into liquid droplets directly on contact efficiency for well designed wet collectors of all types is a function of the energy consumed in the air-to-water contact process.

12.10 (a)

Wind Speed $u = 4$ m/s

Stack gas speed $V_s = 8$ m/s

Diameter $d = 1$ m

Heat rise $Q_h = 2000$ kJ/s

We know that effective height of stack,

$$\Delta h = \frac{V_s \times d}{u}\left[1.5 + 0.0096 \times \frac{Q_h}{V_s \times d}\right]$$

$$\Delta h = \frac{8 \times 1}{4}\left[1.5 + 0.0096 \times \frac{2000}{8 \times 1}\right] = 7.8 \text{ m}$$

12.11 (d)

To obtain representative samples, selection of sampling site and the number of sampling point are the most important factors. The gas stream in the stack is normally under turbulent flow condition and any flow disturbance would result in non-uniform and unstable gas flow profiles and non-uniform particle concentration patterns. The other important factor in stack sampling is high temperature and high velocities of the gases encountered inside the stack. Therefore isokinetic technique is adopted for particulate laden gas stream. In this technique a sample is drawn into the probe such that the condition at the tip of the probe is same as those in the free gas stream at the sampling point. It can be carried out in two ways :

1. By computing point velocity of free stream by Pitot tube and adjusting sampling velocity.
2. By using a hull type of sampling probe, in which the static pressures measured between the inner and outer probe walls is balanced.

12.12 (c)

- Carbon monoxide affect human aerobic metabolism by forming carboxyhaemoglobin (COHb).
- Oxides of nitrogen (NO_x) includes – nitric oxide (NO), nitrogen dioxide (NO_2), nitrous oxide (N_2O), nitrogen sesquioxide (N_2O_3), Nitrogen tetroxide (N_2O_4) and nitrogen pentoxide (N_2O_5). NO and NO_2 are of primary concern as air pollutants. NO_2 plays major role in the production of secondary air pollutant ozone (O_3).
- SO_2 is responsible for acid rains as it combines with water vapour to form secondary pollutants like H_2SO_4 which cause acidity.
- CO_2 is major contributor to green house effect (approximately 57%).

12.13 (d)

Aromatic hydrocarbons are biochemically and biologically active, and some are potentially carcinogenic. All aromatics are derived from or related to benzene. Transportation, including incomplete combustion from car engines, along with evaporative emissions from fuel tanks, crankcases and carburettors contributed the largest percentage of hydrocarbons.

12.14 (c)

Carbon monoxide (CO) combines with haemoglobin (Hb) of blood to give carboxy-haemoglobin (COHb), thus reducing the capability of the blood to carry oxygen. Since the affinity of haemoglobin for carbon monoxide is more than 200 times as great as its affinity for oxygen, CO can seriously impair the transport of O_2, even when present at low concentrations.

12.15 (b)

$$\text{Lapse rate} = \frac{\text{Temperature difference}}{\text{Elevation difference}} = \frac{\Delta T}{\Delta h}$$

Curve	ΔT	Δh	$\Delta T/\Delta h$
A	– ive	—	– ive
B	9.8°C	> 1.5 km	< 6.53°C/km
C	9.8°C	1 km	9.8°C/km
D	9.8°C	0.5 km	17.6°C/km

Dry adiabatic lapse rate is 9.8°C/km and wet adiabatic lapse rate is 6°C/km. The actual adiabatic lapse rate is between 6°C/km to 9.8°C/km.

If ambient lapse rate is more than actual adiabatic lapse rate then it is super-adiabatic. If ambient lapse rate is 9.8°C/km then dry adiabatic condition prevail. If ambient lapse rate is less than actual adiabatic lapse rate then it is sub-adiabatic. For negative rate the condition is said to be inversion.

Therefore

A represents	Inversion	(4)
B represents	Sub-adiabatic	(3)
C represents	Dry adiabatic	(2)
D represents	Super adiabatic	(1)

12.16 (b)

Isokinetic conditions exist when the velocity in the stack 'V_s' equals the velocity at the top of the probe nozzle 'V_n' at the sample point. Sampling in these conditions provides a uniform, unbiased sample of the pollutants being emitted by the source.

12.17 (a)

	Name of mechanical device	Minimum particle size that can be removed	Efficiency
(i)	Gravity settling chamber	> 50 μm	< 50%
(ii)	Centrifugal collectors including cyclone collectors	> 5-25 μm	50-90%
(iii)	Electrostatic precipitator	> 1 μm	95-99%
(iv)	Fabric filters	< 1μm	> 99%

12.18 (d)

CO_2 is major greenhouse gas and absorbs long wave radiations like infrared radiation.

12.19 (a)

Looping Plume:

- High degree of convective turbulence
- Superadiabatic lapse rate – strong instabilities.
- Associated with clear daytime conditions accompanied by strong solar heating and light winds.
- High probability of high concentrations sporadically at ground level close to stack.
- Occurs in unstable atmospheric conditions.

12.20 (a)

Photochemically, the NO_x and hydrocarbons are two groups of chemical compounds which are the necessary ingredients for the production of photochemical smog.

$$HC + NO_x \xrightarrow{\text{sunlight}} Smog$$

The end product of photochemical reactions is photochemical smog consisting of air contaminants such as O_3, PAN, aldehydes, ketones, alkyl nitrates and carbon monoxide.

12.21 (d)

From the earth surface atmosphere to the outer periphery of the planet, there are several layers of gaseous clouds which are - Troposphere, Stratosphere, Mesosphere and Thermosphere with increasing height from the surface.

Height (km)	Layer	Sub-layer	Water Vapour	Temperature	Gases
40000	Thermo-sphere		None	Slowly increases with height	Rarified
500		Iono-sphere		Rapidly increases with height	Atomic H, O, N, Ne
80	Meso-sphere		None	Decreases with height	
50	Strato-sphere		Little		Ozone layer
			Moist with vapour	Increases with height	Diatomic O_2
12	Tropo-sphere		Cloud	Decreases with height	N_2, O_2, CO_2, O_2, Kr, He, Ne, Ar

Earth Surface

12.22 (d)

Wind rose diagram represents direction, frequency and velocity of the winds in a particular location. The direction of wind is characteristically from high pressure to low, but the Coriolis force tends to deflect air currents out of these expected patterns. Wind speed is measured by an anemometer, an instrument typically consisting of three or four hemispherical cups arranged on a vertical axis.

A pollution rose is used for plotting the data necessary to determine the source direction of specific air contaminants. The velocity of wind determines the travel time of a particulate to a receptor and also the dispersion rate of air contaminants.

Cyclone is a low pressure system in the northern hemisphere. Horizontal air movement is anticlockwise and the vertical movement is upward. Low pressure systems are usually associated with cloudy skies, gusty winds and atmospheric instability, and the formation of fronts. Under such unstable conditions dispersion of pollutants is likely, and air pollution problems are minimal

Cyclone collectors are centrifugal collectors used for removal of 5–25 μm size particulate matter.

12.23 (b)

Refer Solution 10.1.

- Carbon monoxide affect human aerobic metabolism by forming carboxyhaemoglobin (COHb).
- Oxides of nitrogen (NO_x) includes – nitric oxide (NO), nitrogen dioxide (NO_2), nitrous oxide (N_2O), nitrogen sesquioxide (N_2O_3), Nitrogen tetroxide (N_2O_4) and nitrogen pentoxide (N_2O_5). NO and NO_2 are of primary concern as air pollutants. NO_2 plays major role in the production of secondary air pollutant ozone (O_3).
- SO_2 is responsible for acid rains as it combines with water vapour to form secondary pollutants like H_2SO_4 which cause acidity.
- CO_2 is major contributor to green house effect (approximately 57%).

12.24 (b)

During extreme winter nights, under extreme inversion conditions, the emission spreads horizontally due to extremely stable environment. Such a plume is called fanning plume. The negative environmental lapse rate from the ground causes these extreme inversion conditions.

12.25 (a)

Pneumoconiosis is an occupational lung disease and a restrictive lung disease caused by the inhalation of silica and coal dust, often in mines and from agriculture.

12.27 (c)

An **electrostatic precipitator (ESP)** is a filtration device that removes fine particles, like dust and smoke, from a flowing gas using the force of an induced electrostatic charge minimally impeding the flow of gases through the unit.

12.29 (b)

No at low level is relatively harmless but at high concentrations it may cause asphyxiation and respiratory discomfort.

12.30 (c)

NO_x emitted by 50000 vehicles

$= 2.0 \times 50000$

$= 100000$ g/km

NO_x produced by 50000 vehicles for a distance of 20000 km

$= 100000 \times 20000 \times 10^{-6}$ tonnes

$= 2000$ tonnes

12.31 (b)

Acid rain is caused by a chemical reaction that begins when compounds like sulfur dioxide and nitrogen oxides are released into the air. These substances can rise very high into the atmosphere, where they mix and react with water, oxygen, and other chemicals to form more acidic pollutants, known as acid rain.

12.32 (d)

PAN (peroxyacetyl nitrate) is a secondary pollutant which is formed in the atmosphere by a photochemical reaction or by hydrolysis or oxidation. Alkenes like ethylene in the presence of sunlight react which nitrogen dioxide at high concentrations to form PAN and Ozone (O_3). This ethylene is produced in automobile exhausts and not PAN which is a secondary pollutant.

12.33 (d)

Smog is formed when NO_x and hydrocarbons react photochemically in presence of intense sunlight. The end products of photochemical reactions is photochemical smog consisting of air contaminants such as O_3, PAN, Aldehydes, Ketones etc.

12.34 (b)

Photochemically the NO_x and hydrocarbons are two groups of chemical compounds which are necessary for production of photochemical smog. The end products of photochemical reaction are photochemical smog consisting of air contaminants such as O_3, PAN, Aldehydes, Ketones etc.

12.35 (d)

Photochemically, NO_x and hydrocarbons are two groups of chemical compounds which are the necessary ingredients for the production of photochemical smog.

$$\text{Hydrocarbons} + NO_x \xrightarrow{\text{sunlight}} \text{Smog.}$$

12.36 (d)

Environmental Impact Assessment (EIA) is the process by which the anticipated effects on the environment of a proposed development or project are measured. If the likely effects are unacceptable, design measures or other relevant mitigation measures can be taken to reduce or avoid those effects.

12.37 (a)

The most common form of surface-based inversion is the *radiational inversion*. It occurs whereever, the surface of the earth can become cooler during the night and winters by the thermal radiation of energy. As a result of a decrease in temperature of the ground, the lower atmosphere in contact with it loses sensible heat through conduction, conversion, and more importantly, radiation. Consequently a temperature inversion, is setup between the cool low-level air and the warmer air above. This type of inversion is most pronounced during the late night and early morning hours and is sometimes called a *nocturnal inversion*. The effects of stable low-level air are evident early in the day when the smoke from chimney's and small fires is confined close to the ground.

12.38 (c)

Particulates have irregular shape and size can be determined by an equivalent aerodynamic diameter by comparing with a perfect sphere.

Particulate matter is defined only for sizes less than 2.5 μ (PN 2.5) and between 2.5 m and 10 μ (PM 10). There is no PM above 10 μ. So, statement 3 goes wrong.

The diameter of these particles very between 0.1 μm to 10 μm.

12.39 (b)

Electrostatic precipitators are sensitive to variable dust loading and flow rates.

Hence, statement 2 is wrong.

Its other characteristics are:

1. Particles may be collected wet or dry.
2. 99% and more efficiency can be obtained.
3. Small particles and in large volume can be removed.
4. Maintenance is nominal.
5. They can be operated at high temperatures upto 300-450°C.
6. Power requirement is high

12.40 (c)

Chief source of Carbon monoxide in urban atmosphere is incomplete combustion of fuel, i.e. especially due to automobile exhaust.

12.41 (d)

Sound wave propagates through vibrations. The speed of sound in directly related to the frequency at which there molecular collection occur and that frequency is directly related to density of how there molecules are packed. The density of atmosphere decrease with altitude and hence speed of sound decreases.

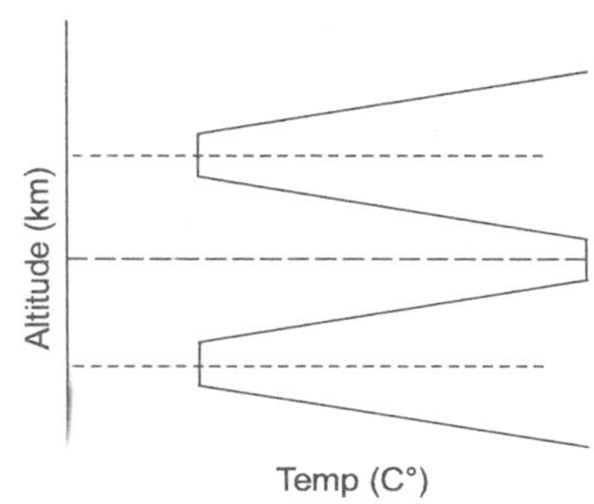

The ozone layer acts as a filler for UV-rays thereby protects people from harmful effects of solar radiation.

12.42 (b)

In electrostatic precipitators , the emitted gas is passed through a highly ionised atmosphere i.e. in high voltage field, and in that zone particulates, get electrically charged with the result that they got separated out from the gaseous stream with the help of electrostatic forces.

12.43 (a)

Electrostatic Precipitators (ESP) can remove particles of very small size (>1 μm). In electrostatic precipitators, the flue gas is made to pass through a highly ionised zone, where the particals get electrically charged and are separated out from the gas, with help of electrostatic forces in the powerful electric field.

They are widely used in thermal power plants, pulp and papers industries, mining and steel plants.

12.44 (b)

Due to high voltage, power requirement is very high in electrostatic precipitator as compared to other devices viz-a-viz cyclone collector, wet scrubber, adsorbers, etc.

12.45 (a)

Ground level ozone (i.e. tropospheric ozone), though less concentrated than ozone aloft, is more of a problem because of its health effects and is also a green house gas.

Ozone is a powerful oxidizing agent readily reacting with other chemical components to make many possibly toxic oxides and hence can cause severe diseases like asthma, chronic bronchitis, etc. Whereas, stratospheric ozone stops UV radiations from reaching the earth.

Ozone is the largest constituent of photochemical smog along with PAN.

12.46 (b)

Necrosis is the name given to unprogrammed death of cells and living tissue.

Chlorosis is a condition in which leaves produce insufficient chlorophyll. As chlorophyll is responsible for the green colour of leaves, chlorotic leaves becomes pale, yellow or yellow-white.

12.49 (c)

- Fluorine is a member of the halogen family.
- It is pale yellow diatomic gas.
- Chlorine, iodine and bromine are member of halogen family but helium is not, it is a noble gas.
- Fluorine in combination with chlorine forms chlorine trifluoride which can burn highly fire-retardant material line sand, asbestos etc.

12.50 (c)

Photochemical smog is formed in the presence of high concentration of hydrocarbons, nitrogen oxides, air stagnation as not to cause disperses of air and abundant sunlight.

12.51 (d)

A **Biome** is a formation of plants and animals that have common characteristics due to similar climates and can be found over a range of continents. Spanning continents, biomes are distinct biological communities that have formed in response to a shared physical climate. Biomes are distinct from habitats, because any biome can comprise a variety of habitats.

An **Ecotone** is a transition area between two biomes. It is where two communities meet and integrate. It may be narrow or wide, and it may be local (the zone between a field and forest) or regional (the transition between forest and grassland ecosystems). An ecotone may appear on the ground as a gradual blending of the two communities across a broad area, or it may manifest itself as a sharp boundary line.

An **Ecological niche** is the role and position a species has in its environment; how it meets its needs for food and shelter, how it survives, and how it reproduces. A species' niche includes all of its interactions with the biotic and abiotic factors of its environment.

12.52 (d)

Barry commoners four laws of ecology.

- Everything is connected to everything else.
- Everything must go somewhere.
- Nature knows best.
- There is no such thing as a free lunch.

12.55 (b)

GHGs absorb a large amount of radiation from the sun thereby increasing the surface temperature. It melts ice increasing the sea level reducing land area. A year of not enough or too much rainfall can affect local crop yield and thus food shortage can take place due to climate change.

12.56 (a)

Atmospheric turbulence in troposphere is a chaotic structure generated by the irregular air movements in which the wind randomly varies in speed and direction. Turbulence is important because it churns and mixes the atmosphere and causes water vapour, smoke, other substances as well as energy to become distributed at al elevations. Its behavior makes the weather (locally).

The changes (chemical) takes place due to energy predominantly provided by the sun.

(Ex. Photochemical smog)

4th statement is incorrect. The average height of troposphere is 10 km and at equator it is about 17 km and at the poles, it is about 9 km.

12.57 (b)

Fanning plume is formed under extreme inversion condition when pollutants are heavier than surrounding air, hence are not able to rise vertically up. It is formed during light turbulence and extremely stable atmosphere below the stack.

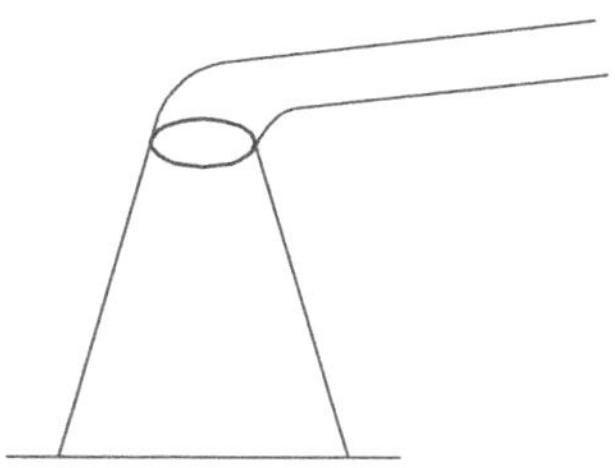

12.58 (b)

Coal burnt per sec

$$= \frac{8 \times 10^6}{60 \times 60} = 2222.22 \text{ gm/sec}$$

$$\text{Sulphur content} = \frac{4.5 \times 2222.22}{100} = 100 \text{ gm/sec}$$

Assuming entire sulphur is converted to SO_2

$$S + O_2 \rightarrow SO_2$$

1 mole of SO_2 is formed from 1 mole of S

64 gm of SO_2 is formed from 32 gm of S

$$SO_2 \text{ emission rate} = \frac{64}{32} \times 100 = 200 \text{ gm/sec}$$

■■■■

13 Noise Pollution

13.1 Which of the following pairs are correctly matched?

1. Reverberation time : Time required to reduce noise by 60 dB
2. NIPTS : Responsible for permanent hearing loss
3. Sound foci : Formed when sound waves are reflected from convex surface
4. TTS : Responsible for temporary hearing loss

Select the correct answer using the codes given below:

(a) 2, 3 and 4 (b) 1, 3 and 4
(c) 1, 2 and 4 (d) 1, 2 and 3

[ESE : 2001]

13.2 Acoustics of an auditorium is considered to be excellent when its reverberation time is between

(a) 0.50 and 1.50 s (b) 1.50 and 2.00 s
(c) 2.00 and 3.00 s (d) 3.00 and 5.00 s

[ESE : 2004]

13.3 Which one of the following is the correct sound intensity expression with usual notations?

(a) $dB = 10\log_{10}(I/I_0)^2$
(b) $dB = 10\log_{10}(I/I_0)$
(c) $dB = 10\log_{10}(I - I_0)^2$
(d) $dB = 10\log_{10}(I - I_0)$

[ESE : 2005]

13.4 What type of noise can be abated by providing lining on walls and ceiling with sound absorbing materials?

(a) Source noise
(b) Reflection noise
(c) Structural noise
(d) Direct air-borne noise

[ESE : 2005]

13.5 A machine in a steel plate fabricating industry is found to be producing a sound level of 50 dB. In the expansion plans one more such machine needs to be added. What will be the combined noise level?

(a) 80 – 100 dB (b) 101 – 150 dB
(c) 51 – 70 dB (d) 40 – 50 dB

[ESE : 2007]

13.6 What will be the resultant decibel level when two sources make noise of equal decibels?

(a) Decibel level will the same
(b) Decibel level will increase by 3 decibels
(c) Decibel level will decrease by 3 decibels
(d) Decibel level will be equal to the sum of decibels of the two sources

[ESE : 2008]

13.7 Noise pollution in a road-side building can be reduced by

(a) providing a ditch around the building and filling it with water
(b) providing a steel mesh around the building
(c) providing a thick bush around the building
(d) planting tall trees around the building and fencing them with barbed wires

[ESE : 2010]

13.8 For noise measurement, formula for sound pressure level (SPL) is $20\log\frac{P}{P_{ref}}$. What will be the resultant noise in dB if P is 0.0002 μ bar ?

(a) 0 (b) 60
(c) 90 (d) 100

[ESE : 2012]

13.9 Consider the following statements:

1. Noise pollution can be reduced using double-glass window panes.
2. Glass absorbs the noise.
3. The air trapped in the double-glass system acts as an insulator and reduces the noise.
4. The noise totally reflects back due to the two layers of glass.

Which of these statements are correct?

(a) 1, 2, 3 and 4 (b) 1, 2 and 3 only
(c) 1 and 3 only (d) 2 and 4 only

[ESE : 2013]

13.10 Consider the following statements:

1. Sound is a form of mechanical energy from a vibrating substance, transmitted by a cyclic series of compressions and rarefactions of the molecules of the material through which it passes.
2. In a pure tone, the wave pattern of the alternating positive and negative sound pressure is of ideal sinusoidal form with fixed wave length, frequency and amplitude.
3. The speed of transmission of sound is a function of the transmitting medium and its temperature.
4. The audible range of 200 to 25000 Hz is considered normal for young adults.

Which of these statements are correct in respect of sound transmission?

(a) 1, 2, 3 and 4 (b) 1, 2 and 4 only
(c) 1, 2 and 3 only (d) 2, 3 and 4 only

[ESE : 2013]

13.11 Consider the following statements related to noise:

1. The range of sound power and sound pressures produced is from 0.0002 μ to 10000 μ bars.
2. Human ears do not respond linearly to increase in sound pressure.
3. Regular exposure to moderate noise makes the human ear more resistant to occasional exposures of high-intensity noise.

Which of the above statements are correct?

(a) 1 and 2 only (b) 1 and 3 only
(c) 2 and 3 only (d) 1, 2 and 3

[ESE : 2015]

13.12 The noise value of sound waves depends upon:

1 The frequency of sound waves.
2. The intensity of sound waves.
3. The time of exposure of sound waves.

(a) 1 and 2 only (b) 1 and 3 only
(c) 2 and 3 only (d) 1, 2 and 3

[ESE : 2020]

13.13 The frequency range for hearing the sound by a human ear is in the range of

(a) 20 Hz - 200 kHz (b) 10 Hz - 20 kHz
(c) 20 Hz - 20 kHz (d) 10 Hz - 20 Hz

[ESE : 2020]

13.14 Physiological response accompanying response and other noise exposures include:

1. A vascular response characteristic by peripheral vasoconstriction, changes in heart beat rate and blood pressure.
2. Various glandular charges such as increased output of adrenaline evidenced by chemical changes in blood.
3. Slow, deep breathing.

(a) 1 and 2 only (b) 1 and 3 only
(c) 2 and 3 only (d) 1, 2 and 3

[ESE : 2020]

■■■■

Answers Noise Pollution

13.1 (c) 13.2 (a) 13.3 (b) 13.4 (b) 13.5 (c) 13.6 (b) 13.7 (c) 13.8 (a) 13.9 (c)

13.10 (a) 13.11 (a) 13.12 (d) 13.13 (c) 13.14 (d)

Explanations Noise Pollution

13.1 (c)

Reverberation time is the time required to reduce the intensity to one-millionth of its initial value.

$$\text{Sound level (dB)} = 10 \log \left(\frac{I_1}{I_2}\right)$$

$$= 10 \log (10^6) = 60 \text{ dB}$$

TTS (Temporary Threshold Shift) is temporary impairment of hearing acuity as indicated by change in the threshold of audibility PTS (Permanent Threshold Shift) is related to permanent hearing loss. The sound foci are formed when sound waves are reflected from concave surface.

13.2 (a)

Reverberation time is the time required to reduce the intensity to one millionth of its initial value. When this time is between 0.5 to 1.5 seconds, acoustics are considered to be excellent.

13.3 (b)

In terms of sound intensity level

$$\text{dB} = 10\log_{10}\left(\frac{I}{I_0}\right)$$

where I_0 = Reference sound intensity = 20 μPa

13.4 (b)

Linings and sound absorbing materials are most commonly used in theatres, concert arena and movie halls because they abate reflection noises.

13.5 (c)

The sound pressure level in dB is given by

$$L_p = 20\log_{10}\left(\frac{p_{rms}}{20\,\mu Pa}\right)$$

Sound level of 1st machine = 50 dB

$$\therefore \quad 50 = 20\log_{10}\left(\frac{p_{rms}}{20\,\mu Pa}\right)$$

$$\Rightarrow \quad \frac{p_{rms}}{20\,\mu Pa} = \text{Antilog}\left(\frac{50}{20}\right)$$

$$\Rightarrow \quad p_{rms} = 6324.55\,\mu\text{Pa}$$

Since the sound level of 2nd machine is also 50 dB, therefore p_{rms} of 2nd machine will also be 6324.55 μPa.

∴ Resultant sound level in rms

$$= \sqrt{(6324.55)^2 + (6324.55)^2}$$

$$= 8944.26\,\mu\text{Pa}$$

$$\therefore \quad L_p = 20\log_{10}\left(\frac{8944.26}{20}\right) = 53.01 \text{ dB}$$

13.6 (b)

Adding of equal loud incoherent sound sources

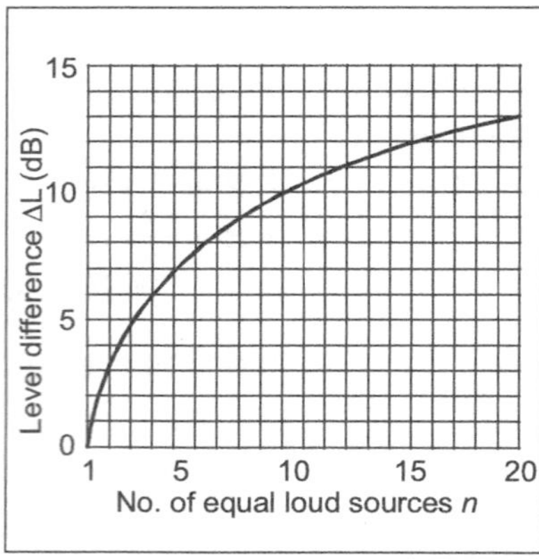

Level increase ΔL for n equal loud sound sources	
No. of n equal loud sound sources	Level increase ΔL in dB
1	0
2	3.0
3	4.8
4	6.0
5	7.0
6	7.8
7	8.5
8	9.0
9	9.5
10	10.0
2	10.8
16	12.0
20	13.0

13.7 (c)

Methods to reduce Noise Pollution:

- Proper Town planning technique
- Raising of obstructions and barriers in between noise sources (automobiles) and residences.
- Raising of thick and high vegetation and growing trees.

13.8 (a)

$$\text{SPL} = 20\log_{10}\left(\frac{P}{P_{ref}}\right)$$

$$P_{ref} = 20\,\mu\text{Pa}$$

$$P = 0.0002\,\mu\text{ bar}$$

$$= 0.0002 \times 10^5\,\mu\text{Pa}$$

$$\therefore \quad \text{SPL} = 20\log_{10}\left(\frac{0.0002 \times 10^5}{20}\right)$$

$$= 20\log_{10}(1) = 0$$

13.9 (c)

Glass doesn't reflect sound waves. Sound waves require a medium to travel and they travel faster in glass than in air. Double glass window panes have air in between which is a poor conductor than glass and hence reduces noise.

13.11 (a)

Most cases of noise-induced hearing loss are caused by repeated exposure to moderate levels of noise over many years, not by a few cases of very loud noise. Wearing hearing protectors can help prevent damage from both moderate and loud noise.

13.14 (d)

Physiological response accompanying response and other noise exposures include:

1. A vascular response characteristic by peripheral vasoconstriction, changes in heart beat rate and blood pressure.
2. Various glandular charges such as increased output of adrenaline evidenced by chemical changes in blood.
3. Slow, deep breathing.
4. A change in the electrical resistance of skin with changes in activity of sweat glands.

■■■■

UNIT

V

Soil Mechanics and Foundation Engineering

Syllabus

Soil Mechanics: Soil exploration - planning & methods, Properties of soil, classification, various tests and inter-relationships; Permeability & Seepage, Compressibility, consolidation and Shearing resistance, Earth pressure theories and stress distribution in soil; Properties and uses of geo-synthetics.

Foundation Engineering: Types of foundations & selection criteria, bearing capacity, settlement analysis, design and testing of shallow & deep foundations; Slope stability analysis, Earthen embankments, Dams and Earth retaining structures: types, analysis and design, Principles of ground modifications.

Contents

■■■■

1 Properties of Soils

1.1 Match **List-I** with **List-II** and select the correct answer using the codes given below the lists:

List-I	List-II
A. Loess	1. Deposited from suspe-nsion in running water
B. Peat	2. Deposits of marine origin
C. Alluvial soil	3. Deposits by wind
D. Marl	4. Organic soil

Codes:

	A	B	C	D
(a)	3	4	2	1
(b)	4	3	1	2
(c)	4	3	2	1
(d)	3	4	1	2

[ESE : 2000]

1.2 The value of porosity of a soil sample in which the total volume of soil grains is equal to twice the total volume of voids would be
(a) 75% (b) 66.66%
(c) 50% (d) 33.33%
[ESE : 2000]

1.3 A soil has a liquid limit of 40% and plasticity index of 20%. The plastic limit of the soil will be
(a) 20% (b) 30%
(c) 40% (d) 60% [ESE : 2000]

1.4 A sample of saturated sand has a dry unit weight of 18 kN/m^3 and a specific gravity of 2.7. If density of water is 10 kN/m^3, the void ratio of the soil sample will be
(a) 0.5 (b) 0.6
(c) 0.4 (d) 0.9 [ESE : 2000]

1.5 **Assertion (A):** Black cotton soils are clays and they exhibit characteristic property of swelling.
Reason (R): These clays contain Montmorillonite which attracts external water into its lattice structure.
(a) both A and R are true and R is the correct explanation of A
(b) both A and R are true but R is not a correct explanation of A
(c) A is true but R is false
(d) A is false but R is true [ESE : 2001]

1.6 A dry soil sample has equal amounts of solids and voids by volume. Its void ratio and porosity will be

	Void ratio	Porosity (%)
(a)	1.0	100%
(b)	0.5	50%
(c)	0.5	100%
(d)	1.0	50%

[ESE : 2001]

1.7 The plasticity index and the percentage of grain size finer than 2 microns of a clay sample are 25 and 15, respectively. Its activity ratio is
(a) 2.5 (b) 1.67
(c) 1.0 (d) 0.6 [ESE : 2001]

1.8 A soil sample having a void ratio of 1.3, water content of 50% and a specific gravity of 2.60, is in a state of
(a) partial saturation
(b) full saturation
(c) over saturation
(d) under saturation [ESE : 2001]

1.9 The natural void ratio of a sand sample is 0.6 and its density index is 0.6. If its void ratio in the loosest state is 0.9, then the void ratio in the densest state will be
(a) 0.2 (b) 0.3
(c) 0.4 (d) 0.5 [ESE : 2001]

1.10 Which one of the following phase diagrams represents a clay at its shrinkage limit?

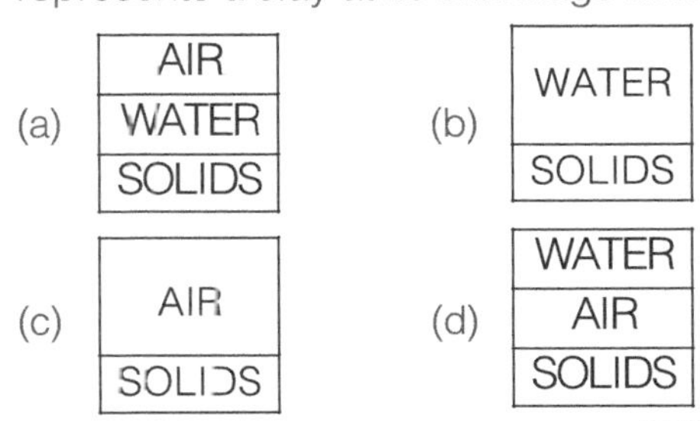

[ESE : 2001]

1.11 Which one of the following correctly represents the dry unit weight of a soil sample which has a bulk unit weight of γ_t at a moisture content of w%?

(a) $\frac{w\gamma_t}{100}$ (b) $\gamma_t\left(1+\frac{w}{100}\right)$

(c) $\gamma_t\left(\frac{100}{100+w}\right)$ (d) $\frac{\gamma_t(100-w)}{100}$

[ESE : 2001]

1.12 Which one the following represents relative density of saturated sand deposit having moisture content of 25%, if maximum and minimum void ratio of sand are 0.95 and 0.45 respectively and specific gravity of sand particles is 2.6?

(a) 40% (b) 50%

(c) 60% (d) 70% [ESE : 2002]

1.13 The correct sequence of plasticity of minerals in soil in an increasing order is

(a) silica, kaolinite, illite, montmorillonite

(b) kaolinite, silica, illite, montmorillonite

(c) silica, kaolinite, montmorillonite, illite

(d) kaolinite, silica, montmorillonite, illite

[ESE : 2002]

1.14 **Assertion (A):** A soil is at its liquid limit if the consistency index of the soil is equal to zero.

Reason (R): The consistency index of a soil is defined as ratio of (liquid limit minus the natural water content) to (natural water content minus plastic limit).

(a) both A and R are true and R is the correct explanation of A

(b) both A and R are true but R is not a correct explanation of A

(c) A is true but R is false

(d) A is false but R is true [ESE : 2002]

1.15 Consider the following statements:

A dispersed structure is formed in clay when

1. the net electrical forces between adjacent soil particles at the time of deposition are repulsive
2. remoulding and compacting of clays have high void ratio
3. there is concentration of dissolved minerals in water
4. platelets have face to face contact in more or less parallel arrays

Which of these statements are correct?

(a) 1, 2 and 3 (b) 2, 3 and 4

(c) 1, 2 and 4 (d) 1, 3 and 4

[ESE : 2003]

1.16 The collapsible soil is associated with

(a) dune sands (b) laterite soils

(c) loess (d) black cotton soils

[ESE : 2003]

1.17 Consistency as applied to cohesive soils is an indicator of its

(a) density (b) moisture content

(c) shear strength (d) porosity

[ESE : 2003]

1.18 While computing the values of limits of consistency and consistency indices, it is found that liquidity index, has a negative value.

Consider the following comments on this value:

1. Liquidity index cannot have a negative value and should be taken as zero.
2. Liquidity index can have a negative value.
3. The soil tested is in semisolid state and stiff.
4. The soil tested is in medium soft state.

Which of these statements are correct?

(a) 1 and 4 (b) 1 and 3

(c) 2 and 4 (d) 2 and 3

[ESE : 2003]

1.19 The predominant mineral responsible for shrinkage and swelling in black cotton soils is

(a) illite (b) kaolinite

(c) mica (d) montmorillonite

[ESE : 2003]

1.20 Two soil sample A and B have porosities $n_A = 40\%$ and $n_B = 60\%$ respectively. What is the ratio of void ratios $e_A : e_B$?

(a) 2 : 3 (b) 3 : 2

(c) 4 : 9 (d) 9 : 4 [ESE : 2004]

1.21 Match **List-I** (Densities) with **List-II** (Expressions) and select the correct answer using the codes given below the lists:

(Symbols G, e, γ_w and S stand for specific gravity of soil grains, void ratio, unit weight of water and degree of saturation respectively)

List-I	List-II
A. Dry density	1. $\gamma_w(G + Se)/(1 + e)$

B. Moist density 2. $\gamma_w G/(1 + e)$
C. Submerged density 3. $\gamma_w(G + e)/(1 + e)$
D. Saturated density 4. $\gamma_w(G - 1)/(1 + e)$

Codes:

	A	B	C	D
(a)	2	1	4	3
(b)	2	3	4	1
(c)	4	1	2	3
(d)	4	3	2	1

[ESE : 2004]

1.22 What are the respective values of void ratio, porosity ratio and saturated density (in kN/m³) for a soil sample which has saturation moisture content of 20% and specific gravity of grains as 2.6? (take density of water as 10 kN/m³)
(a) 0.52, 1.08, 18.07 (b) 0.52, 0.34, 18.07
(c) 0.77, 1.08, 16.64 (d) 0.52, 0.34, 20.14
[ESE : 2004]

1.23 Embankment fill is to be compacted at a density of 18 kN/m³. The soil of the borrow area is at a density of 15 kN/m³. What is the estimated number of trips of 6 m³ capacity truck for hauling the soil required for compacting 100 m³ fill of the embankment? (Assume that the soil in the borrow area and that in the embankment are at the same moisture content)
(a) 14 (b) 18
(c) 20 (d) 23 [ESE : 2004]

1.24 Match **List-I** (Unit/Test) with **List-II** (Purpose) and select the correct answer using the codes given below the lists:

List-I
A. Casagrande's apparatus
B. Hydrometer
C. Plate load test
D. Oedometer

List-II
1. Determination of grain size distribution
2. Consolidation characteristics
3. Determination of consistency limits
4. Determination of safe bearing capacity of soil

Codes:

	A	B	C	D
(a)	1	3	2	4
(b)	1	3	4	2
(c)	3	1	2	4
(d)	3	1	4	2

[ESE : 2004]

1.25 A clay sample has a void ratio 0.54 in dry state. The specific gravity of soil solids is 2.7. What is the shrinkage limit of the soil?
(a) 8.5% (b) 10.0%
(c) 17.0% (d) 20.0%
[ESE : 2005]

1.26 Consider the following clay minerals:
1. Kaolinite
2. Illite
3. Montmorillonite

Which one of the following is the correct sequence of the minerals given above in the increasing order of their grain size?
(a) 3-2-1 (b) 1-3-2
(c) 1-2-3 (d) 3-1-2 [ESE : 2006]

1.27 Consider the following statements:
1. Mica is a clay mineral.
2. Rock dust particles, even of clay size are non-plastic.
3. A particle of kaolinite is electrically charged.

Which of these statements are correct?
(a) 1, 2 and 3 (b) Only 1 and 2
(c) Only 2 (d) Only 1 and 3
[ESE : 2006]

1.28 Which one of the following relation gives the value of degree of saturation S, in terms of unit weight γ, water content w (as ratio) and specific gravity of soil solids G_s, γ_w is unit weight of water?

(a) $S = \dfrac{w}{(\gamma/\gamma_w)(1+w) - 1/G_s}$

(b) $S = \dfrac{w}{(\gamma/\gamma_w)(1+w) + 1/G_s}$

(c) $S = \dfrac{w}{(\gamma_w/\gamma)(1+w) + 1/G_s}$

(d) $S = \dfrac{w}{(\gamma_w/\gamma)(1+w) - 1/G_s}$ [ESE : 2006]

1.29 **Assertion (A):** The water content of inorganic soils is determined by heating the soil in an oven at a temperature of 105° to 110°C.
Reason (R): The free water, absorbed water and structural water are all completely removed from the soil by heating it at 105° to 110°C.

(a) both A and R are true and R is the correct explanation of A
(b) both A and R are true but R is not a correct explanation of A
(c) A is true but R is false
(d) A is false but R is true [ESE : 2007]

1.30 In comparison to Atterberg limits of normal soils, the expansive soils have which of the following?
1. More liquid limit
2. Less plastic limit
3. Less shrinkage limit
4. More volumetric shrinkage

Select the correct answer using the codes given below:
(a) 1, 2, 3 and 4 (b) 1, 3 and 4 only
(c) 2 and 3 only (d) 1, 2 and 4 only
[ESE : 2008]

1.31 For a sandy soil with soil grains spherical in shape and uniform in size, what is the theoretical void ratio ?
(a) 0.61 (b) 0.71
(c) 0.91 (d) 0.81 [ESE : 2009]

1.32 A soil has liquid limit = 35, plastic limit = 20, shrinkage limit = 10 and natural moisture content = 25%. What will be its liquidity index, plasticity index and shrinkage index?
(a) 0.67, 15 and 25 (b) 0.33, 15 and 10
(c) 0.67, 25 and 15 (d) 0.33, 20 and 15
[ESE : 2009]

1.33 A cohesive soil yields a maximum dry density of 16 kN/m^3 during a standard Proctor Compaction test. If the specific gravity is 2.65, what would be its void ratio?
(a) 0.552 (b) 0.624
(c) 0.712 (d) 0.583 [ESE : 2009]

1.34 Consider the following properties for clays *X* and *Y*:

Sl. No.	Properties	Clay X (%)	Clay Y (%)
1.	Liquid limit	42	56
2.	Plastic limit	20	34
3.	Natural water content	30	50

Which of the clays, *X* or *Y*, experiences larger settlement under identical loads; is more plastic; and is softer in consistency?
(a) *X*, *Y* and *X* (b) *Y*, *X* and *X*
(c) *Y*, *X* and *Y* (d) *X*, *X* and *Y*
[ESE : 2009]

1.35 Consider the following clay minerals:
1. Kaolinite
2. Montmorillonite
3. Illite

What is the correct sequence in an increasing order of their plasticity index?
(a) 1–2–3 (b) 3–2–1
(c) 1–3–2 (d) 3–1–2 [ESE : 2009]

1.36 Consider the following statements:
1. Each year, black cotton soil appreciably shrinks during dry season and swells during rainy season. This alternate cycle of shrinking and swelling causes severe stresses in structures supported directly by such soil.
2. Black cotton soil contains predominantly a clay mineral called kaolinite, which is responsible for causing appreciable shrinking and swelling.
3. Shrinking and swelling of black cotton soils are observed only upto a certain depth below the ground level. Below that level, there is neither shrinking nor swelling.

Which of these statements is/are correct?
(a) 1, 2 and 3 (b) 3 only
(c) 2 and 3 only (d) 1 and 3 only
[ESE : 2010]

1.37 In a wet soil mass, air occupies one-sixth of its volume and water occupies one-third of its volume. The void ratio of the soil is
(a) 0.25 (b) 0.50
(c) 0.75 (d) 1.0 [ESE : 2010]

1.38 Consider the following statements:
1. Peat and muck are organic soils.
2. Peat is an inorganic soil whereas muck is an organic soil.
3. Indurated clay is a type of clay which does not soften under prolonged wetting.

Which of these statements is/are correct?
(a) 1, 2 and 3 (b) 2 only
(c) 3 only (d) 1 and 3 only
[ESE : 2010]

1.39 A specimen of clayey silt contains 70% silt size particles. Its liquid limit = 40 and plastic limit = 20. In liquid limit test, at moisture content of 30%, required number of blows was 50. Its plasticity index, activity and consistency index will respectively be

(a) 20, 0.67 and 0.5 (b) 20, 1.5 and 2.0
(c) 30, 1.5 and 0.72 (d) 20, 0.286 and 0.38

[ESE : 2010]

1.40 Consider the following statements:

1. Mica is a clay mineral.
2. The shape of clay particle is usually flaky.
3. A particle of kaolinite is electrically charged.

Which of these statements is/are correct?

(a) 1, 2 and 3 (b) 1 and 3 only
(c) 2 and 3 only (d) 2 only

[ESE : 2010]

1.41 Match **List-I** (Deposit) with **List-II** (Soil structure) and select the correct answer using the codes given below the lists:

List-I	List-II
A. Coarse grained soil	1. Flocculated
B. Silt deposit	2. Cohesive matrix
C. Clay deposit	3. Honeycomb
D. Composite soil	4. Single-grained

Codes:

	A	B	C	D
(a)	2	3	1	4
(b)	4	3	1	2
(c)	2	1	3	4
(d)	4	1	3	2

[ESE : 2010]

1.42 Match **List-I** with **List-II** and select the correct answer using the code given below the lists:

List-I	List-II
A. Plate load	1. Specific gravity
B. Pycnometer	2. Bearing capacity
C. Core cutter	3. Grain size analysis
D. Mechanical sieve	4. Field density

Code:

	A	B	C	D
(a)	3	1	4	2
(b)	2	1	4	3
(c)	3	4	1	2
(d)	2	4	1	3

[ESE : 2011]

1.43 The natural water content of the soil sample was found to be 40%, specific gravity is 2.7 and void ratio 1.2; then the degree of saturation of the soil will be

(a) 100% (b) 69%
(c) 87% (d) 90%

[ESE : 2012]

1.44 The porosity of a soil sample having its void ratio equal to unity would be

(a) 33.34% (b) 50.0%
(c) 66.66% (d) 75.0%

[ESE : 2012]

1.45 The porosity of a certain soil sample was found to be 80% and its specific gravity was 2.7; the critical hydraulic gradient will be estimated as

(a) 0.34 (b) 0.92
(c) 1.0 (d) 1.5

[ESE : 2012]

1.46 The field density and field moisture content of a soil can be determined by

1. Core cutter method
2. Sand replacement method
3. Proctor compaction test
4. Modified proctor compaction test

(a) 1, 2, 3 and 4 (b) 1 and 2 only
(c) 2 and 3 only (d) 2 and 4 only

[ESE : 2012]

1.47 Consider the following statements:

1. Relative density is a meaningful parameter for all types of soils.
2. Relative density is a meaningful parameter only for cohesionless soils.
3. Relative density is a better indicator of the denseness of an in-situ granular soil deposit than the void ratio.

Which of these statements are correct?

(a) 1, 2 and 3 (b) 1 and 2 only
(c) 2 and 3 only (d) 1 and 3 only

[ESE : 2012]

1.48 Consider the following statements:

1. Activity is a property typical of clay soils.
2. An activity value of 7 in a clay soil is indicative of the presence of montmorillonite mineral.
3. An activity value of 7 in a clay soil is indicative of the presence of illite mineral.

Which of these statements are correct?

(a) 1, 2 and 3 (b) 1 and 2 only
(c) 1 and 3 only (d) 2 and 3 only

[ESE : 2013]

1.49 In a wet soil mass, air occupies one-fourth of its volume and water occupies one-half of its volume. The void ratio of this soil is:
(a) 1 (b) 2
(c) 3 (d) 4 [ESE : 2013]

1.50 Consider the following statements:
1. If the soil is not black in colour, it is unlikely to be a swelling soil.
2. The swelling pressure of a fine-grained soil depends on its initial water content and density.
3. The swelling pressure of a fine-grained soil depends on the nature of the pore fluid.

Which of these statements are correct?
(a) 1, 2 and 3 (b) 1 and 2 only
(c) 1 and 3 only (d) 2 and 3 only
[ESE : 2013]

1.51 A soil sample has a void ratio of 0.5; its porosity will be
(a) 50% (b) 66%
(c) 100% (d) 33% [ESE : 2014]

1.52 A saturated specimen of clay was immersed in mercury and displaced volume was 21.8 cc. The weight of the sample was 32.2 gm. After oven drying for 48 hours, weight reduced to 20.2 gm while volume came down to 11.6 cc. The shrinkage limit of the soil is
(a) 7.9% (b) 8.0%
(c) 8.9% (d) 9.8% [ESE : 2014]

1.53 **Statement (I):** Clays exhibit more hygroscopicity than sands.

Statement (II): Clays are colloidal and consequently their specific surface is very high.
(a) Both Statement (I) and Statement (II) are individually true and Statement (II) is the correct explanation of Statement (I).
(b) Both Statement (I) and Statement (II) are individually true but Statement (II) is not the correct explanation of Statement (I).
(c) Statement (I) is true but Statement (II) is false.
(d) Statement (I) is false but Statement (II) is true.
[ESE : 2014]

1.54 The specific gravity of a soil sample is 2.7 and its void ratio is 0.945. When it is fully saturated the moisture content of the soil will be
(a) 25% (b) 30%
(c) 35% (d) 40%
[ESE : 2015]

1.55 A soil sample has shrinkage limit of 6% and the specific gravity of the soil grains is 2.6. The porosity of soil at shrinkage limit is
(a) 7.5% (b) 9.5%
(c) 13.5% (d) 16.5%
[ESE : 2016]

1.56 What is the dry unit weight of a clay soil when the void ratio of a sample there of is 0.50, the degree of saturation is 70% and the specific gravity of soil grains is 2.7? Take the value of γ_w to be 9.81 kN/m^3.
(a) 13.65 kN/m^2 (b) 19.95 kN/m^2
(c) 23.65 kN/m^2 (d) 29.95 kN/m^2
[ESE : 2016]

1.57 A fill having volume of 150000 m^3 is to be constructed at a void ratio of 0.8. The borrow pit solid has a void ratio of 1.4. The volume of soil required to be excavated from the borrow pit will be
(a) 150000 m^3 (b) 200000 m^3
(c) 250000 m^3 (d) 300000 m^3
[ESE : 2016]

1.58 The ratio of dry unit weight to unit weight of water represents
(a) specific gravity of soil solids
(b) specific gravity of soil mass
(c) specific gravity of dry soil
(d) shrinkage ratio
[ESE : 2016]

1.59 Proctor's compaction test for the maximum dry density of a certain soil gave the results as : 1.77 gm/cc and OMC 14.44%. The specific gravity of the clay soil grain was 2.66. What was the saturation degree for this soil?
(a) 44% (b) 55%
(c) 66% (d) 77%
[ESE : 2016]

1.60 The value of porosity of a soil sample in which the total volume of soil grains is equal to twice the total volume of voids would be
(a) 30% (b) 40%
(c) 50% (d) 60%
[ESE : 2016]

1.61 Consider the following statements:

1. Illite the mineral largely responsible for the swelling and shrinkage behaviour of clayey soils.
2. A differential free swell value of 55% indicates a soil with low degree of expansiveness.
3. Higher the plasticity index of a soil, greater its swelling potential.
4. A low shrinkage limit of a soil indicates possibility of swelling at low water content.

Which of the above statements are correct?

(a) 1 and 2 (b) 2 and 3
(c) 1 and 4 (d) 3 and 4

[ESE : 2017]

1.62 Consider the following effects as indicative of complete saturation of a soil sample:

1. Pore water pressure is positive.
2. Volume of water to volume of voids is equal to 1.
3. Relative density is equal to 1.

Which of the above statements are correct?

(a) 1 and 2 (b) 1 and 3
(c) 2 and 3 (d) 1, 2 and 3

[ESE : 2017]

1.63 Statement (I): At shrinkage limit, the soil is fully saturated.

Statement (II): Montmorillonite clay minerals are non-expansive.

(a) Both Statement (I) and Statement (II) are individually true and Statement (II) is the correct explanation of Statement (I)
(b) Both Statement (I) and Statement (II) are individually true but Statement (II) is ***not*** the correct explanation of Statement (I)
(c) Statement (I) is true but Statement (II) is false
(d) Statement (I) is false but Statement (II) is true

[ESE : 2017]

1.64 Consider the following statements:

1. When a soil sample is dried beyond its shrinkage limit, the volume of the soil slowly decreases.
2. Plastic limit is always lower than the liquid limit for any type of soil.
3. At the liquid limit, the soil behaves like a liquid and possesses no shear strength at all.
4. When subjected to drying, the volume of the soil remains unchanged once the water content of the soil goes below its shrinkage limit.

Which of the above statements are CORRECT?

(a) 1 and 3 only (b) 1 and 4 only
(c) 2 and 3 only (d) 2 and 4 only

[ESE : 2018]

1.65 A sand sample has a porosity of 30% and specific gravity of solids as 2.6. What is its degree of saturation at moisture content of 4.94%?

(a) 40% (b) 35%
(c) 30% (d) 25%

[ESE : 2018]

1.66 What will be the unit weight of a fully saturated soil sample having water content of 38% and grain specific gravity of 2.65?

(a) 19.88 kN/m^3 (b) 17.88 kN/m^3
(c) 16.52 kN/m^3 (d) 14.65 kN/m^3

[ESE : 2018]

1.67 How many cubic metres of soil having void ratio of 0.7 can be made from 30 m^3 of soil with void ratio of 1.2?

(a) 36.6 m^3 (b) 30.0 m^3
(c) 25.9 m^3 (d) 23.2 m^3

[ESE : 2018]

1.68 A sample of dry soil is coated with a thin layer of paraffin and has a mass of 460 g. It displaced 300 cc of water when immersed in it. The paraffin is peeled off and its mass was found to be 9 g. If the specific gravity of soil solids and paraffin are 2.65 and 0.9 respectively, the voids ratio of soil is nearly

(a) 0.92 (b) 0.71
(c) 0.59 (d) 0.48

[ESE : 2018]

1.69 The property of clays by virtue of which they regain, if left alone for a time, a part of the strength lost due to remoulding at unaltered moisture content, is known as

(a) Thixotropy (b) Sensitivity
(c) Consistency (d) Activity

[ESE : 2019]

1.70 The plastic limit and liquid limit of a soil are 30% and 42% respectively. The percentage volume change from liquid limit to dry state is 35% of the dry volume. Similarly the percentage volume change from plastic limit to dry state is 22% of the dry volume. The shrinkage ratio will be nearly

(a) 4.2 (b) 3.1
(c) 2.2 (d) 1.1

[ESE : 2019]

1.71 The ratio of a given volume change in a soil, expressed as percentage of the dry volume, to the corresponding change in water content is called

(a) Specific gravity of soil solids
(b) Mass-specific gravity of soils
(c) Shrinkage ratio of soils
(d) Density ratio of soils

[ESE : 2019]

1.72 A soil sample has porosity of 40%, and the specific gravity of solids is 2.70. If the soil is 50% saturated, the unit weight will be nearly

(a) 22 kN/m^3 (b) 20 kN/m^3
(c) 18 kN/m^3 (d) 16 kN/m^3

[ESE : 2020]

1.73 Oven dry mass of a pat of clay is 10.8 gm and mass of mercury displaced on immersion is 84.2 gm. If the specific gravity of solids is 2.72 and the density of the mercury is 13.6 g/cm^3, the shrinkage limit of the soil will be nearly

(a) 12% (b) 15%
(c) 18% (d) 21%

[ESE : 2020]

1.74 The suitability number of a backfill for D_{50} = 1 mm, D_{20} = 0.5 mm and D_{10} = 0.08 mm will be nearly

(a) 16 (b) 18
(c) 20 (d) 22 [ESE : 2020]

1.75 The porosity of a soil *n* is

(a) $\frac{e}{1+e}$ (b) $\frac{e}{1-e}$

(c) $\frac{e+1}{e}$ (d) $\frac{e-1}{e}$

Where e = void ratio

[ESE : 2020]

1.76 A soil has bulk density of 20.1 kN/m^3 and water content 15%. What is the water content if the soil partially dries to a density of 19.4 kN/m^3 and the void ratio remains unchanged?

(a) 10.86% (b) 10.76%
(c) 10.68% (d) 10.66%

[ESE : 2021]

1.77 A fine grained soil is found to have a liquid limit of 90% and a plasticity index of 50. The natural water content is 28% What is the liquidity index?

(a) –0.34 (b) –0.14
(c) –0.24 (d) –2.40

[ESE : 2021]

1.78 A sample of silty clay has a volume of 14.88 cm^3, a total mass of 28.81 gm, a dry mass of 24.83 gm and a specific gravity of solids 2.7. What is the void ratio?

(a) 0.412 (b) 0.521
(c) 0.618 (d) 0.663

[ESE : 2021]

1.79 The void ratio of a clay sample is 0.5 and the degree of saturation is 70%. What is the bulk unit weight of the soil?(Assume G = 2.7)

(a) 10.46 kN/m^3 (b) 14.32 kN/m^3
(c) 17.77 kN/m^3 (d) 19.95 kN/m^3

[ESE : 2021]

1.80 Consistency is a term used to indicate

(a) the quantitative analysis of soils
(b) the degree of firmness of cohesive soils
(c) the fineness of non-cohesive soils
(d) the fineness of clay soils

[ESE : 2022]

1.81 An oven-dried soil having a mass of 200 g is placed in a pycnometer which is then completely filled with water. The total mass of the pycnometer with water and soil inside is 1605g. The pycnometer filled with water alone has a mass of 1480 g. What is the specific gravity of the soil?

(a) 2.21 (b) 2.41
(c) 2.67 (d) 3.32

[ESE : 2022]

1.82 A soil sample has a porosity of 40 percent. The specific gravity of solids is 2.70. What is voids ratio?

(a) 0.467 (b) 0.567
(c) 0.667 (d) 0.743 [ESE : 2022]

1.83 Large parts of northern India lying north of Vindhya-Satpura range in the Indo-Gangetic and Brahmaputra flood plains are covered by
(a) the colluvial soils (b) the aeolian soils
(c) the alluvial soils (d) the talus soils
[ESE : 2022]

1.84 The water remaining in the soil after the removal of gravitational water is called
(a) Capillary water
(b) Gravity water
(c) Hygroscopic water
(d) Field capacity
[ESE : 2024]

1.85 Consider the following statements regarding sedimentation analysis of soil particles based on Stokes' law:
1. The finer soil particles are never perfectly spherical.
2. All the soil grains may have the same specific gravity.
3. Particles constituting to fine soil fraction may carry surface electric charges.

Which of the above statements are correct?
(a) 1 and 2 only (b) 1 and 3 only
(c) 2 and 3 only (d) 1, 2 and 3 only
[ESE : 2024]

1.86 What is meant by Thixotropy?
(a) Property of a material which is manifested by its resistance to flow
(b) The ratio of the unconfined compression strength in the natural or undisturbed state to that in the remoulded state
(c) The compressive stress at failure, giving due allowance to the reduction in area of cross-section
(d) Phenomenon of strength loss- strength gain, with no change in volume or water content
[ESE : 2024]

1.87 Which one of the following types of soils will usually shrink if drained or if subjected to repeated loading?
(a) Non-cohesive soils
(b) Cohesive soils
(c) Peat
(d) Silts [ESE : 2024]

1.88 If γ_s = unit weight of the material of falling sphere in g/cm^3, γ_t = unit weight of the liquid medium in g/cm^3, μ_t = viscosity of the liquid medium in g sec/cm^2, and D = diameter of the spherical particle in cm. According to Stokes' law, what is the formula for the terminal velocity (v) of the spherical particle?

(a) $v = \frac{1}{18}\left[\frac{(\gamma_s - \gamma_t)}{\mu_t}\right]D^2$

(b) $v = \frac{1}{8}\left[\frac{(\gamma_s + \gamma_t)}{\mu_t}\right]D^2$

(c) $v = \frac{1}{12}\left[\frac{(\gamma_s - \gamma_t)}{\mu_t}\right]D^2$

(c) $v = \frac{1}{2}\left[\frac{(\gamma_s + \gamma_t)}{\mu_t}\right]D^2$

[ESE : 2024]

1.89 Which one of the following soils are transported by wind?
(a) Alluvial soils (b) Lacustrine soils
(c) Aeolian soils (d) Marine soils
[ESE : 2024]

1.90 What is the grain specific gravity for humus type soil?
(a) 2.37 (b) 1.37
(c) 4.37 (d) 3.37
[ESE : 2024]

1.91 When the specific gravity of solids is known, which one of the following types of methods is used to determine the water content?
(a) Pycnometer method
(b) Rapid moisture tester method
(c) Oven-drying method
(d) Sand-replacement method
[ESE : 2024]

■■■■

Answers Properties of Soils

1.1 (d)	1.2 (d)	1.3 (a)	1.4 (a)	1.5 (a)	1.6 (d)	1.7 (b)	1.8 (b)	1.9 (c)
1.10 (b)	1.11 (c)	1.12 (c)	1.13 (a)	1.14 (c)	1.15 (c)	1.16 (c)	1.17 (c)	1.18 (d)
1.19 (d)	1.20 (c)	1.21 (a)	1.22 (d)	1.23 (c)	1.24 (d)	1.25 (d)	1.26 (a)	1.27 (c)
1.28 (d)	1.29 (c)	1.30 (a)	1.31 (c)	1.32 (b)	1.33 (b)	1.34 (c)	1.35 (c)	1.36 (d)
1.37 (d)	1.38 (d)	1.39 (a)	1.40 (d)	1.41 (b)	1.42 (b)	1.43 (d)	1.44 (b)	1.45 (a)
1.46 (b)	1.47 (c*)	1.48 (b)	1.49 (c)	1.50 (d)	1.51 (d)	1.52 (c)	1.53 (a)	1.54 (c)
1.55 (c)	1.56 (b)	1.57 (b)	1.58 (d)	1.59 (d)	1.60 (a)	1.61 (d)	1.62 (a)	1.63 (c)
1.64 (d)	1.65 (c)	1.66 (b)	1.67 (d)	1.68 (b)	1.69 (a)	1.70 (d)	1.71 (c)	1.72 (c)
1.73 (d)	1.74 (d)	1.75 (a)	1.76 (a)	1.77 (c)	1.78 (c)	1.79 (d)	1.80 (b)	1.81 (c)
1.82 (c)	1.83 (c)	1.84 (a)	1.85 (b)	1.86 (d)	1.87 (b)	1.88 (a)	1.89 (c)	1.90 (b)
1.91 (a)								

Explanations Properties of Soils

1.1 (d)

Loess is a loosely compacted wind-blown sediment typically in the 20–50 micrometer size range.

Peat is a brown material consisting of partly decomposed vegetable matter (humus) forming a deposit in a water saturated environment and in the absence of oxygen.

Marl is a fine grained calcium-carbonated soil of marine origin.

1.2 (d)

Given, $V_S = 2V_V$

$V_T = V_S + V_V$

Porosity, $n = \dfrac{V_v}{3V_V} = 0.33 = 33\%$

1.3 (a)

Plastic limit, $w_P = w_L - PI$

$= 40 - 20 = 20\%$

1.4 (a)

$$\gamma_d = \frac{G}{1+e}\gamma_w$$

$$\Rightarrow \quad e = \frac{G\gamma_w}{\gamma_d} - 1$$

$$= \frac{2.7 \times 10}{18} = 0.5$$

1.5 (a)

For black cotton soil:

(i) Liquid limit – 40 to 100%

(ii) Plasticity Index – 20 to 60%

(iii) Shrinkage limit – 10 to 15%

(iv) Optimum Moisture content – 25 to 30%

1.6 (d)

Void ratio, $e = \dfrac{V_v}{V_s} = 1$

Porosity, $n = \dfrac{e}{1+e} \times 1 = \dfrac{1}{1+1} \times 100$

$= 50\%$

1.7 (b)

Activity

$$= \frac{\text{Plasticity index}}{\text{Percent of clay particles finer than } 2\mu m}$$

$$= \frac{25}{15} = 1.67$$

As activity is more than 1.25 so it is active soil.

1.8 (b)

$$Gw = Se$$

$$S = \frac{2.60 \times 50}{1.3} = 100\%$$

Therefore soil is fully saturated.

Remember	S ≤ 100% always
	w can be more than 100%
	e can be more than 1.0.

1.9 (c)

$$\text{Density Index} = \frac{e_{max} - e}{e_{max} - e_{min}} = 0.6$$

Void ratio in loosest state,

$$e_{max} = 0.9$$

Void ratio in natural state,

$$e = 0.6$$

$$\therefore \quad 0.9 - e_{min} = \frac{0.9-0.6}{0.6}$$

$$e_{min} = 0.9 - 0.5 = 0.4$$

1.10 (b)

At shrinkage limit, there is no air and degree of saturation is 100%.

1.11 (c)

$$\gamma_d = \frac{\gamma_t}{\left(1+\frac{w}{100}\right)} = \frac{100\gamma_t}{(100+w)}$$

1.12 (c)

At saturated moisture content void ratio is

$$e = \frac{Gw}{S} = \frac{2.6\times 25}{100} = 0.65$$

Relative density,

$$D_r = \frac{e_{max} - e}{e_{max} - e_{min}}$$

$$= \frac{0.95-0.65}{0.95-0.45} = 0.6$$

$$D_r = 60\%$$

1.13 (a)

Silica has least plasticity while Montmo-rillonite has highest plasticity.

1.14 (c)

For $w = w_L$

$$\text{Consistency index} = \frac{w_L - w}{w_L - w_P} = 0$$

1.15 (c)

- Dispersed structure is formed when there is net repulsive force.
- Remoulding and compaction converts the edge to face orientation to face to face orientation (dispersed structure)
- Concentration of dissolved mineral decreases the charge of clay particle, thus stabilizing them to form flocculated structure.

1.16 (c)

Loess is predominantly a silt-sized sediment that is formed by the accumulation of wind-blown dust. It covers about 10% of the Earth's surface. It is usually homogeneous and highly porous and is traversed by vertical capillaries that permit the sediment to fracture and form vertical bluffs.

1.17 (c)

Consistency of soil refers to the resistance offered by it against forces that tend to deform or rupture the soil aggregate. It is related to strength.

1.19 (d)

Montmorillonites expand considerably more than other clays due to water penetrating the interlayer molecular spaces and concomitant adsorption. The amount of expansion is due largely to the type of exchangeable cation contained in the sample. The presence of sodium as the predominant exchangeable cation can result in the clay swelling to several times its original volume. Also the water content of montmorillonite is variable and it increases greatly in volume when it absorbs water.

1.20 (c)

$$e_A = \frac{n_A}{1-n_A} = \frac{0.4}{0.6} = \frac{2}{3}$$

$$e_B = \frac{n_B}{1-n_B} = \frac{0.6}{0.4} = \frac{3}{2}$$

$$\therefore \quad \frac{e_A}{e_B} = \frac{4}{9}$$

1.21 (a)

Moist density,

$$\gamma = \left(\frac{G+Se}{1+e}\right)\gamma_w$$

For dry density put $S = 0$

For saturated density put $S = 1$

Submerged density

= saturated density – density of water

1.22 (d)

Given $w = 20\%$, $G = 2.6$,

$S = 100\%$

$$\therefore \quad e = \frac{Gw}{S} = \frac{2.6 \times 20}{100} = 0.52$$

$$n = \frac{e}{1+e} = 0.34$$

$$\gamma_{sat} = \left(\frac{G+e}{1+e}\right)\gamma_w$$

$$= \frac{2.6+0.52}{1+0.52} \times 10$$

$$= 20.53 \text{ kN/m}^3$$

If density of water is taken as 9.81 kN/m^3

$$\gamma_{sat} = 20.14 \text{ kN/m}^3$$

1.23 (c)

In problems of fill and borrow pit, the volume of soil solids shall be equated.

$$V_s = \frac{V}{1+e} = \frac{V\gamma}{G(1+w)\gamma_w}$$

As the moisture content of borrow area and that of embankment are same.

$$V_1\gamma_1 = V_2\gamma_2$$

For 100 m fill at embankment, the volume required from borrow pit

$$V_2 = \frac{100 \times 18}{15} = 120 \text{ m}^3$$

$$\text{Number of trips of truck} = \frac{120}{6} = 20$$

1.24 (d)

Cassagrande apparatus is used to determine the liquid limit of soils. Liquid Limit is defined as the water content at which the soil changes from liquid state to plastic state.

Hydrometer method is used to determine the particle size distribution of fine-grained soils passing 75 μ sieve. The hydrometer measures the specific gravity of the soil suspension at the centre of its bulb. The specific gravity depends upon the mass of solids present, which in turn depends upon the particle size.

Plate Load Test is a field test for determining the ultimate bearing capacity of soil and the likely settlement under a given load. The Plate Load Test basically consists of loading a steel plate placed at the foundation level and recording the settlements corresponding to each load increment.

An **oedometer test** is a test that measures a soil's consolidation properties. Oedometer tests are performed by applying different loads to a soil sample and measuring the deformation response. The results from these tests are used to predict how a soil in the field will deform in response to a change in effective stress.

1.25 (d)

At shrinkage limit the soil remains fully saturated,

$$w_s = \frac{e_{min}}{G} \times 100$$

$$= \frac{0.54}{2.7} \times 100 = 20\%$$

1.26 (a)

	Thickness (Å)	Lateral dimension (Å)
Kaolinite	100-1000	1000-20,000
Montmorillonite	10-50	1000-5000
Illite	50-500	1000-5000

Thus size of kaolinite > Illite > Montmorillonite

1.27 (c)

Feldspars are the most common rock minerals which account for the abundance of clays derived from the feldspars on the earth's surface. Quartz comes next in order of frequency. Most sands are composed of quartz.

Mica constitute very small fraction of original igneous rocks and thus it can not be said to be clay mineral.

Rock flour (rock dust particles) generally consists of more or less uniform size grains of quartz and they are least plastic varieties.

Kaolinite layer has no net electrical charge.

1.28 (d)

Unit weight,

$$\gamma = \frac{G_s(1+w)}{(1+e)}\gamma_w$$

$$1 + e = \frac{G_s(1+w)\gamma_w}{\gamma}$$

Using $G_s w = Se$

$$1+\frac{G_s w}{S} = \frac{G_s(1+w)\gamma_w}{\gamma}$$

Dividing both sides by G_s and rearranging, we get

$$S = \frac{w}{(\gamma_w/\gamma)(1+w) - 1/G_s}$$

1.29 (c)

The drying of soil is recommended at a temperature of 105°-110°C as temperature higher than 110°C may break the crystalline structure of clay particles and result in the loss of chemically bound water of crystallization.

1.32 (b)

Plasticity index, $I_P = w_L - w_P$
$= 35 - 20 = 15\%$

Liquidity index, $I_L = \frac{w_N - w_P}{I_P} = \frac{25-20}{15} = 0.33$

Shrinkage Index, $I_S = w_P - w_S = 20 - 10 = 10\%$

1.33 (b)

The maximum dry density of the cohesive soil is given by

$$\gamma_d = \frac{G\gamma_w}{1+e}$$

$$\Rightarrow \quad 16 = \frac{2.65\times 9.81}{1+e}$$

$$\Rightarrow \quad e = 0.625$$

1.34 (c)

Plasticity Index (I_P) is the numerical difference between liquid limit and plastic limit. The soil with higher value of I_P will be more plastic.

$$(I_P)_X = 42 - 20 = 22$$
$$(I_P)_Y = 56 - 34 = 22$$

But if plasticy index is same for both soils, then the soil with larger difference in liquid limit and natural water content is more plastic.

Consistency Index,

$$I_C = \frac{\text{Liquid limit} - \text{Natural water content}}{I_P}$$

$$\therefore \quad (I_C)_X = \frac{42-30}{22} = 0.545$$

$$(I_C)_Y = \frac{56-50}{22} = 0.273$$

Soil which has low value of consistency index is softer. Thus the softer soil will undergo large settlements under identical loads.

1.35 (c)

The values of plasticity index of kaolinite, Illite and montmorillonite are 15, 30 and 300 respectively.

1.36 (d)

Black cotton soils contain predominantly a clay mineral called montmorillonite, which is responsible for causing appreciable shrinking and swelling. Clay soils containing other clay minerals do not exhibit the volume change characteristics to the same degree as those containing montmorillonite mineral.

1.37 (d)

If V is the volume of wet soil mass, then

Volume of air, $V_a = \frac{V}{6}$

Volume of water $V_w = \frac{V}{3}$

But void ratio is given by

$$e = \frac{V_v}{V_s}$$

where V_v = Volume of voids

$$= V_a + V_w$$

$$= \frac{V}{6} + \frac{V}{3} = \frac{V}{2}$$

Also, $V_s = V - V_v = V - \frac{V}{2} = \frac{V}{2}$

$$\therefore \quad e = \frac{V/2}{V/2} = 1$$

1.38 (d)

Peat is a highly organic soil, consisting almost entirely of vegetable matter in varying stages of decomposition. It is brown to black in colour and possess a strong organic odour. Peat in general is highly compressible and hence one should be cautious while using it for supporting foundations of structures.

Muck is a mixture of fine particled, inorganic solid and black decomposed organic matter. It is usually found accumulated under conditions of imperfect drainage e.g. in swamps or is deposited by overflowing rivers.

Indurated clay is a soil that had been hardened by extremes of climate.

1.39 (a)

Plasticity Index, $I_P = W_L - W_P$

$= 40 - 20 = 20$

Activity of clay,

$$A_C = \frac{\text{Plasticity index}}{\text{Per cent finer than } 2\mu}$$

$$= \frac{\text{Plasticity index}}{\text{Per cent of clay size particles in the soil}}$$

$$= \frac{20}{100-70} = 0.67$$

Consistency Index,

$$I_C = \frac{W_L - W_N}{I_P} = \frac{40-30}{20} = 0.50$$

1.40 (d)

Clay soils have flaky particles to which water adheres, thus imparting the property of plasticity.

Kaolinite: This clay mineral is the weathering product of feldspars. It has a white, powdery appearance. Because kaolinite is electrically balanced, its ability of adsorb ions is less than that of other clay minerals.

1.41 (b)

Single grained structure is formed by the settlement of coarse grained soils in suspension in water. A flocculent structure is formed by the deposition of fine soil fraction (like clay particles) in water. The particles oriented in flocculent structure will have edge to face contact.

Honeycomb structure is formed by the disintegration of flocculent structure under superimposed load. The particles oriented in honeycomb structure will have face to face contact.

1.42 (b)

Allowable bearing pressure is determined by conducting a plate load test at the site. It is also used for determining the settlement for a given intensity of loading.

Pycnometer can be used to determine specific gravity as well as water content of soil sample.

Core cutter method is a field method for determination of mass density.

1.43 (d)

$$e = \frac{wG_S}{S}$$

∴ Degree of saturation,

$$S = \frac{wG_S}{e} = \frac{0.4 \times 2.7}{1.2} = 0.9 = 90\%$$

1.44 (b)

$$n = \frac{e}{1+e} = \frac{1}{2} = 50\%$$

1.45 (a)

$$i_c = \frac{G-1}{1+e}$$

Given, $n = 80\%$, using $n = \frac{e}{1+e}$

We get, $e = 4$

$$\therefore \quad i_c = \frac{2.7-1}{1+4} = 0.34$$

1.46 (b)

Core cutter method: One method of examining the in-situ density of the soil is the core cutter method. This method involves pushing steel rings in the soil to determine the wet and dry volume weight including the water content.

Sand Replacement Method: Determination of field density of cohesion less soil is not possible by core cutter method, because it is not possible to obtain a core sample. In such situation, the sand replacement method is employed to determine the unit weight.

Proctor test: The Proctor test is used to determine the maximum density that can be practically achieved for a soil. The Proctor soil compaction test is performed by measuring the density of the soil being tested at different moisture content points. The aim of the Proctor test is to determine the optimum moisture content for the soil.

1.47 (c*)

The term relative density is used to express the relative compactness of a natural cohesionless soil deposit. Relative density is the ratio of the difference between the voids ratio of soil in its loosest state e_{max}, and its natural state, e to the difference between the void ratios in the loosest and densest states.

$$I_D = \frac{e_{max} - e}{e_{max} - e_{min}}$$

This term is used for cohesionless soils only. This term is not applicable to cohesive soil because of uncertainities in the laboratory determination of the voids ratio in the loosest state of the soil (e_{max}).

Note: In point no. 2 soil mentioned should be cohesionless.

1.49 (c)

$$V_a = V/4, \quad V_w = V/2$$

$$\therefore \quad V_v = \frac{V}{4} + \frac{V}{2} = \frac{3V}{4}$$

$$\therefore \quad e = \frac{V_v}{V_s} = \frac{\frac{3V}{4}}{V - \frac{3}{4}V} = \frac{\frac{3V}{4}}{\frac{V}{4}} = 3$$

1.50 (d)

There are a lot of swelling soils which are brown or light coloured, even though most expansive soils are dark in colour.

Swelling pressure of a soil depends on lots of factors such as initial dry density, method of compaction, moisture content, height of specimen, surcharge pressure and nature of pore fluid.

1.51 (d)

$$\text{Porosity,} \quad n = \frac{e}{1+e} = \frac{0.5}{1+0.5} = 33\%$$

1.52 (c)

Mass of water in initial condition

$$= 32.2 - 20.2 = 12 \text{ gm}$$

Loss of water from original to shrinkage limit

$$= (21.8 - 11.6) \times 1 = 10.2 \text{ gm}$$

Shrinkage limit

$$= \frac{32.2 - 20.2 - 10.2}{20.2} \times 100 = 8.91 \simeq 8.9\%$$

1.53 (a)

Since clays are colloidal in nature and consequently their specific surface is very high. Due to very high specific surfaces of clay, the hygroscopicity of clay is more than sands.

1.54 (c)

$$wG = Se$$

$$w \times 2.7 = 1 \times 0.945$$

$$w = 0.35 \text{ i.e. } 35\%$$

1.55 (c)

$$w_s = 6\%; \quad G = 2.6$$

⇒ At shrinkage limit, soil is fully saturated.

$$se = wG$$

$$1.e = 0.06 \times 2.6$$

$$\Rightarrow \quad e = 0.156$$

$$n = \frac{e}{1+e} = \frac{0.156}{1+0.156}$$

$$= 0.1349 = 13.49\%$$

1.56 (b)

Bulk unit weight

$$\gamma_b = \frac{(G+Se)\gamma_w}{1+e}$$

$$= \frac{2.7 + 0.4 \times 0.5}{1+0.5} \times 9.81$$

$$= 19\,947 \text{ kN/m}^3$$

$$\Rightarrow \quad Se = wG$$

water content

$$w = \frac{0.7 \times 0.5}{2.7} = 0.12962$$

$$\gamma_d = \frac{\gamma_b}{1-w} = \frac{19.947}{1+0.12962}$$

$$= 17.658 \text{ kN/m}^3$$

***Note:** Dry unit weight has been asked but no answer is matching. But bulk unit weight is matching.*

1.57 (b)

Volume of solids will remain same.

$$(V_s)_{file} = (V_s)_{Borrowpit}$$

$$\left(\frac{V_T}{1+e}\right)_{fill} = \left(\frac{V_T}{1+e}\right)_{Borrowpit}$$

$$\frac{150000}{1+0.8} = 200000 \text{ m}^3$$

1.58 (d)

$$\text{Shrinkage ratio, } R = \frac{\gamma_d}{\gamma_w}$$

1.59 (d)

$$(\gamma_d)_{max} = 1.77 \text{ g/cc}$$

$$\text{OMC} = 14.44\%$$

$$G = 2.66$$

$$\gamma_d = \frac{G \cdot \gamma_w}{1+e}$$

$$\Rightarrow \quad 1.77 = \frac{2.66 \times 1}{1+e}$$

$$e = 0.5028$$

$$\Rightarrow \quad Se = wG$$

$$S = \frac{wG}{e} = \frac{0.1444 \times 2.66}{0.5028}$$

$$= 0.7639 = 76.39\%$$

1.60 (a)

Given, $V_S = 2V_V$

$V_T = V_S + V_V$

Porosity, $n = \frac{V_v}{3V_V} = 0.33 = 33\%$

1.61 (d)

- Montmorillonite is the mineral largely responsible for swelling and shrinkage. Activity of montmorillonite is 1 to 7 and activity of Illite is 0.5 – 1.
- Differential free swell value more then 50 indicates highly expansiveness.

1.62 (a)

Degree of saturation of saturated soil is 1.

$$S = \frac{V_w}{V_v} = 1$$

Hence, $V_w = V_v$

Relative density $I_D = \frac{e_{max} - e}{e_{max} - e_{min}} = 1$

i.e., $e = e_{min.}$

Hence soil is in densest condition.

1.63 (c)

Montmorillonite clay mineral is expansive having activity 1 to 7. Soil containing high Montmorillonite show more swelling and shrinkage.

1.64 (d)

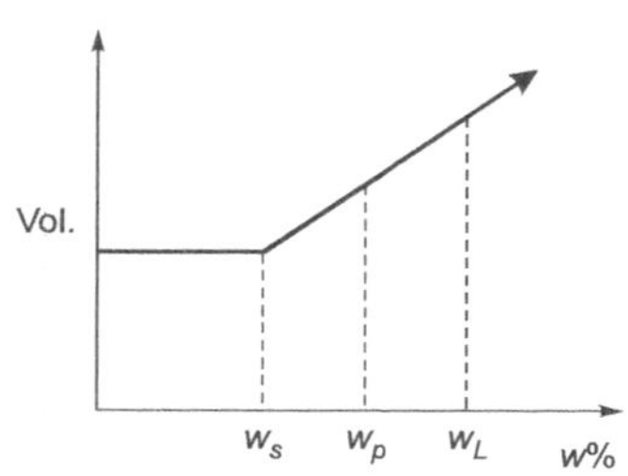

When soil is dried below shrinkage limit volume remains constant.

$w_s < w_p < w_L$

At liquid limit all the soils have almost same and negligible shear strength $\simeq$ 2.7 kN/m²

1.65 (c)

$$e = \frac{n}{1-n} = \frac{0.3}{1-0.3} = 0.428$$

$$Se = wG$$

$$s \times 0.428 = 4.94 \times 2.6$$

$$s = 30\%$$

1.66 (b)

$$Se = wG$$

$$1 \times e = 1.38 \times 2.65$$

$$e = 1.007$$

$$\gamma_{sat} = \frac{(G+e)\gamma w}{1+e}$$

$$= \frac{(2.65+1.007) \times 9.81}{1+1.007}$$

$$= 17.87 \text{ kN/m}^3$$

1.67 (d)

For soil-1 $e = 1.2$

$V = V_s(1 + e)$

$30 = V_s(1 + 1.2)$; $V_s = 13.63 \text{ m}^3$

For soil-2 $e = 0.7$

$V = V_s(1 + e)$

$V = 13.63\ (1 + 0.7) = 23.2 \text{ m}^3$

1.68 (b)

Volume of wax $= \frac{9}{G} = \frac{9}{0.9} = 10$cc

Weight of dry soil=wt. of solid $= w_s = 451$ gm

Volume of soil = 300 – 10 = 290 cc

Volume of solid $= V_s = \frac{W_s}{G} = \frac{451}{2.65} = 170.18$cc

Volume of air = Volume of void

$= V_v = 290 - 170.18$

$V_v = 119.81$cc

$$e = \frac{V_v}{V_s} = \frac{119.81}{170.18} = 0.70$$

$$e = 0.70 \simeq 0.71$$

1.69 (a)

Thixotropy is that property of soil due to which loss in shear strength due to remoulding at unaltered moisture content may be regained. This increase in strength is due to the gradual reorientation of the adsorbed molecules of water.

1.70 (d)

Given data:

$w_P = 30\%$

$w_L = 42\%$

$V_L - V_d = 0.35\ V_d$

$V_P - V_d = 0.22\ V_d$

$V_L = 1.35\ V_d$

$V_P = 1.22\ V_d$

$$R = \frac{\left(\frac{V_1 - V_2}{V_d}\right)}{w_1 - w_2} \times 100 = \frac{\frac{V_L - V_P}{V_d}}{w_L - w_P} \times 100$$

$$= \frac{\left(\frac{1.35V_d - 1.22V_d}{V_d}\right)}{42 - 30}$$

$$R = \frac{0.13}{12} \times 100 = 1.083 \simeq 1.1$$

1.71 (c)

$$\text{Shrinkage ratio } R = \frac{\frac{V_1 - V_2}{V_d}}{w_1 - w_2} \times 100$$

1.72 (c)

$$\text{Bulk unit weight, } \gamma_b = \frac{(G + Se)\gamma_w}{1 + e}$$

$$\text{Void ratio, } e = \frac{n}{1-n} = \frac{0.4}{1-0.4} = \frac{2}{3}$$

$$\therefore \quad \gamma = \left\{\frac{2.7 + 0.5 \times \frac{2}{3}}{1 + \frac{2}{3}}\right\} 9.81$$

$$= 17.85 \text{ kN/m}^3$$

1.73 (d)

Dry mass of clay

$= 10.8$ gm

Volume of mercury displaced

= Total volume of soil mass

$$= \frac{\text{Mass of mercury}}{\text{Unit weight of mercury}}$$

$$= \frac{84.2}{13.6} = 6.1911 \text{ cc}$$

Dry unit weight of soil,

$$\rho_d = \frac{M_{dry}}{V_T} = \frac{10.8}{6.1911} = 1.744$$

Shrinkage ratio,

$$R = \frac{\rho_d}{\rho_w} = \frac{1.744}{1} = 1.744$$

Shrinkage limit,

$$w_s = \frac{1}{R} - \frac{1}{G}$$

$$w_s = \frac{1}{1.744} - \frac{1}{2.72} = 0.2056$$

$= 20.56\%$

Alternatively

Weight of solids,

$W_s = 10.8$ gm

$G_s = 2.72$

Density of mercurym,

$\rho_m = 13.6$ g/cc

Volume of mercury displaced,

$$V = \frac{84.2}{13.6} = 6.19 \text{ cc}$$

$\therefore$Volume of soil,

$V = 6.19$ cc

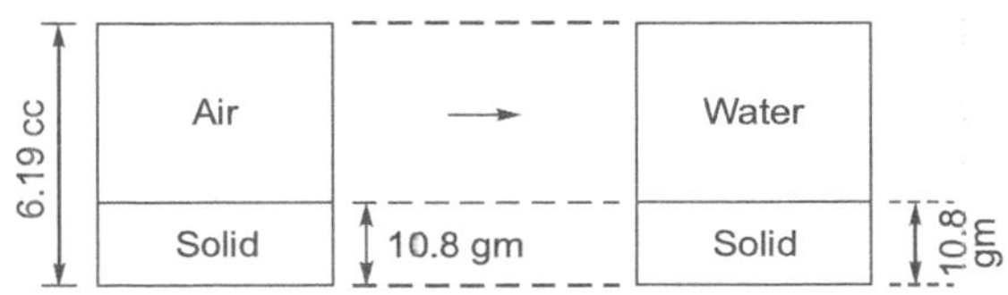

At shrinkage limit soil is just saturated

$\therefore$Volume of water

= Volume of void in dry state

$$= 6.19 - \frac{10.8}{2.72} = 2.22 \text{ cc}$$

So shrinkage limit,

$$w_s = \frac{W_w}{W_s} = \frac{2.22 \times 1}{10.8} \times 100\%$$

$= 20.55\% \simeq 21\%$

1.74 (d)

Suitability Number

$$= 1.7\sqrt{\frac{3}{(D_{50})^2} + \frac{3}{(D_{20})^2} + \frac{1}{(D_{10})^2}}$$

$$= 1.7\sqrt{\frac{3}{(1)^2} + \frac{3}{(0.5)^2} + \frac{1}{(0.08)^2}}$$

$= 21.72$

1.75 (a)

$$n = \frac{e}{1+e}$$

$$n = \frac{V_V}{V} = \frac{\frac{V_V}{V_S}}{\frac{(V_S+V_S)}{V_S}} \quad \ldots\ldots \quad \left(e = \frac{V_V}{V_S}\right)$$

$$n = \frac{e}{1+e} \text{ proved}$$

1.76 (a)

Case-1:

Bulk density, $\gamma = 20.1 \text{ kN/m}^3$

Water content, $w = 15\%$

Case-2:

When soil partially dries to

$$\gamma = 19.4 \text{ kN/m}^3$$

$$w = ?$$

$$\gamma = \left(\frac{G_S + Se}{1+e}\right)\gamma_w$$

$$\gamma = \frac{G_S\gamma_w(1+w)}{1+e}$$

Since, void ratio remain same

$$\therefore \quad \frac{\gamma_1}{\gamma_2} = \frac{1+w_1}{1+w_2}$$

($\because$ G_s and γ_w also same)

$$\frac{20.1}{19.4} = \frac{1+0.15}{1+w_2}$$

$$w_2 = \frac{1.15\times 19.4}{20.1} - 1$$

$$= 0.1099 \simeq 10.99\%$$

No option is matching, answer can be taken approximately as 'a' because it is the closest.

1.77 (c)

Given:

Liquid limit, $w_l = 90\%$

Plasticity index, $I_p = 50\%$

Natural moisture content,

$$w_n = 28\%$$

$$I_P = w_l(\%) - w_p(\%)$$

$$50 = 90 - w_p$$

$$\therefore \quad w_p = 40\%$$

Liquidity index, $$I_L = \frac{w_n - w_p}{I_P} = \frac{28-40}{50}$$

$$= -\frac{12}{50} = -0.24$$

1.78 (c)

For a sample of silty clay. The data given can be represented in phase diagram as:

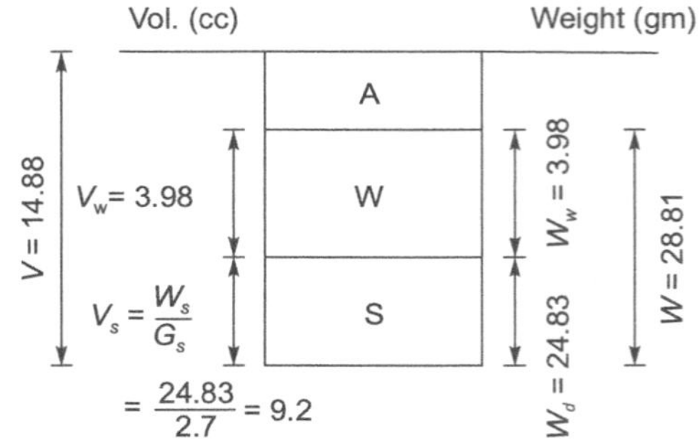

$$G_s = 2.7$$

For the data given, all value are shown in phase diagram.

Volume of air,

$$V_a = 14.88 - (3.98 + 9.2) = 1.7 \text{ cc}$$

Volume of void,

$$\therefore \quad V_v = V_a + V_w$$

$$= 1.7 + 3.98 = 5.68 \text{ cc}$$

$$\text{Void ratio, } e = \frac{V_V}{V_S} = \frac{5.68}{9.2} = 0.618$$

1.79 (d)

Given: Void ratio,

$$e = 0.5$$

Degree of saturation,

$$s = 70\%$$

Sp. gr. of soil solids,

$$G_s = 2.7$$

Bulk unit weight,

$$\gamma_t = \frac{(G_S + Se)\gamma_w}{1+e}$$

$$= \frac{(2.7 + 0.7\times 0.5)\times 9.81}{1+0.5}$$

$$= 19.95 \text{ kN/m}^3$$

1.81 (c)

Specific gravity of soil,

$$G = \frac{W_2 - W_1}{(W_4 - W_3) + (W_2 - W_1)}$$

$$= \frac{W_d}{(W_4 - W_3) + W_d}$$

$$\Rightarrow \quad = \frac{200}{1480 - 1605 + 200}$$

$$= \frac{200}{75} = \frac{8}{3} = 2.667$$

1.82 (c)

Porocity, $n = 40\% = 0.4$

Void ratio, $e = \dfrac{n}{1-n} = \dfrac{0.4}{1-0.4}$

$= \dfrac{2}{3} = 0.667$

1.84 (a)

Capillary water: The water that is stored in the soil after the gravitational water is drained out is called capillary water. They are stored in the capillary spaced around the soil granules against the force of gravity by the capillary force in them.They are available to the plant root absorption.

NOTE: The closest answer is capillary water (a).

Field capacity: The maximum amount of capillary water remaining in the soil after the removal of gravitational water (is called its field capacity). In general it represents the water held at a tension of 1/3 atmosphere.

1.86 (d)

Thixotropy is the phenomenon of strength loss-strength gain with no change in volume or water content.

1.87 (b)

If cohesive soil is drained then it will show the property of swelling and shrinkage.

1.90 (b)

As the humus soil is organic soil and specific gravity of organic soil is less than 2, So its specific gravity is 1.37.

191 (a)

Water content, $w = \left\{\dfrac{W_2 - W_1}{W_3 - W_4}\left(\dfrac{G-1}{G}\right) - 1\right\}$

Where G is specific gravity. So if 'G' is known, then this method can be used.

■■■■

2 Classification of Soils

2.1 Match **List-I** (Soils) with **List-II** (Group symbols) and select the correct answer using the codes given below the lists:

	List-I		List-II
A.	Clayey gravel	**1.**	SM
B.	Clayey sand	**2.**	OH
C.	Organic clay	**3.**	SC
D.	Silty sand	**4.**	GC

Codes:

	A	B	C	D
(a)	3	4	2	1
(b)	4	3	1	2
(c)	4	3	2	1
(d)	3	4	1	2

[ESE : 2000]

2.2 Based on grain distribution analysis, the D_{10}, D_{30} and D_{60} values of a given soil are 0.23 mm, 0.3 mm and 0.41 mm respectively. As per **IS code**, the soil classification will be

(a) SW (b) SP
(c) SM (d) SC [ESE : 2001]

2.3 Given that coefficient of curvature = 1.4,

D_{30} = 3 mm, D_{10} = 0.6 mm.

Based on this information of particle size distribution for use as subgrade, this soil will be taken to be

(a) uniformly-graded sand
(b) well-graded sand
(c) very fine sand
(d) poorly-graded sand [ESE : 2001]

2.4 In a soil specimen, 70% of particles are passing through 4.75 mm IS sieve and 40% of particles are passing through 75 μ **IS** sieve. Its uniformity coefficient is 8 and coefficient of curvature is 2. As per **IS** classification, this soil is classified as

(a) SP (b) GP
(c) SW (d) GW

[ESE : 2002]

2.5 The laboratory tests on a sample yielded the following results:

Plasticity index : 32%
Liquidity index : 0.15
Activity number : 1.58

Which of the following inferences can be drawn?

1. The soil is very stiff.
2. The soil is medium soft.
3. The soil is highly plastic.
4. The soil is medium plastic.
5. The soil is active.

Select the correct answer using the codes given below :

(a) 1, 3 and 5 (b) 1, 3, and 4
(c) 2, 3 and 5 (d) 1, 2 and 4

[ESE : 2005]

2.6 Match **List-I** (Soil classification symbol) with **List-II** (Soil property) and select the correct answer using the codes given below the lists:

	List-I		List-II
A.	GW	**1.**	Soil having uniformity coefficient > 6
B.	SW	**2.**	Soil having uniformity coefficient > 4
C.	ML	**3.**	Soil have low plasticity
D.	CL	**4.**	Soil have low compressibility

Codes:

	A	B	C	D
(a)	1	2	4	3
(b)	2	1	3	4
(c)	2	1	4	3
(d)	1	2	3	4

[ESE : 2008]

2.7 Match **List-I** (Soil) with **List-II** (Type) and select the correct answer using the codes given below the lists:

	List-I		List-II
A.	Fine sand	**1.**	Expansive soil
B.	Silt	**2.**	Coarse grained soil
C.	Peat	**3.**	Fine grained
		4.	Organic soil

Codes:

	A	B	C
(a)	3	2	1
(b)	4	3	1
(c)	3	1	2
(d)	2	3	4

[ESE : 2008]

2.8 Match **List-I** (Symbol) with **List-II** (Soil) and select the correct answer using the codes given below the lists:

List-I	List-II
A. *ML*	1. Silty sand
B. *SM*	2. Inorganic silt with large compressibility
C. *Pt*	3. Inorganic silt with small compressibility
D. *MH*	4. Soil with high organic content with high compressibility

Codes:

	A	B	C	D
(a)	3	2	4	1
(b)	4	1	3	2
(c)	3	1	4	2
(d)	4	2	3	1

[ESE : 2008]

2.9 Match **List-I** (Measuring device) with **List-II** (Soil parameter) and select the correct answer using the codes given below the lists:

List-I	List-II
A. Pycnometer	1. Compressibility
B. Hydrometer	2. Permeability
C. Oedometer	3. Specific gravity
D. Permeameter	4. Particle size analysis

Codes:

	A	B	C	D
(a)	2	4	1	3
(b)	3	1	4	2
(c)	2	1	4	3
(d)	3	4	1	2

[ESE : 2010]

2.10 Consider the following statements:

1. The minimum value of group index for a soil can be taken as zero.
2. The maximum possible value of group index for a soil is twenty.

Which of these statements is/are correct?

(a) Both 1 and 2 (b) 1 only
(c) 2 only (d) Neither 1 nor 2

[ESE : 2010]

2.11 Consider the following statements:

1. For a well graded sand, the coefficient of curvature should lie between 1 and 3.
2. A soil having uniformity coefficient smaller than about 2.0 is considered uniform.

Which of these statements is/are correct?

(a) Both 1 and 2 (b) Neither 1 nor 2
(c) 1 only (d) 2 only

[ESE : 2010]

2.12 On analysis of particle size distribution of a soil, it is found that D_{10} = 0.1 mm, D_{30} = 0.3 mm and D_{60} = 0.8 mm. The uniformity coefficient and coefficient of curvature, as given by the particle size distribution curve, are respectively

(a) 3 and 3 (b) 2.67 and 1.125
(c) 2.67 and 3 (d) 8 and 1.125

[ESE : 2011]

2.13 Which one of the following statements is correct?

(a) Grain size is the primary criterion for classification of coarse, as well as fine-grained soil
(b) Grain size is the primary criterion for classification of coarse-grained soils
(c) Plasticity curve classifies coarse grained soils
(d) Plasticity characteristics relate to classification of coarse-grained soils

[ESE : 2013]

2.14 Consider the following statements:

1. A conspicuous break in the continuity of a grain size distribution curve indicates a mixture of soil from two different layers.
2. A steep grain size distribution curve indicates prevalence of nearly uniform grain size.
3. A flat grain size distribution curve indicates certain range of missing grain sizes.

Which of these statements are correct?

(a) 1, 2 and 3 (b) 2 and 3 only
(c) 1 and 3 only (d) 1 and 2 only

[ESE : 2013]

2.15 **Statement (I):** A semi-log plot is used to represent the grain size distribution of a soilsample.

Statement (II): In a semi-log plot, a wide range of grain sizes can be accommodated.

(a) Both Statement (I) and Statement (II) are

individually true and Statement (II) is the correct explanation of Statement (I)
(b) Both Statement (I) and Statement (II) are individually true but Statement (II) is NOT the correct explanation of Statement (I)
(c) Statement (I) is true but Statement (II) is false
(d) Statement (I) is false but Statement (II) is true

[ESE : 2014]

2.16 The standard plasticity chart by Casagrande to classify fine-grained soils is shown in the figure.

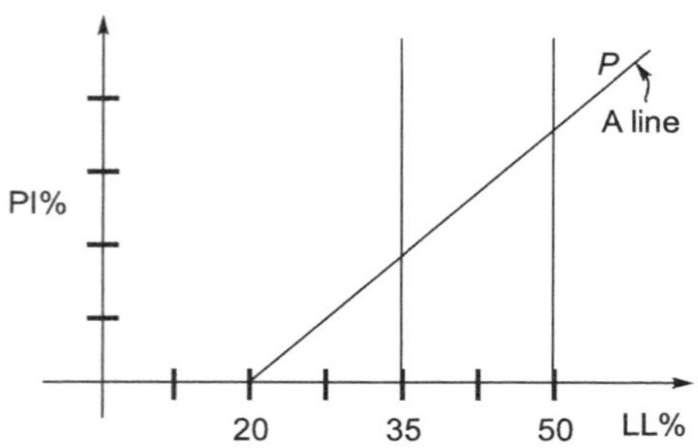

The area marked *P* represents
(a) Inorganic clays of high plasticity
(b) Organic clays and highly plastic organic silts
(c) Organic and inorganic silts and silt-clays
(d) Clays

[ESE : 2016]

2.17 Consider a soil sample, for which test yield the following results:
Passing 75 micron sieve 62%
Liquid limit 35%
Plasticity index 14%
As per the group index classification of soil, what is the soil condition of the above soil sample?
(a) Poor (b) Fair
(c) Good (d) Excellent

[ESE : 2016]

2.18 According to Highway Research Board (HRB) classification system, which one of the following is NOT relevant for dependency of group index of soil?
(a) The amount of material passing the 75-micron IS sieve
(b) The liquid limit
(c) The plastic limit
(d) The shrinkage limit

[ESE : 2022]

2.19 The unified classification system is based on the recognition of the type and predominance of the constituents considering
(a) grain length (b) gradation
(c) elasticity (d) non-compressibility

[ESE : 2023]

Answers Classification of Soils

2.1 (c) 2.2 (b) 2.3 (b) 2.4 (d) 2.5 (a) 2.6 (b) 2.7 (d) 2.8 (c) 2.9 (d)
2.10 (a) 2.11 (a) 2.12 (d) 2.13 (b) 2.14 (a) 2.15 (a) 2.16 (a) 2.17 (a) 2.18 (d)
2.19 (b)

Explanations Classification of Soils

2.1 (c)

The Unified Soil Classification System (USCS) is a soil classification system used to describe the texture and grain size of a soil. The classification system can be applied to most unconsolidated materials, and is represented by a two-letter symbol. Each letter is described below (with the exception of Pt):

First and / or Second Letters

Letter	Definition
G	gravel
S	sand
M	silt
C	clay
O	organic

Second Letters

Letter	Definition
P	poorly graded (uniform particle sizes)
W	well-graded (diversified particle sizes)
H	high plasticity
L	low plasticity

2.2 (b)

Coefficient of uniformity,

$$C_u = \frac{D_{60}}{D_{10}} = \frac{0.41}{0.23} = 1.78$$

Coefficient of curvature,

$$C_c = \frac{D_{30}^2}{D_{60} \times D_{10}} = \frac{0.3^2}{0.41 \times 0.23} = 0.95$$

For well graded sand $1 < C_c < 3$ and $C_u > 6$ as per IS classification. So it is a poorly graded sand.

2.3 (b)

Coefficient of curvature,

$$C_c = \frac{D_{30}^2}{D_{60} \times D_{10}}$$

$$\Rightarrow \quad D_{60} = \frac{D_{30}^2}{C_c \times D_{10}}$$

Coefficient of uniformity,

$$C_u = \frac{D_{30}^2}{C_c \times D_{10}^2} = \left(\frac{3}{0.6}\right)^2 \times \frac{1}{1.4} = 17.9$$

As $1 < C_c < 3$ and $C_u > 6$ so it is well graded sand.

2.4 (d)

As less than 50% are passing through 75 μ IS Sieve, it is coarse grained soil. As % Sand (30%) = % Gravel (30%) willbe classified as Gravelly soil.

Also C_u (8) > 6 and C_c (2) is between 1 and 3 so it is well graded (W). So symbol will be GW as per IS.

2.5 (a)

Consistency index (CI)	Liquidity index (LI)	Consistency
1.00 - 0.75	0.00 - 0.25	Stiff
0.75 - 0.50	0.25 - 0.50	Medium soft
0.50 - 0.25	0.50 - 0.75	Soft
0.25 - 0.00	0.75 - 1.00	Very soft

$$CI = \frac{w_L - w}{PI}$$

and $$LI = \frac{w - w_P}{PI}$$

where w is natural water content
w_L is liquid limit
w_p is plastic limit
PI is plasticity index

$$CI + LI = 1.00$$

If $CI > 1$ or $LI < 0$, then soil is very stiff.
If $CI < 0$ or $LI > 1$ soil behaves as liquid.

Activity number	Classification
0.75	Inactive
0.75 - 1.25	Normal
> 1.25	Active

So soil is active

Liquid limit	Plasticity
0 - 30 %	Low
35 - 50%	Medium
> 50%	High

Plasticity Index	Plasticity
0	Non plastic
1 - 5	Slight
5 - 10	Low
10 - 20	Medium
20 - 40	High
> 40	Very high

So soil is high plastic

2.6 (b)

GW is well graded gravel for which coefficient of uniformity (C_u) > 4.
SW is well graded sand for which coefficient of uniformity (C_u) > 6
ML is silt with low plasticity (< 35%)
CL is clay with low plasticity (< 35%). It also possess low compressibility.

2.7 (d)

Sand (both fine and coarse – Coarse grained soil
Silt – Fine grained soil
Peat – Organic soil

2.9 (d)

The **Pycnometer** is used for determination of both the specific gravity of soil particles and water content of soil. The specific gravity of soil is determined using the relation:

$$G = \frac{M_2 - M_1}{(M_2 - M_1) - (M_3 - M_4)}$$

Water content is determined by using the relation

$$w = \left[\frac{(M_2 - M_1)}{(M_3 - M_4)} \times \left(\frac{G-1}{G}\right) - 1\right] \times 100$$

Where M_1 = mass of empty Pycnometer,
M_2 = mass of the Pycnometer with dry soil
M_3 = mass of the Pycnometer and soil and water
M_4 = mass of Pycnometer filled with water only.
G = Specific gravity of soils.

Hydrometer method is used to determine the particle size distribution of fine-grained soils passing 75 μ sieve. The hydrometer measures the specific gravity of the soil suspension at the centre of its bulb. The specific gravity depends upon the mass of solids present, which in turn depends upon the particle size.

An **oedometer test** is a test that measures a soil's consolidation properties. Oedometer tests are performed by applying different loads to a soil sample and measuring the deformation response. The results from these tests are used to predict how a soil in the field will deform in response to a change in effective stress.

The constant head **permeameter** is a test used to measure the permeability wherein the head causing flow through the specimen remains constant throughout the test.

2.10 (a)

For qualitative evaluation of the desirability of a soil as a highway subgrade material, a number referred to as the group index has been developed. It varies from a value of 0 to 20, lower the value higher is the quality of the sub-grade and greater the value, poor is the sub-grade.

2.11 (a)

For a soil to be well graded coefficient of curvature (C_c) must lie between 1 and 3. In addition to this coefficient of uniformity (C_u) must be greater than 4 for gravels and greater than 6 for sands.

A soil is said to be very uniform, if $C_u < 5$; it is said to be of medium uniformity, if $C_u = 5$ to 15; and it is said to be very non-uniform or well graded, if $C_u > 15$

2.12 (d)

$$C_u = \frac{D_{60}}{D_{10}} = \frac{0.8}{0.1} = 8$$

$$C_c = \frac{(D_{30})^2}{D_{60} \times D_{10}} = \frac{(0.3)^2}{0.8 \times 0.1} = 1.125$$

2.13 (b)

Coarse grained soils are classified based on grain size and fine grained soils are classified based on Atterberg's limits.

2.14 (a)

Both the position and the shape of the grading curve for a soil can aid in establishing its identity and description. Some typical grading curves are shown.

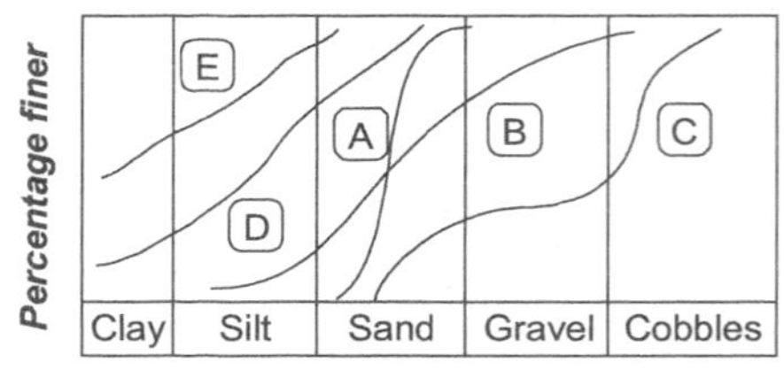

Curve A - a poorly-graded medium SAND
Curve B - a well-graded GRAVEL-SAND (i.e. having equal amounts of gravel and sand)
Curve C - a gap-graded COBBLES-SAND
Curve D - a sandy SILT
Curve E - a silty CLAY (i.e. having little amount of sand)

2.15 (a)

In view of large range of particles sizes, a logarithmic scale becomes necessary. Further a semilogarithmic plot has the merit that soils of equal uniformity exhibit the same shape of grain size curve, irrespective of the particles size frictions present in the soil.

2.16 (a)

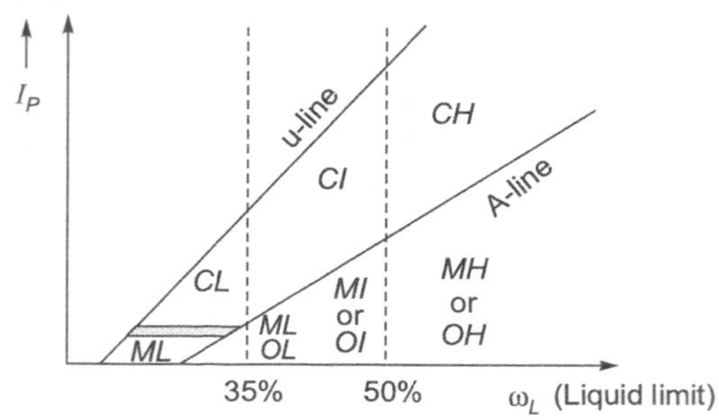

Plasticity curve

CH = High plastic inorganic clay

2.17 (a)

$$p = 62\%$$
$$LL = 35\%$$
$$I_P = 14\%$$
$$GI = 0.2a + 0.005ac + 0.01bd$$
$$a = p - 35 = 62 - 35 = 27 \not> 40 = 27$$
$$b = p - 15 = 62 - 15 = 47 \not> 40 = 40$$
$$c = LL - 40 = 0$$
$$d = I_P - 10 = 14 - 10 = 4$$
$$GI = 0.2 \times 27 + 0.005 \times 27 \times 0 + 0.01 \times 40 \times 4 = 5.4 + 1.6 = 7$$

General Evaluation of Subgrade	Group Index Range of Subgrade
Excellent	
Good	0-1
Fair	2-4
Poor	5-9
Very Poor	10-20

2.18 (d)

Group India of soil,

$$GI = 0.2\,a + 0.005\,ac + 0.01\,bd$$

Where, a = % finer passing 75μ sieve $= p - 35 \not> 40$

b = % finer passing 75μ sieve $= p - 15 \not> 40$

$c = w_L - 40 \not> 20$

$d = I_P - 10 \not> 20$

w_L = Liquid limit

I_P = Plasticity Index

2.19 (b)

Unified classification system is based on recognition of the type and predominance of the constituents considering grain size, gradation, plasticity and compressibility.

■■■■

3 Effective Stresses and Permeability

3.1 Which one of the following diagrams represents the effective pressure distribution for a saturated soil mass of depth 'z' submerged under water of height 'z_1' above its top level (γ' = submerged density of soil, γ_{sat} = saturated density of soil and γ_w = unit weight of water)?

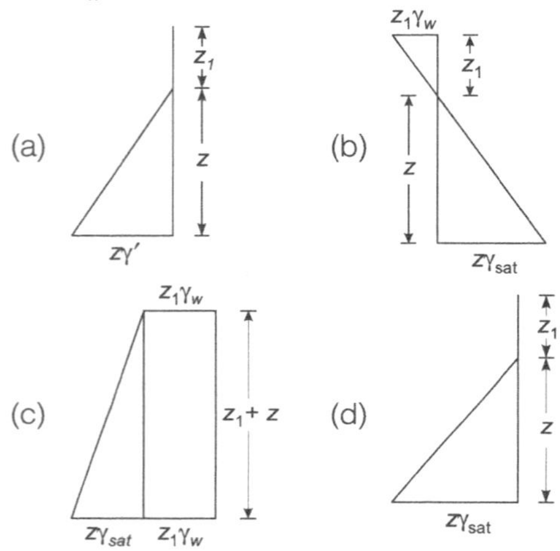

[ESE : 2000]

3.2 In a two-layer soil system, the top soil and bottom soil are of same thickness but the coefficient of permeability of the top soil is twice that of the bottom soil of coefficient of permeability 'k'. When horizontal flow occurs, the equivalent coefficient of permeability of the system will be

(a) 2 k (b) 1.5 k

(c) 1.25 k (d) 1.2 k

[ESE : 2000]

Directions: The following items consists of two statements; one labelled as **'Assertion (A)'** and the other as **'Reason (R)'**. You are to examine these two statements carefully and select the answers to these items using the codes given below:

Codes:

(a) both A and R are true and R is the correct explanation of A

(b) both A and R are true but R is not a correct explanation of A

(c) A is true but R is false

(d) A is false but R is true

3.3 **Assertion (A):** Effective vertical stress at some depth below a river bed is unaffected by the water depth in the river.

Reason (R): Equal amounts of increase in total stress and pore pressure will not change the effective stress.

[ESE : 2001]

3.4 **Assertion (A):** Permanent lowering of GWT results in settlements.

Reason (R): Increase in effective stress results in settlement of soils.

[ESE : 2001]

3.5 The following data were obtained when a sample of medium sand was tested in a constant head permeameter:

Cross-section area of sample	:	100 cm^2
Hydraulic gradient	:	10
Discharge collected	:	10 cc/s

The coefficient of permeability of the sand is

(a) 0.1 m/s (b) 0.01 m/s

(c) 1×10^{-4} m/s (d) 1×10^{-8} m/s

[ESE : 2001]

3.6 The total, neutral and effective vertical stresses (in t/m^2) at a depth of 5 m below the surface of a fully saturated soil deposit with a saturated density of 2 t/m^3 would, respectively, be

(a) 5, 5 and 10 (b) 5, 10 and 5

(c) 10, 5 and 10 (d) 10, 5 and 5

[ESE : 2001]

3.7 A stratified soil deposit has three layers of thicknesses : $z_1 = 4$, $z_2 = 1$, $z_3 = 2$ units and the corresponding permeabilities of $k_1 = 2$, $k_2 = 1$ and $k_3 = 4$ units, respectively. The average permeability perpendicular to the bedding planes will be

(a) 4 (b) 2

(c) 8 (d) 16 [ESE : 2003]

3.8 Match **List-I** (Soil description) with **List-II** (Coefficient of permeability, mm/s) and select the correct answer using the codes given below the lists:

List-I	List-II
A. Gravel	1. > 1
B. Clay silt admixtures	2. 10^{-2} to 10^{-4}
C. Loess	3. < 10^{-6}
D. Homogeneous clays	4. 10^{-4} to 10^{-6}

Codes:

	A	B	C	D
(a)	4	1	3	2
(b)	1	4	3	2
(c)	4	1	2	3
(d)	1	4	2	3

[ESE : 2003]

3.9 Consider the following statements:
1. Coarse sand is more than a million times permeable than a high plasticity clay.
2. The permeability depends on the nature of soil and not on properties of liquid flowing through soil.
3. If a sample of sand and a sample of clay have the same void ratio, both samples will exhibit the same permeability.
4. Permeability of soil decreases as the effective stress acting on the soil increases.

Which of these statements are correct?

(a) 1 and 2 (b) 1 and 3
(c) 1 and 4 (d) 2 and 3

[ESE : 2004]

3.10 In a falling head permeability test on a soil, the time taken for the head to fall from h_0 to h_1 is t. The test is repeated with same initial head h_0, the final head h' is noted in time $t/2$. Which one of the following equations gives the relation between h', h_0 and h_1?

(a) $h' = h_0/h_1$ (b) $h' = \sqrt{h_0 / h_1}$
(c) $h' = h_0h_1$ (d) $h' = \sqrt{h_0h_1}$

[ESE : 2004]

3.11 Assertion (A): The drained tests using triaxial test apparatus are useful in the study of drainage of water through the soil sample and hence the permeability of the soil sample.

Reason (R): The permeability of soil is an important property useful in estimation of loss of imporded water through permeable soil layers below the earth dams.

(a) both A and R are true and R is the correct explanation of A
(b) both A and R are true but R is not a correct explanation of A
(c) A is true but R is false
(d) A is false but R is true

[ESE : 2004]

3.12 A soil deposit has three layers having same thickness each but the permeabilities of the layers are in the ratio of 1 : 2 : 4 from top to bottom. What is the ratio of average permeability in the horizontal direction to that in the vertical direction?

(a) 7 : 2 (b) 14 : 6
(c) 28 : 24 (d) 49 : 36

[ESE : 2005]

3.13 Consider the following statements:
1. Organic matter increases the permeability of a soil.
2. Entrapped air decreases the permeability of a soil.

Which of these statements is/are correct?

(a) 1 only (b) 2 only
(c) Both 1 and 2 (d) Neither 1 nor 2

[ESE : 2007]

3.14 Consider the following statements:
1. Permeability of a soil decreases as the effective stress acting on the soil increases.
2. The presence of organic matter in the soil increases its permeability.
3. Entrapped air decreases the permeability of a soil.

Which of these statements is/are correct?

(a) 1 only (b) 1 and 2
(c) 2 and 3 (d) 1 and 3

[ESE : 2009]

3.15 Consider the following statements
The coefficient of permeability k depends upon
1. void ratio of the soil.
2. duration of flow.

3. equivalent diameter of the soil grains.
4. shape of the particle.

Which of these statements are correct?

(a) 1, 2, 3 and 4 (b) 2 and 3 only
(c) 1, 3 and 4 only (d) 3 and 4 only

[ESE : 2010]

3.16 A soil has discharge velocity of 5×10^{-7} m/s and a void ratio of 0.50. Its seepage velocity will be

(a) 15×10^{-7} m/s (b) 10×10^{-7} m/s
(c) 20×10^{-7} m/s (d) 30×10^{-7} m/s

[ESE : 2010]

3.17 Consider the following statements:

1. Organic matter decreases the permeability of a soil.
2. Entrapped air decreases the permeability of a soil.

Which of these statements are correct?

(a) 1 only (b) 2 only
(c) Both 1 and 2 (d) Neither 1 nor 2

[ESE : 2012]

3.18 Two observation wells penetrating into a confined aquifer are located 1500 m apart in the direction of flow. Heads of 50 m and 25 m are indicated at these two observation wells. If the coefficient of permeability for the aquifer is 30 m/day and its porosity is 0.25, the time of travel of an inert tracer from one well to another is

(a) 75 days (b) 750 days
(c) 1200 days (d) 3000 days

[ESE : 2012]

3.19 Consider the following statements:

1. Knowledge about moisture content is necessary to determine permeability.
2. Some typical marine clay may have moisture content as high as 400%.
3. Sand-bath method of moisture content determination is not suitable for soils containing organic matter.

Which of the above statements are correct?

(a) 1, 2 and 3 (b) 1 and 2 only
(c) 2 and 3 only (d) 1 and 3 only

[ESE : 2014]

3.20 The void ratio of a given soil *A* is twice that of another soil *B*, while the effective size of particles of soil *A* is one-third of that of soil *B*. The ratio of height of capillary rise of water in soil *A* to that in soil *B* will be

(a) 0.67 (b) 1.0
(c) 1.5 (d) 2.0

[ESE : 2014]

3.21 A stratum of soil consists of three layers of equal thickness. The permeability of both the top and the bottom layers is 10^{-4} cm/s; and that of the middle layer is 10^{-3} cm/s; then the value of the horizontal coefficient of permeability for the entire composite of the soil layers is

(a) 2×10^{-4} cm/s (b) 3×10^{-4} cm/s
(c) 4×10^{-4} cm/s (d) 5×10^{-4} cm/s

[ESE : 2015]

3.22 **Assertion (A):** Permanent lowering of ground water table results in settlement of foundations.

Reason (R): Increase in effective stress does not result in settlement of strata.

(a) Both A and R are true and R is the correct explanation of A
(b) Both A and R are true but R is not a correct explanation of A
(c) A is true but R is false
(d) A is false but R is true

[ESE : 2015]

3.23 In a three-layered soil system, the thicknesses of the top and bottom layers each are half the thickness of the middle layer. The coefficients of permeability of top and bottom layers each are double the coefficients of permeability *k* of the middle layer. When horizontal flow occurs, the equivalent coefficient of permeability of the system will be

(a) 1.5 *k* (b) 3.0 *k*
(c) 4.5 *k* (d) 6.0 *k*

[ESE : 2016]

3.24 An extended layer of soil with homogeneous rounded grains has 10% of the material finer than 0.07 mm. The constant to be adopted to determine its permeability has been recommended as 750. What is its permeability?

(a) 2.583 m/day (b) 2.857 m/day
(c) 3.244 m/day (d) 3.675 m/day

[ESE : 2017]

3.25 Consider the following statements in the context of capillary pressure in soils:

1. Pore water pressure is negative in capillary

zone.

2. Water is in tension in capillary zone.
3. Capillary pressure is more in coarse grained soils.

Which of the above statements are correct?

(a) 1 and 2 only (b) 1 and 3 only
(c) 2 and 3 only (d) 1, 2 and 3

[ESE : 2017]

3.26 Statement (I): Fine-grained soils are difficult to drain.

Statement (II): Capillary forces act on pore water.

(a) Both Statement (I) and Statement (II) are individually true and Statement (II) is the correct explanation of Statement (I)
(b) Both Statement (I) and Statement (II) are individually true but Statement (II) is ***not*** the correct explanation of Statement (I)
(c) Statement (I) is true but Statement (II) is false
(d) Statement (I) is false but Statement (II) is true

[ESE : 2017]

3.27 A constant head permeability test is carried out on a cylindrical sample of sand 10 cm diameter and 15 cm height 160 cm^3 of water is collected in 1.75 minutes, under a head of 30 cm. What is the coefficient of permeability in m/year?

(a) 1257 m/year (b) 2111 m/year
(c) 3060 m/year (d) 3332 m/year

[ESE : 2021]

3.28 Which one of the following does NOT affect the permeability of soils?

(a) Void ratio (b) Soil strength
(c) Grain size (d) Temperature

[ESE : 2022]

3.29 Which of the following is/are the laboratory methods of determination of coefficients of permeability of soils?

1. Constant head permeability method
2. Falling head permeability method

Select the correct answer using the codes given below:

(a) 1 only (b) 2 only
(c) Both 1 and 2 (d) Neither 1 nor 2

[ESE : 2022]

3.30 For non-homogeneous clays, the coefficient of permeability in (mm/s) should be ranges between

(a) 10^{-1} to 10^{-2} (b) 10^{-2} to 10^{-3}
(c) 10^{-3} to 10^{-4} (d) 10^{-4} to 10^{-6}

[ESE : 2022]

3.31 What is the value of coefficient of permeability for coarse gravel?

(a) Greater than 100 mm/s
(b) Greater than 10 mm/s
(c) Greater than 1 mm/s
(d) Less than 0.1 mm/s

[ESE : 2024]

■■■■

Answers Effective Stresses and Permeability

3.1 (a)	3.2 (b)	3.3 (a)	3.4 (a)	3.5 (c)	3.6 (d)	3.7 (b)	3.8 (d)	3.9 (c)
3.10 (d)	3.11 (b)	3.12 (d)	3.13 (b)	3.14 (d)	3.15 (c)	3.16 (a)	3.17 (c)	3.18 (b)
3.19 (a)	3.20 (c)	3.21 (c)	3.22 (c)	3.23 (a)	3.24 (d)	3.25 (a)	3.26 (a)	3.27 (c)
3.28 (b)	3.29 (c)	3.30 (d)	3.31 (a)					

Explanations Effective Stresses and Permeability

3.1 (a)

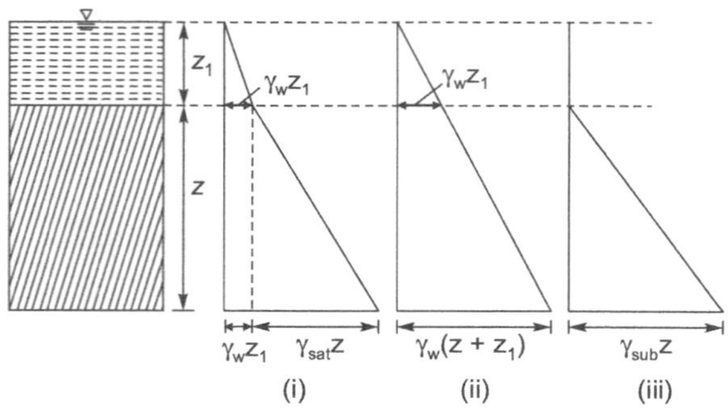

(i) σ(total stress)

(ii) u(pore water pressure)

(iii) $\bar{\sigma}$(effective stress)

3.2 (b)

Equivalent horizontal permeability,

$$k_{ex} = \frac{k_1 h_1 + k_2 h_2}{h_1 + h_2} = \frac{2kh + kh}{h + h} = 1.5\text{ k}$$

3.3 (a)

Effective stress = Total stress – pore pressure

Effective stress remains unaffected by water depth in river.

3.4 (a)

Permanent lowering of Ground Water Table (GWT) reduces pore pressure and thereby increases effective stress so settlement occurs.

3.5 (c)

$$k = \frac{Q}{iA} = \frac{10}{100 \times 10} \times 10^{-2} = 1 \times 10^{-4}\text{ m/s}$$

3.6 (d)

Total stress, σ = $2 \times 5 = 10\text{ t/m}^2$

Neutral stress, u = $1 \times 5 = 5\text{ t/m}^2$

Effective stress, $\sigma' = \sigma - u = 10 - 5 = 5\text{ t/m}^2$

3.7 (b)

$$k_{ev} = \frac{z_1 + z_2 + z_3}{\frac{z_1}{k_1} + \frac{z_2}{k_2} + \frac{z_3}{k_3}}$$

$$= \frac{4 + 1 + 2}{\frac{4}{2} + \frac{1}{1} + \frac{2}{4}} = \frac{7}{3.5} = 2$$

3.8 (d)

Gravel has maximum permeability and clays have minimum permeability. Loess deposit has a loose structure and permeability is more than clay silt admixture.

3.9 (c)

Coarse sand is more than million times permeable than a high plasticity clay.

$$k = CD_s^2 \frac{e^3}{(1+e)} \frac{\gamma}{\mu}$$

Permeability depends upon

(i) Grain size

(ii) Properties of pore fluid

(iii) Void ratio of soil

(iv) Structural arrangement of the soil particles

(v) Entrapped air and foreign matter

(vi) Adsorbed water in clayey soils.

It means that a sample of sand and clay having same void ratio may not have same permeability due to structural arrangement of particles and adsorbed water.

Increase in effective stress causes compression of soil and rearrangement of particles. So permeability decrease.

3.10 (d)

In falling head permeability test

$$k = 2.3 \frac{aL}{At} \log_{10}\left(\frac{h_0}{h_1}\right)$$

$$= 2.3 \frac{aL}{A(t/2)} \log_{10}\left(\frac{h_0}{h'}\right)$$

$$\therefore \quad \log_{10}\left(\frac{h_0}{h_1}\right) = 2\log\left(\frac{h_0}{h'}\right)$$

$$\Rightarrow \quad \frac{h_0}{h_1} = \left(\frac{h_0}{h'}\right)^2$$

$$\Rightarrow \quad h' = \sqrt{h_0 h_1}$$

3.11 (b)

Triaxial test is used to measure the permeability of low permeable soils. Test involves the establishment of a steady-state flow condition in a cylindrical condition housed in triaxial permeability cell.

3.12 (d)

Average permeability in horizontal direction,

$$k_{ex} = \frac{k_1h_1 + k_2h_2 + k_3h_3}{h_1 + h_2 + h_3}$$

Average permeability in vertical direction,

$$k_{ev} = \frac{h_1 + h_2 + h_3}{\frac{h_1}{k_1} + \frac{h_2}{k_2} + \frac{h_3}{k_3}} = \frac{12}{7}$$

$$\frac{k_{ex}}{k_{ev}} = \frac{49}{36}$$

3.13 (b)

Organic matter and entrapped air obstruct the flow in the soil voids, and hence reduce the permeability.

3.14 (d)

- As the effective stress increases, void ratio decreases. Hence permeability decreases.
- Entrapped air decreases the permeability of soil.
- Organic matter has the tendency to move towards flow channel and choke them. Thus the permeability is reduced.

3.15 (c)

Since permeability is the property governing the ease with which a fluid flows through the soil, it depends on the characteristics of the fluid or permeant as well as those of the soil. Mathematically,

$$k = CD_s^2 \cdot \frac{\gamma}{\mu} \cdot \frac{e^3}{1+e}$$

where k is coefficient of permeability, D_s is effective particle size, γ is unit weight of permeant, μ is viscosity of permeant, e is void ratio and C is shape factor.

3.16 (a)

Porosity is expressed as

$$n = \frac{e}{1+e} = \frac{0.5}{1+0.5} = \frac{1}{3}$$

Seepage velocity,

$$V_s = \frac{V}{n} = \frac{5\times10^{-7}}{1/3}$$

$$= 15 \times 10^{-7} \text{ m/s}$$

3.17 (c)

The permeability is greatly reduced if air is entrapped in the voids thus reducing its degree of saturation. Organic foreign matter also has the tendency to move towards critical flow channels and choke them up, thus decreasing the permeability.

3.18 (b)

Let the inert tracer travel by a distance 'dx' in time dt

$$\therefore \quad V_s = \frac{dx}{dt} = \text{Actual vel. of flow}$$

$$V_s = \frac{V}{n} = \frac{dx}{dt}$$

Now, $V = ki$

$$\Rightarrow \quad V_s . n = ki$$

$$\Rightarrow \quad \frac{dx}{dt}.n = ki$$

$$\Rightarrow \frac{1500}{t}0.25 = 30\times\frac{(50-25)}{1500}$$

$$\Rightarrow \quad t = 750 \text{ days}$$

3.19 (a)

Permeaibility depends upon degree of saturation hence on water content.

3.20 (c)

There empirical formula for capillary height is

$$h = \frac{C}{eD_{10}}$$

$$\Rightarrow \quad \frac{h_A}{h_B} = \frac{e_B(D_{10})_B}{e_A(D_{10})_A} = \left(\frac{e_B}{e_A}\right)\times\left(\frac{D_{10B}}{D_{10A}}\right)$$

$$= \frac{1}{2}\times3 = 1.5$$

3.21 (c)

$$k_H = \frac{k_1H + k_2H + k_3H}{3H} = \frac{10^{-4} + 10^{-3} + 10^{-4}}{3}$$

$$= \frac{12\times10^{-4}}{3} = 4 \times 10^{-4} \text{ cm/s}$$

3.22 (c)

Permanent lowering of ground water table results in settlement of foundation and it is due to increase in effective stress in settlement strata.

3.23 (a)

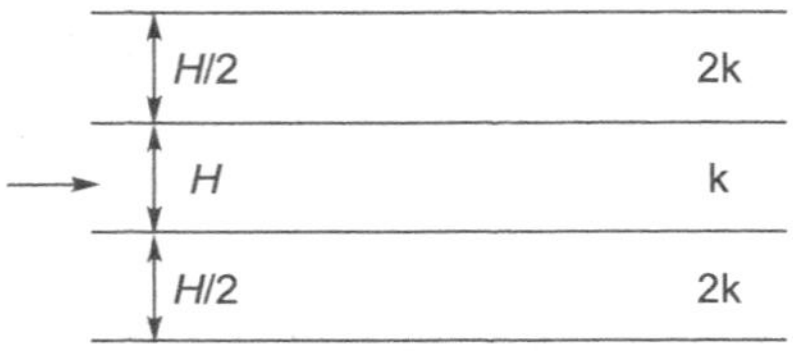

$$K_{avg} = \frac{\Sigma k_i z_i}{\Sigma z_i} = \frac{2k\,H/2 + k \cdot H + 2k\,H/2}{\frac{H}{2} + H + \frac{H}{2}}$$

$$= \frac{3kH}{2H} = 1.5\,k$$

3.24 (d)

$k = CD_{10}^2 = 750 \times (0.07)^2 = 3.675$ m/day

Note: As per Allen Hazen

$k = CD_{10}^2$ where $C = 100$, D_{10} is in cm and k is in cm/sec

Similarly if we solve this equation

then $k = 750 \times (0.007\text{ cm})^2$

$= 0.03675$ cm/sec

$= 0.03675 \times 10^{-2}$ m/day $\times 24 \times 3600$

$= 31.752$ m/day

Answer is not matching hence option 'd' is correct.

3.26 (a)

Fine grained soils are more difficult to drain than coarse grained soils because the voids in the former are so small that the contained water is strongly held by capillary attraction, while in soil with large grains this force does not act so strongly.

3.27 (c)

Given: For a constant head permeability test

Dia. of soil sample, $d = 10$ cm

Height of soil sample, $L = 15$ cm

Constant head, $\Delta h = 30$ cm

$\forall = 160\text{ cm}^3$

$t = 1.75$ min

Coefficient of permeability

$$k = \frac{\forall L}{t(\Delta h)A}$$

$$= \frac{(160 \times 10^{-6}) \times 15 \times 60 \times 24 \times 365}{1.75 \times 30 \times \frac{\pi}{4} \times (10)^2 \times 10^{-4}}$$

$= 3059.267$ m/year $\simeq 3060$ m/year

3.28 (b)

$$k = CD_{10}^2 \frac{e^3}{1+e} \frac{\gamma_w}{\mu}$$

k depends on D_{10}, e, μ, γ_w (μ, γ_w depends temperature)

3.30 (d)

Clay has least coefficient of permeability.

4 Seepage Analysis

4.1 A strata of 3.5 m thick fine sand has a void ratio of 0.7 and *G* of 2.7. For a quick sand condition to develop in this strata, the water flowing in upward direction would require a head of

(a) 7 m (b) 5.56 m
(c) 5 m (d) 3.5 m

[ESE : 2000]

4.2 A flownet of a coffer dam foundation has 6 flow channels and 18 equipotential drops. The head of water lost during seepage is 6 m. If the coefficient of permeability of foundation is 4×10^{-5} m/min, then the seepage loss per metre length of dam will be

(a) 2.16×10^{-2} m^3/day
(b) 6.48×10^{-2} m^3/day
(c) 11.52×10^{-2} m^3/day
(d) 34.56×10^{-2} m^3/day [ESE : 2000]

4.3 **Assertion (A):** The possibility of quicksand condition occurring is more on the downstream of a weir on permeable foundation.

Reason (R): Seepage lines are directed upwards at the downstream of such a weir.

(a) both A and R are true and R is the correct explanation of A
(b) both A and R are true but R is not a correct explanation of A
(c) A is true but R is false
(d) A is false but R is true

[ESE : 2001]

4.4 Match **List-I** (Protective measure) with **List-II** (Purpose) and select the correct answer using the codes given below the lists:

List-I
A. Inverted filter
B. Riprap
C. Clay core
D. Sand columns

List-II
1. To protect river banks from erosion by river flow
2. To prevent escape of fine soils by seepage water
3. To reduce camage due to liquefaction of saturated granular soils during earthquakes
4. To reduce seepage of water through the body of the earth dams

Codes:

	A	B	C	D
(a)	4	3	2	1
(b)	2	3	4	1
(c)	4	1	2	3
(d)	2	1	4	3

[ESE : 2004]

4.5 Match **List-I** (Unit) with **List-II** (Purpose) and select the correct answer using the codes given below the lists:

List-I
A. Graded filter
B. Lime treatment
C. Impervious clay core
D. Curtain grouting

List-II
1. To reduce seepage of water through body of earth cam
2. To reduce water seepage through foundation below dam
3. To stabilize black cotton soils
4. To drain water without losing fines from the soil

Codes:

	A	B	C	D
(a)	4	3	1	2
(b)	1	2	4	3
(c)	4	2	1	3
(d)	1	3	4	2

[ESE : 2005]

4.6 A soil mass under seepage has a downward flow of water. Which of the following statements are correct with regard to stresses at any point in the soil mass?

1. Effective stress is decreased by an amount equal to the seepage force
2. Effective stress is increased by an amount equal to the seepage force
3. Total stress will change
4. Total stress will be unaltered

Select the correct answer using the codes given below:

(a) 1 and 3 (b) 2 and 3
(c) 1 and 4 (d) 2 and 4

[ESE : 2006]

4.7 From a flow net which of the following information can be obtained?

1. Rate of flow 2. Pore water pressure
3. Exit gradient 4. Permeability

Select the correct answer using the codes given below:

(a) 1, 2, 3 and 4 (b) 1, 2 and 3 only
(c) 2, 3 and 4 only (d) 1 only

[ESE : 2007]

4.8 Consider the following statements:

1. Quicksand is a special variety of sand.
2. Quicksand is not a material but a hydraulic condition.
3. In nature, quicksand condition is observed usually in coarse silt or fine sand.

Which of the above statements are correct?

(a) 1, 2 and 3 (b) 1 and 2 only
(c) 2 and 3 only (d) 1 and 3 only

[ESE : 2009]

4.9 In a 6 m thick stratum of fine sand having submerged density of 11 kN/m^3, quicksand condition occurred at a depth of 4.2 m of excavation. What is the depth of lowering of groundwater table required for making an excavation 5 m deep ? Take density of water as 10 kN/m^3.

(a) 3.85 m (b) 1.68 m
(c) 1.1 m (d) 0.897 m

[ESE : 2009]

4.10 For a sheet pile wall constructed in a soil having effective grain size = 0.1 mm, the difference of the upstream and downstream water levels is 3 m. If the flow net drawn for the problem yields 2 as the ratio of number of head drops to number of flow channels, then what is the discharge in unit of m^3/s/m length of sheet pile wall?

(a) 3.0×10^{-4} (b) 3.0×10^{-2}
(c) 1.5×10^{-4} (d) 1.5×10^{-2}

[ESE : 2009]

4.11 A homogeneous earth dam is 43 m high with 3 m as free board and a 30 m long horizontal filter at the downstream end. The flownet drawn in the dam section consists of 5 flow channels and 15 potential drops. If the dam is 500 m long and the permeability of the material of the dam is uniformly 3×10^{-3} m/s, the total discharge permeating through the body of the dam will be

(a) 0.0004 m^3/s (b) 0.0036 m^3/s
(c) 0.0039 m^3/s (d) 0.20 m^3/s

[ESE : 2010]

4.12 Consider the following statements:

1. The hydraulic head at a point in the soil includes piezometric head as well as datum head.
2. Piping in soil occurs when effective pressure becomes equal to zero.
3. Piping in soil occurs when soil is highly porous.

Which of these statements is/are correct?

(a) 1, 2 and 3 (b) 1 and 2 only
(c) 2 and 3 only (d) 2 only

[ESE : 2010]

Directions: The following items consists of two statements; one labelled as **'Assertion (A)'** and the other as **'Reason (R)'**. You are to examine these two statements carefully and select the answers to these items using the codes given below:

Codes:

(a) both A and R are true and R is the correct explanation of A
(b) both A and R are true but R is not a correct explanation of A
(c) A is true but R is false
(d) A is false but R is true

4.13 **Assertion (A):** The possibility of piping failure in earth dams is more if black cotton soil is the foundation material.

Reason (R): The highly expansive black cotton soils are the most common soils wherever basalt rock is present.

[ESE : 2010]

4.14 **Assertion (A):** Flow net is dependent on the permeability of soil through which flow is taking place.

Reason (R): The flow net is useful in finding the discharge.

[ESE : 2010]

4.15 A flow net for seepage under a sheet pile wall has $n_f = 4$, $n_d = 8$ and the permeabilities of the soil in the horizontal and vertical directions are: $K_h = 8 \times 10^{-5}$ m/sec and $K_v = 2 \times 10^{-5}$ m/sec. If the head loss through the soil is 2 m, the quantity of seepage per meter length of the wall will be

(a) 2×10^{-5} m^3/sec (b) 4×10^{-5} m^3/sec
(c) 8×10^{-5} m^3/sec (d) 16×10^{-5} m^3/sec

[ESE : 2011]

4.16 **Statement-I :** In a flownet, each field must be a (curvilinear) square:

Statement-II : Each flow channel in a flownet has the same rate of flow.

(a) Both Statement (I) and Statement (II) are individually true and Statement (II) is the correct explanation of Statement (I)
(b) Both Statement (I) and Statement (II) are individually true but Statement (II) is NOT the correct explanation of Statement (I)
(c) Statement (I) is true but Statement (II) is false
(d) Statement (I) is false but Statement (II) is true

[ESE : 2012]

4.17 Consider the following statements:

1. Quantity of seepage in each flow channel of a flow-net is independent of the size of field.
2. Drop in head between adjacent equipotential lines of a flow-net is dependent on upstream head.
3. With increase in the length of a flow path, the corresponding exit gradient will decrease.

Which of these statements are correct?

(a) 1, 2 and 3 (b) 1 and 2 only
(c) 1 and 3 only (d) 2 and 3 only

[ESE : 2013]

4.18 A sand sample has a bulk density of 20 kN/m^3 and a degree of saturation of 70%. If the specific gravity of soil grains is 2.65, the value of critical hydraulic gradient for the soil will be

(a) 1.02 (b) 1.05
(c) 1.10 (d) 1.15

[ESE : 2014]

4.19 **Statement (I):** The possibility of quicksand condition occurring is more on the downstream of a weir on a permeable foundation than on the upstream end with an upward component of seepage velocity.

Statement (II): Seepage lines end with an upward component of seepage velocity at the downstream reaches of such a weir.

(a) Both Statement (I) and Statement (II) are individually true and Statement (II) is the correct explanation of Statement (I)
(b) Both Statement (I) and Statement (II) are individually true but Statement (II) is NOT the correct explanation of Statement (I)
(c) Statement (I) is true but Statement (II) is false
(d) Statement (I) is false but Statement (II) is true

[ESE : 2016]

4.20 A uniform collapsible sand stratum, 2.5 m thick, has specific gravity of its sand as 2.65, with a natural void ratio of 0.65. The hydraulic head required to cause quick collapsible sand condition is

(a) 2.50 m (b) 2.75 m
(c) 3.25 m (d) 3.50 m

[ESE : 2016]

4.21 Consider the following statements regarding flow net:

1. It helps determine the quantity of seepage.
2. It helps determine the upward lift below a hydraulic structure.
3. It is applicable to rotational flow only.

Which of the above statements are correct?

(a) 1 and 2 (b) 1 and 3
(c) 2 and 3 (d) 1, 2 and 3

[ESE : 2017]

4.22 A soil sample has an average grain diameter as 0.03 mm. The size of interstices is one-eighth of the mean grain diameter. Considering σ of water as 0.075 g/cm, the water will rise in the clay to a height of

(a) 2.4 m (b) 3.0 m
(c) 3.6 m (d) 4.0 m

[ESE : 2018]

4.23 A masonry is founded on pervious sand. A factor of safety of 4 is required against boiling. For the sand, n = 45% and G_s = 2.65. The maximum permissible upward hydraulic gradient will be nearly

(a) 0.18 (b) 0.23
(c) 0.28 (d) 0.33

[ESE : 2019]

4.24 A coarse-grained soil has a void ratio of 0.78 and specific gravity as 2.67. The critical gradient at which a quick sand condition occurs will be

(a) 0.62 (b) 0.74
(c) 0.82 (d) 0.94

[ESE : 2020]

4.25 What one of the following problems is required to be studied in the design of earth dams?

(a) The prediction of the position of the line of seepage in the longitudinal section.
(b) The computation of seepage loss.
(c) The seepage line should cut the downstream slope.
(d) The seepage loss through the dam should be maximum.

[ESE : 2021]

4.26 Consider the following statements regarding energy heads:

1. The velocity head in soils is negligible.
2. Direction of flow is determined by the difference in total head.
3. Negative pore pressure cannot exist.

Which of the above statements are correct?

(a) 1 and 2 only (b) 1 and 3 only
(c) 2 and 3 only (d) 1, 2 and 3

[ESE : 2024]

Answers Seepage Analysis

4.1 (d)	4.2 (c)	4.3 (a)	4.4 (d)	4.5 (a)	4.6 (d)	4.7 (b)	4.8 (c)	4.9 (b)
4.10 (c)	4.11 (*)	4.12 (d)	4.13 (d)	4.14 (d)	4.15 (b)	4.16 (d)	4.17 (a)	4.18 (c)
4.19 (a)	4.20 (a)	4.21 (a)	4.22 (d)	4.23 (b)	4.24 (d)	4.25 (b)	4.26 (a)	

Explanations Seepage Analysis

4.1 (d)

The critical hydraulic gradient,

$$i_c = \frac{G-1}{1+e} = 1.0$$

For quick sand condition

$$i = i_c = \frac{h}{L}$$

$$h = 1.0 \times 3.5 = 3.5 \text{ m}$$

4.2 (c)

Seepage loss per metre length,

$$q = k \times \frac{N_f}{N_d} \times H$$

Given, $k = 4 \times 10^{-5}$ m/min

$$= 4 \times 10^{-5} \times 24 \times 60 \text{ m/day}$$

$$= 0.0576 \text{ m/day}$$

$$\therefore \quad q = 0.0576 \times \frac{6}{18} \times 6$$

$$= 0.1152 \text{ m}^3/\text{day}$$

$$= 11.52 \times 10^{-2} \text{ m}^3/\text{day}$$

N_f is number of flow channels = 6.0

N_d is number of equipotential drops = 18

4.3 (a)

On the upstream side the seepage lines (streamlines) are directed downwards and increase the effective stress. While on the downstream side seepage line are directed upwards and reduce the effective stress.

4.4 (d)

Sand columns increase the permeability and therefore drainage of water occurs easily and faster. During earthquake conditions, the excess pore pressure will be quickly dissipated and thus liquefaction of sands can be reduced.

Riprap protect river banks from erosion.

4.5 (a)

Lime treatment reduces liquid limit slightly and increases plastic limit immensely. So there is reduction in plasticity index.

4.6 (d)

In the case of downward flow of water the pore pressure is decreased and effective stress increased. Total stress remains unaltered in the flow.

4.7 (b)

Seepage pressures, uplift pressures, exit gradient and pore-water pressure can be obtained from a flownet.

4.8 (c)

Quicksand is a condition and not a soil type. This condition is created in saturated thick layers of loose fine sandy soils when disturbed either due to vibration, such as, from pile driving in the neighbourhood, or due to pressure of flowing water.

4.9 (b)

Let x be the depth of ground water table initially. Total upward water force on the sand stratum at the bottom of excavation

$$= (6 - x) \times \gamma_w$$

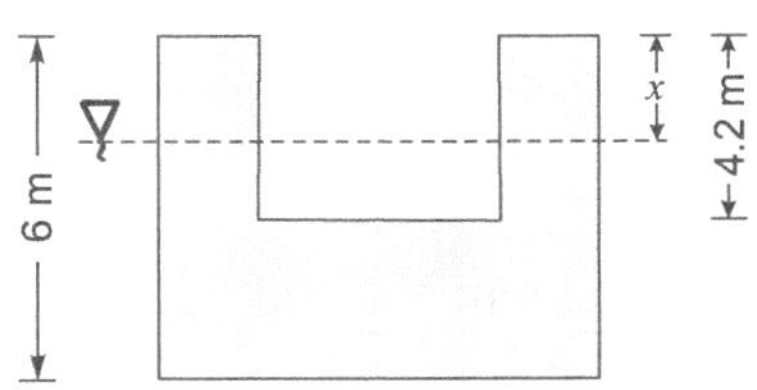

Total downward force at the bottom of excavation

$$= \text{Weight of soil in saturated condition}$$
$$= \gamma_{sat} \times (6 - 4.2)$$
$$= (\gamma_{sub} + \gamma_w) \times 1.8$$

When quicksand condition occurs, then total upward water force becomes equal to total downward force i e.

$$(6 - x) \times \gamma_w = (\gamma_{sub} + \gamma_w) \times 1.8$$
$$\Rightarrow (6 - x) \times 10 = (11 + 10) \times 1.8$$
$$\Rightarrow x = 6 - \frac{21 \times 1.8}{10}$$
$$\Rightarrow x = 2.22 \text{ m}$$

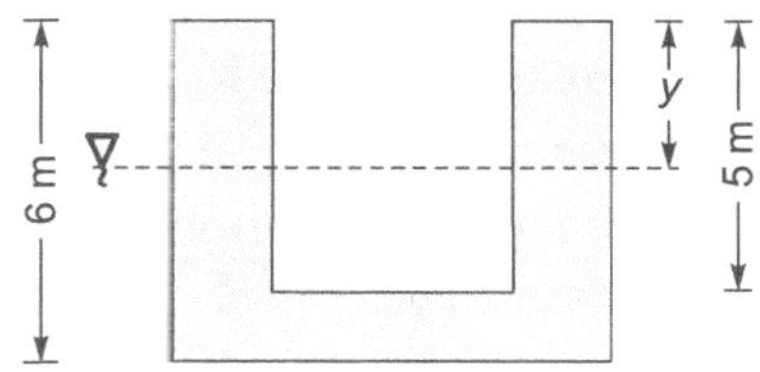

Similarly, if the depth of displaced ground water table is y from the surface, then

$$(6 - y) \times \gamma_w = (6 - 5) \times (\gamma_{sub} + \gamma_w)$$
$$\Rightarrow (6 - y) \times 10 = 1 \times (11 + 10)$$
$$\Rightarrow y = 6 - 2.1$$
$$\Rightarrow y = 3.9 \text{ m}$$

$\therefore$ Lowering of water table $= y - x$

$$= 3.9 - 2.2 = 1.68 \text{ m}$$

4.10 (c)

The permeability of a soil as per Hazen equation is given by

$$k = 100\, D_{10}^2 \text{ in cm/s}$$

where D_{10} is Hazen's effective grain size in cm.

$$\therefore \quad k = 100 \times \left(\frac{0.1}{10}\right)^2$$
$$= 0.01 \text{ cm/s}$$
$$= 10^{-4} \text{ m/s}$$

The discharge per unit length is given by

$$q = k \times H \times \frac{n_f}{n_d}$$
$$= 10^{-4} \times 3 \times \frac{1}{2}$$
$$= 1.5 \times 10^{-4} \text{ m}^3\text{/s/m}$$

4.11 (*)

The discharge per unit length of the dam is given by

$$q = k \times \frac{N_f}{N_d} \times H$$
$$= 3 \times 10^{-3} \times \frac{5}{15} \times (43 - 3)$$
$$= 40 \times 10^{-3} \text{ m}^2\text{/sec}$$

∴ Total discharge permeating through the body of the dam is given by

$$Q = q \times L$$
$$= 40 \times 10^{-3} \times 500$$
$$= 20 \text{ m}^3/\text{s}$$

None of the options is correct.

4.12 (d)

- Piezometric head = Pressure Head + Datum Head. Piezometric head already includes datum head.
- When effective pressure becomes zero this leads to erosion of the soil, this begins near the downstream toe but may lengthen progressively towards the upstream. This phenomenon is known as piping, because erosion occurs in the form of a pipe or closed channel underneath the structure.
- Highly porous refers to gravel in which large discharge is required for piping. But piping can also occur in sands which are comparatively lesser porous and require lesser discharge.

4.13 (d)

Black cotton soils are held together even at critical hydraulic gradient due to inherent cohesion that holds the grains together. Possibility of piping failure is less because the cohesion component of shear strength is an independent quantity.

4.14 (d)

Flow net depends upon boundary condition, and not on the permeability of soil. Flow net is useful in finding the seepage discharge, uplift pressure, exit gradient etc.

4.15 (b)

For a sheet pile the quantity of seepage per unit length of wall is given by

$$q = K' \cdot H \cdot \frac{n_f}{n_d}$$
$$k' = \sqrt{K_h \times K_v} = 4 \times 10^{-5}$$
$$\Rightarrow \quad q = \left(4 \times 10^{-5}\right) \times 2 \times \frac{4}{8}$$
$$= 4 \times 10^{-5} \text{ m}^3/\text{s}$$

4.16 (d)

All fields in a flow net are drawn as "elementary squares", meaning that average distances between opposite sides are equal. Therefore each flow channel in a flow net has same rate of flow.

4.17 (a)

$$\text{Drop between equipotential line} = \frac{H}{N_d}$$

Where, H = Upstream head

4.18 (c)

$$\text{Hydraulic Gradient, } i_c = \frac{G-1}{1+e}$$
$$\text{Bulk density} = \frac{(G+Se)\gamma_w}{1+e}$$
$$20 = \frac{(2.65+0.7e)}{1+e} \times 10$$
$$20 + 20e = 26.5 + 7e$$
$$e = 0.5$$
$$\Rightarrow \quad 6.5 = 13e$$
$$\text{So,} \quad i_c = \frac{2.65-1}{1+0.5} = \frac{1.65}{1.5}$$
$$i_c = 1.1$$

4.19 (a)

Upward seepage reduces effective stress and downward seepage increases effective stresses.

4.20 (a)

Critical hydraulic gradient,

$$i_{cr} = \frac{G-1}{1+e} = \frac{2.65-1}{1+0.65} = 1$$

Hydraulic gradient,

$$i_{cr} = \frac{\text{Head}}{\text{Length}} = \frac{H}{2.5} = 1$$

Hydraulic head

$$= 2.50 \text{ m}$$

4.21 (a)

4.22 (d)

$$d = \frac{0.03 \text{ mm}}{8} = \frac{0.03}{8} \times 10^{-3} \text{ m}$$
$$h = \frac{4\sigma}{\gamma_w d} = \frac{4 \times 0.075 \times 10^{-3} \times 9.81 \times 10^2}{9.81 \times 10^3 \times \frac{0.03}{8} \times 10^{-3}} = 8 \text{ m}$$

Note: Since no option is matching by assuming interstices as diameter.

Assuming interstices as radius the answer is 4 m. Which is given as option (d).

4.23 (b)

$$n = 45\%, \; G_s = 2.65$$

$$\text{F.O.S} = 4$$

$$e = \frac{n}{1-n} = \frac{0.45}{1-0.45} = \frac{0.45}{0.55} = 0.818$$

$$\text{F.O.S} = \frac{i_c}{i}$$

$$\Rightarrow \quad i = \frac{i_c}{\text{F.O.S}} = \frac{\left(\frac{G-1}{1+e}\right)}{\text{F.O.S}}$$

$$i = \frac{\left(\frac{2.65-1}{1+0.818}\right)}{4} = \frac{0.9075}{4}$$

$$= 0.22689 \simeq 0.23$$

4.24 (d)

Critical hydraulic gradient,

$$i_c = \frac{G-1}{1+e} = \frac{2.67-1}{1+0.78} = 0.938$$

4.25 (b)

Controlled seepage or limited uniform seepage is inevitable in all earth dams, and ordinally it does not produce any harm. However, uncontrolled or concentrated seepage through the dam body or through its foundation may lead to piping or sloughing and subsequent failure of the dam. The following problem must be studied in the design of earth dams.

- The computation of seepage loss.
- The prediction of the position of the line of seepage in the cross-section.
- The seepage line should not cut the downstream slope.
- The seepage loss through the dam should be minimum.

4.26 (a)

Negative pore pressure can exist in soil in case of capillary pores.

■■■■

5 Compaction of Soil

5.1 **Assertion (A):** For a given soil, the optimum moisture content increases with the increase in compactive effort.
Reason (R): Higher the compactive effort, higher is the dry density at the same moisture content.
[ESE : 2000]

5.2 Match **List-I** (Roller type) with **List-II** (Soil type) and select the correct answer using the codes given below the lists:

List-I
A. Pneumatic roller
B. Smooth wheeled roller
C. Sheep foot roller
D. Vibratory roller

List-II
1. Cohesive and granular soils
2. Plastic soils of moderate cohesion
3. Cohesionless soils
4. Silty soils of low plasticity

Codes:

	A	B	C	D
(a)	4	2	1	3
(b)	3	1	2	4
(c)	4	1	2	3
(d)	3	2	1	4

[ESE : 2002]

5.3 In a compaction test if the compacting effort is increased, it will result in
(a) increase in maximum dry density and OMC
(b) increase in maximum dry density but OMC remains unchanged
(c) increase in maximum dry density and decrease in OMC
(d) no change in maximum dry density but decrease in OMC
[ESE : 2003]

5.4 Match **List-I** (Equipment) with **List-II** (Use) and select the correct answer using the codes given below the lists:

List-I
A. Vibratory rollers
B. Sheep foot rollers
C. Frog hammers
D. Vibrofloats

List-II
1. To compact soils in confined areas and at corners
2. To compact road and railway embank-ments of sandy soils
3. To densify sandy soils over a large area and to a larger depth
4. To compact clayey soil fills

Codes:

	A	B	C	D
(a)	4	2	1	3
(b)	4	2	3	1
(c)	2	4	1	3
(d)	2	4	3	1

[ESE : 2004]

5.5 Soil is compacted at which one of the following when a higher compactive effort produces highest increase in dry density?
(a) Optimum water content
(b) Dry side of the optimum moisture content
(c) Wet side of the optimum moisture content
(d) Saturation moisture content
[ESE : 2004]

5.6 **Assertion (A):** The clay core of an earth dam is usually compacted on the wet side of OMC.
Reason (R): Compaction on the wet side of OMC reduces permeability and prevents cracking in core.
(a) both A and R are true and R is the correct explanation of A
(b) both A and R are true but R is not a correct explanation of A
(c) A is true but R is false
(d) A is false but R is true
[ESE : 2005]

5.7 Match **List-I** (Equipment) with **List-II** (Purpose) and select the correct answer using the codes given below the lists:

List-I

A. Sheep foot roller
B. Frog hammer
C. Vibratory roller

List-II

1. To densify cohesionless soils to relatively larger depths
2. To compact lumpy cohesive soil fills
3. To compact soils at corners and places which bigger rollers cannot access
4. To compact cohesionless soils of shallow depth

Codes:

	A	B	C
(a)	4	1	2
(b)	2	1	4
(c)	4	3	2
(d)	2	3	4

[ESE : 2005]

5.8 Match **List-I** with **List-II** and select the correct answer using the codes given below the lists:

List-I

A. Smooth wheel rollers
B. Sheep foot rollers
C. Pneumatic tyred rollers
D. Rammers

List-II

1. Most suitable for compacting cohesive soils
2. Most suited for compacting coarse grained soils
3. Used for compacting soils in confined places
4. Suitable for both cohesionless and cohesive soils

Codes:

	A	B	C	D
(a)	4	3	2	1
(b)	2	1	4	3
(c)	4	1	2	3
(d)	2	3	4	1

[ESE : 2006]

5.9 Match **List-I** (Type of soil) with **List-II** (Compaction equipment) and select the correct answer using the codes given below the lists:

List-I

A. Wet clays and silts
B. Crushed rock
C. Fill soils
D. Sands

List-II

1. Smooth wheel rollers
2. Vibrators
3. Pneumatic tyred rollers
4. Grid rollers

Codes:

	A	B	C	D
(a)	4	2	3	1
(b)	3	1	4	2
(c)	4	1	3	2
(d)	3	2	4	1

[ESE : 2006]

5.10 The following soils are compacted at the same compactive effort in the field. Which one of the following is the correct sequence in the increasing order of their maximum dry density?

(a) Silty clay – Clay – Sand – Gravel sand clay mixture
(b) Sand – Gravel sand clay mixture – Silty clay – Clay
(c) Clay – Silty clay – Sand – Gravel sand clay mixture
(d) Sand – Gravel sand clay mixture – Clay – Silty clay

[ESE : 2007]

5.11 Consider the following statements:
On addition of lime to a clay soil, generally

1. MDD and strength both increase
2. MDD decreases but strength increases
3. MDD and OMC both increase

Which of these statements is correct ?

(a) 1 only
(b) 2 only
(c) 3 only
(d) None of these

[ESE : 2007]

5.12 An increase in compaction effort will lead to which of the following?

(a) Decrease in both the optimum moisture content (OMC) and maximum dry density.
(b) Decrease in the optimum moisture content (OMC) and increase in the maximum dry density.
(c) Increase in the optimum moisture content (OMC) and decreases in the maximum dry density.
(d) Increase in both the optimum moisture content (OMC) and maximum dry density.

[ESE : 2007]

5.13 In laboratory compaction tests, the optimum moisture content of soil decreases

(a) with increase of compaction energy and with decrease of coarse grains in the soil
(b) with decrease of compaction energy and with increase of coarse grains in the soil
(c) with increase of both compaction energy and coarse grains in the soil
(d) with decrease of both compaction energy and coarse grains in the soil

[ESE : 2008]

5.14 Consider the following:

1. Increase in shear strength and bearing capacity
2. Increase in slope stability
3. Decrease in settlement of soil
4. Decrease in permeability

Which of the above with respect to compaction of soil is/are correct?

(a) 1 only (b) 1 and 2 only
(c) 2 and 3 only (d) 1, 2, 3 and 4

[ESE : 2009]

5.15 Consider the following statements:

1. In clay soils, the maximum dry unit weight tends to decrease as plasticity increases.
2. In clay soils, the maximum dry unit weight tends to increase as plasticity increases.
3. Heavy clays with high plasticity have the minimum dry unit weight and high OMC.

Which of these statements are correct?

(a) 1, 2 and 3 (b) 1 and 2 only
(c) 2 and 3 only (d) 1 and 3 only

[ESE : 2010]

5.16 Consider the following statements:

1. Poorly graded or uniform sands compact to low dry unit weights.
2. Heavy clays with high plasticity have very low maximum dry unit weight.
3. In clay soils, the maximum dry weight tends to decrease as plasticity increases.

Which of these statements is/are correct?

(a) 1, 2 and 3 (b) 1 and 2 only
(c) 2 and 3 only (d) 3 only

[ESE : 2010]

5.17 When the compaction energy increases the compaction of soils:

(a) Both of OMC and maximum dry density decrease
(b) Both of OMC and maximum dry density increase
(c) OMC decreases but maximum dry density increases
(d) OMC increases but maximum dry density decreases

[ESE : 2011]

Directions: The following items consists of two statements; one labelled as **'Assertion (A)'** and the other as **'Reason (R)'**. You are to examine these two statements carefully and select the answers to these items using the codes given below:

Codes:

(a) both A and R are true and R is the correct explanation of A
(b) both A and R are true but R is not a correct explanation of A
(c) A is true but R is false
(d) A is false but R is true

5.18 **Assertion (A):** The efficiency of the sheeps foot roller depends on the weight of the roller and the number of 'feet' in contact with the ground at a time.
Reason (R): Sheep foot rollers are considered most suitable for compacting clayey soils.

[ESE : 2011]

5.19 **Assertion (A):** In a compaction test at $\gamma_{d_{max}}$ and OMC, the degree of saturation is never 100%.
Reason (R): It is not possible to expel all the air entrapped in soil by compaction.

[ESE : 2011]

5.20 Consider the following statements:

1. Swelling is greater and shrinkage is less for clay compacted on the dry side of optimum.
2. Clays compacted on the dry side of optimum are characterized by larger strength.
3. An increase in organic content of clay soils causes an increase in optimum moisture content value.

Which of these statements are correct?

(a) 1, 2 and 3 (b) 1 and 2 only
(c) 1 and 3 only (d) 2 and 3 only

[ESE : 2013]

5.21 Consider the following statements:

1. Cohesionless soils experience relatively quicker compression under load.
2. Clays compacted dry-of-optimum compress more rapidly.
3. Clays compacted wet-of-optimum consolidate more under low stress range.

Which of these statements are correct?

(a) 1, 2 and 3 (b) 1 and 2 only
(c) 1 and 3 only (d) 2 and 3 only

[ESE : 2013]

5.22 A sample of sand has a volume of 1000 ml in its natural state. Its minimum volume when compacted is 750 ml. When gently poured into a measuring cylinder, its possible maximum volume is 1320 ml. What is the relative density?

(a) 56 (b) 52
(c) 58 (d) 60

[ESE : 2014]

Directions: The following items consists of two statements; one labelled as **'Statement (I)'** and the other as **'Statement (II)'**. You are to examine these two statements carefully and select the answers to these items using the codes given below:

Codes:

(a) Both Statement (I) and Statement (II) are individually true and Statement (II) is the correct explanation of Statement (I)
(b) Both Statement (I) and Statement (II) are individually true but Statement (II) is NOT the correct explanation of Statement (I)
(c) Statement (I) is true but Statement (II) is false
(d) Statement (I) is false but Statement (II) is true

5.23 **Statement (I):** The zero-air void curve is non-linear

Statement II: The dry density at 100% saturation is a non-linear function of void ratio.

[ESE : 2014]

5.24 **Statement (I):** For a given soil, optimum moisture content increases with the increase in compactive effort.

Statement (II): Higher the compactive effort higher is the dry density at the same moisture content.

[ESE : 2016]

5.25 A soil yielded a maximum dry unit weight of 18 kN/m^3 at a moisture content of 16% during a Standard Proctor Test. What is the degree of saturation of the soil if its specific gravity is 2.65?

(a) 98.42% (b) 95.50%
(c) 84.32% (d) 75.71%

[ESE : 2018]

5.26 **Statement (I)** : Coarser the particles, less is optimum moisture content.

Statement (II) : The specific surface area of coarser particle is less.

(a) Both Statement (I) and Statement (II) are individually true; and Statement (II) is the correct explanation of Statement (I)
(b) Both Statement (I) and Statement (II) are individually tru'e; but Statement (II) is NOT the correct explanation of Statement (I)
(c) Statement (I) is true; but Statement (II) is false
(d) Statement (I) is false; but Statement (II) is true

[ESE : 2020]

■■■■

Answers Compaction of Soil

5.1 (d)	5.2 (c)	5.3 (c)	5.4 (c)	5.5 (b)	5.6 (a)	5.7 (d)	5.8 (b)	5.9 (c)
5.10 (c)	5.11 (b)	5.12 (b)	5.13 (c)	5.14 (d)	5.15 (d)	5.16 (a)	5.17 (c)	5.18 (b)
5.19 (a)	5.20 (a)	5.21 (a)	5.22 (a)	5.23 (a)	5.24 (d)	5.25 (b)	5.26 (a)	

Explanations Compaction of Soil

5.1 (d)

The optimum moisture content decreases and maximum dry density increases with increase in compactive effort. At the same time dry density increases at any moisture content with increase in compactive effort. This is due to increase in the dispersed orientation of particles at same moisture content.

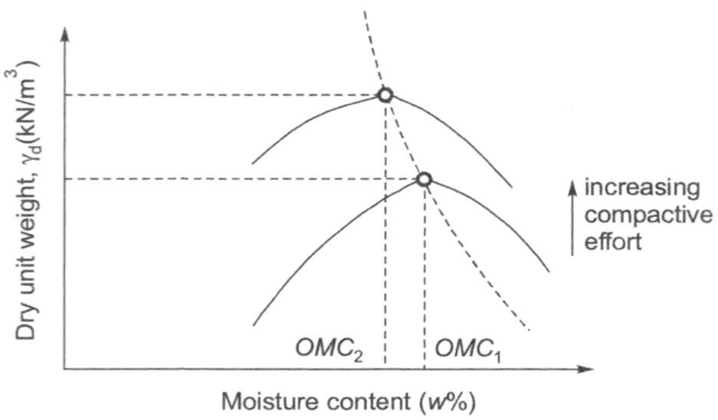

5.2 (c)

Pneumatic tyred rollers are also called as rubber tyred rollers. These rollers are used for compaction of coarse grained soils with some fines. These rollers are least suitable for uniform coarse soils and rocks. Generally pneumatic tyred rollers are used in pavement subgrade works both earthwork and bituminous works.

Smooth wheeled roller are most suitable for well graded sand, gravel, crushed rock, asphalt etc. where crushing is required. These are used on soils which does not require great pressure for compaction. These rollers are generally used for finishing the upper surface of the soil. These roller are not used for compaction of uniform sands.

Sheepsfoot rollers are used for compacting fine grained soils such as heavy clays and silty clays. Sheepsfoot rollers are used for compaction of soils in dams, embankments, subgrade layers in pavements and rail road construction projects.

Vibratory rollers compact the soil through impact forces. These forces result in compaction greater than the equivalent static load. Vibratory rollers are suitable for coarse grained soil. Vibratory rollers can be used for all types of soil except for plastic clays.

5.3 (c)

The effect of increasing the amount of compactive effort is to increase the maximum dry density and to decrease the optimum water content (OMC).

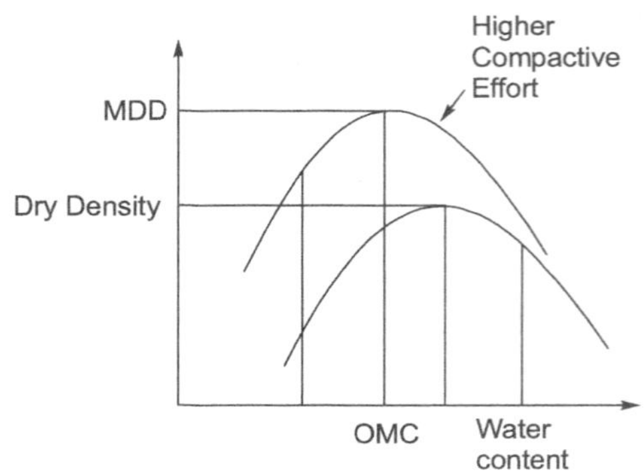

5.4 (c)

Frog hammers are used to compact soils in confined areas and at corners e.g., back of retaining wall, etc.

Soil densification techniques are used for reducing liquefaction hazards is to avoid large increases in pore water pressure during earthquake shaking. **Vibrofloatation** involves the use of a vibrating probe that can penetrate granular soil to depths of over100 feet. The vibrations of the probe cause the grain structure to collapse thereby densifying the soil surrounding the probe. To treat an area of potentially liquefiable soil, the vibroflot is raised and lowered in a grid pattern.

5.5 (b)

With increase in compactive effort, the optimum moisture content decrease and dry density increases. Thus soil is to be compacted on the dry side of the optimum moisture content of original (lower compactive effort) Proctor curve.

5.6 (a)

To avoid large expansions and swelling pressure under pavements and the floors, cohesive soils are generally compacted at a water content more than the optimum moisture content (OMC) with the resulting dry density less than the maximum dry density. The clayey soil in the impervious core of an earth dam is also compacted to the wet side of the optimum to reduce swelling pressure and to achieve low permeability. On the other hand, the highway embankment of cohesive soils are generally compacted of a water content lower than OMC to achieve higher shear strength and low compressibility. Likewise, the soil in the outer shells of the earth dams is compacted on dry side of the optimum to obtain higher shear strength, high permeability and low pore pressure.

5.7 (d)

- By sheep foot roller the compaction is achieved by a combination of tamping and kneading. It is suitable for cohesive soils.
- Frog hammers are pneumatic rammers. They are used for cohesionless soils, especially in small restricted and confined areas such as beds of drainage trenches and backfills of bridge abutments.
- Vibratory rollers or drum have a separate motor drive i.e., an arrangement of eccentric weights so as to produce high frequency, low amplitude vertical oscillation. Layers of the order of 1 m deep could be compacted to high densities.
- Vibrofloat can be used for compacting cohesionless soils up to 12 m depth (i.e., relatively larger depths.)

5.8 (b)

- **Rammers** are used for compacting small areas by providing impact load to the soil. This equipment is light and can be hand or machine operated. Rammers are suitable for compacting cohesive soils as well as other soils. This machine is used in areas with difficulty of access.
- **Smooth wheeled roller:** Most suited for compacting coarse grained soil.
- **Sheep foot roller:** Only cohesive soil.
- **Pneumetic tyred rollers:** Both cohesive and cohesionless soil.

5.9 (c)

Grid rollers are used for compaction of weathered rocks, well graded coarse soils. These rollers are not suitable for clayey soils, silty clays and uniform soils. The main use of these rollers are in subgrade and sub-base in road constructions.

5.10 (c)

A well graded sand attains a much higher dry density than a poorly graded sand for same compactive effort.

Cohesive soil has high air voids. This soil attains a relatively lower maximum dry density as compared to the cohesionless soil. It also requires more water than cohesionless soil and so the optimum water content is high. Heavy clay of very high plasticity has very low dry density and very high optimum water content.

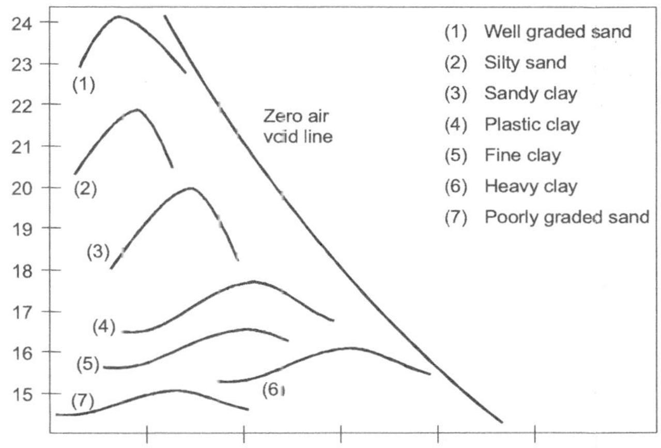

Typical proctor compaction curve for different soils

5.11 (b)

- Lime decreases the plasticity index of highly plastic soils and make them friable and more easily handled in the field.
- It increases the optimum moisture content and decreases the maximum compacted density however there will be an increase in strength.
- About 5-10% lime is required for cohesive soils.

5.12 (b)

For a given type of compaction, the higher the compactive effort, the higher the maximum dry unit weight and lower the optimum moisture content.

5.13 (c)

As the grain size increases, specific surface area of solids decreases so less water is required to make lubricating water layer around soil particles hence OMC decreases. As compactive effort increases the OMC decreases.

5.14 (d)

Compaction is the application of mechanical

energy to a soil so as to rearrange its particles and reduce the void ratio.

It is applied to improve the properties of an existing soil or in the process of placing fill such as in the construction of embankments, road bases, runways, earth dams, and reinforced earth walls. Compaction is also used to prepare a level surface during construction of buildings. There is usually no change in the water content and in the size of the individual soil particles.

The objectives of compaction are:

- To increase soil shear strength and therefore its bearing capacity.
- To reduce subsequent settlement under working loads.
- To reduce soil permeability making it more difficult for water to flow through.

5.15 (d)

As plasticity increases in clay soils, the maximum dry unit weight decreases. Heavy clays of very high plasticity have very low dry unit weight and a very high OMC.

5.16 (a)

Poorly graded or uniform sand compact to low dry unit weights while well graded sand compact to high dry weights because in well graded soil, smaller particle can be fitted in voids of large size particle, so voids are less and this results into a dense mass.

As the plasticity increases, smaller size clay particles increases. This results in decrease in maximum dry unit weight and increase in OMC.

5.17 (c)

The effect of increasing the amount of compactive effort is to increase the maximum dry density and to decrease the optimum water content. At a water content less than the optimum, the effect of increased compaction is more predominant. At a water content more than the optimum, the volume of air voids becomes almost constant and the effect of increased compaction is not significant.

But, the maximum dry density does not go on increasing with an increase in the compactive effort.

5.18 (b)

The sheep-foot roller consists of a hollow drum with a large number of small projections (known as feet) on its surface. These projections penetrate the soil layers during the rolling operations and cause compaction.

The sheep-foot rollers are ideally suited for compaction of cohesive soils. The rollers compact the soil by a combination of tamping and kneading action. When the roller is passed for the first time, the projections penetrate the soil layer and the lower portion of the layer is compacted. In successive passes, compaction is obtained in the middle and the top portion of the layer. This continually rising effect of the compaction is called walking-out of the roller.

5.19 (a)

The water content corresponding to the maximum dry density is known as the optimum water content (O.W.C.) or the optimum moisture content (O.M.C.).

For a given water content, theoretical maximum density, $(\rho_d)_{theomax}$, is obtained corresponding to the condition when there are no air voids (i.e. degree of saturation is equal to 100%). The theoretical maximum dry density is also known as saturated dry density $(\rho_d)_{sat}$. In this condition, the soil becomes saturated by reduction in air voids to zero but with no change in water content. It may be mentioned that compaction methods cannot remove all the air voids, and, therefore, the soil never becomes fully saturated. Thus, the theoretical maximum density is only hypothetical. The line indicating the theoretical maximum density can be plotted along with the compaction curve. It is also known as zero air void line or 100% saturation line.

5.21 (a)

- In cohesionless soils, compression is relatively quicker because settlement is mainly immediate.
- Due to large permeability, clays are compacted more rapidly thus leading to quicker consolidation process on dry side.
- Clays compacted wet of optimum is more compressible than the flocculated soil on dry

side of optimum, at relatively low stress levels. This is because the strong interparticle bonds in a flocculated structure do not allow the particles to be displaced.

5.22 (a)

$$V = 1000 \text{ ml},$$
$$V_{min} = 750 \text{ ml},$$
$$V_{max} = 1320 \text{ ml}$$

$$e = \text{Void ratio} = \frac{V_V}{V_S}$$

Hence, V_s = constant for all condition only V_V changes.

Let, $V_s = x$

$$e = \frac{V_V}{V_s} = \frac{V - V_s}{V_s} = \frac{V}{V_s} - 1$$

For natural state, $e = \frac{1000}{x} - 1$

For maximum compacted state,

$$e_{max} = \frac{1320}{x} - 1$$

For minimum compacted,

$$e_{min} = \frac{750}{x} - 1$$

$\Rightarrow$ Relative density $= \frac{e_{max} - e}{e_{max} - e_{min}} \times 100\,\%$

$$= \frac{\left(\frac{1320}{x} - 1\right) - \left(\frac{1000}{x} - 1\right)}{\left(\frac{1320}{x} - 1\right) - \left(\frac{750}{x} - 1\right)} \times 100$$

$$= \frac{\frac{320}{x}}{\frac{570}{x}} \times 100 = \frac{320}{570} \times 100 = 56.14\%$$

5.24 (d)

Optimum moisture content decreases with compactive effort.

5.25 (b)

$$\gamma_{d\,max} = 18 \text{ kN/m}^3$$
$$w = 16\%$$
$$G = 2.65$$

$$\gamma_{d\,max} = \frac{G\gamma_w}{1+e}$$

$$18 = \frac{2.65 \times 9.81}{1+e}$$

$$\Rightarrow \quad e = 0.444$$
$$Se = wG$$
$$S \times 0.444 = 16 \times 2.65$$
$$S = 95.5\%$$

5.26 (a)

Coarser the grains, the lesser is the specific surface area, so lesser is the amount of water required to make the soil grains wet at maximum dry density.

6 Compressibility and Consolidation

6.1 Match **List-I** with **List-II** and select the correct answer using the codes given below the lists (notations have their usual meaning):

List-I	List-II
A. Coefficient of compressibility	1. m_v
B. Compression index	2. $C_v t/H^2$
C. Time factor	3. a_v
D. Coefficient of volume compressibility	4. C_c

Codes:

	A	B	C	D
(a)	3	2	4	1
(b)	1	2	4	3
(c)	1	4	2	3
(d)	3	4	2	1

[ESE : 2000]

6.2 Reduction in volume of soil primarily due to squeezing out of water from the voids is called
(a) primary consolidation
(b) plastic flow
(c) creep
(d) secondary consolidation [ESE : 2000]

6.3 The settlement analysis for a clay layer draining from top and bottom shows a settlement of 2.5 cm in 4 years and an ultimate settlement of 10 cm. However detailed subsurface investigation reveals that there is no drainage at the bottom. The ultimate settlement in this condition will be
(a) 2.5 cm (b) 5 cm
(c) 10 cm (d) 20 cm
[ESE : 2000]

6.4 **Assertion (A):** The rates of settlement of building structures on sandy clay soils are faster compared to those of buildings constructed on clayey soils.
Reason (R): The rate of consolidation is dependent on the permeability of soils and the permeability of sandy clay is more than that of clayey soil.
(a) both A and R are true and R is the correct explanation of A
(b) both A and R are true but R is not a correct explanation of A
(c) A is true but R is false
(d) A is false but R is true [ESE : 2000]

6.5 The figure given shows the state of a sample of clay before and after consolidation. Based on these figures, the settlement of a clay layer of initial thickness H will be

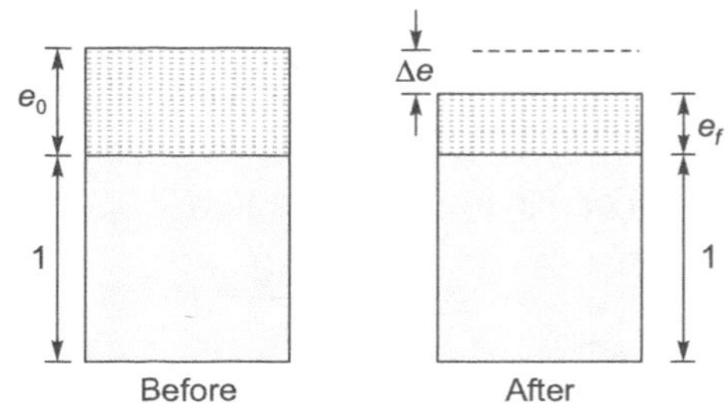

(a) $\dfrac{a_v}{1+e_0}$ (b) $\dfrac{H\Delta e}{(1+e_f)}$
(c) $\dfrac{\Delta e}{(1+e_f)}$ (d) $\dfrac{H\Delta e}{(1+e_0)}$
[ESE : 2001]

6.6 Excess pore pressure distributions within the thickness of a soil sample tested in oedometer sometime after loading is shown in the figure below labelled 1, 2, 3 and 4. Which one of these figures, refers to a situation where the operator forgot to put on the porous stones at the top and bottom of the sample before the test?

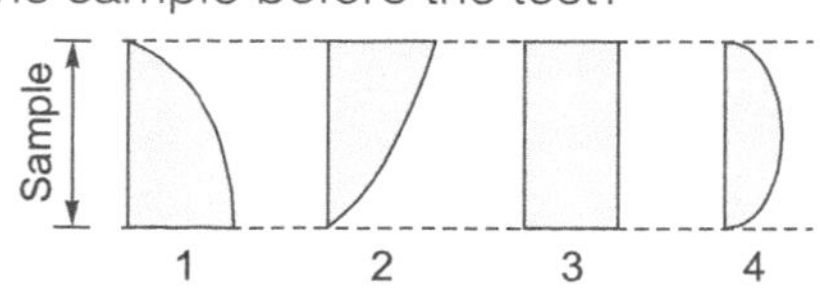

(a) 1 (b) 2
(c) 3 (d) 4 [ESE : 2001]

6.7 In soil consolidation process, the following events take place after loading:
1. Decrease in excess pore pressure.
2. Increase in total stress.
3. Development of excess pore pressure.
4. Increase in effective stress.

The correct sequence of these events is
(a) 3, 2, 1, 4 (b) 2, 3, 1, 4
(c) 2, 3, 4, 1 (d) 3, 2, 4, 1
[ESE : 2001]

6.8 The e-p curve for a soil is shown in the figure below. The coefficient of compressibility (in m²/kN) of the soils is

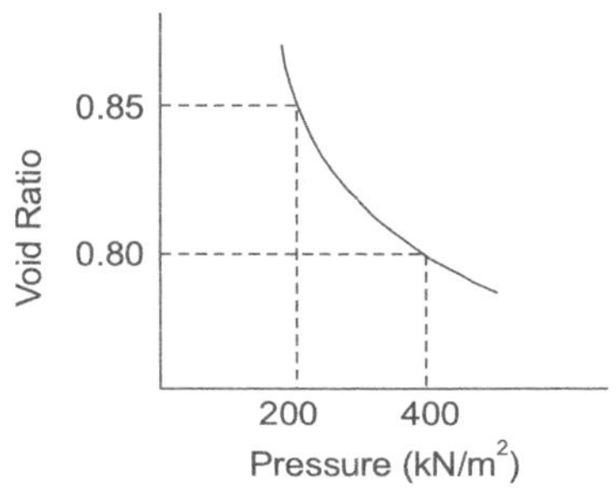

(a) 4000 (b) 2000
(c) 2.5×10^{-4} (d) 1.25×10^{-4}
[ESE : 2001]

6.9 A saturated clay layer with double drainage takes 5 years to attain 90% degree of consolidation under a structure. If the same layer were to be single drained, what would be the time (in years) required to attain the same consolidation under the same loading conditions?
(a) 10 (b) 15
(c) 20 (d) 25 [ESE : 2001]

6.10 In the phase diagrams given the change due to initial state changing into final state is shown due to consolidation. Depth of soil layer undergoing consolidation is H; e_0 is initial void ratio; e_f is final void ratio; Δe is change in void ratio.

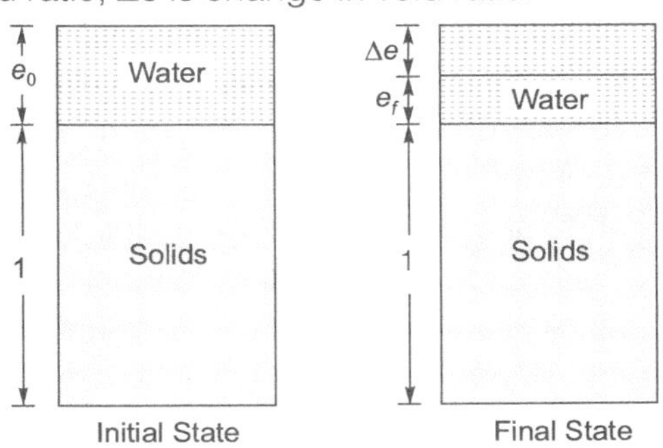

Indicate which of the following expressions gives settlement of the layer.

(a) $H\log_{10}\left(\frac{\Delta e}{1+e_0}\right)$ (b) $\log_{10}\left(H\frac{\Delta e}{1+e_0}\right)$

(c) $\frac{\Delta e}{1+e_0}$ (d) $H\frac{\Delta e}{1+e_0}$

[ESE : 2002]

6.11 Which one of the following statements regarding coefficient of consolidation C_v is correct?
(a) $C_v \propto k$ (b) $C_v \propto \frac{1}{k}$
(c) $C_v \propto m_v$ (d) $C_v \propto a_v$ [ESE : 2002]

6.12 Assertion (A): A clay layer of thickness. H, with single drainage suffers an ultimate settlement of S under increase in pressure of Δp. When a thin sand layer is present at its mid depth, the layer, under otherwise identical conditions, will have no change in ultimate settlements.

Reason (R): The path of travel of water for the clay layer is H under single drainage. When a thin sand layer is present at mid depth of clay layer, the path of travel in lower half of the layer gets reduced to $H/2$ and that in upper half of the layer gets reduced to $H/4$. This considerably reduces the time required to realize ultimate settlement.
(a) both A and R are true and R is the correct explanation of A
(b) both A and R are true but R is not a correct explanation of A
(c) A is true but R is false
(c) A is false but R is true [ESE : 2002]

6.13 Consider the following statements:
Theory of consolidation predicts settlement due to primary consolidation; it cannot include settlement due to initial compression nor due to secondary consolidation. This happens because of the following assumptions made by the theory:
1. Soil grains and water are incompressible.
2. Soil is fully saturated.
3. Compression takes place in vertical direction only.
4. Time lag in consolidation is entirely due to low permeability of soil.

Which of these statements are correct?

(a) 1, 2 and 3 (b) 2 and 3
(c) 3 and 4 (d) 1, 2 and 4

[ESE : 2002]

6.14 For a certain loading condition, a saturated clay layer undergoes 40% consolidation in a period of 178 days. What would be the additional time required for further 20% consolidation to occur?

(a) 89 days (b) 222.5 days
(c) 329.5 days (d) 400.5 days

[ESE : 2004]

6.15 The one dimensional theory of consolidation proposed by Terzaghi derives its name due to the fact that

(a) only vertical dimension of the soil sample is used for consolidation test and lateral dimensions are neglected
(b) water in the soil sample in conventional consolidometer escapes in the lateral directions resulting into settlements only in vertical direction
(c) normal stress on the sample is applied in only one (vertical) direction
(d) lateral movements of soil grains are not permitted across any vertical boundary resulting into only vertical settlements to account for the decrease in volume due to escape of water from soil sample

[ESE : 2004]

6.16 If the time required for 60% consolidation of a remoulded soil sample of clay with single drainage is 't', then what is the time required to consolidate the same sample of clay with the same degree of consolidation but with double drainage?

(a) $4t$ (b) $2t$
(c) $t/2$ (d) $t/4$ [ESE : 2005]

6.17 If Δp is increment of pressure on a normally consolidated saturated soil mass, as per Terzaghi's theory at the instant of application of pressure increment i.e., when time $t = 0$, what is the pore pressure developed in the soil mass?

(a) Zero
(b) Very much less than Δp
(c) Equal to Δp
(d) Greater than Δp [ESE : 2005]

6.18 The void ratio-pressure diagram is shown below. What is the coefficient of compressibility?

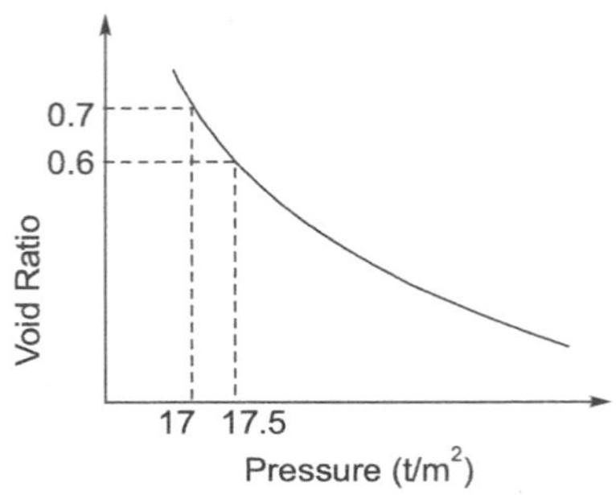

(a) 0.5 m^2/t (b) 0.73 m^2/t
(c) 0.20 m^2/t (d) 0.25 m^2/t

[ESE : 2005]

6.19 Match **List-I** (Term) with **List-II** (Definition) and select the correct answer using the codes given below the lists (where 'e' is void ratio, 'p' is effective vertical stress, 'k' is permeability and γ_w is unit weight of water):

List-I
A. Compression index, C_c
B. Coefficient of volume change, m_v
C. Coefficient of compressibility, a_v
D. Coefficient of consolidation, C_v

List -II
1. Slope of the e-p curve defined as $\Delta e/\Delta p\,(1+e_0)$
2. Slope of e-log p curve given by $[-\Delta e/\Delta(\log p)]$
3. Term defined as $k/(m_v\gamma_w)$
4. Slope of the e-p curve defined as $-\Delta e/\Delta p$

Codes:

	A	B	C	D
(a)	2	1	4	3
(b)	4	3	2	1
(c)	2	3	4	1
(d)	4	1	2	3

[ESE : 2005]

6.20 A clay layer 3 m deep is sandwiched by a pervious material. The soil reaches 90% consolidation in 3 years. At another site, the same clay layer is bounded by impervious boundary at the bottom and sand at the top. What is the time taken by this layer to reach 90% settlement?

(a) 9 years (b) 12 years
(c) 15 years (d) 21 years

[ESE : 2005]

6.21 In a consolidation test void ratio decreased from 0.80 to 0.70 when the load was changed from 40 kN/m^2 to 80 kN/m^2. What is the compression index?

(a) 0.14 (b) 0.16
(c) 0.33 (d) 0.66 [ESE : 2006]

6.22 In a consolidation test the sample tested has height H; water content is w; specific gravity of solids G. After increasing the loading by an increment Δp, the height decrease is ΔH. Which one of the following expresses the corresponding change in void ratio Δe?

(a) $\Delta e = \dfrac{\Delta H}{H(1+wG)}$ (b) $\Delta e = \dfrac{\Delta H(1+wG)}{H}$

(c) $\Delta e = \dfrac{H(1+wG)}{\Delta H}$ (d) $\Delta e = \dfrac{H}{\Delta H(1+wG)}$

[ESE : 2006]

6.23 The virgin compression curve for a soil is shown in figure below. What is the compression index of the soil?

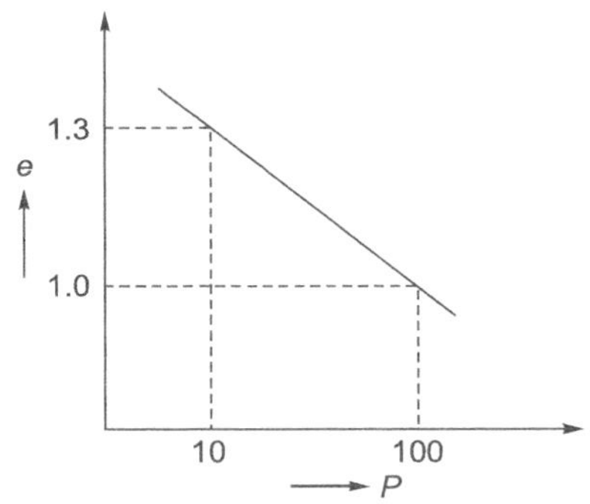

(a) 0.3 (b) 0.5
(c) 1.5 (d) 15 [ESE : 2007]

6.24 Consider the following statements:

1. Coefficient of consolidation normally increases with decreasing liquid limit of clay.
2. The larger the value of coefficient of consolidation, the longer it takes for full consolidation to occur.

Which of these statements is/are correct?

(a) 1 only (b) 2 only
(c) Both 1 and 2 (d) Neither 1 nor 2

[ESE : 2008]

6.25 During the consolidation of a clay layer, if instead of double drainage, the drainage is allowed at single face only, the rate of compression will be

(a) 4 times slower (b) 4 times faster
(c) 2 times slower (d) 2 times faster

[ESE : 2008]

6.26 A clay sample, originally 26 mm thick at a void ratio of 1.22, was subjected to a compressive load. After the c ay sample was completely consolidated, its thickness was measured to be 24 mm. What is the final void ratio?

(a) 1.322 (b) 1.421
(c) 1.311 (d) 1.050

[ESE : 2009]

6.27 A 1 m thick layer of saturated clay, drained at both faces, settles by 10 cm in one year. If a thin layer of perv ous soil is introduced in the middle of this layer, then what will be the period during which the settlement of 10 cm will be completed?

(a) 4 years (b) 0.5 year
(c) 0.25 year (d) 2 years [ESE : 2009]

6.28 **Assertion (A):** At the same voids ratio desiccated clay is stronger than saturated clay.

Reason (R): Desiccation impacts (induces) pre-compressive forces in the soil structure.

(a) both A and R are true and R is the correct explanation of A
(b) both A and R are true but R is not a correct explanation of A
(c) A is true but R is false
(d) A is false but R is true [ESE : 2010]

6.29 During consolidation process of clayey soils, indicate the sequence of occurrence of the following events in the order from first to last

1. Load being taken up by the pore-water
2. Load being taken up by the soil grains
3. Drainage of water from the pores of the soil

(a) 1, 2 and 3 (b) 2, 3 and 1
(c) 1, 3 and 2 (d) 2, 1 and 3

[ESE : 2011]

6.30 If, instead of single drainage, the number of drainage faces is increased to two in responding soils, the rate of compression will be

(a) 4 times slower (b) 2 times slower
(c) 4 times faster (d) 2 times faster

[ESE : 2011]

6.31 Settlement due to creep in soils is contingent on
(a) Primary consolidation
(b) Secondary consolidation
(c) Initial settlement
(d) Compaction settlement [ESE : 2011]

6.32 **Assertion (A):** In non-cohesive soils, settlement occurs immediately after application of load.
Reason (R): The settlement is attributed to volume changes caused by lateral yielding or shear strains occurring in the soil.
(a) both A and R are true and R is the correct explanation of A
(b) both A and R are true but R is not a correct explanation of A
(c) A is true but R is false
(d) A is false but R is true [ESE : 2011]

6.33 Consider the following statements:
1. Mathematically speaking, the time taken for 100% consolidation is infinite.
2. The time factor for a particular average degree of consolidation depends upon the distribution of initial excess hydrostatic pressure.
3. Secondary consolidation obeys Terzaghi's one-dimensional theory of consolidation.

Which of these statements are correct?
(a) 1, 2 and 3 (b) 1 and 2 only
(c) 2 and 3 only (d) 1 and 3 only
[ESE : 2012]

6.34 Consider the following statements:
1. Consolidation time increases with increasing compressibility.
2. Consolidation time decreases with increasing permeability.
3. Consolidation time is dependent on the magnitude of stress increase.

Which of these statements are correct?
(a) 1, 2 and 3 (b) 1 and 2 only
(c) 2 and 3 only (d) 1 and 3 only
[ESE : 2012]

6.35 Consider the following statements:
1. The maximum possible over-consolidation ratio of normally consolidated soil is unity.
2. The compressibility of a field deposit is slightly greater than that shown by a laboratory sample.
3. In Terzaghi's theory of one-dimensional consolidation, only hydrodynamic lag is considered whereas plastic lag is ignored.

Which of these statements are correct?
(a) 1, 2 and 3 (b) 1 and 2 only
(c) 1 and 3 only (d) 2 and 3 only
[ESE : 2013]

6.36 Compression index developed by Casagrande is
(a) $C_c = 0.009(LL + 10\%)$
(b) $C_c = 0.009(LL - 10\%)$
(c) $C_c = 0.0009(LL + 10\%)$
(d) $C_c = 0.0009(LL - 10\%)$ [ESE : 2014]

6.37 When a structural load is applied on a soil stratum, which of the following soil types will have the minimum settlement?
(a) Over-consolidated clay stratum
(b) Clayey silt stratum
(c) Normally consolidated clay stratum
(d) Sandy clay stratum [ESE : 2014]

6.38 **Statement (I):** In a consolidation test, pressure on the soil specimen is doubled at each step.
Statement (II): It is intended that the soil always remains in a normally consolidated condition.
(a) Both Statement (I) and Statement (II) are individually true and Statement (II) is the correct explanation of Statement (I).
(b) Both Statement (I) and Statement (II) are individually true but Statement (II) is not the correct explanation of Statement (I).
(c) Statement (I) is true but Statement (II) is false.
(d) Statement (I) is false but Statement (II) is true.
[ESE : 2014]

6.39 If the coefficient of permeability is doubled and the coefficient of volume compressibility in simultaneously halved, the coefficient of consolidation
(a) increase by 2 times
(b) decrease by 2 times
(c) increase by 4 times
(d) decrease by 4 times [ESE : 2015]

6.40 In a 7 m thick soil stratum, with its initial void ratio of 0.40, the void ratio decreases to 0.30 when the effective pressure on the stratum is increased by 1.0 kg/cm^2. The consolidation settlement of the stratum will be

(a) 5 cm (b) 50 cm
(c) 100 cm (d) 150 cm [ESE : 2015]

6.41 The virgin compression curve for a particular soil is as shown in the below figure on the standard graphical format. The compression index of the soil is

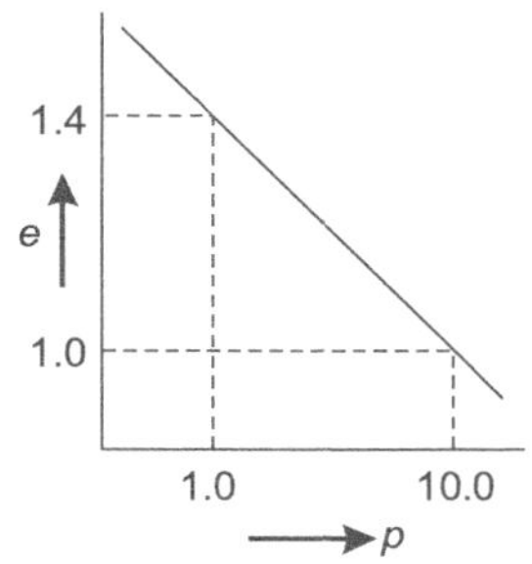

(a) 0.3 (b) 0.4
(c) 0.5 (d) 0.6

[ESE : 2015]

6.42 The time taken to construct a building was from April 1992 to September 1993. In September 1996, the average settlement was found to be 5.16 cm. If the ultimate settlement is estimated to be 25 cm, then the settlement in January 1997 would have been

(a) 6 cm (b) 7 cm
(c) 8 cm (d) 9 cm

[ESE : 2015]

6.43 A soil deposit has a void ratio of 1.0. If the void ratio is reduced to 0.60 by compaction, the percentage volume loss is

(a) 10% (b) 20%
(c) 30% (d) 40% [ESE : 2015]

6.44 Consider the following statements:

1. Effective stress in a sand layer below a lake with standing water does not alter as the water level fluctuates.
2. Regarding water table below the ground surface, any rise in the water table causes equal changes in both pore pressure and effective stress.
3. Capillary saturation will cause the effective stress in increase.

Which of the above statements are correct?

(a) 1, 2 and 3 (b) 1 and 2 only
(c) 2 and 3 only (d) 1 and 3 only

[ESE : 2015]

6.45

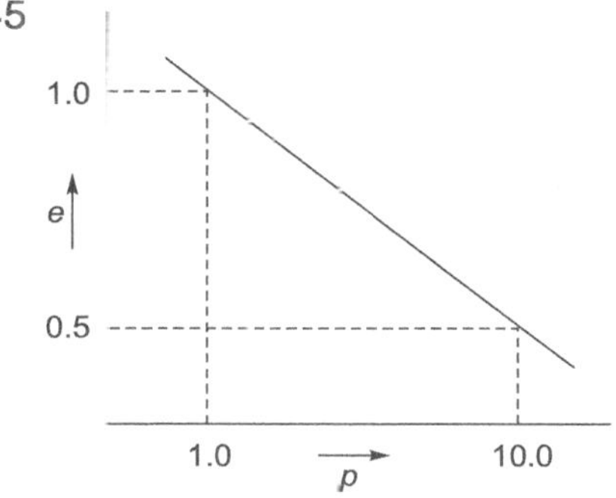

The virgin compression curve with axes adopted as per convention in this regard for a clay soil is shown in the figure. The compression index of the soil is

(a) 0.25 (b) 0.50
(c) 1.00 (d) 1.25 [ESE : 2016]

Directions: Each of the next items consists of two statements, one labelled as the **'Statement (I)'** and the other as **'Statement (II)'**. Examine these two statements carefully and select the answers to these items using the codes given below :

Codes:

(a) Both Statement (I) and Statement (II) are individually true and Statement (II) is the correct explanation of Statement (I)
(b) Both Statement (I) and Statement (II) are individually true but Statement (II) is ***not*** the correct explanation of Statement (I)
(c) Statement (I) is true but Statement (II) is false
(d) Statement (I) is false but Statement (II) is true

6.46 **Statement (I)** : Rate of settlement of a consolidating layer depends upon its coefficient of consolidation, which is directly proportional the permeability and number of drainage paths available.

Statement (II): The excess hydrostatic pore pressure is relieved fast in soil of higher permeability, in turn, depending on the number of drainage paths available in the consolidating layer.

[ESE : 2016]

6.47 **Statement (I):** Consolidation is the process of expulsion of pore water by applying steady load for a long time.

Statement (II): Volume change occurs immediately after the action of the load in case of consolidation tests.

[ESE : 2017]

6.48 Consider the following statements:

1. Secondary consolidation of soil follows Terzaghi's one-dimensional theory of consolidation.
2. Consolidation is a function of total stress.
3. Even after complete dissipation of excess pore pressure, the soil undergoes a little more consolidation.

Which of the above statements is/are correct?

(a) 3 only (b) 1 and 2
(c) 2 and 3 (d) 1 only

[ESE : 2017]

6.49 Consider the following statements:

1. Immediate settlement takes place as soon as the load is placed.
2. Secondary settlement is significant in the case of organic soil.
3. Secondary settlement is estimated based on the 'void ratio versus time curve' for a particular load under consolidation test.

Which of the above statements are correct?

(a) 1 and 2 (b) 1, 2 and 3
(c) 2 and 3 (d) 1 and 3

[ESE : 2017]

6.50 Consider the following statements regarding permeability of soils:

1. Permeability of coarse grained soil is inversely proportional to the specific surface at a given porosity.
2. Direct measurement of permeability of a soil specimen at any stage of loading in odometer test can be made only in fixed-ring type oedometer.
3. The permeability of an aquifer increased with decreases in temperature of water moving through it.

Which of the above statements is/are correct?

(a) 1 and 2 (b) 1 and 3
(c) 2 and 3 (d) 1, 2 and 3

[ESE : 2017]

6.51 The theory of a consolidation predicts settlement due to primary consolidation; it cannot include settlement due to initial compression nor due to secondary consolidation. This happens because of the following assumptions made in developing the theory:

1. Soil grains and water are incompressible.
2. Soil is fully saturated.
3. Compression takes place in the vertical direction only.
4. Time lag in consolidation is entirely due to low permeability of soil.

Which of the above statements are CORRECT?

(a) 1, 2 and 3 only (b) 1, 2 and 4 only
(c) 3 and 4 only (d) 1, 2, 3 and 4

[ESE : 2018]

6.52 Consider the following statements:

1. Secondary consolidation results due to prolonged dissipation of excess hydrostatic pressure.
2. Primary consolidation happens under expulsion of both air and water from voids in early stages.
3. Initial consolidation in the case of fully saturated soils is mainly due to compression of solid particles.
4. Primary consolidation happens more quickly in coarse-grained soils than in fine-grained soils.

Which of the above statements are CORRECT?

(a) 1 and 2 only (b) 2 and 3 only
(c) 3 and 4 only (d) 1 and 4 only

[ESE : 2018]

6.53 The settlement due to secondary compression is predominant in

(a) Granular soils (b) Inorganic clays
(c) Organics clays (d) Very fine sand & silts

[ESE : 2019]

6.54 A bed consists of compressible clay of 4 m thickness with pervious sand on top and impervious rock at the bottom. In a consolidation test on an undisturbed specimen of clay from this deposit 90% settlement was reached in 4 hours. The specimen was 20 mm thick. The time for the building founded over this deposit to reach 90% of its final settlement will be

(a) 91 years (b) 82 years
(c) 73 years (d) 64 years

[ESE : 2020]

6.55 What is the coefficient of volume change (using change in void ratio method) for pressure range 100 kN/m^2 to 200 kN/m^2? (Consider $\sigma_0' = 100$ kN/m^2, $e_0 = 1.121$, $\sigma' =$ 200 kN/m^2, $e_0 = 0.964$, $\Delta\sigma' = 100$ kN/m^2 and $\Delta e = -0.157$)

(a) 0.25 m^2/MN (b) 0.48 m^2/MN
(c) 0.69 m^2/MN (d) 0.74 m^2/MN

[ESE : 2021]

6.56 A footing 3 m × 1.5 m in plan transmits a pressure of 160 kN/m^2 on a cohesive soil having $E = 8 \times 10^4$ kN/m^2 and $\mu = 0.48$. What is the immediate settlement at the centre, assuming the footing to be flexible? (Take $I_w = 1.52$)

(a) 2.11 mm (b) 3.21 mm
(c) 3.51 mm (d) 4.62 mm

[ESE : 2023]

6.57 Which one of the following is the part of assumptions made by Terzaghi while developing the mathematical statement of the consolidation process?

(a) The soil is non-homogeneous
(b) The soil particles and water are incompressible
(c) The partial deformation of soil is due to partial change in volume
(d) Coefficient of permeability is variable during consolidation

[ESE : 2023]

■■■■

Answers Compressibility and Consolidation

6.1 (d) 6.2 (a) 6.3 (c) 6.4 (a) 6.5 (d) 6.6 (c) 6.7 (b) 6.8 (c) 6.9 (c)
6.10 (d) 6.11 (a) 6.12 (b) 6.13 (d) 6.14 (b) 6.15 (d) 6.16 (d) 6.17 (c) 6.18 (c)
6.19 (a) 6.20 (b) 6.21 (c) 6.22 (b) 6.23 (a) 6.24 (a) 6.25 (a) 6.26 (d) 6.27 (c)
6.28 (a) 6.29 (c) 6.30 (c) 6.31 (b) 6.32 (c) 6.33 (b) 6.34 (a) 6.35 (a) 6.36 (b)
6.37 (a) 6.38 (b) 6.39 (c) 6.40 (b) 6.41 (b) 6.42 (a) 6.43 (c) 6.44 (d) 6.45 (b)
6.46 (a) 6.47 (c) 6.48 (a) 6.49 (b) 6.50 (a) 6.51 (b) 6.52 (c) 6.53 (c) 6.54 (c)
6.55 (d) 6.56 (c) 6.57 (b)

Explanations Compressibility and Consolidation

6.2 (a)

After initial consolidation, further reduction in volume occurs due to expulsion of water from the voids. When a saturated soil is subjected to a pressure, initially all the applied pressure is taken up by water as an excess pore water pressure. A hydraulic gradient will develop and the water starts flowing out and a decrease in volume occurs. This reduction in volume is called as the primary consolidation of soil.

6.3 (c)

Ultimate settlement does not depend upon drainage condition.

6.4 (a)

Rate of settlement depends upon permeability, drainage conditions (single drained or double drained) and coefficient of volume change as

$$\text{Time factor, } T_v = \frac{C_v t}{d^2}$$

6.5 (d)

$$\text{Settlement, } \Delta H = \frac{H\Delta e}{1+e_0}$$

6.6 (c)

	Top plate	Bottom plate
Sample-1	Yes	No
Sample-2	No	Yes
Sample-3	No	No
Sample-4	Yes	Yes

6.7 (b)

- **Initial Consolidation :** When a load is applied to a partially saturated soil, a decrease in volume occurs due to expulsion and compression of air in the voids. A small decrease in volume occurs due to compression of solid particles. The reduction in volume of the soil just after the application of the load is known as initial consolidation or initial compression. For saturated soils, the initial consolidation is mainly due to compression of solid particles.
- **Primary Consolidation :** After initial consolidation, further reduction in volume occurs due to expulsion of water from the voids. When a saturated soil is subjected to a pressure, initially all the applied pressure is taken up by water as an excess pore water pressure. A hydraulic gradient will develop and the water starts flowing out and a decrease in volume occurs. This reduction in volume is called as the primary consolidation of soil.
- **Secondary Consolidation :** The reduction in volume continues at a very slow rate even after the excess hydrostatic pressure developed by the applied pressure is fully dissipated and the primary consolidation is complete. The additional reduction in the volume is called as the secondary consolidation.

6.8 (c)

Coefficient of compressibility,

$$a_v = -\frac{\Delta e}{\Delta \sigma}$$

$$= -\left(\frac{0.8-0.85}{400-200}\right) = \frac{0.05}{200}$$

$$= 2.5 \times 10^{-4} \text{ m}^2/\text{kN}$$

6.9 (c)

$$\frac{t_2}{t_1} = \left(\frac{d_2}{d_1}\right)^2$$

But $\quad d_2 = 2d_1$

$$\therefore \quad \frac{t_2}{t_1} = 4$$

d_2 is drainage path for single drained

d_1 is drainage path for double drained

$$\therefore \quad t_2 = 5 \times 4 = 20 \text{ years}$$

6.10 (d)

Settlement, $\quad \Delta H = H\left(\frac{e_0 - e_f}{1+e_0}\right) = \frac{H\Delta e}{1+e_0}$

6.11 (a)

Coefficient of consolidation,

$$C_v = \frac{k}{m_v \gamma_w}$$

6.12 (b)

The ultimate settlement is given by

$$S = m_v \Delta p\, H$$

It does not depend upon type of drainage (single or double drainage) conditions. Therefore A is correct. The presence of thin sand layer reduces drainage path and settlement becomes fast.

6.13 (d)

Since the assumption of compression in vertical direction only is required to compute settlement due to volumetric change so it has no relation with primary and secondary compression.

6.14 (b)

Time factor, $\quad T = \frac{\pi}{4}U^2$ for $U < 60\%$

For same clay layer at different times t_1 and t_2 the relationship reduces to

$$\frac{t_2}{t_1} = \left(\frac{U_2}{U_1}\right)^2$$

$$\Rightarrow \quad t_2 = \left(\frac{60}{40}\right)^2 \times 178$$

$$= 400.5 \text{ days}$$

Thus additional time needed $= t_2 - t_1$

$$= 400.5 - 178 = 222.5 \text{ days}$$

6.15 (d)

Assumptions of Terzaghi's Theory:

1. The soil is homogenous (uniform in composition throughout) and isotropic(show same physical property in each direction).

2. The soil is fully saturated (zero air voids due to water content being so high).
3. The solid particles and water are incompressible.
4. Compression and flow are one-dimensional.
5. Strains in the soil are relatively small.
6. Darcy's Law is valid for all hydraulic gradients.
7. The coefficient of permeability and the coefficient of volume compressibility remain constant throughout the process.
8. There is a unique relationship, independent of time, between the void ratio and effective stress

6.16 (d)

For same degree of consolidation,

$$\frac{t_1}{t_2} = \left(\frac{d_1}{d_2}\right)^2$$

Single drainage $t_1 = t$; $d_1 = H$

Double drainage $t_2 = ?$; $d_2 = H/2$

$$\therefore \quad t_2 = t \times \left(\frac{H/2}{H}\right)^2 = \frac{t}{4}$$

6.17 (c)

A $t = 0$ the pore pressure increment or pore pressure developed will be Δp. At any time $t > 0$ excess pore pressure dissipates and reduces.

6.18 (c)

Coefficient of compressibility,

$$a_v = -\frac{\Delta e}{\Delta \sigma} = -\left(\frac{0.6 - 0.7}{17.5 - 17}\right) = 0.20 \text{ m}^2/\text{t}$$

6.19 (a)

6.20 (b)

For same degree of consolidation,

$$t_2 = t_1\left(\frac{d_2}{d_1}\right)^2 = 4t_1 = 4 \times 3 = 12 \text{ years}$$

6.21 (c)

Compression Index

$$C_c = \frac{e_0 - e_f}{\log(\sigma_f / \sigma_0)} = \frac{0.80 - 0.70}{\log(80/40)} = 0.33$$

6.22 (b)

$$\frac{\Delta H}{H} = \frac{\Delta e}{1 + e_0}$$

Using $e_0 = G\,w$ (for full saturation)

$$\Delta e = \frac{\Delta H(1 + Gw)}{H}$$

6.23 (a)

Compression index is given by the equation,

$$C_c = \frac{e_1 - e_2}{\log_{10}(\bar{\sigma}_2/\bar{\sigma}_1)}$$

$$= \frac{1.3 - 1.0}{\log_{10}(100/10)} = 0.3$$

6.24 (a)

From Taylor's formula, we have

$$T_v = \frac{C_v t}{d^2}$$

For full consolidation i.e. $U = 100\%$. T_v will more or less be a constant.

$$\therefore \quad C_v t = \text{constant}$$

$$\Rightarrow \quad C_v \propto \frac{1}{t}$$

Thus, larger the value of coefficient of consolidation, smaller it takes for full consolidation to occur.

6.25 (a)

From Taylor's formula, we have

$$T_v = \frac{C_v t}{d^2}$$

$$\Rightarrow \quad t \propto d^2$$

$$\Rightarrow \quad \frac{t_d}{t_s} = \left(\frac{d_d}{d_s}\right)^2 = \left(\frac{H/2}{H}\right)^2$$

$$\Rightarrow \quad t_s = 4\,t_d$$

Thus the time taken by the clay in single drainage is four times more than in double drainage.

6.26 (d)

$$\frac{\Delta H}{H} = \frac{\Delta e}{1 + e}$$

$$\Rightarrow \quad \frac{26 - 24}{26} = \frac{\Delta e}{1 + 1.22}$$

$$\Delta e = 0.171$$

But $\quad e_1 - e_2 = \Delta e$

$$\Rightarrow \quad e_2 = 1.22 - 0.171$$

$$\Rightarrow \quad e_2 = 1.05$$

6.27 (c)

For a particular soil

$$t \propto d^2$$

$$\Rightarrow \quad \frac{t_1}{t_2} = \left(\frac{d_1}{d_2}\right)^2$$

$$\Rightarrow \quad \frac{1}{t_2} = \left(\frac{1}{0.5}\right)^2$$

$$\Rightarrow \quad t_2 = 0.25 \text{ years}$$

6.28 (a)

Desiccation in clay soils can result in shrinkage of the soil and subsidence of the ground. As the soil re-hydrates, it can swell, resulting in ground heave. One of the reasons of preconsolidation is desiccation of the clay deposit. During drying due to rise in temperature, tension develops in pore water and effective stress increases.

6.29 (c)

Consolidation of soil deposit has three stages:

1. **Initial Consolidation:** When a load is applied to a partially saturated soil, a decrease in volume occurs due to expulsion and compression of air in the voids.
2. **Primary Consolidation:** After initial consolidation, further reduction in volume occurs due to expulsion of water from voids. When a saturated soil is subjected to a pressure, initially all the applied pressure is taken up by water as an excess pore water pressure, as water is almost incompressible as compared with solid particles. A hydraulic gradient develops and the water starts flowing out and a decrease in volume occurs. The decrease depends upon the permeability of the soil and is, therefore, time dependent. This reduction in volume is called primary consolidation.
3. **Secondary consolidation:** It occurs due to the plastic readjustment of the solid particles and the adsorbed water to the new stress system.

6.30 (c)

Coefficient of consolidation,

$$C_v = \frac{T_v d^2}{t}$$

for two drainage path, rate of compression becomes 4 times faster.

6.31 (b)

Creep is prolonged deformation under the action of constant applied stress. In secondary consolidation drainage of water from the pores of the soil due to development of pore water pressure causes compression of the soil layer.

6.32 (c)

In non-cohesive soils, immediate settlement occurs due to expulsion and compression of pore air and expulsion of pore water after the load is applied. Thus, volume of soil also reduces.

6.33 (b)

Secondary consolidation continues much beyond primary consolidation and occurs at a much slower rate. It occurs at a constant effective stress and is not associated with the dissipation of excess pore water pressure. It does not obey Terzaghi's 1 – *D* theory.

6.34 (a)

$$T_V = \frac{C_V t}{D^2}$$

$$t \propto \frac{1}{C_v} \propto \frac{m_V}{k}$$

$$\therefore \quad \left(t \propto \frac{1}{k}\right)$$

$$k = C_v m_V \gamma_w$$

$$m_v = \frac{a_v}{1+e_0} = \frac{\Delta e}{\Delta\bar{\sigma}(1+e_0)}$$

From above, it is clear that:

- Consolidation time increase with increasing compressibility.
- Consolidation time decreases with increasing permeability.
- Consolidation time is dependent on the magnitude of stress increase.

6.35 (a)

The **over consolidation ratio** or OCR is defined as the highest stress experienced divided by the current stress. A soil which is currently experiencing its highest stress is said to be normally consolidated and to have an OCR of one. Even though the consolidation is 100% complete due to the phenomenon of **hydrodynamic lag** that means the entire load increment that has been applied to the soil has been transferred to the soil grains, yet the settlement of soil continues. This phenomenon is due to a number of reasons but primarily due to what is known as **plastic lag** that these solid particles go on adjusting themselves even at constant stresses they continue to undergo some deformation even at constant stress and readjust themselves. And this gradual movement of the particle and readjustment even under the constant stress is what is called **secondary compression behavior** or **secondary consolidation.**

6.36 (b)

The compression index, C_c is sometimes approximately estimated by empirical equations relating C_c to some simple index property of the soil.

One of the most widely used equations with ±30 percent error, is the one proposed by Terzaghi and Peck (1967) for undisturbed clays of low to medium sensitivity.

$$C_c = 0.009\,(W_L - 10)$$

in which W_L is the liquid limit expressed in percent.

This equation can be used only for clays which are not very sensitive ($s_t < 4$)

For remoulded clays, the following equations is used

$$C_c = 0.007(W_L - 7)$$

Some of the other equations that have been proposed are:

$$C_c = 1.15(e_0 - 0.35)$$

for all clays

$$C_c = 0.30\,(e_0 - 0.27)$$

for inogranic cohesive soil

$$C_c = 1.15 \times 10^{-12}\, W_n$$

for organic soils

Where, W_n is the natural water content of the soil.

6.37 (a)

The compression index for the over consolidated clay stratum (called recompression index) is smaller as compared to other given soils , thus it will have minimum settlement.

6.38 (b)

In consolidation test, pressure on the soil specimen is doubled at each step. The pressure commonly used are 25, 50, 100, 200, 400 and 800 kN/m^2. In this test, it is intended that the soil always remains in a normally consolidated condition.

6.39 (c)

$$C_v = \frac{k}{\gamma_w m_v}$$

$$\frac{C_{v_1}}{C_{v_2}} = \frac{k_1}{k_2} \times \frac{m_{v_2}}{m_{v_1}} = \frac{1}{2} \times \frac{1}{2}$$

$$C_{v_2} = 4C_{v_1}$$

6.40 (b)

$$\Delta H_c = H_0 m_v \Delta\sigma$$

$$= H_o \times \left(\frac{\dfrac{\Delta e}{1 - e_o}}{\Delta\sigma} \right) \times \Delta\sigma$$

$$= 700 \times \left(\frac{0.40 - 0.30}{1 + 0.40} \right)$$

$$= 700 \times \frac{0.1}{1.4} = 50 \text{ cm}$$

6.41 (b)

Compression index,

$$C_c = \frac{e_1 - e_2}{\log_{10}\left(\dfrac{\bar{\sigma}_2}{\bar{\sigma}_1}\right)} = \frac{1.4 - 1}{\log\left(\dfrac{10}{1}\right)} = 0.4$$

6.42 (a)

From September 1993 to September 1996 i.e. in 3 years, the average settlement is 5.16 cm

i.e., $t_1 = 3$ years

$\Delta h_1 = 5.16$ cm

Then settlement in January 1997 i.e., in 3.25 year will be Δh_2

$$\text{Consolidation in 3 year} = \frac{5.16}{25} = 0.2064$$

$$\therefore \quad T_V = \frac{\pi}{4} U^2 = 0.0334$$

$$\therefore \quad \frac{C_V t}{H^2} = T_V = 0.0334$$

$$\frac{C_V}{H^2} = \frac{0.0334}{3} = 0.0113$$

Now in 3.25 year

$$T_V = \frac{\pi}{4} U^2 = \frac{C_v t}{H^2}$$

$$= 0.0113 \times 3.25 = 0.03618$$

$$U = 0.2146$$

$$\therefore \quad U = \frac{\Delta h_2}{H}$$

$$0.2146 = \frac{\Delta h_2}{25}$$

$$\therefore \quad \Delta h_2 = 5.37 \text{ cm}$$

6.43 (b)

Volume of soil (V_s) remains constant

$$V_s = \frac{V}{1+e} = \text{Constant}$$

$$\therefore \quad \frac{V_1}{1+e_1} = \frac{V_2}{1+e_2}$$

$$\frac{V_1}{1+1} = \frac{V_2}{1+0.6}$$

$$\frac{V_2}{V_1} = \frac{1.6}{2} = \frac{4}{5}$$

% volume loss

$$= \left(\frac{V_1 - V_2}{V_1}\right) \times 100\% = \left(1 - \frac{V_2}{V_1}\right) \times 100$$

$$= \left(1 - \frac{4}{5}\right) \times 100 = 20\%$$

6.44 (d)

When water level fluctuation takes place above a lake, total stress (σ) and pore water pressure (u) changes with same value in a sand layer below a lake, hence effective stress does not change.

But when the WT is below the ground level, any rise or fall in the WT causes different and unequal changes in both σ & u.

Capillary saturation leads to development of negative pore pressure (i.e. $-u$)

Effective stress, $\bar{\sigma} = \sigma - (-u) = \sigma + u$

$\therefore$ Increment is observed in $\bar{\sigma}$.

6.45 (b)

Compression index,

$$C_c = \frac{\Delta e}{\log\left(\frac{\sigma_2}{\sigma_1}\right)} = \frac{1.0 - 0.5}{\log\left(\frac{10}{1}\right)} = 0.50$$

6.46 (a)

$$T_V = \frac{C_v \cdot t}{d^2} = \frac{k}{m_v d_w} \cdot \frac{t}{d^2}$$

6.48 (a)

- Terzaghi's one dimensional theory of consolidation is valid for primary consolidation.
- Consolidation is function of effective stress.
- After complete dissipation of excess pore pressure, secondary consolidation may occur.

6.49 (b)

Secondary consolidation becomes important for certain types of soil, such as peat, soft organic clay and some loosely deposited clays.

Secondary compression index (C_s) can be calculated as

$$C_s = \frac{\Delta e}{\Delta \log(t)}$$

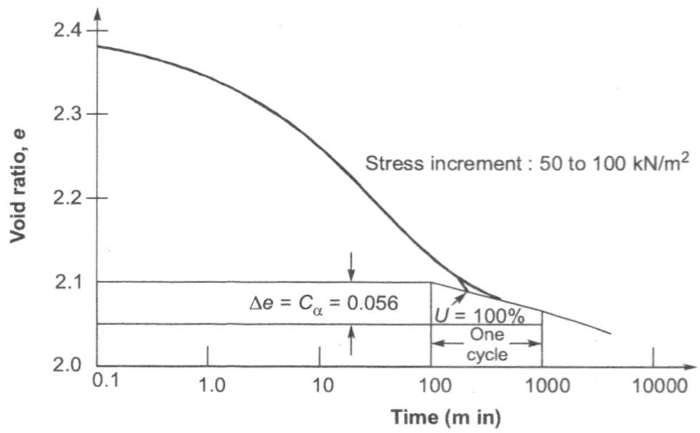

Determination of the rate of secondary compression, C_α

6.50 (a)

As per Kozney-Carman equation

$$k = \left(\frac{\gamma_w}{\mu}\right) \frac{e^3}{1+e} \times \frac{1}{S_s^2} \propto \frac{1}{S_s^2}$$

Dynamic viscosity (μ) is inversely proportional to temperature hence permeability is directly proportional to the temperature.

In consolidation test, two types of consolidation cell, the floating ring cell and the fixed ring cell are commonly used. A falling head permeability test can be performed on the consolidation specimen attaching a stand pipe to fixed ring consolidometer, hence the coefficient of permeability can be determined only in fixed ring test.

6.51 (b)

- Terzaghi theory of consolidation is only for primary consolidation. It does not include compaction and secondary consolidation.
- This theory is one-dimensional theory, it means change in volume of soil is due to change in depth only and expulsion of water is in vertical dimension only.

- Third statement does not justify that, why Terzaghi considered only primary consolidation because in compaction (air) also compression takes place in vertical direction.

6.52 (c)

Secondary consolidation is due to readjustment of soil particles.

Primary consolidation happens due to expulsion of excess pore water.

6.53 (c)

Organic clays and highly plastic clays have secondary compression.

6.54 (c)

By taking single drainage in field slab,

In lab, $$(T_V)_{90} = C_V \frac{t}{d^2}$$

$$\frac{(T_V)_{90}}{C_V} = \frac{4 \text{ hrs}}{(20 \times 10^{-3})^2}$$

In field, $$(T_V)_{90} = C_V \frac{t}{d^2}$$

$$t = \frac{(T_V)_{90}}{C_V} \times d^2$$

$$= \frac{4}{(20 \times 10^{-3})^2} \times 4^2$$

$$t = 18.26 \text{ years}$$

Note: Answer is not matching, hence try assuming double drainage in laboratory.

In lab, $$(T_V)_{90} = C_V \frac{t}{d^2}$$

$$= C_V \times \frac{4 \text{ hrs}}{(10 \times 10^{-3})^2}$$

In field, $$(T_V)_{90} = C_V \frac{t}{(4\text{m})^2}$$

$$t = \frac{(T_V)_{90}}{C_V} \times (4 \text{ m})^2$$

$$= \frac{4}{(10 \times 10^{-3})^2} \times 4^2$$

$$= 73.059 \text{ years}$$

Alternatively,

$$t \propto H^2$$

$$\frac{t_{field}}{t_{lab}} = \left(\frac{H_{field}}{H_{lab}}\right)^2$$

$$H_{lab} = \frac{20}{2} = 10 \text{ mm}$$

$$\therefore \quad t_{field} = \left(\frac{4 \times 10^3}{10}\right)^2 \times \frac{4}{24 \times 365}$$

$$= 73 \text{ years}$$

6.55 (d)

m_v = Coefficient of volume change or coefficient of volume compressibility

$$m_v = \frac{a_v}{1+e_0} = -\left(\frac{\Delta e}{\Delta \bar{\sigma}}\right) \times \frac{1}{1+e_0}$$

$$= -\frac{(-0.157)}{100} \times \frac{1}{1+1.121}$$

$$= \frac{0.157}{100 \times (1+1.121)} \times 10^3 \text{ m}^2/\text{MN}$$

$$= 0.74 \text{ m}^2/\text{MN}$$

6.56 (c)

mmediate settlement or elastic settement is given as,

$$S = \frac{qB(1-\mu^2)I_w}{E}$$

$$= \frac{160 \times 1.5(1-0.48^2) \times 1.52}{8 \times 10^4}$$

$$= 3.51 \times 10^{-3} \text{ m} = 3.51 \text{ mm}$$

6.57 (b)

Assumptions of Terzaghi's theory for 1-D consolidation:

- Soil is homogeneous.
- Soil is fully saturated.
- Solid particles are incompressible.
- Compression and flow are one dimensional.
- Strains in the soil are relatively small.
- Darcy's law is valid.
- Coefficient of permeability and coefficient of volume compressibility remains constant.

7 Stress Distribution in the Soil

7.1 In the case of stratified soil layers, the best equation that can be adopted for computing the pressure distribution is

(a) Prandtl's (b) Skempton's
(c) Westergaard's (d) Boussinesq's

[ESE : 2000]

7.2 In a Newmark's influence chart for stress distribution, there are 10 concentric circles and 50 radial lines. The influence factor of the chart is

(a) 0.0002 (b) 0.002
(c) 0.02 (d) 0.2 [ESE : 2001]

7.3 σ_z is the vertical stress at a depth equal to Z in the soil mass due to a surface point load Q. The vertical stress at depth equal to $2Z$ will be

(a) $0.25\sigma_z$ (b) $0.50\sigma_z$
(c) $1.0\sigma_z$ (d) $2.0\sigma_z$

[ESE : 2002]

7.4 The change in the vertical stress in the soil mass estimated by Boussinesq's equation when Poisson's ratio of soil changes from 0.3 to 0.5 will be

(a) reduction by 30% (b) increase by 50%
(c) reduction by 20% (d) no change

[ESE : 2002]

7.5 In a Newmark's chart for stress distribution, there are 10 concentric circles and 20 radial lines. The influence factor for the chart would be

(a) 0.1 (b) 0.05
(c) 0.01 (d) 0.005

[ESE : 2003]

7.6 A footing of 3 m × 3 m size transmits a load of 1800 kN. The angle of load dispersion in soil $\alpha = \tan^{-1} 0.5$. What is the stress created by the footing load at a depth of 5 m?

(a) 26.12 kN/m^2 (b) 27.12 kN/m^2
(c) 28.12 kN/m^2 (d) 29.12 kN/m^2

[ESE : 2004]

7.7 Stresses obtained from Boussinesq's theory are considered reasonably satisfactory in foundation engineering because

(a) they represent stress distribution in inhomogeneous soils below loaded area
(b) they account for anisotropy of soil property
(c) they give due regard to plastic behaviour of soils, particularly for settlement analysis
(d) they consider elastic soil medium, and the intensity of allowable stresses below foundations in most cases are quite small and justify elastic solutions

[ESE : 2005]

7.8 Westergaard's formula for vertical stress gives greater value of stress than that by the Boussinesq's formula, when r/z exceeds

(a) 1.5 (b) 2.5
(c) 3.5 (d) 4.0 [ESE : 2005]

7.9 A point load of 650 kN is applied on the surface of a thick layer of clay. Using Boussinesq's elastic analysis, what is approximate value of the estimated vertical stress at a depth 2 m and a radial distance of 1.0 m from the point of application of load?

(a) 55 kN/m^2 (b) 44 kN/m^2
(c) 41 kN/m^2 (d) 37 kN/m^2

[ESE : 2006]

7.10 A line load of infinite length has an intensity q per unit length. What is the vertical stress σ_z at a depth z below the earth at the centre of the load?

(a) $\sigma_z = \frac{2qz}{\pi}$ (b) $\sigma_z = \frac{2q}{\pi z}$
(c) $\sigma_z = \frac{2qz^2}{\pi}$ (d) $\sigma_z = \frac{2q}{\pi z^2}$

[ESE : 2008]

7.11 Consider the following statements:

1. Westergaard's theory of stress distribution is

more appropriate for soil deposits which exhibit large lateral strain.
2. Newmark's influence chart can be used for the determination of vertical stress under any shape of a loaded area.

Which of these statements is/are correct?
(a) 1 only (b) 2 only
(c) Both 1 and 2 (d) Neither 1 nor 2
[ESE : 2010]

7.12 An isobar is a line which connects all points below the ground surface at which
(a) The local ground elevation is same
(b) The settlement is same
(c) The vertical stress is the same
(d) The ground elevation is varying
[ESE : 2012]

7.13 Critical shear stress of cohesive sediment
(a) Decreases with the void ratio for a given plasticity index.
(b) Increases with the plasticity index for a given void ratio
(c) Increases with shear strength for a given clay content
(d) All of the above
[ESE : 2012]

7.14 A concentrated load of 50 kN acts on the surface of ground. The increase in vertical stress directly below the load at a depth of 4 m will be (Value of influence factor is 0.48)
(a) 1.5 kN/m^2 (b) 15 kN/m^2
(c) 150 kN/m^2 (d) 0.15 kN/m^2
[ESE : 2013]

7.15 Consider the following statements:
1. The bulb of pressure under strip-footing forms in the lengthwise direction.
2. The bulb of pressure under a strip-footing forms in the width-wise direction.
3. Contact pressure at the edges of a rigid footing in a saturated clay is theoretically infinite.

Which of these statements are correct?
(a) 1, 2 and 3 (b) 2 and 3 only
(c) 1 and 2 only (d) 1 and 3 only
[ESE : 2013]

7.16 **Assertion (A):** Boussinesq equation is not suitable for sedimentary deposits.
Reason (R): Sedimentary deposits do not represent an iso-tropic-cum-homogeneous system.
(a) Both A and R are true and R is the correct explanation of A
(b) Both A and R are true but R is not a correct explanation of A
(c) A is true but R is false
(d) A is false but R is true [ESE : 2015]

7.17 Which of the following statements are correct?
1. Stress Isobar can be prepared using Boussinesq's stress distribution theory.
2. Equivalent point load method yield accurate results.
3. Newmark's method relates the vertical stress with the help of influence chart.
4. Westergaard's method helps in determination of stress distribution for layered soils.

Select the correct answer using the codes given below:
(a) 1, 2 and 3 (b) 1, 3 and 4
(c) 1, 2 and 4 (d) 2, 3 and 4
[ESE : 2017]

7.18 **Statement (I):** The stresses and strains in a soil mass depend on the stress-deformation characteristics, anisotropy and non-homogeneity of the soil and also on the boundary conditions.
Statement (II): Boussinesq's theory of stress distribution in soils deals with layered soils only.
(a) Both Statement (I) and Statement (II) are individually true and Statement (II) is the correct explanation of Statement (I)
(b) Both Statement (I) and Statement (II) are individually true but Statement (II) is ***not*** the correct explanation of Statement (I)
(c) Statement (I) is true but Statement (II) is false
(d) Statement (I) is false but Statement (II) is true
[ESE : 2017]

7.19 What is the Boussinesq's vertical stress at a point 6 m directly below a concentrated load of 2000 kN applied at the ground surface?
(a) 53.1 kN/m^2 (b) 26.5 kN/m^2
(c) 11.8 kN/m^2 (d) 8.8 kN/m^2
[ESE : 2018]

7.20 The vertical stress at any point at a radial distance r and at depth z as determined by using Boussinesq's influence factor K_B and Westergaard's influence factor K_W would be almost same for $\left(\frac{r}{z}\right)$ ratios equal to or greater than

(a) 2.0 (b) 1.8
(c) 1.5 (d) 1.2

[ESE : 2019]

7.21 A concentrated load of 2000 kN is applied at the ground surface. What is the vertical stress at a point 6 m directly below the load?

(a) 16.42 kN/m^2 (b) 26.53 kN/m^2
(c) 36.12 kN/m^2 (d) 40.51 kN/m^2

[ESE : 2021]

7.22 By means of Boussinesq's stress distribution theory, which one of the following pressure distribution diagrams can be prepared?

(a) Strain isobar
(b) Vertical pressure distribution on a horizontal plane
(c) Vertical pressure distribution on an inclined plane
(d) Pressure distribution diagram on an oblique plane

[ESE : 2023]

Answers Stress Distribution in the Soil

7.1 (c)	7.2 (b)	7.3 (a)	7.4 (d)	7.5 (d)	7.6 (c)	7.7 (d)	7.8 (a)	7.9 (b)
7.10 (b)	7.11 (b)	7.12 (c)	7.13 (d)	7.14 (a)	7.15 (b)	7.16 (a)	7.17 (b)	7.18 (c)
7.19 (b)	7.20 (c)	7.21 (b)	7.22 (b)					

Explanations Stress Distribution in the Soil

7.1 (c)

Westergaard's equation is applicable for non-isotropic condition commonly encountered in sedimentary stratified deposits.

7.2 (b)

$$\text{Influence factor} = \frac{1}{10 \times 50} = 0.002$$

7.3 (a)

Under the point load, the vertical stress is given by

$$\sigma_z = \frac{0.4775Q}{z^2}$$

(From Boussinesq's equation at $r/z = 0$)

$$\therefore \quad \sigma_{2z} = \frac{0.4775Q}{4z^2} = 0.25\sigma_z$$

The depth at which $\sigma_z = 0.1$ per unit area is

$z = \sqrt{4.775} = 2.185$ units under the load.

Thus isobar extends up to 2.185 units of depth under the load to dissipate 90% of stresses.

7.4 (d)

Vertical stress at a point due to a point load Q is

$$\therefore \quad \sigma_z = \frac{3Q}{2\pi z^2}\left[\frac{1}{\left(1+(x/z)^2\right)}\right]^{5/2}$$

It depends only on r/z and is independent of poisson's ratio. So, there wont be any change in vertical stress.

7.5 (d)

Influence factor

$$= \frac{1}{\text{No. of concentric circles} \times \text{No. of radial lines}}$$

$$= \frac{1}{10 \times 20} = 0.005$$

7.6 (c)

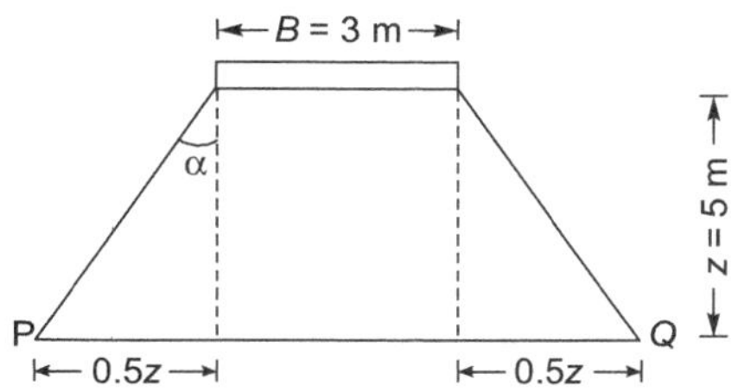

$$PQ = B + z = 3 + 5 = 8 \text{ m}$$

Stress at $\quad z = 5 \text{ m is } \sigma = \dfrac{1800}{8 \times 8}$

$$= 28.125 \text{ kN/m}^2$$

7.7 (d)

Boussinesq's theory is based on the following assumptions :

(i) The soil medium is elastic, homogeneous, isotropic and semi infinite medium.
(ii) The medium obeys Hooke's law.
(iii) The self weight of the soil is ignored.
(iv) The soil is initially unstressed.
(v) The change in volume of the soil upon application of the loads on to it is neglected i.e., it is non-dilatant.
(vi) The top surface of the soil medium is free from shear stresses and is subjected to only point loads.
(vii) The stresses are distributed symme-trically with respect to vertical axis.

7.8 (a)

The Boussinesq's equation gives vertical stress

$$\sigma_z = K_B \frac{Q}{z^2}$$

where

$$K_B = \frac{3/2\pi}{\left[1+(r/z)^2\right]^{5/2}} \quad \ldots\text{(i)}$$

For $\mu = 0$, the Westergaard's solution is

$$\sigma_z = K_W \frac{Q}{z^2}$$

$$K_W = \frac{1/\pi}{\left[1+2(r/z)^2\right]^{3/2}} \quad \ldots\text{(ii)}$$

Equating (i) and (ii), we get,

$$r/z = 1.524$$

for $\quad r/z < 1.524;\ K_W < K_B$

Westergaard's formula gives lesser value of vertical stress. For $r/z > 1.524$, $K_W > K_B$ i.e., Westergaard's formula gives greater value of stress.

7.9 (b)

According to Boussinesq's theory,

$$\sigma_z = \frac{3Q}{2\pi z^2} \frac{1}{\left[1+(r/z)^2\right]^{5/2}}$$

$$= \frac{3 \times 650}{2\pi \times 2^2 \left[1+\left(\frac{1}{2}\right)^2\right]^{5/2}}$$

$$= 44.4 \text{ kN/m}^2$$

7.10 (b)

Vertical stress at a point P due to a line load of intensity q/unit length.

$$\therefore \quad \sigma_z = \frac{2q}{\pi z}\left[\frac{1}{\left(1+(x/z)^2\right)}\right]^2$$

where,
x = distance of point from centreline of load
z = depth of point below the centre of load

7.11 (b)

- Westergaard assumed an elastic soil to be laterally reinforced by numerous closely spaced, horizontal sheets of negligible thickness but of infinite rigidity that allow only vertical movement but prevent the mass as a whole from undergoing any lateral strain.
- Newmark has constructed a circular influence chart to compute the vertical stress at any point at a depth (z) under a circular footing, carrying a uniform load of intensity q. This chart incidentally can be used not only for circular areas but for all other types of footings like square, rectangular or any other shape.

7.12 (c)

An isobar or pressure bulb is a stress contour or a line which connects all points below the ground surface at which the vertical pressure is the same. In fact, an isobar is a spatial curved surface and resembles a bulb in shape; this is because the vertical pressure at all points in a horizontal plane at equal radial distances from the load is the same.

7.14 (a)

$$\sigma_z = \frac{Q}{Z^2}.I_B = \frac{Q}{Z^2}(0.48)$$

$$= \frac{50}{4^2}(0.48) = 1.5 \text{ kN/m}^2$$

7.15 (b)

At a given depth and for some distance from center line of footing, pressure remains constant along the length of strip footing, hence pressure bulb forms in the width wise direction and along the length.

According to theory of elasticity, contact pressure at the edges of a rigid footing on saturated clay assuming clay to be perfectly elastic and homogenous and extending to a great depth increases from a value less than 0.7 q at the center line to an infinite value at the edges.

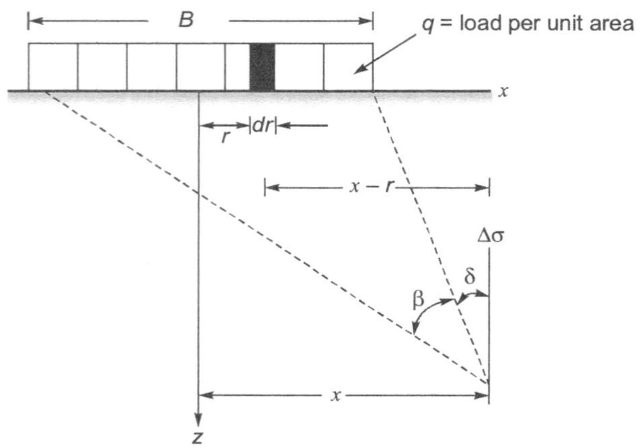

$$\Delta\sigma = \frac{q}{\pi}[\beta + \sin\beta\cos(\beta + 2\delta)]$$

7.16 (a)

In Boussinesq's Equation, it is assumed that the soil must be elastic, isotropic and also a homogeneous mass extending infinitely on all sides from the level surface.

7.17 (b)

Equivalent point load method is in approximate method.

7.18 (c)

Boussinesq's assumed soil as isotropic medium and layered soils is a non-isotropic medium hence Boussinesq's is not valid for layered soil.

7.19 (b)

$$\sigma_v = 0.4775 \times \frac{Q}{z^2}$$

$$= 0.4775 \times \frac{2000}{(6)^2} = 26.5 \text{ kN/m}^2$$

7.20 (c)

For $\frac{r}{z} = 1.5$

K_B and K_W would be almost same.

7.21 (b)

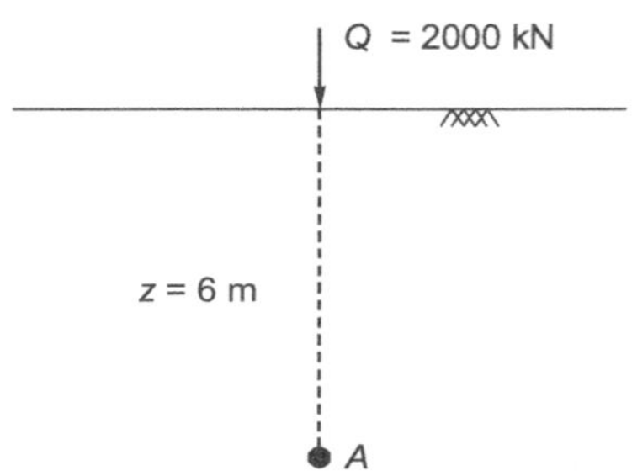

From Boussinesq's equation

Vertical stress at A:

$$\sigma_A = \frac{Q}{z^2}K_B$$

$$= \frac{2000}{(6)^2} \times \frac{3}{2\pi}\left\{\frac{1}{1+\left(\frac{r}{z}\right)^2}\right\}^{5/2}$$

$$\because \quad \frac{r}{z} = 0$$

$$\therefore \quad \sigma_A = \frac{2000}{(6)^2} \times \frac{3}{2\pi} = 26.53 \text{ kN/m}^2$$

7.22 (b)

Boussinesq's stress distribution theory is used for vertical pressure distribution on horizontal plane, vertical pressure distribution on vertical plane and stress isobar.

■■■■

8 Shear Strength of Soil

8.1 **Assertion (A):** Stress paths can be plotted for stress conditions during triaxial test.

Reason (R): It is not possible to control drainage in a triaxial test.

(a) both A and R are true and R is the correct explanation of A
(b) both A and R are true but R is not a correct explanation of A
(c) A is true but R is false
(d) A is false but R is true

[ESE : 2000]

8.2 Consider the following features of direct shear test:

1. Failure takes place on the predetermined plane.
2. It is a quick test.
3. Drainage conditions cannot be changed.
4. Failure of the sample is progressive.

Which of these are the disadvantages of direct shear test?

(a) 1, 2 and 3 (b) 1, 3 and 4
(c) 1, 2 and 4 (d) 3 and 4

[ESE : 2000]

8.3 Which one of the following soil types is most likely to be subjected to liquefaction under seismic forces?

(a) Soft saturated clays
(b) Loose saturated sands
(c) Murum
(d) Fractured rocky strata [ESE : 2000]

8.4 A soil sample tested in a triaxial compression apparatus failed when the total maximum and minimum principal stresses were100 kPa and 40 kPa, respectively. The pore pressure measured at failure was 10 kPa. The effective principal stress ratio at failure is

(a) 2.5 (b) 3.0
(c) 2.75 (d) 2.0 [ESE : 2001]

8.5 In a triaxial test at failure, major principal stress was 180 kPa, minor principal stress was 100 kPa, and pore pressure was 20 kPa. The tangent of the angle of shearing resistance of the sandy soil tested is

(a) 1/3 (b) 2/7
(c) 1/2 (d) 1/6 [ESE : 2002]

8.6 Laboratory vane shear test can also be used to determine

(a) shear parameters of silty sand
(b) shear parameters of sandy clay
(c) liquid limit of silty clay
(d) plastic limit of clayey silt [ESE : 2002]

8.7 A wooden bridge in Assam failed and was observed to have arched up in the middle after the earthquake. The right abutment settled 40 cm and left abutment 30 cm. The bridge was supported on wooden piles, which floated up during the earthquake. The most probable cause of failure is

(a) shear failure of soil below abutments.
(b) excessive settlement below abutments due to increased forces.
(c) liquefaction of foundation soil below abutments and piles.
(d) failure of abutments due to dynamic earth pressure behind abutments.

[ESE : 2002]

8.8 Consider the following statements:

1. Quick condition and liquefaction of saturated sands are based on similar phenomenon.
2. Quick condition is associated with only earth dams.
3. Liquefaction is possible in dry sand.
4. Liquefaction is associated with increase in pore water pressure due to vibrations.

Which of these statements are correct?

(a) 2 and 4 (b) 1 and 4

(c) 1 and 2 (d) 1, 3 and 4

[ESE : 2002]

8.9 **Assertion (A):** The ultimate strength of soil material is determined by the stresses in the potential failure plane.

Reason (R): The critical shear stress causing failure depends upon the properties of soil as well as on the normal stress in the failure plane.

(a) both A and R are true and R is the correct explanation of A

(b) both A and R are true but R is not a correct explanation of A

(c) A is true but R is false

(d) A is false but R is true [ESE : 2003]

8.10 A *CD* triaxial test was conducted on a granular soil. At failure $\frac{\sigma_1'}{\sigma_3'}$ was 3.0. The effective minor principal stress of failure was 75 kPa. The principal stress difference at failure will be

(a) 75 kPa (b) 150 kPa

(c) 225 kPa (d) 300 kPa

[ESE : 2003]

8.11 On which of the following factors does the behaviour of sand mass to cause liquefaction during an earthquake depend largely?

1. The number of stress cycles
2. The frequency and amplitude of vibrations of the earthquake shock
3. Angle of internal friction of sand
4. Relative density of sand

Select the correct answer using the codes given below:

(a) 1, 2, 3 and 4 (b) 2, 3 and 4

(c) 1 and 3 (d) 4 only [ESE : 2003]

8.12 Consider the following statements:

1. All soils can experience liquefaction under vibrations.
2. Liquefaction is generally associated with sandy soils.
3. Liquefaction is not possible in normal clays.
4. Highly sensitive clays may undergo liquefaction under vibrations.

Which of these statements are correct?

(a) 1 and 3 (b) 2 and 4

(c) 2 and 3 (d) 2, 3 and 4

[ESE : 2003]

8.13 A shear test was conducted on a soil sample. At failure the ratio of $\frac{\sigma_1 - \sigma_3}{2}$ to $\frac{\sigma_1 + \sigma_3}{2}$ is equal to unity. Which one of the following shear tests represents this condition?

(a) Drained triaxial compression test

(b) Undrained triaxial compression test

(c) Undrained triaxial compression test

(d) Unconfined compression test [ESE : 2004]

8.14 Which one of the following expressions represents the shear strength of soil at point *X*? Angle of shear resistance of soil is ϕ and symbols γ_w, γ_{sat}, γ_{sub}, γ_{dry} stand for unit weights of water, saturated soil, submerged soil and dry soil respectively.

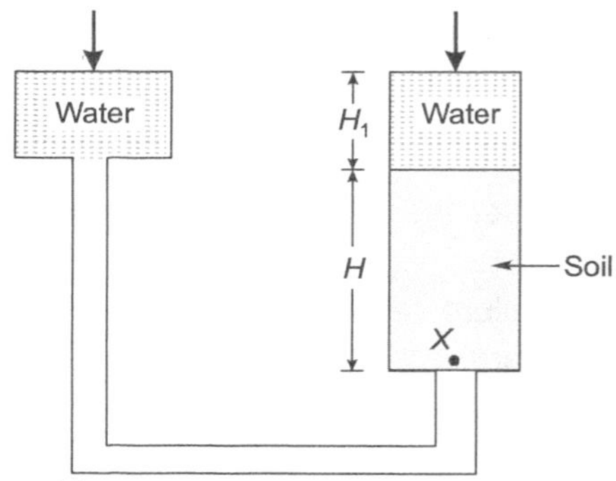

(a) $\gamma_{dry} H \tan\phi$

(b) $[\gamma_{sat} H - (H + H_1)\gamma_w]\tan\phi$

(c) $\gamma_{sub} H \tan\phi$

(d) $(\gamma_{dry} H - H_1 \gamma_w)\tan\phi$ [ESE : 2004]

8.15 Match **List-I** (Situation) with **List-II** (Stress Path) and select the correct answer using the codes given below the lists:

List-I

A. Embankment construction

B. Excavation of a pit

C. Hydrostatic loading

D. Lateral expansion of a backfill

List-II

1. Stress path moves in left side upward direction
2. Stress path moves in left side downward direction
3. Stress path moves in right side upward direction
4. Stress path moves along the horizontal axis

Codes:

	A	B	C	D
(a)	3	2	1	4
(b)	3	2	4	1
(c)	2	3	4	1
(d)	2	3	1	4

[ESE : 2004]

8.16 Which one of the following is the appropriate triaxial test to assess the immediate stability of an unloading problem, such as an excavation of a clay slope?

(a) UU test
(b) CU test
(c) CD test
(d) Unconsolidated drained tests [ESE : 2005]

8.17 Well-graded dense saturated sands have high shear strength because

(a) such sands have a better grade (superior type) of sand grains resulting in higher strength
(b) such sands have lower water content, which increases shear strength
(c) such sands have better interlocking of grains, higher inter-particle contacts and higher inter-particle frictional resistance resulting in higher strength
(d) presence of water in such sands induces capillary pressure generating higher inter granular stresses, which generate apparent cohesion and hence higher shear strength

[ESE : 2005]

8.18 A footing is resting on a fully saturated clayey strata. For checking the initial stability, shear parameters are used from which one of the following?

(a) Consolidated non-drained test
(b) Unconsolidated drained test
(c) Unconsolidated non-drained test
(d) Unconsolidated non-drained test with pore pressure measurement [ESE : 2005]

8.19 In a shear test on cohesionless soils, if the initial void ratio is less than critical void ratio, the sample will

(a) increase in volume
(b) initially increase in volume and then remain constant
(c) decrease in volume
(d) initially decrease and then increase in volume

[ESE : 2005]

8.20 As the state of strain of an element of dense sand changes from plain strain to triaxial strain condition, the effective angle of internal friction

(a) increases
(b) decreases
(c) remains constant
(d) first increases and then remains constant

[ESE : 2005]

8.21 Which one of the following factors are associated with the behaviour of sand mass during earthquake to cause liquefaction?

1. Number of stress cycles
2. The frequency and amplitude of vibration of waves generated by an earthquake.
3. Characteristics of sand
4. Relative density

Select the correct answer using the codes given below:

(a) 1, 2 and 3 (b) 1, 2, 3 and 4
(c) 2 and 4 (d) 3 and 4

[ESE : 2005]

8.22 Which one of the following different types of submerged soils is susceptible to liquefaction under earthquake shocks?

(a) Dense sand (b) Soft clay
(c) Loose silt (d) Fissured clay

[ESE : 2005]

8.23 Consider the following statements:

1. A sand with its void ratio higher than its critical void ratio increases in volume when sheared.
2. A sand with its void ratio less than its critical void ratio increases in volume when sheared.
3. For a sand at critical void ratio, the volume change during shear is minimum.

Which of these statements are correct?

(a) 1, 2 and 3 (b) Only 1 and 2
(c) Only 2 and 3 (d) Only 1 and 3

[ESE : 2006]

8.24 Consider the following statements about the shearing resistance as a function of effective stress:

1. Effective stress on the failure plane governs the shearing resistance and not the total stress.
2. Two soils equally dense, consolidated to same effective stress will show different shear resistance with drainage and undrained condition.
3. The post peak drop off in shearing resistance is less pronounced in over consolidated clays and more in normally consolidated clays at same effective stress.

Which of these statements are correct?

(a) 1, 2 and 3 (b) Only 1 and 3
(c) Only 1 and 2 (d) Only 2 and 3

[ESE : 2006]

8.25 Consider the following statements:

1. Stress path is a locus of stress points developed by stress changes in the soil and can be obtained from Mohr's Stress Circle.
2. Stress path can be used to determine the intensity of stress at a point due to the application of uniformly applied circular loaded area.
3. Stress path has a value in giving insight into probable soil response-particularly if a part of the previous history stress path can be reproduced.

Which of these statements are correct?

(a) 1, 2 and 3 (b) Only 1 and 2
(c) Only 2 and 3 (d) Only 1 and 3

[ESE : 2006]

8.26 What does the confining pressure used in triaxial compression tests on an undisturbed soil sample represent?

(a) The in-situ total normal stress
(b) The in-situ total lateral stress
(c) The in-situ effective principal stress
(d) The in-situ shear stress

[ESE : 2007]

8.27 Consider the following statements:

Liquefaction is a phenomenon

1. observed in fine sands
2. associated with development of positive pore pressure

Which of these statements is/are correct?

(a) 1 only (b) 2 only
(c) Both 1 and 2 (d) Neither 1 nor 2

[ESE : 2007]

8.28 **Assertion (A):** Shear strength parameters of sand can be estimated by conducting unconfined compression test.

Reason (R): The effective angle of shearing resistance of sand is nearly the same for dry and saturated sands, in drained condition.

(a) both A and R are true and R is the correct explanation of A
(b) both A and R are true but R is not a correct explanation of A
(c) A is true but R is false
(d) A is false but R is true

[ESE : 2007]

8.29 Consider the following statements:

1. Pore pressure parameter A is a constant for a soil.
2. The shear strength of soil is a function of the effective stress in the soil and not of the total stress in the soil.

Which of these statements is/are correct?

(a) 1 only (b) 2 only
(c) Both 1 and 2 (d) Neither 1 nor 2

[ESE : 2008]

8.30 Which one of the following conditions is valid in case of unconfined compression test in comparison to triaxial test?

(a) Minor principal stress = 0
(b) Minor principal stress = 0.5 × major principal stress
(c) Minor principal stress = major principal stress
(d) Major principal stress = 3 × minor principal stress

[ESE : 2009]

8.31 In an unconfined compression test on stiff clay, if the failure plane made an angle of 52° to the horizontal, what would be the angle of shearing resistance?

(a) 16° (b) 14°
(c) 12° (d) 13°

[ESE : 2009]

8.32 Considerable loss of shear strength due to shock or disturbance is exhibited by
(a) under-consolidated clays
(b) normally consolidated clays
(c) over-consolidated clays
(d) organic soil

[ESE : 2010]

8.33 Consider the following statements:
1. Sensitivity of a natural soil deposit cannot be less than 1.0.
2. A saturated loose sand deposit liquefies when water flows through it in upward direction under critical hydraulic gradient.
3. A quick clay has very high sensitivity.

Which of these statements are correct?
(a) 1, 2 and 3 (b) 1 and 2 only
(c) 1 and 3 only (d) 2 and 3 only

[ESE : 2010]

8.34 Liquefaction of foundation soil during an earthquake shall not be the reason for cracking of
(a) only floors in the building
(b) walls and roof in the building
(c) beams and columns in the building
(d) only balcony in the building

[ESE : 2010]

8.35 Consider the following statements:
Apparent cohesion in sands is exhibited mainly due to
1. reduction in density.
2. increase in density.
3. capillary moisture in pores.

Which of these statements is/are correct?
(a) 1, 2 and 3 (b) 1 and 3 only
(c) 2 and 3 only (d) 2 only

[ESE : 2010]

8.36 Unconfined compression test is most suitable for determining the
1. Sensitivity of clays
2. Settlement of embankments
3. 'Strength' of partially saturated clay sample
4. 'Strength' of fully saturated clay sample

(a) 1, 2, 3 and 4 (b) 2 and 3 only
(c) 3 and 4 only (d) 1 and 4 only

[ESE : 2011]

8.37 Consider the following statements:
1. For a saturated soil, Skempton's B-parameter is nearly equal to unity.
2. For an undisturbed sensitive clay, the stress-strain curve shows a peak.
3. Interlocking contributes significantly to the shearing strength in case of dense sand.

Which of these statements are correct?
(a) 1, 2 and 3 (b) 1 and 2 only
(c) 2 and 3 only (d) 1 and 3 only

[ESE : 2012]

8.38 For the determination of shear strength parameters, c and ϕ, of soil in the laboratory, the test to be conducted will be
(a) Triaxial compression test
(b) Sieve analysis
(c) Compaction test
(d) Relative density test

[ESE : 2012]

8.39 Consider the following statements:
For pure clay, the shear strength parameters will be
1. Cohesion $c = 0$; and angle of internal friction ϕ will be maximum.
2. Cohesion c is maximum; and angle of internal friction ϕ is also maximum.
3. Angle of internal friction ϕ is zero, with some value of cohesion c.

Which of these statements is/are correct?
(a) 1, 2 and 3 (b) 1 only
(c) 3 only (d) 2 only

[ESE : 2012]

8.40 **Statement-I :** A curved, or straight, line connecting the relevant stress points is called the stress path.
Statement-II : All the total stress paths and the effective stress paths for the drained tests are straight lines at a slope of 45°.
(a) Both Statement (I) and Statement (II) are individually true and Statement (II) is the correct explanation of Statement (I)
(b) Both Statement (I) and Statement (II) are individually true but Statement (II) is NOT the correct explanation of Statement (I)
(c) Statement (I) is true but Statement (II) is false
(d) Statement (I) is false but Statement (II) is true

[ESE : 2012]

8.41 Consider the following statements:

1. Presence of moisture in sand does not affect the magnitude of friction angle.
2. Moisture introduces capillary effect in a sandy soil leading to apparent increase in angle of internal friction.
3. Moisture introduces capillary effect in a sandy soil leading to development of apparent cohesion.
4. The capillary film is broken by drying or submergence leading to loss in the apparent cohesion.

Which of these statements are correct?

(a) 1, 2, 3 and 4 (b) 2 and 3 only
(c) 2 and 4 only (d) 3 and 4 only

[ESE : 2013]

8.42 Consider the following statements:

1. Brittle behaviour of soils can be obtained when the soil is heavily over-consolidated clay.
2. Remoulded cohesive clays show a tendency towards progressive failure.
3. Undisturbed sensitive clays show a tendency towards progressive failure.

Which of these statements are correct?

(a) 1, 2 and 3 (b) 1 and 3 only
(c) 1 and 2 only (d) 2 and 3 only

[ESE : 2013]

8.43 Consider the following statements:

1. The water content of a soil remains unchanged during the entire duration of a 'quick' test.
2. Effective stress cum shear parameters of clay can be obtained from drained triaxial shear test.
3. Strain-controlled shear test is a good option since this test procedure maintains better control on the rate of loading than other tests.

Which of the above statements is/are correct?

(a) 1, 2 and 3 (b) 1 only
(c) 1 and 3 only (d) 2 and 3 only

[ESE : 2014]

8.44 Arrange the following soils with respect to increasing order of realizable friction ratio:

1. Loose gravel fill
2. Sands or gravels
3. Clay sand mixtures and silts
4. Clays and peats

(a) 1, 2, 3 and 4 (b) 4, 2, 3 and 1
(c) 1, 3, 2 and 4 (d) 4, 3, 2 and 1

[ESE : 2015]

8.45 **Statement (I) :** Mohr's theory is based on logical arrangement of experimental results.

Statement (II) : Mohr's theory generalizes Coulomb's theory.

(a) Both Statement (I) and Statement (II) are individually true and Statement (II) is the correct explanation of Statement (I)
(b) Both Statement (I) and Statement (II) are individually true but Statement (II) is NOT the correct explanation of Statement (I)
(c) Statement (I) is true but Statement (II) is false
(d) Statement (I) is false but Statement (II) is true

[ESE : 2014]

8.46 In a triaxial compression test, the major principal stress was 90 kPa and the minor principal stress was 30 kPa, at failure. The pore pressure at failure was observed to be 10 kPa. The tangent of the angle of shearing resistance of the sandy soil that was tested was

(a) 1/2 (b) 1/3
(c) 2/3 (d) 3/4 [ESE : 2015]

8.47 A cylindrical soil specimen of saturated clay, 3.50 cm diameter and 8 cm length, is tested in an unconfined compression testing machine. The specimen failed under a vertical load of 50 kg together with an accompanying additional deformation of 8 mm. What is the unconfined compressive strength of this clay?

(a) 4.67 kg/cm^2 (b) 5.0 kg/cm^2
(c) 5.5 kg/cm^2 (d) 6.0 kg/cm^2

[ESE : 2015]

8.48 An undrained triaxial compression test is carried out on a saturated clay sample under a cell pressure of 50 kN/m^2. The sample failed at a deviator stress of 100 kN/m^2. The cohesion of this clay sample would be

(a) 25 kN/m^2 (b) 50 kN/m^2
(c) 75 kN/m^2 (d) 100 kN/m^2

[ESE : 2017]

Directions: The following items consists of two statements; one labelled as **'Statement (I)'** and the other as **'Statement (II)'**. You are to examine these two statements carefully and select the answers to these items using the codes given below:

Codes:

(a) Both Statement (I) and Statement (II) are individually true and Statement (II) is the correct explanation of Statement (I)

(b) Both Statement (I) and Statement (II) are individually true but Statement (II) is NOT the correct explanation of Statement (I)

(c) Statement (I) is true but Statement (II) is false

(d) Statement (I) is false but Statement (II) is true

8.49 **Statement (I):** Cohesion and angle of internal friction are shear strength parameters of soils.

Statement (II): Cohesion is zero for pure sand and angle of internal friction is zero for pure clay.

[ESE : 2017]

8.50 **Statement (I):** Saturated fine, as well as medium, sands of uniform particle size are most susceptible to liquefaction.

Statement (II): Fine particles reduce the permeability which is a prime factor for liquefaction.

[ESE : 2017]

8.51 A dry sand specimen is put through a tri-axial test. The cell pressure is 50 kPa and the deviator stress at failure is 100 kPa. The angle of internal friction for the sand specimen is

(a) 15° (b) 30°
(c) 45° (d) 55°

[ESE : 2018]

8.52 Consider the following statements with regard to Soil Testing:

1. The origin and pole are at the same point in a Mohr's circle.
2. The shear stress is maximum on the failure plane.
3. Mohr's circle drawn with data from an unconfined compression test passes through the origin.
4. Maximum shear stress occurs on a plane inclined at 45° to the principal plane.

Which of the above statements are CORRECT?

(a) 1 and 2 only (b) 2 and 3 only
(c) 3 and 4 only (d) 1 and 4 only

[ESE : 2018]

8.53 The representative liquid limit and plastic limit values of a saturated consolidate clay deposit are 60% and 30%, respectively. The saturated unit weight of the soil is 19 kN/m^3. The water table is at 3 m below ground level. Ata depth of 10 m from the ground surface, the undrained shear strength of the soil will be nearly

(a) 37.7 kN/m^2 (b) 33.5 kN/m^2
(c) 29.3 kN/m^2 (d) 25.1 kN/m^2

[ESE : 2019]

8.54 **Statement (I)** : The inclination of the resultant stress with normal can exceed the angle of repose (adopting old terminology).

Statement (II) : The ratio of the difference between greatest and least intensities of pressure to their sum cannot exceed the sine of the angle of repose (adopting old terminology).

(a) Both Statement (I) and Statement (II) are individually true, and Statement (II) is the correct explanation of Statement (I)

(b) Both Statement (I) and Statement (II) are individually true, but Statement (II) is NOT the correct explanation of Statement (I)

(c) Statement (I) is true, but Statement (II) is false

(d) Statement (I) is false, but Statement (II) is true

[ESE : 2019]

■■■■

Answers Shear Strength of Soil

8.1 (c)	8.2 (b)	8.3 (b)	8.4 (b)	8.5 (a)	8.6 (c)	8.7 (c)	8.8 (b)	8.9 (a)
8.10 (b)	8.11 (a)	8.12 (d)	8.13 (d)	8.14 (c)	8.15 (b)	8.16 (b)	8.17 (c)	8.18 (d)
8.19 (d)	8.20 (b)	8.21 (b)	8.22 (c)	8.23 (c)	8.24 (c)	8.25 (d)	8.26 (b)	8.27 (c)
8.28 (d)	8.29 (b)	8.30 (a)	8.31 (b)	8.32 (a)	8.33 (a)	8.34 (d)	8.35 (b)	8.36 (d)
8.37 (a)	8.38 (a)	8.39 (c)	8.40 (c)	8.41 (d)	8.42 (c)	8.43 (a)	8.44 (d)	8.45 (c)
8.46 (d)	8.47 (a)	8.48 (b)	8.49 (b)	8.50 (a)	8.51 (b)	8.52 (c)	8.53 (a)	8.54 (d)

Explanations Shear Strength of Soil

8.1 (c)

The triaxial compression test consists of two stages:

- ***First stage:*** In this, a soil sample is set in the triaxial cell and confining pressure is then applied.
- ***Second stage:*** In this, additional axial stress (also called deviator stress) is applied which induces shear stresses in the sample. The axial stress is continuously increased until the sample fails.

During both the stages, the applied stresses, axial strain, and pore water pressure or change in sample volume can be measured. Drainage can be very much controlled in Triaxial Test.

8.2 (b)

The most important limitation of direct shear test is that the failure plane is predetermined and this may not be the weakest plane.

There is virtually no control of drainage of soil specimen as the water content of saturated soil changes rapidly with stress.

The stress conditions across the soil sample are very complex because of nonuniform distribution of normal and shear stress on the potential surface of sliding. The stress is more at edge and less at centre. Due to this there is progressive failure of the specimen i.e., the entire strength of soil is not mobilised simultaneously.

8.3 (b)

Soil liquefaction describes a phenomenon whereby a saturated or partially saturated soil substantially loses strength and stiffness in response to an applied stress, usually earthquake shaking or other sudden change in stress condition, causing it to behave like a liquid. It is more likely to occur in sandy or non-plastic silty soils.

8.4 (b)

Effective principal stress ratio

$$= \frac{\sigma_1 - u}{\sigma_3 - u} = \frac{100-10}{40-10} = 3.0$$

8.5 (a)

Major principal stress:

$\bar{\sigma}_1 = \sigma_1 - u = 180 - 20 = 160$ kPa

Minor principal stress:

$\bar{\sigma}_3 = \sigma_3 - u = 100 - 20 = 80$ kPa

For sand,

$c = 0$

$$\sigma_1 = \sigma_3 \tan^2\left(45° + \frac{\phi}{2}\right)$$

Thus on solving $\phi = 19.47°$

And $\tan\phi = 0.353$

So, nearest option is (a)

8.6 (c)

Vane shear test is used to determine the cohesion or shear strength of sensitive soft clays because their cylindrical specimen cannot be easily prepared. It is necessary that the soil mass should be saturated completely because the angle of shearing resistance will not be zero for partially saturated soils. This method should be used when:

(i) Clay is deep, normally consolidated and sensitive

(ii) Only the undrained shear is required.

The test cannot be conducted on the fissured clay or the clay containing sand or silt laminations.
It can be also be used for determination of liquid limit of soils because the shear strength at liquid limit is about 2.7 kN/m^2.

8.7 (c)
The differential settlement of abutments and subsequent rotation results in arching up of the bridge. It is given that wooden piles floated up during the earthquake. So liquefaction is the main cause of settlement and failure.

8.8 (b)
When the effective stress is zero soil looses its strength and behave as a liquid. The phenomenon of quick condition is due to flow of water through the soil. While liquefaction occurs due to vibration caused by earthquake, explosion, volcanic eruption etc. However, in both cases effective stress becomes zero.

8.9 (a)

$$\tau = c + \sigma \tan \phi$$

where σ is the normal stress on the failure plane.

8.10 (b)

$$\sigma_3' = 75 \text{ kPa}; \ \sigma_1' = 3\sigma_3'$$

Principal stress difference

$$= \sigma_1' - \sigma_3' = 2\sigma_3' = 2 \times 75 = 150 \text{ kPa}$$

8.11 (a)
Although all the factors are responsible for liquefaction of soils. However, relative density is the most important factor as liquefaction can occur in loose sandy soils.

8.12 (d)
Cohesive soils generally do not show liquefaction due to cohesion between particles. However highly sensitive clays when subjected to vibration may change their structure and start flowing like a liquid.

8.13 (d)
When ratio of $\frac{\sigma_1 - \sigma_3}{2}$ to $\frac{\sigma_1 + \sigma_3}{2}$ is equal to unity

$$\frac{\sigma_1 - \sigma_3}{2} = \frac{\sigma_1 + \sigma_3}{2}$$

$\sigma_3 = 0$ (Confining pressure zero)
This case is represented by unconfined compression test.

8.14 (c)
As the free water surface on both sides is same so there is no flow occurring through soil. Thus effective stress at X,

$$\sigma' = \gamma_w H_1 + \gamma_{sat} H - \gamma_w (H + H_1) = \gamma_{sub} H$$

$\therefore$ Shear strength,

$$\tau = \sigma' \tan\phi = \gamma_{sub} H \tan\phi$$

8.15 (b)
In hydrostatic loading $\Delta\sigma_h = \Delta\sigma_v$. So stress path moves along a horizontal axis. Lateral expansion of backfill means $\Delta\sigma_h < 0$ and $\Delta\sigma_v = 0$ so stress path moves in left side upward direction. During embankment construction $\Delta\sigma_h = 0$ and $\Delta\sigma_v > 0$ so stress path moves in right side upward direction. During excavation of a pit $\Delta\sigma_h = 0$ and $\Delta\sigma_v < 0$ so stress path moves in left side downward direction.

8.17 (c)
Well grade sands have higher shearing resistance as they have high inter-particle contacts. Dense sands have better interlocking of particles. Saturated sands have higher frictional resistance. Also, loose, saturated, uniformly graded fine grain sands are very susceptible to liquefaction.

8.18 (d)
UU (unconsolidated undrained) test: In this, cell pressure is applied without allowing drainage. Then keeping cell pressure constant, deviator stress is increased to failure without drainage.
CU (consolidated undrained) test: In this, drainage is allowed during cell pressure application. Then without allowing further drainage, deviator stress is increased keeping cell pressure constant.
CD (consolidated drained) test: This is similar to **CU test** except that as deviator stress is increased, drainage is permitted. The rate of loading must be slow enough to ensure no excess pore water pressure develops.

8.19 (d)
If initial void ratio is less than critical void ratio then the sample is dense sand. So in this case the volume will decrease initially to get fully compacted and then to achieve critical void ratio the volume will increase.
If initial void ratio is more than critical void ratio

then the sample is loose sand. In this case volume initially decreases and then remains constant.

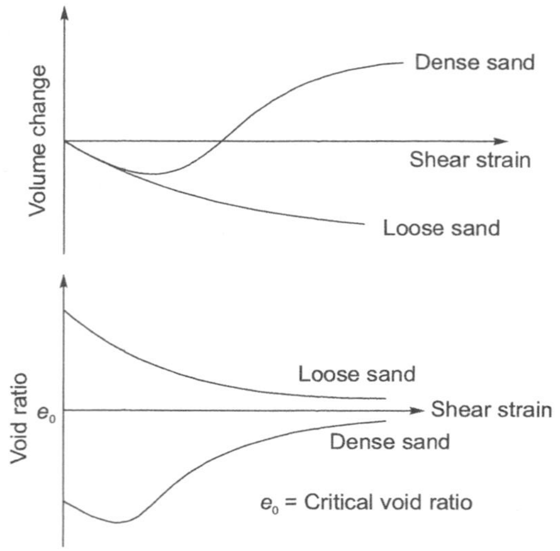

8.20 (b)

Trivial strain condition means zero strain in all directions. This condition gives higher angle of internal friction compared to plane strain condition.

8.21 (b)

The phenomenon when the sand loses its shear strength due to oscillatory motion is known as liquefaction. The structures resting on such soils sink. The soils most susceptible to liquefaction are the saturated, fine and medium sands of uniform particle size. When such a deposit has void ratio greater than the critical void ratio i.e. loose sand is subjected to sudden shearing stresses, the volume decreases and pore pressure increases. The effective stresses in soil decrease and sand behaves as a dense liquid. If the deposits are compacted to a void ratio. Smaller than the critical void ratio, the chances of liquefaction reduces.

The liquefaction can occur due to explosion, heavy vibration of waves generated by an earthquake and number of stress cycles.

8.22 (c)

In loose silt, under earthquake shocks the volume decreases and positive pore water pressure is generated due to which effective stress decreases. When effective stress becomes zero, soil looses all of its shear strength and this phenomenon is called liquefaction.

8.23 (c)

Loose sands have initial void ratio more than critical void ratio. When this sand is sheared the final void ratio reaches critical void ratio i.e., the void ratio decreases and therefore volume of soil also decreases. Dense sands have initial void ratio less than critical void ratio when this sand is sheared, the void ratio decreases first to achieve 100% relative density (minimum void ratio) and then volume increases to achieve critical void ratio. The sand at critical void ratio have minimum change in volume during shearing. The critical void ratio at ultimate or critical state is a function of either effective all round pressure or normal stress.

8.24 (c)

The effective stress principle given by Terzaghi states that:

(a) The effective stress is equal to total stress minus pore pressure

(b) The effective stress controls certain aspects of soil behaviour, notably compressibility and shear strength.

Thus according to Terzaghi, it is the effective stress on the failure plane that governs the shearing strength and not the total stress.

Further undrained conditions reduce the effective stress due to development of pore pressure. Thus two soils equally dense consolidated to same effective stress in the confining pressure stage and sheared under drained and undrained conditions will show different shear resistance.

Over consolidated clays behave similar to dense sands and they show pronounced peak during shearing. Normally consolidated clays show less pronounced peak or no peak during shearing. Both the soils finally show same strength at large value of strain due to rolling and translation of bulky particles. This final strength is called residual strength and can be determined by ring shear apparatus.

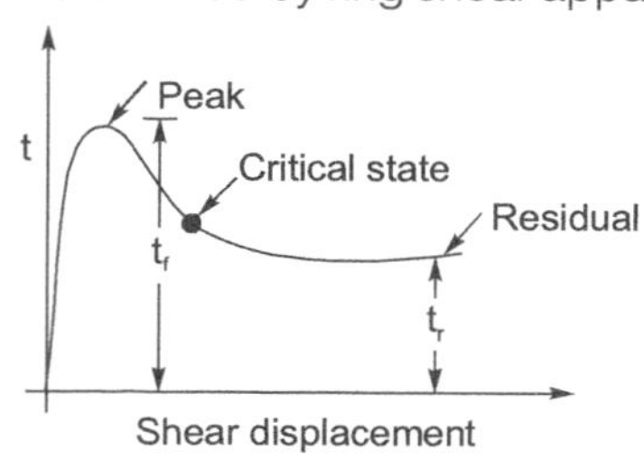

(Ring shear test)

8.25 (d)

The successive states of stress in a test specimen or an in-situ element of soil can be represented by a series of Mohr's circles or in a less confusing way, by series of stress points (p, q).

$$p = \frac{\sigma_1 + \sigma_3}{2} \text{ and}$$

$$q = \frac{\sigma_1 - \sigma_3}{2}$$

The curve or straight line connecting the relevant stress points is called the stress path, giving a clear representation of states of stress. The horizontal distance between effective and total stress path is the value of pore pressure at any particular state. In general the horizontal distance between the two stress paths is the sum of the pore water pressure due to change in total stress and the static pore water pressure.

8.26 (b)

The triaxial cell contains soil sample which is subjected to an all round pressure. This is called the cell pressure or confining pressure. It acts radially on the vertical surface of the specimen and axially at the top and bottom.

8.27 (c)

Due to development of excessive pore water pressure, the effective stress of soil becomes zero and it tends to behave like liquid and loses its shear strength.

8.28 (d)

Unconfined compression test is a special case of triaxial test carried out at zero cell pressure. This test is applicable only to soils for which $\phi_u = 0$ i.e. fully saturated, non-fissured clays.

8.29 (b)

Pore water parameters are empirical coefficients which are used to express the response of pore pressure to changes in total stress under undrained conditions. Pore pressure parameter A is not a constant and is different for different soils and stress conditions. For a given soil, A depends upon the strain, anisotropy, sample disturbance and the over consolidation ratio.

8.30 (a)

Unconfined compression test is a special case of triaxial test in which the confining pressure is zero. This test may be conducted on undisturbed or remoulded cohesive soils. It cannot be conducted on coarse-grained soils such as sands and gravels as these cannot stand without lateral support.

8.31 (b)

If the failure plane makes an angle α with the horizontal, then

$$\alpha = 45° + \frac{\phi}{2}$$

where ϕ is the angle of shearing resistance

$$\therefore \quad 52° = 45° + \frac{\phi}{2}$$

$$\Rightarrow \quad \phi = 14°$$

8.32 (a)

Under-consolidated clays when undergo shock or disturbance, remoulding occurs which leads to considerable loss of shear strength, because excess pore pressure developed will be more in U.C. soil.

8.33 (a)

Natural soils generally have a sensitivity value of 1 or more than 1. However, stiff clays having fissures and cracks sometimes exhibits a sensitivity value of less than 1, since these weaknesses i.e. fissures and cracks present in an undisturbed sample, will be removed upon remoulding.

$$\text{Sensitivity,} \quad S_t = \frac{(q_u)\text{ undisturbed}}{(q_u)\text{ remoulded}}$$

Soils which are yet to consolidate under the existing overburden pressure, as in case of a recently placed fill, are sometimes referred to as under consolidated soil.

8.35 (b)

Bulking in sands is manifestation of capillary moisture. The capillary menisci surrounding the soil grains produces apparent cohesion which holds the particle together in clusters. The increase in volume known as bulking occurs only in moist sands, thereby decreasing the density of sand.

8.36 (d)

The unconfined compression test is a special form of a triaxial test in which the confining pressure is zero. The test can be conducted only on clayey soils which can stand without confinement.

8.37 (a)

Dense sands derive their shear strength from interlocking and friction whereas clays derive their shear strength from cohesion. For a saturated soil, skempton's *B* parameter = 1.

8.38 (a)

The triaxial test is the most versatile of all shear testing method, to determine c and ϕ of the soil. Drainage conditions can be controlled, whatever be the type of soil. In the triaxial test, pore water pressure measurements can be made accurately. Volume changes can also be measured. The specimens can fail on any weak plane. The stress distribution on the failure plane is fairly uniform.

8.39 (c)

For pure clay, failure envelope (ϕ = 0)

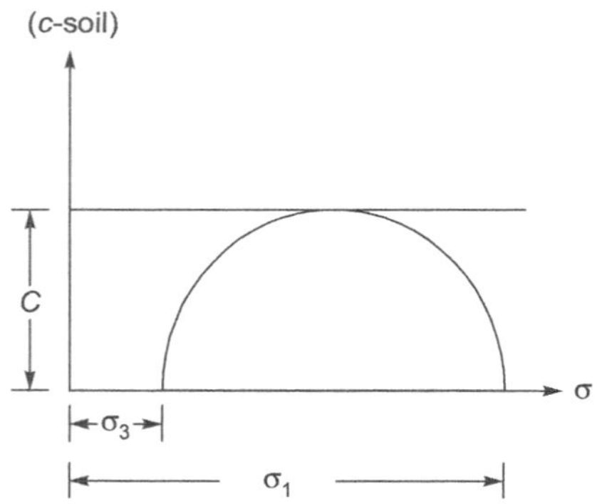

8.40 (c)

Progressive changes in the state of a particular load application can be represented by a series of Mohr circles. The locus of points on the Mohr diagram whose coordinates represent the maximum shear stress and the associated principal stress for the entire stress history is defined as a stress path. For given principal stresses σ_1 and σ_3, the coordinates of a point on the stress path are

$$p = \frac{\sigma_1 + \sigma_3}{2}$$

and
$$q = \frac{\sigma_1 - \sigma_3}{2}$$

Such a plot is referred to as p-q diagram.

The stress path for σ_3 = constant and σ_1 increasing is a 45° line. Figure shows stress paths for different variations of σ_1 and σ_3. A stress path diagram may be constructed for total or effective stress conditions.

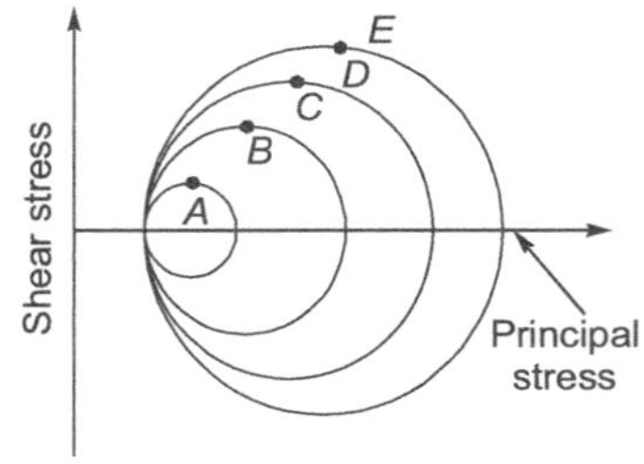

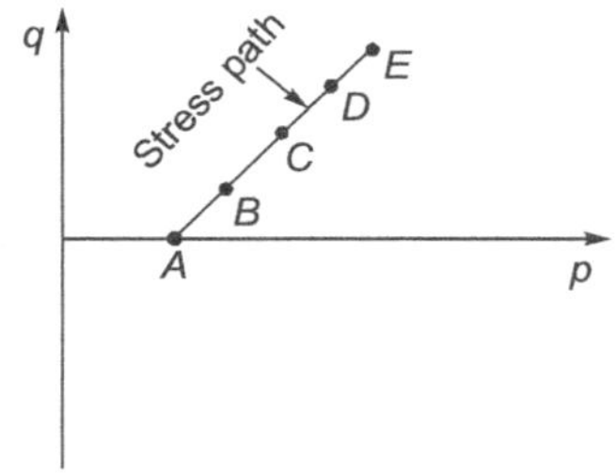

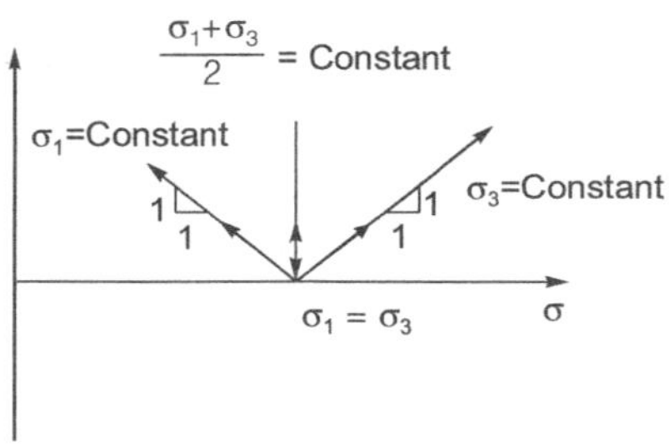

8.41 (d)

Presence of moisture in sand does have some effect on friction angle. However, this is not usually taken into account in practice.

Moisture introduces capillary effects in sand, leading to development of apparent cohesion and capillary film is broken by drying or submergence leading to loss of apparent cohesion.

8.42 (c)

- In case of heavily over consolidated clays, the stress strain curve has a sharp peak. Hence it shows brittle behaviour.
- Remoulded clays or soft clays show constant failure at similar values of stress hence it has a plastic failure. This is the reason for progressive failure.
- Undisturbed sensitive clay shows a sharp peak which occurs at relatively low strain in stress-strain curve.

8.46 (d)

$$\sin\phi = \frac{\bar{\sigma}_1 - \bar{\sigma}_3}{\bar{\sigma}_1 + \bar{\sigma}_3}$$

$$= \frac{(90-10)-(30-10)}{(90-10)+(30-10)} = \frac{60}{100}$$

$$\sin\phi = \frac{3}{5}$$

$$\therefore \quad \tan\phi = \frac{3}{4}$$

8.47 (a)

$$A_0 = \frac{\pi}{4}\times(3.50)^2 = 9.621\text{ cm}^2$$

$$A_f = \frac{A_o}{1-\epsilon_0} = \frac{9.62}{1-\frac{0.8}{8}}$$

$$= 10.69\text{ cm}^2$$

$$\therefore \quad q_u = \frac{50}{10.69} = 4.677\text{ kg/cm}^2$$

8.48 (b)

Undrained triaxial compression

saturated clays ($\phi = 0$)

$\sigma_3 = 50$ kN/m^2

$\sigma_d = 100$ kN/m^2

$\sigma_1 = \sigma_3 + \sigma_d = 50 + 100$

$= 150$ kN/m^2

From stress relationship

$$\sigma_1 = \sigma_3\tan^2\left(45+\frac{\phi}{2}\right) + 2c\tan\left(45+\frac{\phi}{2}\right)$$

$$150 = 50 + 2c$$

$$c = \frac{100}{2} = 50\text{ kN/m}^2$$

8.50 (a)

Prime factor for liquefaction is "Decrease in effective stress, due to development of excess pore pressure which is caused due to dynamic or seismic loading.

8.51 (b)

$\sigma_c = \sigma_3 = 50$ kpa

$\sigma = \sigma_3 + \sigma_d = 50 + 100 = 150$ kpa

Sand $\quad C = 0, \quad \phi = ?$

$\sigma_1 = \sigma_3 \tan^2(45 + \phi/2)$

$150 = 50\tan^2(45 + \phi/2)$

$o = 30°$

8.52 (c)

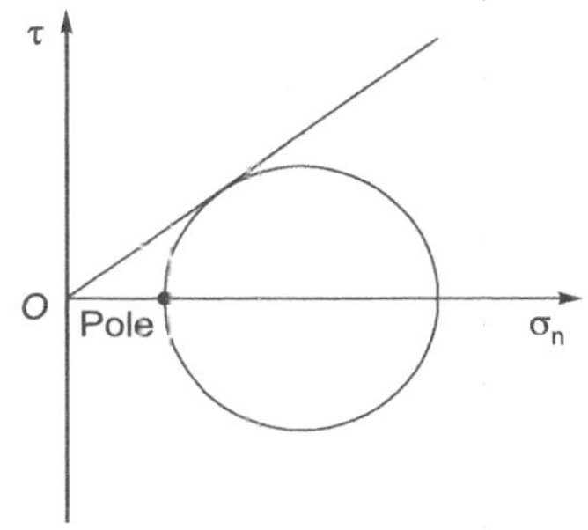

- In unconfined compression test

  $\sigma_c = \sigma_3 = 0$

  Hence Mohr circle passes through origin.
- Shear stress is maximum on the plane inclined 45° to major principal plane because

$$\tau = \left(\frac{\sigma_1-\sigma_3}{2}\right)\sin 2\theta$$

and τ is maximum when $\theta = 45°$

$$\tau_{max} = \left(\frac{\sigma_1-\sigma_3}{2}\right)$$

At failure plane $\tau < \tau_{max}$

8.53 (a)

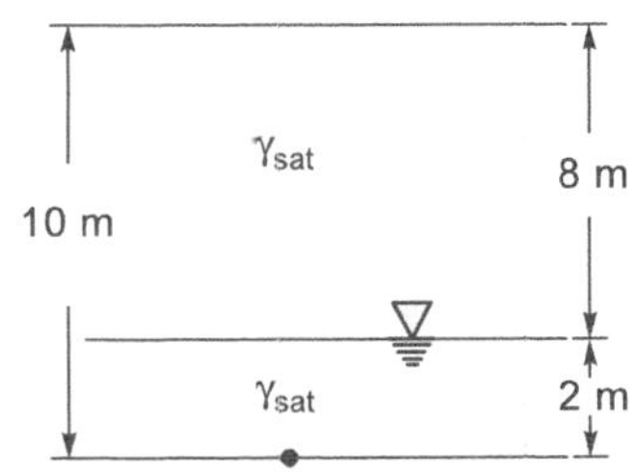

$\bar{\sigma}_{n@base} = \sigma - u$

$\bar{\sigma}_n = 10\,\gamma_{sat} - 2\gamma_w$

$\bar{\sigma}_n = 10 \times 19 - 2 \times 9.81$

$= 190 - 19.62$

$= 170.38$ kN/m^2

$$\frac{c_u}{\bar{\sigma}_n} = 0.11 + 0.0037\,I_p\%$$

$$\frac{c_u}{170.38} = 0.11 + (0.0037 \times 30)$$

$$\frac{c_u}{170.38} = 0.221$$

$c_u = 37.7$ kN/m^2

8.54 (d)

The maximum slope at which a material, dropped down through a natural process, remains stable, is called angle of repose.

The angle of repose of a granular soil can be determined by pouring the material on a level surface from a small height and measuring the angle between the sloping surface and the horizontal.

The angle of repose is approximately equal to the angle of internal friction of the soil. At the failure plane angle of obliquity is maximum, which is also termed as internal angle of friction.

At failure plane, resultant stress with normal stress makes maximum angle of obliquity.

$$\Rightarrow \quad \sigma_1 = \sigma_3 \tan^2\left(45+\frac{\phi}{2}\right) + 2c\tan\left(45+\frac{\phi}{2}\right)$$

$$\sigma_1 = \sigma_3\left(\frac{1+\sin\phi}{1-\sin\phi}\right)$$

$$\sin\phi = \frac{\sigma_1-\sigma_3}{\sigma_1+\sigma_3}$$

■■■■

9 Retaining Wall/ Earth Pressure Theories

9.1 The correct sequence of the given parameters in descending order of earth pressure intensity is
(a) active, passive, at rest
(b) passive, active, at rest
(c) passive, at rest, active
(d) at rest, passive, active

[ESE : 2000]

9.2 Match **List-I** with **List-II** and select the correct answer using the codes given below the lists:

List-I
A. Dense sand
B. Loose sand
C. Plastic clay

List-II

1. 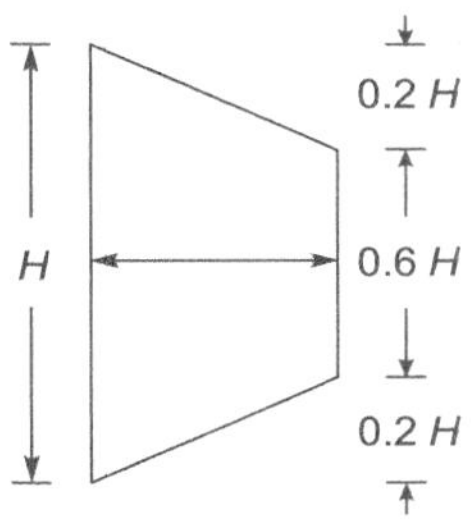

(Terzaghi-Peck, 1948)

2. 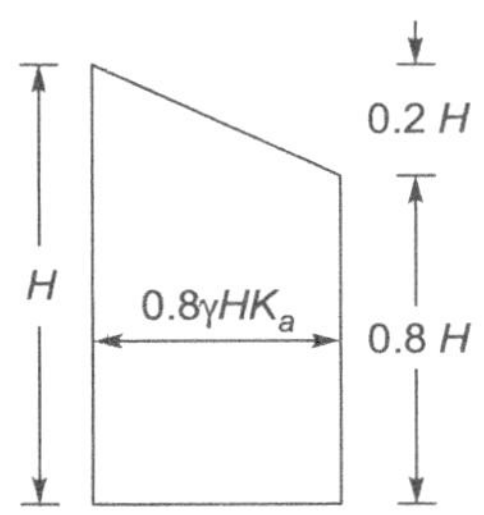

(Terzaghi-Peck, 1948)

3. 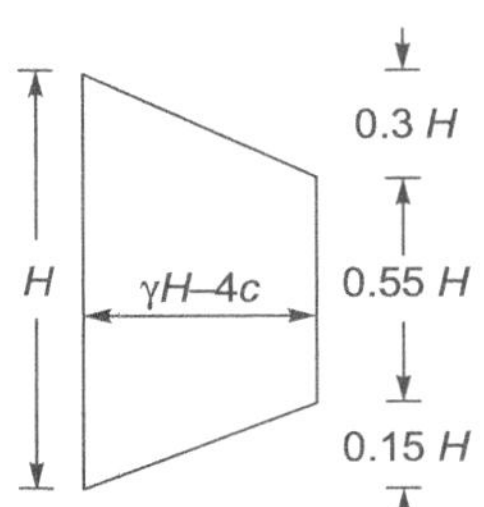

(Peck, 1943)

Codes:

	A	B	C
(a)	3	2	1
(b)	2	1	3
(c)	1	3	2
(d)	1	2	3

[ESE : 2000]

9.3 An earth-retaining structure may be subjected to the following lateral earth pressures:
1. Earth pressure at rest
2. Passive earth pressure
3. Active earth pressure

The correct sequence of the increasing order of the magnitudes of these pressures is
(a) 3, 2, 1 (b) 1, 3, 2
(c) 1, 2, 3 (d) 3, 1, 2

[ESE : 2001]

9.4 For a sand having an internal friction of 30°, the ratio of passive to active lateral earth pressure will be
(a) 1 (b) 3
(c) 6 (d) 9

[ESE : 2001]

9.5 When movement of a wall under the earth pressures from the backfill was prevented the coefficient of earth pressure was recorded as 0.5. The ratio of the coefficient of passive and active earth pressure of the backfill is
(a) 1/3 (b) 3
(c) 1/9 (d) 9

[ESE : 2002]

9.6 **Assertion (A):** Passive earth pressure is always greater than the earth pressure at rest and active earth pressure.
Reason (R): In passive state the structure becomes the actuating element and soil becomes the resisting element to maintain the stability.

(a) both A and R are true and R is the correct explanation of A
(b) both A and R are true but R is not a correct explanation of A
(c) A is true but R is false
(d) A is false but R is true

[ESE : 2003]

9.7 The wall friction of the retaining wall
(a) decreases active earth pressure but increases passive earth pressure
(b) decreases passive earth pressure but increases active earth pressure
(c) decrease both active and passive earth pressure
(d) increase both active and passive earth pressure

[ESE : 2005]

Directions: The following items consists of two statements; one labelled as **'Assertion (A)'** and the other as **'Reason (R)'**. You are to examine these two statements carefully and select the answers to these items using the codes given below:

Codes:
(a) both A and R are true and R is the correct explanation of A
(b) both A and R are true but R is not a correct explanation of A
(c) A is true but R is false
(d) A is false but R is true

9.8 **Assertion (A):** Basement walls and bridge abutments are usually designed for earth pressure at rest.
Reason (R): These are usually not restrained by the floor slab and by the deck structure at their bases.

[ESE : 2005]

9.9 **Assertion (A):** The earth pressure at rest is greater than active pressure but less than passive earth pressure.
Reason (R): When the wall moves away from backfill from the rest condition, the mobilization of the internal resistance of soil occurs; on the other hand, if the wall moves towards the fill, the shearing resistance build up.

[ESE : 2005]

9.10 **Assertion (A):** The state of earth pressure at rest is the state of equilibrium with zero strain condition.
Reason (R): In rest condition neither the wall nor the soil moves.

[ESE : 2006]

9.11 In a braced vertical excavation of 5 m height and 2.50 m width in a cohesive soil having undrained cohesion equal to 20 kN/m^2 and bulk unit weight of 20 kN/m^3, what is the factor of safety against heave failure at its base?
(a) 1.0 (b) 1.20
(c) 1.40 (d) 2.0

[ESE : 2006]

9.12 If the coefficient of active earth pressure is 1/3, then what is the value of the coefficient of passive earth pressure?
(a) 1/9 (b) 1/3
(c) 3 (d) 1

[ESE : 2007]

9.13 Why are weep holes provided at the back of retaining walls?
(a) To reduce the active earth pressure on the walls.
(b) To reduce the build-up of hydrostatic pressure.
(c) To provide better compaction.
(d) To increase the passive earth pressure.

[ESE : 2007]

9.14 **Assertion (A):** Rankine's earth pressure theory should not be used for concrete retaining walls and Coulomb's theory should not be used for estimating passive earth pressures.
Reason (R): Rankine assumed that the retaining wall has a vertical back and Coulomb assumed that the resultant reaction due to earth pressure acts at one-third the height of the wall.
(a) both A and R are true and R is the correct explanation of A
(b) both A and R are true but R is not a correct explanation of A
(c) A is true but R is false
(d) A is false but R is true

[ESE : 2008]

9.15 A vertical cut is to be made in saturated clay with $c = 15$ kN/m^2, $\phi = 0°$, and $\gamma = 20$ kN/m^3. What is the theoretical depth to which the clay can be excavated without side collapse?

(a) 6 m (b) 2 m
(c) 2.5 m (d) 3 m

[ESE : 2008]

9.16 What is the intensity of active earth pressure at a depth of 10.0 m in dry sand with an angle of shearing resistance of 30° and unit weight of 18 kN/m^3?

(a) 50 kN/m^2 (b) 60 kN/m^2
(c) 70 kN/m^2 (d) 80 kN/m^2

[ESE : 2009]

9.17 When a vertical face excavation was made in a clayey silt, having density of 20 kN/m^3, it failed at a depth of excavation of 4 m. What is the cohesive strength (in kN/m^2) of the soil, if its angle of internal friction was 30°?

(a) 23.1 (b) 20.0
(c) 11.6 (d) 10.2

[ESE : 2009]

9.18 Consider the following statements:

1. The yield of a retaining wall required to reach plastic equilibrium in active case is more than that in positive case.
2. Culman's graphical method is simplified version of the more general trial wedge method.
3. For a masonry gravity retaining wall coulomb's theory of earth pressure is preferred for designing.

Which of these statements is/are correct?

(a) 1, 2 and 3 (b) 1 and 2 only
(c) 2 and 3 only (d) 3 only

[ESE : 2010]

9.19 The lateral earth pressure coefficient of a soil, K_a for active state, K_p for passive state and K_0 for at-rest condition, compare as

(a) $K_0 < K_a < K_p$ (b) $K_a < K_0 < K_p$
(c) $K_a < K_p < K_0$ (d) $K_p < K_0 < K_a$

[ESE : 2011]

9.20 Consider the following statements regarding Coulomb's theory of earth pressure:

1. It is based on wedge theory of earth pressure.
2. It assumes the wall surface to be rough.
3. It may or may not satisfy the static equilibrium condition occurring in nature.

Which of these statements are correct?

(a) 1, 2 and 3 (b) 1 and 2 only
(c) 2 and 3 only (d) 1 and 3 only

[ESE : 2012]

9.21 A wall with smooth vertical back and 10 meters height retains cohesionless material with a horizontal surface. The cohesionless material weighs 4.91 kN/m^3 and has an angle of internal friction of 30°. The total active earth pressure is

(a) 81.585 kN/m length of wall
(b) 91.585 kN/m length of wall
(c) 40.743 kN/m^2
(d) 3.158 kN/m^2

[ESE : 2012]

9.22 The earth pressure behind a bridge abutment is:

(a) Active
(b) Passive
(c) At rest
(d) Constant always and everywhere

[ESE : 2013]

9.23 Consider the following statements:

1. According to Terzaghi, the average angle of wall friction may be taken equal to two-thirds of the respective angle of internal friction.
2. Wall friction increases the active earth pressure and also decreases the passive earth resistance of the soil.
3. Wall friction decreases the active earth pressure and increases the passive earth resistance of the soil.

Which of the above statements is/are correct?

(a) 1,2 and 3 (b) 1 and 2 only
(c) 3 only (d) 1 and 3 only

[ESE : 2014]

9.24 Consider the following statements:

1. Uniform surcharge increases active earth pressure even as it decreases passive resistance.
2. For analyzing the conditions at a masonry gravity wall, application of Coulomb's theory of earth pressure is preferred.
3. Culmann's method is applicable to a stratified backfill; whereas Poncelet's method is not.

Which of the above statements are correct ?

(a) 1 and 2 (b) 1 and 3
(c) 2 and 3 (d) I, 2 and 3

[ESE : 2014]

9.25 **Statement (I):** For a cantilever retaining wall, Rankine's theory of earth pressure may be used.

Statement (II): For a retaining wall with smooth vertical back, Rankine's theory is appropriate.

(a) Both Statement (I) and Statement (II) are individually true and Statement (II) is the correct explanation of Statement (I)
(b) Both Statement (I) and Statement (II) are individually true but Statement (II) is NOT the correct explanation of Statement (I)
(c) Statement (I) is true but Statement (II) is false
(d) Statement (I) is false but Statement (II) is true

[ESE : 2014]

9.26 Consider the following statements:

1. The required yield of a retaining wall to reach equilibrium in the active case in less than for the passive case.
2. The active pressure caused by a cohensionless backfill on a smooth vertical retaining wall may be reduced by compacting the backfill.
3. Given a choice, one should prefer a cohesive soil for a backfill vis-a-vis a non-cohesive soil.

Which of the above statements is/are correct?

(a) 1,2 and 3 (b) 1 and 2 only
(c) 1 and 3 only (d) 2 and 3 only

[ESE : 2015]

9.27 A rigid retaining wall of 6 m height has a saturated backfill of soft clay soil. What is the critical height when the properties of the clay soil are γ_{sat} = 17.56 kN/m^3 and cohesion c = 18 kN/m^2

(a) 1.1 m (b) 2.1 m
(c) 3.1 m (d) 4.1 m [ESE : 2016]

9.28 In a certain sea shore, the height of a retaining wall with smooth vertical back is 4.4 m. The foundation is over an expansive collapsible soil and has a horizontal surface at the level of the top of the wall and carries a UDL of 197 kPa. The unit weight and angle of internal friction are 19 kN/m^3 and 30°, respectively. What is the nearest magnitude of the total active pressure per metre length of this sea shore wall?

(a) 270 kN/m (b) 360 kN/m
(c) 450 kN/m (d) 640 kN/m

[ESE : 2016]

9.29 Consider the following assumptions regarding Coulomb's Wedge Theory:

1. There is equilibrium of every element within the soil mass of the material.
2. There is equilibrium of the whole of the material.
3. Backfill is wet, cohesive, and ideally elastic.
4. The wall surface is rough.

Which of the above assumptions are CORRECT?

(a) 1 and 3 only (b) 1 and 4 only
(c) 2 and 3 only (d) 2 and 4 only

[ESE : 2018]

9.30 A 6 m high retaining wall with a vertical back has a backfill of silty sand with a slope of 10° for thw backfill. With values of K_H = 760 kg/m^2/m and K_V = 100 kg/m^2/m, the total active earth pressure will approximately be

(a) 128 kN/m (b) 134 kN/m
(c) 138 kN/m (d) 142 kN/m

[ESE : 2019]

9.31 Which of the following assumptions of the Rankine theory of lateral earth pressure are correct?

1. The soil mass is semi-infinite, homogeneous, dry and cohesionless.
2. The ground surface is a plane which may be horizontal or inclined.
3. The wall yields about the base and thus satisfies the deformation condition for plastic equilibrium.

(a) 1 and 2 only (b) 1 and 3 only
(c) 1, 2 and 3 (d) 2 and 3 only

[ESE : 2020]

9.32 The ratio of the horizontal stress to the vertical stress is called coefficient of

(a) Active earth pressure
(b) Passive earth pressure
(c) Earth pressure
(d) Plastic earth pressure

[ESE : 2020]

9.33 Which one of the following is the correct assumption of Rankines theory?
(a) The soil mass is infinite.
(b) The soil mass is non homogeneous
(c) The soil mass is cohesive.
(d) The ground surface is a plane which may be horizontal or inclined.
[ESE : 2021]

9.34 If a retaining wall 5 m high is restrained from yielding, what is the at-rest earth pressure per meter length of wall? (Consider the backfill is cohesionless soil having ϕ = 30° and γ = 18 kN/m^3)
(a) 108 kN/m (b) 112.5 kN/m
(c) 115 kN/m (d) 124 kN/m
[ESE : 2021]

9.35 Which one of the following coffer dams is suitable for small heights?
(a) Cantilever sheet pile coffer dam
(b) Braced coffer dam
(c) Embankment protected coffer dam
(d) Double wall coffer dam
[ESE : 2023]

9.36 Which one of the following is an assumption made by Coulomb in the development of his earth pressure theory for sand and for active state?
(a) The rupture surface is non-uniform surface
(b) There is a wall friction on the pressure surface
(c) Failure is three dimensional
(d) The soil is cohesive
[ESE : 2023]

9.37 Which one of the following is a correct general consideration for a mechanically stabilized earth wall failure?
(a) Compression in reinforcement
(b) Bearing capacity failure
(c) Shearing in the wall
(d) Partly sliding of wall soil
[ESE : 2023]

■■■■

Answers Retaining Wall/Earth Pressure Theories

9.1 (c)	9.2 (d)	9.3 (d)	9.4 (d)	9.5 (d)	9.6 (a)	9.7 (a)	9.8 (c)	9.9 (a)
9.10 (a)	9.11 (c)	9.12 (c)	9.13 (b)	9.14 (b)	9.15 (d)	9.16 (b)	9.17 (c)	9.18 (c)
9.19 (b)	9.20 (a)	9.21 (a)	9.22 (c)	9.23 (d)	9.24 (c)	9.25 (a)	9.26 (b)	9.27 (d)
9.28 (b)	9.29 (d)	9.30 (c)	9.31 (c)	9.32 (c)	9.33 (d)	9.34 (b)	9.35 (a)	9.36 (b)
9.37 (b)								

Explanations Retaining Wall/Earth Pressure Theories

9.1 **(c)**

Passive pressure is maximum.
Active pressure is minimum.

9.2 **(d)**

List-II gives the lateral pressures on braced sheet pile walls. The pressure distribution on the wall is non-hydrostatic. Based on the measurements on actual job sites, empirical pressure diagram have been proposed by Terzaghi and Peck, and Tschebotarioff depending upon the type of soil.

9.3 **(d)**

Pressure at rest: When the wall is at rest and the material is in its natural state then the pressure applied by material is known as Earth Pressure at Rest.

Active earth pressure: When the wall moves away from the backfill, there is a decrease in the pressure on the wall and this decrease continues until a minimum value is reached after which there is no reduction in the pressure and the value will become constant. This kind of pressure is known as active earth pressure.

Passive earth pressure: When the wall moves towards the backfill, there is an increase in the pressure on the wall and this increase continues until a maximum value is reached after which there is no increase in the pressure and the value will become constant. This kind of pressure is known as passive earth pressure.

9.4 (d)

$\frac{K_p}{K_a}$ will be the ratio of passive to active earth pressure.

$$\frac{K_p}{K_a} = \left(\frac{1+\sin\phi}{1-\sin\phi}\right)^2 = \left(\frac{1+\sin 30^\circ}{1-\sin 30^\circ}\right)^2 = 9$$

9.5 (d)

Coefficient of earth pressure at rest,

$$1-\sin\phi = K_0$$
$$\sin\phi = 1-0.5$$
$$\sin\phi = 0.5$$
$$\therefore \quad \frac{K_p}{K_a} = \left(\frac{1+\sin\phi}{1-\sin\phi}\right)^2 = \left(\frac{1.5}{0.5}\right)^2 = 9$$

9.6 (a)

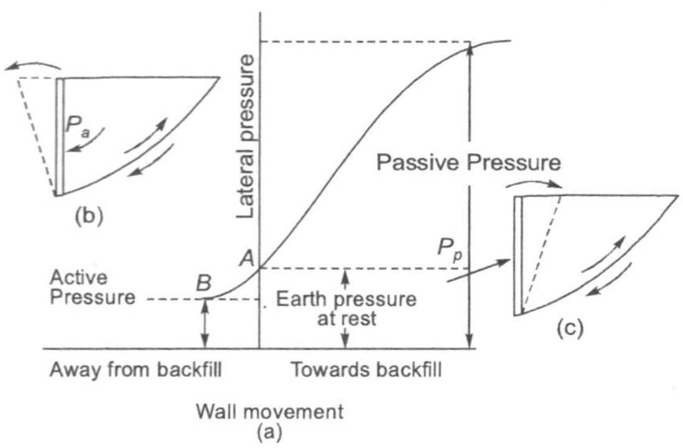

9.7 (a)

The angle of friction (δ) between the soil and the wall can be determined from direct shear test. The value of δ may range from $\frac{1}{2}\phi$ to $\frac{3}{4}\phi$ in most cases and it can be assumed as $\frac{2}{3}\phi$ in the absence of precise data.

The wall friction decreases active earth pressure and increases passive earth pressure.

From Coulomb's theory,

$$K_p = \frac{\sin^2(\alpha-\phi)}{\sin^2\alpha\sin(\alpha+\delta)\left[1-\sqrt{\frac{\sin(\phi+\delta)\sin(\phi+\beta)}{\sin(\phi+\delta)\sin(\phi+\beta)}}\right]}$$

$$K_a = \frac{\sin^2(\alpha+\phi)}{\sin^2\alpha\sin(\alpha-\delta)\left[1+\sqrt{\frac{\sin(\phi+\delta)\sin(\phi+\beta)}{\sin(\phi-\delta)\sin(\phi+\beta)}}\right]^2}$$

9.8 (c)

Both basement walls and bridge abutments are designed for earth pressure at rest.

9.9 (a)

$$K_a < K_0 < K_p$$

9.10 (a)

$$K_0 = \frac{\mu}{1-\mu}$$

Where μ is the Poisson's ratio $\simeq 1-\sin\phi'$ where ϕ' is the angle of shearing resistance

9.11 (c)

Factor of safety,

$$F = \frac{C_u N_c}{\gamma H}$$

Given $C_u = 20$ kN/m², $\gamma = 20$ kN/m³

Bearing capacity factor N_c from Skempton's analysis is

$$N_c = 5\left(1+0.2\frac{D_f}{B}\right),$$

for $\frac{D_f}{B} < 2.5$

$$= 7.5 \text{ for } \frac{D_f}{B} > 2.5$$

Thus $N_c = 5\times\left(1+0.2\times\frac{5}{2.5}\right) = 7.0$

$$\therefore \quad F = \frac{20\times 7}{20\times 5} = 1.4$$

9.12 (c)

The relation between K_a and K_p is given by

$$K_a K_p = 1$$
$$\Rightarrow \quad \frac{1}{3}\times K_p = 1$$
$$\Rightarrow \quad K_p = 3$$

9.13 (b)

Weep holes at the base of retaining walls are intended to bleed off excess moisture that collects behind the wall. Due to rise of water table, the pore pressure develops and it causes excessive hydrostatic pressure on the wall. To reduce the development of excessive lateral pressure, weep holes are provided.

9.14 (b)

The **Rankine's theory** assumes that there is no wall friction, the ground and failure surfaces are straight planes, and that the resultant force acts parallel to the backfill slope

For higher values of δ, the error becomes unacceptable because of pronounced curvature of slip surface.

9.15 (d)

The critical (theoretical) depth of vertical cut in a cohesive soil is given by

$$H_c = \frac{4c}{\gamma\sqrt{K_a}}$$

$$K_a = \frac{1-\sin\phi}{1+\sin\phi} = \frac{1-\sin 0^\circ}{1+\sin 0^\circ} = 1$$

$$\therefore \quad H_c = \frac{4\times 15}{20\times\sqrt{1}} = 3 \text{ m}$$

9.16 (b)

The intensity of active earth pressure is given by

$$p_a = K_a \times \gamma \times H$$

$$= \left(\frac{1-\sin 30^\circ}{1+\sin 30^\circ}\right)\times 18\times 10$$

$$= 60 \text{ kN/m}^2$$

9.17 (c)

$$H_c = \frac{4c}{\gamma\sqrt{K_a}}$$

But $$K_a = \frac{1-\sin 30^\circ}{1+\sin 30^\circ} = \frac{1}{3}$$

$$\therefore \quad 4 = \frac{4\times c\times\sqrt{3}}{20}$$

$$\Rightarrow \quad c = 11.6 \text{ kN/m}^2$$

9.18 (c)

The yield of a retaining wall required to reach plastic equilibrium in active case is less than that in passive case. Active pressures are accompanied by movements of the retaining wall directed away from the soil and passive resistances are accompanied by movements towards the soil.

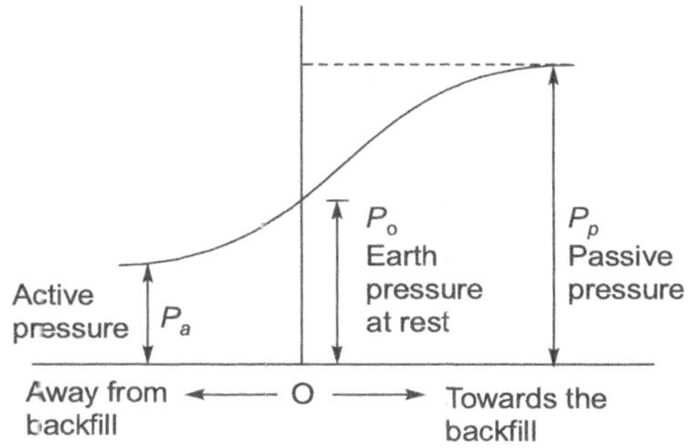

Culmann gave the graphical method to evaluate the earth pressure from Coulomb's trial wedge method. Culmann's method permits one to determine graphically the magnitude of the earth pressure and to locate the most dangerous rupture surface according to Coulomb's trail wedge method. Culmann's method is in fact a simplified version of the more general trial wedge method. It may be conveniently used for ground surface of any shape, for different types of surcharge loads, and for layered backfill with different unit weights for different layers.

For cantilever and counterfort walls, Rankine's theory is used; for gravity and semi-gravity retaining walls, Coulomb's theory is preferred.

9.19 (b)

$$K_a = \frac{1-\sin\phi}{1+\sin\phi}$$

$$K_p = \frac{1+\sin\phi}{1-\sin\phi}$$

and $$K_o = 1-\sin\phi$$

$$K_p > K_o > K_a$$

9.20 (a)

One of the main deficiencies in Coulomb's theory is that, in general, it does not satisfy the static equilibrium condition that occurs in nature. The three forces, weight of sliding wedge, soil reaction and the earth pressure acting on the sliding wedge, do not meet at a common point when the sliding surface is assumed to be planer.

9.21 (a)

$$P_a = \frac{1}{2}K_a\gamma h^2$$

$$K_a = \frac{1-\sin\phi}{1+\sin\phi}$$

$$= \frac{1-\sin 30^\circ}{1+\sin 30^\circ} = \frac{1}{3}$$

$$\therefore \quad P_a = \frac{1}{2}\times\frac{1}{3}\times 4.91\times 10^2$$

$$= 81.83 \text{ kN/m}$$

9.22 (c)

In bridge abutment, there is no tendency of wall movement.

⇒ At rest earth pressure.

9.24 (c)

Uniform surcharge increases both active and passive earth pressure.

9.25 (a)

Retaining walls are designed for active earth pressure. Rankine's earth pressure theory over estimates the active earth pressure. So it is safe to design the retaining wall using this theory.

9.26 (b)

On compaction, of cohesionless backfill, active earth pressure decreases and passive earth pressure increases.

Granular (non-cohesive) backfill offers the benefit of good drainage, easy compaction and increased sliding resistance.

9.27 (d)

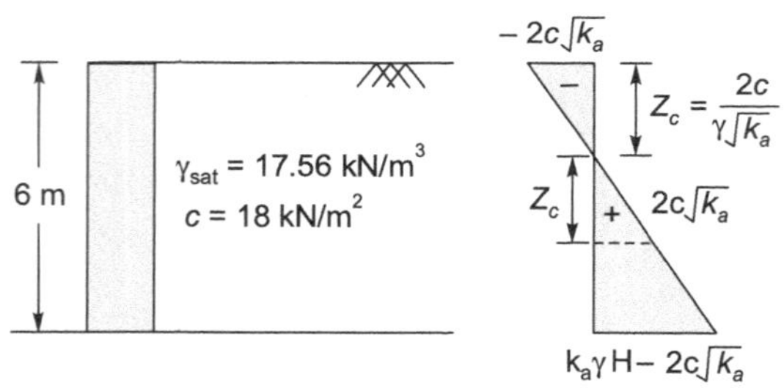

Critical height of unsupported cut

$$H_c = \frac{4c}{\gamma\sqrt{k_a}}$$

$$k_a = \frac{1-\sin 0^\circ}{1+\sin 0^\circ} = 1$$

$$H_c = \frac{4c}{\gamma} = \frac{4\times 18}{17.56} = 4.1 \text{ m}$$

9.28 (b)

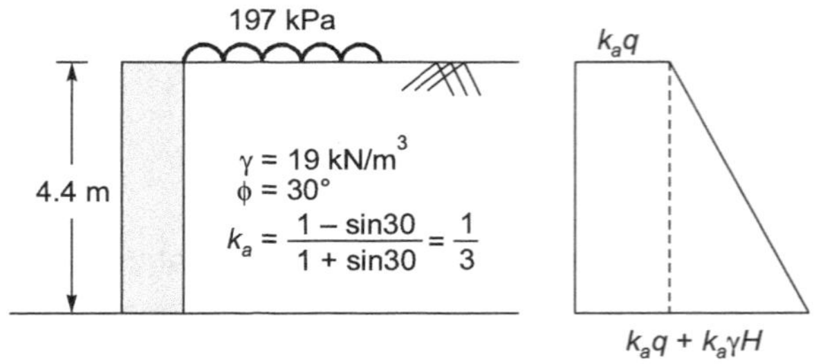

$$P_a = k_a q\cdot H + \frac{1}{2}k_a\gamma H^2$$

$$= \frac{1}{3}\times 197\times 4.4 + \frac{1}{2}\times\frac{1}{3}\times 19\times 4.4^2$$

$$= 350.24 \text{ kN/m}$$

9.29 (d)

Coulomb's theory of earth pressure is based on the following assumptions:

(i) The backfill is dry, cohesionless, homogeneous, isotropic and ideally plastic material.

(ii) The slip surface is a plane surface which passes through the heel of the wall.

(iii) The surface of wall is rough.

(iv) The sliding wedge itself acts as a rigid body.

9.30 (c)

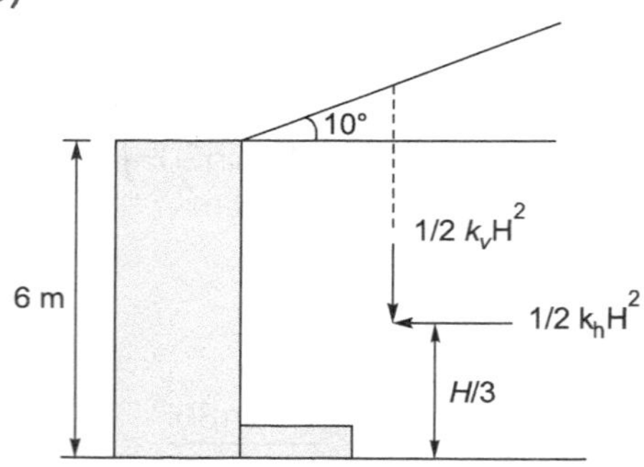

$K_H = 760 \text{ kg/m}^2\text{/m}$

$K_V = 100 \text{ kg/m}^2\text{/m}$

By Peck Hensen and Thornburn method

$$P = \sqrt{\left(\frac{1}{2}K_V H^2\right)^2 + \left(\frac{1}{2}K_H H^2\right)^2}$$

$$P = \sqrt{\left(\frac{1}{2}\times 100\times 6^2\right)^2 + \left(\frac{1}{2}\times 760\times 6^2\right)^2}$$

$$= \sqrt{(1800)^2 + (13680)^2}$$

$$= 13797.9 \text{ kg/m}$$

$$= 13797.9 \text{ kg/m} \times 9.81$$

$$= 135.357 \text{ kN/m}$$

By taking g = 10 m/s^2

P = 13797.9 kg/m × 10

= 137.979 kN/m

9.32 (c)

$$\frac{\text{Horizontal stress}}{\text{Vertical stress}} = \frac{\sigma h}{\sigma V} = K$$

K = Earth pressure coefficient.

9.33 (d)

Rankine's theory: Rankine's theory considered stress in soil mass when it attains plastic equilibrium

Assumption in Rankine's Theory:

1. Soil mass is semi-infinite, homogenous, isotropic, dry and cohesionless.
2. Soil mass is the state of plastic equilibrium condition at the time of active and passive earth pressure generation.
3. Backfill surface is planner which may be horizontal or inclined.
4. Back of wall is vertical and smooth.
5. Rupture surface is planner surface.

9.34 (b)

H = 5 m, ϕ = 30°, γ = 18 kN/m^3

k_o = earth pressure at rest coefficient

= 1 − sin ϕ = (1 − sin 30°)

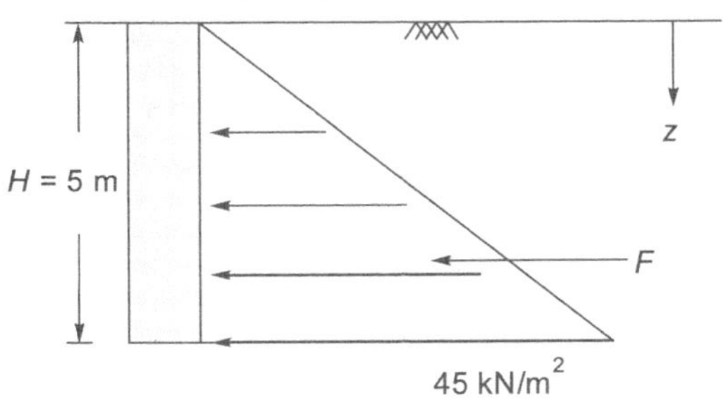

Earth pressure at rest

$$p_0 = k_0\bar{\sigma}_z$$

At z = 0; $\bar{\sigma}_z$ = 0

∴ p_0 = 0

At z = 5 m, $\bar{\sigma}_z$ = $\gamma z = 18 \times 5 = 90$ kN/m

$p_0 = k_0\bar{\sigma}_z = (1 - \sin 30°) \times 90$

= (1 − sin30°)18 × 5

= 45 kN/m^2

F = Thrust due to earth pressure at rest

= Area under earth pressure diagram

$$F = \frac{1}{2} \times 45 \times 5$$

= 112.5 kN/m

9.35 (a)

A coffer dam may be made of earth materials, timber or steel sheet piling. Cantilever sheet pile coffer dams are suitable for small heights, since these are susceptible to large leakage and flood damage.

9.36 (b)

Assumptions of Coulomb's theory:

1. Soil is homogeneous, isotropic, semi infinite, dry and cohesionless.
2. Face of wall in contact with backfill is vertical or inclined and is rough.
3. The failure wedge acts as a rigid body and stresses over it are uniform distributed.
4. The failure is essentially 2D and rupture surface is planer which passes from heel of the wall.
5. The location and direction of resultant thrust is known which passes from lower third point of the wall by assuming triangular distribution of earth pressure.
6. Wedge failure is considered which is under equilibrium.

■■■■

10 Stability Analysis of Slopes

10.1 Consider the following statements associated with stability of slope:

1. Stability number is inversely proportional to cohesion and directly proportional to height.
2. Swedish method of analysis is based on circular failure surfaces.
3. The Culmann method assumes that rupture will occur in a plane.

Which of these statements are correct?

(a) 2 and 3 (b) 1 and 3
(c) 1 and 2 (d) 1, 2 and 3

[ESE : 2000]

10.2 Taylor's stability number S_n is given by which one of the following expressions? (c is cohesion, F_c is factor of safety, γ is density of soil and H, the height of the slope).

(a) $\frac{c}{F_c \gamma}$ (b) $\frac{c}{\gamma H}$

(c) $\frac{c}{F_c \gamma H}$ (d) $\frac{c}{F_c(\gamma + H)}$

[ESE : 2002]

10.3 Consider the following statements:
Dewatering increases the slope stability of a cohesionless soil mainly because

1. it causes change in pH
2. it reduces pore water pressure

Which of the statements is/are correct?

(a) 1 only (b) 2 only
(c) Both 1 and 2 (d) Neither 1 nor 2

[ESE : 2007]

10.4 For stability analysis of slopes of purely cohesive soils, the critical centre is taken to lie at the intersection of

(a) The perpendicular bisector of the slope and the locus of the centre.
(b) The perpendicular drawn at the one-third slope from the toe and the locus of the centre.
(c) The perpendicular drawn at the two-third slope from the toe and the locus of the centre.
(d) Directional angles.

[ESE : 2018]

10.5 In a clayey soil having 50 kN/m^2 as unit cohesion and 18 kN/m^3 as unit weight, an excavation is made with a vertical face. Taking Taylor's stability number as 0.261, what is the maximum depth of excavation so that the vertical face remains stable?

(a) 5.30 m (b) 7.06 m
(c) 10.6 m (d) 12.4 m

[ESE : 2018]

10.6 An excavation is made with a vertical face in a clay soil which has C_u = 50 kN/m^2, γ_t = 18 kN/m^3 and s_n = 0.261. The maximum depth of a stable excavation will be nearly

(a) 10.6 m (b) 12.4 m
(c) 14.2 m (d) 16.0 m

[ESE : 2019]

10.7 A canal of 4 m deep has side slopes of 1 : 1. The properties of the soil are c = 15 kN/m^2, ϕ = 15°, e = 0.76 and G = 2.7. Taylor's stability number for that sudden drawdown = 0.136. The factor of safety with respect to cohesion in the case of sudden drawdown will be

(a) 0.64 (b) 1.43
(c) 2.22 (d) 3.01

[ESE : 2020]

Answers Stability Analysis of Slopes

10.1 (a) 10.2 (c) 10.3 (b) 10.4 (d) 10.5 (c) 10.6 (a) 10.7 (b)

Explanations Stability Analysis of Slopes

10.1 (a)

Taylor's stability number, $S = \dfrac{C_m}{\gamma H}$

where, C_m is unit cohesion mobilised, γ is unit weight of soil, H is height of slope

Thus, stability number is directly proportional to cohesion and inversely proportional to height. The culmann's graphical method of earth pressure determination is based on Coulomb's wedge theory. It is better than Ponclet's method and can be used for ground surface of any shape, for different types of surcharge loads, and for layered backfill with different unit weights for different layers.

10.2 (c)

A dimensionless parameter called a Taylor's stability number is often useful for the analysis of slopes of c-ϕ soils and can be defined by the following equation:

$$S_n = \frac{C_m}{\gamma H} = \frac{C}{F_c \gamma H} = \cos^2 i(\tan i - \tan\phi)$$

10.3 (b)

Dewatering in cohesionless soil reduces pore water pressure. This leads to increase in effective stress and which in result increases the shear strength of soil. Hence slope stability increases.

10.5 (c)

$$S_n = \frac{C}{\text{FOS}\times\gamma\times H}$$

$$0.261 = \frac{50}{1\times 18\times H}$$

$$H = 10.6 \text{ m}$$

10.6 (a)

Taylor's stability number,

$$s_n = \frac{c}{\gamma H_c} = \frac{50}{18\times H_c} = 0.261$$

$$H_c = 10.6428 \text{ m}$$

10.7 (b)

In case of sudden drawdown γ_{sat} should be used in stability No.

$$\gamma_{sat} = \left(\frac{G+e}{1+e}\right)\gamma_w = \left(\frac{2.7+0.76}{1+0.76}\right)9.81$$

$$= 19.28 \text{ kN/m}^2$$

$$s_n = \frac{C}{\gamma H_c} = \frac{C}{\gamma_{sat}(\text{FOS})H}$$

$$0.136 = \frac{15}{19.28(\text{FOS})4}$$

$$\text{FOS} = 1.43$$

11 Shallow Foundation and Bearing Capacity

11.1 Match **List-I** (In-situ test) with **List-II** (Measurement) and select the correct answer using the codes given below the lists:

List-I

A. SPT test
B. Plate load test
C. Field vane shear test
D. CPT test

List-II

1. Penetration resistance (N value)
2. Load settlement data
3. Point resistance and skin friction
4. In situ shear strength

Codes:

	A	B	C	D
(a)	1	2	4	3
(b)	1	2	3	4
(c)	2	1	3	4
(d)	2	1	4	3

[ESE : 2000]

11.2 Match **List-I** (Contact pressure distribution diagrams) with **List-II** (Description of footings) and select the correct answer using the codes given below the lists:

List-I

A.
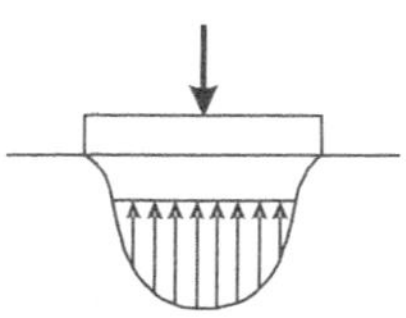

B.
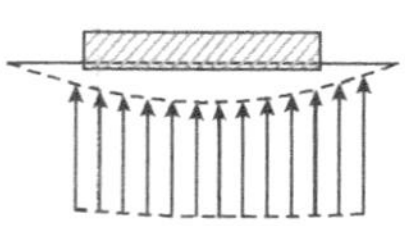

C.
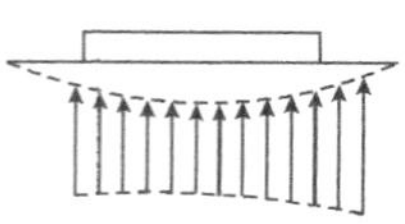

D.
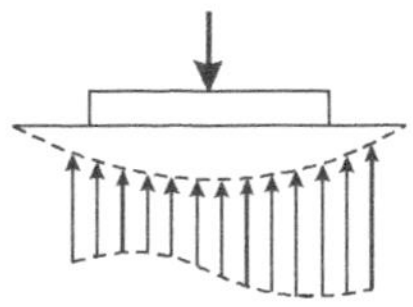

List-II

1. Rigid footing on cohesive soil
2. Flexible footing on cohesive soil
3. Rigid footing on cohesion-less soil at ground level
4. Flexible footing on cohesionless soil at ground level

Codes:

	A	B	C	D
(a)	3	1	4	2
(b)	4	2	3	1
(c)	3	2	4	1
(d)	4	1	3	2

[ESE : 2000]

11.3 **Assertion (A):** Larger footings settle more than the smaller footings under the same load intensity.

Reason (R): Size of the pressure bulb depends upon the size of the footing.

(a) both A and R are true and R is the correct explanation of A
(b) both A and R are true but R is not a correct explanation of A
(c) A is true but R is false
(d) A is false but R is true

[ESE : 2000]

11.4 Depth of foundation depends upon

(a) scour depth, minimum grip length and Rankine depth
(b) scour depth, minimum grip length and depth of bearing stratum

(c) scour depth, Rankine depth and depth of bearing stratum
(d) minimum grip length, Rankine depth and depth of bearing stratum

[ESE : 2000]

11.5 Match **List-I** (Test) with **List-II** (Property) and select the correct answer using the codes given below the lists:

List-I	List-II
A. Proctor Test	1. Grain Size Analysis
B. Vane Test	2. Shear Strength
C. Penetration Test	3. Bearing Capacity
D. Hydrometer Test	4. Compaction

Codes:

	A	B	C	D
(a)	2	4	1	3
(b)	4	2	1	3
(c)	4	2	3	1
(d)	2	4	3	1

[ESE : 2001]

11.6 Two circular footings of diameters D_1 and D_2 are resting on the surface of a purely cohesive soil. The ratio $D_1/D_2 = 2$. If the ultimate load carrying capacity of the footing of diameter D_1 is 200 kN/m^2, then the ultimate bearing capacity (in kN/m^2) of the footing of diameter D_2 will be
(a) 100 (b) 200
(c) 314 (d) 571

[ESE : 2001]

11.7 The minimum bearing capacity of a soil under a given footing occurs when the groundwater table at the location is at
(a) the base of the footing
(b) the ground level
(c) a depth equal to one-half the width of footing
(d) a depth equal to the width of footing

[ESE : 2001]

11.8 Match **List-I** (Field test) with **List-II** (Property) and select the correct answer using the codes given below the lists:

List-I	
A. Pumping test	1. Bearing capacity
B. Plate load test	2. Load carrying capacity
C. Pile load test	3. Permeability

Codes:

	A	B	C
(a)	1	2	3
(b)	3	1	2
(c)	2	3	1
(d)	3	2	1

[ESE : 2001]

11.9 A rigid strip footing subjected to vertical central load fails under general shear failure. The rupture surface developed is
(a) circular one sided rotational failure
(b) plane surface of failure originating at the edge of footing and extending downwards with orientation of Rankine passive state equilibrium
(c) symmetrical failure surface consisting of segment of spiral and Rankine's passive state failure surface
(d) symmetrical failure surface consisting of segments of log spiral and plane failure surface as per Rankine active state of equilibrium

[ESE : 2002]

11.10 In a plate load test on sandy soil, the test plate of 60 cm × 60 cm undergoes a settlement of 5 mm at a pressure of 12 × 10^4 N/m^2. What will be the expected settlement of 3 m × 3 m footing under same pressure?
(a) 25 mm (b) 20 mm
(c) 15 mm (d) 9 mm [ESE : 2002]

11.11 **Assertion (A):** Terzaghi's bearing capacity theory overestimates the bearing capacity of footings.
Reason (R): Terzaghi's theory neglects the shearing resistance of rupture surface in the soil above the foundation level.
(a) both A and R are true and R is the correct explanation of A
(b) both A and R are true but R is not a correct explanation of A
(c) A is true but R is false
(d) A is false but R is true [ESE : 2002]

11.12 If two foundations, one narrow and another wide, are resting on a bed of sand carrying the same intensity of load per unit area, then which one is likely to fail early?
(a) Narrow foundation
(b) Wider foundation

(c) Both will fail simultaneously
(d) Difficult to judge since other conditions are unknown [ESE : 2003]

11.13 The contact pressure distribution under a rigid footing on a cohesionless soil would be
(a) uniform throughout
(b) zero at centre and maximum at edges
(c) zero at edges and maximum at centre
(d) maximum at edges and minimum at centre
[ESE : 2003]

11.14 A strip footing having 1.5 m width founded at a depth of 3 m below ground level in a clay soil having c = 20 kN/m^2. ϕ = 0° and unit weight γ = 20 kN/m^3. What is the net ultimate bearing capacity using Skempton's analysis?
(a) 30 kN/m^2 (b) 60 kN/m^2
(c) 100 kN/m^2 (d) 140 kN/m^2
[ESE : 2004]

11.15 When a load test was conducted by putting a 60 cm square plate on top of a sandy deposit, the ultimate bearing capacity was observed as 60 kN/m^2. What is the ultimate bearing capacity for a strip footing of 1.2 m width to be placed on the surface of the same soil?
(a) 75 kN/m^2 (b) 120 kN/m^2
(c) 150 kN/m^2 (d) 160 kN/m^2
[ESE : 2004]

11.16 A plate load test is conducted on a cohesionless soil with a test plate having width B_p(cm) and settlement of this plate S_p(cm) is obtained at the same load intensity as a foundation. A footing having a width B_f(cm) is to be constructed as foundation. What is the settlement S_f(cm) experienced by this footing?
(a) $S_f = S_p\{[B_f(B_p + 30)]/[B_p(B_f + 30)]\}^2$
(b) $S_f = S_p\{[B_p(B_f + 30)]/[B_f(B_p + 30)]\}^2$
(c) $S_f = S_p\,[B_f/B_p]$
(d) $S_f = S_p\,[B_p/B_f]$
[ESE : 2004]

11.17 Match **List-I** (Type of strata below foundation) with **List-II** (Type of foundation movement) and select the correct answer using the codes given below the lists:

List-I
A. Sand
B. Heterogeneous landfill
C. Black cotton soil
D. Hard rock

List-II
1. Practically no movements
2. Immediate settlements
3. Large relative settlements
4. Heaving of foundations

Codes:

	A	B	C	D
(a)	2	3	1	4
(b)	2	3	4	1
(c)	3	2	1	4
(d)	3	2	4	1

[ESE : 2004]

11.18 A strap footing is to be provided for the two columns shown in the diagrams given below.

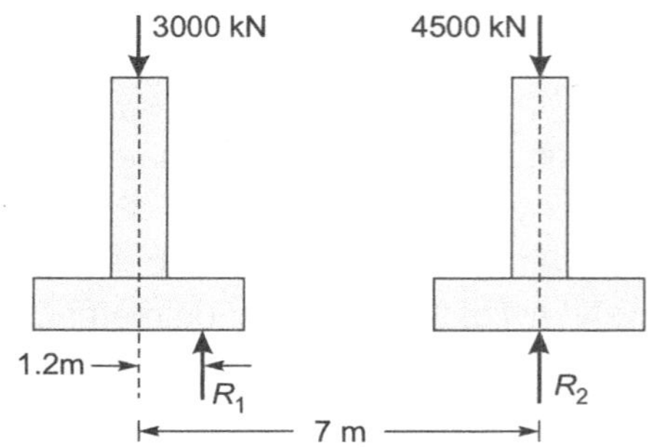

What is the length L of the left footing? Take width of both the footings as 3.5 m and safe bearing capacity of soil as 350 kN/m^2.
(a) 2.0 m (b) 2.5 m
(c) 3.5 m (d) 2.95 m [ESE : 2004]

Directions: The following items consists of two statements; one labelled as **'Assertion (A)'** and the other as **'Reason (R)'**. You are to examine these two statements carefully and select the answers to these items using the codes given below:

Codes:
(a) both A and R are true and R is the correct explanation of A
(b) both A and R are true but R is not a correct explanation of A
(c) A is true but R is false
(d) A is false but R is true

11.19 **Assertion (A):** Raft foundations are the best type of shallow foundations to support heavy structures.

Reason (R): Raft foundations has the ability to redistribute load coming on weak pockets of soil below raft to the adjacent soils where the stresses are less severe for the soil at that place. This reduces the chances of differential settlements and increases the bearing capacity.

[ESE : 2004]

11.20 **Assertion (A):** Plate load test carried out at the site gives field test data which is useful in evaluation of bearing capacity and settlements. It is normally conducted at the level of the proposed foundation.

Reason (R): Plate load test is reliable because it reflects the true behaviour of foundation stratum below the proposed level of foundation and extending up to large depth below.

[ESE : 2004]

11.21 Consider the following statements:

1. The standard penetration test is a reliable method for measuring the relative density of granular soils.
2. For a sand having the same relative density, *N*-values remain the same at all depths.
3. For a sand having the same relative density, *N*-values are different at different depths.

Which of these statements is/are correct?

(a) 1, 2 and 3 (b) 1 and 2
(c) 3 only (d) 1 and 3

[ESE : 2005]

11.22 Which one of the following states of field compaction of sand deposit truly represents the corrected standard penetration test value: *N* (corrected) = 27?

(a) Loose (b) Medium dense
(c) Dense (d) Very dense

[ESE : 2005]

11.23 In case of footing on the surface or shallow depth in very dense sand which one of the following types of failure is likely to occur?

(a) Punching shear failure
(b) Local shear failure
(c) General shear failure
(d) Any of the above three

[ESE : 2005]

11.24 The bearing capacity factors N_c, N_q and N_γ are functions of

(a) width and depth of footing
(b) density of soil
(c) cohesion of soil
(d) angle of internal friction of soil

[ESE : 2005]

11.25 **Assertion (A):** The bearing capacity of a footing always gets affected by the location of ground water table.

Reason (R): Water in soil affects the shear strength parameters as well as the unit weight.

(a) both A and R are true and R is the correct explanation of A
(b) both A and R are true but R is not a correct explanation of A
(c) A is true but R is false
(d) A is false but R is true

[ESE : 2005]

11.26 In a plate load test, how is the ultimate load estimated from the load settlement curve on a log-log graph?

(a) Directly
(b) By drawing tangents to the curve at the initial and final points
(c) By the secant method
(d) At 0.2 per cent of the maximum settlement

[ESE : 2005]

11.27 In which one of the following zones is a logarithmic spiral shape of failure surface assumed in the case of bearing capacity analysis of c-ϕ soils?

(a) Active zone
(b) Passive zone
(c) Radial shear zone
(d) Surcharge zone

[ESE : 2005]

11.28 A test plate 30 cm square, settles by 12 mm under a load of 4.5 kN in a sandy soil. By how much will a footing 2 m × 2 m subjected to a load of 200 kN settle?

(a) 36.3 mm (b) 20.87 mm
(c) 75.75 mm (d) 18.15 mm

[ESE : 2006]

11.29 The contact pressure distribution below rigid footing on the surface of a clay soil is
(a) uniform for the full width
(b) maximum at the centre and minimum at the edges
(c) maximum at the edges and minimum at the centre
(d) of an irregular shape

[ESE : 2007]

11.30 Consider the following statements:
1. The bearing capacity of a footing on clay does not significantly get altered by the presence of water table.
2. The bearing capacity of a footing on saturated clay ($\phi = 0$) is a function of its size.

Which of these statements is/are correct?
(a) 1 only (b) 2 only
(c) Both 1 and 2 (d) Neither 1 nor 2

[ESE : 2007]

11.31 Consider the following statements:
Criteria for satisfactory performance of footings are that the
1. soil supporting the footing must be safe against shear failure
2. footing must not settle more than a pre-specified value
3. footing must be rigid
4. footing should be above water table.

Which of these statements are correct?
(a) 3 and 4 (b) 1 and 2
(c) 1 and 3 (d) 2 and 4

[ESE : 2007]

11.32 The observed *N*-value from a standard penetration test conducted in saturated sandy strata is 30; the *N*-value corrected for dilatancy may be taken as
(a) 18 (b) 23
(c) 28 (d) 36 [ESE : 2007]

11.33 **Assertion (A):** Estimation of settlement of foundations on sandy soils can be done by using SPT values .
Reason (R): Sampling in cohesionless soils without disturbing the structure is difficult.
(a) both A and R are true and R is the correct explanation of A
(b) both A and R are true but R is not a correct explanation of A
(c) A is true but R is false
(d) A is false but R is true [ESE : 2007]

11.34 Match **List-I** (Equipments) with **List-II** (Use) and select the correct answer using the codes given below the lists:

List-I
A. Hydrometer
B. Plate load test set up
C. Pycnometer
D. Direct shear apparatus

List-II
1. Determination of shear parameters
2. Determination of specific gravity
3. Determination of bearing capacity of soils
4. Grain size distribution tests for clays

Codes:

	A	B	C	D
(a)	2	1	4	3
(b)	4	3	2	1
(c)	2	3	4	1
(d)	4	1	2	3

[ESE : 2008]

11.35 Consider the following statements in respect of static cone penetration test:
1. The cone used has an apex angle of 60° and base area of 10 cm^2.
2. This test give a continuous record of cone resistance.

Which of these statements is/are correct?
(a) 1 only (b) 2 only
(c) Both 1 and 2 (d) Neither 1 nor 2

[ESE : 2008]

11.36 Which factors influence the bearing capacity of a purely cohesionless soil?
1. Relative density of soil
2. Width and depth of footing
3. Unit weight of soil

Select the correct answer using the codes given:
(a) 1 and 2 only (b) 2 and 3 only
(c) 1 and 3 only (d) 1, 2 and 3

[ESE : 2008]

11.37 Match **List-I** (Field test) with **List-II** (Suitability) and select the correct answer using the codes given below the lists:

List-I
A. Plate load test
B. Standard penetration test
C. Vane shear test
D. Dilatancy test
List-II
1. To estimate bearing capacity of granular soil
2. To estimate in situ strength of soft clay
3. To identify silt from clay
4. To estimate bearing capacity for permissible settlement in clays

Codes:

	A	B	C	D
(a)	4	3	2	1
(b)	2	1	4	3
(c)	4	1	2	3
(d)	2	3	4	1

[ESE : 2008]

11.38 Match **List-I** (Test) with **List-II** (Utility) and select the correct answer using the codes given below the lists:

List-I
A. Field density test
B. Plate load test
C. CBR
List-II
1. Stress deformation characteristics
2. Compaction characteristics
3. Design of pavement
4. Safe load bearing capacity of soil

Codes:

	A	B	C
(a)	2	4	1
(b)	4	2	1
(c)	4	2	3
(d)	2	4	3

[ESE : 2008]

11.39 For a proposed building, raft foundation, isolated footings and combined footings are being considered. These foundations are to be listed in the decreasing order of preference in terms of performance. Which one of the following is the correct order of listing?
(a) Raft foundation – Combined footings –Isolated footings
(b) Isolated footings – Raft foundation –Combined footings
(c) Combined footings – Raft foundation –Isolated footings
(d) Combined footings – Isolated footings –Raft foundation

[ESE : 2008]

11.40 Consider the following statements:
Foundation design in expansive soil can be done by
1. isolating the foundation from the swelling soil
2. taking measures to prevent the swelling
3. employing measures to make the structure withstand the movement

Which of these statements is/are correct?
(a) 1 and 2 only (b) 2 and 3 only
(c) 1 and 3 only (d) 1, 2 and 3

[ESE : 2008]

11.41 Assertion (A): Foundation on expansive soils must be designed for as high a bearing pressure as possible consistent with the bearing capacity and settlement requirement.
Reason (R): Lightly loaded single and two storeyed buildings experience maximum damage when built on expansive soils.
(a) both A and R are true and R is the correct explanation of A
(b) both A and R are true but R is not a correct explanation of A
(c) A is true but R is false
(d) A is false but R is true [ESE : 2008]

11.42 On which of the following soils is the standard penetration test useful?
1. Cohesionless soils
2. Medium clays
3. Gravelly soils
4. Very stiff clays

Select the correct answer using the codes given below:
(a) 1 only (b) 1 and 3
(c) 1 and 2 (d) 3 and 4

[ESE : 2009]

11.43 Consider the following statements:
1. Standard penetration test (SPT) is conducted by pushing a cone into soil at the rate of 2 cm/s.
2. Standard penetration test results are unreliable in deposits containing large number of boulders.

3. Dutch cone is a static penetrometer.

Which of these statements is/are correct:

(a) 1 only (b) 1 and 2

(c) 2 and 3 (d) 3 only [ESE : 2009]

11.44 Consider the following statements:

Increasing width of footing results in

1. increase in settlement of a consolidating clay layer
2. Increase in bearing capacity of sandy soils
3. decrease in bearing capacity of clays

Which of these statements is/are correct?

(a) 1 only (b) 1 and 2 only

(c) 2 and 3 only (d) 1, 2 and 3

[ESE : 2009]

11.45 If an SPT test gave the average blow count of 32 in fine sand below water table, then what is the corrected value of blow count?

(a) 22.1 (b) 23.5

(c) 24.2 (d) 24.8 [ESE : 2009]

11.46 The standard penetration resistance value obtained in a deep deposit of sand at a depth of 6.0 m was 28. The unit weight of sand is 18.0 kN/m^3. What is the corrected value of number of blows for overburden pressure?

(a) 60 (b) 57

(c) 59 (d) 55 [ESE : 2009]

11.47 The net ultimate bearing capacity of a purely cohesive soil

(a) depends on the width of the footing and is independent of the depth of the footing

(b) depends on the width as well as the depth of the footing

(c) depends on the depth, but is independent of the width, of the footing

(d) is independent of both the width and the depth of the footing

[ESE : 2009]

11.48 A soil has a low allowable bearing capacity. The soil deposit contains compressible loess. A foundation is to be provided for a structure carrying a heavy load. Which one of the following foundation types is to be adopted?

(a) Strap footing

(b) Continuous footing

(c) Raft foundation

(d) Combined spread foundation

[ESE : 2009]

11.49 The observed *N*-value from a standard penetration test conducted in saturated sandy strata is 30; the *N*-value corrected for dilatancy may be taken as

(a) 15 (b) 20

(c) 23 (d) 39 [ESE : 2011]

11.50 Two footings, one circular and the other square, are founded on the surface of a purely cohesionless soil. The diameter of the circular footing is the same as that of the side of the square footing. The ratio between their ultimate bearing capacities will be

(a) 1.0 (b) 1.3

(c) 1.33 (d) 0.75 [ESE : 2011]

11.51 Consider the following statements:

1. Standard penetration test is commonly used for cohesionless soils.
2. Standard penetration test results in respect of a cohesionless soil are correlated to its density index and friction angle.
3. Use of *N*-value not corrected for overburden pressure leads to highly conservative design of footings at shallow depths.

Which of these statements are correct?

(a) 1, 2 and 3 (b) 1 and 2 only

(c) 2 and 3 only (d) 1 and 3 only

[ESE : 2012]

11.52 If water table is encountered in the standard pit while conducting plate load test

(a) The load test should be abandoned

(b) The pit is considered unsafe

(c) Test should be conducted with complete dewatering continuously throughout the test duration.

(d) The bearing capacity of soil cannot be determined in this condition

[ESE : 2012]

11.53 Consider the following statements relating to foundations on expansive soils:

1. Strength should be improved and compressibility should be reduced.
2. Compressibility should be increased.

3. No stabilization should be done.

Which of these statements is/are correct?

(a) 1, 2 and 3 (b) 2 only
(c) 1 only (d) 3 only [ESE : 2012]

11.54 Consider the following statements

1. The benefit of surcharge and depth of foundation is only marginal in case of footings on purely cohesive soils.
2. The bearing capacity of a footing in pure clay increases with increase in size of the footing.
3. Size effects in plate load tests are more important in case of cohesionless soils.

Which of these statements are correct?

(a) 1, 2 and 3 (b) 1 and 2 only
(c) 2 and 3 only (d) 1 and 3 only

[ESE : 2012]

11.55 The mean unconfined compressive strength of a purely cohesive soil was found to be 50 kN/m^2. The ultimate bearing capacity of a square footing calculated by Terzaghi's concept (bearing capacity factor $NC = 5.7$) will be

(a) 185.25 kN/m^2 (b) 390.5 kN/m^2
(c) 285 kN/m^2 (d) 142.5 kN/m^2

[ESE : 2012]

11.56 The ultimate bearing capacity of a square footing on surface of a saturated clay having unconfined compression strength of 50 kN/m^2 (using Skempton's equation) is

(a) 250 kN/m^2 (b) 180 kN/m^2
(c) 150 kN/m^2 (d) 125 kN/m^2

[ESE : 2013]

11.57 Consider the following statements:

1. In the case of footings on purely cohesive soils, the benefit due to surcharge or depth of foundation is only marginal.
2. The bearing capacity of a footing in pure clay may be increased by increasing its size.
3. Factor of safety should be applied only to the net ultimate bearing capacity.

Which of the above statements are correct?

(a) 1, 2 and 3 (b) 1 and 2 only
(c) 2 and 3 only (d) 1 and 3

[ESE : 2014]

11.58 The gross bearing capacity of a footing is 450 kN/m^2. If the footing is 1.5 m wide and is at a depth of 1 m in a clayey soil which has a unit weight of 20 kN/m^3, then the net bearing capacity is

(a) 410 kN/m^2 (b) 420 kN/m^2
(c) 430 kN/m^2 (d) 440 kN/m^2

[ESE : 2014]

11.59 A footing 1 m × 1 m in size rests on the surface of an infinite layer of soil. It is subjected to a load of 600 kN. What is the immediate settlement of the soil by considering $Eu = 2.0$ MPa, $N = 0.5$ and influence factor = 0.95?

(a) 22.5 mm (b) 25.5 mm
(c) 27.5 mm (d) 30.0 mm

[ESE : 2015]

11.60 Consider the following statements:

1. The proportioning of footing in sand is more often governed by settlement rather than by bearing capacity.
2. The pressure bulb profiles under a strip footing form as co-axially imaginable bulbs under its length.
3. Friction piles are also called floating piles.

Which of the above statements are correct?

(a) 1, 2 and 3 (b) 1 and 2 only
(c) 1 and 3 only (d) 2 and 3 only

[ESE : 2015]

11.61 Which of the following factors affect the bearing capacity of cohesive soils?

1. Density of the soil
2. Angle of shearing resistance of the soil
3. Depth of the footing
4. Width of the footing

(a) 1, 2 and 3 only (b) 1, 2 and 4 only
(c) 2, 3 and 4 only (d) 1, 2, 3 and 4

[ESE : 2015]

11.62 Consider the following statement:

1. The ultimate bearing capacity of a footing on sand increases with an increase in its width
2. The settlement of the footing on sand increases with increase in its width.

Which of the above statements are correct?

(a) 1 only (b) Both 1 and 2
(c) 2 only (d) Neither 1 nor 2

[ESE : 2015]

11.63 **Assertion (A):** In cohesive soils, the ultimate bearing capacity is independent of foundation width.

Reason (R): The ultimate bearing capacity of cohesive soils increases with depth below ground level.

(a) Both A and R are true and R is the correct explanation of A
(b) Both A and R are true but R is not a correct explanation of A
(c) A is true but R is false
(d) A is false but R is true

[ESE : 2015]

11.64 In a plate load test on a soil, at a particular magnitudes of the settlement, it was observed that the bearing pressure beneath the footing is 100 kN/m^2 and the perimeter shear is 25 kN/m^2. Correspondingly, the load capacity of a 2 m square footing at the same settlement will be

(a) 200 kN (b) 300 kN
(c) 400 kN (d) 600 kN [ESE : 2017]

11.65 Consider the following statements:

1. According to Terzaghi, a foundation is shallow if its depth is equal to or less than its width.
2. Spread footing, strap footing and raft footing are types of shallow foundations.
3. Combined footing may be trapezoidal if the two columns carry unequal loads; and rectangular if both columns carry equal loads.
4. For water tanks, providing raft foundations will avoid unequal settlements.

Which of the above statements are correct?

(a) 1, 2, 3 and 4 (b) 1, 2 and 3 only
(c) 1, 2 and 4 only (d) 3 and 4 only

[ESE : 2017]

11.66 Consider the following statements:

For a rigid footing placed at the ground surface on sand, the contact pressure

1. is maximum at the edges.
2. is zero at the edges.
3. distribution is parabolic.
4. is uniform throughout the base of the footing.

Which of the above statements are correct?

(a) 1 and 3 only (b) 1 and 4 only
(c) 2 and 3 only (d) 2 and 4 only

[ESE : 2017]

11.67 Consider the following statements as suggestive of the bearing capacity of soil:

1. The maximum net loading intensity at which neither the soil fails in shear nor is there excessive settlement detrimental to the structure.
2. The maximum net pressure which the soil can carry without shear failure.
3. The net ultimate bearing capacity of the soil divided by a factor of safety.

Which of the above statements is/are correct?

(a) 1, 2 and 3 (b) 1 only
(c) 2 only (d) 3 only [ESE : 2017]

11.68 A strip footing 2 m in width, with its base at a depth of 1.5 m below ground surface, rests on a saturated clay soil with γ_{sat} = 20 kN/m^3, c_u = 40 kN/m^2, ϕ_u = 0, c' = 10 kN/m^3, and ϕ' = 20°. The natural water table is at 1 m depth below ground level. As per IS : 6403 – 1981, the ultimate bearing capacity of this footing will be (taking the relevant N_c as 5.14)

(a) 327 kN/m^2 (b) 285 kN/m^2
(c) 253 kN/m^2 (d) 231 kN/m^2

[ESE : 2019]

11.69 A raft foundation 10 m wide and 12 m long is to be constructed in a clayey soil having shear strength of 12 kN/m^2. Unit weight of soil is 16 kN/m^3. The ground surface carries a surcharge of 20 kN/m^2, the factor of safety is 1.2 and the value of N_c = 5.7. The safe depth of foundation will be nearly

(a) 8.2 m (b) 7.3 m
(c) 6.4 m (d) 5.5 m

[ESE : 2019]

11.70 A 30 cm square bearing plate settles by 8 mm in the plate load test on cohesionless soil when the intensity of loading is 180 kN/m^2. The settlement of a shallow foundation of 1.5 m square under the same intensity of loading will be nearly

(a) 30 mm (b) 26 mm
(c) 22 mm (d) 18 mm

[ESE : 2020]

11.71 When the observed value of N exceeds 15, the corrected penetration number N_c as per Terzaghi and Peck recommendation in the silty fine sands will be

(a) $15-\frac{1}{2}(N_R-15)$ (b) $15-\frac{1}{2}(N_R+15)$

(c) $15+\frac{1}{2}(N_R-15)$ (d) $15+\frac{1}{2}(N_R+15)$

where : N = Penetration number, and

N_R = Recorded value [ESE : 2020]

11.72 Which one of the following is a characteristic of local shear failure?

(a) Failure pattern is not clearly defined.
(b) Failure surfaces reach ground surfaces.
(c) There is no bulging of soil around the footing
(d) Failure is not sudden and there is no tilting of footing

[ESE : 2021]

11.73 According to Terzaghi, a foundation is shallow if its

(a) depth is equal to or less than its width
(b) depth is twice the width
(c) width is thrice the depth
(d) width is one fourth of depth

[ESE : 2023]

11.74 The ultimate bearing capacity is

(a) the minimum net pressure intensity causing shear failure of soil
(b) the minimum gross pressure at the lose of the foundation at which the soils fails in shear.
(c) the difference in intensities of the gross pressure after the construction of the structure.
(d) the total pressure at the base of the footing due to the weight of the superstructure.

[ESE : 2023]

11.75 Consider the following statements related to ultimate bearing capacity of footing based on the SPT and CPT values:

1. The empirical correlations established in the USA between N and soil properties indicate the value of N conforms to certain standard energy ratios.
2. The value of N to be used for getting ϕ is the corrected values for standard energy.
3. The angle ϕ obtained by this method can't be used for obtaining the bearing capacity factors.

Which of the above statements are correct?

(a) 1 and 2 only (b) 2 and 3 only
(c) 1 and 3 only (d) 1, 2 and 3

[ESE : 2023]

11.76 A reinforced concrete foundation, of dimensions 18 m × 36 m exerts a uniform pressure of 180 kN/m^2 on a soil mass, with E-value 45 MN/m^2. What is the value of immediate settlement under the foundation?

(a) 1 m (b) 1.54 m
(c) 18 mm (d) 54 mm

[ESE : 2024]

11.77 Which one of the following terms is defined as "Maximum pressure which a foundation can withstand without the occurrence of shear failure of the foundation"?

(a) Gross bearing capacity
(b) Bearing capacity
(c) Ultimate bearing capacity
(d) Safe bearing capacity

[ESE : 2024]

11.78 If S_i = immediate settlement at a corner of a rectangular flexible foundation of size L × B, B = Width of the foundation, q = Uniform pressure on the foundation, E_S = Modulus of elasticity of the soil beneath the foundation, ν = Poisson's ratio of the soil, and I_t = Influence value, which is dependent on L/B. What is the immediate settlement of a flexible foundation?

(a) $S_i = q.B\left(\frac{1+\nu^2}{E_S}\right)\cdot I_t$

(b) $S_i = q.B\left(\frac{1-\nu}{E_S}\right)\cdot I_t$

(c) $S_i = q\left(\frac{1+\nu^2}{BE_S}\right)\cdot I_t$

(d) $S_i = q\cdot B\left(\frac{1-\nu^2}{E_S}\right)\cdot I_t$

[ESE : 2024]

11.79 Consider the following statements regarding bearing capacity values specified in building codes:

1. The codes tacitly assume that the allowable bearing capacity is dependent only on the soil type.
2. The codes assume that the bearing capacity is dependent of the size, shape and depth of foundation.
3. Building codes are usually not up-to-date

Which of the above statements are correct?

(a) 1 and 2 only (b) 1 and 3 only
(c) 2 and 3 only (d) 1, 2 and 3

[ESE : 2024]

11.80 Which one of the following comprises two or more footings connected by a beam called strap?

(a) Continuous footing
(b) Spread footing
(c) Combined footing
(d) Cantilever footing

[ESE : 2024]

Answers Shallow Foundation and Bearing Capacity

11.1 (a)	11.2 (c)	11.3 (a)	11.4 (a)	11.5 (c)	11.6 (b)	11.7 (b)	11.8 (b)
11.9 (c)	11.10 (d)	11.11 (d)	11.12 (a)	11.13 (c)	11.14 (d)	11.15 (c)	11.16 (a)
11.17 (b)	11.18 (d)	11.19 (a)	11.20 (c)	11.21 (c)	11.22 (b)	11.23 (c)	11.24 (d)
11.25 (d)	11.26 (a)	11.27 (c)	11.28 (a)	11.29 (c)	11.30 (a)	11.31 (b)	11.32 (b)
11.33 (b)	11.34 (b)	11.35 (c)	11.36 (d)	11.37 (c)	11.38 (d)	11.39 (a)	11.40 (d)
11.41 (a)	11.42 (a)	11.43 (c)	11.44 (b)	11.45 (b)	11.46 (d)	11.47 (d)	11.48 (c)
11.49 (c)	11.50 (d)	11.51 (a)	11.52 (c)	11.53 (c)	11.54 (d)	11.55 (a)	11.56 (c)
11.57 (d)	11.58 (c)	11.59 (*)	11.60 (a)	11.61 (a)	11.62 (b)	11.63 (b)	11.64 (d)
11.65 (a)	11.66 (c)	11.67 (c)	11.68 (d)	11.69 (d)	11.70 (c)	11.71 (c)	11.72 (d)
11.73 (a)	11.74 (b)	11.75 (a)	11.76 (c)	11.77 (c)	11.78 (d)	11.79 (a)	11.80 (b)

Explanations Shallow Foundation and Bearing Capacity

11.1 (a)

The **vane shear test** is an in-situ geotechnical testing methods used to estimate the undrained shear strength of fully saturated clays without disturbance.

11.2 (c)

Factor Influencing Contact Pressure:

The actual distribution of contact pressure depends upon a number of factors such as

- Elastic properties of footing
- Elastic properties of soil
- Thickness of footing

(i) Flexible footing over clayey soil

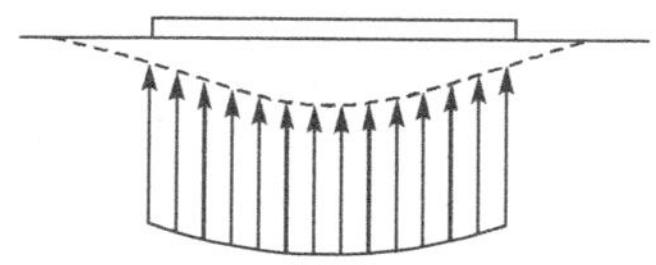

(ii) Rigid footing over clayey soil

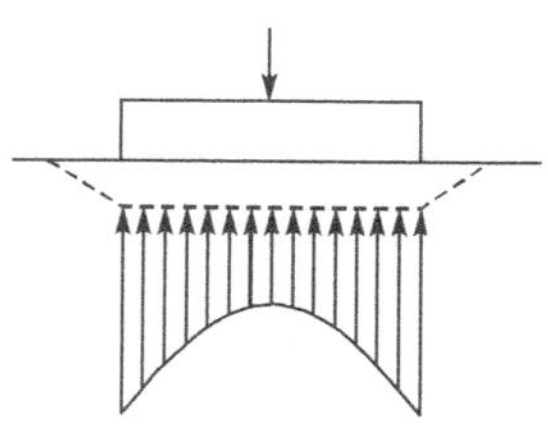

(iii) Flexible footing over Granular soil

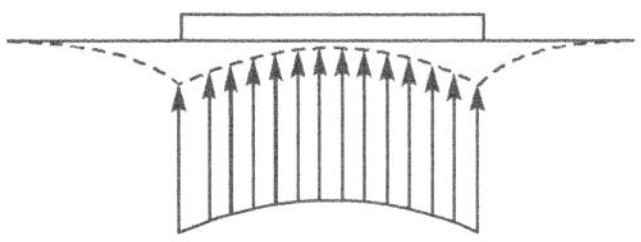

(iv) Rigid footing over granular soil

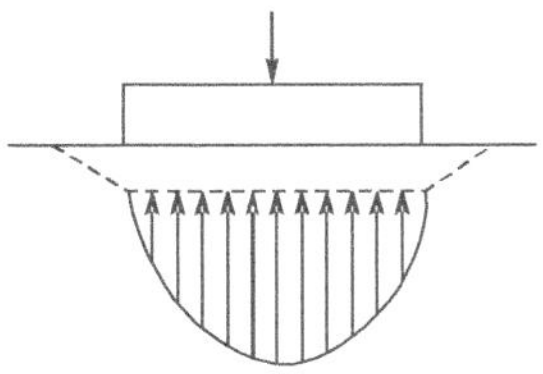

11.3 (a)

The region within the 0.20 q - isobar is called by Terzaghi the "seat of settlement". The depth below centre of footing in active zone (i.e., in "seat of settlement") is 1.5B. For a larger footing more area will be affected so the settlement will be more.

11.4 (a)

For well foundations, the depth of foundation depends upon

(i) maximum scour level

(ii) grip length

For shallow foundations, the minimum depth of footing is

$$D_{f,\min} = \frac{q_f}{\gamma}\left(\frac{1-\sin\phi}{1+\sin\phi}\right)^2$$

$D_{f,\min}$ is called Rankine depth

11.5 (c)

The **Proctor compaction test** is a laboratory method of experimentally determining the optimum moisture content at which a given soil type will become most dense and achieve its maximum dry density.

Standard Penetration Test is a field test which is basically meant of determine the relative density of the sandy grounds; but has been empirically extended to be used for determining the allowable bearing pressure for a given settlement, and also to approximate shear parameters of the ground.

In case of fine grained soil, sieve analysis test does not give reliable test result. This because a fine grained soil consist of different sizes of particles starting from 0.075 mm to 0.0002 mm. and it is not practicable to design sieve having so smaller screen size. Also there is a chance of lost of sample during sieving. Therefore **hydrometer analysis** is done for grain size analysis of fine grained soils.

11.6 (b)

Terzaghi bearing capacity equation for purely cohesive soil

$$q_u = 5.7c + \gamma D_f$$

For footing on ground surface ($D_f = 0$)

$$\therefore \quad q_u = 5.7c$$

Hence, for footings on purely cohesive soils, the bearing capacity does not depend upon the size of the footing. So the bearing capacity for diameter D_2 will be 200 kN/m^2.

11.7 (b)

When the water table is at ground level,the surcharge depth D_f and soil below foundation base will be submerged. Thus minimum bearing capacity will occur for this condition.

11.8 (b)

A field **permeability test** consists of **pumping** out water from a main well and observing the resulting drawdown surface of the original horizontal water table from at least two observation wells. When a steady state of flow is reached, the flow quantity and the levels in the observation wells are noted.

Plate Load Test is a field test for determining the ultimate bearing capacity of soil and the likely settlement under a given load. The Plate Load Test basically consists of loading a steel plate placed at the foundation level and recording the settlements corresponding to each load increment.

11.9 (c)

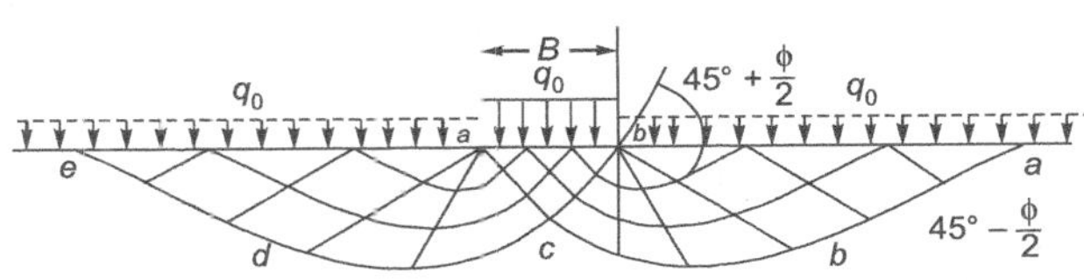

Failure under a strip footing

11.10 (d)

$S_p = 5.0$ mm; $B_p = 0.6$ m;
$B_f = 3$ m

$$S_f = S_p \times \left(\frac{B_f(B_p+0.3)}{B_p(B_f+0.3)}\right)^2$$

$$\therefore \quad S_f = 5 \times \left(\frac{3(0.6+0.3)}{0.6(3+0.3)}\right)^2 = 9.3 \text{ mm}$$

11.11 (d)

Terzaghi's theory neglects the shearing resistance of rupture surface in the soil above the foundation level so this theory underestimates the bearing capacity of footing.

Terzaghi considered a continuous footing of width B placed at a depth of D below the ground surface. In the derivation of the equation the following assumptions were made.

- The soil is homogeneous, isotropic and Coulumb's law of shear strength is valid.
- The footing is continuous and has a rough base.
- Failure zone does not extend above the base of the foundation.
- Shear resistance of the soil above the base of the foundation is neglected.
- The soil above the base of the foundation is replaced by a uniform surcharge.
- Principle of superposition holds good.

11.12 (a)

For foundation on sand, the bearing capacity increases with width of the foundation. So narrow foundation will have lesser bearing capacity and it may fail.

11.13 (c)

A rigid footing settles uniformly. So contact pressure will be maximum at centre and zero at edge in cohesionless soils. The distribution is close to parabolic.

Rigid footing over granular soil

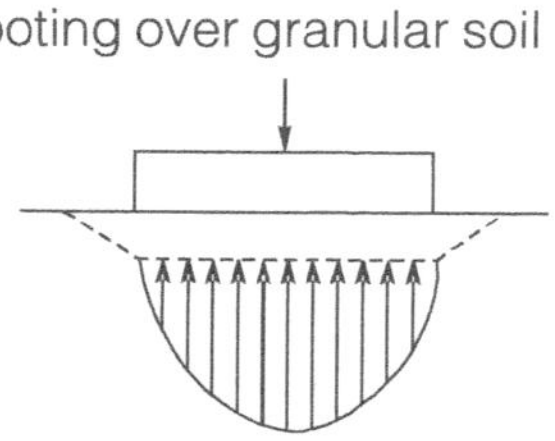

11.14 (d)

Net ultimate bearing capacity,

$$q_{nu} = c\,N_c$$

For strip footing

$$N_c = 5(1 + 0.2\,D_f/b)$$
$$= 5(1 + 0.2 \times 3/1.5) = 7$$
$$q_{nu} = 20 \times 7 = 140 \text{ kN/m}^2$$

Note: For square footing or circular footing,

$$N_c = 6(1 + 0.2\,D_f/b)$$

For rectangular footing,

$$N_c = 5(1 + 0.2\,D_f/b)\,(1 + 0.2\,b/L)$$

11.15 (c)

According to Terzaghi,

For strip footing

$$q_{nu} = 0.5\gamma B\,N_\gamma$$

For square footing

$$q_{nu} = 0.4\,\gamma B\,N_\gamma$$

Using the above equations,

$$\gamma N_\gamma = \frac{60}{0.4 \times 0.6} = 250 \text{ kN/m}^3$$

For strip footing,

$$q_{nu} = 0.5 \times 1.2 \times 250$$
$$= 150 \text{ kN/m}^2$$

11.16 (a)

The size effect of footing has been considered by empirically evolving the equation:

$$\frac{S_f}{S_p} = \left(\frac{B_f(B_p+30)}{B_p(B_f+30)}\right)^2$$

This equation is applicable for dense sands having general shear failure.

11.17 (b)

Sand shows immediate settlements. **Heterogeneous landfill** shows uneven or relative settlements. **Black cotton soil** is highly compressible (shows large volume change) so it can lead to heaving of foundation. **Hard rock** shows practically no movements/settlements.

11.18 (d)

Taking moment about R_2

$$\text{Reaction } R_1 = \frac{3000 \times 7}{(7-1.2)} = 3620.69 \text{ kN}$$

Thus $R_2 = 3000 + 4500 - R_1 = 3879.31$ kN

$$\text{For left footing,} \quad L_1 = \frac{3620.69}{350 \times 3.5} = 2.95 \text{ m}$$

For right footing, $L_2 = \dfrac{3879.31}{350 \times 3.5} = 3.17\text{ m}$

11.19 (a)

A raft or mat foundation is a thick reinforced concrete slab which supports all the load bearing walls and column loads of a structure or a large portion of structure. A mat is required where load is heavy and the soil is very weak or highly compressible.

11.20 (c)

The plate load test reflect the characteristics of soil located only within a depth of about twice the width of the plate.

11.21 (c)

SPT is not considered to be a refined and completely reliable method of investigation. The *N* values give useful information about relative density of cohesionless soils. Thus it does not measure relative density rather correlate *N* values with relative density.

N values increases with depth due to overburden pressure for sand having same relative density.

Therefore overburden pressure correction are applied.

11.22 (b)

According to Terzaghi and Peck, the following correlation between correct *N* value and compactness are useful:

N-value	Compactness
0 – 4	Very loose
4 – 10	Loose
10 – 30	Medium dense
30 – 50	Dense
> 50	Very dense

$N = 27$ corresponds to medium dense sands.

11.23 (c)

At shallow depth or surface

Relative Density	Type of failure
< 35	Punching shear
35 – 70	Local shear
> 70	General shear

Very dense sands have high relative density so general shear failure is most likely.

11.24 (d)

Terzaghi's bearing capacity factors N_c, N_q and N_γ are function of angle of friction only.

$N_c > N_q > N_\gamma$	for $\phi < 30°$
$N_c > N_\gamma > N_q$	for $30° < \phi < 40°$
$N_\gamma > N_c > N_q$	for $\phi < 40°$

11.25 (d)

The position of ground water has a significant effect on the bearing capacity of soil. Only if it is present at a depth less than the width of the foundation from the foundation bottom. At depths greater than this, there is no effect on bearing capacity.

11.26 (a)

The log-log graph of load versus settlement distinctly shows, the failure point. The pressure corresponding to the point is taken as the bearing capacity.

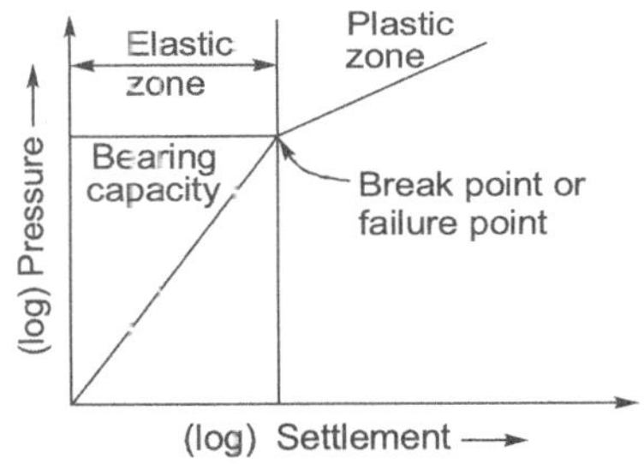

11.27 (c)

According to Terzaghi there are three zones:

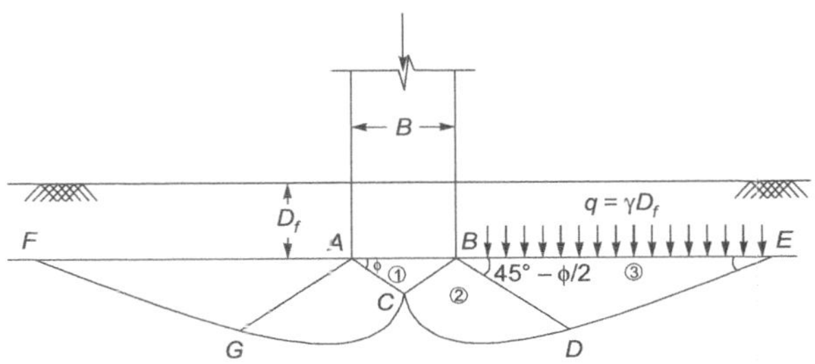

Terzaghi system for ideal soil, rough base and surcharge

Zone-1: Elastic zone of wedge shape located beneath the loaded strip wherein the major principal stress are vertical. This remains in elastic equilibrium and behaves as if it were a part of footing itself and $\angle CAB = \angle CBA = \phi$.

Zone-2: Radial shear zone where arc *CD* is logarithmic spiral. (Circular when $\phi = 0$).

Zone-3: It is called Rankine passive zone. An overburden pressure $q = \gamma D_f$ acts as a surcharge on this zone.

11.28 (a)

Load intensity for 4.5 kN load on 30 cm square plate will be

$$q = \frac{4.5}{(0.3)^2} = 50 \text{ kN/m}^2$$

Load intensity for 200 kN load on 2 m × 2 m footing will be

$$q' = \frac{200}{4} = 50 \text{ kN/m}^2$$

Since load intensity is same so,

$$S_f = S_p \times \left(\frac{B_f(B_p + 0.3)}{B_p(B_f + 0.3)}\right)^2$$

$$= 12 \times \left(\frac{2 \times 0.6}{0.3 \times 2.3}\right)^2 = 36.3 \text{ mm}$$

11.29 (c)

Rigid footing on surface of clay, contact pressure is maximum at edges and minimum at centre.

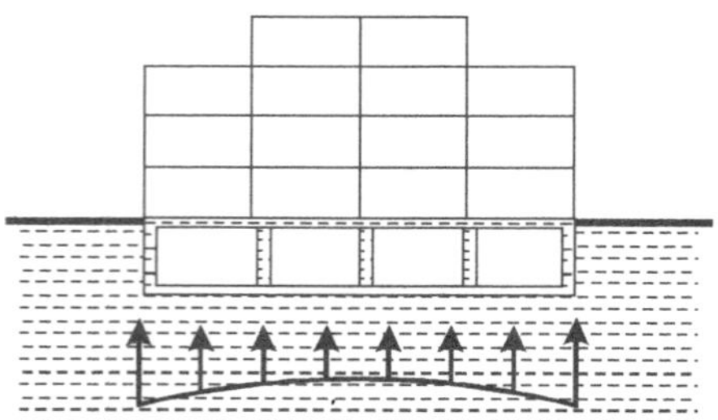

11.30 (a)

The bearing capacity of a footing on saturated clay ($\phi = 0$) is given by

$$q_u = 5.7\, C_u + \gamma D_f$$

11.31 (b)

If footing is satisfying shear criteria and settlement criteria then it does not matter if footing is flexible or rigid, below or above the water table.

11.33 (b)

Sampling in cohensionless soils without disturbing the structure is difficult because there is no cohesion to keep individual grains together and grains may read just among themselves due to slight impacts also while extracting.

11.35 (c)

Static Cone Penetration Test – Also known as Dutch Cone for CPT, it gives a continuous record of variation of both cone resistance and friction resistance with depth. CPT however does not yield any sample. It is suitable for soft clay and silts and fine to medium sand deposits.

11.36 (d)

The bearing capacity of soil is influenced by various factors. The physical features of foundation such as type of foundation, size of foundation, depth of foundation and shape of foundation significantly affect the bearing capacity. The amount of total and differential settlement is one of the main controlling factors for the bearing capacity of the soil. The relative density in the case of granular soil and consistency in the case of cohesive soil play a decisive role in influencing the bearing capacity. The physical as well as engineering properties of soils such as density, cohesion and friction, position of water table and original stresses are the other factors governing the soil bearing capacity.

11.37 (c)

The distinction between silt & clay cannot be based on particle size because the significant physical properties of the two materials are related only indirectly to the size of particles. Furthermore, since both are microscopic, physical properties other than particle size must be used as criteria for field identification. Thus **Dilatancy or Shaking test** is used to distinguish between silts and clays.

11.38 (d)

Field density Test is used to determine in situ density of natural or compacted soils. The in situ density of natural soil is needed for the determination of bearing capacity of soils, for the purpose of stability analysis of slopes, for the determination of pressures on underlying strata for the calculation of settlement and the design of underground structures.

The **california bearing ratio test** is penetration test meant for the evaluation of subgrade strength of roads and pavements. The results obtained by these tests are used with the empirical curves to determine the thickness of pavement and its component layers.

11.39 (a)

- Isolated foundation is provided to support an individual column.
- Combined footing is used when two columns are so close to each other that their individual footings would overlap each other.
- Raft foundation is a large slab supporting a number of columns and walls under the entire structure or a large part of the structure.

11.40 (d)

Expansive soil is a very common cause of foundation problems. Depending upon the supply of moisture in the ground, shrink-swell soils will experience changes in volume of up to thirty percent or more. Foundation soils which are expansive will "heave" and can cause lifting of a building or other structure during periods of high moisture. Conversely during periods of falling soil moisture, expansive soil will "collapse" and can result in building settlement. Either way, damage can be extensive.

In order for expansive soil to cause foundation problems, there must be fluctuations in the amount of moisture contained in the foundation soils. If the moisture content of the foundation soils can be stabilized, foundation problems can often be avoided.

11.41 (a)

Foundation on expansive soils must be designed for as high bearing pressure as possible so that they can counteract the swelling pressure generated by the swelling of the expansive soil. However it must be consistent with the bearing capacity and settlement requirement.

11.42 (a)

Vane shear test is more useful for clays.

11.43 (c)

Standard Penetration Test (SPT) is conducted by pushing a split spoon sampler into the soil through a borehole 55 to 150 mm in diameter at the desired depth.

11.44 (b)

Changing the width of footing does not affect the bearing capacity of clays.

11.45 (b)

Corrected SPT value

$$N = 15 + \frac{1}{2}(N' - 15)$$

$$= 15 + \frac{1}{2}(32 - 15)$$

$$= 15 + 8.5 = 23.5$$

11.46 (d)

The effect of overburden pressure on SPT value may be approximated by the equation

$$N = N' \times \left(\frac{350}{\sigma + 70}\right)$$

where σ is effective overburden pressure in kN/m^2 not exceeding 280 kN/m^2

But $\sigma = \gamma H = 18 \times 6 = 108$ kN/m^2

$$\therefore \quad N = 28 \times \left(\frac{350}{108 + 70}\right) = 55$$

11.47 (d)

As per Terzaghi's bearing capacity equation,

$$q_{ult} = cN_c + \gamma D_f N_q + 0.5\,\gamma B N_\gamma$$

For a purely cohesive soil, $N_q = 1$ and $N_\gamma = 0$

$$\therefore \quad q_{ult} = cN_c + \gamma D_f$$

But $q_{net,ult} = q_{ult} - \gamma D_f$

$$\Rightarrow \quad q_{net,ult} = cN_c + \gamma D_f - \gamma D_f$$

$$\Rightarrow \quad q_{net,ult} = cN_c$$

11.48 (c)

Raft or mat foundation is a combined footing which covers the entire area beneath of a structure and supports all the walls and columns. This type of foundation is most appropriate and suitable when the allowable soil pressure is low, or the loading is heavy.

Also, when the soil contains lenses of compressible strata which are likely to cause considerable differential settlement, a raft foundation is well suited.

11.49 (c)

$$N_c = 15 + \frac{1}{2}(N_R - 15)$$

$$= 15 + \frac{1}{2}(30 - 15) = 22.5 \approx 23$$

11.50 (d)

Ultimate bearing capacity is given by

Circular footing:

$$Q_{uc} = 1.3\,cN_c + \gamma D_f N_q + 0.3\,b\gamma N_\gamma$$

Square footing:

$$Q_{us} = 1.3\,cN_c + \gamma D_f N_q + 0.4\,b\gamma N_\gamma$$

For cohesionless soil :

$$c = 0$$

For footing on surface of soil:

$$D_f = 0$$

$$\therefore \quad \frac{Q_{uc}}{Q_{us}} = \frac{0.3}{0.4} = 0.75$$

11.51 (a)

The N-value in a granular soil is influenced by the over burden pressure. If two granular soil possesing the same relative density but having different confining pressure are tested, one with a higher confining pressure gives a higher N value. Since the confining pressure increases with depth, the N values at shallow depths are underestimated and the N values at larger depths are overestimated.

11.52 (c)

If water table is encountered while conducting plate load test, it has to be lowered by pumping before placing the plate. Throughout the test duration complete dewatering is to be carried out.

11.53 (c)

Expansive soils have high compressibility and low strength. Hence, strength should be improved and compressibility should be reduced.

11.54 (d)

$$q_u = cN_C + qN_q + 0.5\,\gamma BN_\gamma$$

For pure clay,

$$N_C = 5.7,\ N_q = 1 \text{ and } N_\gamma = 0$$

$$\therefore \quad q_u = 5.7\,c + q$$

∴ Bearing capacity of a footing in pure clay is independent of size of the footing.

11.55 (a)

For a square footing

$$q_{nu} = 1.3\,cN_C$$

where $c = \dfrac{50}{2} = 25\text{ kN/m}^2$

$$\therefore \quad q_{nu} = (1.3 \times 25 \times 5.7)\text{ kN/m}^2 = 185.25\text{ kN/m}^2$$

11.56 (c)

$$q_{nu} = c_u N_c$$

where $c_u = \dfrac{50}{2} = 25\text{ kN/m}^2$

$$N_c = 6(1 + 0.2\,D_f/b)$$

for $D_f = 0$, we get $N_c = 6$

$$\therefore \quad q_{nu} = 25 \times 6 = 150\text{ kN/m}^2$$

11.57 (d)

In case of footings on purely cohesive soils

$$q_{nu} = CN_C$$

Therefore the benefit due to surcharge or depth of foundation is only marginal. Moreover, the bearing capacity is independent of the size of footing.

11.58 (c)

Gross bearing capacity

$$q_g = 450\text{ kN/m}^2$$

Net bearing capacity $= q_g - \gamma \times D_f$

$$= 450 - 20 \times 1 \;\; 430\text{ kN/m}^2$$

11.59 (*)

$B = 1000$ mm, $E = 2$ N/mm^2, $I_f = 0.95$, $\mu = 0.5$

$$q = \frac{600 \times 10^3}{(1000)^2} = 0.6\text{ N/mm}^2$$

Immediate settlement

$$S_i = \frac{qB}{E}(1-\mu^2)I_f$$

$$= \frac{0.6 \times 1000}{2}(1-0.5^2) \times 0.95$$

$$= 213.75\text{ mm}$$

11.61 (a)

The bearing capacity of cohesive soil is given by

$$q_u = cN_c + \gamma D_f$$

So, the bearing capacity of cohesive soil will be affected by depth of footing, density of soil, angle of shearing resistance of the soil and not affected by width of footing.

11.62 (b)

Following equation is used from Plate load test data for sands.

$$q_f = q_p \frac{B_f}{B}$$

Settlement of footing (S_f) on sand increases with increase in its width (B_f), can be explained as

$$\frac{S_{f_1}}{S_p} = \left[\frac{B_{f_1}(B_p + 0.3)}{B_p(B_{f_1} + 0.3)}\right]^2 \text{ (for 1st footing)}$$

$$\frac{S_{f_2}}{S_p} = \left[\frac{B_{f_1}(B_p + 0.3)}{B_p(B_{f_2} + 0.3)}\right]^2 \text{ (for 2nd footing)}$$

$$\therefore \quad \frac{S_{f_1}}{S_{f_2}} = \left[\frac{B_{f_1}(B_{f_2} + 0.3)}{B_{f_2}(B_{f_1} + 0.3)}\right]^2 = \left(\frac{1 + \frac{0.3}{B_{f_2}}}{1 + \frac{0.3}{B_{f_1}}}\right)^2$$

If $B_{f_2} > B_{f_1}$; $\frac{S_{f_1}}{S_{f_2}} < 1$

$\therefore \quad S_{f_2} > S_{f_1}$

Hence settlement increases.

11.63 (b)

In cohesive soil, the ultimate bearing capacity q_u is defined as

$$q_u = cN_c + \gamma D_f N_q$$

So, bearing capacity of cohesive soil is dependent on angle of shearing friction, density of soil and depth of footing only.

11.64 (d)

By an empirical equation given by Housel ,

$$Q = Aq + pS$$

Q = Total load on bearing area

q = Bearing pressure beneath the footing

S = Perimeter shear

A = Area of plate (2 × 2 m^2)

p = Perimeter of plate (2 × 4 = 8 m)

$$Q = (2)^2\,(100) + (8)\,(25)$$
$$= 400 + 200 = 600 \text{ kN}$$

11.65 (a)

As per Terzaghi, footing is shallow if $\frac{D_f}{B} \le 1$.

11.66 (c)

Rigid footing over granular soil

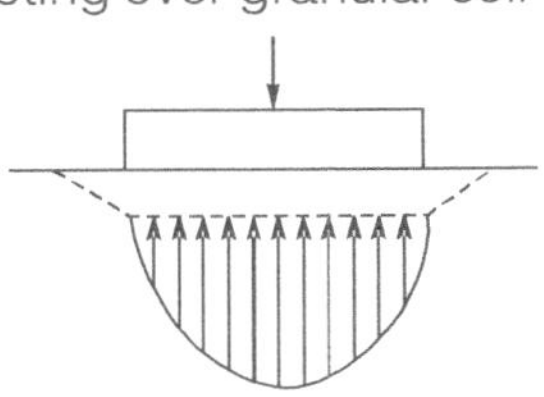

11.67 (c)

Two types of failure in foundation:

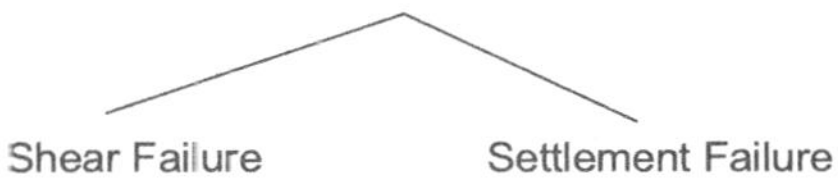

Foundation should be safe in both.

(a) Safe load against shear failure
= Safe bearing pressure × Footing area

(b) Safe load against settlement failure
= Safe bearing pressure × Footing area

11.68 (c)

By assuming depth factor = 1

Shape factor = 1

Inclination factor = 1

$$q_{nu} = CN_C\,(s_c d_c i_c) + q(N_q - 1)$$
$$s_q d_q i_q + 0.5\,B\gamma N_r (s_r d_r i_r)\,W$$

$N_c = 5.14$, $N_q = 1$, $N_\gamma = 0$

$$q_{nu} = CN_C = 5.14 \times 40$$

$$q_u = q_{nu} + \bar{\sigma} = 5.14 \times 40 + (1 \times \gamma + 0.5\gamma')$$
$$= 5.14 \times 40 + [1 \times 20 + 0.5 \times (20 - 10)] = 230.6$$

11.69 (d)

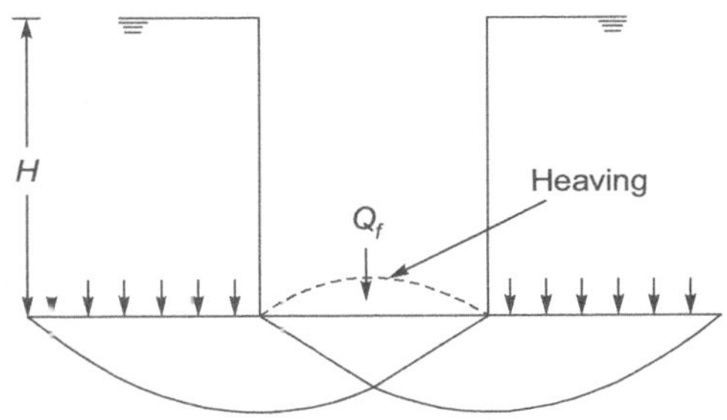

Given: q = 20 kN/m^2, $\tau_u = C_u$ = 12 kN/m^2

γ = 16 kN/m^3, B = 10 m, L = 12 m

Bearing capacity of soil for rectangular raft footing in cohesive soil is given by

$$Q_f = \left(1 + 0.3\frac{B}{L}\right) CN_C + \gamma D_f + q$$

$$= \left(1 + 0.3 \times \frac{10}{12}\right) 12 \times 5.7 + 16D_f + 20$$

$$= 105.5 + 16D$$

Base failure will occur when $Q_f = 0$

Hence,

$$D = -\frac{10.5}{6} = -6.59 \text{ m}$$

(–ve sign indicates that it is excavation)

$$\text{Safe depth} = \frac{D}{\text{FOS}} = \frac{6.59}{1.2} = 5.49 \text{ m}$$

11.70 (c)

$B_f = 1.5$ m. $B_P = 0.3$ m, $S_P = 8$ mm

For cohesionless soil,

$$\frac{S_f}{S_p} = \left\{\frac{B_f(B_P+0.3)}{B_P(B_f+0.3)}\right\}^2$$

$$S_f = 8\left\{\frac{1.5(0.3+0.3)}{0.3(1.5+0.3)}\right\}^2$$

$$= 22.22 \text{ mm}$$

11.72 (d)

Local shear failure

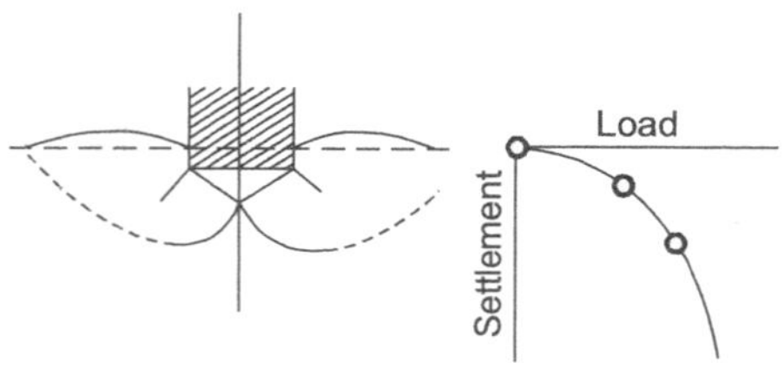

- Local shear failure is an intermediate failure mode characterized by well defined slip surface immediately below the footing but extending only a short distance into the soil mass, as shown in figure.
- Since, the stress zone does not extend upto ground level. Therefore, only little bulging of soil around the footing.
- Main characteristics of local shear failure are :
  (a) Occurs in loose sands and soft clays in shallow foundations.
  (b) Well defined wedge and slip surface only below the footing.
  (c) Slip surface not visible beyond the edges of the foundation.
  (d) Soil below footing is more stressed as compared to adjacent soil.

11.73 (a)

According to Terzaghi,

For shallow foundation, $\frac{D_f}{B} \le 1$

For deep foundation, $\frac{D_f}{B} > 1$

11.74 (b)

Ultimate bearing capacity is the minimum gross pressure intensity at the base of foundation at which the soil fails in shear.

OR

Ultimate bearing capacity is the maximum gross pressure the soil can support without shear failure.

11.75 (a)

The angle ϕ obtained by using N values can be used to find out the values of bearing capacity factor, N_c, N_q and N_r.

11.76 (c)

$$S_i = \frac{qB(1-\mu^2)I_f}{E_s} = \frac{180 \times 18 \times (1-0.5^2) \times (1.2)}{45 \times 10^3}$$

$$I_f = 1.2, \mu = 0.5$$

$$= 0.0648 \text{ mm} = 64.8 \text{ mm}$$

However if $I_f = 1$ then $S_i = 54$ mm

Closest answer is (d).

11.77 (c)

Maximum pressure which a foundation can withstand without the occurrence of shear failure of the foundation is the definition of ultimate bearing capacity.

11.80 (b)

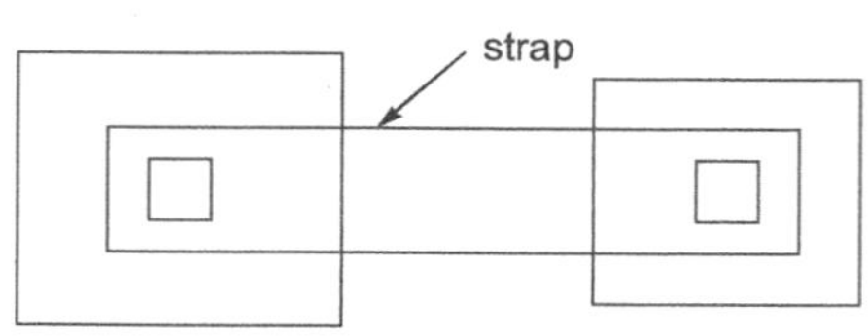

Spread footing

■■■■

12 Deep Foundation, Sheet Pile Walls and Machine Foundation

12.1 **Assertion (A):** Pile driving formulae have very limited use in the case of cohesive soils.
Reason (R): Pile driving in cohesive soils results in liquefaction of clays.
(a) both A and R are true and R is the correct explanation of A
(b) both A and R are true but R is not a correct explanation of A
(c) A is true but R is false
(d) A is false but R is true
[ESE : 2000]

12.2 Match **List-I** (Suitable condition) with **List-II** (Foundations) and select the correct answer using the codes given below the lists:
List-I
A. When structural load is uniform and soil is soft clay, made up of marshy land.
B. When structural load is heavy and/or soil having low bearing capacity for considerable depth.
C. When soil is having good bearing capacity at shallow depth and structural load is within permissible limit.
D. When structural load of bridge is to be transferred through sandy soil to bed rock.
List-II
1. Footings
2. Piles
3. Rafts
4. Wells or piers

Codes:

	A	B	C	D
(a)	3	1	2	4
(b)	4	1	2	3
(c)	4	2	1	3
(d)	3	2	1	4

[ESE : 2000]

12.3 The nature of earth pressure above dredge line behind a cantilever sheet pile wall is
(a) active
(b) passive
(c) at rest
(d) active and passive
[ESE : 2000]

12.4 A machine foundation weighs 39.24 kN and has a spring constant $k = 10000$ kN/m. The system will be critically damped if the damping coefficient C_c is (acceleration due to gravity, $g = 9.81$ m^2/s)
(a) 100 kN-s/m (b) 200 kN-s/m
(c) 400 kN-s/m (d) 800 kN-s/m
[ESE : 2000]

12.5 In the Engineering News Record formula for determining the safe load carrying capacity of a pile, the factor of safety used is
(a) 2.5 (b) 3
(c) 4 (d) 6 [ESE : 2001]

12.6 In the case of a pile foundation, negative skin friction may occur at a load which is
(a) lower than the designed load
(b) higher than the designed load
(c) equal to the designed load
(d) of any magnitude [ESE : 2001]

12.7 In under-reamed pile construction, the ratio of shaft diameter to bulb diameter is
(a) 1/1.5 (b) 1/2
(c) 1/2.5 (d) 1/4 [ESE : 2001]

12.8 A single pile, 50 cm in diameter and 15 m long is driven in clay having an average un-confined compressive strength of 100 kN/m^2. The ultimate bearing capacity of the pile, neglecting end bearing, if any, and assuming shear mobilization factor of 0.8 around the pile is

(a) 942 kN (b) 1884 kN
(c) 1177.5 kN (d) 1334.5 kN

[ESE : 2002]

12.9 Match **List-I** (Method of estimating pile capacity) with **List-II** (Parameter to be estimated) and select the correct answer using the codes given below the lists:

List-I
A. Dynamic formulae
B. Static formulae
C. Pile load test
D. Cyclic pile load test

List-II
1. Bearing capacity of cast-in-situ concrete pile
2. Separating end-bearing and friction-bearing powers of a pile
3. Bearing capacity of a timber pile
4. Settlement of a friction bearing pile

Codes:

	A	B	C	D
(a)	3	1	4	2
(b)	4	2	3	1
(c)	3	2	4	1
(d)	4	1	3	2

[ESE : 2002]

12.10 For designing end bearing piles of square cross-section in clays having average unconfined compressive strength of 6 t/m^2, the net ultimate bearing capacity may be taken as

(a) 15 t/m^2 (b) 18 t/m^2
(c) 20 t/m^2 (d) 27 t/m^2

[ESE : 2002]

12.11 **Assertion (A):** In a machine foundation subjected to vibrations, frequency of vibrations with damping is less than the natural frequency of vibrations of the system.

Reason (R): The frequency of vibrations with damping ω_d and natural frequency of vibrations ω_n are related by expression $\omega_d = \omega_n(1 - d^2)$ where d is the damping ratio.

(a) both A and R are true and R is the correct explanation of A
(b) both A and R are true but R is not a correct explanation of A
(c) A is true but R is false
(d) A is false but R is true

[ESE : 2003]

12.12 Skin frictional capacities of a 40 cm diameter driven concrete pile for the portions *A*, *B* and *C* are 17 kN, 63 kN and 503 kN respectively, and point load capacity is 11000 kN/m^2. Total pile load capacity will be

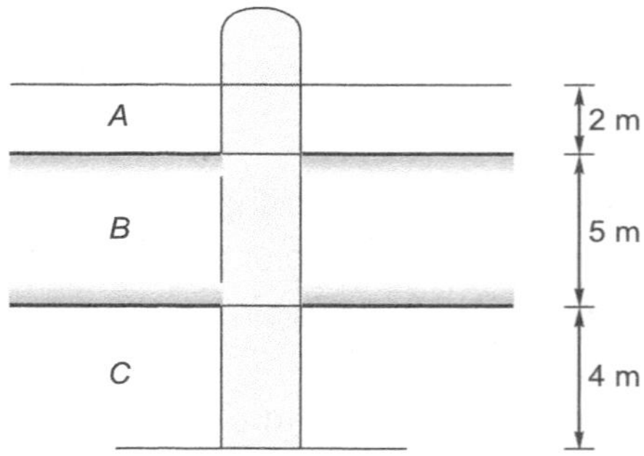

(a) 3743 kN (b) 2864 kN
(c) 1965 kN (d) 1529 kN

[ESE : 2003]

12.13 Sinking effort in well foundation is the ratio of weight of well steining to that of skin friction developed on the sides and should preferably be

(a) < 1.0 (b) = 1.0
(c) > 1.0 (d) > 2.0

[ESE : 2003]

12.14 Match **List-I** (Type of Foundation) with **List-II** (Type of Soil) and select the correct answer using the codes given below the lists:

List-I
A. Floating piles
B. Micro Piles
C. Combined footing
D. Under-reamed piles

List-II
1. Closed spaced columns resting on compressible soil
2. Expansive soils
3. Deep soft clays
4. Loose sands

Codes:

	A	B	C	D
(a)	2	1	4	3
(b)	2	4	1	3
(c)	3	1	4	2
(d)	3	4	1	2

[ESE : 2004]

12.15 A square pile of section 30 cm × 30 cm and length 10 m penetrates a deposit of clay having c = 5 kN/m^2 and the mobilizing factor m = 0.8. What is the load carried by the pile by skin friction only?

(a) 192 kN (b) 75 kN
(c) 60 kN (d) 48 kN [ESE : 2004]

12.16 Consider the following statements:

1. Pile foundations are usually provided when loads coming on the foundation are quite large. Such piles may often extend up to a large depth below ground level.
2. Precast piles inserted into the holes bored at the site do not get damaged while they are driven into the ground.

Which of these statements is/are correct?

(a) 1 only (b) 2 only
(c) Both 1 and 2 (d) Neither 1 nor 2

[ESE : 2004]

12.17 Match **List-I** (Type of foundation) with **List-II** (Use of the foundation) and select the correct answer using the codes given below the lists:

List-I

A. Point bearing piles
B. Sheet piles
C. Compaction piles
D. Batter piles

List-II

1. To retain soil filling
2. To transfer heavy loads to strong stratum below a weak stratum
3. To resist lateral loads
4. To densify loose soils

Codes:

	A	B	C	D
(a)	3	1	4	2
(b)	3	4	1	2
(c)	2	1	4	3
(d)	2	4	1	3

[ESE : 2004]

12.18 The installation of sand drains in clayey soil causes the soil adjacent to the sand drains to undergo which one of the following?

(a) Increase in porosity
(b) Increase in compressibility
(c) Decrease in horizontal permeability
(d) Decrease in shear strength [ESE : 2004]

12.19 The load carrying capacity of an individual friction pile is 200 kN. What is the total load carrying capacity of a group of 9 such piles with group efficiency factor of 0.8?

(a) 1800 kN (b) 1640 kN
(c) 1440 kN (d) 900 kN

[ESE : 2005]

12.20 Match **List-I** (Type of soil) with **List-II** (Suitable foundation) and select the correct answer using the codes given below the lists:

List-I

A. Strong soil in surface layer
B. Weak surface layer followed by rock at shallow depth below ground
C. Swelling soil in surface layer extending up to a few meters below ground level
D. Weak heterogenous surface soil layer

List-II

1. Raft foundation
2. Isolated footing
3. End bearing pile
4. Under-reamed piles

Codes:

	A	B	C	D
(a)	1	4	3	2
(b)	2	3	4	1
(c)	1	3	4	2
(d)	2	4	3	1

[ESE : 2005]

12.21 Match **List-I** (Type of pile) with **List-II** (Situation for use) and select the correct answer using the codes given below the lists:

List-I

A. End bearing pile
B. Pedestal pile
C. Friction pile
D. Sand piles

List-II

1. When weak foundation soil is to be compacted
2. When foundation soil is granular
3. When foundation soil is relatively weak
4. When hard formation or rock is at a shallow depth

Codes:

	A	B	C	D
(a)	4	2	3	1
(b)	1	3	2	4
(c)	4	3	2	1
(d)	1	2	3	4

[ESE : 2005]

12.22 While driving a large number of piles in loose sand

(a) it is advantageous to follow a sequence of pile driving such that the inner piles are driven first and then proceed outwards

(b) it is advantageous to follow a sequence of pile driving such that the piles near the periphery are driven first and inner piles are driven later

(c) it is advantageous to follow a sequence of pile driving such that alternately inner and outer piles are driven

(d) driving of piles can be done in any random order

[ESE : 2005]

12.23 Match **List-I** (Type of foundations) with **List-II** (Suitability) and select the correct answer using the codes given below the lists:

List-I

A. Spread footings
B. Under reamed piles
C. Raft foundation
D. Deep foundation

List-II

1. Soft clay for 20 m followed by hard rock stratum
2. Up to 3 m black cotton soil followed by medium dense sand
3. Compact sand deposit extending to great depth
4. Loose sand extending to great depth

Codes:

	A	B	C	D
(a)	4	1	3	2
(b)	3	2	4	1
(c)	4	2	3	1
(d)	3	1	4	2

[ESE : 2006]

12.24 Consider the following statements:

1. The proportioning of a footing is more often governed by its bearing capacity.
2. Friction piles are also called 'floating piles'.

Which of these statements is/are correct?

(a) Only 1 (b) Only 2
(c) Both 1 and 2 (d) Neither 1 nor 2

[ESE : 2006]

12.25 Consider the following statements:

1. Initial load tests and routine tests are carried out on test piles and working piles, respectively.
2. A cyclic load test is performed to determine a pile's skin resistance and base resistance separately.
3. In a pile load test, the safe load is taken as half the final load at which the settlement equals 10% of pile-diameter.

Which of these statements are correct?

(a) 1, 2 and 3 (b) Only 1 and 2
(c) Only 2 and 3 (d) Only 1 and 3

[ESE : 2006]

12.26 A machine and its foundation weigh 981 kN and has a spring constant k = 10,000 kN/m. What is the value of the damping coefficient C_c if the system is to be critically damped?
(Acceleration due to gravity g = 9.81 m/s^2)

(a) 1000 kN-s/m (b) 4000 kN-s/m
(c) 2000 kN-s/m (d) 8000 kN-s/m

[ESE : 2007]

12.27 Which of the following types of piles is likely to have the highest load capacity in compression?

(a) Driven pre-cast concrete pile
(b) Pre-cast pile placed in a pre-drilled bore
(c) Driven steel pipe pile
(d) Steel pipe pile placed in a pre-drilled bore

[ESE : 2007]

12.28 What is the value of negative skin friction for a group of piles of 30 cm diameter, 5 m long and spaced at 80 cm c/c and having cohesive strength of soil as 25 kN/m^2? (Neglect bottom contribution in bearing capacity)

(a) 925 kN (b) 950 kN
(c) 975 kN (d) 1000 kN

[ESE : 2007]

12.29 A static cone penetration test is usually conducted when the structure is likely to be founded on which of the following?

(a) Shallow foundations
(b) Pile foundations
(c) Drier foundation
(d) Improved ground [ESE : 2007]

12.30 Match **List-I** (Foundation) with **List-II** (Demanding situation) and select the correct answer using the codes given below the lists:

List-I
A. Under-reamed piles
B. Floating rafts
C. Combined footing
D. Strap footing

List-II
1. To restrict settlement of soft clays/silts
2. To transfer the moment in between two adjacent footings
3. To restrict damages due to volume changes of swelling soils
4. When columns are very close to the property line

Codes:

	A	B	C	D
(a)	3	2	4	1
(b)	4	1	3	2
(c)	4	2	3	1
(d)	3	1	4	2

[ESE : 2008]

12.31 Match **List-I** (Method) with **List-II** (Suitability) and select the correct answer using the codes given below the lists:

List-I
A. Static formulae
B. Dynamic formulae
C. Static pile load test
D. Cyclic pile load test

List-II
1. To estimate allowable load for permis-sible settlement for a pile
2. To separate point bearing and friction bearing capacities of a pile
3. To estimate allowable load for driven piles in granular soils
4. To estimate allowable load for bored piles in clays

Codes:

	A	B	C	D
(a)	4	2	1	3
(b)	1	3	4	2
(c)	4	3	1	2
(d)	1	2	4	3

[ESE : 2008]

12.32 Consider the following statements about the under-reamed pile in swelling soils:

1. Its bulb provides anchor against movement due to volume changes of soil.
2. It is a driven pile.
3. Its bulb diameter is 2.5 times its shaft diameter.

Which of these statements are correct?

(a) 1, 2 and 3 (b) 1 and 2 only
(c) 2 and 3 only (d) 1 and 3 only

[ESE : 2008]

12.33 Consider the following statements:

1. Piles are provided in groups, which are connected together by a pile cap. The structure rests on top of the pile cap. The pile cap is situated below the ground level.
2. For situations where pile is subjected to upward pull, pedestal piles or under-reamed piles are more suitable type of foundations.

Which of these statements is/are correct?

(a) 1 only (b) 2 only
(c) Both 1 and 2 (d) Neither 1 nor 2

[ESE : 2008]

Directions: The following items consists of two statements; one labelled as **'Assertion (A)'** and the other as **'Reason (R)'**. You are to examine these two statements carefully and select the answers to these items using the codes given below:

Codes:
(a) both A and R are true and R is the correct explanation of A
(b) both A and R are true but R is not a correct explanation of A
(c) A is true but R is false
(d) A is false but R is true

12.34 **Assertion (A):** Generally driven piles are adopted in granular soils and not in clays.
Reason (R): Vibratory loading helps in densification of sands but it has adverse effects in clays.

[ESE : 2008]

12.35 **Assertion (A):** Under-reamed piles are suitable for loose filled up sites and black-cotton soils.

Reason (R): Black cotton soils have expansive montmorillonite clay minerals.

[ESE : 2008]

12.36 Consider the following statements with respect to an anchored sheet pile wall:

1. Failure occurs due to bulging.
2. Failure occurs due to rotation.
3. Large yield is observed above the anchor rod.
4. Large yield is observed below the anchor rod and above the dredge line.

Which of these statements is/are correct?

(a) 1 only (b) 2 and 3
(c) 1 and 4 (d) 3 and 4

[ESE : 2009]

12.37 Consider the following statements:

1. Underreamed piles are precast piles with one or more underreams in each pile.
2. The ratio of pile shaft size to bulb size in an underreamed pile may be 0.33 to 0.50.
3. In a multibulb underreamed pile, the load-carrying capacity is a function of the area of cross-section of the lowest bulb.

Which of these statements is/are correct?

(a) 1 only (b) 1 and 2 only
(c) 2 and 3 only (d) 1, 2 and 3

[ESE : 2009]

12.38 Consider the following statements:

1. Underreamed piles are designed as bearing piles.
2. In multiple-bulb underreamed piles, the bulbs are spaced at 1.5 to 2.0 times the diameter of the underream, the centre of the first underream being at a minimum depth of 1.75 m.
3. The length of traditional undereamed pile ranges from 3 m to 4 m.

Which of these statements is/are correct?

(a) 2 and 3 (b) 1 and 2
(c) 2 only (d) 1 and 3 [ESE : 2009]

12.39 **Assertion (A):** Batter piles are provided to resist lateral load coming onto structures.

Reason (R): The batter of batter piles is helpful in converting a part of the applied lateral load into axial compressive load on the batter piles.

(a) both A and R are true and R is the correct explanation of A
(b) both A and R are true but R is not a correct explanation of A
(c) A is true but R is false
(d) A is false but R is true [ESE : 2011]

12.40 Match **List-I** with **List-II** and select the correct answer using the code given below the lists:

List-I

A. Large diameter piles
B. Drilled pier
C. Open caisson
D. Box caisson

List-II

1. Heavy loads in water structures, but foundation strata at shallow depth
2. Heavy loads in water structures with horizontal loads
3. Heavy but isolated loads
4. Very heavy loads

Code:

	A	B	C	D
(a)	3	2	4	1
(b)	1	2	4	3
(c)	3	4	2	1
(d)	1	4	2	3

[ESE : 2011]

12.41 Consider the following statements:

1. Friction piles are also called floating piles
2. Minimum number of piles to qualify as a pile group is three
3. The group efficiency of a pile group may be either less than 100% or more than 100%.

Which of these statements are correct?

(a) 1, 2 and 3 (b) 1 and 2 only
(c) 2 and 3 only (d) 1 and 3 only

[ESE : 2012]

12.42 **Statement-I** : Different types of piles are used in construction work depending on the type of load to be carried, the sub-soil conditions and the ground water table.

Statement-II: The load transfer mechanism from a pile to the soil is selfsame in all cases.

(a) Both Statement (I) and Statement (II) are individually true and Statement (II) is the correct explanation of Statement (I)
(b) Both Statement (I) and Statement (II) are individually true but Statement (II) is NOT the correct explanation of Statement (I)
(c) Statement (I) is true but Statement (II) is false
(d) Statement (I) is false but Statement (II) is true [ESE : 2012]

12.43 A single pile, 50 cm in diameter and 16 m long, is driven in clay having an average unconfined compressive strength of 100 kN/m^2. The ultimate load carrying capacity of the pile, neglecting end bearing if any, and assuming shear mobilization of 0.75 around the pile, is nearly:
(a) 840 kN (b) 900 kN
(c) 940 kN (d) 1000 kN **[ESE : 2013]**

12.44 A drop hammer is used to drive a wooden pile. The hammer weight is 25 kN and its free falling height is 0.8 m. The penetration in the last blow is 12 mm. What is the nearest approximation to the load carrying capacity of the pile according to the Engineering News Formula?
(a) 125 kN (b) 110 kN
(c) 3000 kN (d) 90 kN **[ESE : 2014]**

12.45 Consider the following statements:
1. Initial load test and routine load tests are carried out on test piles and working piles, respectively.
2. In a pile load test, the allowable load can be taken equal to half the ultimate load at which the total settlement amounts to one-tenth of the pile diameter.
3. In a pile load test, the allowable load can be taken equal to two-thirds of the total load which causes a plastic settlement of 6 mm.

Which of the above statements are correct ?
(a) 1, 2 and 3 (b) 1 and 2 only
(c) 1 and 3 only (d) 2 and 3 only
[ESE : 2014]

12.46 Consider the following statements:
Problems regarding foundations to be constructed in expansive soils are solved by
1. Resorting to light foundation pressure.
2. Using under-reamed piles for light loads.
3. Making the structure rigid enough so that settlement and uplift would not affect them.
4. Providing a well-designed basement with the foundation below the neutral point.

Which of the above statements are correct?
(a) 1 and 2 (b) 2 and 4
(c) 2 and 4 (d) 2 and 3 **[ESE : 2015]**

12.47 Consider the following statements:
1. Dynamic resistance of a soil is not much different from its static resistance.
2. The most comprehensive pile driving formula is Hiley formula.
3. Pile driving formulae are more useful if the subsoil consists of coarse grained soils.

Which of the above statements are correct?
(a) 1 and 2 only (b) 1 and 3 only
(c) 2 and 3 only (d) 1, 2 and 3
[ESE : 2015]

12.48 The Engineering News Record Formula, $Q_a = \dfrac{Wh}{6(s+0.25)}$, is used for the case of
1. Drop hammer
2. Single-acting hammer
3. Double-acting hammer

Which of the above is/are correct?
(a) 1 only (b) 2 only
(c) 3 only (d) 1, 2 and 3
[ESE : 2016]

12.49 **Statement (I):** Negative skin friction will act on the piles of a group in filled-up reclaimed soils or peat soil.
Statement (II): The filled-up or peat soils are not fully consolidated but start consolidating under their own overburden pressure, developing a drag on the surface of the piles.
(a) Both Statement (I) and Statement (II) are individually true and Statement (II) is the correct explanation of Statement (I)
(b) Both Statement (I) and Statement (II) are individually true but Statement (II) is ***not*** the correct explanation of Statement (I)
(c) Statement (I) is true but Statement (II) is false
(d) Statement (I) is false but Statement (II) is true **[ESE : 2016]**

12.50 Consider the following statements?

1. A braced cofferdam is used in shallow trench excavation as well as in deep excavation exceeding 6 m in depth.
2. Cofferdams, braced or un-braced, are temporary structures either on land or in water bodies.
3. When sheet piling is used for retaining soil, or soil and water, without any bracing, it is called a bulkhead.

Which of the above statements are correct?

(a) 1, 2 and 3 (b) 1 and 2 only
(c) 1 and 3 only (d) 2 and 3 only

[ESE : 2017]

12.51 The skin frictional resistance of a pile driven in sand does not depend on

(a) Lateral earth pressure coefficient
(b) Angle of friction between pile and soil
(c) Pile material
(d) Total stress analysis [ESE : 2019]

12.52 The maximum test load on a working pile should not exceed

(a) 250 kN
(b) 180 kN
(c) two and a half times the design load
(d) one and a half times the design load

[ESE : 2022]

Answers Deep Foundation, Sheet Pile Walls and Machine Foundation

12.1 (c)	12.2 (d)	12.3 (a)	12.4 (c)	12.5 (d)	12.6 (d)	12.7 (c)	12.8 (a)
12.9 (a)	12.10 (d)	12.11 (c)	12.12 (c)	12.13 (d)	12.14 (d)	12.15 (d)	12.16 (a)
12.17 (c)	12.18 (c)	12.19 (c)	12.20 (b)	12.21 (c)	12.22 (a)	12.23 (b)	12.24 (b)
12.25 (a)	12.26 (c)	12.27 (a)	12.28 (*)	12.29 (b)	12.30 (d)	12.31 (c)	12.32 (d)
12.33 (b)	12.34 (a)	12.35 (a)	12.36 (c)	12.37 (c)	12.38 (a)	12.39 (a)	12.40 (c)
12.41 (a)	12.42 (c)	12.43 (c)	12.44 (d)	12.45 (b)	12.46 (b)	12.47 (c)	12.48 (b)
12.49 (a)	12.50 (d)	12.51 (d)	12.52 (d)				

Explanations Deep Foundation, Sheet Pile Walls and Machine Foundation

12.1 (c)

In general, the dynamic pile driving formulae appear to be more appropriate to piles driven in cohesionless soils. If the pile is driven into saturated loose silt and sand, liquefaction might reduce the pile capacity. Dynamic pile driving formulae are not considered to be applicable to piles driven into cohesive soils. This is because there may be apparent increase in driving resistance due to development of pore pressures, while there may be a tendency for it to decrease later due to dissipation of pore pressure. Pile driving in cohesive soil disturbs the soil structure and consequently the resistance tends to decrease due to remoulding in sensitive clays although there might be some regaining of strength with passage of time, due to thixotropy.

12.3 (a)

The pile tends to point of rotation, deflecting away from the backfill. Above the point of rotation, the sheet pile wall deflects away from the backfill, thus generating active conditions on the back of the wall. At the same time, between the dredge line and the point of rotation, as the wall tends to move towards the soil in front of the wall, passive conditions are generated in this side.

12.4 (c)

For critical damping $C_c = 2\sqrt{kM}$

Mass (M) should be in kg.

$$\therefore\ C_c = 2\times\sqrt{\frac{10{,}000\times 39.24}{9.81}}$$

$$= 400 \text{ kN-s/m}$$

12.5 (d)

The Engineering News formula is largely an empirically developed formula. It has a factor of safety of six. The intent of the factor of safety was to produce a load that could be used as an applied load of the structure the pile supported.

12.6 (d)

Negative skin friction is related to the relative rate of settlement of pile and soil surrounding the pile. It has no relation with load.

12.7 (c)

Under reamed piles are bored cast-in-situ concrete piles having one or more number of bulbs formed by enlarging the pile stem. These piles are best suited in soils where considerable ground movements occur due to seasonal variations, filled up grounds or in soft soil strata.

The ratio of bulb diameter to the pile shaft diameter is generally between 2 to 3; usually a value of 2.5 is used.

12.8 (a)

The skin friction resistance,

$$f_s = \alpha c_u$$

$$c_u = 100/2 = 50 \text{ kN/m}^2$$

Ultimate bearing capacity $= \alpha\, c_u\, PL$

$$= 0.8 \times 50 \times (\pi \times 0.5) \times 15$$

$$= 942.5 \text{ kN}$$

12.10 (d)

End bearing resistance,

$$q_b = 9c_u$$

$$c_u = \frac{6}{3} = 3 \text{ t/m}^2$$

$$\therefore \quad q_b = 9 \times 3 = 27 \text{ t/m}^2$$

12.11 (c)

The correct relationship is

$$\omega_{dn} = \omega_n\sqrt{1-D^2}$$

12.12 (c)

Total pile capacity

$$= 17 + 63 + 503 + 11{,}000 \times \left(\frac{\pi \times 0.4^2}{4}\right)$$

$$= 1965.3 \text{ kN}$$

12.14 (d)

The pile resisting load by skin friction only is called floating pile.

The swelling pressure in expansive soils like black cotton soils can be resisted by under-reamed pile. Micro pile resists the vertical load largely by end bearing resistance. Therefore they will be suitable in loose sands.

12.15 (d)

The load carried by skin-friction only

$$= m \times c \times P \times L$$

$$= 0.8 \times 5 \times 0.3 \times 4 \times 10 = 48 \text{ kN}$$

12.16 (a)

The precast concrete piles are generally used for a maximum in load of 80 tonnes approximately. They must be reinforced to withstand handling stresses. They are driven into the ground without any holes bored at the site.

12.17 (c)

Point Bearing Piles - If a bedrock or rocklike material is present at a site within a reasonable depth, piles can be extended to the rock surface. In this case, the ultimate bearing capacity of the pile depends entirely on the underlying material; thus the piles are called end or point bearing piles. In most of these cases the necessary length of the pile can be fairly well established.

Batter pile shall be used to transfer inclined load and horizontal forces. Free standing batter piles are subjected to bending moment due to their own weight, or external forces from other sources. Batter pile in loose fill or consolidating deposits may become laterally loaded due to settlement of the surrounding soil. In consolidating clay, special precautions, like provision of permanent casing, shall be taken.

Sheet piling is an earth retention and excavation support technique that retains soil, using steel sheet sections with interlocking edges. **Sheet piles** are installed in sequence to design depth along the planned excavation perimeter or seawall alignment. The interlocked sheet piles form a wall for permanent or temporary lateral earth support with reduced groundwater inflow. Anchors can be included to provide additional lateral support if required.

12.18 (c)

The installation of sand drains causes peripheral smear (i.e., wiping action caused by casing or hollow mandrel used to form the well as it is driven down into the soil and then pulled out after it has been filled with sand). For a soil originally having a greater permeability in the horizontal than that in the vertical direction, the smeared zone forms a barrier to the horizontal flow of water, thereby slowing down the process of consolidation considerably.

12.19 (c)

Load carrying capacity of pile group

= Group efficiency factor × Number of piles × Individual pile capacity

= 0.8 × 9 × 200 = 1440 kN

12.20 (b)

Raft foundations (sometimes known as Mat Foundations) are a large concrete slab which can support a number of columns and walls. The slab is spread out under the entire building or at least a large part of it which lowers the contact pressure compared to the traditionally used strip or trench footing.

Isolated footings are independent footings which are provided for each column. This type of footing is used when. Columns are not closely spaced. Loads on footings are less. The Safe bearing capacity of soil is generally high.

In **End bearing piles,** the bottom end of the pile rests on a layer of especially strong soil or rock. The load of the building is transferred through the pile onto the strong layer. In a sense, this pile acts like a column.

12.21 (c)

Pedestal pile is used to bypass weak soil stratum below which thin bearing stratum is reached with reasonable depth. The pedestal of pile gives the effect of spread footing on this comparatively thin bearing.

12.22 (a)

For piles driven in loose and medium dense cohesionless soils, the group efficiency is high. The soil around and between the piles is compacted due to vibration caused during the driving operation. For better results, it is essential to start driving the piles at the centre and then proceed outwards.

12.23 (b)

In the case of loose sand there may be loose pockets causing differential settlements. Raft foundations reduce the differential settlements and higher allowable soil pressure may be used for design of rafts on loose sands.

Under-reamed piles are suitable to bear the swelling pressure caused by black cotton soil.

12.24 (b)

The proportioning of footing is governed by 'Service load' criterion to ensure equal settlement of footing for various columns. For ordinary buildings,

Service load = Dead load + 0.5 × Live Load.

Friction piles are called floating piles also.

12.25 (a)

Initial load tests can be carried out on test piles or working piles. The routine tests are carried out on working piles only.

12.26 (c)

Damping coefficient is given by

$$C_c = 2\sqrt{km}$$

$$= 2\sqrt{10000 \times \frac{981}{9.81}}$$

$$= 2000 \text{ kN-s/m}$$

12.27 (a)

Driven pile has more load carrying capacity than bored pile because value of lateral earth pressure coefficient is more in case of driven pile, so skin friction is more.

12.28 (*)

Diameter of pile = 30 cm

Spacing = 80 cm c/c

For group action of piles, minimum spacing required = 2.5D = 2.5 × 30 = 75 cm

Hence negative skin friction of group is to be calculated

Assuming 4 number of piles

Neglecting point bearing capacity

$Q_g(u) = \alpha . C . P_g . D$

For cohesive soil, assuming $\alpha = 1$

Given, $C = 25$ kN/m^2

P_g = Perimeter of block = 4 × 110 cm
Depth of block = 5 m

$$Q_g(u) = 1.0 \times 25 \times \left(\frac{4\times110}{100}\times 5\right) = 550 \text{ kN}$$

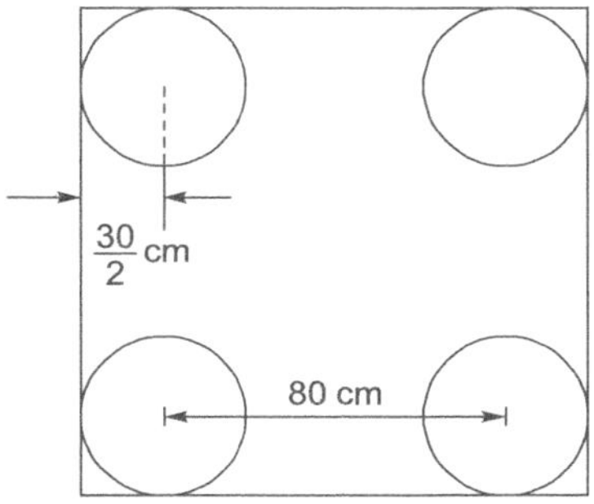

12.29 (b)

Static Cone Penetration Test: Also known as Dutch Cone for CPT, it gives a continuous record of variation of both cone resistance and friction resistance with depth. CPT however does not yield any sample. It is suitable for soft clay and silts & fine to medium sand deposits.

12.30 (d)

- **Under reamed piles** are used to restrict damages due to volume change of swelling soils (as it causes uplift pressure).
- **Combined** footing is used when a foundation is built close to an existing building or the property line, there may not be sufficient space for equal projections on the sides of the exterior column.
- **Strap footing** consists of two spread (isolated) footings joined by a rigid beam known as strap. Its main function is to transfer the moment from the exterior footing to the interior footing.
- An exact balance of weight removed against weight imposed, the result is zero settlement of the building. This is the principle of **floating foundations**. Such a foundation is very useful for weak soils.

12.31 (c)

Initial Tests on piles:

This test is performed to confirm the design load calculations and to provide guidelines for setting up the limits of acceptance for routine tests. It also gives an idea of the suitability of the piling system. Initial Test on piles are to be carried out at one or more locations depending on the number of piles required.

Load applied for the initial (cyclic) load test is 2.5 times the safe carrying capacity of the pile. Loading for Initial Tests is conducted as per Appendix 'A' Clause 6.3of IS-2911 Part IV.

Routine Tests on piles:

Selection of piles for the Routine Test is done based on number of piles required subject to maximum of 0.5% of total number of piles required. The number of tests may be increased to 2% depending on the nature / type of structure. The test load applied is 1½ times the safe carrying capacity of the pile. **The Maintained load method** as described in Clause 6.2 of **IS-2911 (Part IV) – 1985** shall be followed for loading for the Routine Tests.

This test will be performed for the following purposes:

(a) To ensure the safe load capacity of piles

(b) Detection of any unusual performance contrary to the findings of the Initial Test.

12.32 (d)

Under-reamed pile is a special type of bored pile which is provided with a bulb/pedestal at the end. The usual size of such piles are 150 to 200 mm shaft diameter, 3 to 4 m long. The diameter of under-reamed portion is usually 2 to 3 times the shaft diameter.

12.33 (b)

A pile cap may or may not be situated below the ground level. In case of free standing pile group, the pile cap stands above the ground. In case it rests on the soil, partially or fully buried below the ground it is called piled foundation. So statement 1 is incorrect.

12.34 (a)

Driven piles may be of concrete, steel or timber. These piles are driven into the soil strata by the impact of a hammer. Generally boring is not used in these cases. When pile is driven into granular soils it densifies the soil and increases stiffness (strength) of soil. But in case of clays, driven piles remould the soil to a distance of about two times diameter of pile.

12.36 (c)

In anchored sheet pile wall large yield is observed below the anchor rod and above the dredge line, not above the anchor rod.

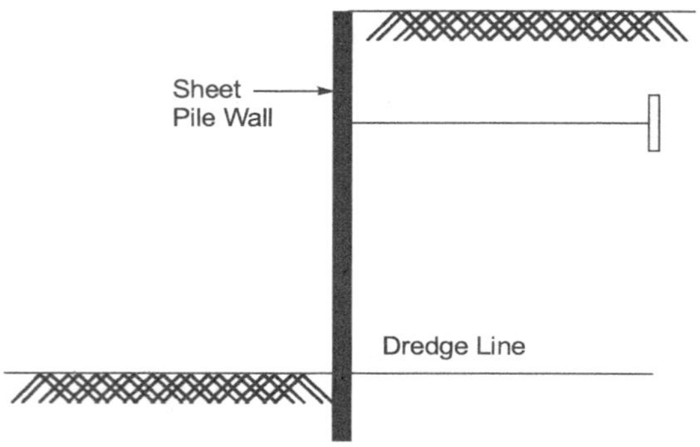

12.37 (c)

Under-reamed piles are cast-in-situ piles which may be installed both in sandy and in clayey soils.

12.38 (a)

Under Reamed piles are designed for both skin friction resistance and point bearing resistance.

12.39 (a)

Piles are sometimes subjected to lateral loads due to wind pressure, water pressure, earth pressure, earthquakes, etc.

If the horizontal load is large, inclined piles, known as raking piles or batter piles, are provided to take the horizontal load. These piles have a high resistance to lateral loads, as a large portion of the horizontal component of the load is carried axially by the pile. Batter piles, along with vertical piles, are provided in situations where the horizontal loads are significant, such as wharves, jetties, bridge piers, trestles, retaining wall and tall chimneys.

12.40 (c)

Drilled Pier: A drilled pier is a large diameter concrete cylinder built in the ground. A drilled pier is a type of deep foundation constructed to transfer heavy axial or lateral loads to a deep stratum below the ground surface.

Caisson: A caisson is a type of foundation of the shape of a hollow prismatic box, which is built above the ground level and then sunk to the required depth as a single unit. It is a watertight chamber used for laying foundations under water, as in rivers, lakes, harbours, etc. The caissons are of three types: (i) Open caissons, (ii) Pneumatic caissons, and (iii) Floating caissons.

Open caissons are hollow chambers, open both at the top and the bottom. The bottom of the caisson has a cutting edge. The caisson is sunk into place by removing the soil from the inside of the shaft (chamber) until the bearing stratum is reached. Well foundations are special type of open caissons.

Pneumatic caissons are closed at the top, but open at the bottom. A pneumatic caisson has a working chamber at its bottom in which compressed air is maintained at the required pressure to prevent entry of water into the chamber.

Floating caissons are open at the top but closed at the bottom. These caissons are constructed on land and then transported to the site, and floated to the place where these are to be finally installed.

12.41 (a)

The ratio of the ultimate load capacity of the pile group, Q_{ug} to the sum of the individual load capacities of the piles in the group, is called group efficiency η.

$$\eta = \frac{Q_{ug}}{nQ_u}$$

Disturbance of soil during installation of piles and overlap of stresses between adjacent piles may cause the group capacity become less than the sum of the individual capacities, that is $\eta < 1$. Generally, for smaller spacings between piles, $\eta < 1$. For larger spacings, the effect of pile interaction diminishes and η approaches unity. In driven piles where the soil around the piles gets densified, as in loose to medium sand, η may be even more than 1.

12.42 (c)

In case of friction piles, load is a transferred through friction along the length of pile. In case of end bearing piles load is transferred through bearing at the end of pile.

12.43 (c)

$$P = \alpha(\pi dL)C_u$$

$$= (0.75)\pi(0.5)16\left(\frac{100}{2}\right)$$

$$= 942 \text{ kN} \approx 940 \text{ kN}$$

12.44 (d)

As per Engineering News Formula

$$Q_{\text{allowable}} = \frac{W.h}{\text{FOS}(S+C)}$$

Where, $C = 2.5$ cm for drop hammer

weight, $W = 25$ kN

$h = 0.8$ m = 80 cm

$S = 12$ mm = 1.2 cm

FOS = 6

$$\therefore \quad Q_{\text{allowable}} = \frac{25 \times 80}{6 \times (1.2 + 2.5)}$$

$$= 90.090 = 90 \text{ kN}$$

12.45 (b)

As per IS – 2911, Part IV the allowable load on a single pile will be minimum of the following two cases :

1. $\frac{2}{3}$rd of the load at which total settlement is 12 mm.
2. 50% of the ultimate load at which total settlement is equal to the $\frac{1}{10}$ of the diameter of pile.
3. $\frac{1}{2}$ to $\frac{2}{3}$rd of the load which gives a net settlement of 6 mm.

12.47 (c)

Due to various energy losses occurring during dynamic resistance test, the static resistance and dynamic resistance of soil is different.

12.48 (b)

For drop hammer

$$Q_a = \frac{Wh}{6(s+2.5)}$$

For single acting steam hammer

$$Q_a = \frac{Wh}{(s+0.25)}$$

12.50 (d)

- Braced cofferdams are economical for small to moderate height only for earth embankment type
- In cofferdam there is no height limitation.
- Sheet piles are used as bulkhead or as impervious cutoff to reduce seepage and uplift under hydraulic structure.

12.51 (d)

Skin frictional resistance of a pile driven in sand

$$Q_{sf} = q_s A_s = \left(\frac{1}{2}k\gamma L \tan\delta\right) A_s$$

In the presence of GWT,

$$Q_{sf} = \left(\frac{0 + k\bar{\sigma}_v}{2}\right) \tan\delta A_s$$

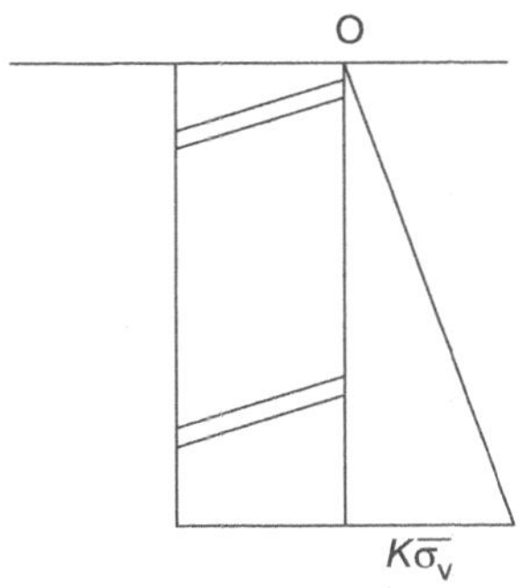

For piles in granular soil, the design is based on an effective stress analysis. In clays, it is common to use a "total stress" analysis in which the load. Capacity is related to the undrained shear strength, C_u.

12.52 (d)

The maximum test load on a working pile should not exceed one and a half times the design load.

13 Soil Stabilization and Soil Exploration

13.1 A wet, cohesive subgrade is most effectively stabilized by the addition of

(a) cement (b) fly ash
(c) bitumen (d) lime

[ESE : 2000]

13.2 A soil sampler has inner and outer radii of 25 mm and 30 mm, respectively. The area ratio of the sampler is

(a) 24% (b) 34%
(c) 44% (d) 54%

[ESE : 2001]

13.3 The correct sequence of the increasing order of the disturbance to soil samples obtained from chunk, piston, split spoon and remoulded sampler is

(a) piston sampler, chunk sampler, split spoon sampler, remoulded sampler
(b) chunk sampler, piston sampler, split spoon sampler, remoulded sampler
(c) piston sampler, chunk sampler, remoulded sampler, split spoon sampler
(d) chunk sampler, piston sampler, remoulded sampler, split spoon sampler

[ESE : 2002]

13.4 Which one of the expressions given below correctly relates natural frequency f_n, spring stiffness 'k', mass 'm' of the machine foundation vibrating system with C_u, the coefficient of elastic uniform compression of the foundation soil?

(a) $f_n = \frac{1}{2\pi}\sqrt{\frac{k}{m}} = \sqrt{\frac{C_u A}{m}}$

(b) $f_n = \sqrt{\frac{k}{m}} = 2\pi\sqrt{\frac{C_u A}{m}}$

(c) $f_n = \sqrt{\frac{k}{m}} = \frac{1}{2\pi}\sqrt{\frac{C_u A}{m}}$

(d) $f_n = \frac{1}{2\pi}\sqrt{\frac{k}{m}} = \frac{1}{2\pi}\sqrt{\frac{C_u A}{m}}$

[ESE : 2002]

13.5 Match **List-I** (Sampler) with **List-II** (Use) and select the correct answer using the codes given below the lists:

List-I

A. Split spoon sampler
B. Stationary piston sampler
C. Rotary sampler
D. Compressed air sampler

List-II

1. To obtain representative samples in all types of soil
2. To obtain undisturbed samples of sands below water table
3. To obtain undisturbed samples in clay and silts
4. To obtain approximately undisturbed samples of hard cemented cohesive soils

Codes:

	A	B	C	D
(a)	1	3	2	4
(b)	3	1	4	2
(c)	1	3	4	2
(d)	3	1	2	4

[ESE : 2003]

13.6 Consider the following statements:
Prevention or elimination of swelling can be brought by

1. providing an impervious apron around the structure
2. pre-wetting the ground to a moisture content equal to equilibrium moisture content
3. packing downward loads large enough to exceed swelling pressures
4. chemically stabilizing the soil with lime

Which of these statements are correct?

(a) 1 and 2 (b) 1 and 3
(c) 2 and 3 (d) 1, 2, 3 and 4

[ESE : 2004]

13.7 Match **List-I** (Type of exploration) with **List-II** (Soil profile) and select the correct answer using the codes given below the lists:

List-I

A. Diamond core drilling
B. Uncased wash boring
C. Open pit excavation
D. Cased boring

List-II

1. Medium strong cohesive soils
2. Rocky formation
3. Soft cohesive soils and cohesionless soils. Exploration up to relatively large depths
4. Exploration up to a shallow depth below ground level

Codes:

	A	B	C	D
(a)	4	3	2	1
(b)	2	1	4	3
(c)	4	1	2	3
(d)	2	3	4	1

[ESE : 2005]

13.8 Consider the following statements:

1. Obtaining reliable properties of soil is very important in geotechnical engineering. Chunk samples obtained from the field are the best for this purpose.
2. Chunk samples may easily be obtained at all depths below ground level and for all types of soils.
3. Undisturbed soil samples are enclosed in rugged containers. As such, they do not get disturbed easily during transportation and handling.

Which of these statements is/are correct?

(a) 1, 2 and 3 (b) 1 only
(c) 2 and 3 (d) 1 and 3

[ESE : 2005]

13.9 Consider the following statements:

1. A recovery ratio of less than 1 implies that the soil has compressed.
2. A recovery ratio greater than 1 implies that the soil has swelled.
3. A recovery ratio of less than 1 implies that the soil has swelled
4. A recovery ratio greater than 1 implies that the soil has compressed.

Which of these statements is/are correct?

(a) 1 and 2 (b) 1 only
(c) 3 and 4 (d) 4 only

[ESE : 2005]

13.10 A sampling tube with a cutting edge is used for extracting the samples. The sampling tube has the following dimensions:

Inner diameter of cutting edge = D_c
Outer diameter of cutting edge = D_w
Inner diameter of the sampling tube = D_s
Outer diameter of the sampling tube = D_t

What is the area ratio A_r of the sampling tube?

(a) $A_r = \dfrac{D_w^2 - D_c^2}{D_c^2} \times 100\%$

(b) $A_r = \dfrac{D_t^2 - D_s^2}{D_c^2} \times 100\%$

(c) $A_r = \dfrac{D_t^2 - D_w^2}{D_w^2} \times 100\%$

(d) $A_r = \dfrac{D_t^2 - D_s^2}{D_s^2} \times 100\%$

[ESE : 2006]

13.11 During a sampling operation, the drive sampler is advanced 600 mm and the length of the sample recovered is 525 mm. What is the recovery ratio of the sample?

(a) 0.125 (b) 0.140
(c) 0.875 (d) 0.143

[ESE : 2006]

13.12 A differential free swell test on a soil gives a value of differential free swell of 40%. What is the degree of swelling?

(a) Low (b) Medium
(c) High (d) Very high

[ESE : 2008]

13.13 Consider the following statements:

On addition of lime to swelling soils,

1. their liquid limit increases
2. their plastic limit increases
3. their shrinkage limit increases
4. their swelling potential decreases

Which of these statements are correct?

(a) 1 and 3 only (b) 1, 2 and 4 only
(c) 2, 3 and 4 only (d) 1, 2, 3 and 4

[ESE : 2008]

13.14 In seismic exploration method, velocities V_1 and V_2 were 600 m/s and 4000 m/s. The distance corresponding to the break point of velocities was 30 m. Based on this detail, what is the depth of overburden?

(a) 11.5 m (b) 12.5 m
(c) 12.9 m (d) 13.2 m

[ESE : 2008]

13.15 Consider the following statements:

1. The soil obtained from wash boring is a representative sample.
2. Recovery ratio will be high during drilling in sound rock.
3. Hollow stem augers are sometimes used to drill holes in silty sand.

Which of these statements is/are correct?

(a) 1 only (b) 1 and 2
(c) 2 and 3 (d) 3 only

[ESE : 2009]

13.16 Consider the following statements:

1. A differential free swell value of 40% indicates a soil with a high degree of expansiveness.
2. A swelling pressure of less than 20 kN/m^2 is not of much consequence.
3. The swelling pressure is a unique parameter for a swelling soil and is not influenced by other factors.

Which of these statements is/are correct?

(a) 1, 2 and 3 (b) 1 and 2 only
(c) 2 and 3 only (d) 1 only

[ESE : 2010]

13.17 Match **List-I** with **List-II** and select the correct answer using the code given below the lists:

List-I

A. Geophysical methods
B. SPT
C. DCPT
D. Piston-type sampler

List-II

1. Primarily for cohesive soils
2. Clays and silts
3. Reconnaissance covering large area and large depth
4. Suitable for sandy soils

Code:

	A	B	C	D
(a)	2	1	4	3
(b)	3	1	4	2
(c)	2	4	1	3
(d)	3	4	1	2

[ESE : 2011]

13.18 Consider the following statements:

1. Buried service lines should be avoided in an expansive soil region.
2. A swelling pressure less than 20 kN/m^2 is not of much consequence.
3. If soil is not black in colour, it is unlikely to be an expansive soil.

Which of these statements are correct?

(a) 1, 2 and 3 (b) 1 and 2 only
(c) 2 and 3 only (d) 1 and 3 only

[ESE : 2012]

13.19 Consider the following statements:

1. Undisturbed samples may be obtained with the help of augers.
2. Auger drilling is most effective in clayey soils.
3. Hollow stem augers are sometimes used to drill holes in silty sand.

Which of these statements are correct?

(a) 1, 2 and 3 (b) 1 and 2 only
(c) 2 and 3 only (d) 1 and 3 only

[ESE : 2013]

13.20 Consider the following statements:

1. The static cone penetration test is unsuitable for layered deposits of sands, silts and clays.
2. The results of groundwater investigation are recorded as water-table contours over the site.
3. Closed piezometers are used to measure pore water pressure in soils having low permeability.

Which of the above statements are correct?

(a) 1, 2 and 3 (b) 1 and 2 only
(c) 2 and 3 only (d) 1 and 3 only

[ESE : 2014]

13.21 Consider the following statements:
Lime stabilization of soil leads to
1. Decrease in shrinkage limit.
2. Increase in plastic limit.
3. Decrease in liquid limit.
4. Flocculation of clay particles

Which of the above statements are correct?
(a) 1, 2 and 3 (b) 1, 2 and 3
(c) 1, 3 and 4 (d) 2, 3 and 4
[ESE : 2015]

13.22 Consider the following statements related to the properties of a good quality soil sample:
1. Area ratio should be low
2. Cutting edge should be thick
3. Inside clearance should be low
4. Outside clearance should be low

Which of the above statements are correct?
(a) 1 and 2 (b) 2 and 3
(c) 3 and 4 (d) 1 and 4
[ESE : 2015]

13.23 Which of the following tests are essential for designing a foundation on expansive soils?
1. Swelling pressure test
2. Free swell test
3. Estimation of differential free swell
4. Shrinkage limit test

Which of the above statements are correct?
(a) 1, 2 and 3 only (b) 1, 2 and 4 only
(c) 1, 2, 3 and 4 (d) 2, 3 and 4 only
[ESE : 2015]

13.24 Which technique of stabilization for the sub-base is preferred for a heavy plastic soil?
(a) Cement stabilization
(b) Mechanical stabilization
(c) Lime stabilization
(d) Bitumen stabilization [ESE : 2016]

13.25 Consider the following statements:
1. Functions of reinforcement in reinforced soil and in reinforced concrete are comparable.
2. The design of a geotextile reinforced wall is similar in principle to that of a reinforced earth wall.

Which of the above statements is/are correct?
(a) 1 only (b) 2 only
(c) Both 1 and 2 (d) Neither 1 nor 2
[ESE : 2017]

13.26 Consider the following statements :
1. All soils can be identified in the field by visual examination
2. Fine-grained soils can be identified in the field by visual examination and touch
3. Fine-grained soils can be identified in the field by dilatancy test
4. By visual examination, only coarse grained soils can be identified

Which of the above statements are correct?
(a) 1 and 2 only (b) 2 and 3 only
(c) 3 and 4 only (d) 1 and 4 only
[ESE : 2018]

13.27 Which one of the following methods is not the category of Geophysical methods of sub-surface investigation?
(a) Electrical resistivity method
(b) Electric logging
(c) Gamma-ray logging
(d) Electrical response surveying
[ESE : 2020]

13.28 The stability or shear strength of fine grained soils can be increased by draining them with the passage of direct current through them. This process is known as
(a) Electro-osmosis
(b) Zeta potential
(c) Electro-chemical hardening
(d) Consolidation
[ESE : 2020]

13.29 Consider the following steps related to construction with the use of geotextiles:
1. Starts with an adequate working surface and staging areas.
2. Lay a geotextile sheet of proper width on the ground surface.
3. Construction equipment must work from the soil backfill and be kept off the unprotected geotextile.

Which of the above steps are correct?
(a) 1 and 2 only (b) 2 and 3 only
(c) 1 and 3 only (d) 1, 2 and 3
[ESE : 2021]

13.30 The primary function of geogrids is
(a) connecting two layers
(b) separators
(c) reinforcement
(d) protection from corrosion [ESE : 2022]

13.31 Which one of the following characteristics is NOT measured by geophysical method of soil exploration?
(a) Magnetism
(b) Density
(c) Electrical resistivity
(d) Plasticity [ESE : 2022]

13.32 Which one of the following is NOT the common nomenclature of geosynthetics?
(a) Geotextiles (b) Geogrids
(c) Geogrouts (d) Geonets
[ESE : 2023]

13.33 Which one of the following is an advantage when geotextiles is used in road works?
(a) Prior stripping of the natural soil
(b) Economy of aggregate
(c) Increase of different settings
(d) Slower consolidation of fills
[ESE : 2023]

13.34 Consider the following objectives related to the sub-surface investigation:
1. To establish the geological profile.
2. To establish Hydrogeological conditions.
3. To monitor future changes in ground conditions through instrumentation.

Which of the above objectives are correct?
(a) 1 and 2 only (b) 2 and 3 only
(c) 1 and 3 only (d) 1, 2 and 3
[ESE : 2023]

13.35 Which one of the following statements is NOT correct?
(a) The porosity of rock depends upon size and shape of the rock grains
(b) Specific gravity is useful for calculating the rock over burden stress
(c) The porosity does not depend on rock mechanical processes
(d) The porosity of spherical rock grains is high in case of cubic packing
[ESE : 2023]

■■■■

Answers Soil Stabilization and Soil Exploration

13.1 (d)	13.2 (c)	13.3 (b)	13.4 (d)	13.5 (c)	13.6 (d)	13.7 (b)	13.8 (c)
13.9 (a)	13.10 (a)	13.11 (c)	13.12 (c)	13.13 (c)	13.14 (c)	13.15 (c)	13.16 (b)
13.17 (d)	13.18 (b)	13.19 (c)	13.20 (a)	13.21 (a)	13.22 (d)	13.23 (c)	13.24 (c)
13.25 (c)	13.26 (c)	13.27 (d)	13.28 (a)	13.29 (d)	13.30 (c)	13.31 (d)	13.32 (c)
13.33 (b)	13.34 (d)	13.35 (c)					

Explanations Soil Stabilization and Soil Exploration

13.1 (d)

Granular soils with sufficient fines are ideally suited for cement stabilization. Such soils can be easily pulverized and mixed. They require least amount of cement. Lime stabilization is useful for clayey soils. The plasticity of the soil decreases and material is more suitable as a subgrade material. Bitumen stabilization is very effective in stabilizing sandy soils having little or no fines. It can be used for desert soils, (Loess).

13.2 (c)

$$\text{Area ratio} = \frac{D_0^2 - D_i^2}{D_i^2} \times 100$$

$$= \frac{(30)^2 - (25)^2}{(25)^2} \times 100 = 44\%$$

13.3 (b)

Piston sampler is a thin type sampler used in case of soft sensitive clays for obtaining very fine undisturbed samples. **Chunk samples** are considered equivalent to undisturbed samples.

Split Spoon sampler gives disturbed but representative samples and can be used in all soil types.

Stationary Piston Sampler is ideal in case of soft and saturated soils.

13.4 (d)

Natural circular frequency of the system

$$\omega_n = \sqrt{k/m}$$

where m is mass of the system

$$f_n = \frac{\omega_n}{2\pi}$$

and $$k = C_u A$$

13.5 (c)

Rotary samplers are double walled samplers. It consists of a cutting bit in its outer tube which is used in case of hard cemented cohesive soils and rocks.

13.6 (d)

Treatment of expansive soils

- Pre-wetting a site to increase the moisture content can eliminate an expansive soil problem if the high moisture content can be maintained.
- Vertical or horizontal moisture-loss barriers are installed in only a few cases in the past, but their use is becoming more frequent. The combined effect of a moistened sub-grade and moisture barriers, when applied correctly, can be very successful. The application of moisture treatment coupled with proper degrees of compaction has been utilized extensively.

13.7 (b)

In **diamond core drilling**, the drill uses a diamond encrusted drill bit to drill through the rock.

Cased boring is done for larger depths in cohesive or cohesionless soils. Casing prevents the falling of soil into bore.

13.8 (c)

Chunk samples fall in the category of undisturbed samples. They are obtained from bottom of the test pits for all type of soils. Including those containing pebbles and cobbles. Thus reliable properties of soil can be obtained by testing chunk samples.

However, the depth of test pit is limited. Chunk samples can be obtained from exposed soil in a shaft or tunnel also. Thus they can be obtained from all the depths.

Chunk samples are not the best for obtaining the engineering properties. Piston sampler and compressed air sampler cause least disturbance to sensitive soils. So rugged containers are used to enclose samples. Thus 2 and 3 are correct and 1 is partially correct only.

13.9 (a)

$$\text{Recovery ratio}, R_r = \frac{L_a}{L_t}$$

L_a = actual length of sample in tube

L_t = total length of the sampling tube driven below the bottom of the bore hole

$R_r = 1$ indicates no disturbance to the sample

$R_r < 1$ indicates that the sample is compressed

$R_r > 1$ indicates that there is expansion of sample within the tube.

13.10 (a)

$$A_r = \frac{(D_w^2 - D_c^2)}{D_c^2} \times 100$$

If $A_r < 10\%$, the sample disturbance is considered to be small.

13.11 (c)

Recovery ratio is defined as

$$R_r = \frac{\text{Actual length of sample in the tube}}{\text{Total length of the sampling tube driven below the bottom of the bore hole}}$$

$$R_r = \frac{525}{600} = 0.875 < 1$$

This indicates ($R_r < 1$) that the sample is compressed.

13.12 (c)

Degree of swelling for different differential free swell (DFS) are tabulated below:

Degree of swelling	DFS, per cent
Low	Less than 20
Moderate	20 to 35
High	35 to 50
Very high	Greater than 50

13.13 (c)

Lime has been found to be the most effective and economical of all additives. Addition of lime to expansive soils reduces swell potential and increases workability and strength. Liquid Limit is decreased and Plastic Limit is increased with increasing lime content resulting in decreased Plasticity Index. Flocculation takes place when lime is mixed with clays.

13.14 (c)

The depth of the boundary between the two strata can be given by

$$D = \frac{d}{2}\sqrt{\frac{V_2 - V_1}{V_2 + V_1}}$$

$$= \frac{30}{2}\sqrt{\frac{4000-600}{4000+600}} = 12.9 \text{ m}$$

13.15 (c)

Soil samples obtained from auger borings and wash borings are non-representative samples. Non-representative samples consist of mixture of materials from various soil or rock strata or are samples from which some mineral constituents have been lost or got mixed up.

13.16 (b)

Degree of expansiveness and differential free swell are tabulated below:

Degree of expansiveness	DFS percent
Low	< 20
Moderate	20 to 35
High	35 to 50
Very high	> 50

The swelling pressure is not a unique parameter for a swelling soil. It is significantly influenced by several factors, the more important of them being the initial moisture content (moulding water content for remoulded soils), initial dry density, method of compaction, confining surcharge and the height of the soil specimen. For a given dry density, lower the initial moisture content, greater the water thirst and higher the swelling pressure.

13.17 (d)

Piston samplers are used for getting undisturbed soil samples from soft and sensitive clays and silts. SPT (Standard penetration test) is an in-situ test used especially for cohesionless soils e.g. sandy soils. The test is useful for determining the relative density and the angle of shearing resistance of cohesionless soils. It can also be used for unconfined compressive strength test of cohesive soils.Geophysical method is used for soil exploration and testing in large areas and large depth.

13.18 (b)

There are a lot of swelling soils which are brown or light coloured, even though most expansive soils are dark in colour. Swelling pressure of a soil depends on lots of factors such as initial dry density, method of compaction, moisture content, height of specimen, surcharge pressure and nature of pore fluid.

A swelling pressure of less than 20 kN/m^2 is not of much consequence.

13.19 (c)

Auger obtained samples are highly disturbed. Auger boring is most effective in soft soils which can stand unsupported and have some stickiness so that they do not fall back into the hole during lifting operation. Thus it is most effective in case of silts and clays.

13.20 (a)

Water table measurements that are taken at the same time of year can be used to develop a water table contour map to show the direction of ground water flow.

13.21 (a)

Lime stabilization is used to improve the strength, stiffness and durability of fine-grained soils.

Addition of lime to soils produces a maximum density under a higher optimum moisture content than in the untreated soil, thus leads to flocculated or aggregated larger particle groups which are fairly stable. Lime produces a decrease in plasticity index and swelling potential and an increase in shrinkage limit. For this reason, lime stabilization has extensively been used for stabilizing highly unstable plastic and swelling clayey soils, such as blackcotton soils.

13.22 (d)

For a good quality soil sample (undisturbed samples), the sampler should have low value of area ratio (less than 10% or so). In order to keep A_r less, thin cutting edge is therefore preferred in a soil sampler.

Inside clearance is meant to reduce the friction between the soil sample and the sampler when the soil enters the tube, by allowing for the elastic expansion of the soil. Thus if inside clearance is high, there will be too much of lateral expansion. IS recommends that the outside clearance should not be much greater than the inside clearance.

13.24 (c)

Hydrated (or slaked) lime is very effective in treating heavy plastic clayey soils.

13.25 (c)

The basic principles of incorporating reinforcement members into a soil mass are much the same as those utilised in the design of reinforced concrete.

The design of reinforced earth wall is based on assumption that "The pressure acting of the wall facing element without the reinforcements is the same as that acts on a rigid vertical retaining wall under active state".

13.27 (d)

- Electric resistivity method : It is based on the difference in the electrical conductivity or the electrical resistivity of different soil.
- Electric logging : This technique can be used in geotechnical investigations to assess the variation with depth of geologic materials and associated fluids.
- Gamma-Ray logging : This is a method of measuring naturally occurring gamma radiation to characterize the rock or sediment in a borehole or drill hole.

13.28 (a)

Electro-osmosis is a method of drainage of cohesive soi in which a Direct Current (DC) is passed through a saturated soil between a positive electrode (anode) and a negative electrode (cathode), as a result pore water migrates to cathode.

The cathode is a well point which collects the water drained from the soil.

As the water is drained out, the consolidation process can be upto a hundred times faster rate than mechanical consolidation that is why substantial improvement in the shear strength of soil is achieved quickly.

13.29 (d)

The construction procedure with the use of geotextile is explained in following figures. The geotextile serve both as reinforcement and also as facing unit. The procedure is described below.

1. Start with adequate working surface and staging area.
2. Lay a geotextile sheet of proper width on the ground surface with 4 to 7 feet at the wall face draped over temporary wooden form.
3. Backfill over the sheet with soil. Granular soil or soils containing maximum 30 percentage silt and for 5% clay are customary.
4. Construction equipment must work from the soil backfill and be kept of the unprotected geotextile. The spreading equipment should be a wide-tracked bulldozer that exerts little pressure agair the ground on which it rests. Rolling equipment likewise should be of relatively light weight.
5. When the first layer has been loaded over the process should be repeated for the second layer with the temporary facing from being extended from the original ground surface or wall being stepped back about 6 inch so that the form can be supported from the first layer. In the later case the support stakes must penetrate the fabric.

6. This process is continued until the wall reaches its intended height.
7. The production against ultraviolet light and safety again vandalism the faces of such walls must be protected. Both shortcrete and gunite have been used for this purpose.

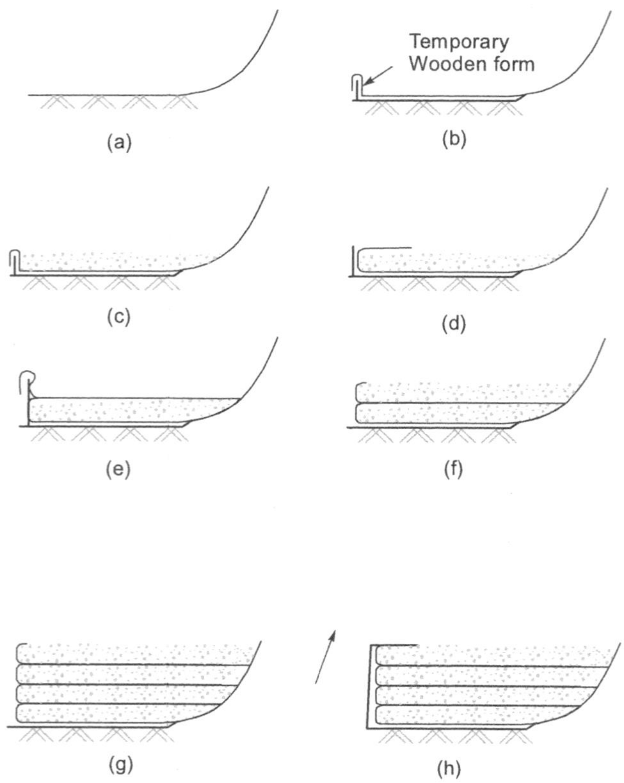

General construction procedures for using geotextiles in fabric wall construction

13.30 (c)

Geogrid is geosynthetic material used to reinforce soil and similar material.

13.32 (c)

Geosynthetics are the synthetic fabrics used in various geotechnical applications such as road and railway embankments, earth dikes and coastal protection structures designed to perform one or more basic functions such as filtration, drainage, separation of soil layers, reinforcement or stablizations. Various Geosynthetic are Geotextiles, Geogrids, Geonets, Geosynthetic clay linear, Geomembrane, Geocomposite and Geofoams.

13.33 (b)

The most frequent role of geotextiles is road construction is as a separator between the subgrade and subbase.This prevents the subgrade material from intruding into sub-base due to repeated traffic loading.

The savings in sub-base materials, which would otherwise be lost due to mixing with the subgrade.

13.34 (d)

Objectives of sub-surface investigation:

1. To establish the geological profile.
2. To determine engineering properties for the various unit within the eventual ground model.
3. To establish hydrogeological conditions.
4. To monitor future changes in ground condition through instrumentation.

13.35 (c)

Porosity of a rock depends upon grain size distribution, shape, the presence of cement between the grains developed to different degrees.

The porosity of spherical rock grain is high is case of cubic packing.

■■■■

UNIT

VI

Surveying and Geology

Syllabus

Surveying: Classification of surveys, various methodologies, instruments & analysis of measurement of distances, elevation and directions; Field astronomy, Global Positioning System; Map preparation; Photogrammetry; Remote sensing concepts; Survey Layout for culverts, canals, bridges, road/railway alignment and buildings, Setting out of Curves.

Geology: Basic knowledge of Engineering geology & its application in projects.

Contents

1 Fundamental Concepts of Surveying & Linear Measurement

1.1 Match **List-I** (Corrections) with **List-II** (Name) and select the correct answer using the codes given below the lists:

List-I

A. $\pm L\left(1-\dfrac{h}{R}\right)$

B. $-\dfrac{l}{24}\left(\dfrac{W}{P}\right)^2$

C. $\pm\alpha(T_f - T_s)L$

D. $\pm\dfrac{(P_f - P_s)L}{AE}$

List-II

1. Sag correction
2. Pull correction
3. Temperature correction
4. Mean sea level correction

Codes:

	A	B	C	D
(a)	4	1	3	2
(b)	1	4	3	2
(c)	4	1	2	3
(d)	1	4	2	3

[ESE : 2002]

1.2 The object of chain and cross-staff survey is to

1. locate the boundaries of an area
2. plot the figure to a scale
3. find the area of the plot
4. find the reduced levels of the plot

Which of these statements is/are correct?

(a) 1, 2, 3 and 4 (b) 1, 2 and 3
(c) 1 and 2 (d) 4 alone

[ESE : 2002]

1.3 Which one of the following figures indicates the principle of traversing?

(a) 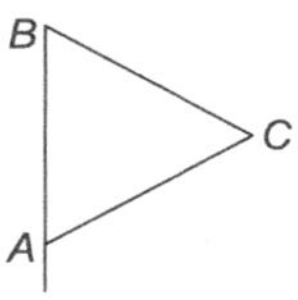

(b) 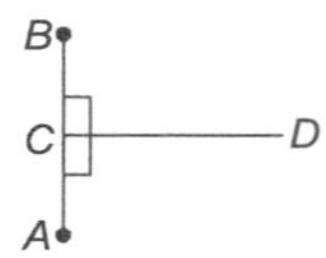

(c) 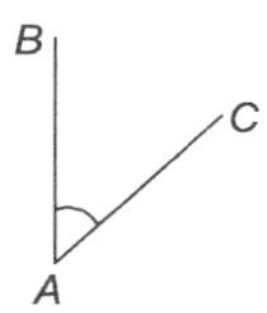

(d) 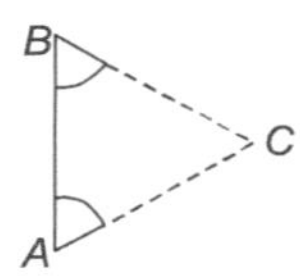

[ESE : 2003]

1.4 Match **List-I** (Type of survey) with **List-II** (Purpose) and select the correct answer using the codes given below the lists:

List-I

A. Topographical survey
B. Reconnaissance survey
C. Cadastral survey
D. Archaeological survey

List-II

1. To determine boundaries of fields, houses, etc.
2. To find relics of antiquity
3. To determine natural features of a country
4. To determine possibility and rough cost of the surveying system to be adopted

Codes:

	A	B	C	D
(a)	3	4	1	2
(b)	3	1	4	2
(c)	2	4	1	3
(d)	2	1	4	3

[ESE : 2004]

1.5 Which of the following figures are equal to one acre?

1. 43560 sq.ft.
2. 40 gunthas
3. 10 sq. Gunter's chain
4. 4850 sq. yds.

Select the correct answer using the codes given below:

(a) 1, 2 and 3 (b) 2, 3 and 4
(c) 1, 2 and 4 (d) 1, 3 and 4

[ESE : 2006]

1.6 What is the slope correction for a length of 30.0 m along a gradient of 1 in 20?

(a) 3.75 cm (b) 0.375 cm
(c) 37.5 m (d) 0.0375 cm

[ESE : 2006]

1.7 A 30 m metric chain is found to be 10 cm too short throughout a measurement. If the distance measured is recorded as 300 m, what is the actual distance?

(a) 300.1 m (b) 301.0 m
(c) 299.0 m (d) 310.0 m

[ESE : 2006]

1.8 A 20 m chain was found to be 10 cm too long after chaining a distance of 2000 m. It was found to be 18 cm too long at the end of the day's work after chaining a total distance of 4000 m. What is the true distance if the chain was correct before the commencement of the day's work?

(a) 3962 m (b) 4019 m
(c) 3981 m (d) 4038 m [ESE : 2007]

1.9 A scale of 1 inch = 50 ft. is mentioned on an old map. What is the corresponding equivalent scale?

(a) 1 cm = 5 m (b) 1 cm = 6 m
(c) 1 cm = 10 m (d) 1 cm = 12 m

[ESE : 2008]

1.10 An invar tape, 50 m in length, standardized at 20°C temperature and 10 kg pull, is used to measure a base line. The correction per tape length, if at the time of measurement the temperature was 30° and the coefficient of linear expansion of the tape was 1×10^{-6} per °C, will be

(a) 0.0200 m (b) 0.0050 m
(c) 0.0005 m (d) 0.0001 m [ESE : 2011]

1.11 A rectangular plot of 16 km^2 in area is shown on a map by a similar rectangular area of 1 cm^2. R.F. of the scale to measure a distance of 40 km will be

(a) $\frac{1}{1600}$ (b) $\frac{1}{400000}$
(c) $\frac{1}{400}$ (d) $\frac{1}{16000}$ [ESE : 2011]

1.12 The magnitude of 'sag correction' during measurement of lengths by taping is proportional to the

(a) Cube of the weight of the tape, in kg per m run
(b) Cube root of the weight of the tape, in kg per m run
(c) Square of the weight of the tape in kg per m run
(d) Square root of the weight of the tape, in kg per m run

[ESE : 2011]

1.13 Working from the whole to the part is followed as the fundamental principle of surveying so as to

1. Distribute errors
2. Improve ease of working
3. Prevent accumulation of errors
4. Compensate errors in a way
5. Refer to a common datum, say MSL

(a) 1, 2 and 4 (b) 1, 3 and 5
(c) 3 and 4 (d) 2 and 5

[ESE : 2011]

1.14 If L is the length of the chain, W is the weight of the chain and T is the tension, the sag correction for the chain line is

(a) $\frac{W^2L^2}{24T^3}$ (b) $\frac{W^2L}{24T^2}$
(c) $\frac{W^2L^2}{24T^2}$ (d) $\frac{W^2L^3}{24T^3}$ [ESE : 2012]

1.15 In an inclined terrain, if the elevation difference between the two ends of a line is h and the inclined length of the line is L, the correction for slope is

(a) $\frac{h^2}{L^2}$ (b) $\frac{h^2}{2L^2}$
(c) $\frac{2h^2}{L^2}$ (d) $\frac{h^2}{2L}$ [ESE : 2012]

1.16 Which one of the following statements is correct?

(a) In a retrograde vernier, $(n - 1)$ divisions on the primary scale are divided into n divisions on the vernier scale.
(b) A double vernier consists of two simple verniers placed end-to-end forming one scale with the zero in the centre.

(c) In an extended vernier, $(2n + 1)$ primary divisions are divided into n divisions on the vernier.
(d) In a direct vernier, $(n + 1)$ primary divisions are divided into n equal divisions on the vernier scale.

[ESE : 2013]

1.17 The length of a line measured with a 30 m chain was found to be 734.6 m. It was afterwards found that the chain was 0.05 m too long. The true length of the line was:

(a) 630.82 m (b) 680.82 m
(c) 735.82 m (d) 780.92 m

[ESE : 2013]

1.18 The length of a survey line when measured with a chain of 20 m nominal length was found to be 841.5 m. If the chain used is 0.1 m too long, the correct length of the measured line is

(a) 845.7 m (b) 837.39 m
(c) 843.6 m (d) 839.4 m

[ESE : 2014]

1.19 Consider the following statements:
The general principles of surveying are

1. To work from part to whole.
2. To locate a new station by measurements from at least two fixed reference points already established and/or identifiable.

Which of the above statements are correct?

(a) 1 only (b) 2 only
(c) Both 1 and 2 (d) Neither 1 nor 2

[ESE : 2015]

1.20 Hypotensuall allowance is given by the expression (adopting standard conventions)

(a) $(1 - \sec\theta) \times$ measured distance
(b) $(1 - \cos\theta) \times$ measured distance
(c) $(\sec\theta - 1) \times$ measured distance
(d) $(\cos\theta - 1) \times$ measured distance

[ESE : 2016]

1.21 Survey of India was publishing toposheets using a scale of

(a) 1 : 1000 (b) 1 : 5000
(c) 1 : 10000 (d) 1 : 50000

[ESE : 2016]

1.22 **Statement (I):** The fundamental principle of surveying is 'to work from the whole to the part'.
Statement (II): Working from the whole to the part ensures prevention of accumulation of possible errors in survey work over large areas.

(a) Both Statement (I) and Statement (II) are individually true and Statement (II) is the correct explanation of Statement (I)
(b) Both Statement (I) and Statement (II) are individually true but Statement (II) is NOT the correct explanation of Statement (I)
(c) Statement (I) is true but Statement (II) is false
(d) Statement (I) is false but Statement (II) is true

[ESE : 2018]

1.23 Reconnaissance survey for determining feasibility and estimation of scheme falls under the classification based on the

(a) Nature of the field survey
(b) Object of surveying
(c) Instruments used
(d) Method employed

[ESE : 2019]

1.24 A 100 m tape is suspended between the ends under a pull of 200 N. If the weight of the tape is 30 N, the correct distance between the tape ends will be nearly

(a) 100.5 m (b) 100.3 m
(c) 100.1 m (d) 99.9 m

[ESE : 2020]

Directions: The following items consists of two statements; one labelled as **'Statement (I)'** and the other as **'Statement (II)'**. You are to examine these two statements carefully and select the answers to these items using the codes given below:

Codes:

(a) Both Statement (I) and Statement (II) are individually true and Statement (II) is the correct explanation of Statement (I)
(b) Both Statement (I) and Statement (II) are individually true but Statement (II) is NOT the correct explanation of Statement (I)
(c) Statement (I) is true but Statement (II) is false
(d) Statement (I) is false but Statement (II) is true

1.25 **Statement (I) :** Geodetic survey cannot be done for works requiring high precision.

Statement (II) : The curvature of earth is accounted for measurements in Geodetic survey.

[ESE : 2020]

1.26 **Statement (I):** Chain surveying is that type of surveying which only linear measurements are made in the field.

Statement (II): Traversing is that type of survey in which number of connected survey lines from framework and the directions and length the survey lines are measured with the help of and angle measuring instrument and a respectively.

[ESE : 2021]

1.27 Consider the following statements related to the classification based upon the object of survey:

1. Archaeological surveys for unearthing relics of antiquity.
2. Geological surveys for determining different strata in the Earth's crust.
3. Mine surveys for exploring mineral wealth such as gold, coal, etc.

Which of the above statements are correct?

(a) 1 and 2 only (b) 2 and 3 only
(c) 1 and 3 only (d) 1, 2 and 3

[ESE : 2021]

1.28 Which one of the following is used for determining different strata in the earth's crust?

(a) Mine survey
(b) Topographic survey
(c) Archaeological survey
(d) Geological survey

[ESE : 2022]

1.29 Which one of the following tapes is generally used for work of the highest precision?

(a) Linen tape
(b) Metric woven metallic tape
(c) Metric steel tape
(d) Invar tape

[ESE : 2022]

1.30 Which one of the following consists of horizontal and vertical location of certain points by linear and angular measurements and is made to determine the natural features of a country?

(a) Cadastral survey
(b) Topographical survey
(c) Astronomical survey
(d) Military survey

[ESE : 2023]

1.31 A line AB between the stations A and B was measured as 348.28 m using a 20 m tape, too short by 0.05 m What is the correct length of line AB?

(a) 349.15 m (b) 348.41 m
(c) 347.41 m (d) 346.15 m

[ESE : 2023]

1.32 The length of a ine originally 100 mm long on a map plotted to a scale of 1/1000, was found to be 96 mm due to shrinkage of the map. The map prepared using a tape of length 20 m was later found to be actually 20.03 m. If a certain area on the map, measured using a planimeter, is 282 mm^2, what is the correct area on the ground?

(a) 237 m^2 (b) 307 m^2
(c) 347 m^2 (d) 397 m^2

[ESE : 2023]

1.33 Match the following lists:

List-I (Correction)	**List-II** (Formula)
P. Absolute length (c_a)	1. $\frac{1}{24}\left(\frac{W}{P}\right)^2 L$
Q. Sag (c_g)	2. $\frac{h^2}{2L}$
R. Alignment (c_m)	3. $\frac{c}{l}L$

Select the correct answer using the code given below:

	P	Q	R
(a)	2	1	3
(b)	3	2	1
(c)	3	1	2
(d)	2	3	1

[ESE : 2024]

■■■■

Answers Fundamental Concepts of Surveying and Linear Measurement

1.1 (a)	1.2 (b)	1.3 (a)	1.4 (a)	1.5 (a)	1.6 (a)	1.7 (c)	1.8 (b)	1.9 (b)
1.10 (c)	1.11 (b)	1.12 (c)	1.13 (c)	1.14 (b)	1.15 (d)	1.16 (b)	1.17 (c)	1.18 (a)
1.19 (b)	1.20 (c)	1.21 (d)	1.22 (a)	1.23 (b)	1.24 (d)	1.25 (d)	1.26 (b)	1.27 (d)
1.28 (d)	1.29 (d)	1.30 (b)	1.31 (c)	1.32 (b)	1.33 (c)			

Explanations Fundamental Concepts of Surveying and Linear Measurement

1.2 (b)

Chain surveying is used for securing data for exact description and marking of the boundaries of a piece of land or for preparing the maps of the area to show various details. It is generally used for plans of estates, fields, etc. on a large scale when the area is small in extent and the ground is fairly level and open.

The cross-staff survey is a special type of chain survey conducted to locate the boundaries of a field for the purpose of determining the area of the field.

1.3 (a)

The principles of traversing say that, a traverse should consist of a series of straight lines connected together to form an open or a closed polygon. The points defining the ends of the traverse lines are called traverse stations or traverse points.

1.4 (a)

Topographic Surveys are used to identify and map the contours of the ground and existing features on the surface of the earth or slightly above or below the earth's surface (i.e. trees, buildings, streets, walkways, manholes, utility poles, retaining walls, etc. It helps determine natural features of a country.

The **reconnaissance survey** is an study of an entire area that might be used for a road or airfield. It helps determine possibility and rough cost of surveying system to be adopted.

Cadastral surveying is concerned with land management, and more specifically with issues of land ownership, measurement and delineation of property boundaries.

In an **archaeological survey**, the archaeologists collect information about the location, distribution and organization of past human cultures and about old and relic structures.

1.5 (a)

1 acre = 4840 sq. Yard

1 sq. Yard = 9 sq feet

1 acre = 43560 sq ft.

1 acre = 10 x (Gunter's chain)^2

1 acre = 40 Guntha

1.6 (a)

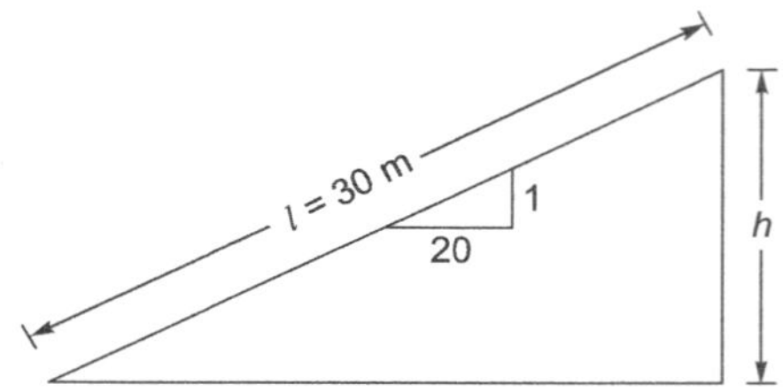

$$\text{Slope correction} = \frac{h^2}{2l}$$

Difference in elevation between points,

$$h = \frac{1}{\sqrt{(20)^2+1}} \times 30 \simeq 1.50 \text{ m}$$

$$\therefore \text{Slope correction} = \frac{(1.50)^2}{2\times 30} = 0.0375 \text{ m} = 3.75 \text{ cm}$$

1.7 (c)

$$\text{Actual distance} = \frac{(30-0.1)}{30} \times 300 = 299.0 \text{ m}$$

1.8 (b)

For the first 2000 m, average error is

$$e = \frac{0+10}{2} = 5 \text{ cm} = 0.05 \text{ m}$$

$\therefore$ Incorrect length of chain,

$$L' = 20 + 0.05 = 20.05 \text{ m}$$

Measured length, l' = 2000 m

$$\therefore \quad \text{True length, } l_1 = \left(\frac{L'}{L}\right) \times l'$$

$$= \left(\frac{20.05}{20}\right) \times 2000$$
$$= 2005 \text{ m}$$

For the next 2000 m, average error is

$$e = \frac{10 + 18}{2} = 14 \text{ cm} = 0.14 \text{ m}$$

$$\therefore \quad L' = 20 + 0.14 = 20.14 \text{ m}$$
$$l' = 2000 \text{ m}$$

$$\therefore \text{ True length, } l_2 = \left(\frac{L'}{L}\right) \times l'$$
$$= \left(\frac{20.14}{20}\right) \times 2000 = 2014 \text{ m}$$

Hence, true distance,

$$l = l_1 + l_2$$
$$= 2005 + 2014 = 4019 \text{ m}$$

1.9 (b)

1 inch = 2.54 cm

1 feet = 0.3048 m

Given that 1 inch = 50 ft.

$$\Rightarrow \quad 2.54 \text{ cm} = 50 \times 0.3048 \text{ m}$$
$$\Rightarrow \quad 1 \text{ cm} = \frac{15.24}{2.54} \text{ m}$$
$$\Rightarrow \quad 1 \text{ cm} = 6 \text{ m}$$

1.10 (c)

Correction for temperature.

$$C_t = \alpha(T_m - T_0)L$$

where α = coefficient of thermal expansion

T_m = mean temperature in the field during measurement

T_0 = the standard temperature for the tape

L = measured length

$$C_t = 1 \times 10^{-6}(30 - 20) \times 50$$
$$= 0.0005 \text{ m}$$

1.11 (b)

R.F. is ratio of two measurements with common unit.

$$\text{R.F.} = \left[\frac{1 \text{ cm}^2}{16 \times (10^3)^2 \times (10^2)^2 \text{cm}^2}\right]^{1/2}$$
$$= \frac{1}{400000}$$

1.12 (c)

The sag correction C_s is given by

$$C_s = \frac{l(wl)^2}{24 P^2}$$

where l = the length of the tape (in metres) suspended between the supports

P = pull applied in kg or N

w = weight of the tape in kg or N per metre run

C_s = sag correction in metres for length l

wl = weight of tape suspended between the supports.

1.13 (c)

Two basic principles of surveying are:

1. Location of a point by measurement from two points of reference.
2. Working from whole to part: The second ruling principle of surveying, whether plane or geodetic, is to work from whole to part.

   The idea of working in this way is to prevent the accumulation of errors and to control and localise minor errors which, otherwise, would expand to greater magnitudes if the reverse process is followed, thus making the work uncontrollable at the end.

1.14 (b)

When a tape is suspended from each end, it forms a curve called a catenary . Because the curve is longer than the straight-line distance between the ends of the tape, the tape always reads longer than the actual distance. It is corrected by sag correction.

1.15 (d)

Correction due to slope

$$C_s = \frac{h^2}{2L} \text{ (Subtractive)}$$

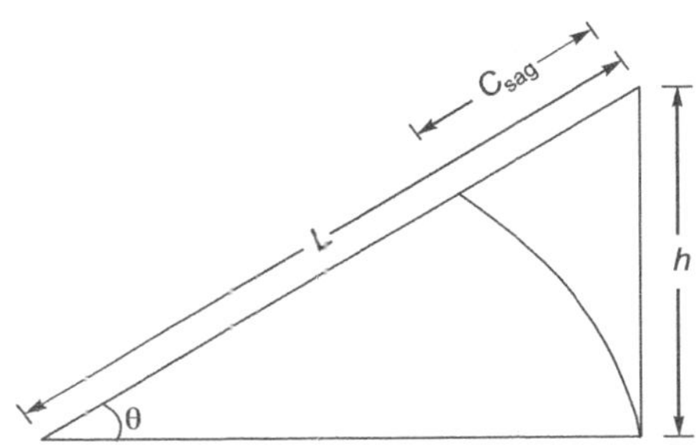

where,

h = difference in elevation between the ends

L = Inclined length measured

l = Horizontal length

C_s = Correction due to sag

1.16 (b)

1. For Direct Vernier
   $(n-1)$ division on main scale = n division on vernier.
2. For retrograde vernier
   $(n+1)$ division on main scale = n division on vernier.
3. For double vernier
   Combination of the direct vernier
4. For inextended vernier
   $(2n-1)$ division on main scale = n division on vernier.

1.17 (c)

Correct length of chain

$$= 30 + 0.05 \text{ m} = 30.05 \text{ m}$$

∴ True length of the line

$$= \frac{734.6 \times 30.05}{30} = 735.82 \text{ m}$$

1.18 (a)

$$L \times l = L_1 \times l_1$$

$$\Rightarrow \quad L = 841.5 \text{ m}, l = 20.1 \text{ m}, l_1 = 20 \text{ m}$$

$$\Rightarrow \quad L_1 = \frac{841.5 \times 20.1}{20}$$

$$L_1 = 845.7 \text{ m}$$

1.19 (b)

The general principle of surveying is to work from whole to part.

1.20 (c)

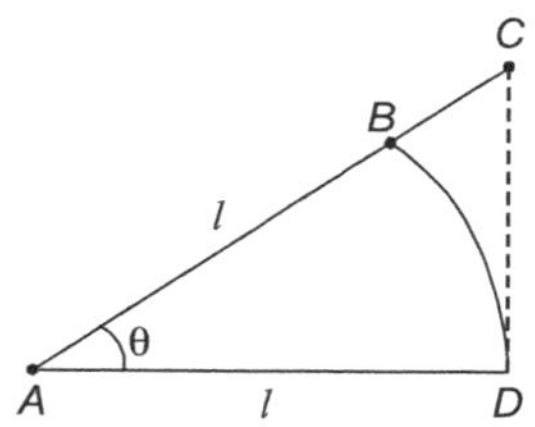

$$c_h = BC = AC - AB$$

$$= (l\sec\theta - l) = l(\sec\theta - 1)$$

1.21 (d)

Topographical maps were printed on 1: 2,50,000 1: 50,000 and 1: 25,000 scales by survey of India before the adoption of nation mapping policy in 2005.

1.23 (b)

On the basis of object of survey the classified can be given as below:

1. **Engineering survey:** To collect requisite data for planning, design and execution of engineering projects. Three broad steps are:
   (a) *Reconnaissance survey:* For determining feasibility and estimation of scheme.
   (b) *Preliminary survey:* Data collection for preparing map/plan for design and planning
   (c) *Location survey:* To set out work on ground for actual execution of the project.
2. **Military survey:** For detemining points of strategic importance.
3. **Mine survey:** For exploring mineral wealth.
4. **Geological survey:** For determining different strata is earth's crust.
5. **Archaeological survey:** For unearthing relics of antiquity.

1.24 (d)

Sag correction,

$$C_s = \frac{w^2 l}{24P_m^2} = \frac{30^2 \times 100}{24 \times 200^2}$$

$$= \frac{30 \times 30 \times 100}{24 \times 200 \times 200} = \frac{9}{24 \times 2 \times 2} = \frac{9}{48 \times 2}$$

$$= 0.09375 \approx 0.1 \text{ (negative)}$$

Measured length of tape = 100 m

Correct length of tape

$$= 100 - 0.1 = 99.9 \text{ m}$$

1.25 (d)

Geodetic survey is done for high precision works.

1.26 (b)

Chain surveying: This is the simplest type of surveying in which only linear measurements are taken with a chain or tape.

Traverse: A traverse is a series of connected lines whose lengths and directions are measured in the field.

Both the statements are individually correct.

1.27 (d)

Classification based on the object of survey:

1. **Engineering survey:** This is undertaken for the determination of quantities or to afford sufficient data for the designing of engineering works such as roads and reservoirs, or those connected with sewage disposal or water supply.
2. **Military survey:** This is used for determining points of strategic importance.
3. **Mine survey:** This is used for exploring mineral earth.
4. **Geological survey:** This is used for determining different strata in the earth's crust.
5. **Archaeological survey:** This is used for unearthing relics of antiquity.

1.28 (d)

Geological survey is used to determine different strata in earth crust.

- Mine survey is used to explore minerals.
- Archaeological survey is used to discover and be certain of the locations of ancient artifacts.
- Topographic survey is used to determine natural features of country such as river, lake, steam lake, wood, hill etc.

1.29 (d)

Invar tape has highest precision.

1.30 (b)

Topographical survey consists of horizontal and vertical location of certain points by linear and angular measurement and is made to determine the natural features of a country such as rivers, streams, lakes hills etc. and artificial features as roads, railways, canals, towns and villages.

1.31 (c)

Given,

Designated length of tape, $L = 20$ m

Incorrect length of tape,

$L' = 20 - 0.05 = 19.95$ m

Measured length of line AB, $l' = 348.28$ m

Let, True length of $AB = l$

As we know, $L \times l = L' \times l'$

$\Rightarrow \quad 20 \times l = 19.95 \times 348.28$

$\Rightarrow \quad l = 347.41$ m

1.32 (b)

Shrunk scale = Shrinkage factor × Original scale

$$= \left(\frac{96}{100}\right) \times \frac{1}{1000}$$

$$= \frac{96}{100} \times \frac{1\text{ mm}}{1\text{ m}} = \frac{1\text{ mm}}{\left(\frac{1}{0.96}\right)\text{m}}$$

$$1\text{ mm} = \left(\frac{1}{0.96}\right)\text{m}$$

$$\therefore \text{Scale for area} = 1\text{ mm}^2 = \left[\frac{1}{0.96}\right]^2 \text{m}^2$$

So, measured area of the ground,

$$\Rightarrow A' = 282 \times \left(\frac{1}{0.96}\right)^2 = 305.98958\text{ m}^2$$

Designated length of tape,

$L = 20$ m

Incorrect length of tape, $L' = 20.03$ m

$$\text{So, true area} = \left[\frac{L'}{L}\right]^2 \times \text{measured area}$$

$$= \left[\frac{20.03}{20}\right]^2 \times 305.98958$$

$$= 306.91\text{ m}^2 \simeq 307\text{ m}^2$$

1.33 (c)

$$\text{Correction for absolute length } (C_c) = \frac{C}{l}L$$

$$\text{Correction for sag } (C_g) = \frac{1}{24}\left(\frac{W}{P}\right)^2 L$$

$$\text{Correction for alignment } (C_m) = \frac{h^2}{2L}$$

■■■■

2 Compass Surveying and Traverse Surveying

2.1 If the declination is 5°40′*W*, which one of the following magnetic bearing would represent the true bearing of *S*25°20′*E*?

(a) *S*19° 20′*E* (b) *S*31° 0′*E*
(c) *S*20° 0′*E* (d) *S*19° 20′*W*

[ESE : 2000]

2.2 Consider the following steps:

1. Calculation of ΣL and ΣD.
2. Correction of latitudes and departures.
3. Calculation of bearings.
4. Calculation of interior angles.
5. Calculation of independent angles.

The correct sequence of these steps in Gale's traverse table calculations is

(a) 3, 4, 5, 2, 1 (b) 4, 3, 1, 2, 5
(c) 2, 1, 3, 4, 5 (d) 4, 3, 5, 2, 1

[ESE : 2001]

2.3 In a closed traverse with five sides, the error found from the fore bearing and back bearing of the last line is + 2°. The correction to the third line will be

(a) 0°24′ (b) 0°48′
(c) 1°12′ (d) 1°36′

[ESE : 2001]

2.4 **Assertion (A):** Whole circle bearing of a line is preferred to a quadrantal bearing.

Reason (R): Bearing is completely specified by an angle.

(a) both A and R are true and R is the correct explanation of A
(b) both A and R are true but R is not a correct explanation of A
(c) A is true but R is false
(d) A is false but R is true

[ESE : 2001]

2.5 The magnetic needle in a prismatic compass is placed

(a) at the bottom of the graduated aluminium ring
(b) above the graduated aluminium ring
(c) below the brass box
(d) below the needle lifter, but above the bottom inside the compass

[ESE : 2003]

2.6 **Assertion (A):** In a theodolite if the lower clamp is not properly clamped or the instrument is not firmly tightened on the tripod head, the error introduced is known as 'slip'.

Reason (R): If the shifting head is loose, the error 'slip' will be introduced.

(a) both A and R are true and R is the correct explanation of A
(b) both A and R are true but R is not a correct explanation of A
(c) A is true but R is false
(d) A is false but R is true [ESE : 2003]

2.7 Which one of the following pairs is not correctly matched?

(a) Declination : Horizontal angle between magnetic meridian and true meridian
(b) Bowditch's rule : Employed to adjust closing error of a closed traverse
(c) Deflection angle : Measured in case of open traverse instead of measuring included angle
(d) Reconnaissance survey : Employed for detailed and precise survey

[ESE : 2005]

2.8 The magnetic bearing of a line *AB* is *S* 30°*E*. If the declination is 6° West, then what is the true bearing?

(a) S 36° E (b) N 36° W
(c) S 24° E (d) N 24° W

[ESE : 2006]

2.9 Diurnal variation is greater
(a) in winter than in summer
(b) at smaller latitudes than at higher latitudes
(c) at magnetic equator points
(d) in summer than in winter

[ESE : 2006]

2.10 There are ten instrument stations occupied in succession during a traverse survey. An observer makes equal error in each station, the magnitude of which is $\delta\theta$ in each instance at all the stations. What is the probable error of the final bearing at the end of the traverse?
(a) $\pm 10\,\delta\theta$ (b) $+ 100\,(\delta\theta)^2$
(c) $\pm 10\sqrt{\delta\theta}$ (d) $+ 10\sqrt{\delta\theta}$

[ESE : 2007]

2.11 The whole circle bearings of lines *AB* and *BC* are 30° 15′ and 120° 30′. What is the included angle *ABC* between the lines *AB* and *BC*?
(a) 229° 45′ (b) 89° 45′
(c) 269° 45′ (d) 90° 15′

[ESE : 2008]

2.12 What is $\angle ABC$ if *FB* of line *AB* is 40° and *BB* of line *BC* is 280°?
(a) 90° (b) 120°
(c) 240° (d) 320°

[ESE : 2008]

2.13 Which one of the following is carried out by two theodolite method?
(a) Circular curve ranging
(b) Tacheometric survey
(c) Geodetic survey
(d) Astronomical survey

[ESE : 2008]

2.14 The whole circle bearings of lines *OP* and *OQ* are 18° 15′ and 335° 45′ respectively. What is the value of the included angle *QOP*?
(a) 307° 30′ (b) 42° 30′
(c) 354° 00′ (d) 177° 00′

[ESE : 2009]

2.15 If L is the perimeter of a closed traverse, ΔD is the closing error in departure, the correction for the departure of a traverse side of length L_1, according to Bowditch rule, is
(a) $\Delta D \frac{L}{L_1}$ (b) $\Delta D \frac{L_1^2}{L}$
(c) $\frac{LL_1}{\Delta D}$ (d) $\Delta D \frac{L_1}{L}$

[ESE : 2010]

2.16 In a transit theodolite, error due to eccentricity of verniers is eliminated by reading
(a) both verniers
(b) both right swing and left swing
(c) right and left faces
(d) different parts of main scale

[ESE : 2010]

2.17 Consider the following pairs in the context of a theodolite:

1. Plunging	:	The process of turning the telescope over its supporting axis through 180° in a vertical plane.
2. Face left	:	When the vertical circle of the theodolite is on the left of the observer while taking the reading.
3. Telescope normal	:	It implies 'bubble up' and face of the vertical circle left.
4. Swinging the telescope	:	It implies turning the telescope in a vertical plane.

Which of these pairs are correctly matched in case of theodolite?
(a) 1, 2 and 4 (b) 1, 3 and 4
(c) 2, 3 and 4 (d) 1, 2 and 3

[ESE : 2010]

2.18 Consider the following in the context of variations in magnetic declination:
1. Secular 2. Diurnal
3. Annual 4. Regular
Which of these are relevant?
(a) 1, 2 and 3 (b) 1, 2 and 4
(c) 2 and 3 only (d) 3 and 4 only

[ESE : 2010]

2.19 Consider the following:
1. Line ranger
2. Reciprocal ranging
3. Random line method
4. Optical square

Which of these are the correct methods of ranging employed to solve the problem of vision obstructed but with chaining free?

(a) 1, 2, 3 and 4 (b) 2 and 3 only
(c) 2 and 4 only (d) 3 and 4 only

[ESE : 2010]

2.20 In a closed traverse *ABC*, following readings were taken

Line	Fore Bearing	Back Bearing
AB	20°	201°
BC	101°	278°
CA	278°	50°

Station *A* is free from local attraction. Correct bearing of *CB* is

(a) 275° (b) 276°
(c) 281° (d) 280°

[ESE : 2011]

2.21 **Assertion (A):** The angle made by the lines of the magnetic force with the earth's surface is called dip

Reason (R): In the northern hemisphere, the south end of the needle dips downwards; and in the southern hemisphere, the north end of the needle dips downwards.

(a) both A and R are true and R is the correct explanation of A
(b) both A and R are true but R is not a correct explanation of A
(c) A is true but R is false
(d) A is false but R is true

[ESE : 2011]

2.22 **Assertion (A):** The diurnal variation is the variation of the declination in a year from the mean position during the year.

Reason (R): The diurnal variation is greater in summer than in winter.

(a) both A and R are true and R is the correct explanation of A
(b) both A and R are true but R is not a correct explanation of A
(c) A is true but R is false
(d) A is false but R is true [ESE : 2011]

2.23 If the whole circle bearing is 315° 20′, its quadrantal bearing would be

(a) *S* 36° 30′ *W* (b) *N* 44° 40′ *W*
(c) *N* 57° 24′ *W* (d) *S* 60° 40′ *W*

[ESE : 2012]

2.24 If the observed fore bearing of a line xy is 16° 26′, the back bearing of this line is

(a) 103° 26′ (b) 118° 36′
(c) 196° 26′ (d) 206° 26′

[ESE : 2012]

2.25 The needle of a magnetic compass is generally supported on a

(a) Bush bearing (b) Ball bearing
(c) Needle bearing (d) Jewel bearing

[ESE : 2012]

2.26 The bearing of line *AB* is 150° and the angle *ABC* is 124°. Bearing of line *BC* is

(a) 94° (b) 98°
(c) 198° (d) 90° [ESE : 2013]

2.27 Which of the following set of terms does not relate to operation of a theodolite?

(a) Transiting and inverting
(b) Face left and face right
(c) Right swing and left swing
(d) Gauging and sounding

[ESE : 2013]

2.28 The magnetic bearing of a line is 55° 30' and the magnetic declination is 4° 30' east. The true bearing of a line will be

(a) 60° (b) 34°30'
(c) 49°30' (d) 51° [ESE : 2014]

2.29 In a transit theodolite, any incidental error due to eccentricity of verniers is primarily counteracted by

(a) Reading both the verniers
(b) Reading different parts of main scale
(c) Reading right and left faces
(d) Taking both right swing readings

[ESE : 2014]

2.30 The Whole Circle Bearing of line *AB* is 50° and of line *BC* is 120°. The deflection angle at *B* at *AB* to *BC* is

(a) 50° (b) 70°
(c) 110° (d) 120°

[ESE : 2015]

2.31 Regarding a Prismatic Compass, which one of the following statements in correct?
(a) The object is sighted first. The observer then moves to the side of the object vane to take the reading.
(b) Sighting and reading are done simultaneously.
(c) The readings are taken from the north end.
(d) The compass has an edge bar needle.
[ESE : 2015]

2.32 For minor adjustments of horizontal angles measured using a theodolite, the tangential screw is adjusted after
(a) both the plates are unclamped
(b) the lower plate in clamped and the upper plate is unclamped
(c) the upper plate is clamped and the lower plate is unclamped
(d) both the plates are clamped [ESE : 2015]

2.33 Consider the following statements:
1. The component of the distance between two points measured in the north-south direction is called the latitude of the line, between the points.
2. The component of the distance between two points measured in the east-west direction is called the departure of the line, between the points.
3. The latitude is considered as positive when reckoned southward.
4. The departure is considered as negative when reckoned westward.

Which of the above statements are correct?
(a) 1, 2 and 3 only (b) 2, 3 and 4
(c) 1, 2 and 4 only (d) 1, 2, 3 and 4
[ESE : 2015]

2.34 The magnetic bearing of a line (on full-circle mensuration basis) is 55° 30′ and the magnetic declination is 5° 30′ East. The true magnetic bearing of the line will be
(a) 61°00′ (b) 55°30′
(c) 40°00′ (d) 50°00′ [ESE : 2016]

2.35 In any closed traverse, if the survey work is error free, then
1. The algebraic sum of all the latitudes should be equal to zero.
2. The algebraic sum of all the departures should be equal to zero.
3. The sum of the northings should be equal to the sum of the southings.

Which of the above statements are correct?
(a) 1 and 2 only (b) 1 and 3 only
(c) 2 and 3 only (d) 1, 2 and 3
[ESE : 2016]

2.36 Turning of the theodolite telescope in vertical plane by 180° about the horizontal axis is known as
(a) Setting (b) Centering
(c) Transiting (d) Swinging
[ESE : 2016]

2.37 Which of the following are among the fundamental lines of a theodolite?
1. The vertical and horizontal axes.
2. The lines of collimation and axis of the plate levels.
3. The bubble line of the altitude level.

(a) 1 and 2 only (b) 1 and 3 only
(c) 2 and 3 only (d) 1, 2 and 3 only
[ESE : 2016]

2.38 A line PQ in an old map had magnetic bearing of N 57° E when the local magnetic declination was 2° E. If the magnetic declination is now 4° W, what will be the magnetic bearing of the line PQ now
(a) N 52° W (b) N 63° E
(c) N 54° E (d) N 52° E
[ESE : 2017]

2.39 **Statement (I):** Compass survey is still used by Geologists to locate the magnetic ores.
Statement (II): Local attraction causes errors in compass survey due to terrestrial features - either natural or man-made.
(a) Both Statement (I) and Statement (II) are individually true and Statement (II) is the correct explanation of Statement (I)
(b) Both Statement (I) and Statement (II) are individually true but Statement (II) is NOT the correct explanation of Statement (I)
(c) Statement (I) is true but Statement (II) is false
(d) Statement (I) is false but Statement (II) is true
[ESE : 2018]

2.40 In an old map, a line AB was drawn to a magnetic bearing of 5°30′, the magnetic declination at the time being 1° East. If the present magnetic declination is 8°30′ East, the line should be set to a magnetic bearing of

(a) 358° (b) 2°
(c) 6°30′ (d) 357°

[ESE : 2019]

2.41 Which of the following statements with reference to isogonic line are correct in magnetic declination?

1. It is drawn through the points of same declination.
2. It does not form complete great circle.
3. It radiates from north and south magnetic regions and follow irregular paths.

(a) 1 and 2 only (b) 1 and 3 only
(c) 2 and 3 only (d) 1, 2 and 3

[ESE : 2020]

2.42 Which one of the following is correct for Prismatic Compass?

(a) The graduated ring rotates with line sight.
(b) Instrument cannot be used without tripod.
(c) The graduation are engraved inverted.
(d) The readings can directly be taken of seeing through the top of the glass.

[ESE : 2021]

2.43 Magnetic declination at a place is the horizontal angle between:

(a) the true meridian and the arbitrary meridian
(b) the magnetic meridian and the arbitrary meridian
(c) the true bearing and the magnetic bearing
(d) the true meridian and the magnetic meridian

[ESE : 2021]

2.44 The magnetic bearing of a line AB is S28°30′E. What is the true bearing of line AB if the magnetic declination is 7°30′ toward west?

(a) S36°E (b) N21°W
(c) S21°E (d) N36°W

[ESE : 2021]

2.45 What is the difference of longitude between two places C and D from the following longitudes?

1. Longitude of C = 46° W
2. Longitude of D = 64° W

(a) 18° (b) 36°
(c) 110° (d) 220°

[ESE : 2022]

■■■■

Answers Compass Surveying and Traverse Surveying

2.1 (b)	2.2 (b)	2.3 (c)	2.4 (a)	2.5 (a)	2.6 (b)	2.7 (d)	2.8 (a)	2.9 (d)
2.10 (a)	2.11 (b)	2.12 (b)	2.13 (a)	2.14 (b)	2.15 (d)	2.16 (a)	2.17 (d)	2.18 (a)
2.19 (b)	2.20 (d)	2.21 (c)	2.22 (d)	2.23 (b)	2.24 (c)	2.25 (d)	2.26 (a)	2.27 (d)
2.28 (a)	2.29 (a)	2.30 (b)	2.31 (b)	2.32 (d)	2.33 (c)	2.34 (a)	2.35 (d)	2.36 (c)
2.37 (d)	2.38 (b)	2.39 (d)	2.40 (a)	2.41 (d)	2.42 (c)	2.43 (d)	2.44 (a)	2.45 (a)

Explanations Compass Surveying and Traverse Surveying

2.1 (b)

Note that negative (W) declination should be added to quadrantal bearing in second and fourth quadrant i.e. $N\ \theta W$ or $S\ \theta E$ and it should be subtracted from quadrantal bearing in first and third quadrant i.e., $N\theta E$ and $S\theta W$. Reverse should be done for positive (E) declination.

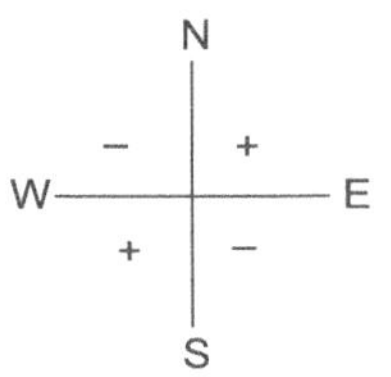

For Positive Declination

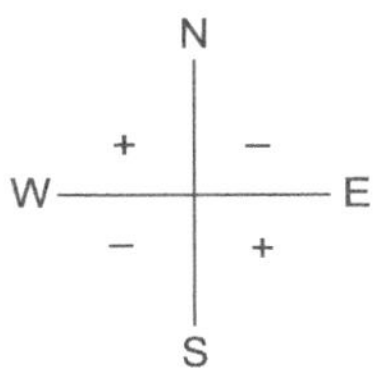

For Negative Declination

$$\text{True bearing} = S\,25°\,20'\,E + 5°\,40'$$
$$= S\,31°\,0'\,E$$

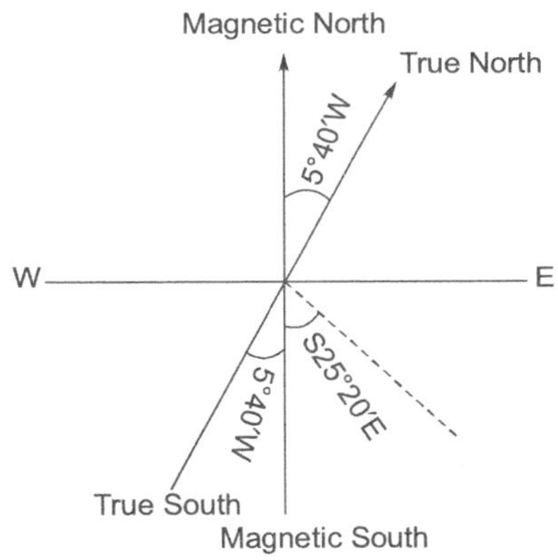

2.2 (b)

The sequence of steps in Gale's traverse table are:

(i) Writing names of traverse station.
(ii) Calculation of interior angles of traverse.
(iii) Correction of angle of disclosure (if any) to obtain correct interior angles.
(iv) Calculation of bearing of all lines with help of initial line bearing and corrected angles.
(v) calculation of latitude and departure of each line.
(vi) Calculation of closing error.
(vii) Determination of correct latitude and departure after applying correction for closing error.
(viii) Calculation of independent coordinates.

2.3 (c)

Number of sides, $n = 5$

Correction to the first line = $e/n = 0°24'$

Correction to the second line = $2e/n = 0°48'$

Correction to the third line = $3e/n = 1°12'$

Correction to the fourth line = $4e/n = 1°36'$

Correction to the fifth line = $5e/n = 2°0'$

2.4 (a)

Whole circle bearing's value ranges from 0° to 360° and is simple to measure. Therefore it is preffered to quadrantal bearing.

2.5 (a)

The *magnetic needle* is attached to a *graduated aluminium ring. The ring is graduated from* 0° to 360°. Degrees are subdivided into half degrees.

2.6 (b)

Slip is an *error which* is introduced *if* the lower *clamp* is *not* properly *clamped,* the shifting head is loose and the instrument is *not* firmly *tightened* on the tripod head.

2.7 (d)

Reconnaissance survey is done to examine the general character of the area for the purpose of determining the most feasible arrangements. It is preliminary inspection, done to have idea of terrain and features of ground. It is not detailed and precise.

2.8 (a)

$$\text{True bearing} = S\,30°\,E + 6°\,W = S\,36°\,E$$

2.9 (d)

Diurnal variation of magnetic declination is due to

(i) geographical position of the place (lesser near equator and increases towards the pole)
(ii) the time of day (more in day)
(iii) season of the year (more in summer)
(iv) the year of the cycle of the secular variation

2.11 (b)

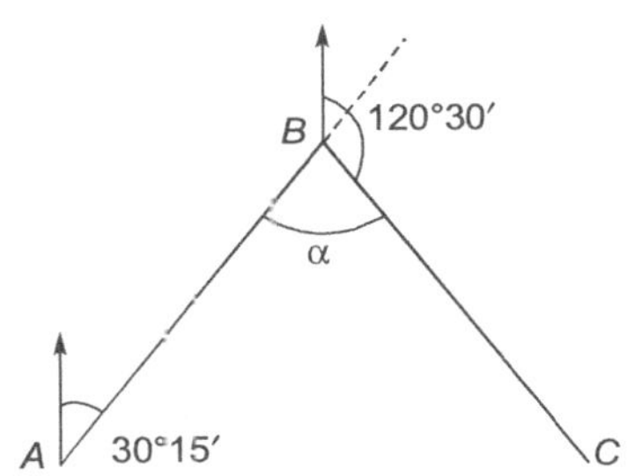

The included angle

$$\alpha = BB \text{ of } AB - FB \text{ of } BC$$
$$= (FB \text{ of } AB + 180°) - FB \text{ of } BC$$
$$= (30°15' + 180°) - 120°30' = 89°45'$$

2.12 (b)

$$BB \text{ of } BC = FB \text{ of } BC + 180°$$
$$\Rightarrow \quad FB \text{ of } BC = 280° - 180°$$
$$\Rightarrow \quad FB \text{ of } BC = 100°$$

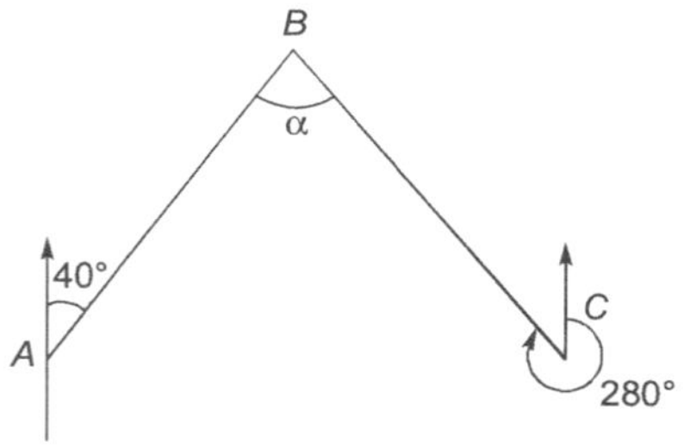

$$BB \text{ of } AB = FB \text{ of } AB + 180°$$
$$= 40° + 180° = 220°$$

∴ Included angle, α

$$= BB \text{ of } AB - FB \text{ of } BC = 220° - 100° = 120°$$

2.13 (a)

Circular curve ranging is carried out by two theodolite method. Only angular measurements (no linear measurements) are done in two theodolite method. This gives very accurate results even in rough terrains.

2.14 (b)

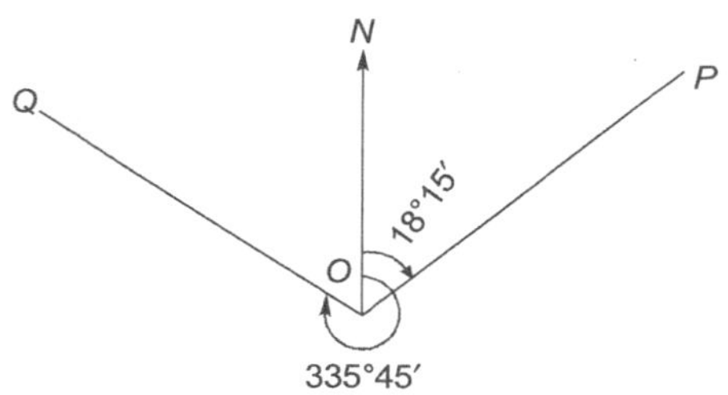

$$\angle NOQ = 360° - 335°45'$$
$$= 24°15'$$

∴ Included angle,

$$QOP = 24°15' + 18°15' = 42°30'$$

2.15 (d)

According to Bowditch rule

Correction for departure of length = Closing error in departure × $\dfrac{\text{Length of traverse side}}{\text{Perimeter of transverse}}$

$$= \Delta D \times \frac{L_1}{L}$$

2.16 (a)

If the zeros of the verniers *A* and *B* are not at the ends of the same diameter and exactly 180° apart, the error occurs. In that case the difference between the two vernier readings will be constant but slightly different from 180°. The error can be eliminated by taking the readings of both the verniers and taking the mean of the two readings. It may be noted that if the vernier is eccentric, no error is introduced in the angle so long as the same vernier is used for the initial and final reading. It is clear from the above discussions that by taking the mean of the two vernier readings, the error due to eccentricity of the centres and that due to eccentricity of verniers both are eliminated.

2.17 (d)

Swinging the telescope is the process of turning the telescope about the vertical axis in a horizontal plane. The swing is called right swing when the telescope is turned clockwise from the previous position. The swing is called left swing when the telescope is turned counter-clockwise.

2.18 (a)

Magnetic declination can be categorized as:

- Daily *declination*
- Annual *declination*
- Secular *declination*
- Irregular declination

2.20 (d)

In a closed traverse with no local attraction

$$FB \sim BB = 180°.$$

Since station *A* is free of local attraction thus

$$FB_{AB} = 20°$$

But, $BB_{AB} - FB_{AB} = 180°$

∴ *A* correction of –1° is applied at station *B*.

$$\therefore \quad FB_{BC} = 101° - 1° = 100°$$

But, $BB_{BC} - FB_{BC} = 178°$.

Therefore a correction of +2° is applied at station *C*.

∴ Correct bearing of *CB* is 280°.

2.21 (c)

The lines of force of earth's magnetic field run generally from south to North. Near the equator, they are parallel to the earth's surface. The horizontal projections of the lines of force define the magnetic meridian. The angle which these lines of force make with the surface of the earth is called the angle of dip or simply the dip of the needle. In elevation, these lines of force (i.e. the North end of the needle), are inclined downward towards the north in the Northern hemisphere and downward towards South in Southern hemisphere. At a place near 70° North latitude and 90° West longitude, it will dip 90°. This area is called North magnetic pole. A similar area in Southern hemisphere is called the South magnetic pole. Since the lines of force are parallel to the surface of the earth only at equator, the dip of the needle will be zero at equator and the needle will remain horizontal.

2.22 (d)

Types of variation in declination are:

(a) Durnal Variation: The diurnal variation or daily variation is the systematic departure of the declination from its mean value during a period of 24 hours. The extent of daily variations depend upon the following factors:
   (i) The locality: More at magnetic poles and less at equator.
   (ii) Season of the year: Considerably more in summer than in winter.
   (iii) Time: More in day and less in night. The rate of variation during 24 hours is variable.
   (iv) The amount of daily variation changes from year to year.

(b) Annual Variation: The variation which has a yearly period is known as annual variation.

(c) Secular variation follows a sine-curve pattern. It swings like a pendulum. For a given place, the compass needle after moving continuously for a period of years in one direction with respect to the true North, gradually comes to a stand still and then begins to move in opposite direction.

(d) Irregular Variation: The irregular variations are due to what are known as 'magnetic strorms', earthquakes and other solar influences.

2.23 (b)

The given line lies in IVth quadrant

∴ quadrantal bearing

$$= 360° - 315° 20'$$
$$= 44°40' = N44° 40' W$$

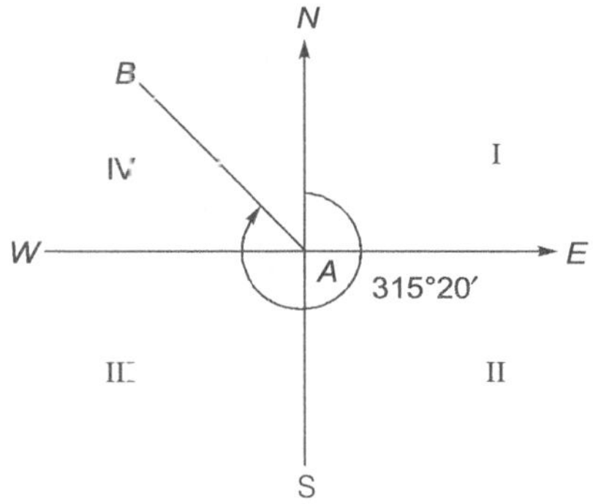

2.24 (c)

$$B.B. = F.B. \pm 180°$$

The + sign is used when $F.B. < 180°$

$$\therefore \quad B.B. = 16° 26' + 180°$$
$$= 196° 26'$$

2.25 (d)

The compass needle is made of a slender symmetrical bar of magnetised steel or iron. It is hung from a conical jewel bearing supported on a sharp, hardened steel pivot.

2.26 (a)

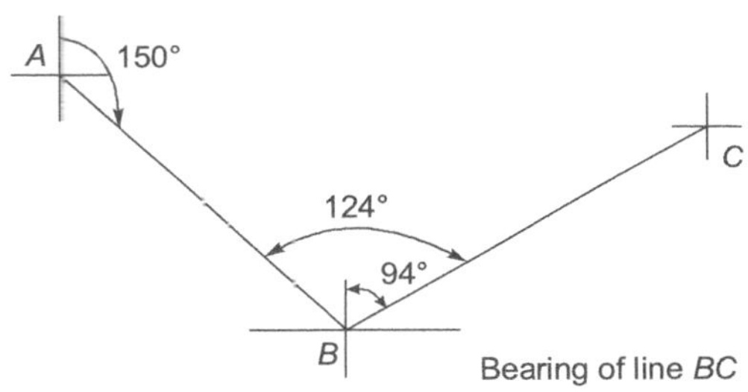

Bearing of line *BC*

2.27 (d)

Transiting is the process of rotating the telescope over the horizontal axis through 180 in the vertical plane. The directions of the objective end and eyepiece are reversed, and the telescope points exactly in the opposite direction. It is also known as **plunging** or reversing.

If the vertical circle is on the right side of observer, the thodolite is in face right condition. If the vertical circle is on the left side of observer, the thodolite is in face left condition.

Swinging is the process of turning the telescope about the vertical axis in a horizontal plane.

The swing is called right swing when the telescope is turned clockwise from the previous position. The swing is called left swing when the telescope is turned anticlockwise from the previous position.

Sounding is the process of determination of bed depth of water body in hydrographic surveying.

Gauges are instruments used to determine the water level and variation with time.

2.28 (a)

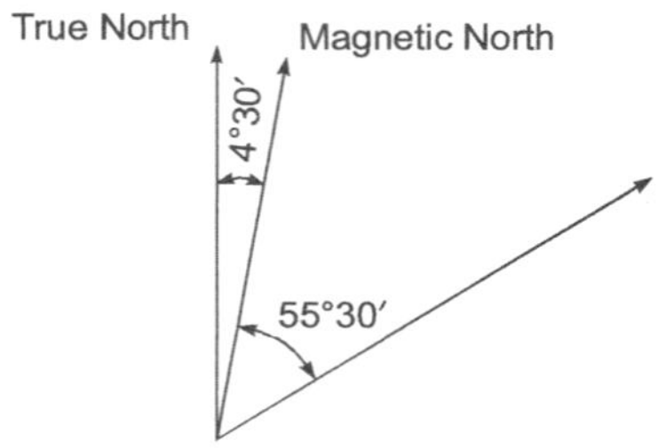

∴ True bearing

= Magnetic bearing + Magnetic declination (East)

= 55°30' +4° 30' = 60°

2.29 (a)

In transit theodolite, any incidental error due to eccentricity of verniers is primarily countracted by reading both the verniers.

2.30 (b)

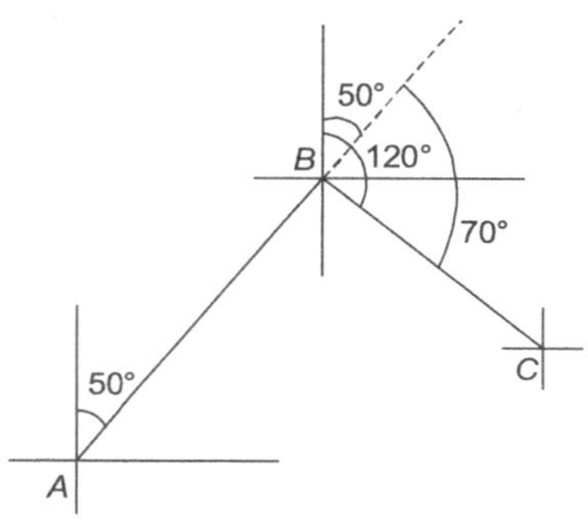

∴ Deflection angle = 120° – 50° = 70°

2.31 (b)

In a prismatic compass, the sighting of the object and the reading of the dial are done simultaneously, whereas in a surveyor's compass, first the object is sighted and then reading of the ring is taken by moving round and looking down the glass cover.

2.32 (d)

For minor adjustment of horizontal angles, measured using a theodolite, the tangential screw is adjusted after both the plates are clamped.

2.33 (c)

The latitude is considered as positive when reckoned northward and negative when reckoned southward. Similarly, the departure is considered positive and negative when reckoned on eastward and westward respectively.

2.34 (a)

$\delta_E = 5° \ 30'E$

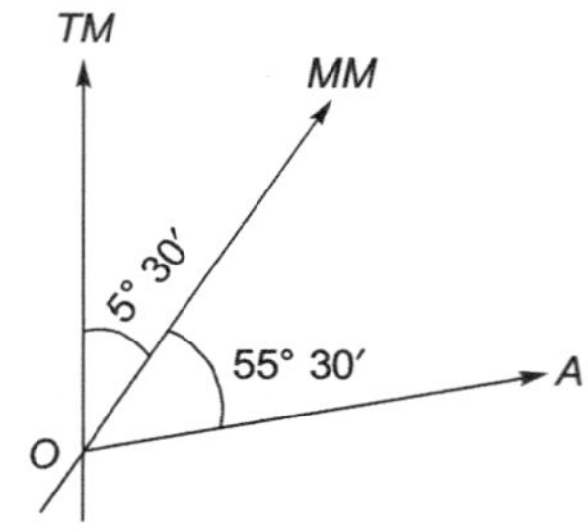

$MB = 55° \ 30'$

$TB = MB + \delta_E$

$= 55° \ 30' + 5° \ 30' = 61° \ 00'$

Note: Here answer given by UPSC is 'd' but the correct answer is 'a', as per Surveying Vol.-I by B.C. Punmia.

2.36 (c)

Transiting the Telescope : It is the process of rotating the telescope over the horizontal axis through 180° in the vertical plane. The directions of the objective end and eyepiece are reversed, and the telescope points exactly in the opposite direction. It is also known as **plunging** or reversing.

Swinging the Telescope : It is the process of turning the telescope about the vertical axis in a horizontal plane. The swing is called right swing when the telescope is turned clockwise from the previous position. The swing is called left swing when the telescope is turned anticlockwise from the previous position

2.37 (d)

Fundamental lines of theodolite and as follows:

(i) Vertical axis

(ii) Axis of plate levels

(iii) Axis of telescope

(iv) Line of collimation

(v) Horizontal axis

(vi) Axis of altitude bubble

2.38 (b)

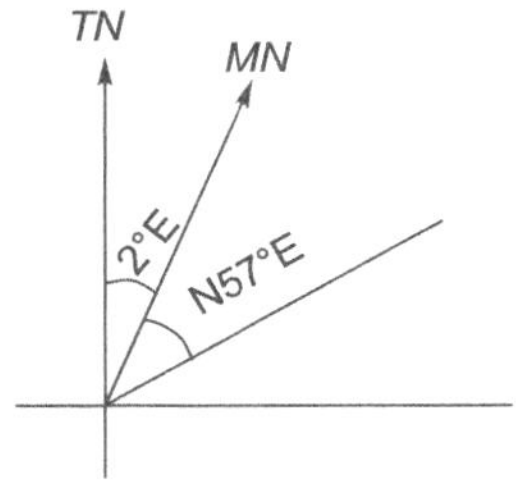

As per old map

$$TB = MB + 2°E$$
$$= 57° + 2° = N\,59°E$$

When magnetic declination becomes 4°W,

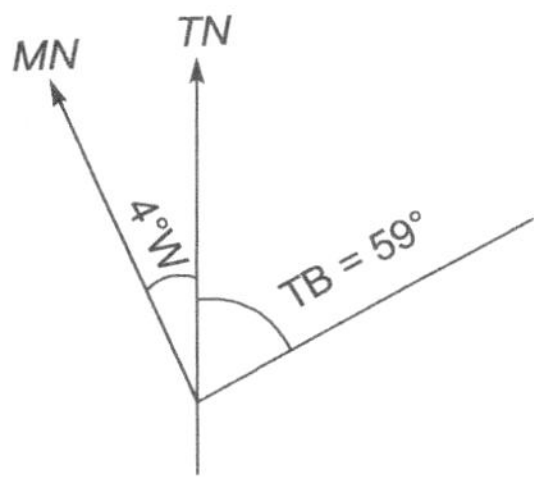

$$TB = MB - 4°W$$
$$\Rightarrow \quad MB = TB + 4°W$$
$$= 59° + 4° = N63°E$$

2.40 (a)

Old Map $MB = 5°30'$

$$\delta_E = 1°$$
$$TB = MB + \delta_E$$
$$= 5°30' + 1° = 6°30'$$

Present: $TB = 6°30'$

$$\delta_E(\text{present}) = 8°30'$$
$$TB = MB + \delta_E$$
$$6°30' = MB + 8°30'$$
$$MB = 6°30' - 8°30' = -2°$$
$$= (360° - 2°)$$
$$= 358°$$

2.42 (c)

Surveyor's compass v/s Prismatic compass

S. No.	Characteristic	Surveyor's compass	Prismatic compass
1.	Magnetic needle	Edge bar needle is used and it acts as an index	Board needle is used which is hidden below the alumirum ring. It does not act as an index.
2.	Graduation	Graduated ring is attached to the box and ring rotates with the box. Graduations are embrossed from 0° to 90° in the four quadrants. Zero points are marked with *N* and *S*.	Graduated ring is attached to the needle and remains stationary when the box is roted. Graduations are embrossed from 0° to 360° clockwise from zero at the south end.
3.	Sighting vanes	Object vane consists of a metal frame with vertical hair. Eye vane also consists of a metal frame with a fine slit. Readings are taken by directly looking down through the glass and reading the north end of the needle.	Object vane consists of a metal frame with vertical hair. Eye vane also consists of a small frame with a fine slit near the prism. Readings are taken through the vertical side of the prism provided at the eye vane.
4.	Reading	Sighting and reading are done separately and that too from different positions. After sighting the object, the observer moves around and takes the reading at the north end of the needle.	Sighting and reading are done simultaneously from the same position of the observer.
5.	Support requirement	This cannot be used without a support like tripod etc.	This can be held in hand while taking the observation but it is better to use tripod with it.

2.43 (d)

- It is quite a common phenomenon that the magnetic meridian and the true meridian at a place generally do not coincide.
- The horizontal angle which the magnetic meridian makes with the true meridian is called as **magnetic declination** (also sometimes referred to as **declination**).

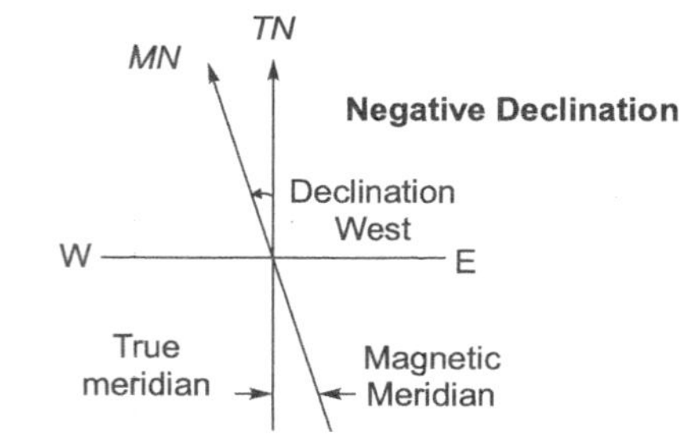

(a)

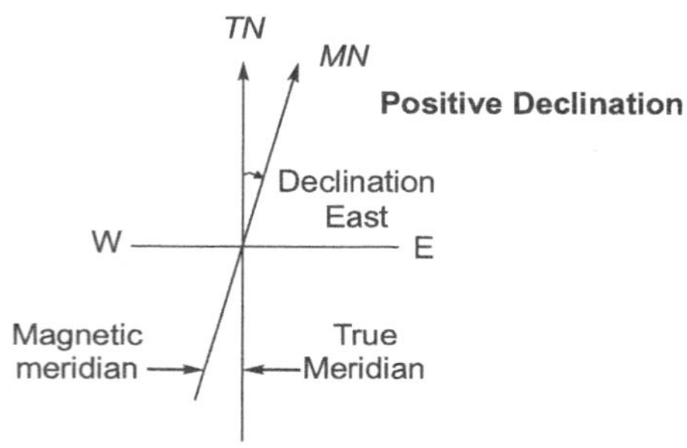

(b)

2.44 (a)

Method 1:

Magnetic bearing,

$$MB = S28°30'E$$

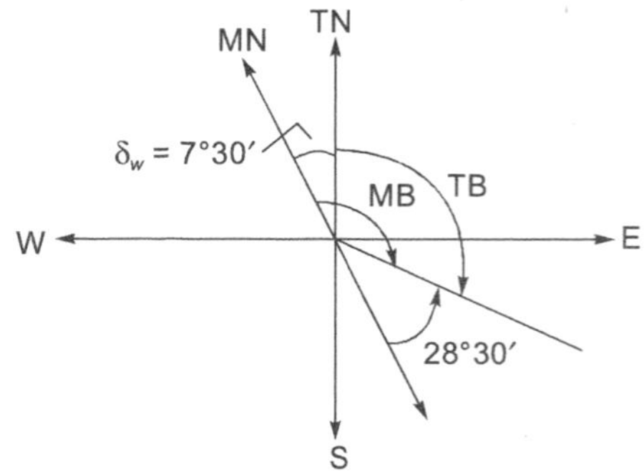

In WCB, $MB = 180° - 28°30' = 151°30'$

Declination, $\delta_w = 7°30'$ W

$\therefore$ True bearing,

$$TB = MB - \delta_w$$
$$= 151°30' - 7°30'$$
$$= 144° \text{ or } S36°E$$

Method 2:

In QB system,

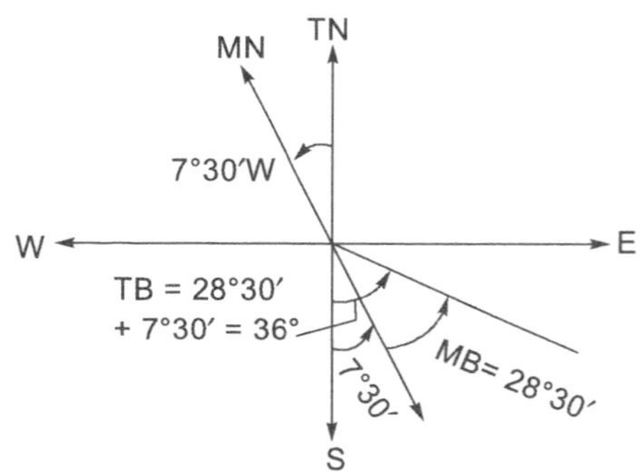

$\therefore$ TB = S36°E

2.45 (a)

Difference in longitude = 64° – 46° = 18°

3 Levelling and Contouring

3.1 The arithmetical check for the computation of *RL* by "rise and fall" method is given by

(a) $\Sigma FS - \Sigma BS$ = *RL* of last station point – *RL* of first station point = ΣFall – ΣRise

(b) $\Sigma BS - \Sigma FS$ = *RL* of first station point – *RL* of last station point = ΣRise – ΣFall

(c) $\Sigma BS - \Sigma FS$ = *RL* of last station point – *RL* of first station point = ΣRise – ΣFall

(d) $\Sigma BS - \Sigma FS$ = *RL* of first station point – *RL* of last station point = ΣRise – ΣFall

[ESE : 2000]

3.2 Match **List-I** with **List-II** and select the correct answer using the codes given below the lists:

List-I

A. Vertical cliff
B. Steep slope
C. Hill
D. Overhanging cliff

List-II

1. Contour lines of different elevations unite to form one line
2. Contour lines of different elevations cross one another
3. Contour lines are closely spaced
4. Closed contour lines with higher values inside them

Codes:

	A	B	C	D
(a)	4	3	1	2
(b)	1	3	4	2
(c)	1	2	4	3
(d)	4	2	1	3

[ESE : 2000]

3.3 The method of orienting a plane table with two inaccessible points is known as

(a) intersection (b) resection
(c) back sighting (d) two-point problem

[ESE : 2000]

3.4 Theory of errors and adjustments deals with minimizing the effects of

(a) instrumental errors
(b) mistakes
(c) systematic errors
(d) personal and accidental errors

[ESE : 2001]

3.5 Which one of the following surveys is employed for collecting sufficient data in connection with sewage disposal and water supply works?

(a) Topographic survey
(b) Cadastral survey
(c) Geodetic survey
(d) Cross-sectioning and profile levelling

[ESE : 2002]

3.6 The following readings were taken with a Dumpy level and a 4 m levelling staff on a continuously sloping ground at 30 m intervals: 0.680 m, 1.455 m, 1.855 m, 2.330 m, 2.885 m, 3.380 m, 1.055 m. The *RL* of the fourth point was calculated to be 79.100 m. The *RL* of the point that was read 0.680 m is

(a) 80.750 m (b) 79.780 m
(c) 78.420 m (d) 77.740 m

[ESE : 2002]

3.7 Match **List-I** (Terms) with **List-II** (Description) and select the correct answer using the codes given below the lists:

List-I

A. Contour
B. Line of collimation
C. Tie line
D. Magnetic meridian

List-II

1. Line joining magnetic North and South
2. Line joining subsidiary station on the main line

3. Line joining points of same elevation
4. Line joining optical centre of the objective lens with point of intersection of cross-wires

Codes:

	A	B	C	D
(a)	3	4	2	1
(b)	4	3	2	1
(c)	3	4	1	2
(d)	4	3	1	2

[ESE : 2003]

3.8 Which one of the following methods of levelling eliminates the error due to curvature and refraction?
(a) Fly levelling
(b) Levelling by equalizing the distances of back sight and foresight
(c) Check levelling
(d) Precise levelling

[ESE : 2003]

3.9 Consider the following statements:
Reciprocal levelling eliminates the effect of
1. errors due to earth's curvature
2. errors due to atmospheric refraction
3. mistakes in taking levelling staff readings
4. errors due to line of collimation

Which of these statements are correct?
(a) 1, 2 and 3 (b) 1, 3 and 4
(c) 2, 3 and 4 (d) 1, 2 and 4

[ESE : 2004]

3.10 Assertion (A): A series of closed contours indicate either a valley or a hill without any outlet, when their elevations, respectively, increase or decrease towards the centre.
Reason (R): Contour lines of different elevations can unite to form one line only at a vertical cliff.
(a) both A and R are true and R is the correct explanation of A
(b) both A and R are true but R is not a correct explanation of A
(c) A is true but R is false
(d) A is false but R is true

[ESE : 2004]

3.11 Match **List-I** with **List-II** and select the correct answer using the codes given below the lists:

List-I
A. Satellite station
B. Gales traverse-method
C. Invar steel tape
D. Intervisibility of adjacent stations

List-II
1. Measurement of baseline
2. Characteristic of triangulation station
3. Auxiliary theodolite station near an inaccessible main triangulation station
4. Computation of co-ordinates of traverse/ triangulation stations.

Codes:

	A	B	C	D
(a)	3	1	4	2
(b)	2	1	4	3
(c)	3	4	1	2
(d)	2	4	1	3

[ESE : 2004]

3.12 How high should a helicopter pilot rise at a point A just to see the horizon at point B, if the distance AB is 40 km?
(a) 101.75 m (b) 110.50 m
(c) 107.75 m (d) 105.50 m

[ESE : 2005]

3.13 What is the angle of intersection of a contour and a ridge line?
(a) 30° (b) 0°
(c) 180° (d) 90°

[ESE : 2006]

3.14 A level when set up 25 m from peg *A* and 50 m from peg *B* reads 2.847 m on staff held on *A* and 3.462 m on staff held on *B*, keeping bubble at its centre while reading. If the reduced levels of *A* and *B* are 283.665 m and 284.295 m respectively, what is the collimation error per 100.0 m?
(a) 0.015 m (b) 0.30 m
(c) 0.045 m (d) 0.060 m

[ESE : 2006]

3.15 A sailor, standing on the deck of a ship, just sees the light beam from a lighthouse on the shore. If the height of the sailor's eye and of the light beam at the lighthouse, above the sea level, are 9 m and 25 m respectively, what is the distance between the sailor and the lighthouse?
(a) 29.8 km (b) 31.1 km
(c) 31.9 km (d) 33.2 km

[ESE : 2007]

3.16 Which one of the following statements is correct?

(a) The axis of plate level should be parallel to the vertical axis.
(b) The axis of striding level must be parallel to the horizontal axis.
(c) The axis of the altitude level must be perpendicular to the line of collimation.
(d) The line of collimation must be perpendicular to the plate level axis.

[ESE : 2007]

3.17 Mean sea level at Indian ports and harbours has generally been established based on the analysis of tidal sea level fluctuations over which period?

(a) 10 years (b) 16 years
(c) 19 years (d) 25 years

[ESE : 2008]

3.18 Which one of the following statements is **not** correct?

(a) Change point is a point denoting shifting of level.
(b) For levelling work both centering and levelling of a dumpy level are prerequisite.
(c) Bench mark is a point whose RL is always known.
(d) None of the above

[ESE : 2008]

3.19 Consider the following statements about the characteristics of contours:

1. Closed contour lines with higher values inside show a lake.
2. Contour is an imaginary line joining points of equal elevations.
3. Closely spaced contours indicate steep slope.
4. Contour lines can cross each other in case an overhanging cliff.

Which of these statements are correct?

(a) 2,3 and 4 (b) 1 and 2 only
(c) 1 and 4 (d) 1, 2 and 3

[ESE : 2008]

3.20 Consider the following statements:

In surveying operations, the word 'reciprocal' can be associated with

1. ranging
2. levelling
3. contouring

Which of these statements is/are correct?

(a) 1 only (b) 1 and 2 only
(c) 2 and 3 only (d) 1, 2 and 3

[ESE : 2009]

3.21 Which of the following sights will be applicable for a change point?

(a) Back sight
(b) Intermediate sight and fore sight
(c) Fore sight
(d) Back sight and fore sight

[ESE : 2009]

3.22 Consider the following statements:

Reciprocal levelling is a method of levelling adopted when

1. the difference of levels between two points at a considerable distance apart is to be determined with great precision.
2. it is not possible to set up the level midway between two points as in the case of a deep valley or a river
3. error due to improper centering of level is to be eliminated

Which of these statements is/are correct?

(a) 1 only (b) 1 and 2
(c) 2 and 3 (d) 1 and 3

[ESE : 2009]

3.23 Consider the following statements:

Errors eliminating by taking both face observations are those due to

1. horizontal axis not being perpendicular to the vertical axis
2. non-parallelism of the axis of telescope level and line of collimation
3. imperfect adjustment of vertical circle vernier

Which of these statements are correct?

(a) 1, 2 and 3 (b) 1 and 2 only
(c) 2 and 3 only (d) 1 and 3 only

[ESE : 2009]

3.24 A light-house is visible just above the horizon at a certain station at the sea level. Distance between the station and the light-house is 60 km. The height of the light-house is

(a) 243.5 m (b) 4.0 m
(c) 287.5 m (d) 5.4 m

[ESE : 2010]

3.25 Error due to inclination of line of collimation in levelling across a river can be eliminated by
(a) reversion
(b) reciprocal ranging
(c) reciprocal levelling
(d) keeping level in middle

[ESE : 2010]

3.26 **Assertion (A):** In plane table photogrammetry, the areas to be mapped are taken from either end of a base line.
Reason (R): The position of the detail point with reference to the base line is obtained by intersection of rays drawn to it from each end of the base.
(a) both A and R are true and R is the correct explanation of A
(b) both A and R are true but R is not a correct explanation of A
(c) A is true but R is false
(d) A is false but R is true

[ESE : 2010]

3.27 An object on the top of a hill 100 m high is just visible above the horizon from a station at sea level. The distance between the station and the object is
(a) 38.53 km (b) 3.853 km
(c) 3853 km (d) 385.3 km

[ESE : 2011]

3.28 Which one of the following statements is **not** correct?
(a) Parallax error is eliminated when there is no change in the staff reading when eye is moved up and down
(b) The objective lens is to be focused towards a white or bright background for clear visibility of cross-hairs
(c) Temporary adjustments of the dumpy level are to be performed at every set up
(d) The eyepiece need not be adjusted after the first set up when the same surveyor is taking readings

[ESE : 2011]

3.29 In a solution of the three-point problem in plane table surveying, the converging of error is attained through
(a) Concyclic concept
(b) Bessel's method
(c) Triangle of error
(d) Tracing paper method

[ESE : 2011]

3.30 **Assertion (A):** Reciprocal levelling is adopted to decide the precise difference of level between two points at a considerable distance apart.
Reason (R): Reciprocal levelling eliminates errors due to:
(i) Curvature
(ii) Refraction, and
(iii) Line of collimation not being exactly parallel to bubble line.
(a) both A and R are true and R is the correct explanation of A
(b) both A and R are true but R is not a correct explanation of A
(c) A is true but R is false
(d) A is false but R is true

[ESE : 2011]

3.31 R.L. of floor at a building is 74.4 m, staff reading on the floor is 1.625 and staff reading when it is held inverted with bottom touching the ceiling of a hall is 2.870; then the height of the ceiling above the floor is
(a) 3.593 m (b) 3.953 m
(c) 4.495 m (d) 4.594 m

[ESE : 2012]

3.32 Following observations were taken during a reciprocal levelling:

Instrument near	P	Q
Staff reading at P	1.824	0.928
Staff reading at Q	2.748	1.606

If reduced level of P is 140.815 m, the reduced level of Q is
(a) 138.014 m (b) 139.616 m
(c) 140.014 m (d) 141.616 m

[ESE : 2012]

3.33 A contour may be defined as an imaginary line passing through
(a) Points on the longitudinal section
(b) Points of equal elevation
(c) Point of equal local ground slope
(d) Points of transverse section surveys

[ESE : 2012]

3.34 A closed contour line with two or more higher contours inside it will represent a
(a) Depression (b) Hill
(c) Cave (d) Well [ESE : 2012]

3.35 In case of levelling, backsight is
(a) A fixed point of known elevation
(b) The last staff reading taken before shifting the instrument.
(c) The first staff reading taken after setting the instrument.
(d) Any staff reading taken on a point of unknown elevation
[ESE : 2012]

3.36 In an instrument, the bubble tube with divisions of 1 mm and a radius of 0.9 m has the sensitivity of
(a) $\frac{1}{2}$ (b) $\frac{1}{70}$
(c) $\frac{1}{90}$ (d) $\frac{1}{900}$
[ESE : 2012]

3.37 The following observations were taken during testing of a dumpy level:

Instrument at	Staff reading on	
	P	Q
P	1.475 m	2.205 m
Q	1.440 m	2.060 m

The collimation error in the instrument was:
(a) +0.055 m (b) +0.005 m
(c) –0.055 m (d) –0.005 m
[ESE : 2013]

3.38 R.L. of a floor level is 200.490 m. Staff reading on the floor is 1.695 m. Reading on the staff held upside down against the bottom of the roof is 3.305 m. Height of the ceiling is:
(a) 3.5 m (b) 4.0 m
(c) 5.0 m (d) 6.0 m
[ESE : 2013]

3.39 Which of the following terms related to levelling are correctly defined?
1. Line of collimation — Line joining the intersection of the crosshairs to the optical centre of the object glass and its continuation.
2. Back-sight — First staff reading taken after the level is set up
3. Fore-sight — Last staff reading prior to shifting of level, or termination of the process of levelling
4. Height of instrument — Height of centre of telescope above the ground where the level is set up.

(a) 1, 2, 3 and 4 (b) 1, 2 and 4 only
(c) 1, 2 and 3 only (d) 2, 3 and 4 only
[ESE : 2013]

3.40 Two stations P and Q are on the opposite banks of a river. Following observations were taken in reciprocal levelling.

Instrument near	Staff reading at	
	P	Q
P	1.400	3.500
Q	0.600	2.200

R.L. of *P* is 200.000 m, and then R.L. of *Q* is nearly
(a) 199.3 (b) 201.7
(c) 200.0 (d) 198.2 [ESE : 2013]

3.41 In lavelling between two points *A* and *B* on the opposite sides of a river, the level was first set up near *A* and the staff readings on *A* and *B* were 2.645 m and 2.30 m respectively. The level was then moved near *B* and set up; the respective staff readings then were 1.085 m and 1.665 m on *A* and *B* respectively. What is the true difference of level between *A* and *B*?
(a) *A* and *B* are at same level
(b) *A* is 0.5825 m below *B*
(c) *A* is 0.4825 m below *B*
(d) *B* is 0.5825 m below *A* [ESE : 2014]

3.42 A plane, which is perpendicular to the plumb line through a point and is tangential to the level surface at that point is called a
(a) Tangential plane (b) Vertical plane
(c) Level plane (d) Horizontal plane
[ESE : 2014]

3.43 Staff reading on the floor of a verandah of a school building is 1.815 m and staff reading when held with bottom of staff touching the ceiling over the verandah is 2.870 m R.L. of the floor is 74.500 m. Height of the ceiling above floor is
(a) 4.270 m (b) 4.685 m
(c) 3.955 m (d) 4.920 m
[ESE : 2014]

3.44 A theodolite was set up at P and the angle of elevation to the top of a mobile tower ST was 30°. The staff reading held at a station of RL 110 m was 2.555 m, the telescope being horizontal. The horizontal distance between the foot of the tower and the instrument station is 810 m. The RL of the top of the tower is

(a) 578.25 m (b) 579.50 m
(c) 580.25 m (d) 582.40 m

[ESE : 2014]

3.45 The levelling staff held at a distance of 200 m is read at 4.54 m with the bubble out of centre by 2 divisions towards the observer. If the sensitiveness of the bubble is 25 secs/division, and 1 division = 2 mm, then actual staff reading must have been

(a) 4.5 m (b) 4.492 m
(c) 4.54 m (d) 4.62 m

[ESE : 2015]

3.46 In a levelling survey, the summation of all backsights and the summation of all foresights are 7.475 m and 7.395 m, respectively. The reduction level of the initial benchmark is 1000.000 m. The reduced level of the of point where the staff is held will be

(a) 100.000 m (b) 100.080 m
(c) 107.395 m (d) 107.475 m

[ESE : 2015]

3.47 The rise and fall method for obtaining the reduced levels of points provides is check on

1. Foresight
2. Backsight
3. Intermediate sight

Which of the above statements are correct?

(a) 1 and 2 only (b) 1 and 3 only
(c) 2 and 3 only (d) 1, 2 and 3

[ESE : 2016]

3.48 Consider the following characteristics of contours:

1. A uniform slope is indicated when contour lines are uniformly spaced.
2. Contour lines cannot and anywhere but can close on themselves.
3. A set of closed contours indicates a depression or a summit, according to the lower or higher values being respectively inward.

Which of the above statements are correct?

(a) 1 and 2 only (b) 1, 2 and 3
(c) 1 and 3 only (d) 2 and 3 only

[ESE : 2017]

3.49 The type of levelling operations carried out for laying a railway track are

1. Longitudinal levelling (LS)
2. Fly levelling
3. Barometric levelling
4. Cross-sectioning

Select the correct answer using the codes given below:

(a) 1, 2 and 3 only (b) 1, 3 and 4 only
(c) 2, 3 and 4 only (d) 1, 2 and 4 only

[ESE : 2017]

Directions: Each of the next items consists of two statements, one labelled as the **'Statement (I)'** and the other as **'Statement (II)'**. Examine these two statements carefully and select the answers to these items using the codes given below :

Codes:

(a) Both Statement (I) and Statement (II) are individually true and Statement (II) is the correct explanation of Statement (I)
(b) Both Statement (I) and Statement (II) are individually true but Statement (II) is ***not*** the correct explanation of Statement (I)
(c) Statement (I) is true but Statement (II) is false
(d) Statement (I) is false but Statement (II) is true

3.50 **Statement (I):** Reciprocal levelling is adopted to decide the difference of level between two points, a considerable distance apart, with great precision.

Statement (II): Reciprocal levelling eliminates errors due to curvature, refraction and collimation.

[ESE : 2017]

3.51 **Statement (I):** Curvature correction must be applied when the sights are long.

Statement (II): Line of collimation is not a level line but is tangential to the level line.

[ESE : 2017]

3.52 A temporary bench mark has been established at the soffit of a chajja on a window opening, and its known elevation is 102.405 m above mean sea level. The backsight used to establish the height of the instrument is by an inverted staff reading of 1.80 m. A foresight reading with the same staff, held normally, is 1.215 m on a recently constructed plinth. The elevation of the plinth is

(a) 95.42 m O.D (b) 99.39 m O.D
(c) 102.42 m O.D (d) 105.99 m O.D

[ESE : 2018]

3.53 The combined correction for curvature and refraction for a distance of 3400 m will be nearly

(a) 0.2 m (b) 0.4 m
(c) 0.6 m (d) 0.8 m [ESE : 2020]

3.54 If backsight and foresight distances are balanced

1. The difference in elevation between two points can be directly calculated by taking difference of the two readings.
2. No corrections for the inclination of the line of sight is necessary.

(a) 1 only (b) 2 only
(c) Both 1 and 2 (d) Neither 1 nor 2

[ESE : 2020]

3.55 **Statement (I)** : The counter interval depends upon the nature of the ground – whether it is undulating or flat.

Statement (II) : In a hilly terrain or undulating ground a smaller interval is adopted, otherwise the contours will come too close for plotting due to the steep slope.

[ESE : 2020]

3.56 Which one of the following is a staff reading taken on a point whose elevation is to be determined?

(a) Fore sight
(b) Back sight
(c) Intermediate sight
(d) Line of sight

[ESE : 2022]

3.57 Which one of the following statements is NOT correct?

(a) Trigonometric levelling has never been used in geodetic surveys
(b) The differences in elevation are determined indirectly by trigonometric levelling
(c) The electronic distance measurement devices can be used for measuring the vertical distances
(d) The combined effects of curvature and refraction produce vertical readings that are slightly too short

[ESE : 2023]

3.58 The levelling is carried out between two stations P and Q separated by 1000 m. The Back Sight (BS) reading is noted as 0.70 m on station P, whose BM is 240.00 m. Next, the Fore Sight (FS) reading is taken on an intermediate point and the value is 1.25 m. Then the instrument is shifted to a new location and BS is taken on intermediate point as 0.85 m. Finally, the FS reading is taken on station Q as 1.10 m. What is the Reduced Level of Q?

(a) 239.20 m (b) 240.80 m
(c) 241.25 m (d) 241.80 m

[ESE : 2023]

3.59 If sensitivity of a bubble tube is 30" per 2 mm division what would be the error in staff reading on a vertically held staff at a distance of 200 m, when the bubble is out of centre by 2.5 divisions?

(a) 0.073 m (b) 0.73 m
(c) 0.0073 m (d) 7.3 m

[ESE : 2024]

3.60 If the equality of back sight distance and foresight cannot be maintained, under such condition, which one of the following levelling types is most important part of geodetic surveying?

(a) Spirit levelling
(b) Reciprocal levelling
(c) Trigonometric levelling
(d) Ordinary levelling

[ESE : 2024]

3.61 Accurate transfer of surface alignment down a vertical shaft using two plumbwires can be achieved by

(a) Rise and fall method
(b) Collimation method
(c) Bowditch's method
(d) Weisbach triangle method

[ESE : 2024]

■■■■

Answers Levelling, Contouring and Plane Table Surveying

3.1 (c)	3.2 (b)	3.3 (d)	3.4 (d)	3.5 (d)	3.6 (a)	3.7 (a)	3.8 (b)	3.9 (d)
3.10 (d)	3.11 (c)	3.12 (c)	3.13 (d)	3.14 (d)	3.15 (b)	3.16 (b)	3.17 (c)	3.18 (b)
3.19 (a)	3.20 (b)	3.21 (d)	3.22 (b)	3.23 (b)	3.24 (a)	3.25 (c)	3.26 (a)	3.27 (a)
3.28 (d)	3.29 (c)	3.30 (a)	3.31 (c)	3.32 (c)	3.33 (b)	3.34 (b)	3.35 (c)	3.36 (d)
3.37 (a)	3.38 (c)	3.39 (c)	3.40 (d)	3.41 (c)	3.42 (d)	3.43 (b)	3.44 (c)	3.45 (b)
3.46 (b)	3.47 (d)	3.48 (b)	3.49 (d)	3.50 (a)	3.51 (a)	3.52 (b)	3.53 (d)	3.54 (c)
3.55 (c)	3.56 (a)	3.57 (a)	3.58 (a)	3.59 (a)	3.60 (b)	3.61 (d)		

Explanations Levelling, Contouring and Plane Table Surveying

3.1 (c)

Two arithmetic checks in collimation method.

$\Sigma BS - \Sigma FS$ = Last RL – First RL

Three arithmetic checks in rise and fall method.

$\Sigma BS - \Sigma FS$ = Last RL – First RL

= ΣRise – ΣFall

3.3 (d)

Resection is a method of plane table surveying in which location of station occupied by plane table is unknown and it is determined by sighting any two or three known points or plotted points. It is also called method of orientation and it can be conducted by two field conditions as follows:

- The three-point problem
- The two-point problem

3.4 (d)

Systematic or cumulative errors which occurs from well understood causes can be reduced by adopting suitable methods. It follows some definite mathematical or physical law and a correction can be determined and applied.

Accidental errors are those which remain after mistakes and systematic errors have been eliminated and are caused by a combination of reasons beyond the ability of the observer to control.

Personal errors arise from the limitations of the human senses such as sight, touch and hearing.

Both accidental and personal error represent the limit of precision in the determination of a value. They obey the law of probability and therefore theory of errors and adjustments applies to them.

3.5 (d)

Cross sectioning is done to determine elevation of point in direction perpendicular to centre line and Profile levelling is done to determine elevation of points along the centre line. Cross sectioning and Profile levelling is used for projects related to sewers, pipelines, highways, railways, canals etc.

3.6 (a)

BS	IS	FS	Rise	Fall	RL
0.680					80.750
	1.455			0.775	79.975
	1.855			0.400	79.575
	2.330			0.475	79.100
	2.885			0.555	78.545
1.055		3.380		0.495	78.050

RL of the point that was read as 0.680

= 79.100 + total fall up to 2.330 reading

= 79.100 + (0.475 + 0.400 + 0.775)

= 80.750

3.8 (b)

Levelling by balancing backsight and fore-sight distances can eliminate collimation error.

Error due to curvature and refraction can be eliminated by equalizing FS and BS distances or by reciprocal levelling.

3.9 (d)

Reciprocal levelling is used to determine the correct difference of elevations of two points which are quite apart and when it is not possible to set up the instrument midway between the points to balance the foresight and backsight. It eliminates the error due to curvature of the earth and refraction and the collimation error.

3.10 (d)

A series of closed contour line with one or more higher ones inside represents a hill. Similarly, a series of closed contour line with one or more lower ones inside it indicates a depression without an outlet (i.e. a lake).

Contour lines cross a valley line at right angles. They form sharp curves of V-shape across it with convex side of the curve towards the higher ground.

3.11 (c)

Satellite station is an auxiliary theodolite station taken near an inaccessible main station.

Gale transverse method is used for computation of coordinates of transverse/triangulation station.

Invar steel tape is a precise tape which is used in measurement of length of baseline in traversing.

The main characteristic of adjacent triangulation stations is that they should be inter-visible from each other.

3.12 (c)

The height of rise $= 0.0673\,D^2$

$= 0.0673 \times (40)^2 = 107.68$ m

3.13 (d)

Contour lines cross a ridge line (watershed) valley line at right angles. However, they form curves of U-shape round the ridge line with the concave side of the curve towards the higher ground. Contour lines form sharp curves of V-shape across valley line with convex side of the curve towards the higher around.

3.14 (d)

Let the collimation error per 100 m is e.

True staff reading at A

$$= 2.847 + \frac{25}{100}e = 2.847 + 0.25e$$

True staff reading at B

$$= 3.462 + \frac{50}{100}e$$

$$= 3.462 + 0.5e$$

Difference in elevation

$= 3.462 + 0.5e - (2.847 + 0.25e)$

$= 0.615 + 0.25e$

From RL of A and B difference in elevation

$= 284.295 - 283.665 = 0.630$ m

$\therefore\ 0.615 + 0.25e = 0.630$

$e = 0.060$ m per 100 m

3.15 (b)

Let A be the position of the top of light house and B be the position sailor's eye.

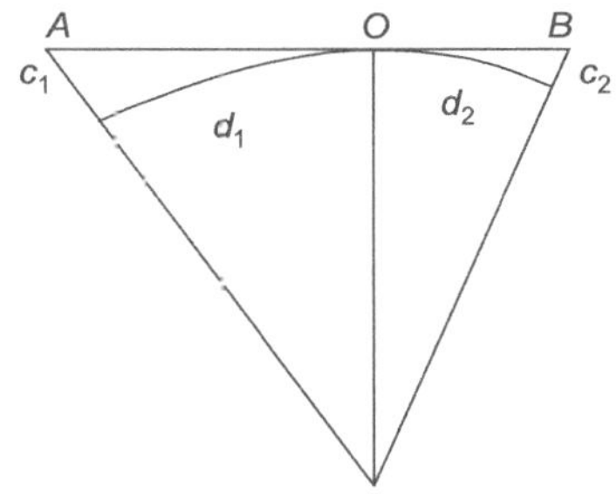

Let AB be tangential to water surface at O.

The distances d_1 and d_2 are given by

$$d_1 = 3.853\sqrt{c_1}\text{ km}$$

$$= 3.853\sqrt{25} = 19.265\text{ km}$$

$$d_2 = 3.853\sqrt{c_2}$$

$$= 3.853\sqrt{9} = 11.559\text{ km}$$

$\therefore$ Distance between A and $B = d_1 + d_2$

$= 19.265 + 11.559 = 30.824$ km

3.16 (b)

The following relationship should exist:

(i) The axis of the plate level should be perpendicular to the vertical axis.

(ii) The axis of the altitude level must be parallel to the line of collimation.

(iii) The axis of the striding level (if provided) must be parallel to the horizontal axis.

(iv) The line of collimation must be parallel to the plate level axis.

(v) The line of collimation must be perpendicular to the horizontal axis at its intersection with the vertical axis.

(vi) The horizontal axis must be perpendicular to the vertical axis.

3.17 (c)

The term Mean Sea Level (MSL) can also refer to a tidal datum, or frame of vertical reference defined by a specific phase of the tide. MSL as a tidal datum is computed as a mean of hourly water level heights observed over 19-years.

3.18 (b)

In levelling, change point or turning point is the point denoting shifting of level. During leveling, temporary adjustments are prerequisite for every setting of instrument.

- Centering
- Levelling
- Focussing

Benchmark is the point whose R.L. is always known.

3.19 (a)

Closed contour lines with higher values inside represents a hill. Closed contour lines with lower values insides represents a depression without an outlet i.e. a lake.

3.20 (b)

Reciprocal Ranging or indirect ranging is used when the end stations are not intervisible due to high ground or a hill or if the ends are too long. In such cases, intermediate points can be fixed on the survey line by a process known as reciprocal ranging.

Differential levelling is method of levelling in which the difference in elevation between two points which are at considerable distance is to be determined accurately and when it is not possible to set up the level midway between the two points. It eliminates the errors due to atmospheric refraction, inclination of line of collimation and the curvature of the earth.

3.21 (d)

A change point is the point selected on the route before shifting the instrument. The change point is also called a turning point. The staff is kept on the turning point and a foresight is taken before shifting the instrument. The staff should not be moved when the instrument is being shifted. After the instrument has been shifted and set up at the new location, a back sight is taken on the staff still held at the change point.

3.22 (b)

Reciprocal levelling is method of levelling in which the difference in elevation between two points which are at considerable distance is to be determined accurately and when it is not possible to set up the level midway between the two points. Reciprocal levelling eliminates the errors due to

- refraction
- inclination of line of collimation
- curvature

3.23 (b)

Errors eliminated by both face readings are:

1. Error due to line of collimation not being perpendicular to the horizontal axis.
2. Error due to horizontal axis not being perpendicular to the vertical axis.
3. Error due to line of collimation not being parallel to axis of altitude level.

3.24 (a)

Height of light house

$$= 0.0673 \times D^2$$

$$= 0.0673 \times (60)^2 = 242.28 \text{ m}$$

3.25 (c)

Reciprocal levelling eliminates :

- refraction error
- inclination of line of collimation
- curvature error

3.27 (a)

Distance,

$$d = \sqrt{\frac{H}{0.06735}} = \sqrt{\frac{100}{0.06735}} \text{ km}$$

$$= 38.53 \text{ km}$$

3.28 (d)

Temporary adjustments or Station adjustments are those which are made at every instrument setting and preparatory to taking observations with the instrument. Permanent adjustments need be made only when the fundamental relations between some parts or lines are disturbed.

3.29 (c)

The trial and error method of three point problem, also known as Lehman's method minimises the triangle of error and converge the error to a point iteratively. The iterative operation consists of drawing of resectors from known points through their plotted position and the adjustment of orientation of the plane table.

3.30 (a)

When it is necessary to carry levelling across a river, ravine or any obstacle requiring a long sight between two points so situated that no place for the level can be found from which the lengths of foresight and backsight will be even approximately equal, special method i.e., reciprocal levelling is used to obtain accuracy and to eliminate the following :

1. Error in instrument adjustment;
2. Comnined effect of earth's curvature and the refraction of the atmosphere and
3. Variations in the average refraction.

3.31 (c)

Given

$$RL = 74.4 \text{ m}, BS = 1.625 \text{ m}$$
$$HI_1 = RL + BS = 74.4 + 1.625$$
$$HI_2 = RL - FS = 74.4 - 2.870$$

$\therefore$ Height of Ceiling = $HI_1 - HI_2$

$$= 1.625 + 2.870 = 4.495 \text{ m}$$

3.32 (c)

$$h_p = 1.824;\ h_Q = 2.748$$
$$h_p' = 0.928;\ h_Q' = 1.606$$

Correct difference of elevation between P and Q is

$$h = \frac{(h_Q' - h_P') + (h_Q - h_P)}{2}$$
$$= \frac{(1.606 - 0.928) + (2.748 - 1.824)}{2}$$
$$= 0.801 \text{ m}$$

$\therefore$ The reduced level of Q

$$= (140.815 - 0.801) = 140.014 \text{ m}$$

3.33 (b)

A contour line is an imaginary line which connects points of equal elevation. Such lines are drawn on the plan of an area after establishing reduced levels of several points in the area. The contour lines in an area are drawn keeping difference in elevation of between two consecutive lines constant.

3.34 (b)

A closed contour line with two or more higher contours inside it will represent a Hill. Similarly a closed contour line with one or more lower ones inside it indicates a depression.

3.35 (c)

The first staff reading taken after setting the instrument is known as backsight. The last staff reading taken after which the position of instrument station is changed is known as foresight. A fixed point of known elevation is called Benchmark.

3.36 (d)

Sensitivity of a bubble tube

$$S = \frac{nl}{R}$$

Now, $n = 1$

$$\therefore \quad S = \frac{1 \times 10^{-3}}{0.9} = \frac{1}{900}$$

3.37 (a)

Using reciprocal levelling,

$$h = \frac{(Q_1 - P_1) + (Q_1 - P_2)}{2}$$
$$= \frac{(2.205 - 1.475) + (2.060 - 1.440)}{2}$$
$$= 0.675 \text{ m}$$

If instrument is kept at P reading at Q will be erroneous

h = (2.205 – e) – 1.475

0.675 m = (2.205 – e) – 1.475

e = 0.055m

3.38 (c)

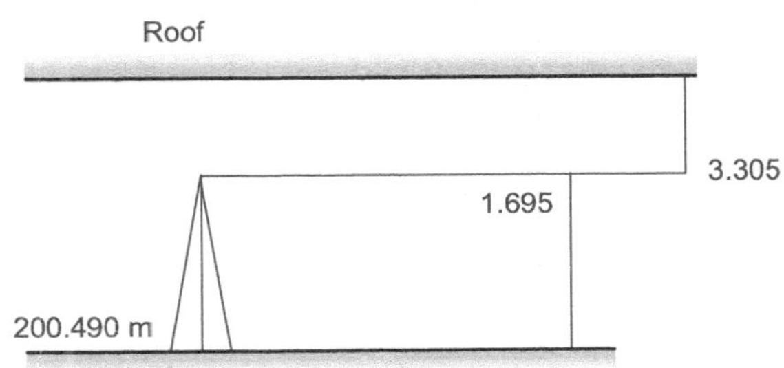

Height of the ceiling

$$= 1.695 + 3.305 = 5.0 \text{ m}$$

3.39 (c)

Line of Collimation: In a surveying telescope, the imaginary line through the optical center of the object glass and the cross-hair intersection in the diaphragm.

Back Sight: short for "back sight reading", the first staff reading taken by the surveyor after the levelling instrument is set up and levelled. B.S is generally taken on the point of known reduced level as on the benchmark or a change point.

Fore sight: short for "fore sight reading", the last staff reading taken before changing the instrument to the other position. It is the staff reading taken on point whose RL is to determined. This sight is considered as negative and deduced from Height of Instrument to determine RL of the point.

Height of Instrument: In survey leveling, it is the height of the line of collimation of the instrument over the station above which it is centered, or above a specified datum level.

3.40 (d)

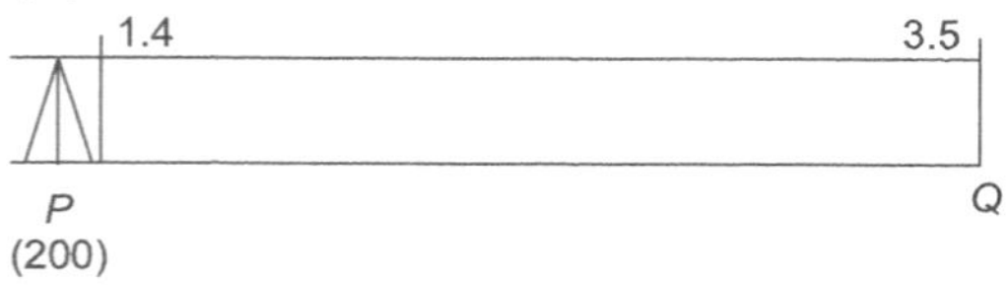

$$200 + 1.4 - 3.5 = RL \text{ of } Q$$

$$\Rightarrow \quad RL \text{ of } Q = 197.9 \text{ m}$$

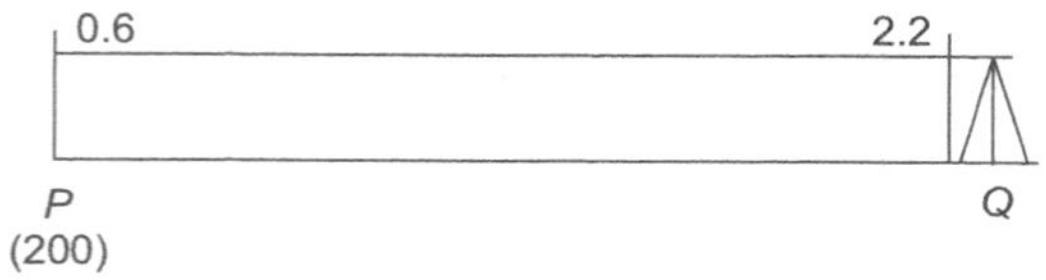

$$RL \text{ of } Q = 200 + 0.6 - 2.2 = 198.4 \text{ m}$$

$$\therefore \quad \text{Avg. } RL \text{ of } Q = \frac{197.9 + 198.4}{2} = 198.15 \approx 198.2 \text{ m}$$

3.41 (c)

True difference of level between A and B

$$= \frac{(2.645 - 2.3) + (-1.085 + 1.665)}{2}$$

$$= 0.4625$$

So, A is below B by 0.4625 m.

Most close option is (c).

3.42 (d)

Horizontal plane: Horizontal plane through a point is a plane tangential to the level surface at that point. It is, therefore, perpendicular to the plumb line through the point.

3.43 (b)

Height of the ceiling above floor

= staff reading on the floor + Reading of inverted staff

= 1.815 + 2.87 = 4.685 m

3.44 (c)

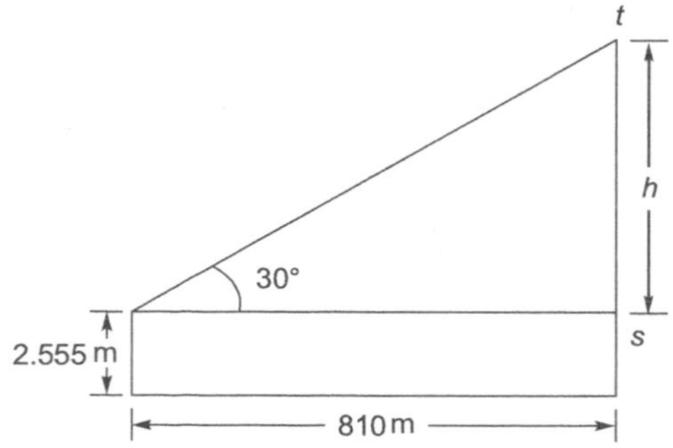

RL of top of tower

$= RL$ of station + staff reading + h

$= 110 + 2.555 + 810 \tan 30°$

$= 580.208$ m

3.45 (b)

Sensitivity,

$$\alpha = \frac{S}{nD} \times 206265$$

$$25'' = \frac{S}{2 \times 200} \times 206265$$

$$S = 0.0484 \text{ m}$$

Now, Staff reading

$$= 4.54 - 0.0484 = 4.492 \text{ m}$$

3.46 (b)

$\because$ ΣBS − ΣFS = RL of last point − RL of initial point

$\Rightarrow$ 7.45 − 7.395 = RL of last point − 100

$\therefore$ RL of last point = 100 + 7.45 − 7.395

= 100.08 m

3.47 (d)

Reduced levels of points on back sight and fore sight is checked along with the intermediate sight in rise and fall method.

3.48 (b)

The principal characteristics of contour lines which help in plotting or reading a contour map are as follows:

1. The variation of vertical distance between any two contour lines is assumed to be uniform.
2. The horizontal distance between any two contour lines indicates the amount of slope and varies inversely on teh amount of slope. Thus contours are spaced equally for uniform slope; closely for steep slope contours and widely for moderate slope.

3. The steepest slope of terrain at any point on a contour is represented along the normal of the contour at that point. They are perpendicular to ridge and valley lines where they cross such lines.
4. Contours do not pass through permanent structures such as buildings.
5. Contours of different elevations cannot cross each other (caves and overhanging cliffs are the exceptions).
6. Contours of different elevations cannot unite to form one contour (vertical cliff is an exception).
7. Contour lines cannot begin or end on the plan.
8. A contour line must close itself but need not be necessarily within the limits of the map.
9. A closed contour line on a map represents either depression or hill. A set of ring contours with higher values inside, depicts a hill whereas the lower value inside, depicts a depression (without an outlet).
10. Contours deflect uphill at valley lines and downhill at ridge lines. Contour lines in U-shape cross a ridge and in V-shape cross a valley at right angles. The concavity in contour lines is towards higher ground in the case of ridge and towards lower ground in the case of valley.
11. Contours do not have sharp turnings.

3.52 (b)

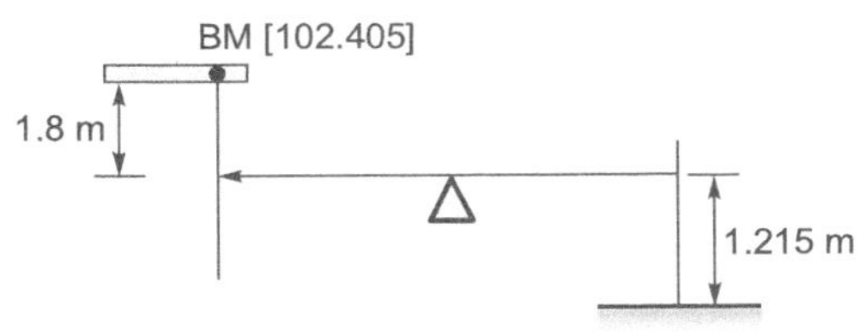

$$HI = RL \text{ of BM} + BS$$
$$= 102.405 + (-1.8) = 100.605 \text{ m}$$
$$RL \text{ of plinth level} = HI - FS$$
$$= 100.605 - 1.215 = 99.39 \text{ m}$$

3.53 (d)

Combined correction,

$$C = -0.06735d^2 = -0.0673 \times 3.4^2$$
$$= 0.77798 \text{ m} \approx 0.8 \text{ m}$$

3.54 (c)

Due to balancing of sight, error due to non-parallism of sight is eliminated.

3.55 (c)

Statement II is wrong. Bigger contour interval is adoptec in hilly terrain.

3.57 (a)

Trigonometric leveling is used in geodetic surveying, particularly in areas where traditional leveling methods are not feasible or practical, such as in mountainous regions or areas with cense vegetation.

3.58 (a)

Point	B.S	F.S.	H.I.	R.L.
P	0.7		240.7	240.0
Intermediate	0.85	1.25	240.3	239.45
Q		1.1		239.2

So, R.L. of Q = 239.2 m

3.59 (a)

$$\alpha = 30'' \quad \theta = 2.5 \times 30''$$
$$l = 2 \text{ mm}$$

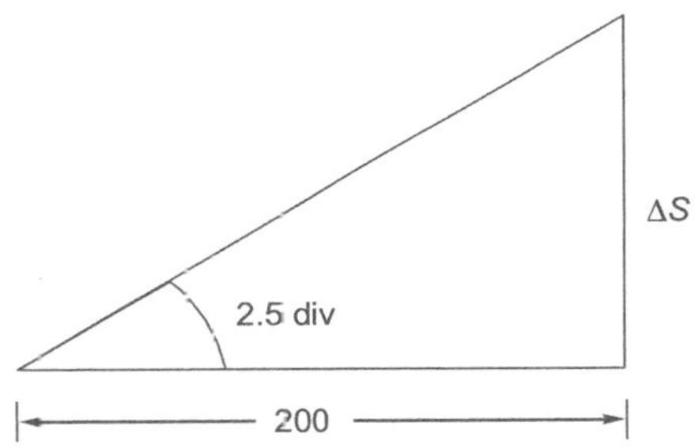

$$\Delta = 200$$
$$\tan(2.5 \times 30'') = \frac{\Delta s}{200}$$
$$\Rightarrow \quad \Delta s = 0.07272 \simeq 0.073$$

3.60 (b)

Reciprocal levelling eliminates all the errors hence is suitable in geodetic surveying.

3.61 (d)

Weisbach triangle method of surface alignment down through a vertical shaft in tunnel is the most accurate and widely excepted method.

■■■■

4 Calculation of Area and Volume

4.1 A two-level section *ABCD* is shown in the given figure with ground slope 1 in r and formation slope 1 in s. *EF* is the centre line (C/L) of the formation. Top width of the formation is $2b$ and widths to the toe are w_1 and w_2. The width 'w_1' will be

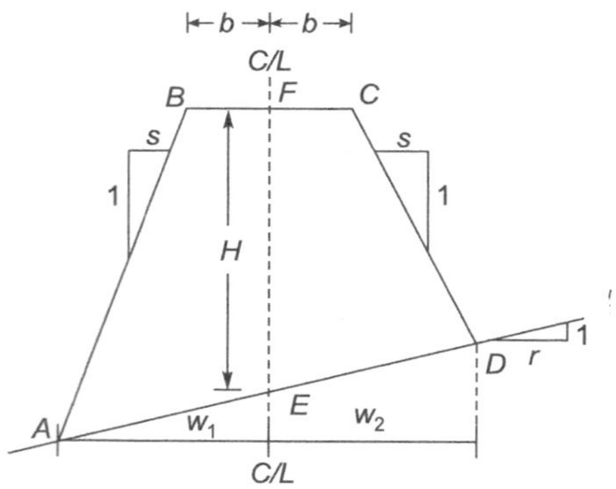

(a) $b+\frac{rs}{r-s}\left(H+\frac{b}{r}\right)$

(b) $b+\frac{rs}{r+s}\left(H-\frac{b}{r}\right)$

(c) $b-\frac{rs}{r-s}\left(H+\frac{b}{r}\right)$

(d) $b+\frac{rs}{r+s}\left(H+\frac{b}{r}\right)$

[ESE : 2000]

4.2 What is the volume of a 6 m deep tank having rectangular shaped top 6 m × 4 m and bottom 4 m × 2 m (computed through the use of prismoidal formula)?

(a) 96 m^3 (b) 94 m^3

(c) 92 m^3 (d) 90 m^3

[ESE : 2005]

4.3 Which one of the following methods of computing area assumes that the short lengths of the boundary between the ordinates are parabolic arcs?

(a) Average ordinate rule

(b) Middle ordinate rule

(c) Simpson's rule

(d) Trapezoidal rule

[ESE : 2009]

4.4 Following offsets were taken from a survey line to a hedge:

Distance (in m)	0	5	10	15	20	30	40
Offsets (in m)	3	4	5.5	5	6	4	4.5

The area between survey line and the hedge is (by trapezoidal method):

(a) 185.5 m^2 (b) 187.5 m^2

(c) 189.5 m^2 (d) 289.5 m^2

[ESE : 2013]

4.5 In a cross staff survey, the perpendicular offsets are taken on right and left of the chain line *AD* as shown in figure – all values are in 'metres'. The area enclosed by *ABCDEFA*, computed by trapezoidal method is

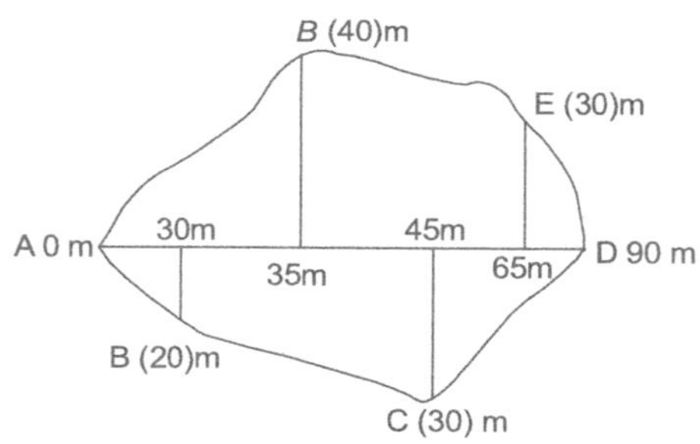

(a) 3650 m^2 (b) 3200 m^2

(c) 3475 m^2 (d) 3500 m^2

[ESE : 2014]

Answers Calculation of Area and Volume

4.1 (a) 4.2 (c) 4.3 (c) 4.4 (b) 4.5 (c)

Explanations Calculation of Area and Volume

4.1 (a)

$$w_1 = b + \left(H + \frac{w_1}{r}\right) \times s$$

After rearranging, we get

$$w_1 = \frac{br}{(r-s)} + H\left(\frac{r.s}{r-s}\right)$$

[After adding and subtracting $\frac{bs}{r-s}$]

$$w_1 = b + \frac{rs}{r-s}\left(H + \frac{b}{r}\right)$$

4.2 (c)

Prismoidal formula is also known as Simpson's rule for volumes.

Given, $2L = 6$ m; $L = 3$ m

$A_1 = 6 \times 4 = 24$ m^2

$A_2 = 4 \times 2 = 8$ m^2

$$A_m = \left(\frac{6+4}{2}\right) \times \left(\frac{4+2}{2}\right) = 15 \text{ m}^2$$

$$\therefore \quad V = \frac{L}{3}(A_1 + 4A_m + A_2)$$

$$= 24 + (4 \times 15) + 8 = 92 \text{ m}^3$$

4.3 (c)

Simpson's one third rule is a numerical method of integration used to determine area of tracts having irregular boundary. It assumes that short lengths of boundary between ordinates are parabolic. *Simpson's rule* is sometimes called as *parabolic rule*.

4.4 (b)

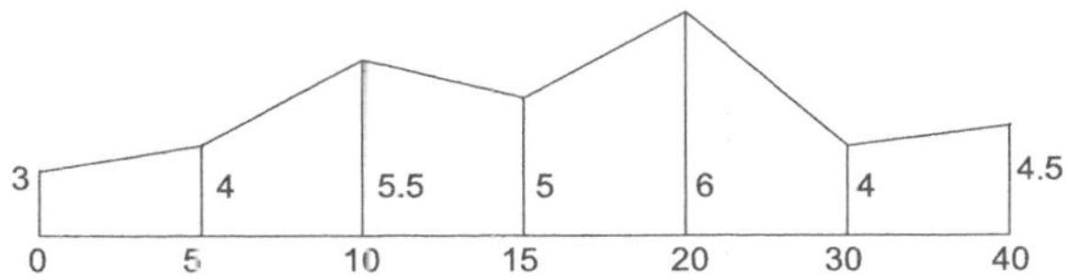

For area as g ven by Trapezoidal method is

$$A = d\left[\left(\frac{O_0 + O_n}{2}\right) + O_1 + O_2 \ldots O_{n-1}\right]$$

Now, since 'd' is not constant

$$\therefore \quad A_1 = 5\left[\frac{3+6}{2} + 4 + 5.5 + 5\right] = 95 \text{ m}^2$$

$$A_2 = 10\left[\frac{6+4.5}{2} + 4\right] = 92.5 \text{ m}^2$$

Total area,

$$A = A_1 + A_2$$

$$= 95 + 95.5 = 187.5 \text{ m}^2$$

4.5 (c)

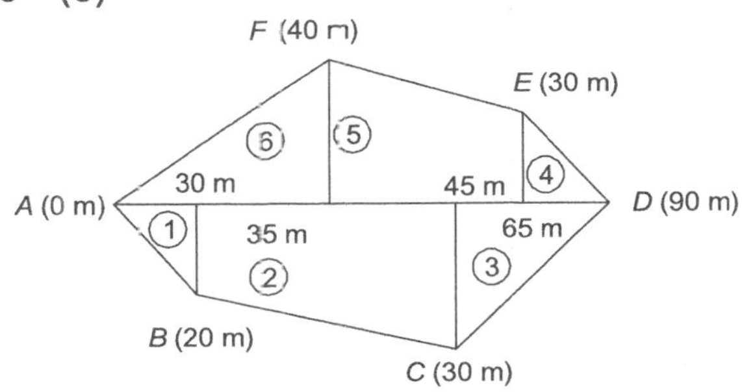

$$\text{Area} = \frac{1}{2}\begin{bmatrix} 20 \times 30 + (20+30) \times 15 + 30 \times 45 \\ +25 \times 30 + (40+30) \times 30 + 30 \times 40 \end{bmatrix}$$

$$= 3475 \text{ m}^2$$

5 Engineering Instruments, Theodolites and Plane Table Surveying

5.1 Which of the following pairs are correctly matched?
1. *Telemeter:* Measurement of distance
2. *Price meter:* Measurement of difference in elevation between points
3. *Sounding sextant :* Measurement of horizontal angles
4. *Clinometer :* Measurement of vertical angles

Select the correct answer using the codes given below:

(a) 2, 3 and 4 (b) 1, 3 and 4
(c) 1, 2 and 4 (d) 1, 2 and 3

[ESE : 2001]

5.2 Match **List-I** (Methods) with **List-II** (Procedures) and select the correct answer using the codes given below the lists:

List-I
- A. Traversing
- B. Resection
- C. Intersection
- D. Radiation

List-II
1. Rays are drawn to locate the station on which the table is set-up
2. At least two rays are drawn from two different stations to the details to be located
3. Rays are drawn in the direction of details through the station point on which the table is set-up
4. Rays are drawn on the map by setting up the table over each of the stations towards the subsequent station

Codes:

	A	B	C	D
(a)	4	3	2	1
(b)	2	1	4	3
(c)	4	1	2	3
(d)	2	3	4	1

[ESE : 2002]

5.3 The process of determining the location of the station (on the map) occupied by the plane table is called as

(a) intersection (b) three-point problem
(c) traversing (d) resection

[ESE : 2002]

5.4 The standard measurement of the Geodimeter 510 is

(a) ± 1 cm (b) ± 5.1 mm
(c) $\sqrt{5.1}$ mm (d) ± 1 mm

[ESE : 2003]

5.5 On which one of the following are the third generation electro-optical instruments based?

(a) Microwave (b) Infra-red
(c) UV light (d) He-Laser light

[ESE : 2004]

5.6 Match **List-I** (Measurements) with **List-II** (Instruments) and select the correct answer using the codes given below the lists:

List-I
- A. Displacement measurement from photograph
- B. Electronic distance measurement
- C. Base line measurement
- D. Horizontal angle measurement

List-II
1. Box sextant
2. Subtense bar
3. Tellurometer
4. Parallax bar

Codes:

	A	B	C	D
(a)	4	3	2	1
(b)	1	3	2	4
(c)	4	2	3	1
(d)	1	2	3	4

[ESE : 2004]

5.7 Which one of the following verniers is employed in Abney Level?
(a) Retrograde vernier
(b) Double vernier
(c) Double folded vernier
(d) Extended vernier

[ESE : 2006]

5.8 For air borne application and materialization of GPS receiver and easy construction, which is the most frequently used antenna?
(a) Microstrip (b) Micropole
(c) Spiral helix (d) Choke ring

[ESE : 2006]

5.9 What is the angle between two plane mirrors of an optical square?
(a) 30° (b) 60°
(c) 45° (d) 90°

[ESE : 2006]

5.10 Which of the following instruments have both horizon glass and index glass?
1. Optical square 2. Line ranger
3. Box sextant 4. Pedometer
Select the correct answer using the codes given below:
(a) 2, 3 and 4 (b) 1, 3 and 4
(c) 1 and 3 only (d) 2 and 4 only

[ESE : 2007]

5.11 Which one of the following instruments can be used as clinometer?
(a) Prism square (b) Line ranger
(c) Abney level (d) Optical square

[ESE : 2007]

5.12 Match **List-I** (Instrument) with **List-II** (Use) and select the correct answer using the code given below the lists:

List-I
A. Subtense bar
B. Sextant
C. Tangent clinometer
D. Range finder

List-II
1. To determine difference in elevation between points
2. To determine horizontal distance
3. To measure angles
4. To establish right angles

Codes:

	A	B	C	D
(a)	2	4	1	3
(b)	1	3	2	4
(c)	2	3	1	4
(d)	1	4	2	3

[ESE : 2007]

5.13 Electronic theodolites of various ranges in which measured angles are displayed originally on display board are based on which one of the following?
(a) Special optical technology
(b) Introduction of Microprocessor technology
(c) Electro-optical technology
(d) Special gearing

[ESE : 2008]

5.14 Which one of the following surveys employs alidade?
(a) Contour survey
(b) Archeological survey
(c) Plane table survey
(d) Reconnaissance survey

[ESE : 2008]

5.15 Which one of the following errors is more severe in plane-table surveying?
(a) Defective sighting
(b) Defective orientation
(c) Movement of board between sights
(d) Non-horizontality of board when points sighted are at large differences of their elevation

[ESE : 2009]

5.16 **Assertion (A):** The solution of a three-point problem in plane-table surveying is aided by Lehmann's rules.
Reason (R): The application of Lehmann's rules reduces the triangle of error and is a controlled trial-and-error technique.
(a) both A and R are true and R is the correct explanation of A
(b) both A and R are true but R is not a correct explanation of A
(c) A is true but R is false
(d) A is false but R is true

[ESE : 2009]

5.17 Consider the following equipments:
1. Tacheometer 2. Odometer
3. Passometer 4. Perambulator
Which of the above equipments can be employed for measurement of horizontal distance?
(a) 1 and 2 only (b) 1 and 3 only
(c) 2 and 3 only (d) 1, 2, 3 and 4
[ESE : 2009]

5.18 Altimetry may be depicted most accurately by
(a) hachures (b) relief shading
(c) layer tinting (d) contour lines
[ESE : 2010]

5.19 For setting out right angles, the instrument used is
(a) Optical square (b) Abney level
(c) Alidade (d) Ceylon ghat tracer
[ESE : 2010]

5.20 Consider the following statements about theodolites:
1. Transit theodolite is a theodolite in which the telescope can be transited.
2. EDMI is a theodolite fitted with a micrometer for measurements.
3. A double reading theodolite is one in which diametrically opposite segments of the graduated circle are brought into view and the readings are averaged.
Which of these statements are correct?
(a) 1, 2 and 3 (b) 1 and 3 only
(c) 1 and 2 only (d) 2 and 3 only
[ESE : 2010]

5.21 The angle between the index glass and the horizon glass of a box sextant is 40°; the horizontal angle between the two points sighted by the instrument is
(a) 20° (b) 60°
(c) 40° (d) 80° [ESE : 2011]

5.22 Which one of the following statements is **not** correct?
(a) A surveyor's compass has two sight vanes
(b) A prismatic compass has an object vane and an eye vane
(c) A trough compass is an accessory to a plane table
(d) In a prismatic compass the graduations on the aluminimum disc rotate and the index remains stationary
[ESE : 2011]

5.23 Regarding plane-table survey, which of the following statements does ***not*** hold?
(a) All the plotting work including contouring can be done in the field.
(b) It is quite suitable for small scale surveys
(c) Less number of control points are required
(d) It can be done in all seasons [ESE : 2012]

5.24 In a plane-table survey, the process of determining the plotted position of a station occupied by the plane-table by means of sights taken towards known points, the locations of which have already been plotted, is known as
(a) Radiation (b) Resection
(c) Intersection (d) Traversing
[ESE : 2012]

5.25 The plotting of inaccessible points in a plane-table survey can be done by the method of
(a) Interpolation (b) Radiation
(c) Intersection (d) Traversing
[ESE : 2012]

5.26 Consider the following pre-conditions for correct use of a theodolite:
1. The vertical axis need not be perpendicular to the plane of the plate level bubble.
2. The line of sight must be perpendicular to the horizontal axis.
3. The axis of the level tube attached to the telescope need not be parallel to the line of sight.
4. The vertical axis, the horizontal axis and the line of sight should all pass through a point known as stadia centre.
Which of these conditions is/are necessary?
(a) 1, 2, 3 and 4 (b) 2 only
(c) 3 only (d) 1 and 4 only
[ESE : 2012]

5.27 Consider the following parameters in respect of RADAR principle:
1. It is an active sensing system.
2. Electromagnetic radiation of wavelength is in centimeter range.
3. It operates in visible region.
4. It comprises Radio detection and ranging.
Which of the above statements are correct ?
(a) 1,2 and 3 (b) 2, 3 and 4
(c) 1, 2 and 4 (d) I, 2, 3 and 4
[ESE : 2014]

5.28 In setting up the plane table at a station *P*, the corresponding point on the plan was not accurately centred above *P*. If the displacement of *P* was 50 cm in a direction perpendicular to the ray, how much on the plan would be the consequent displacement of a point from its true position if the scale was 1 cm = 10 m?

(a) 5 mm (b) 1 mm
(c) 0.5 mm (d) 0.05 mm

[ESE : 2014]

5.29 Which of the following minor instruments are used for setting out right angles in chain surveying?

1. Cross square 2. Optical square
3. Prism square 4. Auto level

(a) 1 and 2 only (b) 2 and 3 only
(c) 1, 2 and 3 (d) 2, 3 and 4 only

[ESE : 2015]

5.30 Which of the following options increase the sensitivity of a bubble tube?

1. Using a liquid of greater surface tension
2. Increasing the diameter of the tube
3. Increasing the length of the tube
4. Decreasing the diameter of the tube

Select the correct answer using the codes given below:

(a) 1 and 2 only (b) 2 and 3 only
(c) 3 and 4 only (d) 1 and 4 only

[ESE : 2017]

5.31 In plane surveying where a graduated staff is observed either with horizontal line of sight or inclined line of sight, the effect of refraction is to

(a) Increase the staff reading
(b) Decrease the staff reading
(c) Neither increase nor decrease the staff reading
(d) Duplicate the staff reading

[ESE : 2019]

5.32 A theodolite is called a transit theodolite, when its telescope can be revolved thrugh a complete revolution about its

(a) Vertical axis in an inclined plane
(b) Horizontal axis in an inclined plane
(c) Vertical axis in a horizontal plane
(d) Horizontal axis in a vertical plane

[ESE : 2020]

5.33 Which one of the following is not an instrument for setting out right angles?

(a) Cross staff (b) Site square
(c) Prism square (d) Optical staff

[ESE : 2021]

5.34 In setting up to plane table at a station P, the corresponding point on the plan was not accurately centered above P. If the displacement of P was 30 cm in a direction at right angle to the ray and scale is 1 cm = 2 m, how much on the plan would be the consequent displacement of point from its true position?

(a) 0.15 mm (b) 6.0 mm
(c) 1.5 mm (d) 0.3 mm

[ESE : 2021]

■■■■

Answers Engineering Instruments

5.1 (b)	5.2 (c)	5.3 (d)	5.4 (b)	5.5 (b)	5.6 (a)	5.7 (d)	5.8 (a)	5.9 (c)
5.10 (c)	5.11 (c)	5.12 (c)	5.13 (c)	5.14 (c)	5.15 (b)	5.16 (a)	5.17 (d)	5.18 (d)
5.19 (a)	5.20 (a)	5.21 (d)	5.22 (d)	5.23 (d)	5.24 (b)	5.25 (c)	5.26 (d)	5.27 (c)
5.28 (c)	5.29 (c)	5.30 (b)	5.31 (c)	5.32 (d)	5.33 (d)	5.34 (c)		

Explanations Engineering Instruments

5.1 (b)

Telemeter is a special device which is attached to the objective end of the telescope to measure directly the horizontal distance.

Price meter is a type of current meter which is used to determine the velocity of water.

Sounding sextant is an instrument used for measurement of horizontal and vertical angels. It is similar to a nautical sextant. However, the index glass of a sounding sextant is generally much larger as compared to that of nautical sextant. It is generally used in hydrographic surveying.

5.4 (b)

Geodimeter is a medium range (less than 25 km) electronic distance measuring equipment. It has a range of 5 km in daylight and about 25 km in the night using a mercury lamp.

5.5 (b)

According to the electromagnetic waves that carries the measuring signal, the EDM (Electronic Distance Measurement) instruments may be classified into two groups:

1. Microwave EDM instruments employing radiowaves.
2. Electro-optical instrument

5.6 (a)

The subtense bar is generally used for the measurement of horizontal distances in the following situations:

(i) In rough country, across water, river or swamps, where direct chaining is very difficult.
(ii) For measurement of the traverse lines in rough country.
(iii) For measurement of base lines of small triangulations.
(iv) To fix control points in aerial photogrammetry.
(v) For measurement of long distance where great accuracy is not required.

The Tellurometer was invented by Dr. T.L. Wadley of the South African National Institute for Telecommunication Research. The Tellurometer was the first instrument which used microwaves in the measurement of distances for surveying purposes.

5.7 (d)

There are five types of verniers in practice:

(i) **Direct Vernier:** It has divisions which are slightly shorter than those of the main scale.
(ii) **Retrograde Vernier:** It has divisions which are slightly longer than those of the main scale.
(iii) **Double Vernier:** It is used when the graduations on the main scale are numbered in both the directions. The double vernier extends in both the directions with the index mark in the centre. Infact a double vernier is a combination of two direct verniers.
(iv) **Double Folded Vernier:** It is a special type of double vernier whose total length is one-half of the combined length of the corresponding double vernier. It is used when it is required to keep the length of the vernier short.
(v) **Extended Vernier:** In this type of vernier n divisions of the vernier scale are equal in length to $(2n-1)$ divisions of the main scale. An abney level has an extended type of vernier with a least count of 5′ or 10′.

5.8 (a)

Microstrip antennas(also known as patch antennas) have added advantage for airborne application, materialization of GPS receiver, easy construction, low fabrication cost. They are mostly used at microwave frequencies and find application in satellite communication, mobile, GPS, military etc.

5.9 (c)

Optical square is an instrument which is used for setting out right angles. It is more convenient and more accurate than a cross-staff. There are two mirrors placed vertically above the base. The mirrors are inclined at 45° to each other. One mirror is called horizon glass and it is unsilvered at the bottom and silvered on the top. Other mirror is called index glass and it is fully silvered. The horizon glass is fixed in position where as index glass can be adjusted.

5.10 (c)

The box sextant is a small pocket instrument used for measuring horizontal and vertical angles, measuring chain angles and locating inaccessible points. By setting the vernier to 90°, it may be used as an optical square. Optical square is an instrument used for setting out a line at right angles to another line. Both these instruments have two sets of mirrors, one being fixed (horizon glass) and other being movable (index glass).

5.11 (c)

Abney level Can be used as clinometers for measuring slope and as hand level for levelling. It is improved modification of hand level.

Line ranger is used for ranging whereas prism square and optical square for setting out offsets at right angles.

5.13 (c)

Liquid crystal diode is used for display in electronic theodolites. It is based on electro-optical technology and is made up of segments filled with liquid.

5.14 (c)

An alidade also called turning board is a device in plane table surveying that allows one to sight a distant object and use the line of sight to perform a task such as to draw a line on a plane table in the direction of the object or to measure the angle to the object from some reference point. Angles measured can be horizontal, vertical or in any chosen plane.

5.15 (b)

If the orientation is defective the plotted work at successive station cannot be connected to plotted work of previous stations.

5.16 (a)

In three point problem, if the orientation of the plane table is not proper, the intersection of the resectors through the three points will not meet at a point but will form a triangle, known as triangle of error. The size of the triangle of error depends upon the amount of angular error in the orientation. The trial and error method of three point problem, also known as Lehman's method minimises the triangle of error to a point iteratively. The iterative operation consists of drawing of resectors from known points through their plotted position and the adjustment of orientation of the plane table.

5.17 (d)

Perambulator Passometer, Pedometer, Tacheometer, measuring wheel, speedometer, Pacing, odometer etc are used in reconnaissance surveying for measuring horizontal distance.

5.18 (d)

The measurement of altitude is called altimetry. Contour lines can be best for altimetry of steep risings, cliffs etc as it gives complete picture of earth's surface.

5.19 (a)

Abney level is an improved version of the hand level. It can be used as a hand level for levelling and as a clinometer for measuring slopes.

An alidade is a straight-edge ruler having some sighting device. It is used for sighting the objects and drawing the lines.

The Ceylon ghat tracer is a simple instrument used for measuring the slopes. It is specially useful for setting out a grade contour on a given gradient in the preliminary survey of a road in hilly area.

5.20 (a)

A theodolite is a precision instrument used for measuring angles both horizontally and vertically. Theodolites can rotate along their horizontal axis as well as their vertical axis. A transit is a surveying instrument that also takes accurate angular measurements. Transits use vernier scales and external graduated metal circles for angular readings.

A double reading theodolite is one in which images of diametrically opposite parts of the graduated circle are brought into the field of view by a suitable optical arrangement, and are read and averaged in one stroke.

5.21 (d)

The sextant is based on the principle that when a ray of light is reflected successively from two mirrors, the angle between the first and last directions of ray is twice the angle between the planes of the two mirror.

∴ The horizontal angle between the two points sighted by the instrument = 2 × 40° = 80°

5.22 (d)

Both prismatic and surveyor's compass has two vanes. The object vane consist of a metal vane with a vertical hair and the eye vane consists of a metal vane with a fine slit.

In prismatic compass the graduated aluminium ring is attached with the needle. The ring does not rotate along with the line of sight.

The compass used with a plane table is a trough compass in which the longer sides of the trough are parallel and flat so that either side can be used as a ruler or lid down to coincide with a straight line drawn on the paper. Trough compass is used for orienting the plane table to magnetic north.

5.23 (d)

Plane table surveying cannot be done in rainy season and wet climatic conditions.

5.25 (c)

Intersection method can be used for plotting inaccessible points in plane table surveying. In this method, the positions of the object on the plan are fixed by the intersection of rays drawn from two instrument stations. The line joining these instrument stations are called baseline.

5.27 (c)

The acronym of radar is radio detection and Ranging, which is used to determine the range, direction or speed of the object. It is an active sensing system.

5.28 (c)

In given scale, 1 cm = 10 m

i.e. 10 m on ground = 1 cm on plan

10 × 100 cm on ground = 1 × 10 mm on plan

$$1 \text{ cm on ground} = \frac{1\times 10}{10\times 100} \text{ mm on plan}$$

∴ 50 cm on ground

$$= \frac{1\times 10}{10\times 100}\times 50 \text{ mm on plan}$$

$$= 0.5 \text{ mm on plan}$$

5.29 (c)

For setting out right angles in chain surveying, the minor instruments used are cross staff, optical square and prism square.

Auto level, a levelling instrument, is an optical instrument used to established or check points in the same horizontal plane. It is used in surveying and building with a vertical staff to measure height difference and so transfer, measure and set heights.

5.30 (b)

The factor which govern the sensitivity of bubble tube are:

1. Viscosity of liquid
2. Radius of circle, larger the radius better will be bubble.
3. Internal fraction, if inside walls are very smooth it will give better sensitivity.

5.31 (c)

Neither increase nor decrease the staff reading.

5.32 (d)

Transiting = Revolution of telescope about horizontal axis (Trunion axis) in vertical plane.

5.33 (d)

There are several types of insruments used to set out a right angle to a chain line, the most common being:

(i) Cross staff (ii) Optical square

(iii) Prism square (iv) Site square

Note: Optical staff is a levelling staff that helps to determine the height difference between two points on a survey.

5.34 (c)

Scale = 1 cm = 2 m

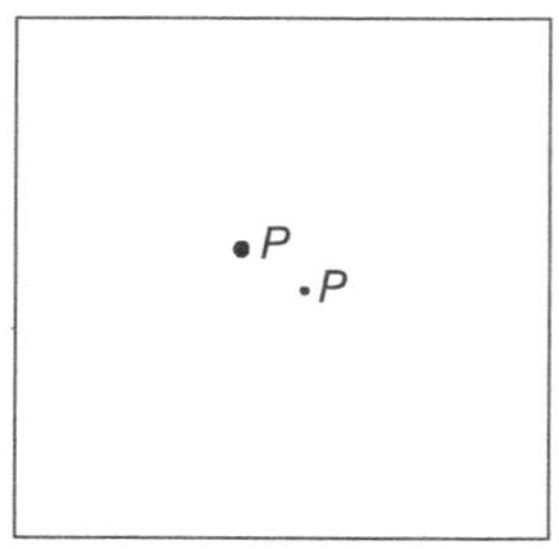

10 mm = 200 cm

$$1 \text{ cm} = \frac{10}{200} = \frac{1}{20} \text{ mm}$$

PP = 30 cm on ground

$$(\text{PP on plan}),\ pp = 30 \text{ cm} = \frac{1}{20}\times 30 = 1.5 \text{ mm}$$

■■■■

6 Tacheometric, Curve, Hydrographic Survey, Tides & Triangulation

6.1 Triangulation station selected close to the main station for avoiding intervening obstruction is called

(a) eccentric station (b) pivot station
(c) satellite station (d) tie station

[ESE : 2000]

6.2 Match **List-I** (Type of transition curves) with **List-II** (Characteristics) and select the correct answer using the codes given below the lists:

List-I

A. Glover's spiral
B. Cubic spiral
C. Froude's transition curve
D. Bernoulli's lemniscate

List-II

1. An autogenous curve of an automobile
2. Radius of curvature at any point varies inversely as the distance from the beginning of the curve
3. $\sin \phi = \phi$
4. $x = l$, $\cos\phi = 1$

Codes:

	A	B	C	D
(a)	2	3	4	1
(b)	3	2	1	4
(c)	2	3	1	4
(d)	3	2	4	1

[ESE : 2003]

6.3 If two triangulation signals of 6.75 m height each, are to be just visible over ground mutually, what is the maximum distance between their locations on the ground surface?

(a) 10 km (b) 20 km
(c) 30 km (d) 40 km

[ESE : 2005]

6.4 Match **List-I** with **List-I** and select the correct answer using the codes given below the lists:

List-I

A. Adjustment of surveying instruments
B. Bowditch rule
C. Triangulation
D. Bessel's method

List-II

1. Bringing the various fixed parts of the instruments into proper relations with one another
2. Solution of three point problem
3. Measuring all the angles and base line
4. Balancing the latitudes and departures

Codes:

	A	B	C	D
(a)	1	2	3	4
(b)	3	4	1	2
(c)	1	4	3	2
(d)	3	2	1	4

[ESE : 2006]

6.5 Match **List-I** (Triangulation station) with **List-II** (Deflection) and select the correct answer using the code given below the lists:

List-I

A. Main stations
B. Subsidiary stations
C. Satellite Stations
D. Pivot Stations

List-II

1. Control points of triangulation network
2. Points not for observation but for continuation of triangulation network
3. Points to provide additional rays to intersected points
4. Points close to main stations to avoid intervening obstructions

Codes:

	A	B	C	D
(a)	1	3	4	2
(b)	1	3	2	4
(c)	3	1	4	2
(d)	3	1	2	4

[ESE : 2007]

6.6 **Assertion (A):** Triangulation networks are to be formed by well conditioned triangles.
Reason (R): Triangulation signals should be conspicuous and centred accurately over the stations.
(a) both A and R are individually true and R is the correct explanation of A
(b) both A and R are individually true but R is not the correct explanation of A
(c) A is true but R is false
(d) A is false but R is true [ESE : 2008]

6.7 Which one of the following linear methods of setting out a circular curve needs reference of the centre of the curve?
(a) Offset from chord produced
(b) Radial offset
(c) Perpendicular offset
(d) Successive bisection of arcs [ESE : 2009]

6.8 In accordance to 'Gauss rule', weights to be assigned are proportional to
(a) 1/(Sum of the residual errors of observations
(b) 1/(Sum of the square of the residual errors of observations)
(c) Sum of the square roots of the residual errors of observations
(d) Sum of the cube roots of the residual errors of observations [ESE : 2010]

6.9 Refraction error is the least in case of
(a) stadia tacheometry
(b) tangential tacheometry
(c) subtense bar tacheometry
(d) omnimeters [ESE : 2010]

6.10 Match **List-I** with **List-II** and select the correct answer using the code given below the lists:

List-I
A. Traverse surveying
B. Geodetic surveying
C. Plane table surveying
D. Hydrographic surveying

List-II
1. Weddel's sounding machine
2. Alidade
3. Chain and compass
4. Theodolite

Code:

	A	B	C	D
(a)	3	4	2	1
(b)	1	4	2	3
(c)	3	2	4	1
(d)	1	2	4	3

[ESE : 2011]

6.11 The purpose of a 'satellite station' in triangulation can be served by
(a) *A* 'Church spire' in order to secure a well-shaped triangle
(b) *A* 'Flag pole' in order to secure a well shaped triangle
(c) *A* 'Steeple' in order to secure a well-shaped triangle
(d) An 'Eccentric station' near the true station whereon the instrument cannot be setup [ESE : 2011]

6.12 The sum of the three interior angles of a triangle, the vertices of which lie on the surface of the earth, covering a vast area of several hundreds of sq kms, is
(a) Less than 180°
(b) Equal to 180°
(c) More than 180° but not less than 270°
(d) More than 180° but not more than 225° [ESE : 2011]

6.13 The subtense tacheometry method is adopted when the ground is
(a) Flat (b) Inclined
(c) Undulating (d) A waterbody [ESE : 2012]

6.14 Following observations were taken with a transit fitted with stadia wires. The line of sight was horizontal and the staff was held vertical.

	Reading on staff (m)
Top hair	1.726
Middle hair	2.278
Bottom hair	2.830

The tacheometric constants *k* and *C* are 100 and

0.4 m respectively. The horizontal distance between staff and instrument is
(a) 90.8 (b) 100.8
(c) 110.8 (d) 120.8 [ESE : 2012]

6.15 If the angle of deflection of a simple curve is θ and its radius is R, then the length of the chord is
(a) $2R \sin \theta$ (b) $2R \sin \frac{\theta}{2}$
(c) $2R \cos \theta$ (d) $2R \tan \frac{\theta}{2}$
[ESE : 2012]

6.16 If the radius of a simple curve is R, then the length of the chord for calculating the offsets by the "method of chords produced" should not exceed.
(a) $R/5$ (b) $R/10$
(c) $R/20$ (d) $R/25$ [ESE : 2012]

6.17 With regard to Trigonometric Levelling, which one of the following statements is correct at its simplest applications?
(a) Determination of the elevations of stations is based on the observed vertical angles and the horizontal distance.
(b) Determination of the horizontal distance is based on the observed vertical angles.
(c) Determination of the vertical angles is based on the observed horizontal distances.
(d) Determination of the horizontal distance
[ESE : 2015]

6.18 For better accuracy in measuring and plotting the sides of a triangle by triangulation, the angles of the triangle
(a) should not be more than 30°
(b) should not be less than 30° or more than 120°
(c) are not restricted in magnitude
(d) should not be less than 120°
[ESE : 2015]

6.19 A survey line BAC crosses a river, A and C being on the near and distant banks respectively. Standing at D, a point 50 m measured perpendiculars to AB from A, the bearings of C and B are 320° and 230° respectively. AB being 25 m. The width of the river will be
(a) 80 m (b) 90 m
(c) 100 m (d) 110 m
[ESE : 2019]

6.20 In triangulation, in order to control the accumulation of errors of length and azimuth subsidiary bases are selected. At certain stations, the astronomical observations for azimuth and longitude are also made. These stations are called
(a) Transportation stations
(b) Bowditch stations
(c) Universe stations
(d) Laplace stations
[ESE : 2019]

6.21 A circular curve has a long chord of 80 m and a versed sine of 4 m. The height and ordinate at a distance of 30 m from the mid-ordinate will be nearly
(a) 3.06 m (b) 2.72 m
(c) 2.24 m (d) 1.76 m
[ESE : 2019]

6.22 In horizontal distance measurement, the basic formula for distance in stadia tacheometry has an additive constant. An anallatic lens is inserted in the tacheometer to make this additive constant zero. This lens is
(a) Convex lens inserted between object glass and diaphragm.
(b) Plano-convex lens between object glass and diaphragm.
(c) Plano-convex lens between diaphragm and eye piece.
(d) convex lens inserted between diaphragm and eye piece.
[ESE : 2020]

6.23 Which one of the following conditions shall be fulfilled when a transition curve is inserted between the tangent and circular curve?
(a) It should not meet the original straight tangentially.
(b) It should not meet the circular curve tangentially.
(c) Its radius at the junction with the circular curve should be the same as that of the circular curve.
(d) The rate of decrease of curvature along the transition curve should be same as that of increase in superelevation
[ESE : 2021]

6.24 Consider the following statements related to set out the curve using two theodolite method:

To set out the curve,

1. set up a theodolite over T_1 and another over T_2.
2. set the vernier of each of the instruments to zero.
3. direct the instrument at T_1 to the ranging rod at the point of intersection B and bisect it.

Which of the above statements are correct?

(a) 1 and 2 only (b) 1 and 3 only
(c) 2 and 3 only (d) 1, 2 and 3

[ESE : 2022]

6.25 In setting out location of piers, the central points of the piers are located by intersection of sights, simultaneous sights being taken from

(a) the ends of a base
(b) the top of the pier
(c) the back of the pier
(d) the front of the pier [ESE : 2022]

Directions for the following TWO (02) items:

Read the following information and answer the two items that follow:

The chainage of the intersection point of two straights is 1060 m, and the angle of intersection is 120°. If radius of a circular curve to be set out is 570 m, and peg interval is 30 m.

6.26 What is the length of the curve?

(a) 190π m (b) 185π m
(c) 180π m (d) 170π m

[ESE : 2023]

6.27 What is the length of the long chord?

(a) 370 m (b) 470 m
(c) 670 m (d) 570 m

[ESE : 2023]

6.28 If R is the radius of the circle and Δ is deflection angle, what is the formula for external distance (E) in circular curve for use in design and setting out?

(a) $E = R\left(\sec\frac{\Delta}{2} - 1\right)$ (b) $E = R\left(1 - \sec\frac{\Delta}{2}\right)$

(c) $E = R\left(\cos\frac{\Delta}{2} - 1\right)$ (d) $E = R\left(1 - \cos\frac{\Delta}{2}\right)$

[ESE : 2024]

■■■■

Answers Tacheometric, Curve, Hydrographic Survey, Tides and Triangulation

6.1 (a,c)	6.2 (a)	6.3 (b)	6.4 (c)	6.5 (a)	6.6 (b)	6.7 (b)	6.8 (b)	6.9 (c)
6.10 (a)	6.11 (d)	6.12 (d)	6.13 (a)	6.14 (c)	6.15 (b)	6.16 (c)	6.17 (a)	6.18 (b)
6.19 (c)	6.20 (d)	6.21 (d)	6.22 (a)	6.23 (c)	6.24 (d)	6.25 (a)	6.26 (a)	6.27 (d)
6.28 (a)								

Explanations Tacheometric, Curve, Hydrographic Survey, Tides and Triangulation

6.1 (a, c)

To secure well conditioned triangles or to have good visibility, objects such as chimneys, church spires, flag poles, towers, etc. are selected as triangulation stations. Such stations can be sighted from other stations, but instruments cannot be set up at such stations for sighting other stations for measuring horizontal angles. In such cases, a subsidiary station is selected near the main stations from the subsidiary station. The subsidiary station is known as the satellite station, false station or eccentric station.

6.2 (a)

Cubic parabola is also known as Froude's transition curve.

$$y = \frac{x^3}{6RL}$$

by putting $x = l$

The assumption is $\cos\phi = 1$ and $\sin\phi = \phi$

Clothoid or Glover's spiral has the property that radius of curvature (R) at any point varies inversely with the distance.

The cubic spiral has equation,

$$y = \frac{x^3}{6RL}$$

The approximation made here is $\sin\phi = \phi$

Cubic spiral is superior to a cubic parabola.

Bernoulli's lemniscate curve is used on highways and it is preferred to the spiral for the following reasons :

(i) Its radius of curvature decreases more gradually.

(ii) Its rate of increase of curvature diminishes towards the transition curve, thus fulfilling an essential condition.

(iii) It corresponds to an autogenous curve of an automobile i.e. the path actually traced by an automobile when turning freely.

6.3 (b)

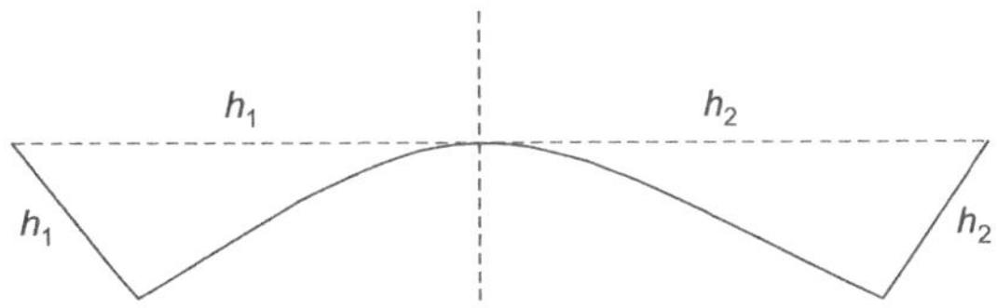

Distance of visible horizon,

$$d = d_1 + d_2$$

$$d = 2 \times 3.855\sqrt{6.75}$$

$$d = 20.03 \text{ km}$$

6.5 (a)

Main stations are the main control points of the network and command boundary of area to be surveyed. The station where observations are not made, but the angles at the station are used in triangulation series, is known as pivot station. When setting up instrument at triangulation station is difficult then satellite station is selected near main station. This helps avoid intervening obstructions. Subsidiary stations provide additional rays to intersected points.

6.6 (b)

Triangulation network should form well conditioned triangles and should be intervisible. Triangulation signals should be conspicuous and clearly visible over ground so that exact position of station can be observed from other station.

6.7 (b)

Radial offsets from tangents require reference of centre of curve for setting out a simple circular curve by linear methods.

6.9 (c)

In subtense bar method, a bar of fixed length, called a subtence bar is placed in horizontal position. The angle subtended by two target points, corresponding to fixed distance on subtense bar, at the instrument station is measured. The distance between the subtence bar and the instrument is computed from the known distance between the targets and the measured horizontal angle. So there will be no refraction error.

6.10 (a)

Sounding Machines are used to measure depth below the water surface. Sounding machines may be hand driven or machine driven. Weddel's sounding machine is hand driven.

A plane table alidade is a straight edge used for sighting with some form of sighting device. Two types of alidade are used: Plain alidade and telescopic alidade.

Geodetic surveying is that type of surveying in which the shape of the earth is taken into account.

The theodolite is the most precise instrument designed for the measurement of horizontal and vertical angles and has wide applicability in surveying especially long distance surveying such as laying off horizontal angles, locating points on line, prolonging survey lines, establishing grades, determining difference in elevation, setting out curves etc.

In traverse surveying compass is used to establish angles between two stations while a chain is used to measure ground distance between stations.

6.11 (d)

Chimneys, church spires, flag poles, towers, lighthouses, etc., are selected as triangulation stations because of their good visibility and forming well-conditioned triangles. Such stations can be sighted from other stations but it is not possible to occupy them directly below such excellent stations for making passivations on other stations. The problem is solved as by taking another station in the vicinity of the main triangulation station from where the other stations are visible. Such stations are called the satellite stations.

6.12 (d)

Properties of spherical triangle:

- If a circle is drawn on a sphere so that the

radius of the circle is the same as the radius of the sphere it is called a great circle. Any other circle is a small circle.

- An infinite number of great circle can pass through one point , but only one great circle can pass through tow points , unless they are diametrically opposite.
- A spherical triangle is a triangle each of whose sides is a great circle.
- The length of the arc of a circle can be measured by the angle which the arc subtends at the center of the circle. The sides of a spherical triangle are measured in degree , minutes , and seconds.
- No sides of a spherical triangle can therefore exceed 180°.
- The three angles of a spherical triangle must together be more than 180° and less than 540°.
- The greater side is opposite the greater angle, if tow sides are equal their opposite angles are equal.
- If one angle of the triangle is 90° it is called a right-angle triangle , and if one side of the triangle 90° it is called a quadrantal triangle .

6.13 (a)

In the subtense bar method, the horizontal angle subtended by two targets fixed on a horizontal bar at a known distance apart is measured at instrument station by theodolite.

6.14 (c)

$$D = kS + C$$
$$S = 2.830 - 1.726 = 1.104$$
$$\therefore \quad D = 100 \times 1.104 + 0.4 = 110.8 \text{ m}$$

6.15 (b)

$L = 2R \sin \theta/2$

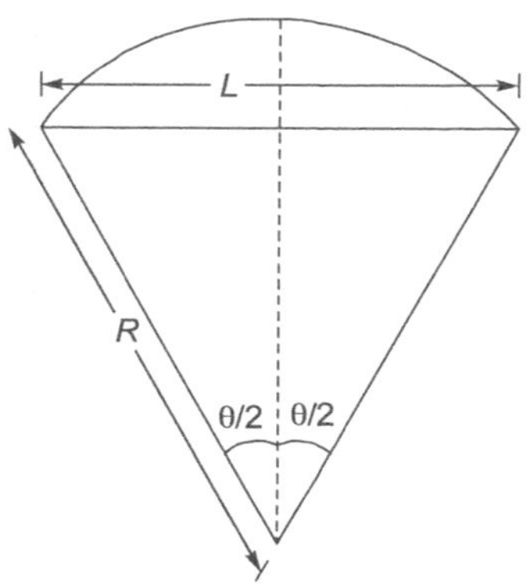

6.16 (c)

Length is selected such that there is not much difference between arc length and chord length. For the error to be within limit the length of full chord should be limited to R/20.

6.17 (a)

In Trigonometric Levelling, determination of the elevations of station is based on the observed vertical angles and the horizontal distances.

6.18 (b)

For better accuracy in measuring and plotting the sides of a triangle by triangulation, the angles of triangle

(i) should not be less than 30°

(ii) should not be more than 120°

6.19 (c)

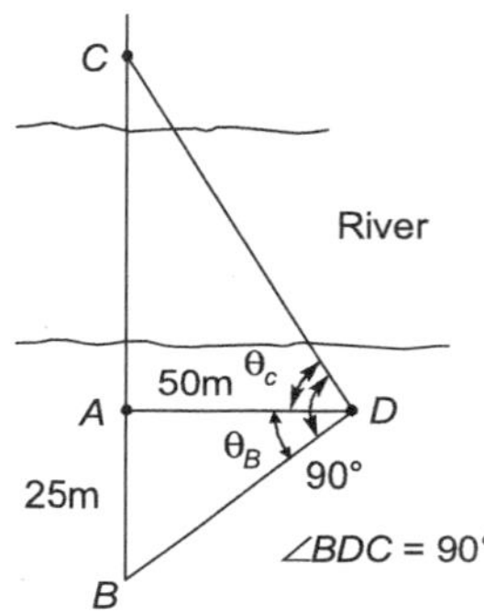

From ΔABD, $\tan \theta_B = \dfrac{25}{50}$

$$\Rightarrow \quad \theta_B = 26°33'54''$$
$$\theta_C = (90 - \theta_B)$$
$$= 90° - 26°33'54'$$
$$= 63°26'6''$$

From ΔACD, $\tan \theta_C = \dfrac{AC}{AD}$

$$\Rightarrow \quad \tan 63°26'6'' = \frac{AC}{50}$$
$$\Rightarrow \quad AC = 100 \text{ m}$$

Width of River = AC = 100 m

6.20 (d)

Laplace station is a geodetic station at which coincident triangulation and astronomic longitude and azimuth determinations are made.

It is defined as a trinagulation or transverse station at which geodetic azimuth is derived from astronomic azimuth by the use of Laplace equation.

6.21 (d)

By ordinate from the long chord

Ordinate at distance x from the mid point of the chord = O_x

$$O_x = \sqrt{R^2 - x^2} - (R_0 - O_0)$$

$$O_x = \sqrt{R^2 - x^2} - (R - O_0)$$

$$O_0 = R - \sqrt{R^2 - \left(\frac{L}{2}\right)^2}$$

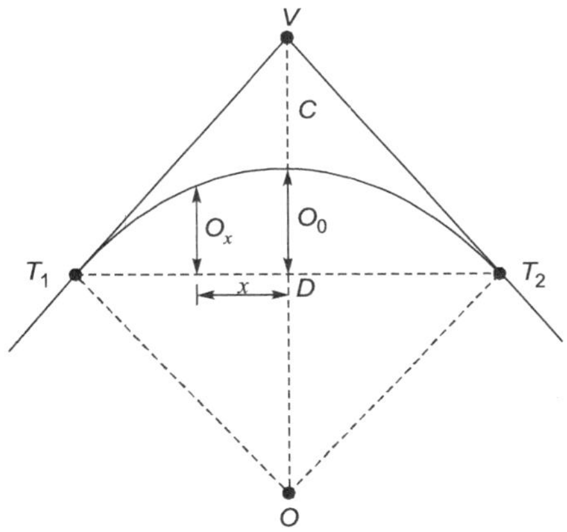

Given data:

O_0 = Versed sine = 4 m

L = Longest chord = 80 m

x = 30 m

Versed Sine (O_0)

$$O_0 = R - \sqrt{R^2 - \left(\frac{L}{2}\right)^2}$$

$$4 \text{ m} = R - \sqrt{R^2 - \left(\frac{80}{2}\right)^2}$$

$$\Rightarrow \quad R = 202 \text{ m}$$

$$O_x = \sqrt{R^2 - x^2} - (R - O_0)$$

$$O_{30} = \sqrt{202^2 - 30^2} - (202 - 4)$$

$$= 1.759 \approx 1.76 \text{ m}$$

6.23 (c)

It should be tangential to the straight line and also meet the circular curve tangentially at the junction.

- It's curvature should be zero (R = ∞) at one end and its curvature should be equal to (1/R) where it meets the circular curve (R = Radius of circular curve).
- The rate of increase of curvature along the transition curve should be equal to rate of increase of superelevation.

Therefore, length of transition curve should be such that full superelevation is achieved at the junction with the circular curve.

6.24 (d)

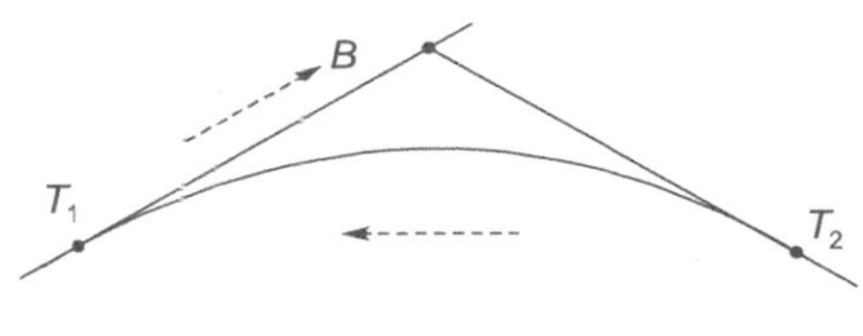

6.25 (a)

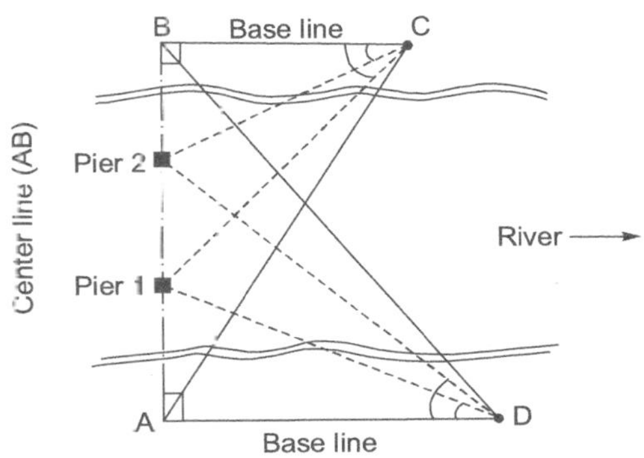

6.26 (a)

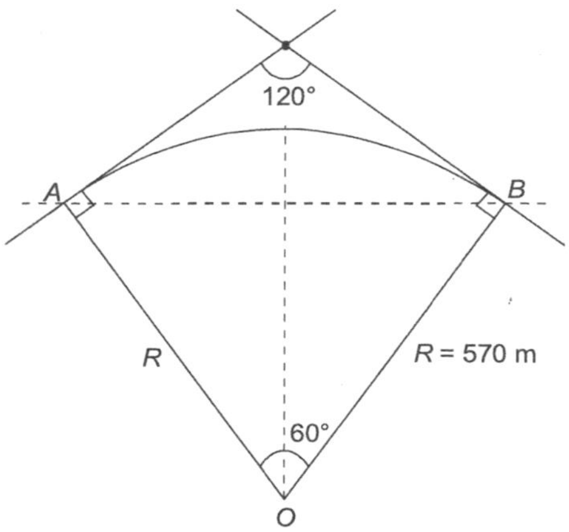

Length of curve = Angle subtended at O × Radius

$$= \frac{\pi}{3} \times 570 = 190\pi \text{ m}$$

6.27 (d)

$$\text{Length of long chord} = AB = 2R\sin\left(\frac{\Delta}{2}\right)$$

$$= 2R\sin\left(\frac{60^\circ}{2}\right) = 2R\sin 30^\circ$$

$$= 2 \times 570 \times \frac{1}{2} = 570 \text{ m} [\because R = 570 \text{ m}]$$

6.28 (a)

External /apex distance(E) is given as

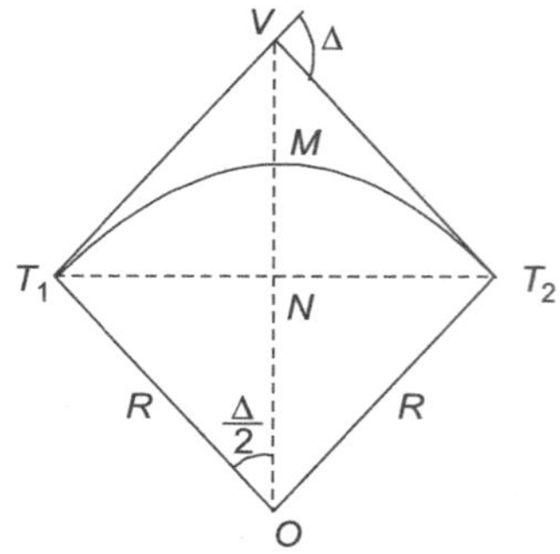

$$E = VM$$

$$= R\left(\sec\frac{\Delta}{2} - 1\right)$$

■■■■

7 Theory of Errors

7.1 The sensitiveness of a bubble tube in a level would decrease if

(a) the radius of curvature of the internal surface of the tube is increased
(b) the diameter of the tube is increased
(c) the length of the vapour bubble is increased
(d) the viscosity of the liquid is increased

[ESE : 2001]

7.2 From the probability equation it is found that the most probable values of a series of errors arising out of observations of equal weightage are those for which the sum of their squares is

(a) zero (b) infinity
(c) minimum (d) maximum

[ESE : 2003]

7.3 If the weight of an angle A (= 40° 24′ 24″ say) is 2, then the weight of the angle $A/3$ (= 13° 28′ 08″) will be

(a) 4 (b) $\pm\sqrt{3}$
(c) 9 (d) 18 [ESE : 2008]

7.4 If everyone of the three angles of a triangle has a probable error of ±1″, then what will be the probable error in the sum of the so-measured internal angles of the triangle?

(a) ±1″ (b) ±3″
(c) ±9″ (d) $\pm\sqrt{3}''$ [ESE : 2009]

7.5 The measured radius of a circle is 80 m with a possible error of 0.05 m in its diameter. The error in the computed area will nearly be

(a) + 6.5 m^2 (b) – 0.65 m^2
(c) ± 12.6 m^2 (d) ± 8.2 m^2

[ESE : 2014]

7.6 The side of a square land was measured as 150 m and is in error by 0.05 m. What is the corresponding error in the computed area of the land?

(a) 5 m^2 (b) 10 m^2
(c) 15 m^2 (d) 20 m^2 [ESE : 2016]

7.7 Which of the following statements are rightly associated with the laws of weights in the theory of errors?

1. If an equation is multiplied by its own weight, then the weight of the resulting equation is equal to the reciprocal of the weight of the original equation.
2. The weight of the algebraic sum of two or more quantities is equal to the reciprocal of the sum of the individual weights.
3. If the quantity of a given weight is multiplied with a factor, then the weight of the resulting quantity is obtained by dividing the original weight by the square root of that factor.
4. If the quantity of a given weight is divided by a factor, then the weight of the resulting quantity is obtained by multiplying the original weight by the square of that factor.

(a) 2 and 3 (b) 2 and 4
(c) 1 and 4 (d) 1 and 3

[ESE : 2016]

7.8 The clogging of chain rings with mud introduces (with 'error' defined in the standard way)

1. Negative cumulative error.
2. Positive cumulative error.
3. Compensating error.

(a) 1 only (b) 2 only
(c) 3 only (d) 1, 2 and 3

[ESE : 2016]

7.9 Which one of the following is the source of error in curve computations and layout?

(a) ability to set on the plates of the theodolite, the required sub-division of a minute for the deflection angles.
(b) use of less than full tape lengths on arc-definition curves.
(c) Carrying out computed elevations to more than 10 mm.
(d) Good intersections between tape line and site line on flat curves. ESE : 2022]

■■■■

Answers Theory of Errors

7.1 (d) 7.2 (c) 7.3 (d) 7.4 (d) 7.5 (c) 7.6 (c) 7.7 (c) 7.8 (b) 7.9 (b)

Explanations Theory of Errors

7.1 (d)

Sensitiveness of a bubble tube depends upon

(i) The radius of curvature of the internal surface i.e. larger the radius greater is the sensitive-ness.
(ii) The diameter of the bubble tube i.e. larger the diameter, greater the sensitiveness.
(iii) The length of the bubble i.e. larger the length greater is the sensitiveness.
(iv) The viscosity and surface tension of the liquid i.e. lesser the viscosity more is its sensitive-ness.
(v) Smoothness of the finish of the internal surface of the tube i.e. greater the smoothness more is the sensitivity.
(vi) Length of one division i.e. greater the length of one division more is the sensitiveness.

7.2 (c)

If the measurements are of equal weight, the most probable value is that which makes the sum of the squares of the residual (v) a minimum. This is called the theory of least squares. Mathematically,

$$\Sigma v^2 = \text{minimum}$$

If the measurements are of unequal weights, the most probable value is that which makes the sum of the products of the weight 'w' and the squares of residuals a minimum. Mathematically,

$$\Sigma w(v)^2 = \text{minimum}$$

7.3 (d)

If a quantity of given weight is divided by a factor, the weight of the result is obtained by multiplying its given weight by the square of that factor.

Angle a has weight of 2

$$\therefore \quad \text{Weight of } \frac{A}{3} = 2 \times (3)^2 = 18$$

7.4 (d)

$$\sigma_A = \pm\sqrt{(\sigma_a)^2 + (\sigma_b)^2 + (\sigma_c)^2}$$

$$= \pm\sqrt{(1)^2 + (1)^2 + (1)^2} = \pm\sqrt{3''}$$

7.5 (c)

$$\text{Area } A = \pi r^2$$

$$\frac{\partial A}{A} = 2\pi r \cdot \frac{\partial r}{r}$$

$$\frac{\partial A}{A} = 2\pi \times 80 \times 0.05$$

$$\text{Error} = \pm 12.56 \text{ m}^2$$

7.6 (c)

$$A = a^2$$

$$\partial A = 2a.\ \partial a = 2 \times 150 \times 0.05 = 15 \text{ m}^2$$

7.7 (c)

Note: Here answer given by UPSC is 'd' but the correct answer is 'c'. As per the Theory of Weights:

Various laws of weight

Let there be quantities A and B with weight W_a and W_b respectively

Operation	Resultant weight
$(A+B)$	$\dfrac{1}{\frac{1}{W_a}+\frac{1}{W_b}}$
$(A-B)$	$\dfrac{1}{\frac{1}{W_a}+\frac{1}{W_b}}$
$k.A.$(k=Constant)	$\dfrac{W_a}{k^2}$
$\dfrac{A}{k}$	$W_a k^2$

7.8 (b)

Length of chain is reduced

Measured length = Less
Noted down value = More
Error = (+)ve
Correction = (–)ve

Positive Cumulative Error

7.9 (b)

Source of errors in curve computations and layout are:

1. Inability to set on plates of theodolite.
2. Use of less than full tape length on arc definition curve.
3. Carrying out computed elevation to less than 3 mm.
4. Poor intersections between tape line and sight line on flat curve.

8 Field Astronomy, Photogrammetic Survey, Remote Sensing, GPS & Geology

8.1 Which one of the following surveys is required in observations of stars?
(a) Astronomical survey
(b) Cadastral survey
(c) Aerial survey
(d) Photogrammetric survey
[ESE : 2000]

8.2 If the equation of time is – 13′28.5″, then the Greenwich Apparent Time corresponding to zero hour Greenwich Mean Time on a day is
(a) 13′ 28.5″
(b) 46′ 31.5″
(c) 23h 46′ 31.5″ of same day
(d) 23h 46′ 31.5″ of previous day
[ESE : 2000]

8.3 A vertical photograph was taken at an altitude of 1500 m above MSL. If the focal length of the camera is 20 m, the scale of photograph for a terrain lying at an elevation of 500 m is
(a) 1 : 50 (b) 1 : 100
(c) 1 : 1000 (d) 1 : 25
[ESE : 2000]

8.4 **Assertion (A):** In a spherical triangle, the sum of the three sides is always less than the circumference of the great circle.
Reason (R): The sum of the three angles of a spherical triangle is greater than two right angles, but less than six right angles.
(a) both A and R are true and R is the correct explanation of A
(b) both A and R are true but R is not a correct explanation of A
(c) A is true but R is false
(d) A is false but R is true [ESE : 2000]

8.5 A road section of length 1 km scales 8 cm on a vertical photograph. The focal length of the camera is 160 mm. If the terrain is fairly level, then the flying height will be
(a) 20 m (b) 2000 m
(c) 20 km (d) 200 km
[ESE : 2001]

8.6 At a given place of observation, the declination of a circumpolar star is
(a) greater than the latitude
(b) equal to the latitude
(c) less than the co-latitude
(d) greater than the co-latitude
[ESE : 2001]

8.7 A tower appears in two successive photo-graphs taken at an altitude of 4000 m above datum. The focal length of the camera is 160 mm. The length of the air base is 300 m. The parallax for the top and bottom of the pole are 72 mm and 63 mm respectively. The height of the top of the tower above its bottom is nearly
(a) 63 m (b) 72 m
(c) 95 m (d) 135 m
[ESE : 2001]

8.8 Which one of the following is the angular distance between the observer's meridian and the vertical circle passing through a star measured along the celestial horizon?
(a) Right ascension (b) Azimuth
(c) Declination (d) Hour angle
[ESE : 2002]

8.9 A vertical photograph of a chimney was taken from an elevation of 500 m above MSL. The elevation of the base of the chimney was 250 m. If the relief displacement of the top of chimney was 51.4 mm and the radial distance of the image of the top of the chimney was 110 mm, the height of the chimney is

(a) 233.64 m (b) 133.2 m
(c) 116.82 m (d) 58.41 m

[ESE : 2002]

8.10 **Assertion (A):** At a given station, a celestial body can be fixed by its co-altitude and co-declination.

Reason (R): In the astronomical triangle ZPS, ZP is constant for the given place and S is the intersection of the vertical circle and declination of circle.

(a) both A and R are true and R is the correct explanation of A
(b) both A and R are true but R is not a correct explanation of A
(c) A is true but R is false
(d) A is false but R is true

[ESE : 2002]

8.11 The relationship between the air-base B, photographic base 'b', flying height H and focal length 'f' of lens in a vertical photograph is given by

(a) $B = \frac{bH}{f}$ (b) $B = \frac{f}{bH}$

(c) $B = \frac{b}{fH}$ (d) $B = \frac{b}{H - f}$

[ESE : 2003]

8.12 Setting off the 'proper principal distance' in the projectors of projection stereoplotters is a procedure for which one of the following?

(a) Finding the focal length of the camera used
(b) Evaluating the digital terrain model
(c) Carrying out the interior orientation
(d) Filling the quotation for the purchase of A_7 stereoplotters

[ESE : 2004]

8.13 Consider the following statements:

IRS series satellites are

1. Low orbiting satellites
2. Geostationary satellites
3. Meteorological satellites
4. Resource survey satellites

Which of these statements are correct?

(a) 1 and 4 (b) 2 and 3
(c) 1, 2 and 4 (d) 2, 3 and 4

[ESE : 2004]

8.14 **Assertion (A):** A mosaic is large photomap assembled from two or more aerial photographs of an area.

Reason (R): Photomap and mosaic are true planimetric representations of the area.

(a) both A and R are true and R is the correct explanation of A
(b) both A and R are true but R is not a correct explanation of A
(c) A is true but R is false
(d) A is false but R is true

[ESE : 2004]

8.15 What is the minimum number of satellites required from which signals can be recorded to enable a global positioning system receiver to determine latitude, longitude and altitude?

(a) One (b) Two
(c) Three (d) Four

[ESE : 2004]

8.16 Which one of the following represents a circumpolar star?

(a) Upper culmination above horizon, lower culmination below horizon
(b) Both upper and lower culminations below horizon
(c) Both upper and lower culminations above horizon
(d) Altitude at upper culmination is minimum

[ESE : 2005]

8.17 In an aerial photogrammetric survey, if the exposure interval is 20 seconds to cover ground distance of 1000 m between exposures, what would be the ground speed of the aircraft?

(a) 90 km/hour (b) 120 km/hour
(c) 150 km/hour (d) 180 km/hour

[ESE : 2005]

8.18 Match **List-I** (Phenomenon) with **List-II** (Method of survey) and select the correct answer using the code given below the lists:

List-I

A. Crab and drift
B. Stadia intercept
C. Culmination and elongation
D. Baseline measurement

List-II

1. Triangulation
2. Astronomical survey

3. Aerial photogrammetry
4. Tacheometric survey

Codes:

	A	B	C	D
(a)	3	1	2	4
(b)	2	4	3	1
(c)	3	4	2	1
(d)	2	1	3	4

[ESE : 2005]

8.19 What is the actual ground area covered by a 20 cm × 20 cm size vertical aerial photograph, at an average scale of 1 cm = 200 m having 60% forward overlap and 30% side overlap?

(a) 1.92 km^2 (b) 4.48 km^2
(c) 6.72 km^2 (d) 2.88 km^2

[ESE : 2005]

8.20 If the mean temperature of Sun's surface 6000 K and λ m of its radiation is 0.5 μm, what is the mean temperature of earth's surface for which λ m is 10.0 μm, according Wien's Displacement Law in Remote Sensing Concept?

(a) 25°C (b) 28°C
(c) 27°C (d) 30°C

[ESE : 2007]

8.21 Match **List-I** (Term) with **List-II** (Definition) and select the correct answer using the code given below the lists:

List-I

A. Apparent solar day
B. Sidereal day
C. Tropical year
D. Sidereal year

List-II

1. The time interval between two successive upper transits of the first point of Aries over the same meridian
2. The interval between two successive lower transits of the centre of the Sun across the same meridian
3. Time interval between two successive passages of the Sun over the meridian of any one of the fixed stars
4. Time interval between two successive vernal equinoxes

Codes:

	A	B	C	D
(a)	1	2	3	4
(b)	2	1	4	3
(c)	2	1	3	4
(d)	1	2	4	3

[ESE : 2007]

8.22 Which of the following can be used as a map substitute?

(a) Terrestrial photographs
(b) Vertical aerial photographs
(c) Oblique aerial photographs
(d) Vertical aerial photo-mosaics

[ESE : 2008]

8.23 Flamsteed gave numbers to stars observed by him in each constellation according to their

(a) Brilliance
(b) Altitudes
(c) Co-declinations
(d) Right ascensions

[ESE : 2008]

8.24 If δ is the declination of a star and ϕ is the latitude of the place of observation, then for a circumpolar star

(a) $\delta \le \phi$ (b) $\delta \ge \phi$
(c) $\delta \le (90° - \phi)$ (d) $\delta \le (90° - 2\phi)$

[ESE : 2008]

8.25 A particular runway measures 6 cm on a vertical photograph, whereas the same runway measures 4 cm on a map plotted on a scale of 1/24000. The scale of the photograph at the runway elevation is

(a) 1/36000 (b) 1/24000
(c) 1/3600 (d) 1/16000

[ESE : 2008]

8.26 Match **List-I** with **List-II** and select the correct answer using the codes given below the lists:

List-I

A. Visual Interpretation
B. Geographical Information System
C. National Remote Sensing Agency
D. Supervised Image Classification

List-II

1. Creation of data bank of multi-information for a project area
2. Visual identification of objects from image characteristics
3. Computer classification of digital image data
4. Supplier of aerial and satellite based remote sensing data products in India

Codes:

	A	B	C	D
(a)	2	1	3	4
(b)	2	1	4	3
(c)	1	3	2	4
(d)	1	4	2	3

[ESE : 2008]

8.27 Which one of the following is **not** strictly a method of remote sensing?
(a) Thermal and multi spectral scanning
(b) Microwave sensing
(c) Earth resource satellite
(d) Stereoscopy
[ESE : 2008]

8.28 Which one of the following conditions requires geodetic surveying?
(a) Horizontal curve ranging
(b) Vertical curve ranging
(c) Survey of a country
(d) Reconnaissance survey [ESE : 2009]

8.29 Which of the following coordinate systems is the most convenient way to specify the position of the star on celestial sphere?
(a) Latitude and longitude
(b) Altitude and azimuth
(c) Declination and right ascension
(d) Declination and hour angle
[ESE : 2009]

8.30 How many sidereal days are there in a solar year?
(a) 365.2840 (b) 366.2422
(c) 360.2500 (d) 365.0000
[ESE : 2009]

8.31 Consider the following statements:
A sidereal year can be defined as the time interval
1. between two successive transits of the sun through the meridian of any of the fixed stars
2. between two successive vernal equinoxes
3. between two successive passages of the sun through perigee

Which of these statements is/are correct?
(a) 3 only (b) 1 and 2
(c) 2 and 3 (d) 1 only
[ESE : 2009]

8.32 Assertion (A): Remote sensing is mostly quantitative in nature.
Reason (R): Remote sensing involves the subjective interpretation of the imagery.
(a) both A and R are true and R is the correct explanation of A
(b) both A and R are true but R is not a correct explanation of A
(c) A is true but R is false
(d) A is false but R is true [ESE : 2009]

8.33 Consider the following statements:
In aerial photogrammetry the 'filter' is placed in front of the lens to
1. reduce the effect of atmospheric haze.
2. protect the lens from dust.
3. provide uniform light distribution over the format

Which of these statements is/are correct?
(a) 1 and 2 only (b) 2 only
(c) 1 and 3 only (d) 1, 2 and 3
[ESE : 2010]

8.34 'Iso-centre' is the point
(a) in which the tilted axis of the camera meets the vertical photograph
(b) in which the bisector of the angle of tilt meets the vertical photograph
(c) in air space, the location of the optical centre of the lens of the camera at the time of exposure
(d) where the perpendicular from the nodal point meets the photograph
[ESE : 2010]

8.35 The declination of a celestial body is the arc of the declination circle intercepted between that body and the
(a) prime vertical through that body
(b) azimuth of the body
(c) equinoxes of the Earth
(d) equator of the Earth [ESE : 2010]

8.36 The difference between the apparent solar time and mean solar time is known as
(a) real time (b) average time
(c) equation of time (d) sidereal time
[ESE : 2010]

8.37 If a 'vertical aerial photograph', (20 cm × 20 cm) in size, on a R.F.1: 10,000 has 60% longitudinal overlap and 40% side overlap, the actual ground length covered by each photograph in the longitudinal direction of the flight will be

(a) 4 km (b) 6 km
(c) 0.8 km (d) 0.4 km [ESE : 2011]

8.38 If the original scale of a negative is 1 : 10,000 the ground resolution, considering that we get nearly 20 lines pair per mm, will be
(a) 50 mm (b) 20 cm
(c) 2 m (d) 25 cm [ESE : 2011]

8.39 When compared with the co-latitude of the place of observation, the declination of a circumpolar star is always.
(a) Lesser (b) Greater
(c) Equal (d) Either lesser or equal
[ESE : 2012]

8.40 When *H* is the flight height, *R* is the appropriate radial measure and *d* is the relief displacement, the vertical height of an object appearing on an aerial photograph is
(a) $\frac{R}{dH}$ (b) $\frac{dH}{R}$
(c) $\frac{H}{dR}$ (d) $\frac{RH}{d}$ [ESE : 2012]

Directions: The following items consists of two statements; one labelled as **'Statement (I)'** and the other as **'Statement (II)'**. You are to examine these two statements carefully and select the answers to these items using the codes given below:
Codes:
(a) Both Statement (I) and Statement (II) are individually true and Statement (II) is the correct explanation of Statement (I)
(b) Both Statement (I) and Statement (II) are individually true but Statement (II) is ***not*** the correct explanation of Statement (I)
(c) Statement (I) is true but Statement (II) is false
(d) Statement (I) is false but Statement (II) is true

8.41 **Statement-I :** Present usage of GPS for positioning includes personal navigation, aircraft, navigation, offshore survey, vessel navigation, etc.
Statement-II : GPS is a satellite navigation system designed to provide information about instantaneous velocity and time almost anywhere on the globe at any time and in any weather.
[ESE : 2012]

8.42 **Statement (I):** Microwave bands are composed of radiation with wavelength between 1 mm and 1 m.
Statement (II): Microwaves are capable of penetrating through the atmosphere under almost all conditions.
[ESE : 2014]

8.43 **Statement (I):** The regions of the electromagnetic spectrum in which the atmosphere is transparent are called atmospheric windows.
Statement (II): The atmosphere is practically transparent in the visible region of the electromagnetic spectrum.
[ESE : 2014]

8.44 **Statement (I):** Scattering is more in lower wavelengths than in higher wavelengths.
Statement (II): Scattering effect decreases the signal value.
[ESE : 2014]

8.45 **Statement (I):** Multistage imaging refers to viewing a given area in several narrow bands.
Statement II: Multistage imaging is also callec spatial resolution.
[ESE : 2014]

8.46 **Statement (I):** Spectral reflectance curves describe the spectral response of a target.
Statement II: Every object on the Earth has its unique spectral reflectance.
[ESE : 2014]

8.47 **Statement (I):** The process by which the geometry of an image area is made planimetric is called rectification.
Statement (II): It is the most precise geometric correction.
[ESE : 2014]

8.48 The observations made over the same area on different dates to monitor ground features like crop growth is called
(a) Temporal resolution
(b) Raciometric resolution
(c) Spatial resolution
(d) Spectral resolution
[ESE : 2014]

8.49 To uniquely determine the position of the user using GPS, one needs to receive signals from at least

(a) 1 satellite (b) 2 satellites
(c) 3 satellite (d) 4 satellites

[ESE : 2015]

8.50 Which one of the following Remote Sensing Systems employs only one detector?

(a) Scanning
(b) Framing
(c) Electromagnetic spectrum
(d) All of the above

[ESE : 2015]

8.51 The magnitude azimuth of one end of a runway is 80° measured clockwise from the magnetic north. The other end of runway will be numbered as

(a) 16 (b) 24
(c) 26 (d) 8

[ESE : 2015]

8.52 Local mean time of a place of longitude of 42°36′ W is 8h 42m 15s AM. The corresponding Greenwich Mean Time is

(a) 10h 32m 40s AM (b) 11h 32m 39s PM
(c) 0h 32m 39s PM (d) 11h 32m 39s AM

[ESE : 2016]

8.53 A vertical photograph of a flat area having an average elevation of 250 m above mean sea level was taken with a camera of focal length 25 cm. A section line AB 300 m long in the area measures 15 cm on the photograph; a tower BP in the area also appears on the photograph. The distance between images of top and bottom of the tower measures 0.5 cm on the photograph. The distance of the image of the top of the tower is 10 cm. The actual height of the tower is

(a) 10 m (b) 15 m
(c) 20 m (d) 25 m [ESE : 2016]

8.54 Cleavage is a mode of folding:

1. It is a process wherein all internal movements are along shear planes which do not change their position during the entire process.
2. It is a process wherein all internal movements are along shear planes which change their position during the process.
3. The process involves dilation in the vertical direction.
4. The process involves compression in a lateral direction.

Which of the above statements are correct?

(a) 1, 2 and 3 only (b) 1, 3 and 4 only
(c) 1, 2 and 4 only (d) 2, 3 and 4 only

[ESE : 2017]

8.55 **Statement (I):** All old systems of surveying can be completely replaced by remote sensing system basing on INSAT data.

Statement (II): Reference data in-situ is essential in interpreting satellite imageries.

(a) Both Statement (I) and Statement (II) are individually true and Statement (II) is the correct explanation of Statement (I)
(b) Both Statement (I) and Statement (II) are individually true but Statement (II) is ***not*** the correct explanation of Statement (I)
(c) Statement (I) is true but Statement (II) is false
(d) Statement (I) is false but Statement (II) is true

[ESE : 2017]

8.56 Consider the following statements regarding Remote Sensing Survey:

1. Information transfer is accomplished by use of electromagnetic radiation
2. Remote Sensing from space is done by satellites
3. Remote Sensing has no application in Earthquake prediction

Which of the above statements are correct ?

(a) 1 and 2 only (b) 1 and 3 only
(c) 2 and 3 only (d) 1,2 and 3

[ESE : 2018]

8.57 Consider the following statements with regard to Global Positioning Systems (GPS):

1. The position of an object can be exactly determined by a single satellite.
2. The position of the observer (moving person or vehicle) on ground is determined by an orbiting satellite.
3. Atomic clocks are fixed in satellites to calculate the positioning of the satellite to aid in determining travel times.
4. Absolute positioning, where accuracy of 1 cm to 5 cm is needed, depends upon the health of the satellite.

Which of the above statements are CORRECT?

(a) 1, 2 and 3 only (b) 1, 2 and 4 only
(c) 1, 3 and 4 only (d) 2, 3 and 4 only

[ESE : 2018]

8.58 A sidereal day is the average time taken by
(a) The Earth to move around the Sun once
(b) The Moon to move around the Earth once
(c) The first point of Aries to cross the same meridian successively
(d) The Earth to move around its own axis once

[ESE : 2019]

8.59 A vertical photograph is taken at an altitude of 1200 m 'above mean sea level' (a.m.s.*l*.) of a terrain lying at an elevation of 80 m a.m.s.*l*. The focal length of camera is 15 cm. The scale of the photograph will be nearly
(a) 1 : 8376 (b) 1 : 7467
(c) 1 : 6558 (d) 1 : 5649

[ESE : 2019]

8.60 Aerial photographs are required to be taken to cover an area of 150 km^2. The longitudinal and side overlaps are to be 60% and 30% respectively. The scale of photograph is 1 cm = 100 m, and the size of each photograph is 20 cm × 20 cm. The minimum required number of photographs will be
(a) 170 (b) 158
(c) 146 (d) 134

[ESE : 2019]

8.61 An unconformity is
(a) A surface of erosion or non-deposition as detected in a sequence of rocks
(b) A layer of boulders and pebbles in a sequence of rocks
(c) A layer of clay or shale in an igneous mass
(d) A type of joint especially associated with folded and faulted rocks

[ESE : 2019]

8.62 If the LMT is $8^h\ 12^m\ 16^s$ AM at 38° 45′ W longitude, the GMT will be
(a) $11^h\ 12^m\ 16^s$ AM
(b) $10^h\ 47^m\ 16^s$ AM
(c) $9^h\ 29^m\ 46^s$ AM
(d) $5^h\ 29^m\ 46^s$ AM

[ESE : 2020]

8.63 A section line *AB* appears to be 10.16 cm on a photograph for which the focal length is 16 cm. The corresponding line measures 2.54 cm on a map, which is to a scale $\frac{1}{50,000}$. The terrain has an average elevation of 200 m above mean sea level. The flying altitude of the aircraft above mean sea level during photograph will be
(a) 1800 m (b) 2000 m
(c) 2200 m (d) 2400 m

[ESE : 2020]

8.64 Stalactites and stalagmites are features of
(a) Stream erosion developed in limestone region by specific chemical reaction.
(b) Ground water deposition in caves formed by precipitation from dripping water rich in calcium carbonate.
(c) Marine erosion and deposition formed along coastal regions by selective erosion followed by deposition by waves.
(d) A centripetal drainage in which streams from different directions flow towards a common central basin.

[ESE : 2020]

8.65 Mountains resulting from the depression or elevation of blocks of the earth crust on a large scale due to faulting and these elevated structures are commonly called
(a) Fault block mountains
(b) Volcanic mountains
(c) Relict mountains
(d) Residual mountains

[ESE : 2020]

8.66 A little gap is left between the head of the glaciated valley and the mass of the glacier ice. This gap is known as
(a) Bergs-chrund (b) Arete
(c) Horn (d) Cirque

[ESE : 2020]

8.67 The Zenith is/are
(a) the point on the upper portion of the celestial sphere marked by plumb line above the observer.
(b) the point on the lower portion of the celestial sphere marked by plumb line below the observer.

(c) the two points in which the Earth's axis of rotation meets the Earth's sphere.
(d) the great circle of the Earth, the plane of which is the right angles to the axis of rotation.

[ESE : 2021]

8.68 Which one of the statements is not correct for remote sensing?
(a) It requires energy source.
(b) It requires propagation of energy through atmosphere.
(c) It requires energy interaction with the Earth's surface features.
(d) It requires absorption of energy by the Earth's surface.

[ESE : 2021]

8.69 Energy in remote sensing deals with which region of electromagnetic spectrum?
(a) Ultraviolet (b) Infrared
(c) X-Ray (d) Gamma Ray

[ESE : 2021]

8.70 A photographic survey is carried out to a scale of 1 : 20000. A camera with a wide angle lens of f = 170 mm was used with 25 cm × 25 cm plane size for a net 65% overlap along the line of flight. What is the error in height given by an error of 0.15 mm in measuring the parallax of the point?
(a) 5.15 m (b) 5.27 m
(c) 5.83 m (d) 6.45 m

[ESE : 2021]

8.71 What is the aeroplane flying height to obtain the average scale of the photograph equal to $\frac{1}{7200}$?
(Ground surface elevations vary from 160 m to 430 m and focal length of the camera lens is 153 mm)
(a) 1021 m (b) 1145 m
(c) 1284 m (d) 1397 m

[ESE : 2021]

8.72 Photographic surveying is suitable for
(a) small-scale mapping of open hilly or mountainous countries
(b) flat or wooded country
(c) roads
(d) transmission lines

[ESE : 2022]

8.73 If the focal length of lens (f), flying height (H) and height of ground above mean sea level (h) are known, then the scale at height 'h' (S_h) is equal to
(a) $\frac{f}{(H-h)}$ (b) $\frac{(H-h)}{f}$
(c) $\frac{(h-H)}{2f}$ (d) $\frac{2f}{(h-H)}$

[ESE : 2022]

8.74 The terrestrial photogrammetry can be divided into how many branches?
(a) Four (b) Three
(c) Two (d) Five

[ESE : 2022]

8.75 Which one of the following is an aerial photograph taken with the camera axis directed intentionally between the horizontal and the vertical?
(a) Tilted photograph
(b) Oblique photograph
(c) Slanting photograph
(d) Vertical photograph

[ESE : 2022]

8.76 A vertical circle which is at right angles to the meridian is also known as
(a) an altitude (b) a co-altitude
(c) a prime vertical (d) an azimuth

[ESE : 2022]

8.77 Dip fault is a fault plane
(a) parallel to the dip of beds
(b) parallel to the strike of a bedding plane
(c) diagonal to the dip of a bedding plane
(d) with no relationship to the bedding plane

[ESE : 2023]

8.78 In Map versus Aerial photograph, due to symbolic representation the clarity of details is
(a) less on map than on a photo
(b) more on map than on a photo
(c) less on a photo than on map
(d) more on a photo than on map

[ESE : 2023]

8.79 In multilevel classification system, Level IV classification is suitable for
(a) Landsat MSS images
(b) High-altitude aerial photographs
(c) Low-altitude aerial photographs
(d) Medium-altitude aerial photographs

[ESE : 2023]

8.80 Consider the following statements regarding aerial photogrammetry:
1. Aerial photogrammetry has the ease with which topography of inaccessible areas can be detailed.
2. In aerial photogrammetry, there is possibility of omitting of few field data.

Which of the above statements is/are correct?
(a) 1 only (b) 2 only
(c) both 1 and 2 (d) Neither 1 nor 2

[ESE : 2024]

8.81 Match the following lists:

	List-I (Region)		List-II (Wavelentgh)
P.	Reflected IR band	1.	Less than 0.03 nm
Q.	X-ray	2.	0.1 – 30 cm
R.	Gamma ray	3.	0.7 – 3.0 μm
S.	Microwave	4.	0.03 – 3.0 nm

Select the correct answer using the code given below:

	P	Q	R	S
(a)	2	3	1	4
(b)	3	4	1	2
(c)	4	1	3	2
(d)	4	1	2	3

[ESE : 2024]

8.82 Which one of the following types of equipment is used in topographic surveys and also for recording the shapes of buildings?
(a) Electromagnetic distance measurement devices
(b) GPS
(c) Satellite camera
(d) Aerial camera

[ESE : 2024]

8.83 When a celestial body crosses the observer's meridian, it is said to be
(a) culminate (b) vernal equinox
(c) obliquity (d) celestial pole

[ESE : 2024]

8.84 Why the observations of field astronomy do not involve the measurement of declination and right ascension?
(a) Because the altitude and azimuth are constantly changed to the motion of the celestial body
(b) Because the stars do not occupy fixed positions on the celestial sphere
(c) Because the distance of the sun from the earth is variable
(d) Because of the obliquity of the ecliptic

[ESE : 2024]

8.85 Consider the following statements regarding global positioning system (GPS):
1. GPS cannot be used in all weather conditions.
2. In GPS surveying inter visibility between stations or points surveyed is not necessary.
3. High cost of GPS surveying has restricted the realization of the full potential of GPS.

Which of the above statements are correct?
(a) 1 and 2 only (b) 2 and 3 only
(c) 1 and 3 only (d) 1, 2 and 3

[ESE : 2024]

8.86 Consider the following statements regarding remote sensing surveys:
1. Different types of land use are distinguishable on images.
2. Most images lack the horizontal perspective.
3. For surveys of small areas, the cost of mobilizing a remote sensing mission may be uneconomical

Which of the above statements is/are correct?
(a) 1 and 2 (b) 2 and 3
(c) 2 only (d) 3 only

[ESE : 2024]

■■■■

Answers Field Astronomy, Photogrammetic Survey, Remote Sensing, GPS & Geology

8.1 (a)	8.2 (d)	8.3 (a)	8.4 (b)	8.5 (b)	8.6 (b)	8.7 (c)	8.8 (b)	8.9 (c)
8.10 (a)	8.11 (a)	8.12 (b)	8.13 (a)	8.14 (c)	8.15 (d)	8.16 (c)	8.17 (d)	8.18 (c)
8.19 (b)	8.20 (c)	8.21 (b)	8.22 (d)	8.23 (d)	8.24 (a)	8.25 (d)	8.26 (b)	8.27 (c)
8.28 (c)	8.29 (b)	8.30 (b)	8.31 (b)	8.32 (b)	8.33 (d)	8.34 (b)	8.35 (d)	8.36 (c)
8.37 (c)	8.38 (*)	8.39 (b)	8.40 (b)	8.41 (a)	8.42 (b)	8.43 (b)	8.44 (c)	8.45 (d)
8.46 (b)	8.47 (b)	8.48 (a)	8.49 (d)	8.50 (a)	8.51 (c)	8.52 (d)	8.53 (d)	8.54 (d)
8.55 (d)	8.56 (a)	8.57 (d)	8.58 (c)	8.59 (b)	8.60 (d)	8.61 (a)	8.62 (b)	8.63 (c)
8.64 (b)	8.65 (a)	8.66 (a)	8.67 (a)	8.68 (d)	8.69 (b)	8.70 (c)	8.71 (d)	8.72 (a)
8.73 (a)	8.74 (c)	8.75 (b)	8.76 (c)	8.77 (a)	8.78 (b)	8.79 (c)	8.80 (c)	8.81 (b)
8.82 (d)	8.83 (a)	8.84 (b)	8.85 (d)	8.86 (b)				

Explanations Field Astronomy, Photogrammetic Survey, Remote Sensing, GPS & Geology

8.1 (a)

Cadastral survey is used to plot the details such as boundaries of fields, houses and property lines in rural and urban areas. These are also known as public land surveys.

Photogrammetry is used to produce a mosaic or map by compiling the photographs. Aerial survey covers photogrammetry and photo-interpretation both.

8.2 (d)

The difference between apparent and mean solar times at any instant is known as the equation of time (*ET*)

$$ET = GAT - GMT$$

where *GAT* is Greenwich Apparent Time

GMT is Greenwich Mean Time

$$ET = -13' \, 28.5''$$

$$GMT = 0 \text{ h on a day}$$

$\therefore$ $GAT = 23$ h $46' \, 31.5''$ of previous day

8.3 (a)

Scale, $S = \dfrac{f}{H-h}$

$$= \frac{20}{1500-500} = \frac{20}{1000} = \frac{1}{50}$$

8.4 (b)

Properties of spherical triangle:

(i) Any angle is less than two right angles or π radians.

(ii) The sum of three angles is less than six right angle or 3π radians and greater than two right angles or π radians.

(iii) The sum of any two sides is greater than the third.

(iv) If the sum of any two sides is equal to two right angles or π radians, the sum of the angles opposite to them is equal to two right angles or π radians.

(v) The smaller angle is opposite to the smaller side and vice versa.

(vi) If two angles are equal the sides opposite to them are also equal.

(vii) The sum of the three sides of a spherical triangle is less than circumference of great circle.

8.5 (b)

Scale of photograph,

$$S = \frac{8\text{ cm}}{1\text{km}} = \frac{80\text{ mm}}{1\text{km}} = \frac{f}{H}$$

Given $f = 160$ mm

$\therefore$ $H = 2$ km $= 2000$ m

8.7 (c)

Parallax for bottom,

$$p_b = \frac{fB}{H - h_b}$$

$$\Rightarrow \quad h_b = H - \frac{fB}{p_b}$$

$$\Rightarrow \quad h_b = 4000 - \frac{160 \times 300}{63}$$

$$\Rightarrow \quad h_b = 3238.09 \text{ m}$$

Parallax difference,

$$\Delta p = \frac{\Delta h \times p_t}{H - h_b}$$

$$\Rightarrow \quad 72 - 63 = \frac{\Delta h \times 72}{4000 - 3238.09}$$

$$\Rightarrow \quad \Delta h = 95.2 \text{ m}$$

8.8 (b)

Azimuth is the direction of a celestial object from the observer, expressed as the angular distance from the north or south point of the horizon to the point at which a vertical circle passing through the object intersects the horizon.

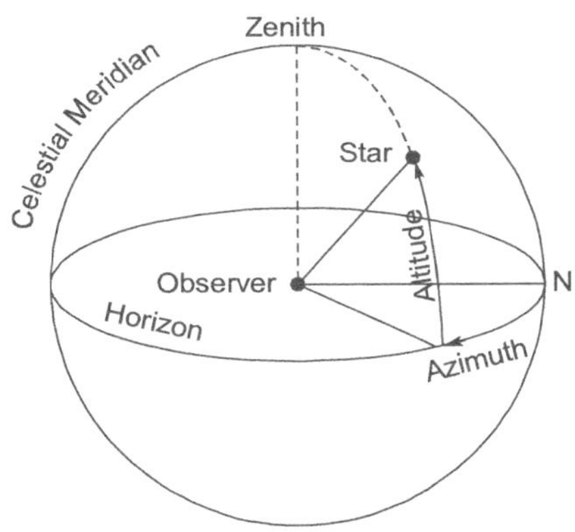

8.9 (c)

Given, $H = 500$ m

$h_1 = 250$ m

If h is the height of chimney then elevation of top of chimney, $h_2 = h_1 + h$

Relief displacement,

$$d = r_2 - r_0 = \frac{r_2 h}{H - h_1}$$

$$\therefore \quad h = \frac{51.4 \times (500 - 250)}{110} = 116.82 \text{ m}$$

8.11 (a)

Airbase is the distance covered by aircraft between two successive exposures

$$B = \frac{bH}{f}$$

8.12 (b)

A stereoplotter, also known as a multiplex, is an optical and mechanical apparatus used for preparing maps of high precision from aerial photographs. Its function is to create a precise three dimensional scale optical model and provide a means to view the spatial model and provide a means for measuring the model and plotting its orthographic or map projection.

8.13 (a)

Sun-synchronous orbit satellites which operates in a near circular polar orbit running nearly north to south at a fixed attitude ranging from 500-1000 km. Every time the satellite passes from north to south, it has a consistent and constant sunlight view of a swath (around 150 km) of earths surface. Indian Remote sensing satellite (IRS) launched by PSLV come under this kind.

8.14 (c)

An assembly of individual overlapping aerial photographs, all on approximately the same scale, which are adjusted along the edges of one another by cutting the overlapping portion is called mosaic. Photographic maps are classified as below:

(i) Index maps in which photograph of the assembled prints arranged in the order of exposure.

(ii) Planimetric map shows natural and cultural features in plan.

(iii) Topographic map shows contour line in addition to the planimetric data.

(iv) Stereotopographic maps.

8.15 (d)

This method is called **trilateration**. The first satellite locates you somewhere on a sphere (top left of Figure). The second satellite narrows your location to a circle created by the intersection of the two satellite spheres (top right). The third satellite reduces the choice to two possible points (bottom left). Finally, the forth satellite helps calculate a timing and location correction and selects one of the remaining two points as your position (bottom right).

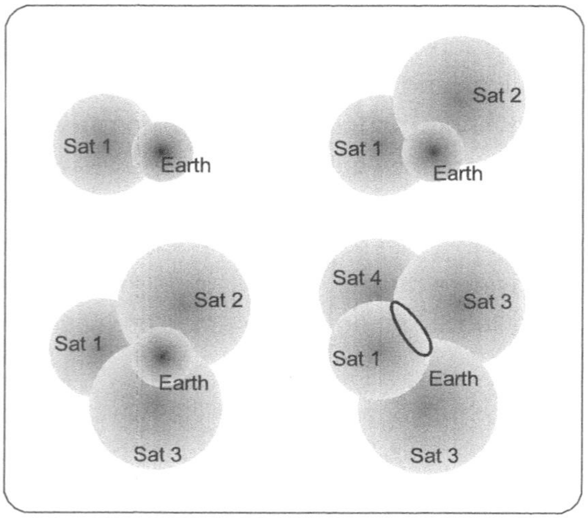

8.16 (c)

The circumpolar stars are always above the horizon and therefore they do not set. At upper culmination their altitude is maximum and at lower culmination their altitude is minimum.

8.17 (d)

Ground speed

$$= \frac{1000}{20} = 50 \text{ m/s} = 50 \times \frac{18}{5} = 180 \text{ km/h}$$

8.18 (c)

Crab of photograph is angle between flight line and edge of photograph in direction of flight. Drift is lateral shifting of photograph.

Stadia survey is a tacheometric form of distance measurement that relies on a fixed-angle intercept made by stadia hair.

Baseline is a line between two points on the earth's surface and the direction and distance between them. In a triangulation network, at least one baseline needs to be measured to calculate the size of the triangles by trigonometry.

In astronomical surveying when the star is on the observer's meridian, it is said to be at culmination. The star is said to be at elongation when it is at its farthest point from the observer's meridian.

8.19 (b)

$$L = (1 - P_l)Sl$$

and $\quad W = (1 - P_w)Sl$

where

l is length of photograph

P_l is forward overlap

P_w is side overlap

$$\therefore \quad L = (1 - 0.6) \times 200 \times 20$$
$$= 1600 \text{ m} = 1.6 \text{ km}$$
$$W = (1 - 0.3) \times 200 \times 20$$
$$= 2800 \text{ m} = 2.8 \text{ km}$$

Area covered by one photograph

$$= LW = 1.6 \times 2.8 = 4.48 \text{ km}^2$$

8.20 (c)

$$\lambda_m \propto \frac{1}{T}$$

$$\Rightarrow \quad \lambda_m = \frac{a}{T}$$

where a is a constant

In case of sun, we have

$$0.5 = \frac{a}{6000}$$

$$\Rightarrow \quad a = 3000 \,\mu\text{m-K}$$

In case of earth, we have

$$\lambda_m = \frac{a}{T}$$

$$\Rightarrow \quad 10 = \frac{3000}{T}$$

$$\Rightarrow \quad T = 300 \text{ K}$$

$\therefore$ Temperature of earth's surface in °C

$$= 300 - 273 = 27°$$

8.22 (d)

Photomaps are the aerial photographs which are used as a substitute for maps. The photomap may consist of one photograph, but usually photomaps are obtained by assembling two or more photographs to form a large map. The large photomaps from two or more photographs are called mosaics. To a varying degree of accuracy, a mosaic is a map substitute. The mosaic has an overall average scale comparable to the scale of a planimetric map.

8.23 (d)

A Flamsteed designation is a combination of a number and constellation name that uniquely identifies stars. *Flamsteed designations* were assigned to 2554 stars. The numbers were originally assigned in order of increasing *right ascension* within each constellation.

8.25 (d)

Scale of photograph

$$= \frac{\text{Photograph distance}}{\text{Map distance}} \times \text{Scale of map}$$

$$= \frac{6}{4} \times \frac{1}{24{,}000} = \frac{1}{16{,}000}$$

8.27 (c)

Remote sensing utilizes satellite and/or airborne based sensors to collect information about a given object or area. Remote sensing data collection methods can be passive or active. Passive sensors (e.g., spectral imagers) detect natural radiation that is emitted or reflected by the object or area being observed. In active remote sensing (e.g., radar) energy is emitted and the resultant signal that is reflected back is measured.

8.28 (c)

Geodetic surveying is the type of surveying in which the curvature of the earth is taken into consideration and a very high standard of accuracy is maintained. It is used for large areas like survey of a country.

8.29 (b)

Altitude and Azimuth are used to describe the location of a star on celestial system as viewed from a particular location at a particular time.

The altitude is the distance an object appears to be above the horizon. The angle is measured up from the closest point on the horizon. The azimuth of an object is the angular distance along the horizon to the location of the object. By convention, azimuth is measured from north towards the east along the horizon.

8.30 (b)

The time interval taken by one apparent revolution of the sun about the earth is called a solar day. The time interval taken by the vernal equinox in one apparent revolution about the earth is called a sidereal day. Because the time period for 366·2422 sidereal days is equal to that for 365·2422 solar days, a sidereal day is shorter than a solar day.

$\because$ 365·2422 solar days = 1 solar year

$\therefore$ 1 solar year = 366·2422 sidereal days

8.31 (b)

It equals 365.25636 SI days. The sidereal year differs from the tropical year, the time interval between vernal equinoxes in successive years, due to the precession of the equinoxes.

8.33 (d)

Lens cone assembly contains the following:

(a) Lens
(b) filter
(c) Shutter and diaphragm

Filter is placed in front of lens and it:

(a) reduces the effect of atmospheric haze,
(b) helps provide uniform light distribution over the entire format
(c) protects the lens from damage and dust.

8.34 (b)

Isocentre is the point on the aerial photo platform that falls on a line half-way between the Principal point and the Nadir point. It is the point at which bisector of angle of tilt meets vertical photograph.

8.35 (d)

Declination of celestial body is angular distance of body above or below celestial equator. Declination is one of the two angles that locate a point on the celestial sphere in the equatorial coordinate system, the other being hour angle. Declination's angle is measured north or south of the celestial equator, along the hour circle passing through the point in question. It is same as latitude of a station on earth.

8.36 (c)

Two types of *solar time* are apparent *solar time* (sundial *time*) and *mean solar time* (clock *time*). The apparent sun is the true sun as seen by an observer on Earth. Apparent solar time or true solar time is based on the apparent motion of the actual Sun. mean solar time is the time as calculated by the motion of the mean sun (such as is shown by an ordinary clock). The equation of time which is this difference between apparent and mean solar time, is cyclical and does not accumulate from year to year.

8.37 (c)

Actual ground length covered

$$L = (1 - P_l)l \times s$$

$$= (1 - 0.6) \times 20 \times \frac{10{,}000}{100 \times 10^3} = 0.8 \text{ km}$$

8.38 (*)

Original scale of negative is 1 : 10000 and we get 20 lines per mm then ground resolution is $\frac{10000}{20}$ mm = 500 mm or 50 cm.

8.41 (a)

The global positioning system is a space based satellite navigation system that provides location and time information in all weather, anywhere or near the earth. It is maintained by the United States Government and is freely accessible to anyone. The GPS program provides critical capabilities to military, civil and commercial users around the world. In addition, GPS is the backbone for modernizing the global air traffic system.

8.42 (b)

Microwaves are a form of electromagnetic radiation with wavelengths ranging from 1 mm to 1 m longer than those of infrared radiation but shorter than that of normal radio wave. They are able to penetrate atmosphere under almost all conditions.

8.43 (b)

Atmospheric windows is that portion of the electromagnetic spectrum that can be transmitted through the *atmosphere* without any distortion or absorption. Light in certain wavelength regions can penetrate the *atmosphere* well. These regions are called *atmospheric windows*. The *atmosphere* is *practically transparent* to a substantial *portion* of the *visible spectrum*.

8.44 (c)

Scattering changes the direction of propagation only not the wavelength. Thus the signal valve will not change.

8.45 (d)

The useful magnification of a light optical microscope is reached by multistage imaging. The total magnification *M* of multistage imaging with single magnification steps M_1, M_2 ... M_n is given by, $M = M_1 M_2 \ldots M_n$.

8.46 (b)

Remote sensing is based on the measurement of reflected or emitted radiation from different bodies. Objects having different surface features reflect or absorb the sun's radiation in different ways. This emitted radiation is known as Spectral Reflectance.

8.47 (b)

Rectification is process by which geometry of an image is made planimetric by referencing to some standard map projection.

Rectification is a precise geometric correction and is used when accurate direction, area or distance measurement is required.

8.48 (a)

Temporal resolution refers to the precision of a measurement with respect to time. Observations are made over same area on different dates. This is useful during crop growth, deforestation, floods etc.

8.49 (d)

This is a high-tech version of triangulation, called trilateration. The first satellite locates you somewhere on a sphere. The second satellite narrows your location to a circle created by the intersection of the two satellite spheres. The third satellite reduces the choice to two possible points. Finally the fourth satellite helps to calculate a timing and location correction and selects one of the remaining two points as your position.

8.50 (a)

Scanning system employs detectors with a narrow field of view which sweeps across the terrain to produce an image. When photons of EM energy radiated or reflected from earth surface feature encounter the detector, an electrical signal is produced that varies in proportion to the number of photons.

8.51 (c)

Numbering of the runway is

$$= \frac{\text{Magnetic azimuth from Magnetic North}}{10}$$

Other end of the runway will be at 180° difference from a given end i.e.

$$80° + 180° = 260°$$

$\therefore$ Number of other end = 26

8.52 (d)

Local mean time difference

$$= \frac{42°36'}{15} = 2 \text{ hr } 50 \text{ min } 24 \text{ sec}$$

Since local place is west of greenwich, LMT is less than GMT.

LMT = GMT − 2 hr 50 min 24 sec

GMT = LMT + 2 hr 50 min 24 sec

= 8 hr 42 min 15 sec + 2 hr 50 min 24 sec

= 11 hr 32 min 39 sec AM

8.53 (d)

$$h_{avg} = 250 \text{ m}, f = 25 \text{ cm}$$

$$\text{Scale} = \frac{15 \text{ cm}}{30000 \text{ cm}} = \frac{1}{2000}$$

$$r = 10 \text{ cm}, d = 0.5 \text{ cm}$$

$$\text{scale} = \frac{f}{H - hav} = \frac{1}{200}$$

$$\frac{0.25}{H-250} = \frac{1}{200}$$

$$H = 750 \text{ m}$$

$$rh = d(h - \text{hav})$$

$$h = \frac{0.5 \times (750-250)}{10} = 25 \text{ m}$$

8.58 (c)

The sidereal day is the interval of time between two successive upper transits of the first point of aries.

8.59 (b)

$H = 1200$ m, $h = 80$ m, $f = 15$ cm

$$\text{Scale} = \frac{f}{H-h} = \frac{15 \times 10^{-2}}{1200-80} = \frac{1}{7466.67}$$

Scale = (1 : 7467)

8.60 (d)

Area of ground = 150 km^2

Overlap: Longitudnal = 60%; Lateral = 30%

Scale : 1 cm = 100 m = 0.1 km

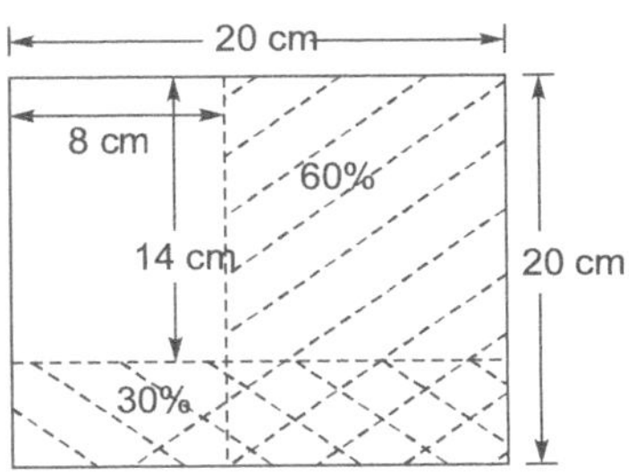

Photosize = 20 cm × 20 cm

Now l = length covered by 1 photo

= 8 × 0.1 = 0.8 km

b = width covered by 1 photo

= 14 × 0.1 = 1.4 km

a = Area covered by 1 photo

= $l \times b$ = 0.8 × 1.4 = 1.12 km^2

Min. no. of photographs

$$= \left(\frac{A}{a}\right) = \left(\frac{150}{1.12}\right) = 133.92$$

≈ 134 photos

8.61 (a)

An unconformity is a buried erosional or non-depositional surface separating two rock masses or strata of different ages, indicating that sediment deposition was not continuous.

8.62 (b)

Local mean time at 38°45′W

= 8 hour 12 minute 16 seconds

Difference of degree from GMT (0°) longitude

= 38° 43′

Time difference

= 38 × 4 minute + 45 × 4 seconds

= 152 min. + 3 min. = 155 min.

= 2 hr. 35 min.

Time at GMT

= LMT + Time difference

= 8 hr. 12 min. 16 sec. + 2 hr. 35 min.

= 10 hr. 47 min. 16 sec.

8.63 (c)

Length of a line on a map = 2.54 cm

$$\text{Scale} = \frac{1}{50{,}000}$$

Actual length on ground = 50,000 × 2.54 cm

Length of same line on photo = 10.16 cm

$$\text{Then, Scale of photo} = \frac{10.16}{50{,}000 \times 2.54}$$

$$= \frac{4}{50{,}000} = \frac{1}{12500}$$

$$\text{Scale of photo} = \frac{f}{H-h}$$

$$\frac{16 \times 10^{-2}}{H-200} = \frac{1}{12500}$$

$$\Rightarrow 16 \times 10^{-2} \times 12500 = H - 200$$

$$\Rightarrow 16 \times 125 + 200 = H$$

$$\Rightarrow 2000 + 200 = H$$

$$\text{Scale of photo} = \frac{f}{H-h}$$

$$\frac{16\times 10^{-2}}{H-200} = \frac{1}{12500}$$

$$\Rightarrow 16 \times 10^{-2} \times 12500 = H - 200$$

$$\Rightarrow \quad 16 \times 125 + 200 = H$$

$$\Rightarrow \quad 2000 + 200 = H$$

$$\Rightarrow \quad H = 2200$$

8.67 (a)

The Zenith and Nadir : The Zenith (Z) is the point on the upper portion of the celestial sphere marked by plumb line above the observer. It is thus the point on the celestial sphere immediately above the observer's station. The Nadir (Z′) is the point on the lower portion of the celestial sphere marked by the plumb line below the observer. It is thus the point on the celestial sphere vertically below the observer's station.

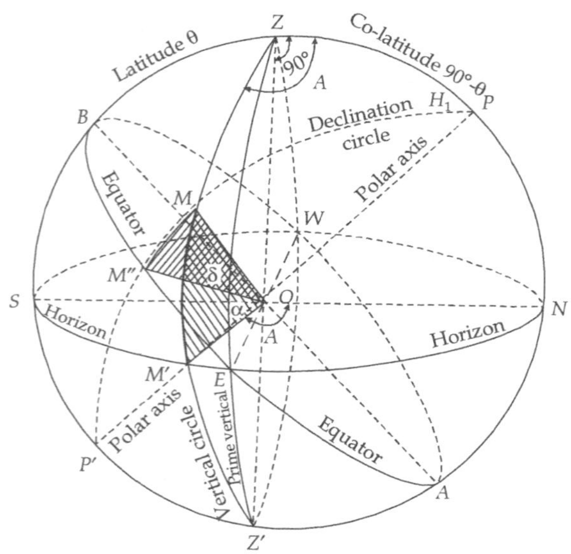

8.68 (d)

An idealised remote sensing system consists of the following stages:

- Energy source.
- Propagation of energy through atmosphere.
- Energy interaction with earth's surface features.
- Airborne/space borne sensors receiving the reflected and emitted energy.
- Transmission of data to earth station and generation of data produce.
- Multiple data users.

8.69 (b)

Remote sensing deals with energy in visible, infrared, thermal and microwage regions.

8.70 (c)

Given: $S = 20000$, $f = 170$ mm

$l = 25$ cm, $b = 25$ cm

$p_l = 65\%$

$\Delta p = 0.15$ mm, $\Delta h = ?$

$$\text{Scale} = \frac{1}{20000} = \frac{f}{H}$$

$$\Rightarrow \quad \frac{1}{20000} = \frac{170/1000}{H}$$

$$H = 3400 \text{ m}$$

Length of air base,

$$B = (1 - p_l)\, ls$$

$$= (1 - 0.65) \times \frac{25}{100} \times 20000$$

$$= 1750 \text{ m}$$

Error in measuring height

$$\Delta h = \frac{(H-h)^2}{Bf} \times \Delta p$$

$$= \frac{(3400-0)^2}{1750 \times 170} \times 0.15$$

$$= 5.83 \text{ m}$$

8.71 (d)

We know average scale of photograph,

$$S_{avg} = \frac{f}{H - h_{avg}}$$

$$\frac{1}{7200} = \frac{0.153}{H - \left(\frac{160+430}{2}\right)}$$

$$H = 1396.6 \text{ m} \simeq 1397 \text{ m}$$

8.72(a)

Photographic surveying is suitable for mountainous and hilly terrain with little vegetation. Also, used for projects demanding higher accuracy, since it provide accurate measurement.

8.73 (a)

Scale of vertical photograph is given by,

$$S = \frac{f}{(H\text{-}h)}$$

8.74 (c)

The terrestrial photogrammetry can be divided into following two branches:

(i)Plane table photogrammetry
(ii)Terrestrial stereo photogrammetry

8.75 (b)

An oblique photograph is an aerial photograph taken with camera axis directed intentionally between the horizontal and vertical.

A vertical photograph is an aerial photograph mode with the camera axis coinciding with direction of gravity.

A tilted photograph is an aerial photograph made with the camera axis unintentionally tilted from vertical by small amount less than 3°.

8.77 (a)

Dip fault runs parallel to the dip of the bed, cut across the strike of the bed. It is a fault that runs perpendicular to the strike of the affected rocks (i.e. parallel to the plane of the angle of dip of the rocks).

8.78 (b)

Differences between Map and Aerial Photograph,

1. The map is an orthographic projection, whereas an aerial photograph is a central projection, i.e. perspective projection.
2. The map has a single constant scale, whereas it varies from point to point depending upon their elevations in an aerial photograph. In the case of the photograph, the terrain area which has higher elevation is closer to the camera and therefore appears larger than the terrains area laying at a lower elevation.
3. The number of details on a map are selective whereas in a aerial photograph there is a large number of details.
4. Due to the symbolic representation, the clarity of details is more on the maps than on photo.

8.79 (c)

Level	System	Image scale
I	Landsat TM and MSS images; AVHRR images	Smaller than 1:250,000
II	High-altitude aerial photographs; TM and SPOT images AVHRR images (with ancillary data)	1:80,000 to 1:250, 000
III	Medium-altitude aerial photographs	1:20,000 to 1:80, 000
IV	Low-altitude aerial photographs	Larger than 1:20,000

8.80 (c)

- If aerial photogrammetry is done at some height then there is a chance to omitting of few field data.
- Aerial photogrammetry has the ease with which topography of inaccessible areas can be detailed.

Hence both the statements are correct.

8.82 (d)

Aerial camera is the most appropriate option.

8.85 (d)

All the statements are correct.

8.86 (b)

Different types of land use are not distringustable very clrearly on the images.

■■■■

UNIT

VII

Highway Engineering

Syllabus

Planning & construction methodology, Alignment and geometric design; Traffic Surveys and Controls; Principles of Flexible and Rigid pavements design.

Contents

■■■■

1 Highway Development & Planning

1.1 The length of National Highways as per 3rd 20 year (Lucknow) road plan is given by
(a) area of the country/75
(b) area of the country/50
(c) area of the country/40
(d) area of the country/25
[ESE : 2000]

1.2 In which one of the following location surveys of the road, soil profile sampling is done up to a depth of 1 m to 3 m below the existing ground level?
(a) Preliminary survey
(b) Final location survey
(c) Construction survey
(d) Material location survey
[ESE : 2003]

1.3 Under the Nagpur Road Plan, which of the following are NOT relevant in planning the road development programme in a backward district?
1. Existing agriculture drainage network of drain canals.
2. Existing number of Panchayat unions.
3. Existing length of mud-track roads.
4. Number of villages with population of 10000 and above.

(a) 1, 2, 3 and 4 (b) 1, 2 and 3 only
(c) 1, 2 and 4 only (d) 2, 3 and 4 only
[ESE : 2013]

1.4 Which one of the following are provided to give access to properties along an import highway with controlled access to expressway freeway?
(a) Lay-bye (b) Frontage roads
(c) Driveways (d) Cycle tracks
[ESE : 2021]

1.5 Which of the following are the design elements in highway embankments?
1. Height
2. Fill material
3. Settlement

Select the correct answer using the codes given below:
(a) 1, 2 and 3 (b) 1 and 3 only
(c) 2 and 3 only (d) 1 and 2 only
[ESE : 2021]

1.6 In which one of the following systems the optimum road length is calculated for an area based on the concept of obtaining maximum utility per unit length of road?
(a) Saturation system
(b) Unsaturated system
(c) Minimum utility system
(d) Average utility system
[ESE : 2023]

1.7 The planning surveys about engineering studies consists of
(a) Income per capita
(b) Living standard
(c) Traffic volume
(d) Road location and alignment studies
[ESE : 2023]

1.8 The survey of the highway construction work for preparation of longitudinal and cross sections, computations of earth work quantities; and other construction material and checking details of geometric design elements is carried out in
(a) Reconnaissance survey
(b) Preliminary survey
(c) Location of final alignment
(d) Detailed survey
[ESE : 2023]

■■■■■

Answers Highway Development & Planning

1.1 (b) 1.2 (a) 1.3 (a) 1.4 (b) 1.5 (a) 1.6 (a) 1.7 (d) 1.8 (b)

Explanations Highway Development & Planning

1.1 (b)

(i) Length of National Highways (NH)
= Area of the country/50

(ii) Length of State Highways (SH)
= Area of the state/25
= 62.5 × number of towns in the state – Area of the state/50

(iii) Length of the Major District Roads (MDR)
= Area of the State/12.5
= 90 × number of towns in the state

(iv) Total length of all categories of roads in a district i.e.
NH + SH + MDR + ODR + VR
= Area of District × 0.82

1.2 (a)

Soil survey is done as a part of preliminary survey to check the suitability of proposed location and details collected are used for comparative study of alternate projects. In this, sample of soil upto depth of 1 to 3 m below ground level or finished road level is taken.

1.3 (a)

Nagpur road plan formula were prepared by radial or star and grid pattern. Villages with population of 501-5000 only were considered. Agricultural area and non agricultural areas were considered. There were no consideration mentioned for existing Panchayat union, existing mud track records or existing drainage network.

1.4 (b)

A frontage road (also known as an access road, outer road, service road, feeder road or parallel road) is a local road running parallel to a higher speed, limited access road.

A frontage road is often used to provide access to private driveways, shops, houses, industries or farms.

Lay-bye: A paved area at the side of a highway designed for drivers to stop in, for emergency parking, or where vehicles can wait, with larger lay-bys possibly having facilities like food vendors or pubic telephones.

Drive way: A driveway is a type of private road for local access to one or a small group of structures and is owned and maintained by an individual or group.

Cycle track: A cycle track, separated bike lane or protected bike lane is an exclusive bikeway that has elements of a separated path and on-road bike lane.

1.5 (a)

The design elements in highway embankments are:

(i) Height
(ii) Fill material
(iii) Settlement
(iv) Stability of foundation
(v) Stability of slope

■■■■

2 Highway Geometric Design

2.1 Total reaction time of a driver does not depend upon
(a) perception time
(b) brake reaction time
(c) condition of mind of the driver
(d) speed of vehicle [ESE : 2000]

2.2 A horizontal curve of 480 m for 7.5 m two-lane road is to be designed for a speed of 80 kmph. The raising of the outer edge of the pavement with respect to the inner edge to cater to the mixed traffic condition is
(a) 0.14 m (b) 0.22 m
(c) 0.24 m (d) 0.27 m [ESE : 2000]

2.3 While driving at a speed of 30 kmph (with available friction 0.4) down the grade, the driver requires a braking distance twice that required for stopping the vehicle when he travels up the same grade. The grade is
(a) 7% (b) 10.6%
(c) 13.3% (d) 33.3% [ESE : 2001]

Directions: The following items consists of two statements; one labelled as **'Assertion (A)'** and the other as **'Reason (R)'**. You are to examine these two statements carefully and select the answers to these items using the codes given below:

Codes:
(a) both A and R are true and R is the correct explanation of A
(b) both A and R are true but R is not a correct explanation of A
(c) A is true but R is false
(d) A is false but R is true

2.4 **Assertion (A):** When a sharp horizontal curve is to be introduced on a road which already has the maximum permissible gradient, the gradient should be decreased.
Reason (R): The gradient should be decreased to compensate for the loss of tractive effort due to the introduction of sharp horizontal curve on the road.
[ESE : 2001]

2.5 **Assertion (A):** The centrifugal ratio decreases along the length of the transition curve.
Reason (R): The superelevation increases along the length of the transition curve.
[ESE : 2001]

2.6 **Assertion (A):** A skilful highway designer 'builds in' speed control at critical locations on horizontal curves rather than increase the superelevation.
Reason (R): A driver slows down on horizontal curve due to feeling of discomfort because of increase in side friction with reduced superelevation.
[ESE : 2002]

2.7 Consider the following factors:
1. Reaction time
2. Speed
3. Coefficient of longitudinal friction
4. Gradient
Which of these factors are taken into account for computing braking distance?
(a) 1 and 3 (b) 1, 2 and 4
(c) 2, 3 and 4 (d) 2 and 3
[ESE : 2002]

2.8 The camber provided on a sloping road is 1 in 48. Which one of the following is the ruling gradient?
(a) 1 in 15 (b) 1 in 20
(c) 1 in 24 (d) 1 in 30
[ESE : 2003]

2.9 Which of the following are the accepted criteria for design of valley curve for highways?
1. Headlight sight distance
2. Passing and non-passing sight distance
3. Aesthetic consideration

4. Motorist comfort
5. Drainage control

Select the correct answer using the codes given below:

(a) 1, 2, 3 and 4 (b) 1, 3, 4 and 5
(c) 2, 3, 4 and 5 (d) 1 and 5

[ESE : 2003]

2.10 If 'g_1' and 'g_2' are the two gradients, 'r' is the rate of change of grade (%) per chain, the length of the vertical curve will be

(a) $\left(\frac{g_1+g_2}{r^2}\right)$ (b) $\left(\frac{g_1-g_2}{\sqrt{r}}\right)$
(c) $\left(\frac{g_1-g_2}{r}\right)$ (d) $\frac{\sqrt{g_1+g_2}}{r^3}$

[ESE : 2003]

2.11 For a given road, safe stopping sight distance is 80 m and passing sight distance is 300 m. What is the intermediate sight distance?

(a) 220 m (b) 190 m
(c) 160 m (d) 150 m [ESE : 2004]

2.12 Consider the following statements:

A transition curve is provided on a circular curve on a highway to provide

1. gradual introduction of centrifugal force
2. minimum stopping sight distance
3. gradual introduction of super elevation
4. comfort and security to passengers

Which of these statements are correct?

(a) 1, 2 and 3 (b) 1, 3 and 4
(c) 2, 3 and 4 (d) 1, 2 and 4

[ESE : 2004]

2.13 What will be the ruling radius of a horizontal curve on a National Highway for a design vehicle speed of 100 km/h, assuming allowable super elevation to be 7% and lateral friction as 0.13?

(a) 405 m (b) 395 m
(c) 385 m (d) 375 m

[ESE : 2004]

2.14 A summit curve is formed at the intersection of a 3% upgrade and a 5% downgrade. What is the length of the summit curve in order to provide a stopping distance of 128 m?

(a) 271 m (b) 298 m
(c) 322 m (d) 340 m [ESE : 2005]

2.15 Which of the following are requirements for the design of a transition curve for a highway system?

1. Rate of change of grade
2. Rate of change of radial acceleration
3. Rate of change of super elevation
4. Rate of change of curvature

Select the correct answer using the codes given below:

(a) 1, 2 and 3 (b) 2, 1 and 4
(c) 1, 3 and 4 (d) 2, 3 and 4

[ESE : 2005]

2.16 What is the minimum length of a transition curve for a design speed of 80 kmph in a horizontal curve of 240 m radius?

(a) 32 m (b) 42 m
(c) 52 m (d) 72 m [ESE : 2006]

2.17 The design speed of a highway is 80 km/hr and the radius of circular curve is 150 m in plain topography. Which one of the following is the minimum length of transition curve?

(a) 115 m (b) 85 m
(c) 64 m (d) 43 m [ESE : 2006]

2.18 A rising gradient of 1 in 50 meets a falling gradient of 1 in 30. Which one of the following is the length of vertical curve if the stopping sight distance is 120 m?

(a) 174 m (b) 158 m
(c) 140 m (d) 120 m [ESE : 2006]

2.19 If a road surface is adequately superelevated on horizontal curve, which one of the following is the proper distribution of pressure on the vehicle wheels?

(a) Pressure on both outer and inner wheels is equal.
(b) Pressure on inner wheels is more than the outer wheels.
(c) Pressure on inner wheels is less than the outer wheels.
(d) Pressure on front wheel is thrice the pressure on rear wheels.

[ESE : 2007]

2.20 Assertion (A): Road camber helps in surface drainage.

Reason (R): In a curved road alignment, super elevation serves the purpose of camber.

(a) both A and R are true and R is the correct explanation of A
(b) both A and R are true but R is not a correct explanation of A
(c) A is true but R is false
(d) A is false but R is true [ESE : 2008]

2.21 If super-elevation is not provided on a horizontal curve of a highway, then on which portion of the road, are pot holes likely to develop?
(a) Outer edge of road
(b) Inner edge of road
(c) Centre of road
(d) Shoulder of road

[ESE : 2010]

2.22 Full amount of super-elevation on a horizontal curve is provided at the
(a) beginning of the transition curve
(b) centre of the circular curve
(c) end of the transition curve
(d) centre of the transition curve

[ESE : 2010]

2.23 With all other relevant conditions remaining the same, the speed of a vehicle negotiating a curve is proportional to
(a) $\sqrt{\text{Weight of the vehicle}}$
(b) Weight of the vehicle
(c) $\dfrac{1}{\text{Weight of the vehicle}}$
(d) $\dfrac{1}{\sqrt{\text{Weight of the vehicle}}}$ [ESE : 2011]

2.24 A 3% downgrade curve is followed by a 1% upgrade curve and rate of change of grade adopted is 0.1% per 20 m length. The length of the respective vertical curve is
(a) 800 m (b) 200 m
(c) 100 m (d) 400 m

[ESE : 2011]

2.25 Match **List-I** with **List-II** and select the correct answer using the code given below the lists:

List-I
A. Lateral friction
B. Cut-off lagoons
C. Skid
D. Sight distance

List-II
1. Disparity between relevant travel distances
2. Vehicle movement on a curve
3. Summit curves
4. Prevention of flooding

Code:

	A	B	C	D
(a)	2	1	4	3
(b)	3	1	4	2
(c)	2	4	1	3
(d)	3	4	1	2

[ESE : 2011]

2.26 **Assertion (A):** IRC has recommended a minimum coefficient of friction in the longitudinal direction on wet pavements after allowing a suitable factor of safety in the range 0.15-0.30.

Reason (R): When the longitudinal coefficient of friction of 0.40 is allowed for stopping the vehicle, the resultant retardation is 3.93 m/sec^2,
(a) Both A and R are true and R is the correct explanation of A
(b) Both A and R are true but R is not a correct explanation of A
(c) A is true but R is false
(d) A is false but R is true [ESE : 2011]

2.27 An ideal horizontal transition curve is a
(a) Parabola (b) Circle
(c) Clothoid spiral (d) Hyperbola

[ESE : 2012]

2.28 The transitional property of a lemniscate curve is disrupted when its deflection angle is around
(a) 30° (b) 45°
(c) 60° (d) 90° [ESE : 2012]

2.29 If R is the radius of the curve and L is the length of the long chord, the shift of the curve is (all in meter units)
(a) $\dfrac{L^2}{R}$ (b) $\dfrac{L^2}{2R}$
(c) $\dfrac{L^2}{24R}$ (d) $\dfrac{L^2}{6R}$ [ESE : 2012]

2.30 The ratio between the adopted centrifugal ratios for roads and railways is
(a) 3 : 1 (b) 4 : 1
(c) 2 : 1 (d) 5 : 1 [ESE : 2012]

2.31 For a bituminous carriageway surface having 2% camber and design speed of 100 kmph, Radius beyond which super-elevation is not essential is nearly:
(a) 1100 m (b) 1500 m
(c) 1800 m (d) 2200 m
[ESE : 2013]

2.32 What is the value of the resultant retardation in m/s^2 when a longitudinal friction coefficient of 0.4 is allowed for stopping the vehicle on road?
(a) 0.98 (b) 1.95
(c) 2.93 (d) 3.93
[ESE : 2013]

2.33 The rate of super-elevation for a horizontal curve of radius 500 m in a national highway for a design speed of 100 kmph is
(a) 0.04 (b) 0.063
(c) 0.07 (d) 0.70 [ESE : 2013]

2.34 If a vehicle travelling at 40 kmph was stopped within 1.8 sec after the application of the brakes, then the average skid resistance coefficient is
(a) 0.63 (b) 0.73
(c) 0.83 (d) 0.93 [ESE : 2014]

2.35 While aligning a hill road with a ruling gradient of 6%, a horizontal curve of 75 m radius is encountered. The compensated gradient at the curve will be
(a) 1% (b) 2%
(c) 3% (d) 5% [ESE : 2014]

2.36 The maximum superelevation to be provided on a road curve is 1 in 15. If the rate of change of superelevation is specified as 1 in 120 and the road width is 10 m, then the minimum length of the transition curve on each end will be
(a) 120 m (b) 100 m
(c) 80 m (d) 180 m [ESE : 2015]

2.37 A four-lane divided highway, with each carriageway being 7.0 m wide, is to constructed in a zone of high rainfall. In this stretch, the highway has a longitudinal slope of 3% and is provided a camber of 2%. What is the hydraulic gradient on this highway in this stretch?
(a) 4.0% (b) 3.6%
(c) 4.5% (d) 3.0% [ESE : 2015]

2.38 In an area of heavy rainfall, a State Highway of high-type bituminous surface with four lanes (14.0 m wide) is to be constructed. What will be the height of the crown of the road relative to the edges for a composite camber (i.e. middle half as parabolic and the rest as straight lines)?
(a) 14 cm (b) 21 cm
(c) 28 cm (d) 7 cm [ESE : 2015]

2.39 A descending gradient of 4% meets an ascending grade of 1 in 40 where a valley curve of length 200 m is to be formed. What will be the distance of the lowest point on the valley curve from its first tangent point?
(a) 100 m (b) 111 m
(c) 125 m (d) 118 m
[ESE : 2015]

2.40 What will be the non-passing sight distance on a highway for a design speed of 100 kmph when its ascending gradient is 2%? Assuming coefficient of friction as 0.7 and brake efficiency as 50%.
(a) 176 m (b) 200 m
(c) 150 m (d) 185 m [ESE : 2015]

2.41 Which one of the following items of hill road construction does not help in the prevention of landslides during the monsoon season?
(a) Breast walls
(b) Hair-pin
(c) Catch-water drains
(d) Retaining walls [ESE : 2015]

2.42 The radius of a horizontal circular curve is 480 m and design speed therein 70 kmph. What will be the equilibrium superelevation for the pressures on the inner and the outer wheels to be equal?
(a) 5% (b) 6%
(c) 7% (d) 8% [ESE : 2015]

2.43 The following purposes served by a transition curve in a highway alignment include:
1. Gradual introduction of the centrifugal force on moving vehicles from zero on the straight alignment to a constant final value on the circular curve.
2. Enabling the gradual introduction of superelevation on the roadway.

Select the correct answer using the codes given below:

(a) 1 only (b) 2 only
(c) Both 1 and 2 (d) Neither 1 nor 2
[ESE : 2017]

2.44 The rate of equilibrium superelevation on a road is

1. Directly proportional to the square of vehicle velocity
2. Inversely proportional to the radius of the horizontal curve
3. Directly proportional to the square of the radius of the horizontal curve

Which of the above statements are correct?

(a) 1 and 2 only (b) 1 and 3 only
(c) 2 and 3 only (d) 1, 2 and 3
[ESE : 2018]

2.45 Which one of the following conditions is **not** correct with respect to the transition curve?

(a) It should be tangential to the straight approaches at the two ends.
(b) It should meet the circular curve tangentially.
(c) Its curvature will necessarily be non-zero at the point of take-off from the straight approaches.
(d) The rate of increase of curvature along the transition reach should match with the increase of cant.

[ESE : 2019]

2.46 Consider two cars approaching from the opposite directions at 90 km/hand 60 km/h. If the reaction time is 2.5 s, coefficient of friction is 0.7 and brake efficiency is 50% in both the cases, the minimum sight distance required to avoid a head-on collision will be nearly

(a) 154 m (b) 188 m
(c) 212 m (d) 236 m [ESE : 2019]

2.47 What is the extra widening required (as nearest magnitude) for a pavement of 7 m width on a horizontal curve of radius 200 m, if the longest wheel of vehicle expected on the road is 6.5 m and the design speed is 65 km/h?

(a) 0.3 m (b) 0.5 m
(c) 0.7 m (d) 0.9 m [ESE : 2019]

2.48 The sight distance available on a road to a driver at any instance depends on

1. Features of the road ahead
2. Height of the driver's eye above the road surface
3. Height of the object above the road surface

(a) 1 and 2 only (b) 1 and 3 only
(c) 2 and 3 only (d) 1, 2 and 3
[ESE : 2020]

2.49 Consider the following data :

Design speed = 96 kmph
Speed of overtaken vehicle = 80 kmph
Reaction time for overtaking = 2 sec
Acceleration = 2.5 kmph/sec
The safe overtaking sight distance on a two-way traffic road will be nearly

(a) 646 m (b) 556 m
(c) 466 m (d) 376 m [ESE : 2020]

2.50 Which one of the following statement is correct?

(a) The radius of transition raises from infinity to a selected minimum in order to attain full super-elevation and curvature gradually.
(b) The compound curve is an arc of circle.
(c) The radius of transition curve is constant for entire length.
(d) The horizontal curves are provided whenever there is a change in gradient.

[ESE : 2020]

2.51 **Statement (I) :** A reverse curve consists of two arcs with their centres of curvature on opposite sides of the curve.

Statement (II) : Superelevation can be provided conveniently at the intersection point of the two arcs.

(a) Both Statement (I) and Statement (II) are individually true; and Statement (II) is the correct explanation of Statement (I)
(b) Both Statement (I) and Statement (II) are individually true; but Statement (II) is NOT the correct explanation of Statement (I)
(c) Statement (I) is true; but Statement (II) is false
(d) Statement (I) is false; but Statement (II) is true

[ESE : 2020]

2.52 What is the value of headlight sight distance for a highway with a design speed of 65 kmph? (Take f = 0.36 and t = 2.5 sec)

(a) 66.5 m (b) 81.3 m
(c) 91.4 m (d) 182.8 m
[ESE : 2021]

2.53 What is the minimum stopping sight distance on a –3.5% grade for a design speed of 110 kmph? (Consider friction coefficient $f = 0.28$, $t = 2.5$ sec and G = 0.035)

(a) 76.4 m (b) 194.4 m
(c) 214.6 m (d) 270.8 m

[ESE : 2021]

2.54 In case of horizontal curves in pavement, the purpose of super-elevation or banking of curve is to

(a) counteract the centripetal acceleration produced as a vehicle rounds a curve
(b) provide proper cross-drainage
(c) prevent vehicle from sliding inwards
(d) make road look good

[ESE : 2022]

2.55 A vehicle has wheel base of 6.5 m. What is the off tracking while negotiating a curved path with a mean radius 32 m?

(a) 1.32 m (b) 1.15 m
(c) 0.86 m (d) 0.66 m

[ESE : 2022]

2.56 The design speed for a two-way traffic on a two way lane road is 50 kmph. What is the value of safe stopping sight distance, if co-efficient of friction is 0.37 and reaction time of driver as 2.5 sec?

(a) 34.8 sec (b) 61.4 sec
(c) 122.8 sec (d) 193.5 sec

[ESE : 2023]

2.57 Match the following lists:

List-I (Median widths at intersections)
P. Four feet or wider
Q. Twentyeight feet or wider
R. Sixteen feet or wider
S. Twenty feet or wider

List-II (Key features)
1. Provides a pedestrian refuge and room for a dual left-turn bay
2. Provides a pedestrian refuge and room for a left-turn bay
3. Provides a pedestrian refuge.
4. Provides refuge for a crossing passenger car.

Select the correct answer using the codes given below.

	P	Q	R	S
(a)	4	1	2	3
(b)	3	1	2	4
(c)	3	4	2	1
(d)	2	3	4	1

[ESE : 2024]

2.58 If V_d is design speed (mph), t_{p-r} is perception-reaction time (sec), a is deceleration rate (ft/sec²) and G is longitudinal grade of the road (%/100). What is the expression for stopping sight distance?

(a) $SSD = 1.468V_d t_{p-r} + \dfrac{V_d^2}{30\left(\left(\dfrac{a}{32.2}\right) \pm G\right)}$

(b) $SSD = 1.468V_d t_{p-r} + \dfrac{V_d^2}{\left(\dfrac{a}{32.2}\right) \pm G}$

(c) $SSD = 1.468V_d t_{p-r} - \dfrac{V_d^2}{30\left(\left(\dfrac{a}{32.2}\right) \pm G\right)}$

(d) $SSD = 1.468V_d t_{p-r} - \dfrac{V_d^2}{\left(\dfrac{a}{32.2}\right) \pm G}$

[ESE : 2024]

2.59 If V_d is design speed (mph), e_{max} is maximum rate of superelevation, f_{max} is coefficient of side friction. What is the expression for minimum radius of curvature (R_{min})?

(a) $R_{min} = \dfrac{V_d^2}{(0.01e_{min} - f_{max})}$

(b) $R_{min} = \dfrac{V_d^2}{15(e_{min} + f_{max})}$

(c) $R_{min} = \dfrac{V_d^2}{15(e_{min} - f_{max})}$

(d) $R_{min} = \dfrac{V_d^2}{15(0.01e_{min} + f_{max})}$

[ESE : 2024]

■■■■

Answers Highway Geometric Design

2.1 (*)	2.2 (*)	2.3 (c)	2.4 (a)	2.5 (d)	2.6 (c)	2.7 (c)	2.8 (d)	2.9 (b)
2.10 (c)	2.11 (c)	2.12 (b)	2.13 (b)	2.14 (b)	2.15 (d)	2.16 (d)	2.17 (a)	2.18 (a)
2.19 (a)	2.20 (b)	2.21 (a)	2.22 (c)	2.23 (d)	2.24 (a)	2.25 (c)	2.26 (d)	2.27 (c)
2.28 (c)	2.29 (c)	2.30 (c)	2.31 (d)	2.32 (d)	2.33 (c)	2.34 (a)	2.35 (d)	2.36 (c)
2.37 (b)	2.38 (a)	2.39 (b)	2.40 (a)	2.41 (b)	2.42 (d)	2.43 (c)	2.44 (a)	2.45 (c)
2.46 (d)	2.47 (c)	2.48 (d)	2.49 (a)	2.50 (a)	2.51 (c)	2.52 (c)	2.53 (d)	2.54 (a)
2.55 (d)	2.56 (b)	2.57 (c)	2.58 (a)	2.59 (d)				

Explanations Highway Geometric Design

2.1 (*)

Total reaction time

= Perception time + Brake reaction time

Perception time varies from driver to driver and also depends upon several other factors such as speed of the vehicle, distance of object and other environmental conditions.

According to PIEV (Perception, Intellection, Emotion and Volition) theory, the total reaction time depends upon several factors such as physical and psychological characteristics of the driver, type of the problem involved, environmental condition and temporary factors (e.g. motive of the trip, travel speed, fatigue, consumption of alcohol, etc.).

So, its also depends on the condition of mind of the driver and the total reaction of a driver depends on the all the four choices given in the question.

2.2 (*)

Equilibrium superelevation for mixed traffic condition,

$$e = \frac{V^2}{225R} = \frac{(80)^2}{225 \times 480}$$

$$= 0.0593 < 0.07$$

Raising of outer edge with respect to inner edge

$$= eW = 0.0593 \times 7.5$$

$$= 0.44 \text{ m}$$

Raising of outer edge with respect to centre

$$= \frac{We}{2} = 0.22 \text{ m}$$

The answer should be 0.44 m, but no such option exists.

2.3 (c)

Braking distance for travelling upgrade,

$$S_1 = \frac{V^2}{254(f + 0.01n)}$$

Braking distance for travelling downgrade

$$S_2 = \frac{V^2}{254(f - 0.01n)}$$

Given, $S_2 = 2S_1$

$$\Rightarrow \frac{1}{(f - 0.01n)} = \frac{2}{(f + 0.01n)}$$

$$\Rightarrow f + 0.01n = 2f - 0.02n$$

$$\therefore n = \frac{f}{0.03} = \frac{0.4}{0.03} = 13.3\%$$

2.4 (a)

For $R > 45$ m, grade compensation $= \frac{75}{R}\%$

The grade compensation in per cent is $\frac{30 + R}{R}$ subject to maximum $75/R$.

In other words,

for $R \leq 45$ m grade compensation

$$= \frac{30 + R}{R}\%$$

2.5 (d)

$$\text{Centrifugal ratio} = \frac{P}{W} = e + f$$

On transition curve 'e' increases from zero to full value so centrifugal ratio increases. The centrifugal force increase on transition curve gradually.

2.6 (d)

Reducing the super-elevation will not affect the side friction. Too much super-elevating friction is dangerous for slow moving vehicles as they can topple so a skilful highway designer will build speed controls at critical locations.

2.7 (c)

Stopping sight distance = Lag distance + Braking distance.

For lag distance, reaction time and speed is taken into consideration.

Lag distance = $0.278\ Vt$

$$\text{Braking distance} = \frac{V^2}{254(f \pm 0.01n)}$$

2.8 (d)

As per IRC 52:1981, ruling gradient values should be

1 in 30: plain and rolling terrain

1 in 20: mountainous terrain

1 in 16.7: Steep terrain

Thus for sloping road it is 1 in 30.

2.10 (c)

Length of vertical curve

$$= \frac{\text{Total change of grade}}{\text{Permissible rate of change of grade per chain length}}$$

Total change of grade = Difference of gradients $= g_1 - g_2$

r = Permisible rate of change of grade/chain length

$$\text{Length of vertical curve} = \frac{g_1 - g_2}{r}$$

2.11 (c)

Safe stopping distance = 80 m

Passing sight distance = 300 m

Intermediate sight distance

$= 2 \times \text{SSD} = 2 \times 80 = 160$

2.12 (b)

Transitional curve is provided between a circular curve and straight road. It provides gradual introduction of centrifugal force, gradual introduction of super elevation and comfort and security to passengers.

2.13 (b)

$$R_{ruling} = \frac{V^2}{127(e+f)}$$

$$= \frac{(100)^2}{127 \times (0.07 + 0.13)} \simeq 395\ \text{m}$$

2.14 (b)

Deviation angle,

$N = 3 + 5 = 8\% = 0.08$

Assuming $L > S$

$$L = \frac{NS^2}{4.4} = \frac{0.08 \times (128)^2}{4.4} \simeq 298\ m$$

2.15 (d)

The function of transition curve is to gradually introduce the centrifugal force between the tangent point and the beginning of the circular curve. So rate of change of centrifugal acceleration (radial acceleration) should be small constant and consistent. This means that radius of the transition curve should consistently decrease from infinity (zero curvature) at tangent point to the radius R {curvature $\left(\frac{1}{R}\right)$} at the beginning of the circular curve.

The other function of transition curve is to introduce the designed superelevation gradually. The extra widening of pavement at the start of circular curve should also be introduced gradually.

2.16 (d)

Rate of change of centrifugal acceleration,

$$C = \frac{80}{75+V} = \frac{80}{75+80} = 0.516\ \text{m/s}^3$$

Length of transition curve,

$$L_s = \frac{0.0215V^3}{CR} = \frac{0.0215 \times (80)^3}{0.516 \times 240}$$

$= 88.9 \simeq 89$ m

For plain and rolling terrain,

$$L_s = \frac{2.7V^2}{R} = 72\ \text{m}$$

For mountainous and steep terrains,

$$L_s = \frac{V^2}{R} \simeq 27\ \text{m}$$

Since 72 m is one option so it should be chosen.

2.17 (a)

Length of transition curve in plain or rolling terrain:

$$L_s = \frac{2.7V^2}{R} = \frac{2.7 \times 80^2}{150} = 115.2\ \text{m}$$

For mountainous and steep terrains,

$$L_s = \frac{V^2}{R}$$

2.18 (a)

Deviation angle,

$$N = \frac{1}{50} + \frac{1}{30} = \frac{4}{75}$$

Assuming $L > S$

$$L = \frac{NS^2}{4.4} = \frac{4}{75} \times \frac{120^2}{4.4}$$

$$= 174.5 \text{ m}$$

2.19 (a)

If the road surface is kept horizontal across the alignment, the pressure on the outer wheels will be higher due to the centrifugal force acting outwards and hence the reaction at the outer wheel would be higher. In order to counteract this centrifugal force and to reduce the tendency of the vehicle to overturn or skid, superelevation is provided.

2.20 (b)

Road camber helps in proper drainage and protects the sub-grade. Super-elevation in curved roads is provided by rotating the inner edge or centre line and helps in drainage, thus serving the purpose of camber.

2.21 (a)

When a vehicle negotiates a horizontal curve on a highway, without super-elevation the pressure on the tyres on outer edge is more than on the inner ones. Thus the chances of appearing of potholes is more on the outer edge of the road.

2.22 (c)

The full amount of super-elevation is attained by the end of transition curve or at the beginning of circular curve.

Superelevation is gradually introduced in the transition curve beginning from tangent point (end of straight line) and the full amount of superelevation is attained by the end of transition curve or the beginning of circular curve.

2.23 (d)

For a vehicle negotiating a curve the relation between weight of vehicle, *W*, centrifugal fore, *P*, velocity, *V*, and radius of the curve, *R*, is given by

$$F = \frac{Wv^2}{gR}$$

∴ With all other relevant conditions remaining same,

$$V = \frac{1}{\sqrt{\text{Weight of vehicle}}}$$

2.24 (a)

$$n = -3\%, \ n = +1\ \%$$

$$\text{Rate of change of grade} = \frac{0.01\%}{20 \text{ m}}$$

$$N = -3 - (+1) = -4\%$$

$$L_c = \frac{N \times 20}{0.1\%} = 800 \text{ m}$$

2.25 (c)

When a vehicle traverses a horizontal curve, the centrifugal force acts horizontally outwards through the centre of gravity of the vehicle. This centrifugal force is counteracted by the transverse frictional resistance developed between the tyres and the pavement which enables the vehicle to change the direction along the curve and to maintain the stability of the vehicle.

Summit curves are designed so as to provide adequate sight distance.

Cut-off lagoons are structures created along the important roadways to protect it from flooding.

Skidding takes place when path travelled along the road is more than circumferential motion of wheels.

2.26 (d)

The coefficient of friction reduces considerably when the pavement surface is smooth or wet. For the calculation of stopping distance, the longitudinal friction coefficient values of 0.35 to 0.40 have been recommended by the Indian Roads Congress, depending upon speed. These values have been suggested keeping in view the minimum coefficient of friction in the longitudinal direction on wet pavements and after allowing a suitable factor of safety. Further, when a longitudinal friction coefficient of 0.40 is allowed for stopping the vehicle, the resultant retardation is 3.93 m/sec² which is not too uncomfortable to the passengers.

In the case of horizontal curve design, the Indian

Roads Congress has recommended the lateral coefficient of friction of 0.15.

2.27 (c)

The Indian Road Congress recommends the use of spiral as transition curve in the horizontal alignment of highways due to the following reasons.

(i) The spiral curve satisfies the requirements of an ideal transition.

(ii) The geometric property of spiral is such that the calculations and setting out the curve in the field is simple and easy.

2.28 (c)

Upto small deviation angle (12°), lemniscate, spiral and cubic parabola are same. Clothoid and lemniscate are identical upto deviation angle of 60°.

2.29 (c)

Shift, $$S = \frac{L^2}{24R}$$

R = Radius of the curve

L = Length of long chord

2.30 (c)

Centrifugal ratio is the ratio of centrifugal force and weight and is given by:

$$\frac{P}{W} = \frac{V^2}{gR}$$

Maximum centrifugal ratio value

For road: $\frac{1}{4}$

For railway: $\frac{1}{8}$

Ration between these two = $\frac{1/4}{1/8} = 2:1$

2.31 (d)

The given camber will serve as superelevation,

$$e = \frac{V^2}{225R}$$

$$\Rightarrow \quad 0.02 = \frac{100^2}{225 \times R}$$

$$\Rightarrow \quad R = 2222.22 \text{ m} \cong 2200 \text{ mm}$$

2.32 (d)

Longitudinal friction coefficient

$$f = \frac{a}{g}$$

$$\Rightarrow \quad 0.4 = \frac{a}{9.81}$$

$$\Rightarrow \quad a = -3.93 \text{ m/s}^2$$

2.33 (c)

$$e = \frac{V^2}{225R} = \frac{(100)^2}{225 \times 500}$$

$$= 0.088$$

Now, limit $e = 0.07$

$$e + f = \frac{V^2}{127R}$$

$$\Rightarrow \quad f = \frac{100^2}{127 \times 500} - 0.07$$

$$= 0.087 \text{ (OK)} < 0.15$$

Provide, $e = 0.07$

2.34 (a)

Data given,

$$u = 40 \text{ kmph},$$

$$V = 0,\ t = 1.8 \text{ sec.}$$

$$\therefore \quad V = u - at$$

$$u = at$$

$$a = \frac{u}{t} = \frac{40 \times \frac{5}{18}}{1.8}$$

$$= 6.1728 \text{ m/sec}^2$$

$\Rightarrow$ Avg. skid resistance,

$$\mu = \frac{6.1728}{9.81} = 0.6292$$

$$\mu \simeq 0.63$$

2.35 (d)

Radius of Horizontal curve = 75 m

Grade compensation = $\frac{30 + R}{R} \le \frac{75}{R}$

$$= \frac{30 + 75}{75} = 1.4$$

and $\frac{75}{R} = \frac{75}{75} = 1\%$

So, Grade compensation = 1%

∴ Compensated gradient

= Ruling gradient − grade compensation

= 6% − 1% = 5%

2.36 (c)

Length of transition curve based on rate of change of superelevation is,

$$L = eNw$$

$$L = \frac{1}{15}\times 120\times 10 = 80 \text{ m}$$

2.38 (a)

Camber to be provided for state highway is 1 in 50.

∴ Height of the crown

$$=\left(\frac{14}{2}\right)\times\frac{1}{50}\times 100 \text{ cm} = 14 \text{ cm}$$

2.39 (b)

$$x = L\left(\frac{n_1}{2N}\right)^{1/2}$$

x = Horizontal distance from end of first tangent point to the lowest point in metres.

n_1 = Natural tangent of first tangent

$$N = n_1 + n_2$$

$$= \left|-\frac{4}{10}-\frac{1}{40}\right| = 0.065$$

$$x = 200\left(\frac{0.04}{2\times 0.065}\right)^{1/2}$$

$$= 110.94 \simeq 111 \text{ m}$$

2.40 (a)

$$\text{SSD} = 0.278Vt + \frac{V^2}{254(kf + S\%)}$$

$$= 0.278\times 100\times 2.5 + \frac{100^2}{254(0.5\times 0.7 + 0.02)}$$

$$= 175.9 \text{ m}$$

2.41 (b)

A hairpin turn or hairpin bend or hairpin corner is a bend in a road with a very acute inner angle, making it necessary for an oncoming vehicle to turn almost 180° to continue on the road. It does not help in prevention of landslides.

2.42 (d)

Equilibrium superelevation

$$e = \frac{V^2}{127R} = \frac{70^2}{127\times 480}$$

$$= 0.08 \text{ i.e. } 8\%$$

2.44 (a)

$$e_{eq} = \frac{V^2}{127R}$$

2.45 (c)

Transition curve should meet straight line and curved posit ons tangentially so it should have same properties as that of straight line and curved positions at their respective junctions.

2.46 (d)

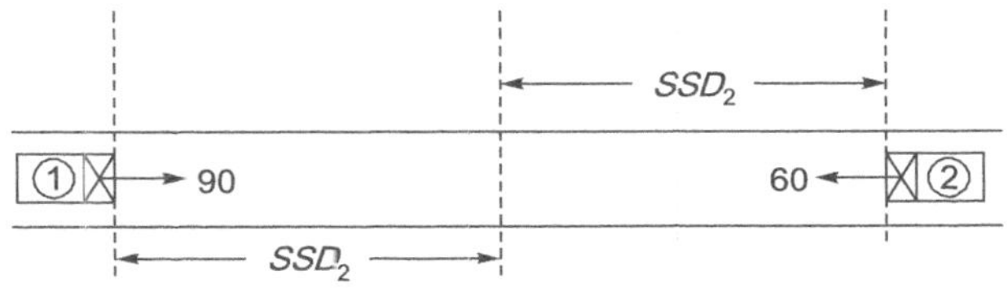

Braking efficiency = 50%, $f = 0.7$

$$SD = SSD_1 + SSD_2$$

$$= \left[0.278V_1t_R + \frac{V_1^2}{254(n_Bf)}\right] + \left[0.278V_2t_R + \frac{V_2^2}{254(n_Bf)}\right]$$

$$= \left[0.278\times 90\times 2.5 + \frac{90^2}{254\times(0.5\times 0.7)}\right] +$$

$$\left[0.278\times 60\times 2.5 + \frac{60^2}{254(0.5\times 0.7)}\right]$$

$$= 153.67 + 82.195 = 235.865 \text{ m} \approx 236 \text{ m}$$

2.47 (c)

Road width = 7 m (i.e., No of lane 2)

Extra widening

$$W_E = \frac{nl^2}{2R} + \frac{V}{9.5\sqrt{R}}$$

(R = 200 m; l = 6.5 m; V = 65 kmph)

$$= \frac{2\times 6.5^2}{2\times 200} + \frac{65}{9.5\sqrt{200}}$$

$$= 0.695 \text{ m} \approx 0.7 \text{ m}$$

2.49 (a)

$$S = (0.2V_B + 6)$$

$$= 0.2\times 80 + 6$$

$$= 22 \text{ m}$$

$$T = \sqrt{\frac{45}{9}} = \sqrt{\frac{4\times 22}{0.694}} = 11.26 \text{ sec}$$

$V_A = V_C = 96$

$V_B = 80$ kmph

$OSD = d_1 + d_2 + d_3$

$$OSD = (0.278 \times 80 \times 2) + (0.278 \times 80 \times 11.26 + 2 \times 22) + (0.278 \times 96 \times 11.26) = 639 \text{ m}$$

2.52 (c)

Given: $V = 65$ kmph, $f = 0.36$, $t_R = 2.5''$

$$HSD = SSD = \left(0.278Vt_R + \frac{V^2}{254f}\right)$$

$$= \left(0.278 \times 65 \times 2.5 + \frac{65^2}{254 \times 0.36}\right)$$

$$= 91.38 \simeq 91.4 \text{ m}$$

2.53 (d)

Given: Falling gradient = 3.5%

Speed, $V = 110$ km/hr

$f = 0.28$, $t_R = 2.5''$

$$S = 0.278Vt_R + \frac{V^2}{254(f - S\%)}$$

$$= 0.278 \times 110 \times 2.5 + \frac{110^2}{254(0.28 - 0.035)}$$

$$= 270.88 \text{ m}$$

2.55 (d)

$l = 6.5$ m, $R = 32$ m

$$\therefore \text{ Offtracking}, O_T = \frac{l^2}{2R} = \frac{6.5^2}{2 \times 32} = 0.66 \text{ m}$$

2.56 (b)

$$SSD = 0.278Vt_R + \frac{V^2}{254f}$$

$$= 0.278 \times 50 \times 2.5 + \frac{50^2}{254 \times 0.37} = 61.4 \text{ m}$$

Note: Stopping sight distance unit should be in meter, not in seconds as given in options.

2.58 (a)

As we know,

$$SSD = V_d t_r + \frac{V_d^2}{2g(f \pm G)}$$

Where V_d is in m/s, t_r in seconds, g is in ms^{-2}.

Now, V_d (in ft/sec) = 1.468 × V_d (in miles/hr)

Now, Deaceleration, a (in ft/sec^2) = μg

$$\therefore \quad f = \frac{a}{g}$$

Now, $g = 32.2$ ft/sec^2

So, $$f = \frac{a}{32.2}$$

Hence, SSD (in feet)

$$= \left[(1.468V_d)t_{p-r}\right] + \left[\frac{(1.468V_d)^2}{2g(f \pm G)}\right]$$

$$= 1.468V_d t_{p-r} + \frac{V_d^2}{\frac{2 \times 32.2}{1.468^2}\left(\frac{a}{32.2} \pm G\right)}$$

$$= 1.468V_d t_{p-r} + \frac{V_d^2}{\frac{2 \times 32.2}{1.468^2}\left(\frac{a}{32.2} \pm G\right)}$$

2.59 (d)

Minimum radius of curvature (R_{min}) is given as

$$R_{min} = \frac{V_d^2}{15[0.01e_{max} + f_{max}]}$$

where, V_d is design speed in miles/hour.

■■■■

3 Traffic Engineering

3.1 Consider the following factors:
1. Length of vehicle
2. Width of vehicle
3. Approach speed
4. Stopping time for approaching vehicle
5. Passing sight distance

Which of these factors are taken into consideration for determining yellow time of a traffic signal at an inter section?
(a) 1, 2 and 5 (b) 2, 3 and 4
(c) 1, 3 and 5 (d) 1, 3 and 4
[ESE : 2000]

3.2 An observer travelling at a constant speed of 70 kmph with the traffic stream over a 5 km stretch is passed by 17 vehicles more than what he passes. When the observer travels against the stream at the same speed, the number of vehicles he meets is 303. The flow of the traffic stream is
(a) 4480 vph (b) 4160 vph
(c) 2240 vph (d) 2002 vph
[ESE : 2001]

3.3 If the total seven-day traffic volume for the week is 3625 and the traffic volume for Monday is 650, then the monthly expansion factor for Monday would be
(a) 5.6 (b) 1.25
(c) 0.8 (d) 0.25 [ESE : 2001]

3.4 Which one of the following is the purpose of divisional island?
(a) To divert the traffic into definite travel path at the intersection.
(b) To reduce the speed of traffic entering the intersection.
(c) To divert traffic from obstacles and expedite the flow of traffic.
(d) To segregate opposing flow of traffic in a multi-lane highway.
[ESE : 2001]

3.5 Based on '30th hour volume' for how much per cent time during the year can the designer willingly tolerate the unfavourable operating conditions?
(a) 30 (b) 29
(c) 2.5 (d) 0.33 [ESE : 2002]

3.6 A journey from work to home made by walking to the bus, travelling by bus to the station and completing the journey by train is regarded as
(a) 4 trips (b) 3 trips
(c) 2 trips (d) 1 trip
[ESE : 2002]

3.7 Which one of the following is taken into consideration for computing traffic capacity per lane of the highway?
(a) Passenger cars and light vehicles
(b) Trucks and buses
(c) Two-wheelers
(d) Equivalent of passenger cars
[ESE : 2003]

3.8 Match **List-I** (Method of traffic volume counts) with **List-II** (Equipment used) and select the correct answer using the codes given below the lists:

List-I
A. Manual count
B. Combination of manual and mechanical methods
C. Automatic devices
D. Photographic method

List-II
1. Video-recorder
2. Pneumatic tube
3. Watch
4. Multiple pen recorder

Codes:

	A	B	C	D
(a)	4	3	1	2
(b)	3	4	2	1

(c)	4	3	2	1
(d)	3	4	1	2

[ESE : 2003]

3.9 Match **List-I** (Type of study) with **List-II** (Data collected) and select the correct answer using the codes given below the lists:

List-I

A. Public transport inventory study
B. Public transport O-D study
C. Public transport usage study
D. Public transport operator's study

List-II

1. Starting and destination of riders
2. Available routing and scheduling
3. Vehicle-km, earning per km
4. Passenger-km, IVTT, walking time, waiting time

Codes:

	A	B	C	D
(a)	1	2	3	4
(b)	2	1	3	4
(c)	1	2	4	3
(d)	2	1	4	3

[ESE : 2003]

3.10 Match **List-I** (Type of traffic signals) with **List-II** (Advantages) and select the correct answer using the codes given below the lists:

List-I

A. Pre-timed signals
B. Vehicle-Actuated signals
C. Semi-vehicle actuated signals
D. Linked traffic signals

List-II

1. Useful for junction of a side street having low traffic volume with a main street having heavy flow
2. Overall optimization of traffic flow
3. Delay is held to a minimum, and maximum lane capacity is achieved
4. Most successfully used in linked system

Codes:

	A	B	C	D
(a)	3	4	1	2
(b)	4	3	1	2
(c)	3	4	2	1
(d)	4	3	2	1

[ESE : 2003]

3.11 Assertion (A): Prohibitory signs are part of regulatory signs which give a definite negative instruction regarding movement prohibition, restriction on weight, or speed of vehicles.

Reason (R): According to the IRC Standards, prohibitory signs are of octagonal shape with side being 900 mm and with a white border and red background for the standard size.

(a) both A and R are true and R is the correct explanation of A
(b) both A and R are true but R is not a correct explanation of A
(c) A is true but R is false
(d) A is false but R is true [ESE : 2003]

3.12 The lost time due to starting delay on a traffic signal approach is noted to be 3 seconds, the actual green time is 25 seconds and amber time is 3 seconds. How much is the effective green time?

(a) 19 sec (b) 25 sec
(c) 27 sec (d) 31 sec [ESE : 2004]

3.13 Match **List-I** (Study) with **List-II** (Purpose) and select the correct answer using the codes given below the lists:

List-I

A. Primary road system
B. Economic studies
C. Engineering studies
D. Road use studies

List-II

1. Population distribution
2. Expressways
3. Traffic volume
4. Topographic details

Codes:

	A	B	C	D
(a)	2	1	4	3
(b)	4	3	2	1
(c)	2	3	4	1
(d)	4	1	2	3

[ESE : 2005]

3.14 Match **List-I** (Traffic survey) with **List-II** (Method) and select the correct answer using the codes given below the lists:

List-I	List-II
A. Spot speed	1. By video tape
B. Traffic volume	2. By road side interview

C. O-D survey 3. By doppler radar
D. Parking survey 4. By pneumatic tube

Codes:

	A	B	C	D
(a)	3	1	2	4
(b)	2	4	3	1
(c)	3	4	2	1
(d)	2	1	3	4

[ESE : 2005]

3.15 Match **List-I** (Unit) with **List-II** (Purpose) and select the correct answer using the codes given below the lists:

List-I
A. Clover leaf interchange
B. Traffic studies
C. Rotary island
D. Road junction approach sign

List-II
1. Informatory sign
2. Traffic control device
3. Traffic flow characteristics
4. Grade -separation

Codes:

	A	B	C	D
(a)	4	3	2	1
(b)	2	1	4	3
(c)	4	1	2	3
(d)	2	3	4	1

[ESE : 2005]

3.16 On a road the free speed was 65 kmph and the space headway at jam density was 6.25 m. What is the maximum flow which could be expected on this road?

(a) 2600 vph (b) 1625 vph
(c) 1300 vph (d) 406 vph

[ESE : 2006]

3.17 Consider the following statements related to interchanges:

1. In diamond interchange there is the possibility of illegal wrong-way turns.
2. Diamond interchange is far superior to cloverleaf design.

Which of these statements is/are correct?

(a) 1 only (b) 2 only
(c) Both 1 and 2 (d) Neither 1 nor 2

[ESE : 2007]

3.18 Match **List-I** (Speed) with **List-II** (Application) and select the correct answer using the code given below the lists:

List-I	List-II
A. Space mean speed	1. Road conditions studies
B. Journey speed	2. Regulatory measures
C. Running speed	3. Traffic flow studies
D. Spot speed	4. Delay studies

Codes:

	A	B	C	D
(a)	3	2	1	4
(b)	1	4	3	2
(c)	3	4	1	2
(d)	1	2	3	4

[ESE : 2007]

3.19 Match **List-I** with **List-II** and select the correct answer using the codes given below the lists:

List-I
A. Traffic Volume
B. Traffic Density
C. Traffic Regulations
D. Rotary Intersection

List-II
1. Number of vehicles occupying a unit length of road at a given instant of time.
2. Number of vehicles passing a given point on road in a given unit of time in a given direction
3. Where all converging vehicles are forced to move in one direction around a large central traffic island
4. Rules covering all aspects of control of vehicles, drivers and all other road users

Codes:

	A	B	C	D
(a)	2	4	1	3
(b)	3	1	4	2
(c)	2	1	4	3
(d)	3	4	1	2

[ESE : 2008]

3.20 Which of the following factors are **not** strictly related to design of traffic rotary intersections?

1 Radius of central island
2. Weaving length
3. Ramps and interchanges
4. Acceleration lanes

Select the correct statements:

(a) 1 and 2 (b) 1 and 4
(c) 2 and 3 (d) 3 and 4 ESE : 2008]

3.21 In urban transportation planning, the 'Modal Split' is the process of which one of the following?
(a) Staggering of working hours
(b) Segregation of fast and slow modes
(c) Separation of traffic streams by flyovers
(d) Deciding the choice for a mode

[ESE : 2008]

3.22 Which one of the following geometric features requires the magnitudes of weaving angle and weaving distance for its design?
(a) Rotary design
(b) Right-angle intersection
(c) Roundabout
(d) Grade-separated junction

[ESE : 2009]

3.23 Which set of traffic studies is needed for functional design as well as for 'highway capacity' design?
(a) Origin and destination studies
(b) Parking and accident studies
(c) Speed and volume studies
(d) Axle load studies

[ESE : 2009]

3.24 Which one of the following traffic survey schemes is most relevant when deciding on locating major 'routes' in a city?
(a) Traffic volume survey
(b) Origin and destination survey
(c) Speed survey
(d) Traffic capacity survey [ESE : 2009]

3.25 Which one of the following equipments is useful in determining spot speed in traffic engineering?
(a) Enoscope (b) Periscope
(c) Radar (d) Tachometer

[ESE : 2009]

3.26 Which one of the following traffic signal systems is useful when there is continuous operation of group of vehicles along the main road?
(a) Simultaneous system
(b) Alternate system
(c) Simple progressive system
(d) Flexible progressive system

[ESE : 2010]

3.27 Match **List-I** (Type of survey) with **List-II** (Method/ Instrument) and select the correct answer using the codes given below the lists:

List-I
A. Traffic volume study
B. Speed and delay study
C. Spot-speed study
D. Multiple character studies

List-II
1. Workspot interview method
2. Doppler radar
3. Floating car method
4. Automatic vehicle counter and classifier
5. Electronic detector

Codes:

	A	B	C	D
(a)	5	3	2	4
(b)	1	3	2	5
(c)	5	2	3	4
(d)	1	2	3	4

[ESE : 2010]

3.28 Light reflecting devices used to guide the driver along the proper alignment are called
(a) rumble strips (b) delineators
(c) attenuators (d) litter bin

[ESE : 2010]

3.29 **Assertion (A):** In an intersection design, the relative speed is dependent on the absolute speed of intersecting vehicles and the angles between them.
Reason (R): When the angle of merging is small, the relative speed will be high.
(a) both A and R are true and R is the correct explanation of A
(b) both A and R are true but R is not a correct explanation of A
(c) A is true but R is false
(d) A is false but R is true

[ESE : 2010]

3.30 Traffic capacity is the:
(a) Ability of roadway to accommodate traffic volume in terms of vehicles/hr
(b) Number of vehicles occupying a unit length of roadway at a given instant expressed as vehicles/km
(c) Capacity of lane to accommodate the vehicles widthwise (across the road)
(d) Maximum attainable speed of vehicles

[ESE : 2013]

3.31 The design speed of a traffic lane is 70 kmph. What is the theoretical capacity per hour taking the total reaction time to be 2 seconds and average length of vehicles as 8 m?
(a) 828 (b) 728
(c) 628 (d) 428 [ESE : 2014]

3.32 The duration of green time of a traffic signal depends on
(a) traffic density (b) traffic volume
(c) traffic speed (d) All of the above
[ESE : 2015]

3.33 What will be the theoretical maximum capacity (to nearest 10 units) for a single lane of highway given that the speed of the traffic stream is 40 kmph?
(a) 3000 veh/h (b) 2860 veh/h
(c) 2010 veh/h (d) 2510 veh/h
[ESE : 2015]

3.34 What will be the initial traffic after construction, in the commercial vehicles per day (CVD) for the following data?
Annual average daily traffic at last count = 400 CVD
Rate of traffic growth per annum = 7%
The road is proposed to be completed in 3 years.
(a) 500 (b) 421
(c) 490 (d) 449
[ESE : 2015]

3.35 Overspeed and delay studies on a preselected section of a Highway are conducted by
(a) Fast moving car method
(b) Enoscope
(c) Radar
(d) Traffic contours [ESE : 2016]

3.36 International Traffic Intelligent Survey Data are related with
(a) Origin and destination studies
(b) Speed and delay studies
(c) Classified traffic volume studies
(d) Accident profiling studies
[ESE : 2016]

3.37 A transportation trip survey was undertaken between private car, and public car transportation. The proportion of those using private cars is 0.45. While using the public transport, the further choices available are Metro Rail and Mono Rail, out of which commuting by a Mono Rail has a proportion of 0.55. In such a situation, the choice of interest in using a Metro Rail, Mono Rail and private car would be respectively
(a) 0.25, 0.3 and 0.45
(b) 0.45, 0.25 and 0.3
(c) 0.25, 0.45 and 0.3
(d) 0.3, 0.25 and 0.45 [ESE : 2016]

3.38 The type of signalling system in which it is possible to vary the length of cycle, cycle division and the time schedule at each signal point is called
(a) Simultaneous system
(b) Alternate system
(c) Simple progressive system
(d) Flexible progressive system [ESE : 2017]

3.39 The normal flows on two approach roads at an intersection are respectively 500 pcu/h and 300 pcu/h. The corresponding saturation flow is 1600 pcu/h on each road. The total lost time per single cycle is 16 s. The optimum cycle time by Webster's method is
(a) 72.5 s (b) 58.0 s
(c) 48.0 s (d) 19.3 s [ESE : 2018]

3.40 A vehicle moving at 40 km/h speed was stopped by applying brake and the length of the skid mark was 12.2 m. If the average skid resistance of the pavement is 0.70, the brake efficiency of the test vehicle will be nearly
(a) 80% (b) 74%
(c) 68% (d) 62% [ESE : 2019]

3.41 The main drawback of automatic counters-cum-classifiers, used for traffic volume studies, is that it is not yet possible to classify and record
(a) Vehicle type (b) Axle spacing
(c) Axle load (d) Speed
[ESE : 2019]

3.42 Which one of the following is **not** a part of 'speed and delay' studies?
(a) Floating car method
(b) Vehicle number method
(c) Interview technique
(d) License number method
[ESE : 2019]

3.43 Which one of the following statements is correct?

(a) The ratio of load on wheel to contact area or area of imprint is called as contact pressure.

(b) The ratio of load on wheel to contact pressure is called as rigidity factor.

(c) The value of rigidity factor is more than three for an average tyre pressure of 7 kg/cm^2.

(d) Rigidity factor does not depend upon the degree of tension developed in walls of tyres.

[ESE : 2020]

3.44 When property designed traffic signals are used, which one of the following is the advantages to traffic signals?

(a) The signals allow crossing of the heavy traffic flow with safety.

(b) The rear-end collision may increase.

(c) Improper design and location of signals may lead to violation of the control system.

(d) Failure of the signal due to electric power failure may cause confusion to the road users.

[ESE : 2021]

3.45 The free mean sped on a roadway is found to be 80 kmph. Under stopped condition, the average spacing between the vehicles is 6.9 m. What is the capacity flow?

(a) 5800 Vehicles/hour (per lane)

(b) 7200 Vehicles/hour (per lane)

(c) 1450 Vehicles/hour (per lane)

(d) 2900 Vehicles/hour (per lane) [ESE : 2021]

3.46 For a street lighting system, having the following conditions:

Street width = 15 m

Mounting height= 7.5 m

Lamp size = 6000 lumen

Luminaire type = II

Coefficient of utilization = 0.44

Maintenance factor = 0.8

What is the spacing between lighting units to produce average Lux = 6?

(a) 18 m (b) 20 m

(c) 23 m (d) 27 m [ESE : 2021]

3.47 In traffic control, the speed at which vehicles are presumed to travel through the coordinated signal system is known as

(a) Signal coordination

(b) Speed of progression

(c) Cycle

(d) Through band [ESE : 2022]

3.48 Space-mean speed represents

(a) the instantaneous speed of a vehicle at a specified section or location.

(b) the effective speed with which a vehicle traverses a particular route between two terminals.

(c) the average speed of vehicle in a certain road length at any time.

(d) the average of instantaneous speed of observed vehicles at the spot.

[ESE : 2022]

3.49 For traffic surveys using origin and destination studies, the most suitable method in case of heavy traffic and absence of skilled or trained personnel is

(a) Road side interview method

(b) License plate method

(c) Work spot or home interview method

(d) Return post card method [ESE : 2022]

3.50 Consider the following aspects of human vision which are important for a traffic engineer:

1. Field of vision
2. Visual acuity
3. Colour perception

Which of the above aspects are correct?

(a) 1 and 2 only (b) 2 and 3 only

(c) 1 and 3 only (d) 1, 2 and 3

[ESE : 2023]

3.51 The distance between successive vehicles moving in the same line measured from head at any instance is

(a) the space headway

(b) the time headway

(c) the one-way stream

(d) the head-way stream [ESE : 2023]

3.52 What is the major difficulty in establishing human surveillance in freeway management systems?

(a) Difficulty in integrating electronic detection with human surveillance.

(b) Infrastructure for providing video providing video to the location of the human operator.

(c) Human operators have superior judgment, but they may lose attention.

(d) Lack of qualified human operators.

[ESE : 2024]

3.53 The weighted average of the damages caused by the individual axle load group with respect to the corresponding volume of the traffic of each group is known as
(a) Lane distribution factor
(b) Load Safety factor
(c) Vehicle damage factor
(d) Heavy-vehicle adjustment factor

[ESE : 2024]

3.54 Which one of the following is a factor that is to be multiplied by the total traffic repetitions in a lane to convert it to equivalent repetitions along the maximum distressed path?
(a) Lane distribution factor
(b) Load Safety factor
(c) Vehicle damage factor
(d) Lateral distribution factor

[ESE : 2024]

Answers Traffic Engineering

3.1 (d)	3.2 (c)	3.3 (b)	3.4 (d)	3.5 (d)	3.6 (d)	3.7 (d)	3.8 (b)	3.9 (d)
3.10 (a)	3.11 (c)	3.12 (b)	3.13 (a)	3.14 (c)	3.15 (a)	3.16 (a)	3.17 (a)	3.18 (c)
3.19 (c)	3.20 (d)	3.21 (d)	3.22 (a)	3.23 (c)	3.24 (b)	3.25 (a)	3.26 (c)	3.27 (a)
3.28 (b)	3.29 (c)	3.30 (a)	3.31 (b)	3.32 (b)	3.33 (b)	3.34 (c)	3.35 (a)	3.36 (c)
3.37 (a)	3.38 (d)	3.39 (b)	3.40 (b)	3.41 (a)	3.42 (b)	3.43 (a)	3.44 (a)	3.45 (d)
3.46 (c)	3.47 (b)	3.48 (c)	3.49 (d)	3.50 (d)	3.51 (a)	3.52 (a)	3.53 (c)	3.54 (a)

Explanations Traffic Engineering

3.1 (d)

Amber or yellow time,

$$Y = \frac{S + W + l}{V}$$

where,

S is safe stopping distance

W is distance from stop line until vehicle (rear) is clear

l is length of the vehicle

V is approach speed

$\frac{S}{V}$ is the stopping time for approaching vehicle

3.2 (c)

This is floating car method for speed and delay studies which gives flow of vehicles,

$$q = \frac{n_a + n_y}{t_a + t_w}$$

n_a is average number of vehicles counted in the direction of the stream when test vehicle travels in the opposite direction

n_y is the average number of vehicles overtaking the test vehicle minus the number of vehicles overtaken when the test vehicle is in the direction of stream

t_w, t_a are average journey times when the test vehicle is travelling with stream and against stream respectively.

Given n_a = 303 vehicle

n_y = 17 vehicles

$$t_a = t_w = \frac{5}{70} \text{hrs}$$

$$\therefore \quad q = \frac{17 + 303}{\frac{5}{70} + \frac{5}{70}} = 2240 \text{ vph}$$

3.3 (b)

$$\text{Monthly expansion factor} = \frac{650 \times 7}{3625} = 1.25$$

3.4 (d)

Divisional Islands are type of traffic islands constructed as channels to divide opposing or same direction traffic streams, usually through movements in multilane. These alert the drivers about the intersection, reduce the night glare and help prevent accidents.

3.5 (d)

Only for 29 hours in a year the traffic volume exceeds the design hourly volume. Therefore, per cent time $= \frac{29 \times 100}{365 \times 24} = 0.33$

3.6 (d)

As per definition of trip, in this case it is a single trip only as the journey is being made to reach the office. It is a one way movement and has a single purpose using different modes of transport.

3.7 (d)

Passenger Car Equivalent (PCE) or Passenger Car Unit (PCU) is a metric used in Transportation Engineering, to assess traffic-flow rate on a highway. A Passenger Car Equivalent is essentially the impact that a mode of transport has on traffic variables (such as headway, speed, density) compared to a single car. For example, typical values of PCE (or PCU) are:

- private car (including taxis or pick-up) 1
- motorcycle 0.5
- bicycle 0.2
- bus, tractor, truck 3.5

Highway capacity is measured in PCE/hour daily

3.8 (b)

Pneumatic tube is automatic device used for counting vehicles. Video recording is a photographic method. Watch measures traffic volume with manual count. Multiple pen recorder is combination of manual and mechanical methods.

3.9 (d)

The O-D survey is an initiative to collect current and reliable information about trip patterns and travel choices of residents which is again classified into the following:

- Home Interview Survey for Internal Movements and internal-external movements.
- Road Side O-D Surveys for through movement, external-Internal movements and also internal-External movement. These are also called non-house hold surveys.

3.10 (a)

Pre-timed signals are fixed signals. Linked system means alternate system.

Semi-actuated signals is usually used when high traffic road intersects low traffic road. Signal time is affected due to detection of vehicles on some but not all approaches.

3.11 (c)

Prohibitory signs are circular in shape and white in colour with a red border.

3.11 (b)

The effective green time denotes that in the given total time (green time and amber time) saturation flow occurs only for small length of time. The initial delay for start of first vehicle and clearance of vehicles in the amber time causes flow to be reduced. Thus

Effective green time, $(G_e) = G + A - L$

Actual green time, $(G) = 25$ sec

Amber time $(A) = 3$ sec

Total lost time, $L =$ Initial lost time (L_i) + Final lost time (L_f)

$L_i = 3$ sec $\quad L_f \leq A$

In the absence of any data on the final lost time, it can be taken to be equal to amber time

$$\therefore \quad G_e = 25 + 3 - (3 + 3) = 22 \text{ sec}$$

But there is no such choice. So (b) can be chosen as the answer.

3.16 (a)

$$\text{Jam density} = \frac{1000}{\text{Space headway(H)}} = \frac{1000}{6.25} = 160 \text{ veh/km}$$

$$\text{Maximum flow} = \frac{\text{Jam density} \times \text{Free speed}}{4} = \frac{160 \times 65}{4} = 2600 \text{ vph}$$

3.17 (a)

Cloverleaf interchange is far superior than diamond interchange or rotary interchange. Of all the interchanges complete clover leaf interchange fulfils all the requirements of turning traffic involving the simplest traffic manoeuvers viz. diverging to the left and merging from the left by providing four indirect ramps.

3.20 (d)

Design factors considered in design of rotary are:

(i) Design speed
(ii) Width of rotary roadway
(iii) Radius and shape of central island
(iv) Weaving length and weaving angle
(v) Camber and super-elevation
(vi) Capacity of rotary.
(vii) Traffic signs and lighting
(viii) Radius of entry and exit curves

3.21 (d)

A modal share (also called mode split, mode-share, or modal split) is the percentage of travellers using a particular type of transportation or number of trips using said type. In freight transportation, this may be measured in mass.

Modal share is an important component in developing sustainable transport within a city or region. In recent years, many cities have set modal share targets for balanced and sustainable transport modes, particularly 30% of non-motorized (cycling and walking) and 30% of public transport. These goals reflect a desire for a modal shift, or a change between modes, and usually encompasses an increase in the proportion of trips made using sustainable modes.

3.22 (a)

Length of rotary roadway between two channelizing islands of adjacent intersecting legs is known as weaving length and angle between path of vehicle entering rotary and path of vehicle leaving rotary is known as weaving angle.

3.23 (c)

Speed and volume studies are used for highway capacity design and functional design. Spot speed studies and speed and delay study etc are used for such design. *Traffic volume studies* are mainly carried out to obtain factual data concerning the movement of vehicles at selected point on the street or highway system.

3.24 (b)

Origin and destination survey are useful:

- in locating major routes of city
- locate intermediate stops of public transport
- to understand travel patterns and characteristics
- provide input to travel demand model development,
- forecasting, and planning for area wide transportation infrastructure needs and services
- monitor progress in implementing transportation policies

3.25 (a)

Enoscope consists of a simple open housing containing a mirror mounted on a tripod at the side of the road in such a way that an observer's line of sight turned through 90°. The observer stands at one end of section at the other end enoscope is placed and measure the time taken by vehicle to cross the section which gives the spot speed.

3.26 (c)

When there is continuous operation of group of vehicles simple progressive system is useful which permit vehicle operation according to a time schedule along the main road at a reasonable speed

3.27 (a)

3.28 (b)

Delineators are light-retro-reflecting devices mounted by the side of the roadway to indicate the roadway alignment, especially for night driving. The three broad types of delineators are roadway indicator, hazard markers and object markers.

3.29 (c)

Relative speed is the vector difference in the velocities of two vehicles in the same flow and is the sum of the speeds of approaching vehicles in the opposite direction. Relative speed is dependent upon the absolute speed of intersecting vehicles and the angles between then. When the angle of merging is small, the relative speed will also be low. If there is a collision between two vehicles at small angle at about the same speed or at low relative speed, the impact would be much less than when vehicle collide at high relative speed. As relative speed increases, the judgement of drivers regarding time and distance is likely to be more inaccurate, thus increasing the possibility and severity of accidents. Thus, in intersection design care has to be taken to keep the relative speed low.

3.30 (a)

Traffic capacity is expressed as the maximum

number of vehicle in a lane or a road that can pass a given point in unit time, usually an hour, i.e., vehicles per hour per lane or roadway. Traffic capacity and traffic volume has same units, difference between the two is that traffic volume represents the actual rate of flow of the traffic and responds to the variation in the traffic demand, while capacity indicates a capability or maximum rate of flow with a certain level of service characteristics that can be carried by the road.

3.31 (b)

Theoretical capacity

$$C = \frac{1000V}{S}, \quad S = SSD + L$$

$$= 0.278 \times V \times t + \frac{(0.278 \times V)}{2\mu g} + L$$

$$= 0.278 \times 70 \times 2 + \frac{(0.278 \times 70)^2}{2 \times 0.4 \times 9.81} + 8$$

(Assume $\mu = 0.4$)

$$= 95.1736 \text{ m}$$

$$\therefore \quad \text{Capacity} = \frac{1000V}{S} = \frac{1000 \times 70}{95.1736}$$

$$= 735.5$$

Note: Here breaking disance have been considered in calculating the space headway so that the answer matches with the given options.

3.33 (b)

Theoretical maximum capacity,

$$C = \frac{1000V}{S} \text{ veh/hr}$$

V = 40 kmph

S = Minimum clear distance between two vehicles

S = 0.2 V + 6

= 0.2 × 40 + 6 = 14 m

$$C = \frac{1000 \times 40}{14} = 2857.14 \text{ Veh/hr}$$

3.34 (c)

Initial traffic,

$$A = P\left(1 + \frac{r}{100}\right)^n$$

P = 400 commercial veh./day

r = Rate of traffic growth = 7%

n = 3 years completion time

$$A = 400\left(1 + \frac{7}{100}\right)^3 = 490 \text{ CVD}$$

3.36 (c)

An Intelligent Traffic Management System aims at managing traffic effectively during emergencies through the use of cutting-edge communication and processing technologies and appropriate intelligent system algorithms. It is related with Classified traffic volume studies.

> **Note: Here answer given by UPSC is 'a' but the correct answer is 'c', as per Traffic Engineering and Transport Planning by L.R. Kadiyali, the intelligent transport system requires data with respect to traffic volume, headway, load carried.**

3.37 (a)

Metro : Mono : Private Car

$(0.55 - 0.55 \times 0.55) : (0.55 \times 0.55) : (0.45)$

0.25 : 0.30 : 0.45

3.38 (d)

Types of co-ordinated signal system:

1. **Simultaneous system:** Under this system, all the signals along a given street always displays the same indication to the same traffic stream at the same time. The division of cycle time is same at the intersections.
2. **Alternate system:** Under this system, consecutive signal installations along a given road show contrary indications at the same time. This permits the vehicles to travel one block in half the cycle time.
3. **Simple progressive system:** With this system, the various signals along a street display green aspects in accordance with a time schedule to permit, as nearly as possible, continues operation of plantoons of vehicles along the street at a planned rate of motion.
4. **Flexible progressive system:** This system is an improvement over the simple progressive system with following provisions:
   (a) It is possible to vary the cycle time and division at each signal depending upon traffic.
   (b) It is possible to introduce flashing of shut-down during offf peak hours.

3.39 (b)

Sum of critical flow ratio

$$Y = y_a + y_b$$

$$= \frac{500}{1600} + \frac{300}{1600}$$

$$= 0.3125 + 0.1875 = 0.5$$

Optimum cycle time

$$C_0 = \frac{1.5L+5}{1-Y} = \frac{1.5 \times 16 + 5}{1-0.5} = 58 \text{ sec}$$

3.40 (b)

u = 40 kmph; L = 12.2 m; f = 0.7

$$f_{obtained} = \frac{u^2}{2gL} = \frac{(0.278 \times 40)^2}{2 \times 9.81 \times 12.2} = 0.5166$$

$$\eta_{Braking} = \left(\frac{f_{obtained}}{f_{max}} \times 100\right)$$

$$= \left(\frac{0.5166}{0.7} \times 100\right) = 73.8\% \approx 74\%$$

3.41 (a)

Automatic counters-cum-classifiers (ATCC) does not help in classifying the type of vehicle.

3.42 (b)

Vehicle number method is not a part of speed and delay studies.

3.44 (a)

Properly designed traffic signals at intersections have the following advantages:

(a) Provide orderly movement of traffic at the intersection.

(b) The quality of traffic flow is improved by forming compact platoons of vehicles, provided all the vehicles move at approximately the same speed.

(c) Reduction in accidents due to crossing conflict, notably the right angled collisions.

(d) Traffic handling capacity is highest amount the different types of intersections at grade.

(e) Provide a chance to traffic of minor road to across the continuous traffic flow os the main road at reasonable intervals of time.

(f) Pedestrians can cross the roads safely at the signalized intersection.

(g) When the signal system is properly coordinated, there is a reasonable speed along the major road traffic.

(h) Automatic traffic signal may work out to be more economical when compared to manual control.

3.45 (d)

Given: Free mean speed,

$$V_f = 80 \text{ kmph}$$

Average spacing between vehicle

$$S = 6.9 \text{ m}$$

$$S = 0.2V + l$$

Since $V = 0,\ S = l$

$$k_J = \frac{1000}{l} = \left(\frac{1000}{6.9}\right)$$

Traffic capacity,

$$q_{max} = \left(\frac{1}{4} v_f k_j\right) = \left(\frac{1}{4} \times 80 \times \frac{1000}{6.9}\right)$$

$$= 2898.55 \text{ veh/hr} \simeq 2900 \text{ veh/hr}$$

3.46 (c)

Street width = 15

Mounting height = 7.5 m

Lamp size = 6000 lumen

$$\frac{\text{Pavement width}}{\text{Mounting height}} = \frac{15}{7.5} = 2.0$$

Coefficient of utilization = 0.44

Maintenance factor = 0.8

Spacing

$$= \frac{\text{Lamp Lumen} \times \text{coefficient of utilisation} \times \text{Maintenance factor}}{\text{Average Lux} \times \text{width of road}}$$

$$= \frac{6000 \times 0.44 \times 0.8}{6 \times 15} = 23.46 \text{ m} \simeq 23 \text{ m}$$

3.48 (c)

The average speed of vehicle in a certain road length at any time.

3.52 (a)

In freeway management systems, the major difficulty is establishing human surveillance with integrated electronic detection.

3.53 (c)

Vehicle damage factor (VDF) is a multiple to convert the number of commercial vehicles of different axle loads and axle configurations to the number of standard axle load repetitions.

■■■■

4 Pavement Design

4.1 Which one of the following is useful in functional evaluation of pavement?
(a) PCU (b) PSI
(c) PIEV (d) Bankelman beam
[ESE : 2000]

4.2 If the modulus of subgrade reaction of a soil is 10 kg/cm^3 when tested with a 30 cm diameter plate, the corrected modulus of subgrade reaction for the standard diameter plate will be
(a) 4 (b) 15
(c) 20 (d) 25
[ESE : 2000]

4.3 **Assertion (A):** The critical combination of stresses on a cement concrete pavement during summer is given by: load stress – warping stress + frictional stress.
Reason (R): The critical combination of stresses in a cement concrete pavement during winter is given by: load stress+ warping stress + frictional stress.
(a) both A and R are true and R is the correct explanation of A
(b) both A and R are true but R is not a correct explanation of A
(c) A is true but R is false
(d) A is false but R is true
[ESE : 2000]

4.4 Modulus of subgrade reaction using 30 cm diameter plate is obtained as 200 N/cm^3. The value of the same (in N/cm^3) using the standard plate will be
(a) 500 (b) 200
(c) 85 (d) 80
[ESE : 2001]

4.5 Consider the following statements regarding soil stabilization:
1. Subgrade should be treated if it has a soaked CBR of 5 or less.
2. Sub-base, after treatment, should have plasticity index not less than 5.
3. Base courses after treatment should have plasticity index of not more than 2 and a CBR in excess of 100.

Which of these statements is/are correct?
(a) 1, 2 and 3 (b) 1 and 3
(c) 2 and 3 (d) 1 alone
[ESE : 2001]

4.6 Consider the following factors:
1. Period of construction, winter/summer
2. Degree of foundation roughness
3. Slab thickness
4. Reinforced/unreinforced

Which of these factors are considered as per IRC for obtaining the maximum expansion joint spacing in rigid pavements?
(a) 1, 2 and 3 (b) 2, 3 and 4
(c) 2 and 3 (d) 1 and 4
[ESE : 2002]

4.7 With addition of lime in soil
(a) *LL* increases and *PL* decreases
(b) Plasticity Index increases
(c) *LL* change very slightly and *PL* increases
(d) *LL* and *PL* both decrease
[ESE : 2002]

4.8 On a National Highway pavement section mean value of Bankelman beam deflection was obtained as 1.45 mm with standard deviation σ of 0.107 mm. Pavement temperature measured was 37°C and moisture correction was obtained as 1.6. What is the value of characteristic deflection considering the value of 2σ?
(a) 2.72 mm (b) 2.69 mm
(c) 2.63 mm (d) 1.94 mm
[ESE : 2002]

4.9 Match **List-I** (Tests) with **List-II** (Properties) and select the correct answer using the codes given below the lists:

List-I

A. CBR test
B. Plate bearing test
C. Triaxial test
D. Stabilometer and Cohesion-meter test

List-II

1. Modulus of sub-grade reaction
2. Arbitrary soil strength
3. Exudation and expansion pressure
4. Shear parameters

Codes:

	A	B	C	D
(a)	1	2	3	4
(b)	2	1	3	4
(c)	1	2	4	3
(d)	2	1	4	3

[ESE : 2003]

4.10 California Bearing Ratio (CBR) is a
(a) measure of soil strength
(b) method of soil identification
(c) measure to indicate the relative strengths of paving materials
(d) measure of shear strength under lateral confinement

[ESE : 2005]

4.11 Which of the following factors have to be considered for the design of the flexible pavement for a highway?
1. Design wheel load
2. Strength of pavement component material
3. Expansion joints
4. Climatic factors

Select the correct answer using the codes given below:
(a) 1, 2 and 3 (b) 2, 3 and 4
(c) 1, 3 and 4 (d) 1, 2 and 4

[ESE : 2005]

4.12 Match **List-I** (Method of design for flexible pavement) with **List-II** (Principle) and select the correct answer using the codes given below the lists:

List-I

A. Group Index Method
B. CBR Method
C. US Navy Method
D. Asphalt Institute Method

List-II

1. Semi-theoretical
2. Quasi-rational
3. Empirical method using soil classifi-cation test
4. Empirical method using soil strength test

Codes:

	A	B	C	D
(a)	3	1	4	2
(b)	2	4	1	3
(c)	3	4	1	2
(d)	2	1	4	3

[ESE : 2006]

4.13 The mix design for pavement concrete is based on the
(a) flexural strength
(b) characteristic compressive strength
(c) shear strength
(d) bond strength

[ESE : 2006]

4.14 What are the maximum value of CBR and minimum value of *GI* of any material, respectively?
(a) 100, 0 (b) 100, 20
(c) 50, 5 (d) 10, 0

[ESE : 2007]

4.15 Which one of the following criteria is used for obtaining the value of modulus of subgrade reaction from the plate bearing test data?
(a) Slope of pressure settlement graph.
(b) Pressure corresponding to the settlement of 1.25 mm.
(c) Pressure corresponding to a pressure of 1.25 kg/cm^2.
(d) Pressure corresponding to the settlement of 1.50 mm.

[ESE : 2007]

4.16 Which of the following factors are used for calculating temperature stress at the critical edge region in rigid pavement design?
1. Maximum temperature difference between summer and winter
2. Coefficient of thermal expansion of concrete
3. Slab length
4. Slab width

Select the correct answer using the codes given below:

(a) 1, 2 and 3 (b) 2, 3 and 4
(c) 1 and 2 only (d) 1 and 3 only

[ESE : 2007]

4.17 IRC code No. 37-1985 deals with which one of the following?
(a) Design of rigid pavements, taking ESWL and CBR into account.
(b) Design of rigid pavements, taking axle load and CBR into account.
(c) Design of flexible pavement, taking ESWL and CBR into account.
(d) Design of flexible pavement taking cumulative axle loads and CBR into account.

[ESE : 2008]

4.18 **Assertion (A):** California Bearing Ratio Test is carried out to evaluate the stability of soil sub grade and other flexible pavement materials.
Reason (R): It is essential that at no time are the soil sub grade as well as other flexible pavement materials over stressed.
(a) both A and R are true and R is the correct explanation of A
(b) both A and R are true but R is not a correct explanation of A
(c) A is true but R is false
(d) A is false but R is true [ESE : 2008]

4.19 Which one of the following methods is used in the design of rigid pavements?
(a) CBR method
(b) Group index method
(c) Westergaard's method
(d) McLeod's method [ESE : 2009]

4.20 Radius of relative stiffness of cement concrete pavement does not depend upon which one of the following?
(a) Modulus of subgrade reaction
(b) Wheel load
(c) Modulus of elasticity of cement concrete
(d) Poisson's ratio of concrete

[ESE : 2009]

4.21 For conditions prevailing in India, at which location in a cement concrete pavement will the combined stresses due to traffic wheel load and temperature have to be critically checked during design?
(a) Corner
(b) Corner and interior
(c) Corner and edge
(d) Corner, edge and interior

[ESE : 2009]

4.22 Which one of the following sets of factors is related to design of thickness rigid pavement by Westergaard method?
(a) CBR value and stiffness index of soil
(b) Deflection factor and traffic index
(c) Swelling index and bulk modulus
(d) Radius of relative stiffness and modulus of subgrade reaction [ESE : 2009]

4.23 Consider the following in relation to group index of soil:
1. Liquid limit
2. Sandy loam
3. Plasticity index
4. Percent passing 75 microns sieve

Which of the above is/are used for estimating the group index?
(a) 1 only (b) 1 and 2
(c) 2 and 3 (d) 1, 3 and 4

[ESE : 2009]

Directions: The following items consists of two statements; one labelled as **'Assertion (A)'** and the other as **'Reason (R)'**. You are to examine these two statements carefully and select the answers to these items using the codes given below:

Codes:
(a) both A and R are true and R is the correct explanation of A
(b) both A and R are true but R is not a correct explanation of A
(c) A is true but R is false
(d) A is false but R is true

4.24 **Assertion (A):** Dowel bars are provided at expansion joints and sometimes also at contraction joints in cement concrete slabs.
Reason (R): Longitudinal joints in cement concrete pavements are constructed with tie bars.

[ESE : 2009]

4.25 **Assertion (A):** Worn out (smooth) tyres offer higher friction factors on dry pavements than new tyres with treads well intact.
Reason (R): Reduced pneumatic pressure is held in tubes which carry smooth tyres over them.

[ESE : 2010]

4.26 **Assertion (A):** Tie bars are used in cement concrete slabs across the longitudinal joints.
Reason (R): Tie bars are designed to act as load transfer devices. [ESE : 2010]

4.27 In a concrete pavement, during summer at and soon after mid-day, the combined stress at the interior of the slab is equal to
(a) Wheel load stress + Temperature warping stress + Sub grade resistant stress
(b) Wheel load stress + Temperature warping stress – Sub grade resistant stress
(c) Wheel load stress – Temperature warping stress + Sub grade resistant stress
(d) Wheel load stress – Temperature warping stress – Sub grade resistant stress
[ESE : 2011]

4.28 Which of the following correspond to the recommendations of IRC for pavement thickness determination by CBR method?
1. CBR tests are to be conducted in-situ
2. Static compression is best adopted
3. The top 50 cm of subgrade should be compacted to as near the proctor density as possible
(a) 1, 2 and 3 (b) 1 and 2 only
(c) 2 and 3 only (d) 1 and 3 only
[ESE : 2011]

4.29 The corrected modulus of sub-grade reaction for standard diameter plate is 6.0 kg/cm^3. What would be the modulus of sub-grade reaction of the soil when tested with a 30 cm diameter plate?
(a) 15 kg/cm^3 (b) 25 kg/cm^3
(c) 30 kg/cm^3 (d) 60 kg/cm^3
[ESE : 2013]

4.30 What is the deflection at the surface of a flexible pavement due to a wheel load of 40 kN and a tyre pressure of 0.5 MPa? The value of E for pavement and subgrade is 20 MPa.
(a) 15 mm (b) 11 mm
(c) 9 mm (d) 6 mm [ESE : 2014]

4.31 In a flexible pavement
(a) Vertical compressive stresses decrease with depth of the layer
(b) The vertical compressive stress is the maximum at the lowest layer
(c) Tensile stresses get developed
(d) Maximum stress induced by a given traffic load is dependent on the location of the load on the pavement surface
[ESE : 2014]

4.32 Bankleman beam deflection method is used for design of
(a) Rigid overlays on rigid pavements
(b) Rigid overlays on flexible pavements
(c) Flexible overlays on flexible pavements
(d) Flexible overlays on rigid pavements
[ESE : 2014]

4.33 A collapsible soil sub-grade sample was tested using Standard California Bearing Ratio apparatus; and the observations are given below:

S.I. No.	Load	Penetration
1.	60.55 kg	2.5 mm
2.	80.55 kg	5.0 mm

Taking the standard assumptions regarding the load penetration curve, CBR value of the sample will be taken as
(a) 3.9% (b) 4.0%
(c) 4.4% (d) 5.5% [ESE : 2016]

4.34 What is the critical thickness of a prestressed concrete pavement (using Westergaard's Corner Load Formula) to support a maximum wheel load of 4200 kg? Allow 10% for impact. Tyre pressure may be taken as 7 kg/cm^2. Assume flexural strength of concrete as 50 kg/cm^2, factor of safety as 2, subgrade reaction for plastic mix road as 6 kg/cm^3. and modulus of elasticity as 3×10^5 kg/cm^2.
(a) 19.6 cm (b) 21.6 cm
(c) 23.6 cm (d) 25.6 cm
[ESE : 2016]

4.35 In revised CBR design method recommended by the IRC for the design of flexible pavement, the total thickness depends upon
(a) Only the CBR value of the soil
(b) The CBR value of the soil and magnitude of wheel load
(c) The CBR value of the soil and cumulative standard axle loads
(d) The CBR value of the soil and number of commercial vehicles passing per day
[ESE : 2017]

4.36 Consider the following statements regarding pavements:

1. Rigid pavements are more suitable than flexible pavements for stage construction.
2. Rigid pavements are more affected by temperature variations than flexible pavements.
3. In a flexible pavement, any deformation in the top layers is transferred to underlaid layers; but, in rigid pavements, there is slab or beam action due to which any deformation is only in the top layer of the concrete slab.

Which of the above statements are correct?

(a) 1 and 2 only (b) 2 and 3 only
(c) 1 and 3 only (d) 1, 2 and 3

[ESE : 2017]

4.37 As per IRC 37 : 2012, the fatigue life of flexible pavement consisting of granular base and sub-base depends upon

1. Resilient Modulus of bituminous layers
2. Horizontal tensile strain at the bottom of bituminous layer
3. Mix design of bitumen
4. Vertical subgrade strain

Which of the above statements are correct ?

(a) 1, 2 and 4 only (b) 1, 3 and 4 only
(c) 1, 2 and 3 only (d) 2, 3 and 4 only

[ESE : 2018]

4.38 Consider the following data with respect to the design of flexible pavement:

Design wheel load = 4200 kg
Tyre pressure = 6.0 kg/cm^2
Elastic modulus = 150 kg/cm^2
Permissible deflection = 0.25 cm

(take $\pi^{1/2} = 1.77$, $\pi^{-1/2} = 0.564$, $\frac{1}{\pi} = 0.318$, and $\pi^2 = 9.87$)

The total thickness of flexible pavement for a single layer elastic theory will be nearly

(a) 42 cm (b) 47 cm
(c) 51 cm (d) 56 cm

[ESE : 2018]

4.39 Which one of the following is **not** the correct type of critical load position in pavement slab design for the load on the pavement surface?

(a) Interior loading
(b) Edge loading
(c) Eccentric loading
(d) Corner loading

[ESE : 2020]

4.40 Which of the following statements are correct regarding Westergaard's concept for temperature stresses?

1. During the day, the top of the pavement slab gets heated under the sun light when the bottom of the slab becomes hot.
2. During summer season as the mean temperature of the slab increases, the concrete pavement expands towards the expansion joints.
3. Due to frictional force at the interface, compressive stress is developed at the bottom of the slab as it tends to expand.

(a) 1 and 2 only (b) 2 and 3 only
(c) 1 and 3 only (d) 1, 2 and 3

[ESE : 2020]

4.41 Consider the following statements related to road pavements:

1. Deflections measured near cracks are normally much lower than the measurements in non-distressed areas.
2. Deflection measurements near longitudinal joints, transverse joints or corners are higher than those measured at mid-slab for concrete pavements.
3. Thermal and moisture gradient in the vertical direction of concrete slabs does not have any influence of deflection measurements.
4. Measurements taken at night or in the early morning are considerably different from those obtained in the afternoon.

Which of the above statements are not correct?

(a) 1 and 2 only (b) 1 and 3 only
(c) 2 and 4 only (d) 1, 2 and 4 only

[ESE : 2021]

4.42 The thickness design of pavement mainly depends on

(a) the pavement materials
(b) the climatic factors
(c) the design wheel road
(d) the subgrade soil

[ESE : 2022]

4.43 A plate load test is carried out on submerged soil using a 300 mm radius rigid plate. A load of 5 Tons resulted in a deflection of 1.20 mm. What is the elastic modulus of the soil by considering the Poisson's ratio as 0.50?

(a) 5216 kPa (b) 521.6 GPa
(c) 52.16 MPa (d) 52160 Pa

[ESE : 2022]

4.44 Consider the following statements related to IRC recommendations for the CBR method of design (IRC: 37-1970):

1. The CBR tests should be performed on remoulded soils on the field.
2. For the design of new roads, the subgrade soil sample should be compacted at OMC to proctor density
3. In new construction, the CBR test samples may be soaked in water for four days period before testing.

Which of the above statements are correct?

(a) 1 and 2 only (b) 2 and 3 only
(c) 1 and 3 only (d) 1, 2 and 3

[ESE : 2022]

4.45 The longitudinal joints in cement concrete pavements are constructed with suitable

(a) tie bars (b) torsion rods
(c) small spacing (d) shear rods

[ESE : 2023]

4.46 Which one of the following tests is carried out using a relatively large diameter plate to evaluate the load supporting capacity of pavement layers?

(a) California bearing ratio test
(b) California resistance value test
(c) Triaxial compression test
(d) Plate bearing test

[ESE : 2023]

Answers Pavement Design

4.1 (b)	4.2 (a)	4.3 (d)	4.4 (d)	4.5 (b)	4.6 (a)	4.7 (c)	4.8 (c)	4.9 (d)
4.10 (c)	4.11 (d)	4.12 (c)	4.13 (a)	4.14 (a)	4.15 (b)	4.16 (b)	4.17 (d)	4.18 (b)
4.19 (c)	4.20 (b)	4.21 (c)	4.22 (d)	4.23 (d)	4.24 (b)	4.25 (b)	4.26 (c)	4.27 (b)
4.28 (c)	4.29 (a)	4.30 (d)	4.31 (a)	4.32 (c)	4.33 (c)	4.34 (a)	4.35 (c)	4.36 (a)
4.37 (c)	4.38 (c)	4.39 (c)	4.40 (b)	4.41 (b)	4.42 (c)	4.43 (c)	4.44 (d)	4.45 (a)
4.46 (d)								

Explanations Pavement Design

4.1 (b)

Present serviceability index (PSI) values are used for functional evaluation of pavement surface conditions. Physical measurements are made on pavement surface and mathematical models

4.2 (a)

Standard plate is of 75 cm diameter.

$$K_1 d_1 = K_2 d_2$$

$$10 \times 30 = K_2 \times 75$$

$$\Rightarrow \quad K_2 = 4 \text{ kg/cm}^3$$

K_1, K_2 are modulus of subgrade reaction for diameter 'd_1' and 'd_2' respectively.

4.3 (d)

During summer: Critical combination of stresses in concrete pavement = Load stress + Warping stress- Frictional stress

During winter: Critical combination of stresses in concrete pavement = Load stress + Warping stress+ Frictional stress

4.4 (d)

$$K_1 d_1 = K_2 d_2$$

$$\therefore \quad K_2 = \frac{30}{75} \times 200 = 80 \text{ N/cm}^3$$

4.5 (b)

Following criteria were given for soil stabilisation in design process of flexible pavements by Dunlop:

(i) Sub-grades are only stabilized if they have a soaked CBR of 5 or less and amount of cement used is limited to keep the final CBR below 50.

(ii) Base courses after treatment should have plasticity index of 2 or less and CBR in excess of 100.

(iii) Sub-bases after treatment should have plasticity index of 5 or less.

4.6 (a)

Expansion joint spacing is designed based on the maximum temperature variation expected and the width of the joint. The width or the gap in expansion joint depends upon the length of the slab. The maximum spacing recommended for 25 mm wide expansion joints is 140 m when the foundation is rough for all slab thicknesses. When the foundation surface is smooth, the maximum spacing may be 90 m for slab thicknesses up to 20 cm and spacing of 120 m for slab thickness up to 25 cm, when the pavement is constructed in summer. However, when the pavement is constructed in winter, the above spacings may be restricted to 50 m and 60 m respectively.

4.7 (c)

With addition of lime to clayey soil there is slight decrease in liquidity limit and considerable increase in plastic limit which reduces plasticity index.

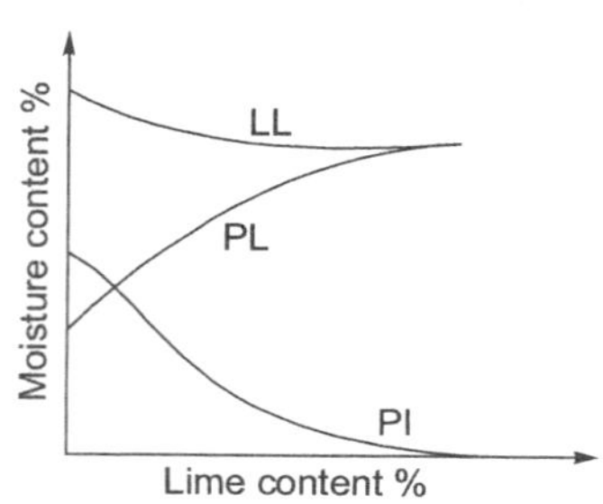

First the rate of increase of plastic limit is rapid which then slows down after lime fixation point.

4.8 (c)

Characteristic deflection

$= \bar{D} + 2\sigma$

$= 1.45 + 2 \times 0.107$

$= 1.664$ mm

Since temperature is above 35°C, so temperature correction

$= (37 - 35) \times 0.0065$

$= 0.013$ mm

Deflection after temperature correction

$= 1.664 - 0.013$

$= 1.651$ mm

Corrected deflection for subgrade moisture

$= 1.651 \times 1.6 = 2.64$ mm

4.10 (c)

CBR test is used to evaluate the stability of soil subgrade and other flexible pavement materials. It is an arbitrary strength test and hence can not be used to evaluate the shear strength parameters. In essence it is a measure to indicate relative strengths of paving materials and not the absolute strength.

4.11 (d)

some factors to be considered for design of pavement are:

(i) Design wheel load

(ii) Strength of pavement component material

(iii) Climatic factors

(iv) Subgrade soil

(v) Environmental factors

4.12 (c)

Various approaches of flexible pavement design are classified as:

(a) Empirical method,

(b) Semi-empirical or semi theoretical methods,

(c) Theoretical methods.

US Navy method is semi-theoretical method based on Burmister approach whereas asphalt in-situ method is quasi rational based on Burmister approach of designing flexible pavements.

4.13 (a)

The mix design for pavement concrete (Highway/

Runway) is based on the flexural strength (f_c) instead of characteristic compressive strength (f_{ck}). Because the soil beneath the concrete pavement may be scoured by flood or some other reasons and the concrete pavement will behave as a beam and the load on pavement is to be limited by flexural strength of concrete. Required flexural strength is converted into corresponding characteristic compressive strength by the expression $f_c = 0.7\sqrt{f_{ck}}$ and the concrete is designed for the required compressive strength as usual.

4.15 (b)

The modulus of sub-grade reaction(K) in plate bearing test may be defined as pressure sustained per unit deformation of sub-grade as specified deformation using specified plate size.

$$K = \frac{P}{0.125} \text{ kg/cm}^3$$

where P is the pressure sustained by the rigid plate of 75 cm diameter at a deflection of 0.125 cm.

IRC specifies that the K value be measured at 1.25 mm settlement.

4.16 (b)

For calculating temperature stress factors considered are:

(i) Maximum temperature difference between top and bottom of slab
(ii) Slab width
(iii) Slab length
(iv) Coefficient of thermal expansion of concrete per degree Celsius
(v) Modulus of elasticity of concrete
(vi) Poisson's ratio

4.19 (c)

CBR method, Group Index method, McLeod method and Burmister method are some of the methods which are used in the design of flexible pavements.

4.20 (b)

Radius of relative stiffness is given by

$$l = \left[\frac{Eh^3}{12K(1-\mu^2)}\right]^{1/4}$$

4.21 (c)

The cumulative effect of different stress gives rise to the following three critical cases :

- **Summer, mid-day:** The critical stress is for edge region given by $\sigma_{critical} = \sigma_e + \sigma_{te} - \sigma_f$ (at bottom)
- **Winter, mid-day:** The critical combination of stress is for the edge region given by $\sigma_{critical} = \sigma_e + \sigma_{te} + \sigma_f$ (at bottom)
- **Mid-nights:** The critical combination of stress is for the corner region given by $\sigma_{critical} = \sigma_e + \sigma_{te}$ (at bottom)

  where; σ_e is stress due to wheel load, σ_{te} is warping stress and σ_f is frictional stresses.

4.22 (d)

While deciding thickness of rigid pavement by Westergaard method factors such as Poisson's ratio, radius of relative stiffness, modulus of sub-grade reaction etc are considered.

4.23 (d)

Group Index is a number assigned to the soil based on its physical properties. It is based on percentage material passing 75 micron sieve, Liquid limit and plasticity index. It varies from a value of 0 to 20, lower the value higher is the quality of the sub-grade and greater the value, poor is the sub-grade.

Group index, GI = 0.2a +0.005ac+ 0.01bd

a= percentage of soil passing 0.075 mm sieve in excess of 35 per cent, not exceeding 75; $0 \le a \le 40$

b= percentage of soil passing 0.075 mm sieve in excess of 15 per cent, not exceeding 55; $0 \le b \le 40$

c= value of liquid limit in excess of 40 and less than 60; $0 \le c \le 20$

d=value of plasticity index exceeding 10 and not more than 30; $0 \le d \le 10$

4.24 (b)

Dowel bars are short steel bars that provide a mechanical connection between slabs without restricting horizontal joint movement. In cement concrete slabs, dowel bars are provided at expansion joints and sometimes at contraction joints.

Longitudinal joints in cement concrete pavements are constructed with tie bars. Tie bars are either deformed steel bars or connectors used to hold the faces of slabs in contact. They are designed to withstand tensile stress and ensure that two slabs remain together.

4.25 (b)

The main factors affecting the friction coefficient are road surface texture and condition and tire composition.

Condition of the roadway refers to the presence of snow, ice, sand and other lubricants. The presence of any of these will make the surface more slippery, which reduces the coefficient of friction.

Texture of the road surface refers to the natural roughness of the asphalt. A relatively new asphalt roadway has a "sharper" surface with a high coefficient of friction. An older well-travelled asphalt roadway has a smoother, slicker surface with a lower coefficient of friction.

Tire composition and wear plays a role in determining the coefficient. Tests show that on a clean, dry surface, bald; worn out tires actually will skid to a stop more quickly than newer, treaded tires. However, on a wet roadway, treaded tires are preferred.

4.26 (c)

Tie bars used across the longitudinal joints of cement concrete pavements. Tie bars ensure two adjacent slabs to remain firmly together. These bars are not designed to act as load transfer devices. Tie bars are thus designed to withstand tensile stresses, the maximum tensile force in tie bars being equal to the force required to overcome frictional force between the bottom of the adjoining pavement slab and the soil subgrade.

4.27 (b)

(i) During Summer: Critical combination of stresses = load stress + warping stress – frictional stress, at edge region.

(ii) During winter: The critical stress combination = load stress + warping sress + frictional stress, at edge region.

(iii) At corner region: The critical stress combination = load stress + warping stress, at corner regions.

4.28 (c)

Some of the important points recommended by the IRC for the CBR method of design (IRC : 37-1970) are:

(a) The CBR tests should be performed on remoulded soils in the laboratory. In-situ tests are not recommended for design purposes. The specimens should be prepared by static compaction wherever possible and otherwise by dynamic compaction. The standard test procedure should be strictly adhered to.

(b) For the design of new roads, the subgrade soil sample should be compacted at OMC to proctor density whenever suitable compaction equipment is available to achieve this density in the field. In the case of existing roads, the sample should be field density of subgrade soil (at OMC or at a field moisture content).

(c) In new constructions the CBR test samples may be soaked in water for four days period before testing.

(d) Atleast three samples should be tested on each type of soil at the same density and moisture content. The specified limits of maximum variation in CBR are 3% for CBR values upto 10% 5% for values 10 to 30 and 10% for values 30 to 60%.

(e) The top 50 cm of subgrade should be compacted atleast upto 95 to 100 percent of Proctor density.

4.29 (a)

When, $k = 6\text{ kg/cm}^3$

$a = 75\text{ cm}$

$$k_1 = \frac{6(75)}{30} = 15\text{ kg/cm}^3$$

4.30 (d)

$$\text{Deflection, } \Delta = \frac{1.5P.a}{E_s} \text{ (For flexible plate)}$$

Where, P = Contact pressure due to wheel load = 0.5 MPa

a = Radius of contact area

Now,

$$\text{Contact area} = \frac{\text{Wheel load}}{\text{Tyre pressure}} = \frac{40 \times 10^3 \text{ N}}{0.5 \text{ N/mm}^2}$$

$$= 80 \times 10^3 \text{ mm}^2$$

and, area,

$$\pi \times a^2 = 80 \times 10^3$$

$$a = \sqrt{\frac{80 \times 10^3}{\pi}} = 159.615 \text{ mm}$$

$$E_s = 250 \text{ MPa}$$

So, $$\Delta = \frac{1.5 \times 0.5 \times 159.617}{20} = 6 \text{ mm}$$

4.32 (c)

The Benkelman Beam Deflection Method is thus widely used for Evaluation of Structural Capacity of Existing Flexible Pavements and also for Estimation and Design of flexible overlays for Strengthening of any weak pavement for Highways. The Benkelman Beam measures the deflection of a flexible pavement under moving wheel loads.

4.33 (c)

$$CBR_{@2.5\text{ mm}} = \frac{60.55}{1370} \times 100 = 4.4\%$$

$$CBR_{@5.0\text{ mm}} = \frac{80.55}{2055} \times 100 = 3.9\%$$

Consider higher value.

4.34 (a)

Assume h = 20 cm and compute l

$$l = \left[\frac{Eh^3}{12k(1-\mu^2)}\right]^{1/4}$$

$$= \left[\frac{3 \times 10^5 \times 20^3}{12 \times 6[1-0.15^2]}\right]^{1/4}$$

$$l = 76.41 \text{ cm}$$

$$S_c = \frac{3P}{h^2}\left[1-\left(\frac{a\sqrt{2}}{l}\right)^{0.6}\right]$$

$$a = \sqrt{\frac{P}{p\pi}} = 13.82 \text{ cm}$$

$$\frac{50}{2} = \frac{3 \times 4200}{h^2}\left[1-\left(\frac{13.82\sqrt{2}}{76.41}\right)^{0.6}\right]$$

$$h = 16.78 \text{ cm}$$

Increase 10% of h to consider impact

$\therefore$ Thicker = 16.78 + 1.678

= 18.45

$\therefore$ Approx answer is h = 19.6 cm

4.35 (c)

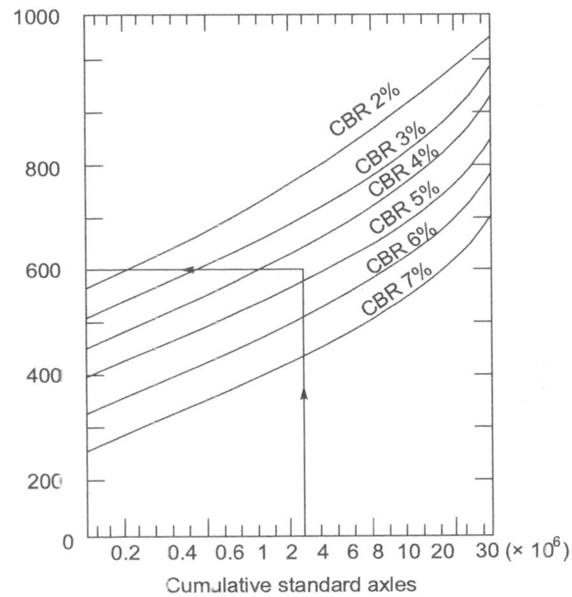

> Note: Here answer given by UPSC is 'd' but the correct answer is 'c'. As per Highway Engineering by Khanna & Justo, (CBR method of pavement design by cumulative standard axle load).

4.36 (a)

> Note: Here answer given by UPSC is 'b' but the correct answer is 'a'. As per Highway Engineering by Khanna & Justo, the flexible pavement layers reflects the deformation of the lower layer on to the surface.

4.37 (c)

As per IRC 37 : 2012

6.2.2 Fatigue Model

$$N_f = 2.21\times10^{-04}\times\left(\frac{1}{\epsilon_t}\right)^{3.89}\times\left[\frac{1}{MR}\right]^{0.854} \quad (80\%)$$

$$= 0.711\times10^{-04}\times\left(\frac{1}{\epsilon_t}\right)^{3.89}\times\left[\frac{1}{MR}\right]^{0.854} \quad (90\%)$$

when

ϵ_t : Max tensile strain at bottom of bituminous layer.

N_f : Fatigue life in number of standard axle.

M_R : Resilient modulus of bituminous layer.

4.38 (c)

P = 4200 kg; p = 6 kg/cm^2;

E = 150 kg/cm^2; Δ = 0.25 cm

$$p = \frac{P}{\pi a^2}$$

$$\Rightarrow \quad 6 = \frac{4200}{\pi\times a^2}$$

$$\Rightarrow \quad a = 14.92 \text{ cm}$$

Radius of contact area

$= a = 14.92$ cm

Thickness of pavement by single layer elastic theory

$$T = \sqrt{\left(\frac{3P}{2\pi E_S\Delta}\right)^2 - a^2}$$

$$= \sqrt{\left(\frac{3\times4200}{2\pi\times150\times0.25}\right)^2 - 14.92^2}$$

$= 51.352$ cm ≈ 51 cm

4.41 (b)

- Deflection measured near cracks are normally much higher than the measurements in non-distressed areas.
- Thermal and moisture gradients in the vertical direction of concrete slabs influence the deflection measurements.

Hence, statement 1 and 3 are incorrect.

4.43 (c)

Radius of rigid plate,

a = 300 mm = 30 cm

Deflection,

Δ = 1.2 mm

Load applied,

P = 5 tons = 5000 kg

Poisson's ratio,

μ = 0.5

$$p = \frac{P}{\pi a^2} = \frac{5000}{\pi\times30^2}$$

$$= 1.768 \text{ kg/cm}^2$$

For rigid plate method [F_2 = 1 for soil subgrade],

$$\Delta = 1.18\frac{p.a}{E_s}\times F_2$$

$$\Rightarrow 1.2 \text{ mm} = 1.18\times\frac{1.768\times300}{E_s}\times1$$

$$\Rightarrow \quad E_s = 521.56 \text{ kg/cm}^2$$

$$= \frac{521.56\times10}{10^{-4}}\text{N/m}^2$$

$$E_s = 52.156 \text{ MPa}$$

■■■■

5 Highway Materials, Maintenance and Properties

5.1 Bitumen grade 80/100 indicates that under the standard test conditions, penetration value of bitumen would vary from
(a) 0.8 mm to 1 mm (b) 8 mm to 10 mm
(c) 8 cm to 10 cm (d) 0.08 mm to 0.1 mm
[ESE : 2000]

5.2 Which one of the following curves illustrates the correct relation between % voids in total mix and % bitumen?

(a)
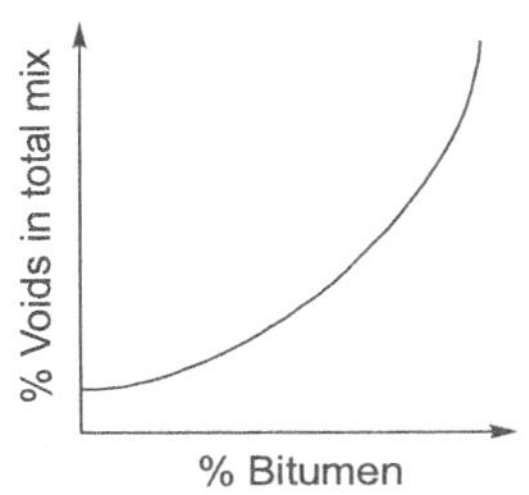

(b)
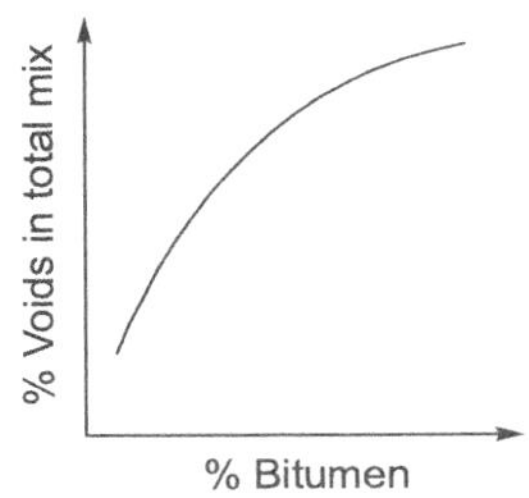

(c)
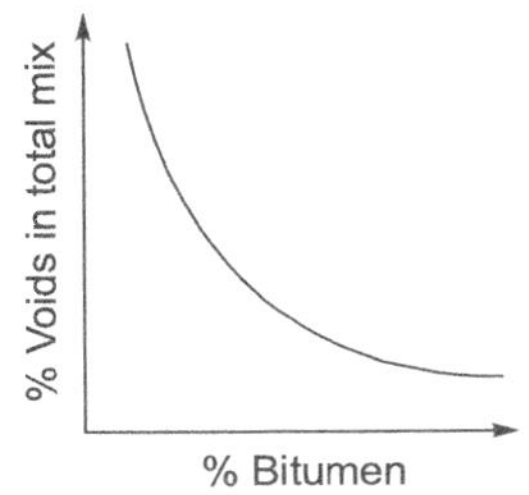

(d)
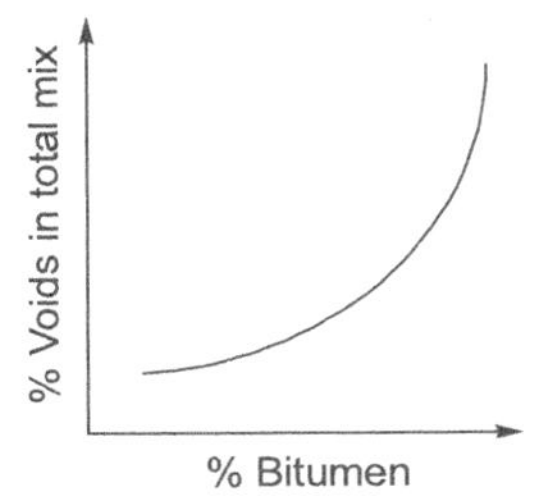

[ESE : 2000]

5.3 **Assertion (A):** The prime coat is an interface bituminous treatment when the existing base course has a pervious texture like water bound macadam.
Reason (R): The primer has to get into the capillary voids in the existing base and it should be of low viscosity. Bituminous emulsion is generally used as a prime coat.
(a) both A and R are true and R is the correct explanation of A
(b) both A and R are true but R is not a correct explanation of A
(c) A is true but R is false
(d) A is false but R is true
[ESE : 2000]

5.4 Match **List-I** (Distress) with **List-II** (Meaning) and select the correct answer using the codes given below the lists:

List-I
A. Settlement
B. Subsidence
C. Depression
D. Distortion

List-II
1. A localized, rather abrupt lowering of the road surface
2. A general lowering of the road surface
3. Irregular deformation of the road
4. Dished localized deformations

Codes:

	A	B	C	D
(a)	2	1	4	3
(b)	1	2	4	3
(c)	2	1	3	4
(d)	1	2	3	4

[ESE : 2001]

5.5 Consider the following stages in the construction of concrete roads:
1. Preparing the subgrade and the base course.
2. Mixing and placing the concrete.
3. Placing the framework and watering the prepared base.
4. Curing.
5. Compaction and floating.

The correct sequence of these stages is
(a) 1, 2, 3, 4, 5 (b) 1, 3, 2, 5, 4
(c) 1, 2, 3, 5, 4 (d) 1, 3, 2, 4, 5

[ESE : 2001]

5.6 Match **List-I** (Type of construction) with **List-II** (% Bitumen content) and select the correct answer using the codes given below the lists:

List-I	List-II
A. Bituminous macadam	1. 3-3.5%
B. Dense bituminous macadam	2. $\nless$ 4%
C. Bituminous concrete	3. 14-17%
D. Bituminous mastic	4. Min. 4.5%

Codes:

	A	B	C	D
(a)	4	2	1	3
(b)	1	2	4	3
(c)	4	3	1	2
(d)	1	3	4	2

[ESE : 2002]

5.7 **Assertion (A):** The Marshall test, being a confined compression test has a correlation with deformation as it occurs on the road.
Reason (R): In road, loaded area is confined by the tyre, the surrounding surfacing and the road base.
(a) both A and R are true and R is the correct explanation of A
(b) both A and R are true but R is not a correct explanation of A
(c) A is true but R is false
(d) A is false but R is true

[ESE : 2002]

5.8 Match **List-I** (Test) with **List-II** (Purpose) and select the correct answer using the codes given below the lists:

List-I	List-II
A. Impact test	1. Bitumen adhesion
B. Los-Angeles abrasion test	2. Toughness
C. Crushing test	3. Hardness
D. Stripping test	4. Strength

Codes:

	A	B	C	D
(a)	2	3	4	1
(b)	4	1	2	3
(c)	4	3	2	1
(d)	2	1	4	3

[ESE : 2004]

5.9 Which one of the following defects indicates progressive disintegration of bituminous premix carpet surfacing by loss of aggregates?
(a) Pot holes (b) Ravelling
(c) Edge breaking (d) Rutting

[ESE : 2004]

5.10 In 500 gm sample of coarse aggregate, there are 100 gm flaky particles and 80 gm elongated particles. What are the flakiness and elongation indices (total) as per IS?
(a) 40% (b) 36%
(c) 18% (d) 4%

[ESE : 2004]

5.11 Match **List-I** (Road drainage system) with **List-II** (Location) and select the correct answer using the codes given below the lists:

List-I
A. Vertical sand drains
B. Causeways and culverts
C. Scuppers and catch water drains
D. Inlets and gratings

List-II
1. Urban road drainage
2. Hill roads
3. High embankments in soft soils
4. Cross drainage in road alignments

Codes:

	A	B	C	D
(a)	3	2	4	1
(b)	1	4	2	3
(c)	3	4	2	1
(d)	1	2	4	3

[ESE : 2004]

5.12 In a shape test of aggregate, which one of the following gives the correct slot for flakiness index for a material passing 50 mm sieve and retained on 40 mm sieve?

(a) 25 mm (b) 27 mm
(c) 81 mm (d) 30 mm

[ESE : 2005]

5.13 Consider the following statements:

1. Mastic asphalt is a mixture of hard grade bitumen or blown bitumen, minerals filler and fine aggregates.
2. % of binder content in the mastic asphalt is 17-20 per cent by weight of the aggregates.

Which of these statements is/are correct?

(a) 1 only (b) 2 only
(c) Both 1 and 2 (d) Neither 1 nor 2

[ESE : 2005]

5.14 Match **List-I** (Type of binder) with **List-II** (Uses) and select the correct answer using the codes given below the lists:

List-I

A. 80/100 penetration grade bitumen
B. 85/25 blown-bitumen
C. MC-70 cutback
D. RT-5 road tar

List-II

1. Mastic-asphalt
2. Bituminous roads
3. Grouting works
4. Prime-coat

Codes:

	A	B	C	D
(a)	3	4	1	2
(b)	2	1	4	3
(c)	3	1	4	2
(d)	2	4	1	3

[ESE : 2006]

5.15 What are the standards for testing of road macadam in Aggregate Impact Test?

(a) 14 kg wt, 38 cm drop, 15 blows
(b) 14 kg wt, 35 cm drop, 20 blows
(c) 18 kg wt, 35 cm drop, 15 blows
(d) 18 kg wt, 30 cm drop, 20 blows

[ESE : 2006]

5.16 Which one of the following is the set of physical requirements of coarse aggregates for construction of WBM roads as per IRC recommendation?

	LAV(%)	AIV(%)	FI(%)
(a)	<50	<40	<15
(b)	<50	<30	<15
(c)	<40	<30	<20
(d)	<40	<30	<15

[ESE : 2006]

5.17 Which one of the following diagrams illustrates the relationship between flow value (FV) and percentage bitumen (%BIT)?

(a)
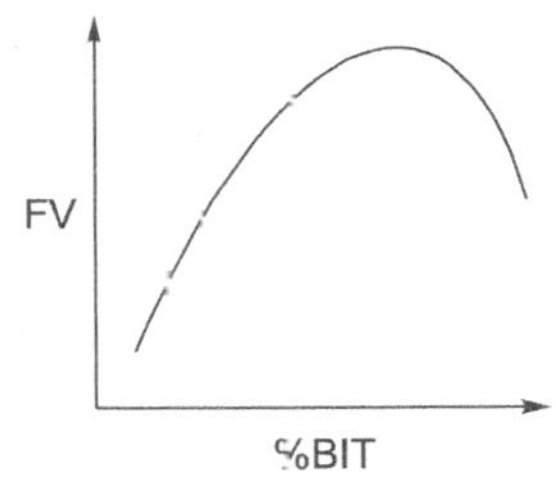

(b)
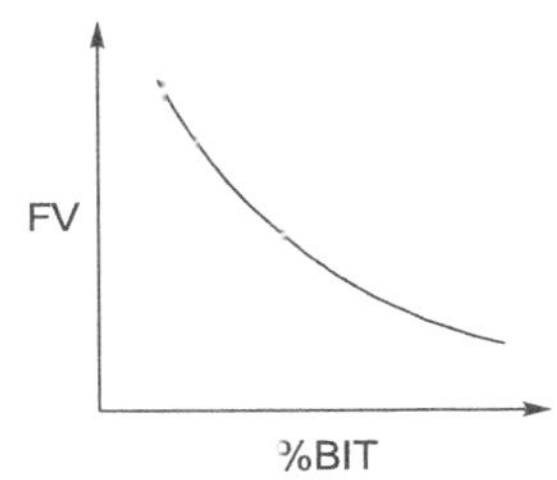

(c)
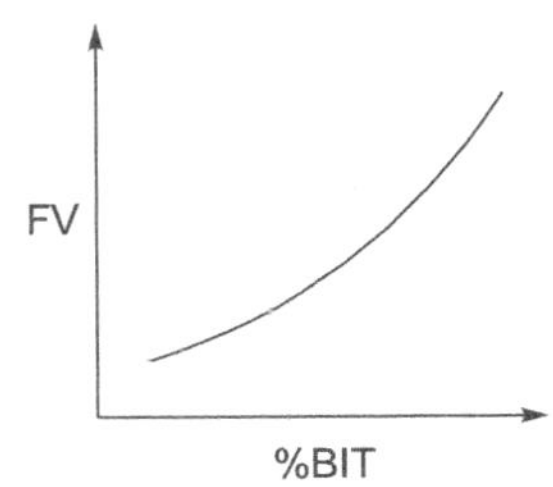

(d)
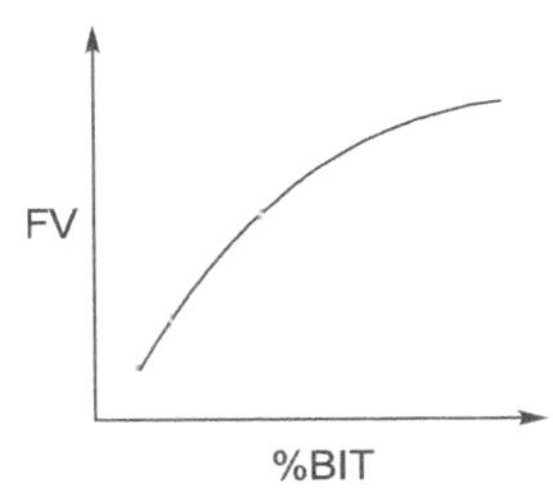

[ESE : 2006]

5.18 Match **List-I** (Type of construction) with **List-II** (Binder content) and select the correct answer using the code given below the lists:

List-I	**List-II**
A. Bituminous macadam	**1.** 8-15%
B. Dense bituminous macadam	**2.** 3-3.5%

C. Bituminous concrete — 3. 4-4.5%
D. Bitumen mastic — 4. 4.5-6.0%

Codes:

	A	B	C	D
(a)	2	1	4	3
(b)	4	3	2	1
(c)	2	3	4	1
(d)	4	1	2	3

[ESE : 2007]

5.19 Based on Fuller's maximum density criterion, for 4 mm maximum size of soil particles what is the percentage of particles between 4 mm and 2 mm by weight?

(a) 80 (b) 50
(c) 30 (d) 20

[ESE : 2007]

5.20 Consider the following statements related to Los Angeles Abrasion test on aggregates:

1. It evaluates hardness of source-rock.
2. It has a coefficient of variation of about 30 per cent.

Which of these statements is/are correct?

(a) 1 only (b) 2 only
(c) Both 1 and 2 (d) Neither 1 nor 2

[ESE : 2007]

5.21 A road surface is corrected by spreading a layer of dry sand in a thickness varying from 5 mm to 10 mm and rolling the surface by heavy rollers. Which one of the following maintenance works does it apply to?

(a) Repair of ruts and patches
(b) Repairing of blow ups
(c) Repair of bleeding surface
(d) Sealing of joints and cracks

[ESE : 2007]

5.22 The weight of aggregate having specific gravity 2.65, completely filled into a cylinder of volume 0.003 m^3 is 5.2 kg. What is the value of the angularity index of aggregate (approximately) as given by Murdock?

(a) 1 (b) 0.34
(c) 0.15 (d) 0.05

[ESE : 2007]

5.23 Match **List-I** (Type of wall) with **List-II** (Feature) and select the correct answer using the code given below the lists:

List-I
A. Parapet wall
B. Check wall
C. Breast wall
D. Gabion wall

List-II
1. Constructed with dry stone masonry encased in wire mesh
2. To add the overall stability to the hill face
3. To buttress the uphill slopes of the road cross-section
4. To give protection to the motorists

Codes:

	A	B	C	D
(a)	2	4	1	3
(b)	4	2	1	3
(c)	4	2	3	1
(d)	2	4	3	1

[ESE : 2007]

5.24 Emulsion is used as a binder in which of the following stages of construction?

1. Surface dressing work
2. Sealing open textured surfacing
3. Filling cracks in pavement
4. Prime coat
5. Pre-coating of aggregates

Select the correct answer the using the codes given below:

(a) 1, 2, 3, and 5 (b) 2, 3, 4 and 5
(c) 1, 2 and 4 (d) 1 and 3 only

[ESE : 2007]

5.25 Consider the following bituminous surfacings:

1. SDBM
2. PMC
3. AC
4. SD
5. Mastic Asphalt (MA)

Which one of the following is the correct sequence in increasing order with respect to their performance and wearing qualities?

(a) 4, 1, 2, 3, 5 (b) 2, 4, 1, 5, 3
(c) 4, 2, 1, 3, 5 (d) 1, 4, 3, 2, 5

[ESE : 2008]

5.26 Consider the following statements with reference to Water Bound Macadam (WBM) and Wet Mix Macadam (WMM):

1. WBM is a road mix and WMM is a plant mix.
2. WBM usually has plastic filler, while WMM

has non-plastic filler.

3. WBM is a modern road mix and WMM is a traditional road mix.

Which of these statements is/are correct?

(a) 1 and 2 (b) 2 and 3
(c) 1 only (d) 2 only

[ESE : 2008]

5.27 Which one of the following tests is performed in the laboratory to determine the extent of weathering of aggregates for roadworks?

(a) Soundness test (b) Crushing test
(c) Impact test (d) Abrasion test

[ESE : 2009]

5.28 Hot bitumen is sprayed over freshly constructed bituminous surface followed by spreading of 6.3 mm coarse aggregates and rolled. Which one of the following is indicated by this type of construction?

(a) Surface dressing (b) Gravel-bitumen mix
(c) Liquid seal coat (d) Seal coat

[ESE : 2009]

5.29 **Assertion (A):** In water-bound macadam construction, grade I has better load dispersion characteristics as compared to grade III aggregates.

Reason (R): The plasticity index of the binding material should be less than 6%.

(a) both A and R are true and R is the correct explanation of A
(b) both A and R are true but R is not a correct explanation of A
(c) A is true but R is false
(d) A is false but R is true

[ESE : 2009]

5.30 Which one of the following is not a desirable property of the subgrade soil as a highway material?

(a) Stability (b) Ease of compaction
(c) Good drainage (d) Bitumen adhesion

[ESE : 2009]

5.31 If aggregate size of 50–40 mm is to be tested for determining the proportion of elongated aggregates, the slot length of the gauge should be

(a) 45 mm (b) 53 mm
(c) 81 mm (d) 90 mm

[ESE : 2014]

5.32 The consistency and flow resistance of a sample of bitumen can be determinated through which of the following tests?

(a) Viscosity test (b) Penetration test
(c) Ductility test (d) Softening point test

[ESE : 2016]

5.33 Flexible concrete is a mix comprising of

(a) Gravel, filler and 30/40 bitumen
(b) Sand, filler and 30/40 bitumen only
(c) Gravel, sand, filler and 60/70 bitumen
(d) Sand, filler and 60/70 bitumen only

[ESE : 2016]

5.34 Consider the following statements regarding ductility of bitumen:

1. Ductility is the property which does not permit bitumen to undergo large deformation without breaking
2. Bitumens with high ductility are generally adhesive but do not have good cementing properties
3. Ductility must be ascertained at two different temperatures in order to pronounce on the suitability of the material

(a) 3 only (b) 2 only
(c) 1 only (d) 1, 2 and 3

[ESE : 2016]

5.35 Consider the following statements:

1. A high aggregate impact value indicates strong aggregates.
2. A low aggregate crushing value indicates high crushing strength of aggregates.
3. Aggregates having elongation index values greater than 15% are generally considered suitable for pavement construction.
4. Flakiness index of aggregates should not be less than 25% for use in road construction.

Which of the above statements are correct?

(a) 2 and 3 only (b) 2 and 4 only
(c) 1 and 3 only (d) 1 and 4 only

[ESE : 2017]

Directions: The following items consists of two statements; one labelled as **'Statement (I)'** and the other as **'Statement (II)'**. You are to examine these two statements carefully and select the answers to these items using the codes given below:

Codes:
(a) Both Statement (I) and Statement (II) are individually true and Statement (II) is the correct explanation of Statement (I)
(b) Both Statement (I) and Statement (II) are individually true but Statement (II) is NOT the correct explanation of Statement (I)
(c) Statement (I) is true but Statement (II) is false
(d) Statement (I) is false but Statement (II) is true

5.36 **Statement (I):** PC A is a preferred raw material for construction of Bituminous pavements in areas of heavy rainfall.
Statement (II): In PCA, no stripping is needed as there is improved binding; and thereby stability is also improved.
[ESE : 2018]

5.37 **Statement (I):** Bituminous roads disintegrate even with light traffic, but such failures are not exclusively attributable to wrong surface treatment.
Statement (II): Improper preparation of the subgrade and the foundation is often responsible for this dis-integration.
[ESE : 2018]

5.38 Consider the following statements related to construction of bituminous pavements:
1. It is not possible to construct relatively thin bituminous pavement layers over a an existing pavement.
2. In India, the bituminous construction is by and large adopted on the surface course.
3. The black top construction is in extensive use in developing nations.

Which of the above statements are correct?
(a) 1 and 2 only (b) 2 and 3 only
(c) 1 and 3 only (d) 1, 2 and 3
[ESE : 2022]

5.39 The held water in subgrade soil forms ice crystals at some spots if the freezing temperatures continue for a certain period. These ice crystals grow further in size if there is a continuous supply of water due to capillary action and the depressed temperature continues. This results in raising of portion of the pavement structure known as
(a) Frost heave
(b) Frost melting
(c) Alternate freeze-thaw cycle
(d) Frost action
[ESE : 2022]

5.40 The roads which are provided with a hard pavement course having at least a water bound macadam (WBM) layer are classified as
(a) Paved roads (b) Unpaved roads
(c) Surface roads (d) Unsurfaced roads
[ESE : 2023]

■■■■

Answers **Highway Materials, Maintenance and Properties**

5.1 (b)	5.2 (c)	5.3 (a)	5.4 (a)	5.5 (b)	5.6 (b)	5.7 (d)	5.8 (a)	5.9 (b)
5.10 (a)	5.11 (c)	5.12 (b)	5.13 (a)	5.14 (b)	5.15 (a)	5.16 (a)	5.17 (c)	5.18 (c)
5.19 (c)	5.20 (a)	5.21 (b)	5.22 (*)	5.23 (c)	5.24 (a)	5.25 (c)	5.26 (a)	5.27 (a)
5.28 (a)	5.29 (d)	5.30 (d)	5.31 (c)	5.32 (a)	5.33 (c)	5.34 (a)	5.35 (a)	5.36 (a)
5.37 (a)	5.38 (b)	5.39 (a)	5.40 (a)					

Explanations Highway Materials, Maintenance and Properties

5.1 (b)

$$\text{Penetration (mm)} = \frac{1}{10} \times \text{reading on scale}$$

5.2 (c)

Curve (a) or (d) → For flow value

Curve (b) → Percent voids filled with bitumen (VFB)

Curve (c) → Percent voids in total mix

5.6 (b)

According to MoRTH specifications bitumen content are as follows:

(i) Bituminous mastic, 14-17%

(ii) Dense bituminous macadam for nominal aggregate size 40 mm minimum 4% and for nominal aggregate size 25 mm minimum 4.5%.

(iii) Bituminous macadam for nominal aggregate size 40 mm, 3.1-3.4% and for nominal aggregate size 19 mm, 3.3-3.5%.

(iv) Bituminous concrete for nominal aggregate size 19 mm, 5-6% and for nominal aggregate size 13mm, 5-7%.

5.7 (d)

Marshall test is unconfined compression test.

5.8 (a)

Impact test evaluates the toughness of stone i.e its ability to absorb energy under repeated impact.

Ductility test or stripping test is used to measure the adhesive property of bitumen and its ability to stretch.

Abrasion test is carried to test the hardness property of aggregates.

Crushing test tests the crushing strength under applied compressive load.

5.9 (b)

Bituminous premixed carpet is a surfacing course. Ravelling is caused due to disintegration of asphalt and dislodgement of aggregate progressively from the surface to downward direction.

5.10 (a)

$$\text{Flakiness index} = \frac{100}{500} \times 100 = 20\%$$

$$\text{Elongation index} = \frac{80}{400} \times 100 = 20\%$$

$$\text{Total} = 20 + 20 = 40\%$$

5.11 (c)

The vertical sand drains increase the effective permeability of soil. This results in faster dissipation of excess pore pressure in high embankment in soft soils.

For urban road surface drainage surface consists of inlets and gratings. The limitation of land width and presence of foot paths, dividing islands and other road facilities necessitates under ground longitudinal drains.

In order to intercept and divert water from the hill slope, catch water drains are provided, running parallel to the roadways. Water from catch water drains is diverted by sloping drain and carried across the road by means of causeways and culverts.

Beside catch water drains, road side drains on one side of hill road are provided. They are angle, saucer, and kerb and channel drain types.

In hill roads where rainfall is heavy, the culverts are needed at 60 to 90 m to facilitate drainage of water across the roads. This may be quite costly. Hence often 8 to 10 scuppers (cross-drainage structures) are provided in addition to the bridges and regular culverts.

5.12 (b)

$$\text{Mean dimension} = \frac{50 + 40}{2} = 45 \text{ mm}$$

$$\text{Slot for flakiness index} = \frac{3}{5} \times 45 = 27 \text{ mm}$$

5.13 (a)

The % binder content in the mastic asphalt is 14-17%.

Mastic asphalt is an intimate homogeneous mixture of selected well graded aggregates as to yield a plastic, voidless and impermeable mass. The paving grade bitumen 30/40, 60/70 and 80/100 may be chosen depending upon the climatic conditions. Hard bitumen 180/200 type are used in surface dressing.

5.14 (b)

80/100 penetration grade bitumen is used for bituminous road construction. Mastic-asphalt is a mixture of blown bitumen, filler and fine aggregates in suitable proportions which yields voidless and impermeable mass. 85/25 bitumin-blow is basically finest quality oxidized bitumen which is known for its high boiling point and accurate composition. Rt-5 road tar is used in grouting works. Cutback MC 70 containing minimum 55% Bitumen is commonly used as a Prime Coating and Maintenance Mixing.

5.15 (a)

Impact test evaluates the toughness of material i.e its ability to absorb energy under repeated impact. A metal hammer of weight 13.5 kg-14 kg having free fall from height of 38 cm is arranged to drop through vertical guides and 15 blows are given to the test specimen.

5.16 (a)

For WBM the prescribed value of LAV, AIV and FI for

(i) Base course are < 50%, < 40% and < 15% respectively.

(ii) Surface course are < 40%, < 30% and < 15% respectively.

where, LAV is Los-Angeles Abrasion Value
AIV is Aggregate Impact Value
FI is Flakiness Index

5.17 (c)

Fig (a) → Marshall stability or unit weight

Fig (b) → Percent voids in total mix (V_v)

Fig (c) → Flow value

Fig (d) → Percent voids filled with bitumen (VFB)

5.18 (c)

Dense bituminous macadam for nominal aggregate size 40 mm -minimum 4%- and for nominal aggregate size 25 mm-minimum 4.5%. Bituminous concrete for nominal aggregate size 19 mm -5 % to 6% and for nominal aggregate size 13 mm-5% to 7%. Bituminous macadam for nominal aggregate size 40 mm -3.1 % to 3.4% and for nominal aggregate size 19 mm-3.3% to 3.5%.

5.19 (c)

According to Fuller's maximum density criterion, the gradation is given by

$$p = 100\left(\frac{d}{D}\right)^n$$

where,

p is per cent finer than diameter 'd' (mm) in the material

D is diameter of largest particle, mm

n is gradation index (taken as 0.5 by Fuller)

$$\therefore \quad p = 100\left(\frac{2}{4}\right)^{0.5} = 70.71$$

Thus, the percentage of particles between 4 mm and 2 mm will be given by

$$100 - 70.71 = 29.29 \simeq 30\%$$

5.20 (a)

The Los Angeles abrasion value of good aggregates acceptable for cement concrete, bituminous concrete and other high quality pavement materials should be less than 30 percent. Values up to 50 percent are allowed in base courses like water bound and bituminous macadam.

5.21 (b)

In the case of blow-ups, road surface is corrected by spreading a layer of dry sand in thickness varying from 5 mm to 10 mm and rolling the surface by heavy rollers.

5.22 (*)

Angularity index depends on the grading of the coarse and fine aggregates as well as on the relative amounts of coarse and fine aggregates in the final mixed concrete. These of course, are the things which are known only after the mix has been proportioned. It is necessarly, therefore, to make preliminary estimates of the probable value of angularity index. Where the tests are carried out, the value of angularity index, F_a can be determined, as given by Murdock as,

$$F_a = 11.05 - \frac{15W}{VG}$$

where, W is weight of aggregate in gm = 5200 gm

V is volume of cylinder = 3000 ml

G is specific gravity = 2.65

$$\text{So, } F_a = 11.05 - \frac{15 \times 5200}{3000 \times 2.56} = 1.239$$

5.23 (c)

Parapet walls which are extensions at the edge are used to give protection to motorcyclists on roads with steep slopes and for protecting vehicles from falling into the valley. Breast walls are constructed to buttress the uphill slope of the road cross section. Check walls are Small retaining structures provided in series to arrest landslides and to add stability to hill face. Gabion walls are provided for retaining fill portion, and constructed with dry stone masonry encased in wire mesh or geo-grid cages.

5.24 (a)

Emulsions are used in bituminous road constructions, especially in maintenance and patch repair works. Three types of bituminous emulsions are prepared viz. Rapid Setting (RS), Medium Setting (MS) and Slow Setting (SS) types. RS type emulsion is suitable for surface dressing and penetration macadam type of construction. MS type is used for premixing with coarse aggregates and SS type emulsion is suitable for fine aggregate mixes.

5.25 (c)

Increasing order of performance and wearing qualities of various bituminous surfacings

Surface dressing < Pre mixed carpet < semi-dense bituminous macadam < asphaltic concrete <Mastic asphalt

5.26 (a)

WBM is a traditional road mix and WMM is a modern road mix.

5.27 (a)

Soundness test is performed in the laboratory to study the resistance of aggregates to weathering action by conducting accelerated weathering test cycle. The chemicals used in the laboratory for conducting this test are sodium sulphate and magnesium sulphate.

5.28 (a)

Surface dressing serves as thin wearing coat over existing pavement. It consists of application of hot bituminous material followed by spreading of aggregate cover and rolling.

5.29 (d)

Aggregates are divided into 3 grades according to their size.

Grade 1 - particles of size 90 mm to 40 mm.

Grade 2 - particles of size 63 to 40 mm.

Grade 3 - particles of size 50 to 20 mm.

Load dispersion of grade 3 > grade 2> grade 1.

5.30 (d)

Bitumen adhesion test is carried out in order to decide the suitability of the road aggregates for use in road construction.

5.31 (c)

Slot length of guage

$$= 1.8 \times (\text{Mean size of agregate})$$

$$= 1.8 \times \frac{50+40}{2} = 81 \text{ mm}$$

5.34 (a)

Bitumen with high ductility generally are adhesive and have good cementing property. Ductility test may be required at two different temperature to obtain some indication of the tendency of the material to change in temperature.

Note: Here answer given by UPSC is 'c' but the correct answer is 'a'. As per Highway Engineering by S.K. Sharma, ductility is a properly which permits bitumen to undergo large deformation without breaking.

5.35 (a)

$$\text{Aggregate crushing value (ACV)} \propto \frac{1}{\text{Strength}}$$

5.36 (a)

Characterisation of PCA bituminous mix:

(a) **Stripping test (IS : 6241-1971)**

The plastic waste coated aggregate bitumen mix prepared by above process was immersed in water. Even after 96 hr. there was no stripping.

This shows that plastic waste coated aggregate bitumen mix has good resistance towards water.

(b) **Marshall stability test:**

It is observed that the Marshall stability values obtained for were generally much higher than Marshall stability value obtained for pure bitumen mix.

5.38 (b)

It is possible to construct relatively thin bituminous pavement layers over an existing pavement.

■■■■

UNIT

VIII

Railway Engineering

Syllabus

Terminology, Planning, designs and maintenance practices; track modernization.

Contents

■■■■

1 Railway Track & it's Components

1.1 **Assertion (A):** Coning of wheels in railways does not cause wear and tear of the wheels.
Reason (R): The rails are tilted at an angle of 1 in 20 to reduce the wear and tear on the rails as well as on the tread of the wheels.
(a) both A and R are true and R is the correct explanation of A
(b) both A and R are true but R is not a correct explanation of A
(c) A is true but R is false
(d) A is false but R is true

[ESE : 2000]

1.2 Squaring of sleepers through packing consists of
(a) adjusting the sleepers to be perpen-dicular to the rails
(b) adjusting the ballast under sleepers to space them parallel to each other
(c) cutting the edges of the sleepers to a square shape
(d) adjusting the rails to be perpendicular to the sleepers

[ESE : 2001]

1.3 Match **List-I** (Device) with **List-II** (Used for) and select the correct answer using the codes given below the lists:

List-I
A. Pandrol clip
B. Densometer
C. Viseur's mire
D. Cotters

List-II
1. Fixing the tie bars with the CI sleepers
2. Measurement of 'packing voids' under the sleeper ends
3. Measurement of unevenness of rail top and rectification of alignment
4. Elastic rail fastening

Codes:

	A	B	C	D
(a)	1	3	2	4
(b)	4	2	3	1
(c)	1	2	3	4
(d)	4	3	2	1

[ESE : 2002]

Directions: The following items consists of two statements; one labelled as **'Assertion (A)'** and the other as **'Reason (R)'**. You are to examine these two statements carefully and select the answers to these items using the codes given below:

Codes:
(a) Both A and R are true and R is the correct explanation of A
(b) Both A and R are true but R is not a correct explanation of A
(c) A is true but R is false
(d) A is false but R is true

1.4 **Assertion (A):** Packing is the method of forcing and packing stone ballast below the sleepers by ramming with a beater-cum-pick axe.
Reason (R): At points and crossings, packing is carried out by lifting the track by means of track jack and requisite quantity of stone chips is evenly spread below the sleepers by shovel and the ships are allowed to consolidate by running trains.

[ESE : 2003]

1.5 **Assertion (A):** CST-9 sleeper is a type of metal sleeper, has been extensively used by Indian Railways due to its uniform strength, economical and satisfactory performance.
Reason (R): Metal sleeper is an inverted channel with folded ends. The folded edges which form a bulb resist the damage caused by the packing of ballast.

[ESE : 2003]

1.6 Breathing length of LWR is the

(a) end portion which gets affected by temperature variation
(b) end portion which does not get affected by temperature variation
(c) central portion which gets affected by temperature variation
(d) central portion which does not get affected by temperature variation

[ESE : 2003]

1.7 Match **List-I** (Track parameter) with **List-II** (Equipments used) and select the correct answer using the codes given below the lists:

List-I	List-II
A. Unevenness	1. Track recording car
B. Gauge	2. Amsler car
C. Superelevation	3. Feeler and spring
D. Alignment	4. Gyroscopic pendulum

Codes:

	A	B	C	D
(a)	2	3	1	4
(b)	3	2	1	4
(c)	2	3	4	1
(d)	3	2	4	1

[ESE : 2003]

1.8 Which one of the following is **not** related to theories of creep of rails?

(a) Wave theory
(b) Percussion theory
(c) Drag theory
(d) Reversal theory

[ESE : 2007]

1.9 Which of the following are related to maintenance of railway track?

1. Jim crow and gauge bar
2. Through packing and boxing
3. Buffer stop and sand hump
4. Creep adjustment

Select the correct answer using the codes given below:

(a) 1, 2 and 3 (b) 1, 3 and 4
(c) 1, 2 and 4 (d) 2, 3 and 4

[ESE : 2008]

1.10 Which of the following factors help in ensuring track modulus, in a railway track?

(a) Gauge and formation width
(b) Track materials and sleeper density
(c) Degree of curvature and super elevation
(d) Length of rail and flange width

[ESE : 2008]

1.11 The 'track modulus' is an index of measure of which of the following?

(a) Resistance due to friction
(b) Resistance due to shear
(c) Resistance due to deformation
(d) Resistance due to rolling

[ESE : 2008]

1.12 **Assertion (A):** The pandrol clip is a fit and forget type of fastening and is made from silicon manganese spring steel bar and heat treated.

Reason (R): The pandrol clip has a point contact and causes indentation on the rail due to heavy toe load and small contact area.

(a) both A and R are true and R is the correct explanation of A
(b) both A and R are true but R is not a correct explanation of A
(c) A is true but R is false
(d) A is false but R is true

[ESE : 2010]

1.13 Match **List-I** with **List-II** and select the correct answer using the code given below the lists:

List-I

A. Rails
B. Sleepers
C. Ballast
D. Fish Plates

List-II

1. Connect one section of rail to next
2. Convert line load into uniformly distributed load
3. Convert point load into uniformly distributed load
4. Convert rolling loads into point load(s)

Code:

	A	B	C	D
(a)	4	3	2	1
(b)	1	2	3	4
(c)	4	2	3	1
(d)	1	3	2	4

[ESE : 2011]

1.14 A 45 kg rail on BG tracks in Indian conditions is expected to carry approximate axle loads of:
(a) 21 t (b) 22 t
(c) 23 t (d) 24 t [ESE : 2013]

1.15 Fish plates are used in rail joints to:
1. Maintain the continuity of rails
2. Provide for any expansion or contraction
3. Transfer the load to the ballast
4. Maintain correct alignment
(a) 1, 2, 3 and 4 (b) 1, 2 and 3 only
(c) 2, 3 and 4 only (d) 1, 2 and 4 only
[ESE : 2013]

1.16 'Composite Sleeper Index' is employed to determine
(a) Sleeper density requirement
(b) Number of fixtures required for a certain type of sleeper
(c) Durability of sleeper
(d) Mechanical strength of wooden sleeper whereby its suitability for use is assessed
[ESE : 2014]

1.17 Which of the following types of Elastic fastenings can be used on all types of sleepers (like Wooden, Cast-Iron or Concrete) on Indian Railways?
(a) Sigma clip (b) IRN-202 clip
(c) LG-20 Lock spike (d) Pandrol clip
[ESE : 2014]

1.18 According to the wave action for creep of rails, the pitch and depth of wave depend upon:
1. Section of rails
2. Track modulus
3. Stiffness of track
4. Stability of foundation
5. Weight of the train
Which of these are correct?
(a) 1 and 2 only (b) 1, 2 and 3
(c) 2, 3 and 4 (d) 3, 4 and 5
[ESE : 2014]

1.19 What will be the optimum depth of ballast cushion required for a *BG* railway track below the sleepers with sleeper density of ($M + 5$) and bottom width of 22.22 cm?
(a) 25 cm (b) 21 cm
(c) 28 cm (d) 30 cm [ESE : 2015]

1.20 'Composite Sleeper Index' is relevant in determining :
1. Required and adoptable sleeper density
2. Durability of s eeper units
3. Mechanical strength of the stock of wooden sleepers
Which of the above statements is /are correct?
(a) 1 and 2 only (b) 2 and 3 only
(c) 1 only (d) 3 only [ESE : 2018]

1.21 Which one of the following types of steel is used in the manufacturing of metro and mono rails?
(a) Mild steel (b) Cast steel
(c) Manganese steel (d) Bessemer steel
[ESE : 2018]

1.22 Which one of the following statement is correct regarding Journal friction?
(a) Caused due to the wave action of rails.
(b) The amount does not depend upon the type of bearing.
(c) For roller bearings, it varies from 0.5 to 1.0 kg per tonne.
(d) For coupled boxes, it lubricates by hard grease from 0.5 kg to 1.0 kg per tonne.
[ESE : 2020]

1.23 For the construction of a 640 m long B.G. railway track by using a sleeper density of $M + 5$, and the length of each rail is 12.8 m, the number of sleepers required will be
(a) 1000 (b) 900
(c) 800 (d) 700 [ESE : 2020]

1.24 Which one of the following statement is correct regarding ballast used for railway tracks?
(a) The minimum depth of ballast for B.G. section is 20 cm – 25 cm.
(b) The quantity of stone ballast required for one metre length of track is 0.53 m^3 for B.G. section.
(c) For M.G. section the width of ballast is 1.83 m.
(d) The minimum depth of ballast for N.G. section is 10 cm. [ESE : 2020]

1.25 Consider the following statements related to the advantages of concrete sleepers:
1. Concrete sleepers can generally be mass produced using local resources.
2. Concrete sleepers are not suitably for beater packing.

3. Concrete sleepers have a very long life-span.
4. Concrete sleepers have no scrap value.

Which of the above statements is/are correct?

(a) 1 only (b) 1 and 3 only
(c) 2 only (d) 2 and 4 only

[ESE : 2021]

1.26 Consider the following statements related to the advantages of uniformly of rail gauges:

1. At transhipping is not required, there is breakage of good.
2. Large sheds to store goods are not required.
3. Labour strikes, etc. do not affect the service and operation of trains.

Which of the above statements are correct?

(a) 1 and 2 only (b) 2 and 3 only
(c) 1 and 3 only (d) 1, 2 and 3

[ESE : 2021]

Answers Railway Track & it's Components

1.1 (d)	1.2 (b)	1.3 (b)	1.4 (b)	1.5 (a)	1.6 (a)	1.7 (c)	1.8 (d)	1.9 (c)
1.10 (b)	1.11 (c)	1.12 (b)	1.13 (a)	1.14 (c)	1.15 (d)	1.16 (d)	1.17 (d)	1.18 (c)
1.19 (a)	1.20 (d)	1.21 (c)	1.22 (c)	1.23 (b)	1.24 (a)	1.25 (b)	1.26 (b)	

Explanations Railway Track & it's Components

1.1 (d)

Tilting is rails done in order to reduce pressure at outer rail which is generating due to effect of centrifugal force.

1.2 (b)

In through packing the sleepers are squared and brought to specified spacing at straight lengths and on curves i.e., the ballast under the sleepers is adjusted to space them parallel to each other.

1.3 (b)

Pandrol clips are elastic fasteners. They are responsible for attaching the rail to the base plate so the rail cannot move vertically or horizontally with respect to the base plate.

Densometer measures the depression of sleeper under a passing train.

Cotters are used to fix the tie bars with the CI sleepers. The different types of cotters are central split cotters, side split cotter, solid end split cotter and bent plate cotter.

Viscur's mire is helpful in measuring of uneveness of the rail and also used for rectification of track alighment.

1.4 (b)

Running trains causes the ballast under the sleepers to get loose and thus the depressions are formed in the plane of rails. Therefore, regular packing of ballast under sleeper is done generally through beater-cum-pick-axe methods of packing of ballast are through packing, scissor packing and shovel packing.

At points and crossings, shovel packing is adopted where the use of beater cum-pick-axe is difficult because of less space.

Beater-pick-axe rammer is used in through packing.

1.5 (a)

The rail seat of a CST-9 has sleeper has narrow bearing tends to reduce the rocking of the sleeper under the wave motion of the rail. The sleeper is designed to provide a firm support to the rail and provides fairly good lateral and longitudinal stability to the rails. Also the inverted channel with folded edges forms a bulb to resist damage caused by packing of ballast. That's why it is one of the most extensively used sleepers by Indian Railways.

1.6 (a)

Due to variation in temperature and inability of the resisting force offered by the ballast and sleeper fastening etc. to the rail, the longitudinal movement of the long welded rail takes place. The portion of the long welded rail at each end, which under goes changes in its length due to temperature variation is called breathing length. For B.G. track in Indian conditions this length is about 125 m.

$$L = \frac{AE\alpha\Delta T}{r}$$

where L is breathing length

A is cross-section area of rail. It is 76.86 cm^2 for 60 kg rail.

E is modulus of elasticity

$= 2.15 \times 10^6$ kg/cm^2

α is coefficient of thermal expansion

$= 1.152 \times 10^{-5}$ per°C

ΔT is temperature change

= 38°C for Zone I, 40°C for Zone II, 45°C for Zone III and 48°C for Zone IV

r is resistance of sleeper, 870 kg for ST sleeper and 640 kg for wooden sleeper.

1.8 (d)

Wave theory states that wave motion is set up by moving of wheels. A vertical reverse curve is formed in the rails ahead of the wheels, resulting from the rail deflection under the load, is the chief cause of creep.

Percussion theory states that the creep is due to impact of wheels at the rail end ahead at joints. Though the creep is very small in single impact but cumulative effect of number of wheels in quick succession results in sufficient creep.

Drag theory states that backward thrust on driving wheels of the locomotive of train has got a tendency to push the sail off the track backward while the other wheels of the locomotive and the vehicles push the rail in the direction of travel. This results in creep of rails in the direction of movement of trains.

1.9 (c)

The dead end of a railway line is provided with a barrier erected across the track to prevent the vehicles running off the track. It is called buffer stop. Sand hump is a method to prevent the vehicles from running off the track at the end of siding. A hump of sand with rising gradient at the dead end of siding is provided.

1.10 (b)

The track modulus depends upon the gauge, the type of rails, the type and density of sleepers, the type of section of ballast and subgrade.

1.11 (c)

Track modulus is a measure of stiffness of the track and is defined as load per unit length of the rail, required to produce a unit depression in the track.

1.13 (a)

The rails on the track can be considered as steel girders for the purpose of carrying axle loads. Rails transmit the loads to sleepers and consequently reduce pressure on ballast and formation below.

Sleepers are members generally laid transverse to the rails on which the rails are supported and fixed, to transfer the loads from rails to the ballast and subgrade below:

Ballast is the granular material usually broken stone or brick, gravel or sand placed and packed below and around the sleepers to transmit load (due to the wheels of the train)from sleepers, to formation and at the same time allowing drainage of the track.

Fish Plates: These plates, resembling in shape to a fish, are used to provide the continuity between the two rails at the rail-joints. They also provide the required gap for expansion and contraction of rails due to temperature variations. They are made of steel.

1.14 (c)

A general rule adopted is to specify a certain constant value of the ratio between the weight of the rail and the locomotive axle load. In India this ratio is 510.

$$\frac{\text{Weight of the rail in tonnes}}{\text{Locomotive axle load in tonnes}} = \frac{1}{510}$$

For 45 kg rail it will be = 45 × 510 = 22.95t

1.15 (d)

Composition of rail steel

(a) For ordinary rail

Carbon - 0.55 – 0.68%

Manganese - 0.65 – 0.90%

Silicon - 0.05% – 0.3%

Sulphur - ≤ 0.05%

Phosphosrus - ≤ 0.06%

(b) For rails at points and crossing
Carbon - 0.5 – 0.6%
Manganese - 0.95 – 1.25%
Silicon - 0.05% – 0.2%
Sulphur - ≤ 0.06%
Phosphosrus - ≤ 0.06%

1.16 (d)

CSI, composite sleeper index measures mechanical strength of timber, derived from its composite properties of strength and hardness.

$$CSI = \frac{S+10H}{20}$$

S = Strength index
H = Hardness index
CSI > 783 for track sleeper
> 1352 for crossing sleeper
> 1455 for bridge sleeper

1.17 (d)

- Pandrol clip PR401 is an elastic rail clip used by Indian railway which can be fixed on wooden, steel, CI and concrete sleeper with help of a base plate and some other ancillary fitting.
- IRN202 clip is an elastic fastening used for two block reinforced cement concrete (RCC) sleeper.
- The lock spike type CG-20 are used for wooden sleeper.

1.18 (c)

Pitch and depth of wave is controlled the wave action. It depends upon:
- Track modulus
- Track stiffness
- Track stability

1.19 (a)

Length of BG rail = 12.8 m
Sleeper density = 12.8 + 5 = 17.8
∴ Spacing,

$$S = \frac{12.8}{17.8} \times 100 \text{ cm} = 71.91 \text{ cm}$$

∴ Optimum depth of Ballast Cushion

$$D_b = \frac{S-W}{2} = \frac{71.91-22.22}{2}$$

= 24.84 cm

1.20 (d)

C.S.I. is used
- To determine suitability of particular timber for use as a sleeper.
- Measures the mechanical strength of timber, derived from its composite properties of strength and hardness.

1.22 (c)

Journal Friction: It is the friction of the locomotive, wagons and compartment itself. Its amount will depend upon:
(i) The type of bearing, (ii) The lubricant used, and (iii) The temperature of atmosphere.
The journal friction for roller bearings varies from 0.5 to 1.0 kg per tonne and for coupled boxes lubricated by hard grease from 1.3 kg to 1.5 kg per tonne.

1.23 (b)

$$\text{No. of sleepers} = \frac{640}{M} \times (M+5)$$

$$= \frac{640}{12.8} \times (12.8+5) = 890 \approx 900$$

1.24 (a)

	BG	MG	NG
Width of Ballast (m)	3.35	2.25	1.83
Depth of Ballast (m)	20.25	15.20	15
Quantity of stone Ballast per m length (m^3)	1.036	0.71	0.53

1.25 (b)

- Concrete sleepers being heavy give more elastic modulus, strength and stability to the track. They are specially suitable for long welded rail track due to great resistance to the buckling of the track.
- Concrete sleepers with elastic fastening maintain better gauge, cross level and alignment. It retains packing also very well.
- Concrete sleepers being flat at bottom are best suited for modern methods of track maintenance like mechanical maintenance and measured should packing (MSP)
- Concrete sleepers can also be used in track circuited areas as concrete is poor conductor of electricity.

- Concrete sleepers neither are susceptible to attack by vermin, corrosion, nor they are inflammable.
- The life of concrete sleepers is very long about 50 to 60 years. Thus rail and sleeper renewals can be matched, which is very big economical advantage.
- Concrete sleeper can be manufactured from local materials also.

1.26 (b)

The uniformity of gauges results in the following advantages:

1. The delay, cost and hardship in transhipping passengers and goods from the vehicles of one gauge to another is avoided.
2. As the transhipping is not required, there is no breakage of goods.
3. Difficulties in loading and unloading are avoided and labour expenses are saved.
4. Possibility of thefts and misplacement, while changing from one vehicle to another, is eliminated.
5. Large sheds to store goods are not required.
6. Labour strikes, etc. do not affect the service and operation of trains.
7. Surplus wagons of one gauge cannot be used on another gauge. This problem will not arise if gauge is uniform.
8. Locomotives can be effectively used on all the tracks if a uniform type of gauge is adopted.
9. Duplication of equipment such as platforms, sanitary arrangements, clocks, etc. is avoided. This saves a lot of extra expenditure.
10. During military movement, no time is wasted in changing personnel and equipment from one vehicle to another if gauge is uniform.

■■■■

2 Track Resistance and Tractive Effort

2.1 What is curve resistance for a 50 tonnes train on a BG track on a 4° curve?
(a) 0.05 tonne (b) 0.06 tonne
(c) 0.08 tonne (d) 0.10 tonne
[ESE : 2005]

2.2 What is the hauling capacity of a railway locomotive, carrying an axle load of 24 tonnes each? (Assume 0.166 as the coefficient of friction)
(a) 16.5 tonnes (b) 16.0 tonnes
(c) 15.5 tonnes (d) 17.0 tonnes
[ESE : 2008]

2.3 The gradient for a B.G. railway line such that grade resistance together with curve resistance due to a 4° curve which will be equivalent to a simple ruling gradient of 1 in 150 is
(a) 1 : 180 (b) 1 : 200
(c) 1 : 300 (d) 1 : 400
[ESE : 2016]

2.4 Which one of the following measures should be taken for maintaining rolling stock?
(a) The different parts of rolling stock need not be cleaned every day.
(b) All axles completed service life need not be replaced.
(c) The parts of rolling stock which get worn out need not be replaced.
(d) Lubrication of all the reciprocating parts and bearings with a suitable lubricant should be done.
[ESE : 2022]

2.5 Consider the following statements related to economic design of modern track:
1. Axle loading, possible weight reduction of the rolling stock and increase in number of axles should be rationalized.
2. Rail to sleeper fastenings for different type of sleepers at high speeds should have definite standards.
3. The weight of the sleeper and the rail may not be related to each other.

Which of the above statements are correct?
(a) 1 and 2 only (b) 2 and 3 only
(c) 1 and 3 only (d) 1, 2 and 3
[ESE : 2023]

■■■■

Answers Track Resistance and Tactive Effort

2.1 (c) 2.2 (b) 2.3 (b) 2.4 (d) 2.5 (a)

Explanations Track Resistance and Tractive Effort

2.1 (c)

$$\text{Curve resistance} = \frac{0.04}{100} \times W \times D$$

$$= \frac{0.04}{100} \times 50 \times 4 = 0.08 \text{ tonne}$$

2.2 (b)

$$\text{Hauling capacity} = \mu \times W \times n$$
$$= 0.166 \times 24 \times 4$$
$$= 15.936$$
$$\approx 16 \text{ tonnes}$$

2.3 (b)

Grade resistance + Curve resistance = Gradient resistance

$$w \tan\theta + 0.0004 \times 4^\circ \times w = \frac{w}{150}$$

$$\tan\theta = \frac{1}{150} - 0.0004 \times 4^\circ = \frac{1}{197}$$

2.4 (d)

- Lubrication of all reciprocating part and bearing with suitable lubricant should be done.
- The different part of rolling stock should be cleaned every day.
- All axle which have completed.

2.5 (a)

A modern track has to be rationalized in view of the following to achieve over-all economy on Indian Railways.

i. Rail requirements, like quality of steel, design of rail-section and manufacturing process, should be rationalized so that the rail section can sustain the expected stresses due to rolling loads.

ii. Wheel diameter in relation to qualities of rail steel and steel of the wheel tyre should also be rationalized so as to keep Hertzian stresses within the endurance limits of the steel.

iii. Axle loading, possible weight reduction of the rolling stock and increase in number of axles should be rationalized.

iv. Rail-to-sleeper fastenings i.e., elastic fastenings for different types of sleepers at high speeds should have definite standards.

v. The weight of the sleeper and the rail should be correctly related to each other.

■■■■

3 Geometric Design of the Railway Track

3.1 **Assertion (A):** Gradient much steeper than the ruling gradient are provided in a hilly terrain and they are negotiated with the help of a pusher engine.

Reason (R): Restricting the gradient to the ruling gradient in hilly terrains would mean increasing the length of the track resulting considerably in heavy excavation.

(a) both A and R are true and R is the correct explanation of A
(b) both A and R are true but R is not a correct explanation of A
(c) A is true but R is false
(d) A is false but R is true [ESE : 2001]

3.2 Grade compensation on curves in Indian Railways for BG is

(a) 0.40% per degree of curve
(b) 0.06% per degree of curve
(c) 0.04% per degree of curve
(d) 0.02% per degree of curve [ESE : 2003]

3.3 Which one of the following is the correct expression for the versine (h) of a curve?
(l = length of rail on the curve and r = radius of curvature)

(a) $h = \frac{l^2}{r}$ (b) $h = \frac{l^2}{2r}$
(c) $h = \frac{l^2}{8r}$ (d) $h = \frac{l^2}{4r}$ [ESE : 2005]

3.4 What is the value of the steepest gradient to be provided on a 2° curve for BG line having ruling gradient of 1 in 200?

(a) 1 in 238 (b) 1 in 227
(c) 1 in 202 (d) 1 in 198 [ESE: 2006]

3.5 The ruling gradient on BG section of railway is 1 in 150 and a 4° curve is also there on it. What is the allowable ruling gradient?

(a) 1 in 146 (b) 1 in 154
(c) 1 in 196 (d) 1 in 232 [ESE: 2006]

3.6 **Assertion (A):** The super elevation provided on a curve of a railway track is less than the equilibrium cant.

Reason (R): Subject to a maximum value depending on speed and gauge, cant deficiency is allowed and therefore actual cant provided is less than the equilibrium cant.

(a) both A and R are true and R is the correct explanation of A
(b) both A and R are true but R is not a correct explanation of A
(c) A is true but R is false
(d) A is false but R is true [ESE : 2006]

3.7 What is the steepest gradient permissible on a 2° Curve for BG line having ruling gradient of 1 in 200?

(a) 1 in 250 (b) 1 in 238
(c) 1 in 209 (d) 1 in 198
[ESE : 2007]

3.8 If the ruling gradient is 1 in 150 on a particular section of a broad gauge track, the allowable ruling gradient on a 4° curve in the track will be

(a) 0.51% (b) 0.53%
(c) 0.61% (d) 0.67% [ESE : 2011]

3.9 For a railway track in hilly terrain, the type of gradient provided is:

(a) Pulling gradient (b) Momentum gradient
(c) Helper gradient (d) Minimum gradient
[ESE : 2013]

3.10 The steepest gradient on a 2° curve on a Broad Gauge line with a stipulated ruling gradient of 1 in 200, given that grade compensation is 0.04% per degree of curve, is

(a) 1 in 200 (b) 1 in 150
(c) 1 in 238 (d) 1 in 283
[ESE : 2014]

3.11 An electric locomotive running at 60 kmph on a curved track of 1.68 m gauge laid at 800 m radius should be provided with superelevation of the rail by an amount of
(a) 50.5 mm (b) 55.5 mm
(c) 59.5 mm (d) 65.5 mm
[ESE : 2014]

3.12 What would be the admissible gradient for a BG track when the grade resistance coupled with a 4° curve resistance shall equal the resistance due to a ruling gradient of 1 in 200?
(a) 0.30% (b) 0.40%
(c) 0.24% (d) 0.34% [ESE : 2015]

3.13 In the layout of an MG track, the versine of a horizontal circular curve is measured over a 11.8 m chord length. What would be the radius of the curve if the value of the versine was 2 cm?
(a) 900 m (b) 800 m
(c) 870 m (d) 850 m [ESE : 2015]

3.14 The maximum speed of a train on B.G. track having a curvature of 3° and a cant of 10 cm with allowable cant deficiency of 76 mm, for conditions obtaining in India, as
(a) 87.6 km/h (b) 99.6 km/h
(c) 76.6 km/h (d) 65.6 km/h
[ESE : 2016]

3.15 Consider the following statements concerning railways:
1. A level stretch equal to the maximum train length is to be provided between the gradient reaches where s rising gradient is followed by a falling gradient.
2. Vertical curves in railway tracks are not set out as parabolas.
3. Diamond crossing can be laid on curves also.
4. Curves should be avoided at the top level segment of bridges.

Which of the above statements are correct?
(a) 1 and 3 only (b) 2 and 3 only
(c) 1 and 4 only (d) 2 and 4 only
[ESE : 2017]

3.16 A transition curve is to be provided for a circular railway curve 300 m radius, the gauge being 1.5 m with the maximum superelevation restricted to 15 cm. What s the length of the transition curve for balancing the centrifugal force?
(a) 72.3 m (b) 78.1 m
(c) 84.2 m (d) 88.3 m
[ESE : 2018]

3.17 A railway curve of 1350 m radius is to be set out to connect two tangents. If the design speed is 110 kmph and the rate of change of acceleration is 0.3 m/s^3, the shift of the circular curve will be nearly
(a) 0.18 m (b) 0.16 m
(c) 0.14 m (d) 0.12 m [ESE : 2020]

3.18 Which one of the following is the correct standard for provision of curves on railway track?
(a) Cant excess on B.G. shall not be allowed to exceed 105 mm.
(b) Minimum radius of vertical curves for group *A*, Broad Gauge track is 4,000 m.
(c) The minimum value of super-elevation according to Railway Board is $\frac{1}{10}$th of gauge.
(d) The speed potential of curve is given by formula $E = \frac{V^2}{127R}$ where, E is superelevation in mm.

[ESE : 2020]

3.19 If the ruling gradient is 1 in 150 on a particular section of broad gauge and at the same time a curve of 4 degree is situated on this ruling gradient, what is the allowable ruling gradient?
(a) 1 in 10 (b) 1 in 72
(c) 1 in 196 (d) 1 in 245
[ESE : 2021]

3.20 The raising of outer rail over inner rail is called
(a) Cant deficiency
(b) Cant
(c) Capacity of the track
(d) Centre bound sleepers [ESE : 2023]

Answers Geometric Design of the Railway Track

3.1 (a)	3.2 (c)	3.3 (c)	3.4 (a)	3.5 (c)	3.6 (*)	3.7 (b)	3.8 (a)	3.9 (c)
3.10 (c)	3.11 (c)	3.12 (d)	3.13 (c)	3.14 (a)	3.15 (a)	3.16 (a)	3.17 (b)	3.18 (b)
3.19 (c)	3.20 (b)							

Explanations Geometric Design of the Railway Track

3.1 (a)

In hilly regions, it becomes uneconomical to provide ruling gradient as the track length will be increased resulting in more excavation.

So pusher gradient is provided for steeper gradient. In that portion, a pusher engine is used for pulling the load.

3.2 (c)

Grade compensation on curves in Indian railways:

- 0.04% per degree of curve - BG track
- 0.03% per degree of curve - MG track
- 0.02% per degree of curve - NG track

3.3 (c)

Versine (h) is perpendicular distance of mid point of a chord from arc of a circle.

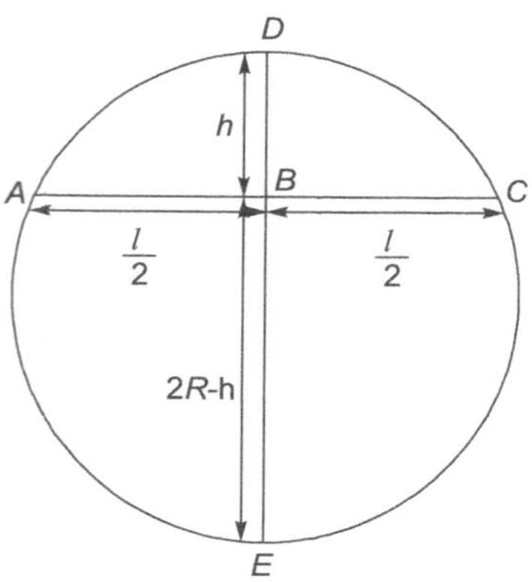

Two chord DE and AC meeting perpendicularly at a common point, hence

$$DB \times BE = AB \times BC$$

$$h \times (2R - h) = \frac{l}{2} \times \frac{l}{2}$$

$$2Rh - h^2 = \frac{l^2}{4}$$

$$\{h^2 \simeq 0\}$$

$$h = \frac{l^2}{8R}$$

3.4 (a)

Grade compensation for 2° curve

$= 0.04 \times 2 = 0.08\%$

Steepest gradient (maximum permissible gradient)

$$= \frac{1}{200} - \frac{8}{10000} = \frac{1}{238}$$

3.5 (c)

$$\text{Compensated grade} = \frac{1}{150} - \frac{0.04 \times 4}{100}$$

$$= \frac{1}{197.36}$$

3.6 (*)

- On the railway curved track equilibrium cant provided at which train is allowed to run at equilibrium speed to balance effect of centrifugal force completely.
- Fast moving train faces deficiency of cant and slow moving train faces excess of cant at provided equilibrium cant.

3.7 (b)

Grade compensation of BG line

= 0.04% per degree of curve

∴ Grade compensation of 2° curve

$= 2 \times 0.04 = 0.08\%$

$$\text{Ruling gradient} = 1 \text{ in } 200 = \frac{1}{200} \times 100 = 0.5\%$$

∴ Maximum allowable gradient

$$= 0.5 - 0.08 = 0.42\%$$

$$= \frac{1}{100/0.42} = \frac{1}{238}$$

3.8 (a)

The ruling gradient is the maximum gradient on a particular section, but if a curve lies on a ruling gradient, the resistance due to gradient is increased by that due to curvature and this further increase the resistance beyong the ruling gradient. In order to avoid resistances beyond the allowable limits, the gradients are reduced on curves and this reduction in gradients is known as grade compensation for curves.

In India compensation for curvature is given at 0.04% per degree of curve for B.G., 0.03% for M.G. and 0.02% for N.G.

∴ Allowable ruling gradient after grade compensation is

$$= \frac{1}{150} - \frac{0.04}{100} \times 4 = 0.0051 = 0.51\%$$

3.9 (c)

Gradient much steeper than the rulling gradient are provided in a hilly terrain and they are negotiated with help of pusher engine.

3.10 (c)

Grade provided

= Ruling gradient -Grade compensation

= 1 in 200 – 0.04% × 2

$$= \frac{21}{5000} = 1 \text{ in } 238$$

3.11 (c)

$$e = \frac{GV^2}{127R}$$

$$= \frac{1.68 \times 60^2}{127 \times 800} \times 1000\,\text{mm}$$

and $e = 59.53$ mm

3.12 (d)

Grade provided

= Ruling gradient – Grade compensation

= 1 in 200 – 0.04% × 4°

$$= \frac{1}{200} - \frac{0.16}{100} = 0.34\%$$

3.13 (c)

Given: Versine = $V = AB = 2$ cm, $a = 11.8$ m

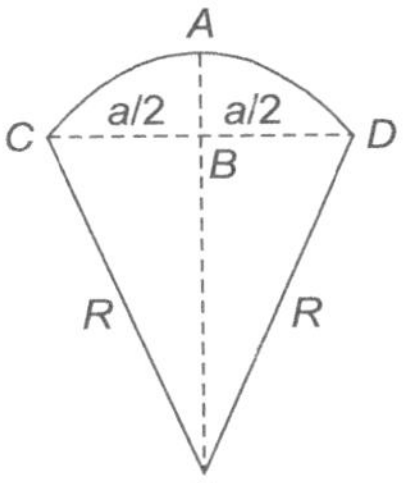

$$AB \times (2AO - AB) = CB \times BD$$

(Property of triangle)

$$V \times (2R - V) = \frac{a}{2} \times \frac{a}{2}$$

$$2RV - V^2 = \frac{a^2}{4}$$

$$2RV = \frac{a^2}{4} \qquad (\because V << R)$$

$$V = \frac{a^2}{8R}$$

$$R = \frac{(11.8)^2}{8 \times 0.02} = 870.25 \text{ m}$$

3.14 (a)

$$e_{th} = e_{act} + CD$$

$$\frac{GV_{max}^2}{127R} = e_{act} + CD$$

$$\frac{1.676 \times V_{max}^2}{127 \times \frac{1720}{3}} = (10 + 7.6) \times 10^{-2}$$

$$V_{max} = 87.44 \text{ kmph}$$

3.15 (a)

Vertical curves in Railway tracks are set out as parabolic curve.

Diamond crossing avoided for curve in order to avoid deflection of wheel to wrong side of nose.

3.16 (a)

Given, $R = 300$ m

$e = 15$ cm

As per Martin's formula,

$$V_{max} = 4.35\sqrt{R - 67}$$

$$= 4.35\sqrt{300 - 67} = 66.4 \text{ kmph}$$

Length of transition curve

$= (0.073)(e)(V_{max})$

$= (0.073) \times (15)(66.4) = 72.71$ m

3.17 (b)

$R = 1350$ m, $V = 110$ kmph

$$L_T = \frac{V^3}{CR} = \left(\frac{110^3 \times \frac{5^3}{18^3}}{0.3 \times 1350}\right)$$

$= 70.41$ m

$$\text{Shift} = \frac{LT^2}{24R} = \frac{70.41^2}{24 \times 1300}$$

$= 0.159 \approx 0.16$ m

3.18 (b)

Minimum radius of vertical curve,

Radius for BG group A is 4000 m.

For BG group B is 3000 m

For BG group C, D, E is 2500 m

For MG all routes is 2500 m.

3.19 (c)

As per recommendation of Indian Railway grade compensation for BG is 0.04 percent per degree of curve.

Then compensation for 4° curve

$$= 0.04 \times 4 = 0.16 \text{ percent}$$

Now, ruling gradient 1 in 150

$$= \frac{1}{150} \times 100 = 0.67 \text{ percent}$$

So, maximum allowable gradient or actual gradient to be provided = 0.67 – 0.16 = 0.51 percent

or $\frac{0.51}{100}$ i.e, 1 in 196.

3.20 (b)

Super elevation or cant can be defined as outer rail raised with respect to inner rail in order to generate centripetal force against centrifugal force which occurs outward in horizontal plane from centre of gravity of vehicle.

■■■■

4 Points and Crossing

4.1 A cross-over of 1 in 10 exists between two broad gauge parallel tracks with centres at 5 m apart, the length of the straight track is
(a) 16.4 m (b) 18.4 m
(c) 19.2 m (d) 19.92 m
[ESE : 2000]

4.2 If the curve lead of a broad gauge railway turnout is 16.76 meters, the angle of crossing of the turnout will be given by
(a) $\tan^{-1} 10$ (b) $\tan^{-1} 5$
(c) $\tan^{-1}\frac{1}{10}$ (d) $\tan^{-1}\frac{1}{5}$ [ESE : 2002]

4.3 The distance between the running faces of the stock-rail and gauge face of tongue rail measured at the heel of the switch is known as
(a) flangeway clearance
(b) throw of switch
(c) heel divergence
(d) flare [ESE : 2003]

4.4 **Assertion (A):** A turn-out is a combination of points and crossings, which enable a branch line to take off from another main line railway track.
Reason (R): Turn-outs are in great use in railway marshalling yards.
(a) both A and R are true and R is the correct explanation of A
(b) both A and R are true but R is not a correct explanation of A
(c) A is true but R is false
(d) A is false but R is true [ESE : 2005]

4.5 The vertical wear on tongue rail should not exceed which one of the following?
(a) 6 mm (b) 10 mm
(c) 12 mm (d) 15 mm
[ESE : 2008]

4.6 **Assertion (A):** When a turnout is taken off from a curved track, it is called gauntlet track.
Reason (R): When a turnout from a curved track turns away in the opposite direction, the curves are said to be of contrary flexure.
(a) both A and R are true and R is the correct explanation of A
(b) both A and R are true but R is not a correct explanation of A
(c) A is true but R is false
(d) A is false but R is true
[ESE : 2009]

4.7 A rail which is tapered to a toe at one end and has a heel at the other end is called as
(a) Stock rail (b) Tongue rail
(c) Wing rail (d) Lead rail
[ESE : 2011]

4.8 The purpose of check rail is:
(a) To aid the flange of the opposite wheel passing through the crossing to steer clear of the nose
(b) To avoid derailment per se
(c) To help control the running speed
(d) To minimize the laying cost of long distances of traces
[ESE : 2013]

4.9 Two parallel rai way tracks are to be connected by a reverse curve, both segments having the same radius. If the centre lines of the tracks are 8 m apart and the maximum adaptable distance between the tangent points is 32 m, the allowable radius for the curve is
(a) 4 m (b) 8 m
(c) 32 m (d) 64 m [ESE : 2014]

4.10 Two parallel railway lines are to be connected by a reverse curve, each section having the same radius. If the lines are 12 m apart and the maximum distance between tangent points measured parallel to the straights is 48 m, then the maximum allowable radius will be

(a) 51.1 m (b) 52.3 m
(c) 53.5 m (d) 54.7 m

[ESE : 2019]

4.11 A cross-over occur between two Metre Gauge parallel tracks of same crossing number 1 in 12 with straight intermediate portion between the reverse curves and the distance between the centres of tracks is 3.5 m. If the value of G is 1 m, the intermediate straight distance will be nearly

(a) 12 m (b) 15 m
(c) 18 m (d) 21 m

[ESE : 2020]

■■■■

Answers Points and Crossing

4.1 (a) 4.2 (d) 4.3 (c) 4.4 (b) 4.5 (b) 4.6 (d) 4.7 (b) 4.8 (a) 4.9 (c)
4.10 (a) 4.11 (c)

Explanations Points and Crossing

4.1 (a)

Length of track, $= DN - GN - G\sqrt{1+N^2}$

$$= 5(10) - 1.676(10) - 1.676\sqrt{1+10^2}$$
$$= 16.4 \text{ m}$$

4.2 (d)

Curve lead, $CL = 2GN$

$$\therefore \quad N = \frac{16.76}{2\times 1.676} = 5 = \cot\alpha$$

$\therefore$ Angle of crossing, $\alpha = \tan^{-1}\left(\frac{1}{5}\right)$

4.3 (c)

Distance between running and gauge face stock and tongue rail at heel is flangeway clearance, whereas heel divergance is either gauge to gauge or running to running face distance of stock rail and tongue rail at heel of switch.

4.4 (b)

Marshalling yards are said to be a "Machine to receive, break up, reform and despatch trains onwards". The important functions of a Marshalling yard are as follows:

(i) Reception (ii) Sorting (iii) Departure

Since there are many sidings, so number of turnouts are provided.

4.7 (b)

Stock Rail: The position of the straight alignment against which the tongue rail fits is known as the stock rail.

Wing Rails: Wing rails help in channelising the wheels in their proper routes and direction meant for their movement. These rails guide the wheel path for movement of trains.

Tongue Rail: A tongue rail is tapered having toe at one end and heel at the other end. It is fixed at heel end and can move or rotate about this point.

Lead Rails: In a turnout, lead rails are the length of rails from the heel of the tongue rail to the toe of the crossing. These rails are of the normal rail sections

4.8 (a)

Purpose of check rail to guide wheel flange and to prevent derailment at level crossing.

4.9 (c)

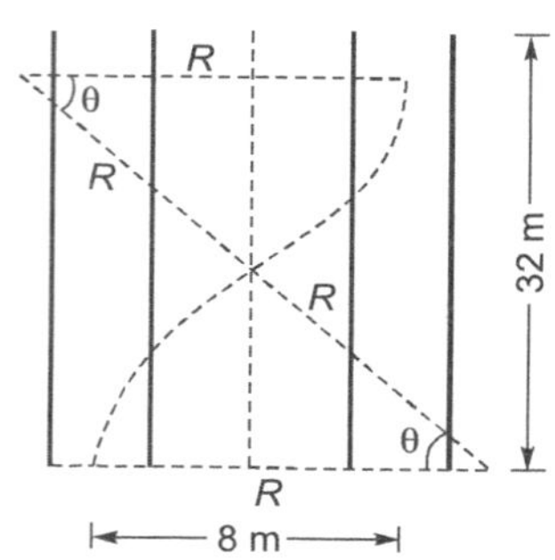

Let the radius of the curves be R.

Now, as shown in the figure above we have,

$$2R\sin\theta = 32 \text{ and}$$

$$2R(1-\cos\theta) = 8$$

$$\Rightarrow \quad \sin\theta = \frac{8}{17}$$

So $$2\times R\times\frac{8}{17} = 32$$

$$\Rightarrow \quad R = 34 \text{ m}$$

4.10 (a)

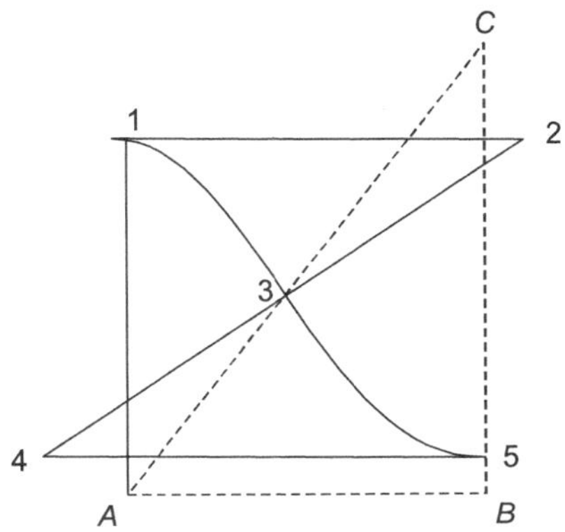

$$D = 12 \text{ m}$$

$$AB = \sqrt{AC^2 + BC^2}$$

$$AB^2 = AC^2 - BC^2$$

$$48^2 = (2R)^2 - (2R - D)^2$$

$$48^2 = 4R^2 - 4R^2 - D^2 + 4RD$$

$$\frac{48^2 + D^2}{4\times D} = R$$

$$\frac{48^2 + 12^2}{4\times 12} = R$$

$$R = 51 \text{ m}$$

4.11 (c)

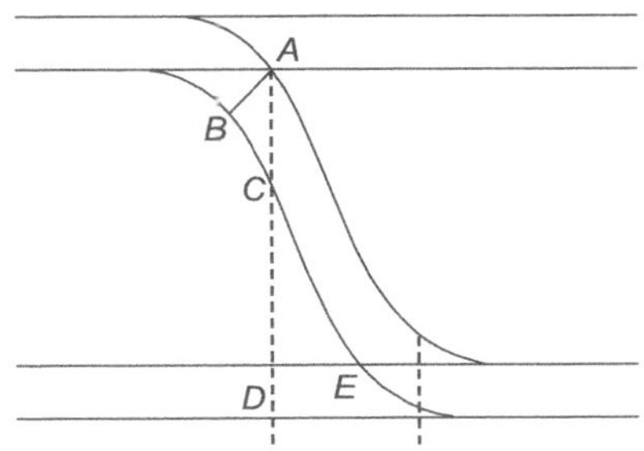

Length of intermediate straight :

$$DE = (D-G)N - G\sqrt{1+N^2}$$

$$= (3.5-1.0)\times 12 - 1\times\sqrt{1+12^2}$$

$$\approx 2.5 \times 12 - 12$$

$$= 1.5 \times 12 = 18 \text{ m}$$

■■■■

5 Miscellaneous

5.1 A railway yard in which wagons are received, sorted and new trains are formed is called a
(a) goods yard (b) station yard
(c) marshalling yard (d) shunting yard
[ESE : 2000]

5.2 Which one of the following signals is provided beyond the trailing points and switches in a railway yard?
(a) Repeater signal (b) Departure signal
(c) Advance starter (d) Routing signal
[ESE : 2000]

5.3 Which one of the following is installed as the second signal on the main signal post for better visibility?
(a) Routing signal (b) Calling on signal
(c) Co-acting signal (d) Repeating signal
[ESE : 2001]

5.4 Match **List-I** with **List-II** and select the correct answer using the codes given below the lists:

List-I
A. Dynamometer car
B. Crossing station
C. Ruling gradient
D. Turn table

List-II
1. Arrangement for turning the direction of an engine
2. Information about condition of railway track
3. Station on a single railway track system where Up and Down trains can pass each other
4. Maximum rising gradient of railway track depending on engine power.

Codes:

	A	B	C	D
(a)	2	3	1	4
(b)	2	3	4	1
(c)	3	2	4	1
(d)	3	2	1	4

[ESE : 2004]

5.5 Match **List-I** (Type of signals) with **List-II** (Characteristics) and select the correct answer using the codes given below the lists:

List-I	List-II
A. Detonating signal	1. Locational
B. Coloured light signal	2. Operational
C. Home signal	3. Special
D. Calling on signal	4. Functional

Codes:

	A	B	C	D
(a)	2	4	3	1
(b)	4	2	1	3
(c)	4	2	3	1
(d)	2	4	1	3

[ESE : 2004]

5.6 Match **List-I** (Signal) with **List-II** (Function) and select the correct answer using the codes given below lists:

List-I	List-II
A. Outer signal	1. Departure signal
B. Starter signal	2. Reception signal
C. Repeater signal	3. Shunting signal
D. Disc signal	4. Co-acting signal

Codes:

	A	B	C	D
(a)	3	4	1	2
(b)	2	1	4	3
(c)	3	1	4	2
(d)	2	4	1	3

[ESE : 2006]

5.7 In which one of the following yards, are reception, sorting and dispatch of railway wagons done?
(a) Loco yard (b) Marshalling yard
(c) Goods yard (d) Passenger yard
[ESE : 2010]

5.8 Marshalling yard in railway system provides facilities for
(a) Maintenance of rolling stock
(b) Safe movements of passengers and coaches
(c) Receiving, breaking up, re-forming and dispatching onwards—of trains
(d) Receiving, loading, unloading and delivery of goods and vehicles, and scheduling their further functioning [ESE : 2018]

Answers Miscellaneous

5.1 (c) 5.2 (c) 5.3 (c) 5.4 (b) 5.5 (d) 5.6 (b) 5.7 (b) 5.8 (c)

Explanations Miscellaneous

5.1 (c)

Station yard is a system of tracks laid usually on a level within defined limit. The types of yards are:

(i) **Goods yard :** The main function is to provide facilities for receiving, loading, unloading and delivery of goods and the movement of goods vehicle.

(ii) **Marshalling yard :** The main function is breakup, reform and despatch of trains onwards. i.e., reception, sorting and departure.

(iii) **Locomotive yard :** Locomotive yard for housing locomotive. All the facilities for oil filing, watering repairing, cleaning, etc. are provided.

(iv) **Passenger bogie yard :** Passenger bogie yard provide facilities for safe movement of passenger and vehicles for the passengers.

5.2 (c)

An advance starter may also be provided (Generally 180 meters or more) beyond the trailing points or switches. The advance starter becomes the LAST STOP SIGNAL at the station where it is provided. In general, advance starter is an indication for the train having left the station and is no more the responsibility of the station master.

5.3 (c)

Routing signals are provided at the point of diversion where a number of lines exist at the station taking off different locations from the main line besides one main Home signal.

Repeating or co-acting signals are provided when the driver's vision is obstructed by (i) an overbridge between the signals or (ii) sharp curvature in the alignment or due to any other reason. It repeats the indication of the signal ahead

Repeating and co-acting signals serve the same purpose of providing visibility. But co-acting signal is provided on the same post at a lower level at which main signal exists.

Calling-on-signals consists of small arms below and parallel to Home signal fixed on the same post When a train stops with the Home signal indication, then a calling on signal permits the train to proceed cautiously by keeping this signal arm at "PROCEED" position.

5.5 (d)

Operating signals: Detonating; Hand and Fixed signals.

Functional signals: Stop or semaphore type; Warner; Shunting; Coloured light signals.

Locational signals – Reception (Outer and Home); Departure (Starter and Advance Starter) signals.

Special signals – Repeater or co-acting; Routing; Calling on; Point indicators; Modified lower quadrant semaphoric signal etc.

5.6 (b)

Reception signal : Outer signal and Home signal

Departure signal: Starter signal and Advance starter signal

Shunting signal: Disc or ground signal

CO-ACTING signal: Repeater signal

5.7 (b)

Marshalling yard is a place where trains are received sorted, reformed and dispatched to their destinations.

Important functions of Marshalling yards are reception, sorting and departure. Places meant for these functions are called sidings. The various sidings are reception siding, sorting siding and departure siding. These sidings are provided in Marshalling yard.

5.8 (c)

Marshalling yard is used to be a machine to receive, break-up, reform and dispatch trains onwards.

■■■■

UNIT

IX

Airport, Dock, Harbour & Tunnelling Engineering

Syllabus

Airport: Layout, planning & design.

Harbour: Terminology, layouts and planning.

Tunnelling: Alignment, methods of construction, disposal of muck, drainage, lighting and ventilation.

Contents

■■■■

1 Airport Engineering

1.1 Match **List-I** with **List-II** with reference to airport engineering and select the correct answer using the codes given below the lists:

	List-I		List-II
A.	PNB	1.	Noise level
B.	LCN	2.	Visual range
C.	STOL	3.	Airport pavement evaluation
		4.	Type of aircraft

Codes:

	A	B	C
(a)	1	3	4
(b)	1	4	3
(c)	4	2	1
(d)	2	3	4

[ESE : 2000]

1.2 For a runway at an elevation of 1000 m above MSL and airport reference temperature of 16°C, the rise in temperature to be taken into account as per ICAO is

(a) 24.5°C (b) 15°C
(c) 7.5°C (d) 6°C

[ESE : 2000]

1.3 **Assertion (A):** Scissor crossing is provided where space does not permit provision of two separate crossovers.

Reason (R): A scissor crossing consists of four pairs of switches, six acute angle crossings and four obtuse angle crossings and check rails.

(a) both A and R are true and R is the correct explanation of A
(b) both A and R are true but R is not a correct explanation of A
(c) A is true but R is false
(d) A is false but R is true

[ESE : 2000]

1.4 A surface longitudinally centered on the extended runway centre line and extending outward and upward is called

(a) primary surface
(b) conical surface
(c) horizontal surface
(d) approach surface

[ESE : 2001]

1.5 Which of the following factors are to be taken into consideration while designing the taxiway system?

1. Volume of air traffic
2. Runway configuration
3. Location of terminal building
4. Location of hangars

Select the correct answer using the codes given below:

(a) 1, 2 and 3 (b) 2 and 3
(c) 1 and 4 (d) 1, 2, 3 and 4

[ESE : 2001]

1.6 In the figure shown below, "cockpit cut-off angle" is the angle labelled

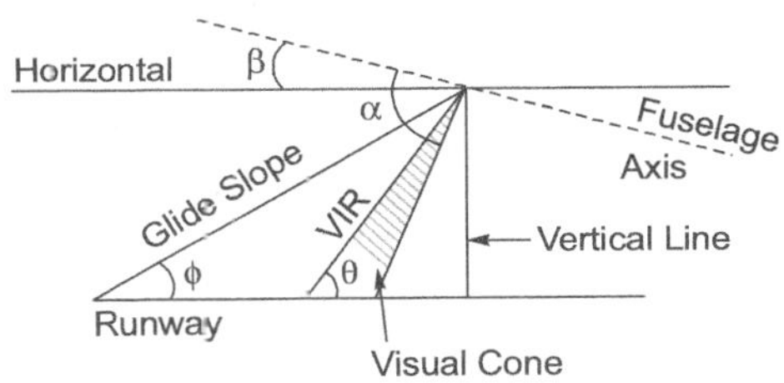

(a) α (b) β
(c) ϕ (d) θ

1.7 Consider the following factors:

1. Air traffic control measures
2. Aircraft traffic composition
3. VFR/ FR operation
4. Runway configuration

Which of these factors affect the capacity of a runway?

(a) 1, 2 and 3 (b) 2 and 3
(c) 1 and 4 (d) 1, 2, 3 and 4

[ESE : 2002]

1.8 Match **List-I** (Imaginary surface) with **List-II** (Characteristic) and select the correct answer using the codes given below the lists:

List-I
A. Conical
B. Approach
C. Primary
D. Transitional

List-II
1. Longitudinally centered on a runway
2. Longitudinally centered on the extended runway
3. Extending outward and upward from the periphery of the horizontal surface
4. Extending outward and upward at right angles to the runway center line

Codes:

	A	B	C	D
(a)	3	1	2	4
(b)	4	2	1	3
(c)	3	2	1	4
(d)	4	1	2	3

[ESE : 2002]

1.9 The monthly mean of maximum daily temperature and monthly mean of average daily temperature of the hottest month of the year are 49°C and 40°C respectively. Then airport reference temperature is
(a) 43°C (b) 69.6°C
(c) 37°C (d) 52°C
[ESE : 2003]

1.10 Which one of the following imaginary surface in airport is circular in plan with centre located at an elevation of 150 m above the airport reference point?
(a) Conical surface
(b) Transitional surface
(c) Inner horizontal surface
(d) Outer horizontal surface
[ESE : 2003]

1.11 Consider the following four components in an Instrumental Landing System:
1. Threshold lights
2. Glide slope antenna
3. 'Z' markers
4. Localizer antenna

These components are installed in the direction of landing in the sequence
(a) 4-2-3-1 (b) 1-4-3-2
(c) 3-1-2-4 (d) 2-3-4-1 [ESE : 2003]

1.12 The orientation of preferential runway in an airport is influenced by
(a) direction of prevailing wind, adequate length, obstruction-free landing and take-off zones
(b) adequate waiting and service facilities
(c) convenience of terminal and control facilities
(d) stable ground and adequate turning space
[ESE : 2004]

1.13 Consider the following aircraft operations:
1. Normal landing
2. Normal take-off with all engines
3. Engine failure at take-off
4. Emergency landing with all engines shut
5. Landing with maximum payload with the help of ILS

Which of the above aircraft operations are taken into consideration in deciding the basic runway length required for an aircraft?
(a) 1, 2 and 3 (b) 2, 3 and 4
(c) 3, 4 and 5 (d) 1 and 5
[ESE : 2004]

1.14 Which one of the following gives the number of gate positions in an airport?

(a) $\frac{\text{Capacity of runway}}{60\times 2}\times$ average gate occupancy time

(b) $\frac{\text{Capacity of apron}}{60\times 2}\times$ number of aircraft movements

(c) $\frac{\text{Capacity of taxiway}}{60\times 2}\times$ average gate occupancy time

(d) $\frac{\text{Capacity of holding apron}}{60\times 2}\times$ average gate occupancy time

[ESE : 2005]

1.15 Match **List-I** (Unit) with **List-II** (Purpose) and select the correct answer using the code given below the lists:

List-I
A. Width and length of safety area of airport
B. Engine failure case
C. Location of exit taxiways
D. Grading of airport size

List-II
1. Basic runway length
2. Runway geometric design

3. Airport drainage
4. Runway capacity

Codes:

	A	B	C	D
(a)	2	4	1	3
(b)	3	1	4	2
(c)	2	1	4	3
(d)	3	4	1	2

[ESE : 2005]

1.16 The percentage of time in a year during which the cross wind component remains within the limit, is

(a) wind coverage (b) head wind
(c) prevailing wind (d) cross wind

[ESE : 2005]

1.17 At a certain station, the mean of the average temperature is 25°C and mean of the maximum daily temperature is 40°C. What is the airport reference temperature (ART)?

(a) 46.6°C (b) 45°C
(c) 35°C (d) 30°C

[ESE : 2005]

1.18 A runway is located 450 m above the mean sea level. If the aeroplane reference field is 1800 m, what is the approximate corrected runway length for elevation?

(a) 1849 m (b) 1889 m
(c) 1987 m (d) 2013 m

[ESE : 2006]

1.19 For design of a runway length, match **List-I** (Factor) with **List-II** (Correction) and select the correct answer using the codes given below the list:

List-I

A. Standard basic length
B. Elevation
C. Temperature
D. Effective gradient

List-II

1. 7% for every 303 metres
2. 0.5%
3. 20% for each 1% gradient
4. Depends upon aircraft and obtained from ICAO Standard Tables

Codes:

	A	B	C	D
(a)	4	1	2	3
(b)	2	3	4	1
(c)	4	3	2	1
(d)	2	-	4	3

[ESE : 2006]

1.20 If M_i = Proportion of ith type of aircraft in the mix,
T_i = "Gate occupancy time" of the ith type of aircraft,
G = Number of gates (considering that all the available gates can be used by all aircrafts).

Which one of the following equations gives the ultimate gate capacity C?

(a) $C = \dfrac{\sum_{i=1}^{n} G \times M_i}{\sum_{i=1}^{n} T_i}$ (b) $C = \dfrac{G \times 60}{\sum_{i=1}^{n} M_i T_i}$

(c) $C = \sum_{i=1}^{n}\left(\dfrac{M_i}{G}\right) \times T_i$ (d) $C = \dfrac{G}{\sum_{i=1}^{n} M_i \times T_i}$

[ESE : 2006]

1.21 The total correction percentage for altitude and temperature, in calculating the runway length from basic runway length, normally does not exceed

(a) 7 (b) 14
(c) 28 (d) 35

[ESE : 2006]

1.22 As the elevation increases, the runway length has to be changed at what rate?

(a) Decreased @ 5% per 300 m rise in elevation above MSL
(b) Increased @ 7% per 300 m rise in elevation above MSL
(c) Decreased @ 9% per 300 m rise in elevation above MSL
(d) Increased @ 15% per 300 m rise in elevation above MSL

[ESE : 2007]

1.23 The orientation of runway is decided by which factors?

1. Maximum wind coverage and least cross winds
2. Landing characteristics of aircrafts
3. Scope for future expansion
4. Obstruction-free approaches

Select the correct answer using the codes given below:

(a) 1 and 2 only (b) 1, 2 and 3
(c) 1, 3 and 4 (d) 2, 3 and 4

[ESE : 2008]

1.24 The size of hangar in airport is guided by which of the following?

(a) Speed and direction of wind and visibility
(b) Weight of aircraft and turning needs
(c) The fuelling capacity and storage space
(d) The length, wingspan and height of aircraft

[ESE : 2008]

1.25 Match **List-I** (Component) with **List-II** (Design feature) and select the correct answer using the codes given below the lists:

List-I
A. Taxiways
B. Control tower
C. Hangar
D. Apron

List-II
1. Large span structures
2. Visibility allround and sound-proof
3. Turning radius
4. Waiting capacity for aircraft

Codes:

	A	B	C	D
(a)	1	4	3	2
(b)	3	2	1	4
(c)	1	2	3	4
(d)	3	4	1	2

[ESE : 2008]

Directions: The following items consists of two statements; one labelled as **'Assertion (A)'** and the other as **'Reason (R)'**. You are to examine these two statements carefully and select the answers to these items using the codes given below:

Codes:
(a) both A and R are true and R is the correct explanation of A
(b) both A and R are true but R is not a correct explanation of A
(c) A is true but R is false
(d) A is false but R is true

1.26 Assertion (A): Wind Rose Diagrams, showing wind direction, duration and intensity are an essential requirement for planning the best direction of main runway of the airport.

Reason (R): Corrections for elevation temperature and gradient have to be applied to determine the length of the main runway of an airport, as per ICAO recommendations.

[ESE : 2008]

1.27 Assertion (A): Runway capacity will be more in airport operating under VFR than IFR.

Reason (R): VFR gives more clear visual reference of airport during good weather conditions.

[ESE : 2008]

1.28 Assertion (A): In case of airports, the innermost portion of approach zone, where any obstruction is to be critically viewed, is known as the clear zone.

Reason (R): The area of airport, other than the approach zone, which is used for turning of aircrafts, is called the turning zone.

[ESE : 2009]

1.29 Assertion (A): Helipads are usually located in close proximity to traffic-generating areas.

Reason (R): A helicopter cannot ascend or descend exactly vertically.

[ESE : 2009]

1.30 Assertion (A): The paved area adjacent to the terminal building and hangers used for loading and unloading, servicing and parking of aircraft is called as apron.

Reason (R): The size of the apron depends on the size and number of gate positions and the way the gate positions are arranged.

[ESE : 2010]

1.31 Wind-rose diagram is useful in deciding on the orientation of

(a) Taxiway (b) Hanger
(c) Apron (d) Runway

[ESE : 2011]

1.32 For safe landing and take-off, the following factors need to be carefully considered:

1. Cross-wind
2. Runway grade
3. Runway width and side clearance
4. Obstructions

(a) 1, 2 and 3 only (b) 1, 2, 3 and 4
(c) 1, 3 and 4 only (d) 2, 3 and 4 only

[ESE : 2011]

1.33 **Statement (I):** The actual payload, particularly in passenger aircraft, is normally less than the maximum structural payload even when the aircraft is completely full.

Statement (II): There are limitations in the use of space when passengers are carried in an aircraft.

(a) Both Statement (I) and Statement (II) are individually true and Statement (II) is the correct explanation of Statement (I)
(b) Both Statement (I) and Statement (II) are individually true but Statement (II) is NOT the correct explanation of Statement (I)
(c) Statement (I) is true but Statement (II) is false
(d) Statement (I) is false but Statement (II) is true

[ESE : 2013]

1.34 The lowest height above the runway where the pilots make the decision to continue the landing manoeuvre to cut it short is called the

(a) Runway height (b) Decision height
(c) Threshold height (d) Runway visual range

[ESE : 2015]

1.35 The runway length for an airport located at 460 m above MSL, corrected for elevation, is 3670 m. The monthly means of maximum and mean daily temperatures for the hottest month of the year are 27°C and 18°C, respectively. What will be the final corrected length of the runway with correction incorporated also due to temperature effects?

(a) 4500 m (b) 4000 m
(c) 3750 m (d) 3400 m

[ESE : 2015]

1.36 What shall be the radius of an exit taxiway with design exit speed of 90 kmph and coefficient of friction 0.13?

(a) 550 m (b) 500 m
(c) 475 m (d) 449 m

[ESE : 2015]

1.37 The section of the tunnel adopted perfectly in lieu of ease of construction and maintenance in hard rock tunnels, where the risk of roof failure or collapse caused by external pressure from water, or from loose or unstable soil conditions on tunnel lining is practically non-existent, is

(a) Circular section
(b) Segmental roof section
(c) Horse-shoe section
(d) Egg-shaped section

[ESE : 2019]

1.38 Consider the following data for designing a taxiway for operating Boeing 707-320 aeroplane:

Wheel base = 17.70 m
Tread of main loading gear = 6.62 m
Turning speed = 40 km/h
Coefficient of friction between tyres and pavement surface = 0.13

The turning radius of the taxiway will be

(a) 98.5 m (b) 94.5 m
(c) 89.5 m (d) 86.5 m

[ESE : 2019]

1.39 Which one of the following instances of performance of aircraft is ***not*** considered for determining basic runway length?

(a) Normal landing case
(b) Normal take-off case
(c) Engine failure case
(d) Emergency landing case

[ESE : 2019]

■■■■

Answers Airport Engineering

1.1	(a)	1.2	(c)	1.3	(c)	1.4	(d)	1.5	(d)	1.6	(a)	1.7	(d)	1.8	(c)	1.9	(a)
1.10	(d)	1.11	(c)	1.12	(a)	1.13	(a)	1.14	(a)	1.15	(c)	1.16	(a)	1.17	(d)	1.18	(c)
1.19	(a)	1.20	(d)	1.21	(d)	1.22	(b)	1.23	(c)	1.24	(d)	1.25	(b)	1.26	(b)	1.27	(a)
1.28	(b)	1.29	(c)	1.30	(b)	1.31	(d)	1.32	(c)	1.33	(b)	1.34	(b)	1.35	(b)	1.36	(b)
1.37	(b)	1.38	(a)	1.39	(d)												

Explanations Airport Engineering

1.1 (a)

PNB: Perceived Noise Level Intensity
LCN: Load Classification Number
STOL: Short Take off and Landing

1.2 (c)

The standard temperature at the airport site can be determined by reducing the standard mean sea level temperature of 15°C at the rate of 6.5°C per thousand metre rise in elevation.

Standard temperature = 15 – 6.5 = 8.5°C
Rise in temperature = 16 – 8.5 = 7.5°C

1.3 (c)

The scissor crossing consists of 4 pair of points (switches), six acute angle crossings; two obtuse angle crossings, check rails and straight lengths.

1.4 (d)

Approach surface is trapezoidal in shape and it is longitudinally centred on the extended centre-line of the runway.

Conical surface extends upwards and outwards from the outer circular edge of the horizontal surface to a point which is at some height above the horizontal surface.

1.5 (d)

A taxiway is a path for aircraft at an airport connecting runways with aprons, hangars, terminals and other facilities. Taxiway is designed considering volume of air traffic, runway configuration, location of terminal building, location of facilities like hanger etc.

1.6 (a)

$\alpha \rightarrow$ Cockpit cutoff angle
$\beta \rightarrow$ Pitch angle
$\phi \rightarrow$ Glide slope angle
$\theta \rightarrow$ Angle formed by visual range with horizontal

1.7 (d)

Air traffic control, environmental factors, characteristic of demand and layout design of runway system affect runway capacity.

1.8 (c)

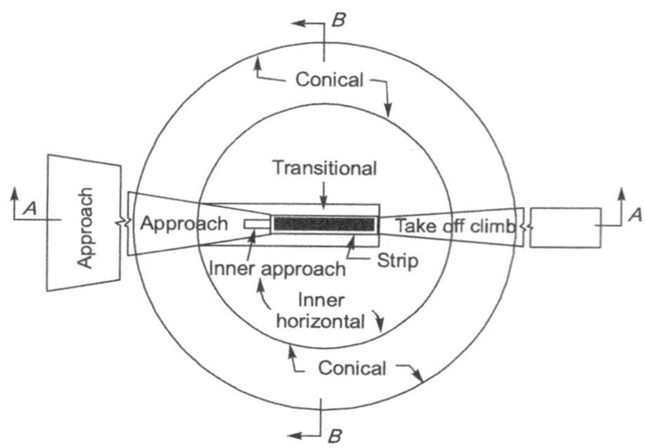

Primary Surface: A surface longitudinally centered on a runway.

Conical Surface: A surface, which extends upward and outward from the outer limits of the Horizontal Surface for a horizontal distance of 4,000 feet. The slope of the conical surface is 20-1 (5 percent) measured in a vertical plane.

Transitional Surface: A surface extending outward and upward, at right angles to the runway centerline and runway centerline extended, from the sides of the Primary Surface and the Approach Surfaces. The slope is 7-1 (14.3 percent) and the surface extends until it intersects the Horizontal or Conical Surface.

Approach Surface: A surface longitudinally

centered, on the extended centerline of the PIR runway, beginning at the end of the Primary Surface and extending outward and upward at a slope of 50:1 (2.0 percent) for a horizontal distance of 10,000 feet and at a slope of 40:1 (2.5 percent) for an additional 40,000 feet.

1.9 (a)

Airport Reference temperature,

$$T = \frac{2}{3} \times T_{av} + \frac{1}{3} \times T_{max}$$

$$= \frac{2}{3} \times 40 + \frac{1}{3} \times 49 = 43°C$$

1.10 (d)

Outer horizontal surface (OHS) is not established for the aerodromes having runway of length less than 900 m. The height of OHS is 150 m above the airport reference point (ARP) elevation.

1.11 (c)

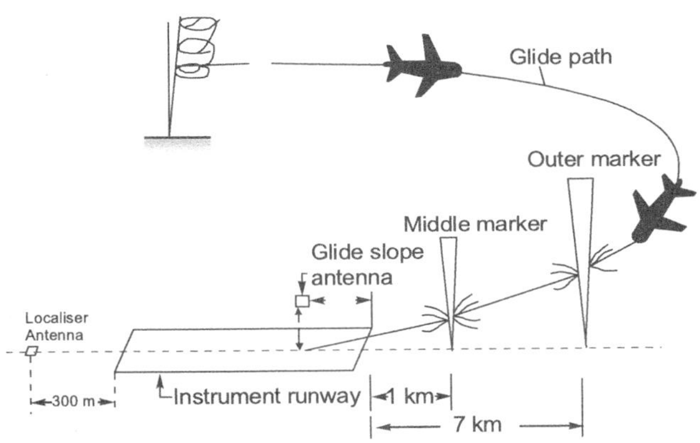

1.12 (a)

Runways are always orientated in the *direction* of the prevailing *winds*, so that the force of the *wind can be utilized* during take-off and *landing* operations. The length of runway should be adequate for landing and takeoff operations and should be free from any obstruction.

1.13 (a)

Factors affecting Basic runway length are:

(i) Normal landing

(ii) Normal take off with all engines.

(iii) Engine failure at take off.

1.14 (a)

The number of gates depend on the following two factors:

(i) Estimated peak hourly volume which should not exceed the capacity of runway for balanced airport design

(ii) Gate occupancy time

∴ Number of Gates

$$= \frac{\text{Capacity of runway}}{60 \times 2} \times \text{Avg. occupancy time}$$

1.15 (c)

The items to be considered in the geometric design of a runway are as follows:

(i) Length

(ii) Longitudinal and effective gradient

(iii) Rate of change of longitudinal gradient

(iv) Safety area

(v) Sight distance

(vi) Transverse gradient

(vii) Width

1.16 (a)

The cross wind component should not exceed 25 kmph. The percentage of time in a year during which the cross wind component remains within the limit of 25 kmph is called wind coverage of the runway. The orientation of the runway should be such that the minimum wind coverage of about 95% is obtained.

Head wind indicates the wind from the opposite direction of the head or nose of the aircraft while it is landing or taking off.

1.17 (d)

Airport reference temperature,

$$T = \frac{2}{3} \times 25 + \frac{1}{3} \times 40 = 30°C$$

1.18 (c)

The elevation correction is 7% per 303 m rise in elevation of airport above MSL.

For 450 m elevation the correction will be

$$= \frac{450}{303} \times 7 = 10.4\%$$

So the corrected runway length will be

$$= 1800 + \frac{10.4}{100} \times 1800 = 1987.2 \text{ m}$$

1.19 (a)

Standard basic length of runway depends upon type of aircraft and are obtained from ICAO tables.

Elevation correction to basic runway length is 7% for every 300 m. Temperature correction is 1% and effective gradient correction is 20% for each 1% gradient.

1.20 (d)

Gate occupancy time × Gate Capacity = Number of Gates (G).

$$\text{Gate occupancy time} = \sum_{i=1}^{n} M_i T_i$$

$$\therefore \quad C = \frac{G}{\sum_{i=1}^{n} M_i T_i}$$

1.21 (d)

As per ICAO's recommendation if the total correction for elevation plus temperature exceeds 35% of basic runway, further checking needs to be done by conducting specific studies through model tests.

1.22 (b)

As per ICAO's recommendation, the runway length should be increased @ 7% per 300 m rise in elevation above the mean sea level. Basic runway length is usually corrected for elevation, gradient and temperature.

1.23 (c)

Generally runways are oriented in direction of prevailing winds which helps in takeoff and landing operation. Obstruction free approaches are essential for orientation of runway. There should for scope for future expansion so orientation of runway is decided accordingly.

1.24 (d)

Hangar in airport useful for servicing, overhauling and repairing aircrafts. Its size depends on turning radius and size of aircraft(its length, height and wingspan)

1.26 (b)

A wind rose is a graphic tool used by meteorologists to give a succinct view of how wind speed and direction are typically distributed at a particular location. Using a polar coordinate system of gridding, the frequency of wind direction, with color bands showing wind speed ranges. The direction of the longest spike shows the wind direction with greatest frequency.

1.27 (a)

Visibility is lower than prescribed in instrumental flight rules(IFR) so rigid traffic control is applied in this. In Visual flight rules (VFR) air traffic control is not practically requires as pilot can maintain separation with visual aids. Runway capacity will be more in airport operating under VRF than IFR. Thus VFR can give more clear visual reference of airport during good weather conditions.

1.29 (c)

The general considerations in the choice of a suitable site for a helipad are:

(i) It should be located in close proximity to the traffic generating areas.

(ii) It should have a convenient access by ground transportation, e.g. roadway or railway.

(iii) The ambient noise level should be considered.

(iv) The site for helipad should provide unobstructed approach departure paths for helicopters.

A helicopter can ascend or descend exactly vertically. However, it is not practicable, within economic limits that helicopter can ascend and descend exactly vertically.

1.30 (b)

The pavement area adjacent to terminal building and hangers used for loading and unloading , servicing and parking of aircraft is called as apron. Apron size depends of:

(i) Size and number of gate positions

(ii) Positioning of gates

(iii) Aircraft parking system

(iv) Size of loading area for aircraft

1.31 (d)

The wind data, i.e., direction, duration and intensity are graphically represented by a diagram called wind rose. The wind data should usually be collected for a period of at least 5 years and preferably of 10 years, so as to obtain an average data with sufficient accuracy. It helps in analysing the wind data and obtaining the most suitable direction of the runway.

1.32 (c)

The runway surface made of asphalt or cement concrete will not have irregularities that would result in the loss of frictional characteristics that would otherwise adversely affect the take-off or landing of an aircraft.

The portion of a strip at the end of a runway is prepared to resist blast erosion, in order to protect a landing airplane from the danger of an exposed edge.

1.33 (b)

Payload is the carrying capacity of an aircraft, usually measured in terms of weight; depending on the nature of the flight, the payload of a vehicle may include cargo, passengers, flight crew, munitions, scientific instruments or experiments, or other equipment. Extra fuel, when optionally carried, is also considered part of the payload. In an airline or air freight carrier, payload may refer only to revenue-generating cargo or paying passengers.

1.34 (b)

Decision height: The lowest height above the runway where the pilots make the decision to continue the landing manoeuvre or to cut is short is called the decision height.

Threshold height: The theoretical height above the runway threshold at which the aircraft's glideslope antenna would be if the aircraft maintained the trajectory established by the mean ILS glideslope or MLS glide path.

Runway visual range: The runway visual range (RVR) is the distance over which a pilot of an aircraft on the centrelne of the runway can see the runway surface markings delineating the runway or identifying its centre line. RVR is normally expressed in feet or meters.

1.35 (b)

Final corrected length of runway

= Given runway length + Length incorporated due to temp. effect

= 3670 + L

$$L = \frac{\text{Runway length}}{100} \times (\text{Airport ref. temp} - \text{Std. temp.})$$

$$\text{Airport ref temp} = T_a + \frac{(T_m - T_a)}{3}$$

$$= 18 + \left(\frac{27-18}{3}\right) = 21°\text{C}$$

$$\text{Std. temp} = 15 - 0.0065 \times 460 = 12.01°\text{C}$$

$$L = \frac{3670}{100}(21 - 12.01) = 329.93 \text{ cm}$$

Final corrected length

$$= 3670 + 329.93$$

$$\simeq 4000 \text{ m}$$

1.36 (b)

Radius of exit taxiway,

$$R = \frac{V^2}{125f} = \frac{90^2}{125 \times 0.13}$$

$$= 498.46 \simeq 500 \text{ m}$$

1.38 (a)

1. $$R = \frac{V^2}{125f} = \frac{40^2}{125 \times 0.13} = 98.46 \text{ m}$$

2. $$R = \frac{0.388w^2}{\frac{T}{2} - S} = \frac{0.388 \times 17.7^2}{\frac{22.5}{2} - \left(\frac{6.62}{2} + 6\right)} = 62.66 \text{ m}$$

1.39 (d)

Basic runway length is determined based on three criteria:

1. Normal landing case
2. Normal takeoff case
3. Engine failure case

2 Dock and Harbour

2.1 The phenomenon of movement and deposition of sand in a zig-zag style due to drifting of sand in the vicinity of coast is called
(a) littoral drift (b) sedimentation
(c) beach drift (d) trough action
[ESE : 2000]

2.2 Match **List-I** (Structure) with **List-II** (Purpose) and select the correct answer using the codes given below the lists:

List-I
A. Breakwater
B. Wharf
C. Fender system
D. Revetments

List-II
1. Separates the land from sea water
2. Lays vessels alongside, receives and discharges cargo and passengers
3. Protects a seashore
4. Absorbs the energy of the moving vessel
5. Retards erosion of an existing beach

Codes:

	A	B	C	D
(a)	5	2	4	3
(b)	3	4	2	1
(c)	3	2	4	1
(d)	1	4	3	5

[ESE : 2000]

2.3 The lowest tide which occurs in half lunar month is called
(a) spring tide (b) neap tide
(c) lunar tide (d) tidal bore
[ESE : 2000]

2.4 In deep water waves, individual particles
(a) are translated
(b) are translated and rotated in a vertical plane about a horizontal axis
(c) are not translated, but they rotate in a vertical plane about a horizontal axis
(d) do not rotate in a vertical plane about a horizontal axis [ESE : 2001]

2.5 **Assertion (A):** The most common approach for control of beach erosion is to build groynes.
Reason (R): Groynes are constructed parallel to the shore.
(a) both A and R are true and R is the correct explanation of A
(b) both A and R are true but R is not a correct explanation of A
(c) A is true but R is false
(d) A is false but R is true [ESE : 2001]

2.6 Consider the following factors:
1. Mass of ships to be berthed
2. Overall transit time
3. Speed of berthing
4. Environmental conditions of the port

Which of these factors are taken into account while selecting the type of fender system?
(a) 1, 2 and 3 (b) 2, 3 and 4
(c) 1, 3 and 4 (d) 1 and 4
[ESE : 2002]

2.7 **Assertion (A):** The principal advantage of the solid fill type of wharf substructure is that its great mass provides adequate resistance to the impact of mooring ships.
Reason (R): Solid fill substructures are inexpensive (except in deep water) and require little maintenance.
(a) both A and R are true and R is the correct explanation of A
(b) both A and R are true but R is not a correct explanation of A
(c) A is true but R is false
(d) A is false but R is true [ESE : 2002]

2.8 **Assertion (A):** Wet docks are enclosed or partially enclosed basins provided with locks and entrance gates to keep the water at fairly constant level for allowing vessels to come in, berth or leave at all times.

Reason (R): Wet docks require expensive arrangements like lock gate and the walls are designed to withstand backfill pressure when the dock is full.

(a) both A and R are true and R is the correct explanation of A
(b) both A and R are true but R is not a correct explanation of A
(c) A is true but R is false
(d) A is false but R is true [ESE : 2003]

Directions: The following items consists of two statements; one labelled as **'Assertion (A)'** and the other as **'Reason (R)'**. You are to examine these two statements carefully and select the answers to these items using the codes given below:

Codes:
(a) both A and R are true and R is the correct explanation of A
(b) both A and R are true but R is not a correct explanation of A
(c) A is true but R is false
(d) A is false but R is true

2.9 **Assertion (A):** Breakwaters are common in natural harbours.

Reason (R): Breakwaters help in controlling wave action and sand movement into the harbour area. [ESE : 2004]

2.10 Which one of the following is the best method for locating soundings to estimate the dredged material form the harbours?
(a) Two angles from shore
(b) Two angles from boat
(c) One angle from shore and other from the boat
(d) Fixed intersecting ranges [ESE : 2005]

2.11 Why are moorings provided?
(a) For anchoring of ships
(b) For towing the ships to the sea
(c) For repair of ships
(d) For washing of ships and ship boards [ESE : 2005]

2.12 What is the use of a station pointer?
(a) For making soundings in water bodies
(b) For plotting of soundings in harbour area
(c) For marking sunken shipping hazards
(d) For making tidal observations [ESE : 2005]

2.13 Consider the following statements:
1. Dock is a marine structure for mooring up vessels, loading and unloading of passengers and/or cargo.
2. Dry dock is generally used only for carrying out repairs, inspections and painting.
3. Wet dock is an enclosed or partially enclosed basin provided with locks and entrance gate to keep the water level at a fairly constant level.

Which of these statements are correct?
(a) 1, 2 and 3 (b) 2 and 3
(c) 1 and 2 (d) 1 and 3 [ESE : 2005]

2.14 Which one of the following is not correct for container ports?
(a) The berth capacity is great
(b) Overall transit time is less
(c) There is minimal damage to cargo
(d) Minimal land is required for the marshalling area [ESE : 2007]

2.15 Consider the following statements associated with groynes:
1. Hydraulic behaviour of a system of groynes is influenced by the characteristics of particles that constitute the littoral drift.
2. Groyne is constructed approximately parallel to shore.

Which of these statements is/are correct?
(a) 1 only (b) 2 only
(c) Both 1 and 2 (d) Neither 1 nor 2 [ESE : 2007]

2.16 What is the wave velocity for a uniform train of wave beyond the storm centre for a wave length of 20 m in 14 m deep water?
(a) 5.5 m/s (b) 11.2 m/s
(c) 4.5 m/s (d) 9 m/s [ESE : 2007]

2.17 Match **List-I** (Harbour structure) with **List-II** (Purpose) and select the correct answer using the codes given below the lists:

List-I
A. Dolphins
B. Groynes
C. Fenders
D. Slipways

List-II
1. To protect the beach or retained earth
2. To cushion to absorb shock of ship during loading and un-loading
3. Mooring structure in combination with loading platform
4. Repairing and building ship

Codes:

	A	B	C	D
(a)	2	4	3	1
(b)	3	1	2	4
(c)	2	1	3	4
(d)	3	4	2	1

[ESE : 2008]

2.18 What is the minimum turning radius for a passenger ship?
(a) 2 × ship length
(b) 3 × ship length
(c) (1.5 × ship length) + water depth
(d) (2 × ship length) + (1.5 × draft)

[ESE : 2008]

2.19 Which of the following pairs are correctly matched?
1. Vishakhapatnam and Plymouth—Seminatural harbour
2. Paradip and Mangalore —Lagoon harbour
3. New York—Natural harbour

Select the correct answer using the codes given below:
(a) 1 and 2 only (b) 2 and 3 only
(c) 1 and 3 only (d) 1, 2 and 3

[ESE : 2008]

2.20 Echo-sounding machine is used for which of the following?
1. Locating objects on sea-shore
2. Measuring sea-surface levels
3. Determining depth of sea-bed below sea surface
4. Locating sunken objects below sea surface

Select the correct answer using the codes given below:
(a) 1 and 3 only (b) 1, 2 and 3
(c) 2 and 4 (d) 3 and 4

[ESE : 2008]

2.21 Match **List-I** (Components) with **List-II** (Function) and select the correct answer using the codes given below the lists:

List-I
A. Keel and Bilge blocks
B. Capstons and Bolards
C. Fenders
D. Buoys

List-II
1. Impact absorbers of quay walls
2. Floating indicators
3. Supports for ships in dry docks
4. Anchorage devices for ships

Codes:

	A	B	C	D
(a)	3	4	1	2
(b)	4	3	1	2
(c)	4	3	2	1
(d)	3	4	2	1

[ESE : 2008]

2.22 Zigzag deposition and movement of sand on sea beach is called
(a) littoral drift (b) beach drift
(c) trough action (d) sedimentation

[ESE : 2010]

2.23 **Assertion (A):** In the overall design and layout of a harbour (or port), differentiation is made between 'entrance' and 'channel' depths.

Reason (R): The 'scend' (or pitching) of a vessel may be larger at the entrance to the channel than within the channel.

[ESE : 2010]

2.24 Which of the following complete sets do not recommend the siting of a harbour layout in that vicinity?
1. Submarine canyon
2. Lee of an island
3. Closely located promontories
4. Indentation coves on the coastline
5. Hooked bays with not-so-rugged rocky bottom

(a) 2, 3, 4 and 5 (b) 1, 4 and 5 only
(c) 2, 3 and 4 only (d) 3, 4 and 5 only

[ESE : 2011]

2.25 Overland flow drainage on and from the tarmac of an airport invokes, in its design principles involving:
(a) Spatially varied flow without hydraulic jumps
(b) Backwater flow
(c) Subcritical flow throughout
(d) Attention to rolling flows within spatially varied flows and possibly to moving hydraulic jumps
[ESE : 2011]

2.26 In the offshore region at a particular harbour facility, an oscillatory wave train approaches with wavelength of 80 m where the mean sea depth is 30 m. What would be the velocity of the individual waves?
(a) 17.15 m/s (b) 16.05 m/s
(c) 15.15 m/s (d) 14.05 m/s
[ESE : 2018]

2.27 For proper planning of harbours, oscillatory waves in the relevant off-shore region must be taken into account. If the sea depth is 30 m and an oscillatory waves train is observed to have wavelength of 50 m, what would be the velocity of the individual waves?
(a) 9.43 m/s (b) 9.21 m/s
(c) 9.08 m/s (d) 8.83 m/s
[ESE : 2018]

2.28 Which one of the following features does *not* pertain to Littoral drift?
(a) It depends on length of wave
(b) It is the process of erosion of deposition by waves
(c) Waves caused by prevailing wind, stir up and move sand particles
(d) Wind tends to carry drifting sand in a zigzag way
[ESE : 2019]

Answers Dock and Harbour

2.1	(a)	2.2	(c)	2.3	(a)	2.4	(d)	2.5	(c)	2.6	(c)	2.7	(b)	2.8	(b)	2.9	(d)
2.10	(d)	2.11	(a)	2.12	(b)	2.13	(a)	2.14	(a)	2.15	(a)	2.16	(b)	2.17	(b)	2.18	(a)
2.19	(d)	2.20	(d)	2.21	(a)	2.22	(a)	2.23	(a)	2.24	(d)	2.25	(a)	2.26	(a)	2.27	(d)
2.28	(d)																

Explanations Dock and Harbour

2.1 (a)
Littoral drift is a geographical process that consists of the transportation of sediments (clay, silt, sand and shingle) along a coast at an angle to the shoreline in a zig-zag style, which is dependent on prevailing wind direction, swash and backwash. This process occurs in the littoral zone, and in or close to the surf zone. The process is also known as longshore drift, longshore current or longshore transport.

2.2 (c)
Breakwaters are structures constructed on coasts as part of coastal defense or to protect an anchorage from the effects of both weather and longshore drift wharf is a structure built on the shore of or projecting into a harbor, stream, etc., so that vessels may be moored alongside and cargos and passengers can be received and discharged.

Fender is a bumper used to absorb the kinetic energy of a boat or vessel berthing against a jetty, quay wall or other vessel.

Revetments are usually built to separate land from sea water, to preserve the existing uses of the shoreline and to protect the slope, as defense against erosion.

2.3 (a)
Spring tides occur usually at new and full moon or rather a day or two after (or twice in each lunar month). At new moon, the sun and moon have the same celestial longitude and cross a meridian of

the earth at the same instant. The sun and moon are on same side of the earth. The high water level of the resulting tide is above the average, whereas the low water level is below the average.

During neap tide, the moon falls behind the sun. High water level is below the average, whereas the low water level is above the average.

2.4 (d)

When a wave is in deep water, the motion of individual particle at the surface follows a circular orbital pattern and orbital radius falls off quickly with depth. Also the particles move forward along with waves so there will be translation as well as rotation.

2.5 (c)

Groynes are constructed perpendicular to the shore to control beach erosion.

2.6 (c)

The fenders are installed:

(i) To prevent direct contact between quay vessel while the vessel is moored alongside the berth (rubbing or protection fenders)

(ii) To reduce the risk of a mooring vessel moving in resonance with waves.

(iii) To absorb berthing impact energy.

The fender system controls the relative motion between the dock and ship caused by wind and waves. The maximum amount of impact depends upon weight of the ship, angle of strike and velocity of approach or speed of berthing.

2.7 (b)

Solid fill type of wharf substructures consist of earth, concrete or rock partly confined by bulkhead or retaining wall. These are heavy so they resist impact of mooring ships. These are inexpensive and require little maintenance except in deep water where the construction cost and maintenance cost rises.

2.8 (b)

Wet docks are the docks in which water is maintained at a level at which a ship is able to float. They are classified into:- Wet docks in tidal basin and Wet docks in mounted/enclosed basin.

Wet docks or harbor docks are enclosed or partially enclosed basins provided with lock and entrance gates to keep water at fairly constant level. This helps keep water level constant and allows vessels to come in and leave. But the arrangements like lock gate and walls which are designed to withstand backfill pressure are quite expensive.

2.9 (d)

They are protective barrier constructed to form artificial harbor. These provide safe accommodation for shipping and protect from effect of sea waves.

2.10 (d)

Locating sounding by intersecting range is best suited for harbours, reservoirs, etc. where it is desired to determine the character of changes in the bed due to scour silting, or to measure the dredged material.

2.11 (a)

Mooring is provided for anchoring a boat or ship. A ship is secured to a mooring to forestall free movement of the ship on the water. An anchor mooring fixes a vessel's position relative to a point on the bottom of a waterway without connecting the vessel to shore. There are various kinds of moorings such as wing mooring, pile mooring etc.

2.12 (b)

Station pointer is a three-armed protractor and consists of a graduated circle with a fixed arm and two movable arms to the either side of the fixed arm. It is used for plotting of soundings.

2.13 (a)

Docks are maritime structures used for loading and unloading of passengers and for mooring up vessels. They can be classified into two types : Dry docks and Wet docks. Dry docks are used for repairs, inspections etc whereas Wet docks or harbor docks are enclosed or partially enclosed basins provided with lock and entrance gates to keep water at fairly constant level.

2.14 (a)

Container ports have limited berth capacity due to which there is minimal damage to cargo and overall transit time is small. Also, they occupy the minimum land for marshalling area.

2.15 (a)

Groyne is connected perpendicular to the shore to prevent erosion. Littoral drift is a geographical process that consists of the transportation of sediments (clay, silt, sand and shingle) along a coast at an angle to the shoreline in a zig-zag style. This movement affects hydraulic behavior of groynes.

2.16 (b)

(i) When the depth of water is great compared to the length of wave, the wave velocity is given by

$$v = 1.25\sqrt{l}$$

where l is length of wave

(ii) In shallow water, if the depth of water is 'd', the wave velocity is given by

$$v = 3.16\sqrt{d}$$
$$= 3.16\sqrt{1.4}$$
$$= 11.82 \text{ m/s}$$

2.18 (a)

Turning radius = 2 × ship length

Minimum dimension of turning basin should not be less than twice the length of largest ship so that ship can turn a complete circle without help of tugs.

2.19 (d)

A natural harbor is a landform where a part of a body of water is protected and deep enough to furnish anchorage. Well known natural harbours are New York City, Rio de Janeiro, San Francisco, Tokyo etc

Semi natural harbour requires artificial protection only at entrance. Vishakhapatnam and Plymouth are semi natural harbours. Lagoon harbor lies where shallow body of water is separated from large body by reefs or barrier island. Paradip and Mangalore are well known lagoon harbours.

2.20 (d)

Echo sounding is a type of SONAR used to determine the depth of water by transmitting should pulses into water. Echo sounding Machines are used comonly to determine depth of sea bed below sea surface, aid in navigation, locating objects below sea surface, etc.

2.22 (a)

Littoral drift is a geographical process that consists of the transportation of sediments (clay, silt, sand and shingle) along a coast at an angle to the shoreline in a zig-zag style, which is dependent on prevailing wind direction, swash and backwash. This process occurs in the littoral zone, and in or close to the surf zone. The process is also known as longshore drift, longshore current or longshore transport.

2.23 (a)

Depth and width required at entrance are more than those required in channel as entrance is more exposed to waves. Due to more exposure to waves, scending or pitching of a vessel may be larger at entrance to the channel than within the channel.

2.26 (a)

Velocity of waves,

$$V = \sqrt{\frac{gl}{2\pi}\tanh\frac{2\pi d}{l}}$$

Given, l = 80 m and d = 30 m

$$\tanh\frac{2\pi d}{l} = 0.982 \simeq 1$$

$$V = \sqrt{\frac{g \times 80}{2\pi}} = 11.17 \text{ m/s}$$

Note: When mean sea depth (h) is very very less than wavelength (L), then the equation applied is

$$V = \sqrt{gh}$$
$$= \sqrt{9.81 \times 30} = 17.15 \text{ m/s}$$

Thus to match the correct option we will take 17.15 m/s as our answer.

2.27 (d)

Wavelength (l) = 50 m

Sea depth (d) = 30 m

Velocity of waves, $V = \sqrt{\frac{gl}{2\pi}\tanh\frac{2\pi d}{l}}$

Considering, d values is more

So, $\tanh\frac{2\pi d}{l} \simeq 1.0$

Then, $V = \sqrt{\frac{gl}{2\pi}} = \sqrt{\frac{9.81\times 50}{2\pi}}$

$V = 8.83$ m/s

2.28 (d)

Littorals, drift is a geological process that consists of the transportation of sediments along a coast parallel to the shoreline, which is dependent on oblique incoming wind direction. In beach drift wind tends to carry drifting sand in a zig-zag way.

■■■■

3 Tunnel Engineering

3.1 Which one of the following methods of tunneling is used in hard rock?
(a) Forepoling method
(b) Needle beam method
(c) Heading and benching method
(d) Shield tunnelling method [ESE : 2000]

3.2 Tunnel alignment is carried out by
(a) surface theodolite traverse
(b) triangulation
(c) compass traverse
(d) aerial photography [ESE : 2001]

3.3 With reference to tunnelling, which of the following factors are to be considered for deciding the size of the shaft?
1. System used for hoisting
2. Size of the muck car
3. Quantity of muck to be lifted
4. Eventual use of the shaft

Select the correct answer using the codes given below:
(a) 1, 2 and 3 (b) 1 and 4
(c) 2, 3 and 4 (d) 1, 2, 3 and 4
[ESE : 2001]

3.4 Which one of these methods of Tunnel Construction is not suitable in rocks?
(a) Full face method
(b) Compressed air method
(c) Heading and benching method
(d) Drift method
[ESE : 2002]

3.5 Match **List-I** (Tunnel construction methods) with **List-II** (Advantages) and select the correct answer using the codes given below the lists:

List-I
A. Full face method
B. Heading and Benching method
C. Drift method
D. Compressed air method

List-II
1. Helps in ventilation of tunnel
2. Useful in soft and water-bearing soil
3. Simultaneous drilling and mucking is possible
4. Tunnelling is continuous

Codes:

	A	B	C	D
(a)	4	3	2	1
(b)	3	4	1	2
(c)	4	3	1	2
(c)	3	4	2	1

[ESE : 2003]

3.6 Match **List-I** (Method) with **List-II** (Suitability) and select the correct answer using the codes given below the lists:

List-I
A. Full face method
B. Heading and Bench method
C. Drift method
D. American method

List-II
1. For tunneling of large cross-section area through soft ground driving a top drift and then widening the sides
2. For rock tunneling carried out first in smaller section of the proposed tunnel and then widened.
3. For driving the top portion in advance of the bottom portion for very large tunnel section in unstable rocks.
4. For tunneling of small cross-section area in stable rock attacking the whole section at a time

Codes:

	A	B	C	D
(a)	2	1	4	3
(b)	4	3	2	1
(c)	4	1	2	3
(d)	2	3	4	1

[ESE : 2004]

Directions: The following items consists of two statements; one labelled as **'Assertion (A)'** and the other as **'Reason (R)'**. You are to examine these two statements carefully and select the answers to these items using the codes given below:

Codes:

(a) both A and R are true and R is the correct explanation of A
(b) both A and R are true but R is not a correct explanation of A
(c) A is true but R is false
(d) A is false but R is true

3.7 **Assertion (A):** Open cut is economical than tunnelling for depths of over burden less than 20 metres.
Reason (R): Heading and Bench method is suitable for tunnelling in soft soils.
[ESE : 2004]

3.8 Consider the following statements with respect to tunnelling methods:
1. Full face excavation is suitable for small size tunnel of short length in competent rock. It is not suitable in urban areas.
2. Heading and benching method is suitable for soft rock tunneling of medium size.
3. Drift method is suitable for large tunnels in difficult or incompetent rock.

Which of these statements are correct?
(a) 1 and 3 (b) 1, 2 and 3
(c) 2 and 3 (d) 1 and 2 [ESE : 2005]

3.9 Which one of the following methods is generally considered the best for tunnel ventilation?
(a) Driving a drift through the tunnel
(b) 'Blow in' method
(c) 'Blow out' method
(d) Combination of 'Blow in' and 'Blow out' methods
[ESE : 2007]

3.10 Which one of the following methods of tunnelling is employed if the strata is sub-aqueous?
(a) Shield tunnelling (b) Drift system
(c) Liner plate method (d) Pilot tunnel method
[ESE : 2008]

3.11 Weisbach Triangle method may be used for which one of the following?
(a) To carry out surface alignment of a tunnel
(b) To transfer levels underground for a tunnel
(c) To connect surface and underground tunnel surveys
(d) To connect two ends of an underground tunnel
[ESE : 2008]

3.12 How can additional access, quick removal of muck and natural ventilation in long tunnels be achieved?
(a) By providing jumbos and pumps
(b) By providing scaffolding and liner plates
(c) By providing adits and shafts
(d) By providing pilot tunnels and well points
[ESE : 2008]

3.13 The minimum possible grade that can be provided in a tunnel and its approaches with providing adequately for proper drainage is
(a) 0.1% (b) 0.2%
(c) 0.3% (d) 0.4% [ESE : 2019]

3.14 Which one of the following methods is adopted for tunneling in soft soils?
(a) Pilot tunnel method
(b) Drift method
(c) Needle beam method
(d) Heading and benching method
[ESE : 2019]

3.15 Which of the following advantages are correct regarding Poly-centric shape tunnel?
1. It can be conveniently used for road and railway traffic.
2. The number of centres and lengths of radii cannot be fixed.
3. It can resist external and internal pressure due to its arch action.

(a) 1 and 2 only (b) 2 and 3 only
(c) 1 and 3 only (d) 1, 2 and 3
[ESE : 2020]

3.16 Which one of the following tunneling methods is adopted of the situations where the metro alignment passage under residential buildings or a canal?
(a) Earth pressure balance tunnelling machine method
(b) Tunnel boring machine method
(c) Tube tunnelling method
(d) Driven shield tunnelling method
[ESE : 2021]

3.17 Which one of the following is not the method of tunnelling is hard rock?
(a) Full-face heading method
(b) Heading and bench method
(c) Drift method
(d) Shaft method [ESE : 2021]

3.18 Which one of the following alignments is obtained by development accompanied by tunneling?
(a) Valley alignment
(b) Cross country alignment
(c) Mountain alignment
(d) Zig-zag alignment [ESE : 2023]

3.19 Consider the following statements regarding tunnel:
1. In hills with soft rocks, a tunnel is cheaper than a cutting.
2. The maintenance cost of a tunnel is considerably higher than that of a bridge.
3. The construction of a tunnel is costly as it requires special construction.

Which of the above statements are correct?
(a) 1 and 2 only (b) 2 and 3 only
(c) 1 and 3 only (d) 1, 2 and 3
[ESE : 2024]

3.20 Match the following lists:
List I (Size of the tunnel)
P. Circular
Q. Elliptical
R. Horseshoe
List II (Purpose of the tunnel)
1. Water and sewage mains
2. Roads and Railways
3. Water and sewage

Select the correct answer using the code given below:

	P	Q	R
(a)	2	1	3
(b)	3	1	2
(c)	3	2	1
(d)	2	3	1

[ESE : 2024]

3.21 Consider the following statements regarding full face method for tunneling in hard rocks:
1. An entire section of the tunnel is tackled at one time.
2. Mucking tracks can be laid on the tunnel floor and extended as the work progresses.
3. It is suitable for unstable rocks.

Which of the above statements are correct?
(a) 1 and 2 only (b) 2 and 3 only
(c) 1 and 3 only (d) 1, 2 and 3
[ESE : 2024]

3.22 Match the following lists:
List-I (Type of soil)
P. Silt
Q. Clay
R. Sand
List-II (Method of tunneling)
1. One or two ports are opened and the material flows continuously into the tunnel.
2. One or two port doors are opened. The material is excavated and deposited at the bottom of the tunnel.
3. Tunneling is of the open type. The material settles on the floor of the shield and it should be continuously removed.

Select the correct answer using the code given below:

	P	Q	R
(a)	2	1	3
(b)	3	1	2
(c)	3	2	1
(d)	2	3	1

[ESE : 2024]

3.23 Which one of the following methods is achieved by drilling a drift through the tunnel from portal to portal?
(a) Blow-in method (b) Exhaust method
(c) Blow-out method (d) Natural method
[ESE : 2024]

3.24 Which one of the following situations is NOT at all suitable for tunneling?
(a) If the beds are parallel to horizontal or have approximately zero degree inclination
(b) If the beds are vertically dipping and the axis of the tunnel is perpendicular to the strike.
(c) If the strike of involved beds is parallel to the axis of the tunnel.
(c) If the dip of involved beds is parallel to the axis of the tunnel.

[ESE : 2024]

■■■■

Answers Tunnel Engineering

3.1	(c)	3.2	(b)	3.3	(d)	3.4	(b)	3.5	(c)	3.6	(b)	3.7	(c)	3.8	(d)	3.9	(d)
3.10	(c)	3.11	(d)	3.12	(c)	3.13	(b)	3.14	(c)	3.15	(c)	3.16	(d)	3.17	(d)	3.18	(c)
3.19	(d)	3.20	(b)	3.21	(a)	3.22	(a)	3.23	(a)	3.24	(c)						

Explanations Tunnel Engineering

3.1 (c)

Methods of tunneling in hard rock

(i) Drift method
(ii) Heading and benching method
(iii) Full face method
(iv) Cantilever car dump method
(v) Pilot tunnel method
(vi) Perimeter method or German method.

Methods of tunneling in soft soil

(i) Forepoling method
(ii) Needle beam method
(iii) Army method or case method
(iv) American method
(v) English method
(vi) Belgian method
(vii) German method
(viii) Austrian method
(ix) Linear plates method
(x) Shield tunnelling method

3.2 (b)

When the length of the tunnel is small, the centre line can be located by theodolite.

When the length of the tunnel is sufficient, elaborate and very accurate methods are used. In this case, micrometer transit theodolite is used and the centre line is located by means of triangulation method.

3.3 (d)

Size of shaft depends upon the following factors:

(i) Quantity of muck to be handled
(ii) Nature of hoisting system
(iii) Nature of equipment used in tunnelling
(iv) Number of labourers needed for tunnelling
(v) Space required to carry pipes and wires
(vi) Eventual use of the shaft.

3.4 (b)

Methods for tunneling in rock areas:

(i) Full face method
(ii) Heading and beaching method
(iii) Drift method

Compressed air method is done for loose and very soft soil.

3.7 (c)

Heading and benching method is used for tunneling in hard rocks.

3.8 (d)

In the drift method, a small tunnel usually of size 300 cm × 300 cm is driven through the full length manually so method cannot be used for difficult or incompetent rocks.

3.9 (d)

'Blow in' method has the advantage of supplying of fresh air right near the working face, but the disadvantage lies in that the foul air, smoke and dust slowly, move out, fogging the atmosphere inside the tunnel, especially in long tunnels. 'Blow out' method has the special advantage of quick removal of dust and smoke from the working face. But the drawback of this method is that the fresh air has to travel a long distance before it reaches the working face and it is quite likely that is may absorb during this time moisture, heat and foul gases of the hauling equipment resulting in an unpleasant working condition at face. Both these methods are used in combination to incorporate their advantages.

3.10 (c)

Liner plate method grants an absolute control of running ground water. It is an excellent and positive method of minimizing ground subsidence in urban areas.

3.11 (d)

Since line joining two underground points gives the direction of centre line, accurate setting can be achieved only if the theodolite can be set out exactly in line with the points. The operation is quite difficult and time consuming so an alternate method known as Weisbach triangle method may be used to connect two ends of an underground tunnel.

3.12 (c)

For the purpose of ventilation, additional access, removal of muck and escape of people from tunnel passage provisions in form of adits and shafts are provided during construction.

3.14 (c)

For tunnelling in soft soil, Needle Beam method is adopted. The needle beam consists of a stout timber beam or a composite flitched beam and form a main temporary support during the excavation.

3.15 (c)

Poly-centric Tunnel:

- This tunnel section consist number of centers.
- The base of tunnel is flat enough for traffic hence can be used for radius and railways.
- The number of centers and length of roads can be fixed as per the local conditions hence required greater skilled persons.
- The tunnel can resist effectively external and internal pressure.

3.17 (d)

Method of tunnelling in hard rock:

Full face method.

Drift method.

Heading and benching method.

Pitot tunnel method.

Perimeter method.

Hence, shaft method is not used for tunnelling is hard rock.

3.18 (c)

The alignment of a tunnel constructed as part of a development project would depend on various factors, such as the topography of the area, the geological conditions, and the tunnel's purpose. A zig-zag alignment would not typically be used in tunnel construction, as it would increase construction costs and complexity without providing significant benefits. A "Mountain Alignment" typically refers to a tunnel alignment that follows the contour of a mountain or hill. This type of alignment is often used when a tunnel is needed to bypass a mountainous area or provide access to resources within a mountain. However, it is important to note that the alignment chosen for a tunneling project would depend on a wide range of factors, such as the area's geology, the presence of other infrastructure, and the project's specific needs.

3.21 (a)

This method is not suitable for unstable rocks.

■■■■